Lecture Notes in Computer Science 15534

Advanced Research in Computing and Software Science
Subline of Lecture Notes in Computer Science

More information about this series at https://link.springer.com/bookseries/558

Marios Mavronicolas · Qi Qi ·
Grant Schoenebeck
Editors

Web and Internet Economics

20th International Conference, WINE 2024
Edinburgh, UK, December 2–5, 2024
Proceedings

Springer

Editors
Marios Mavronicolas
University of Cyprus
Nicosia, Cyprus

Qi Qi
Renmin University of China
Beijing, China

Grant Schoenebeck
University of Michigan
Ann Arbor, MI, USA

ISSN 0302-9743 ISSN 1611-3349 (electronic)
Lecture Notes in Computer Science
ISBN 978-3-032-08559-7 ISBN 978-3-032-08560-3 (eBook)
https://doi.org/10.1007/978-3-032-08560-3

This Springer imprint is published by the registered company Springer Nature Switzerland AG
The registered company address is: Gewerbestrasse 11, 6330 Cham, Switzerland

Preface

This volume contains the papers and extended abstracts presented at the 20th Conference on Web and Internet Economics (WINE 2024), held during December 2–5, 2024, at the University of Edinburgh, UK. The WINE conference series has emerged as a leading forum to present and exchange research results and ideas combining incentives and computational aspects as they appear in a wide and diverse area of Web and Internet applications at the interface of theoretical computer science, artificial intelligence, operations research, and economics. The growing importance of topics in this interface is attested by the fact that for WINE 2024, there was a record number of 248 submissions.

We invited 106 and 37 researchers to be program committee and senior program committee members, respectively. Each submission was reviewed by three program committee members, and an additional meta-review was provided by one senior program committee member. We are very grateful for their thorough, insightful, and fair reviews and discussions, which helped to make the difficult decisions about which papers to accept for the conference. The reviews were done in a double-blind review format where the authors' identities were hidden from the senior program committee and program committee. We ended up selecting 66 papers for presentation at WINE 2024, leading to a very competitive acceptance rate of just above 26%. Of these, 34 appear here as full papers and 32 as abstracts according to the wishes of the authors.

The works accepted for publication in this volume cover a wide range of topics such as computational social choice, fair division, resource allocation, pricing, information design, aggregation, blockchains, auctions, matching markets, contracts, games and Nash equilibria, facility location, and security. To accommodate the publishing traditions of different fields, authors of accepted papers could ask that only a one-page abstract of the paper appear in the proceedings.

Furthermore, due to the generous support by Springer, we were able to provide a Best Paper Award, which we decided to give to the paper "Aggregation of Antagonistic Contingent Preferences: When Is It Possible?" by Xiaotie Deng, Biaoshuai Tao, and Ying Wang, presented in a special session and appearing first in this volume.

Besides the scientific talks on the papers contained in this volume, the conference program included three invited talks by leading researchers in the field: Vittorio Bilò (University of Salento, Italy), Yiling Chen (Harvard University, USA) and Yinyu Ye (Stanford University, USA, Shanghai Jiao Tong University, and the Chinese University of Hong Kong (ShenZhen), China). Furthermore, the conference program featured four tutorial sessions to showcase and explore emerging topics and trends related to incentives and computation.

We would like to thank all the authors for their interest in submitting their work to WINE 2024. We also thank Shengwei Xu for volunteering to help facilitate the electronic organization of these proceedings. Our special thanks go to the steering committee of WINE and to the local organizers.

December 2024

Marios Mavronicolas
Qi Qi
Grant Schoenebeck

Organization

General Chairs

Aris Filos-Ratsikas	University of Edinburgh, UK
Yiannis Giannakopoulos	University of Glasgow, UK

Program Committee Chairs

Marios Mavronicolas	University of Cyprus, Cyprus
Qi Qi	Renmin University of China, China
Grant Schoenebeck	University of Michigan, USA

Steering Committee

Xiaotie Deng	Peking University, China
Paul Goldberg	University of Oxford, UK
Christos Papadimitriou	Columbia University, USA
Paul Spirakis	University of Liverpool, UK
Rakesh Vohra	University of Pennsylvania, USA
Andrew Yao	Tsinghua University, China
Yinyu Ye	Stanford University, USA
Nicole Immorlica (Joint ACM-EC/WINE Executive Committee)	Microsoft Research Lab, USA
Scott Kominers (Joint ACM-EC/WINE Executive Committee)	Harvard University, USA
Katrina Ligett (Joint ACM-EC/WINE Executive Committee)	Hebrew University, Israel

Senior Program Committee

Xiaohui Bei	Nanyang Technological University, Singapore
Vittorio Bilò	University of Salento, Italy

Simina Branzei	Purdue University, USA
Zhigang Cao	Beijing Jiaotong University, China
Jing Chen	Tsinghua University, China
Xujin Chen	University of Massachusetts, USA
George Christodoulou	Aristotle University of Thessaloniki, Greece
Federico Echenique	UC Berkeley, USA
Edith Elkind	University of Oxford, UK
Angelo Fanelli	CNRS, France
Qizhi Fang	Ocean University of China, China
Hu Fu	Shanghai University of Finance and Economics, China
Jie Gao	Rutgers University, USA
Jugal Garg	University of Illinois Urbana-Champaign, USA
Vasilis Gkatzelis	Drexel University, USA
Kristoffer Arnsfelt Hansen	Aarhus University, Denmark
Martin Hoefer	Goethe University Frankfurt, Germany
Zhiyi Huang	Hong Kong University, China
Christos Kaklamanis	University of Patras and CTI "Diophantus", Greece
David Kempe	University of Southern California, USA
Max Klimm	Technische Universität Berlin, Germany
Yuqing Kong	Peking University, China
Bo Li	Hong Kong Polytechnic University, China
Minming Li	City University of Hong Kong, China
Xiao Liu	Tsinghua University, China
Brendan Lucier	Microsoft Research New England, USA
Gianpiero Monaco	University of Chieti, Italy
Rad Niazadeh	University of Chicago, USA
Sigal Oren	Ben-Gurion University of the Negev, Israel
Vijay G. Subramanian	University of Michigan, USA
Biaoshuai Tao	Shanghai Jiao Tong University, China
Alexandros Voudouris	University of Essex, UK
Xintong Wang	Rutgers University, USA
Zihe Wang	Renmin University of China, China
Lirong Xia	Rutgers University, USA
Fang-Yi Yu	George Mason University, USA
Dengji Zhao	Shenghai Tech, China

Program Committee

Yeganeh Alimohammadi	Stanford University, USA
Elliot Anshelevich	Rensselaer Polytechnic Institute, USA
Georgios Birmpas	University of Liverpool, UK
Farzad Bourbabaee	California Institute of Technology, USA
Johannes Brustle	London School of Economics, UK
Martin Bullinger	Technical University of Munich, Germany
Costas Busch	Augusta University, USA
Linda Cai	Princeton University, USA
Bhaskar Ray Chaudhury	University of Illinois Urbana-Champaign, USA
Yilun Chen	Chinese University of Hong Kong, Shenzhen, China
Zhou Chen	Hangzhou City University, China
Yukun Cheng	Jiangnan University, China
Yun Kuen (Marco) Cheung	Australian National University, Australia
Sulagna Dasgupta	University of Chicago, USA
Bart De Keijzer	King's College London, UK
David Delacretaz	University of Manchester, UK
Argyrios Deligkas	Royal Holloway, University of London, UK
Battal Dogan	University of Bristol, UK
Umut Dur	North Carolina State University, USA
Marcin Konrad Dziubiński	University of Warsaw, Poland
Tomer Ezra	Harvard University, USA
Michail Fasoulakis	Foundation for Research and Technology – Hellas, Greece
Zhixuan Fang	Tsinghua University, China
Maximilian Fichtl	Technical University of Munich, Germany
Yiding Feng	Hong Kong University of Science and Technology, China
Jessica Finocchiaro	Boston College, USA
Bailey Flanigan	Carnegie Mellon University, USA
Federico Fusco	Sapienza University of Rome, Italy
Yotam Gafni	Weizmann Institute of Science, Israel
Ling Gai	Shanghai University of Science and Technology, China
Ganesh Ghalme	Indian Institute of Technology Hyderabad, India
Sumit Goel	New York University, USA
Michael L. Hamilton	University of Pittsburgh, USA
Dong Hao	University of Electronic Science and Technology of China, China
Tobias Harks	University of Passau, Germany

Kun He	Renmin University of China, China
Alexandros Hollender	University of Oxford, UK
Anson Kahng	University of Rochester, USA
Panagiotis Kanellopoulos	University of Essex, UK
Gregory Kehne	University of Texas at Austin, USA
Zhonghong Kuang	Renmin University of China, China
Rucha Kulkarni	University of Illinois Urbana-Champaign, USA
Jiangtao Li	Singapore Management University, Singapore
Mengling Li	Xiamen University, China
Weian Li	Shandong University, China
Yingkai Li	National University of Singapore, Singapore
Shengxin Liu	Harbin Institute of Technology, China
Zhengyang Liu	Beijing Institute of Technology, China
Jiaqi Lu	Chinese University of Hong Kong, Shenzhen, China
Junjie Luo	Beijing Jiaotong University, China
Hongtao Lv	Shandong University, China
Simon Mauras	Inria, France
Divyarthi Mohan	Tel Aviv University, Israel
Vishnu V. Narayan	Tel Aviv University, Israel
Lata Narayanan	Concordia University, Canada
Thanh Nguyen	Purdue University, USA
Dario Paccagnan	Imperial College London, UK
Ioannis Panageas	University of California Irvine, USA
Maneesha Papireddygari	University of Colorado at Boulder, USA
Rojin Rezvan	University of Texas at Austin, USA
Daniel Schmand	University of Bremen, Germany
Daniel Schoepflin	Drexel University, USA
Alkmini Sgouritsa	Athens University of Economics and Business, Greece
Weiran Shen	Renmin University of China, China
Yangguang Shi	Shandong University, China
Nian Si	Hong Kong University of Science and Technology, China
Samuel Taggart	Oberlin College, USA
Yixin Tao	Shanghai University of Finance and Economics, China
Yifeng Teng	Google, USA
Nikolas Tsakas	University of Cyprus, Cyprus
Artem Tsikiridis	Centrum Wiskunde en Informatica, The Netherlands
Bertan Turhan	Iowa State University, USA

Cosimo Vinci	University of Salento, Italy
Changjun Wang	Chinese Academy of Sciences, China
Kangning Wang	Rutgers University, USA
Zizhuo Wang	Chinese University of Hong Kong, Shenzhen, China
Xiaowei Wu	University of Macau, China
Zenan Wu	Peking University, China
Dimitrios Xefteris	University of Cyprus, Cyprus
Tao Xiao	Huawei Technologies Ltd., China
Xiang Yan	Huawei Technologies Ltd., China
Yanqing Yang	Beijing University of Posts and Telecommunications, China
Kanye Ye Wang	University of Macau, China
Yuan Yuan	Purdue University, USA
Haifeng Zhang	Chinese Academy of Sciences, China
Hanrui Zhang	Chinese University of Hong Kong, China
Jinshan Zhang	Zhejiang University, China
Meng Zhang	Zhejiang University, China
Qiankun Zhang	Huazhong University of Science and Technology, China
Yichi Zhang	University of Michigan, USA
Jingtong Zhao	Renmin University of China, China
Junyao Zhao	Stanford University, USA
Shuran Zheng	Tsinghua University, China
Juba Ziani	Georgia Institute of Technology, USA

Local Support Team

Jakub Dargaj	University of Edinburgh, UK
Georgios Kalantzis	University of Edinburgh, UK
Charalampos Kokkalis	University of Edinburgh, UK
Kat Molinet	University of Edinburgh, UK

Contents

Extended Abstracts

Best Paper Award

Aggregation of Antagonistic Contingent Preferences: When Is It Possible?

Xiaotie Deng[1], Biaoshuai Tao[2](✉), and Ying Wang[1]

[1] Center on Frontiers of Computing Studies, School of Computer Science, Peking University, Beijing, China

[2] John Hopcroft Center for Computer Science, School of Electronic Information and Electrical Engineering, Shanghai Jiao Tong University, Shanghai, China

bstao@sjtu.edu.cn

Abstract. We study a two-alternative voting game where voters' preferences depend on an unobservable world state and each voter receives a private signal correlated to the true world state. We consider the collective decision when voters can collaborate in a group and have antagonistic preferences—given the revealed world state, voters will support different alternatives. We identify sharp thresholds for the fraction of the majority-type voters necessary for preference aggregation.

We specifically examine the majority vote mechanism (where each voter has one vote, and the alternative with more votes wins) and pinpoint a critical threshold, denoted as $\theta_{\mathtt{maj}}$, for the majority-type proportion. When the fraction of majority-type voters surpasses $\theta_{\mathtt{maj}}$, there is a symmetric strategy for the majority-type that leads to strategic equilibria favoring informed majority decisions. Conversely, when the majority-type proportion falls below $\theta_{\mathtt{maj}}$, equilibrium does not exist, rendering the aggregation of informed majority decisions impossible.

Additionally, we propose an easy-to-implement mechanism that establishes a lower threshold θ^* (with $\theta^* \le \theta_{\mathtt{maj}}$) for both equilibria and informed majority decision aggregation. We demonstrate that θ^* is optimal by proving a general impossibility result: if the majority-type proportion is below θ^*, with mild assumptions, no mechanism can aggregate the preferences, meaning that no equilibrium leads to the informed majority decision for any mechanism.

Keywords: Crowdsourcing · Social Choice · Information Aggregation

1 Introduction

Consider a virtual community with n residents, a decision about whether to impose a housing tax is to be made. The community residents consist of two groups, landlords and tenants, with antagonistic interests with respect to the

Full version: https://arxiv.org/abs/2410.08127. Authors contributed equally and are listed in alphabetical order.

M. Mavronicolas et al. (Eds.): WINE 2024, LNCS 15534, pp. 3–21, 2026.
https://doi.org/10.1007/978-3-032-08560-3_1

rental prices. Landlords favor higher rents, while tenants prefer lower rents. The community decides by the votes of the residents and a desired decision is the alternative supported by more than half of the residents after the housing tax effect is revealed. If the tax effect on the rental prices is known ex-ante, the desired decision can easily be made, for example in the case where the housing tax lowers the rental price. If more than half of the residents are landlords, they reject the housing tax policy. On the other hand, if more than half of the residents are tenants, they will accept the tax policy.

Such a majority vote mechanism provides each resident with two alternatives (either "accept" or "reject") and designates the one with more votes as the winner, successfully revealing the preference of the majority. Also, this majority vote mechanism is strategy-proof for this case, where the tax effect is known ex-ante.

However, the policy effect is uncertain before deployment. In practice, it is usually unclear to the residents whether the housing tax will increase or decrease the rental price. Instead, each resident independently has a hunch of the policy effect, which is obtained based on public information and individual experience. If residents truthfully report their hunches, by the well-known Condorcet Jury Theorem [15], when the majority of agents hold a correct hunch, the policy effect can be correctly aggregated with a high probability via the majority vote.

The real-life situations are even more complex than the idealized scenario described above. First, when the majority of agents hold an incorrect hunch, the majority vote mechanism fails to reveal the true policy effect. Second, as is typical in most social choice settings, voters often have differing preferences. In two alternative cases, the preferences may be antagonistic, as in our housing tax example. While the majority group aims to jointly determine the true effect and vote for their preferred alternative, the minority group with opposing preferences may try to prevent this through strategic behaviors.

Examples of antagonistic preferences abound in the real world: salary adjustment, two-party presidential election, the tradeoff between freedom and security, and more. Often in these cases, the correlation between alternatives and voters' utilities is not immediately clear: the impact of a policy is uncertain before deployment, the performance of a president-elect remains unpredictable before the inauguration, and the consequences of decisions like Brexit were unforeseen by many who abstained from voting against it. Given these complex scenarios, is it still possible to identify a good collective decision?

Specifically, we study the two alternative voting games where voters' preferences are endogenous and depend on an unobservable world state. In our context, (1) voters are *partially informed*: they receive signals correlated to the state, and their preferences are contingent, depending on the unknown state; (2) preferences are *antagonistic*: when the state is revealed, some voters prefer one alternative, while the remaining voters prefer the other alternative. This naturally divides voters into two types: the majority and minority types.

The objective is to uncover the *informed majority decision*: the alternative favored by more than half of the agents if the world state is revealed. In our context, the informed majority decision is the alternative favored by the majority-type voters given the revealed world state.

Our work considers *coalitional behaviors*. The two antagonistic preferences lead to two voter groups. Voters in a group have the same preference for alternatives and may naturally form a coalition to cooperate for better benefits. This corresponds to the solution concept of *strong Nash equilibrium*, where a strategy profile forms an equilibrium if any subset of agents have no incentives to deviate. While the Nash equilibrium ensures no *individual* voter can benefit from changing strategy, the *strong* Nash equilibrium guarantees that no *group* of voters can jointly deviate to improve their outcomes.

In a sentence, we study the following social choice problem. **Can good collective decisions be made for partially informed voters who have antagonistic preferences and can form coalitions?**

1.1 Our Contribution

The two features of voters, *antagonistic preferences* and *coalitional behaviors*, distinguish our work from previous literature.

Antagonistic Preferences vs. Homogeneous Preferences. When voters' preferences are homogeneous rather than antagonistic, previous work has shown that good collective decisions can be achieved across different scenarios. As mentioned, when most agents have a correct hunch on the world state, the informed majority decision can be successfully aggregated based on the Condorcet Jury Theorem [13]. When the majority may be wrong, the surprisingly popular mechanism proposed by Prelec et al. [43] can ensure correct decisions. These methods assume that agents are non-strategic. In another perspective, when agents can act strategically, issues of incentives do not fail the information aggregation. Although truthfully revealing information is not an equilibrium [4], the majority vote mechanism (or its variants) can still lead to the desirable collective decision in many settings, both with and without coalitional behaviors [4,14,16,18–20,23,29,30,33,36,41,47].

In contrast, when voters' preferences are antagonistic, the feasibility of information aggregation and the incentives issue become two distinct challenges. Even if there are some strategy profiles for the community to figure out the true world state, minority-type agents are unlikely to follow such strategies.

Coalitional Agents vs. Non-coalitional Agents. Antagonistic preferences have also been explored in previous studies [1,2,7,8,26], where agents are assumed to be non-coalitional and the solution concept of Nash equilibrium is applied. In contrast, our work considers agents' coalitional behaviors with the strong Nash equilibrium concept.

Voters' reasoning for Nash equilibria and strong Nash equilibria is fundamentally different. For Nash equilibria, a voter reasons his/her strategy as if (s)he is the "pivotal voter". That is, (s)he assumes the remaining voters' votes on the two alternatives are half-half distributed, which is the only case where his/her vote matters. Conditioning on the half-half vote distribution of the remaining voters, a voter forms a posterior belief on the world state, which eventually affects his/her strategy reasoning. This pivotal-voter reasoning inherently disregards possible coalitional behaviors of agents. For strong Nash equilibria, on the other hand, the deviating group size plays an important role in the reasoning, as different group sizes have different powers for influencing the election outcome.

Other related work [3,5,6,9–12,17,21,22,24,25,27,28,31,32,34,35,37–40,42,44,45] can be found in the full version of our paper.

Summary. To summarize, when voters have homogeneous preferences, the informed majority decision can be reached even when considering coalitional agents. On the other hand, for antagonistic preferences, the study of coalitional agents is missing from the previous literature.

Does the majority vote mechanism still work? If not, are there better mechanisms that aggregate agents' preferences and reach the informed majority decision? In this work, we investigate the aggregation of antagonistic preferences in the presence of coalitional behaviors, which is fundamentally more challenging.

Our Results. We find the aggregation of the informed majority decision is not always possible. Deciding informed majority decision in equilibrium becomes impossible for any mechanism when the fractions of the two types of agents (with antagonistic preferences) are close. This outcome is notably unexpected because, a simple majority vote mechanism can effectively elicit the informed majority decision when preferences are antagonistic but predetermined preferences, or contingent but aligned [23]. We reveal sharp thresholds for the fractions that make the aggregation of antagonistic preference possible.

(1) We first study voters' strategic behavior in the majority vote mechanism and identify a sharp required threshold $\theta_{\mathtt{maj}}$ for the proportion of the majority. When the fraction of the majority agents surpasses $\theta_{\mathtt{maj}}$, we show that all ϵ-strong Bayes Nash equilibria are "good", i.e., leading to the informed majority decision with a high probability. We provide a characterization of the symmetric strategy for the majority-type agents to ensure this kind of equilibrium. For the other cases where the majority proportion is below the threshold, we reveal the nonexistence of the strong Bayes Nash equilibrium.
(2) We also identify a sharp required threshold θ^* for all "reasonable" mechanisms. When the fraction of the majority-type agents surpasses θ^*, we present an easy-to-implement mechanism that ensures all equilibria are "good". Specifically, truthful reporting forms an equilibrium in this mechanism. On the other hand, when the majority proportion is below θ^*, our general impossibility result shows that, under some mild assumptions, no mechanism can lead to "good" equilibria.

Our truthful mechanism design is not just a direct application of the revelation principle to the majority vote mechanism. Notably, our mechanism has a lower threshold, i.e., $\theta^* \leq \theta_{\mathtt{maj}}$, with strict inequality in some cases. We explain the distinctions further in Sect. 5.

2 Model

There are two possible world states, $\Omega = \{L, H\}$, where L indicates the imposition of housing tax lowers the rent and H indicates it raises the rent.

Residents are required to vote for one of two alternatives, either accept($\mathbf{A}$) or reject($\mathbf{R}$) the housing tax proposal. Residents do not know the true world state. Instead, each of them has a private judgment, referred to as a random signal. This signal, denoted by a random variable S_i, takes value from $\mathcal{S} = \{l, h\}$. The signal value $S_i = l$ indicates that resident i believes the world is in state L, and similarly, $S_i = h$ implies the belief that state H happens. We assume the signals that residents receive are identically and independently generated, following a distribution that depends on the actual world state $\omega \in \Omega$. All residents share a common knowledge of the joint distribution between world state and signals.

To streamline the expression, we use P_l^L to denote the conditional probability $\Pr[S_i = l \mid \omega = L]$, which is the probability that an agent receives signal l given that the world is in state L. Analogously, we define P_l^H, P_h^L, and P_h^H. By the law of total probability, we have $P_l^H + P_h^H = 1$ and $P_l^L + P_h^L = 1$. Lastly, we use μ to denote the common prior probability that the world is in state H. Correspondingly, $1 - \mu$ is the probability that the true world state is L.

It is important to note that signals may exhibit asymmetrical favorability (one signal, for example, signal l, may have a higher probability than another signal in both world states, i.e., both $P_l^L > P_h^L$ and $P_l^H > P_h^H$ hold). This implies the possibility that most people make mistakes (i.e., the majority of received signals can be inconsistent with the true state). In this work, we consider such biased signals and assume signals are positively correlated with the world states: $P_l^L > P_l^H$ and $P_h^H > P_h^L$. Let $\Delta = P_l^L - P_l^H = P_h^H - P_h^L$ be the difference in signal frequencies across two world states. Our assumption is equivalent to $\Delta > 0$, implying a general tendency for the signals to align with the true state.

Each resident i has a utility function $v_i : \Omega \times \{\mathbf{A}, \mathbf{R}\} \to \{0, 1, \ldots, B\}$, which assigns a payoff value based on the world state and the chosen alternative.[1] In our model, we consider *contingent* agents, whose preferred alternative varies across states. There are two types of agents, i.e., $\mathcal{T} = \{\mathcal{L}, \mathcal{H}\}$:

- An type-$\mathcal{L}$ agent (analogous to a tenant in the example) prefers alternative $\mathbf{A}$ in state L and prefers alternative $\mathbf{R}$ in state H. Formally, $v_i(L, \mathbf{A}) > v_i(L, \mathbf{R})$ and $v_i(H, \mathbf{R}) > v_i(H, \mathbf{A})$ for i such that $t_i = \mathcal{L}$.
- An type-$\mathcal{H}$ agent (analogous to a landlord) prefers alternative $\mathbf{R}$ in state L and prefers $\mathbf{A}$ in state H. Formally, $v_i(L, \mathbf{R}) > v_i(L, \mathbf{A})$ and $v_i(H, \mathbf{A}) > v_i(H, \mathbf{R})$ for i such that $t_i = \mathcal{H}$.

As it is standard in the Bayesian game setting, the types of agents and thus their utility functions are private information.

[1] In many social choice settings, only ordinal preferences are defined for the voters. However, in our setting, cardinal preferences for the voters need to be defined in order to define the equilibrium concepts.

Let $\alpha_{\mathcal{L}}$ denote the approximate fraction of type-$\mathcal{L}$ agents, $\alpha_{\mathcal{H}}$ denote that of type-$\mathcal{H}$. Since there are only two types of agents, $\alpha_{\mathcal{L}} + \alpha_{\mathcal{H}} = 1$. Let $\alpha = \max\{\alpha_{\mathcal{L}}, \alpha_{\mathcal{H}}\}$ denote the proportion of the majority-type in the population. We consider cases with condition $\alpha > \frac{1}{2}$, indicating a clear majority. Formally, given the number of agents n, the number of majority-type agents is $\lfloor \alpha \cdot n \rfloor$.

An *instance* of our model includes:

- the number of agents, n;
- the common prior of the world state, $(\mu, 1-\mu)$, where μ is the probability of state H;
- the signal distributions, (P_h^H, P_l^H) and (P_h^L, P_l^L), where P_s^H and P_s^L denote the conditional probabilities of receiving signal $s \in \mathcal{S}$ in two possible states;
- the utility functions of agents, $\{v_i\}_{i=1}^n$ (assumed to be private);
- and the fraction of each type, $(\alpha_{\mathcal{L}}, \alpha_{\mathcal{H}})$ (consistent with the utilities).

We define a *configuration* of our model as parameters $\{(\alpha_{\mathcal{L}}, \alpha_{\mathcal{H}}), (\mu, P_h^H, P_h^L)\}$. This configuration refers to all instances equipped with the same parameters, where the number of agents n can be arbitrarily large, and the utility functions can be diverse but adhere to the type proportions $(\alpha_{\mathcal{H}}, \alpha_{\mathcal{L}})$.

Our analyses will delve into agents' voting behaviors and the resulting outcome under different configurations. It may be helpful to think of $n \to \infty$ at first and the results for finite n are provided in the appendix.

2.1 Mechanism, Strategy, and Equilibrium

In this section, we will elaborate on the concepts of mechanisms, strategies, and equilibria, which are defined in the standard manner as in game theory.

Mechanism. A *mechanism* $\mathcal{M} : \mathcal{R}^n \to \Delta(\{\mathbf{A}, \mathbf{R}\})$ is a decision-making process that outputs a distribution over alternatives based on the report profile $(r_1, r_2, \ldots, r_n) \in \mathcal{R}^n$ of voters. The *report space* $\mathcal{R}$ is specified by the questions posed to agents, which can be about their received signals, votes on alternatives, preferences in certain states, or beliefs about others' signals. The mechanism only sees agents' reports and does not know any component of the instance.

Strategy. A (mix) *strategy* for an agent is a function $\sigma : \mathcal{T} \times \mathcal{S} \to \Delta(\mathcal{R})$ that maps the type and the signal received by the agent to a distribution on possible reports. A strategy is said to be *truthful* if it honestly answers all questions based on the agent's knowledge and signal. When the type of the agents is clear from the context, we slightly abuse the notation and describe a strategy by the function $\sigma : \mathcal{S} \to \Delta(\mathcal{R})$ with the first function input (i.e., the type) omitted.

Expected Utility Function. Given a strategy profile $\Sigma = (\sigma_1, \sigma_2, \ldots, \sigma_n)$, we denote the expected utility of agent i under mechanism $\mathcal{M}$ as $u_i^{\mathcal{M}}(\Sigma)$, where the expectation is taken over the sampling of agents' signals (with agents' strategies, it decides the report distributions) and the mechanism's outcome. The expected utility is concerned with *ex-ante* outcomes, as opposed to the *ex-post* utility v_i, which is considered after the world state is realized. In this paper, we focus on *ex-ante* utilities when talking about any equilibrium solution concept.

Equilibrium. In a social choice setting with a large number of agents, it is often the case that any single agent's deviation has a negligible impact on the overall outcome. Therefore, rather than considering the typical Bayes Nash equilibrium, we consider a much stronger concept: the strong Bayes Nash equilibrium, where the collective deviations of subsets of agents are taken into account.

Definition 1 (ϵ-Strong Bayes Nash Equilibrium). *A strategy profile $\Sigma = (\sigma_1, \ldots, \sigma_n)$ is an ϵ-strong Bayes Nash equilibrium if no subset of agents D and alternative profile $\Sigma' = (\sigma_1', \ldots, \sigma_n')$ exist such that*

1. *$\sigma_i = \sigma_i'$ for each $i \notin D$,*
2. *$u_i(\Sigma') \geq u_i(\Sigma)$ for each $i \in D$,*
3. *there exist $i \in D$ such that $u_i(\Sigma') > u_i(\Sigma) + \epsilon$.*

2.2 Objective

The objective of our work is to unearth the *informed majority decision*, which is the decision favored by the majority if the true world state ω were known. For example, when type-$\mathcal{H}$ is the majority type, the informed majority decision is **A** if the state is H and **R** otherwise. Conversely, when the majority type is type-$\mathcal{L}$, the informed majority decision is **A** in state L and **R** otherwise.

We identify the necessary and sufficient conditions on parameters for ensuring the informed majority decision in equilibrium, through both the equilibrium analysis of the majority vote mechanism and the mechanism design view.

2.3 Majority Vote Mechanism

The *majority vote mechanism* is a common and natural approach to generating collective decisions. In the majority vote, each voter i votes for either **A** or **R** (the report space $\mathcal{R} = \{\mathbf{A}, \mathbf{R}\}$). The alternative supported by the majority will be the output. In the case of a tie, where exactly half of the agents vote for **A**, both the alternatives have an equal chance of being selected.

In this context, an agent's strategy is a mapping from the signal to a distribution over $\{\mathbf{A}, \mathbf{R}\}$. We use vector (β_l, β_h) to denote strategy, where $0 \leq \beta_s \leq 1$ corresponds to the probability of voting **A** when receiving signal $s \in \mathcal{S}$.

We use alternative notation (δ_l, δ_h) to denote the strategy (β_l, β_h), where $\delta_s \in [-\frac{1}{2}, \frac{1}{2}]$ indicates the deviation of β_s from $\frac{1}{2}$. The deviation direction aligns with the possible shift in the preferred alternative of type-$\mathcal{H}$ agents when receiving signal s. That is, we have $\beta_l = 1/2 - \delta_l$ and $\beta_h = 1/2 + \delta_h$.

We are following using (δ_l, δ_h) rather than (β_l, β_h) to denote strategy for technical simplicity.

It is not clear what strategies are considered truthful: what does it mean by saying that "reporting **A**/**R** truthfully reflects an agent's preference when receiving a signal l/h"? Below we provide some candidates for the "truthful" strategy to familiarize readers with our context.

A natural strategy that can be considered as "truthful" is to report the alternative that exactly corresponds to the signal. For example, for a type-$\mathcal{H}$ agent, this strategy is $(\beta_l, \beta_h) = (0, 1)$, i.e., the agent votes for **A** (with probability 1) when receiving signal h and **R** (with probability 1) when receiving signal l. This is called *the informative strategy.* In our alternative notation, it can be represented as $(\delta_l, \delta_h) = (\frac{1}{2}, \frac{1}{2})$. The informative strategy profile does not always lead to the informed majority decision, even when all the agents have the same type, say, type-$\mathcal{H}$. As we have mentioned, the signals may be biased (e.g., it is possible that both P_h^H and P_h^L are more than $\frac{1}{2}$), voting according to such signals clearly may lead to the wrong outcome.

In order to achieve the informed majority decision, agents need to be more sophisticated than being truthful. Han et al. [23] show that, with common value environment and an infinite number of agents, *a strategy profile leads to the informed majority decision if and only if it is a strong Bayes Nash equilibrium.* Neither the informative strategy profile nor the sincere strategy profile leads to the informed majority decision, so neither of them is a strong Bayes Nash equilibrium. Their work also characterizes the strategy profile that leads to the informed majority decision (or, is a strong Bayes Nash equilibrium) for single-type voters. We briefly go through these, as many intuitions will be useful to the more general case with two types of agents studied in this paper.

We assume all the agents are of type $\mathcal{H}$. According to Han, et al. [23], the symmetric strategy profile $\{(\delta_l, \delta_h)\}$ aggregates the informed majority decision if it satisfies the following two inequalities:

$$\Delta_{\mathbf{A}}^H(\delta_l, \delta_h) := P_h^H \cdot \delta_h - P_l^H \cdot \delta_l > 0 \text{ and } \Delta_{\mathbf{A}}^L(\delta_l, \delta_h) := P_h^L \cdot \delta_h - P_l^L \cdot \delta_l < 0. \quad (1)$$

The amount $\Delta_{\mathbf{A}}^H(\delta_l, \delta_h)$ is the expected increment of **A**'s vote share in state H compared to the midpoint $\frac{1}{2}$, and amount $\Delta_{\mathbf{A}}^L(\delta_l, \delta_h)$ is that in state L.

Why does this work? Notice that the difference in expected vote share across two states is $\Delta_{\mathbf{A}}^H(\delta_l, \delta_h) - \Delta_{\mathbf{A}}^L(\delta_l, \delta_h) = \Delta \cdot (\delta_h + \delta_l)$ (recall that $\Delta = P_h^H - P_h^L = P_l^L - P_l^H$ is the signal frequency difference). The difference in signal frequency, Δ, is utilized to produce the vote share difference, and particularly, the opposite sign between $\Delta_{\mathbf{A}}^H(\delta_l, \delta_h)$ and $\Delta_{\mathbf{A}}^L(\delta_l, \delta_h)$. Such opposite directions ensure the informed majority decision is supported by more than half of the agents in both states. In details, given the strategy profile $\{(\delta_l, \delta_h)\}$, the expected vote share of **A** in state H and the vote share of **R** in state L (denoted by $p_{\mathbf{A}}^H(\delta_l, \delta_h)$ and $p_{\mathbf{R}}^L(\delta_l, \delta_h)$ respectively) can be calculated as

$$p_{\mathbf{A}}^H(\delta_l, \delta_h) = \frac{1}{2} + \Delta_{\mathbf{A}}^H(\delta_l, \delta_h) \text{ and } p_{\mathbf{R}}^L(\delta_l, \delta_h) = \frac{1}{2} - \Delta_{\mathbf{A}}^L(\delta_l, \delta_h). \quad (2)$$

Therefore, (1) implies that the expected vote share of the informed majority decision is more than one half.

By the Law of Large Numbers, the exact vote share concentrates on the expected share as $n \to \infty$, the informed majority decision is aggregated with a probability approaching 1. Such a strategy profile is also a strong Bayes Nash equilibrium for common-interest agents.

The above reasoning holds if all agents have the same type $\mathcal{H}$. The setting with two antagonistic types studied in this paper is much more complex.

3 Equilibrium Analysis Under Majority Vote

In this section, we study the majority vote mechanism with two types of voters. We identify a sharp threshold, $\theta_{\mathtt{maj}}$, for the majority proportion. This threshold determines when the majority vote mechanism can consistently aggregate the informed majority decision.

Theorem 1. *In the majority vote mechanism, a critical threshold exists for the majority proportion, denoted as* $\theta_{maj} = \frac{1}{2M}$ *where*

$$M = \begin{cases} \frac{P_l^L}{P_l^L + P_l^H}, & \text{if } P_h^L + P_h^H \leq 1, \\ \frac{P_h^H}{P_h^L + P_h^H}, & \text{otherwise.} \end{cases}$$

For this threshold, there exist functions $\epsilon(n)$*,* $p(n)$*, and* $\gamma(n)$ *all dependent on the number of agents* n*, such that as* $n \to \infty$*,* $\epsilon(n) \to 0$*,* $p(n) \to 1$ *and* $\gamma(n)$ *does not approach to* 0*. For these functions, the following two statements hold:*

- *If the majority proportion* α *exceeds threshold* θ_{maj}*, i.e.,* $\alpha > \theta_{maj}$*, an* $\epsilon(n)$*-strong Bayes Nash equilibrium exists. Furthermore, any such equilibrium leads to the informed majority decision with a probability of at least* $p(n)$*.*
- *Conversely, if the majority proportion is below the threshold, i.e.,* $\alpha \leq \theta_{maj}$*, there is no* $\gamma(n)$*-strong Bayes Nash equilibrium.*

The remaining part of this section is devoted to demonstrating Theorem 1. In Sect. 3.1, we characterize the best strategy for the majority-type agents that is most robust against the possible adversarial actions from the minority-type agents. In Sect. 3.2, we provide intuitions behind Theorem 1 assuming $n \to \infty$, in which case $\epsilon(n)$ and $p(n)$ in the theorem become just 0 and 1 respectively. The assumption $n \to \infty$ simplifies the proof by avoiding many technical details. The proof of the theorem can be found in the full version of our paper.

3.1 Optimal Strategy for Majority-Type Agents

Without loss of generality, we assume type-$\mathcal{H}$ agents are the majority.

If the majority type-$\mathcal{H}$ agents use a strategy that satisfies inequalities in (1), then a natural choice for the type-$\mathcal{L}$ agents is a strategy that reverses the directions of the two inequalities in (1). Both types of agents would like to choose strategies that maximize the "margins" of vote shares, which are $\Delta_{\mathbf{A}}^H(\delta_l, \delta_h)$ and $\Delta_{\mathbf{A}}^L(\delta_l, \delta_h)$, to guarantee their favored alternatives.

However, the minority type-$\mathcal{L}$ agents are less powerful. They will lose completely if they play in this way, as the margins they create are smaller than the margins created by the majority type-$\mathcal{H}$ agents. To avoid a crushing defeat, type-$\mathcal{L}$ agents need to choose a more aggressive strategy—vote for **A** or **R** with probability 1—to hopefully win in one world state. The majority type-$\mathcal{H}$ agents, on the other hand, would select δ_l and δ_h to maximize both margins $\Delta_{\mathbf{A}}^H(\delta_l, \delta_h)$ and $\Delta_{\mathbf{A}}^L(\delta_l, \delta_h)$ for a greater chance of victory.

We refer to the symmetric strategy (δ_l^*, δ_h^*) that maximizes the margins for the majority type-$\mathcal{H}$ agents as *the optimal strategy*. The following lemma characterizes this strategy.

Lemma 1 (Characterization of Optimal Strategy). *The strategy* (δ_l^*, δ_h^*) *that maximizes the function* $P(\cdot, \cdot)$ *is*

$$\begin{cases} \delta_l^* = \frac{1}{2} \cdot \frac{P_h^L + P_h^H}{P_l^L + P_l^H},\ \delta_h^* = \frac{1}{2} & \text{if } \frac{P_h^L + P_h^H}{2} \le \frac{1}{2}, \\ \delta_l^* = \frac{1}{2},\ \delta_h^* = \frac{1}{2} \cdot \frac{P_l^L + P_l^H}{P_h^L + P_h^H} & \text{otherwise.} \end{cases} \tag{3}$$

where $P(\delta_l, \delta_h) = \min\{p_{\mathbf{A}}^H(\delta_l, \delta_h), p_{\mathbf{R}}^L(\delta_l, \delta_h)\}$ *for* $p_{\mathbf{A}}^H(\delta_l, \delta_h)$ *and* $p_{\mathbf{R}}^L(\delta_l, \delta_h)$ *defined in* (2).

Lemma 1 demonstrates the optimal strategy for the majority type-$\mathcal{H}$ agents. This strategy varies depending on the relative frequency of signals h and l: if signal h is comparatively rare, agents should always vote for **A** upon receiving signal h. However, when they receive signal l, they do not always vote for **R**, but vote for **A** with a certain probability. For the other case where signal l is rare, agents should always vote for **R** when they receive signal l and vote for **R** with a certain probability when receiving signal h.

Table 1. Signal distributions.

	Signal h	Signal l
State H	0.75	0.25
State L	0.5	0.5

The following is an example of the optimal strategy and other strategies satisfying inequalities in (1).

Example 1 (Optimal Strategy and Other Strategies). Consider the signal distributions in Table 1, where signal h is generally more common.

According to inequalities in (1), any strategy (δ_l, δ_h) satisfying $\delta_h < \delta_l < 3\delta_h$ will ensure the wish of type-$\mathcal{H}$ agents when no type-$\mathcal{L}$ agents exist. For example, (δ_l, δ_h) can be $(\frac{1}{6}, \frac{1}{8})$. Then agents vote for **A** with a probability of $\frac{1}{2} + \frac{1}{8} = \frac{5}{8}$ upon signal h and with a probability of $\frac{1}{2} - \frac{1}{6} = \frac{1}{3}$ upon signal l. Such a strategy makes the expected vote share of **A** in state H, and that of **R** in state L be $0.75 \times \frac{5}{8} + 0.25 \times \frac{1}{3} = \frac{53}{96}$ and $0.5 \times \frac{3}{8} + 0.5 \times \frac{2}{3} = \frac{25}{48}$, respectively.

On the other hand, the optimal strategy for type-$\mathcal{H}$ agents is $(\delta_l^*, \delta_h^*) = (\frac{1}{2}, \frac{3}{10})$. Type-$\mathcal{H}$ agents deterministically vote for **R** upon the rarer signal l. The expected vote share of **A** in state H and that of **R** in state L are both $\frac{3}{5}$. □

In this example, the optimal strategy leads to equal vote shares of the informed majority decision in both world states. This is not a coincidence. The equality of shares is a necessary condition for the optimal strategy.

Proposition 1. *The optimal strategy* (δ_l^*, δ_h^*) *satisfies* $p_{\mathbf{A}}^H(\delta_l^*, \delta_h^*) = p_{\mathbf{R}}^L(\delta_l^*, \delta_h^*)$.

We defer the proof of Lemma 1 and Proposition 1 into the full version.

3.2 Equilibrium Analysis for $n \to \infty$

The previous subsection characterizes the optimal strategy for the majority Type-$\mathcal{H}$ agents, which generates the maximum margin for the informed majority decision. Intuitively, when the proportion of minority agents is small, the created margin by the majority is large enough to counter the minority's aggressive strategy, so the informed majority decision will always win. On the other hand, when the minority proportion is relatively large, the margin cannot counter their deterministic voting. Then no equilibrium exists: if the minority agents always vote for **A**, then the majority agents will reduce their probability of voting **A**; if the majority agents play in this way, the minority agents will vote for **R** instead...

Now we provide a threshold on the majority (or minority) proportion for the existence of the strong Bayes Nash equilibrium. We focus on cases that $n \to \infty$, which capture many ideas and intuitions.

We first derive the vote share of the informed majority decision, among the majority type-$\mathcal{H}$ agents, when they use the optimal strategy. This value, denoted as $M = P(\delta_l^*, \delta_h^*)$, can be computed by Equations (2) and (3):

$$M = \begin{cases} \frac{P_l^L}{P_l^L + P_l^H} & \text{if } \frac{P_h^L + P_h^H}{2} \leq \frac{1}{2}, \\ \frac{P_h^H}{P_h^L + P_h^H} & \text{otherwise.} \end{cases} \tag{4}$$

When the Majority Proportion Is Large... In other words, when the expected vote share of the informed majority decision, cast by the majority type agents, is already greater than one-half ($\alpha M > \frac{1}{2}$), the majority agents ensure the informed majority decision. Since there is no room for the majority type agents to improve utilities, any strategy profile of such a case forms a strong Bayes Nash equilibrium.

Lemma 2. *For configuration* $\{(\alpha_{\mathcal{L}}, \alpha_{\mathcal{H}}), (\mu, P_h^H, P_h^L)\}$ *where the majority proportion* $\alpha > \frac{1}{2M}$ *and the number of agents* $n \to \infty$, *any strategy profile* $(\sigma_1, \sigma_2, \ldots, \sigma_n)$ *such that the majority type-*$\mathcal{H}$ *agents adopt the optimal strategy forms a strong Bayes Nash equilibrium in the majority vote mechanism. Such equilibria lead to the informed majority decision with a probability of* 1.

Lemma 2 characterized a specific kind of "good" equilibria that ensure the informed majority decision. One may ask if there exists some "bad" equilibrium. The answer is no. In fact, any strong Bayes Nash equilibrium can ensure the informed majority decision in the majority vote mechanism.

Lemma 3. *For the number of agents $n \to \infty$, any strong Bayes Nash equilibrium in the majority vote mechanism leads to the informed majority decision with a probability of* 1.

Intuitively, for whatever strategies the minority agents play, the majority agents can "fine-tune" the values of δ_l^* and δ_h^* to absorb the impact made by the minority. In this way, the majority can improve the outcome if the informed majority decision is not guaranteed to win. Therefore, no "bad" equilibrium exists.

When the Majority Proportion Is Small... On the other hand, when the population of the majority-type agents is not large enough to create a large margin, the informed majority decision is no longer secured. For whatever strategies the majority agents play, the minority agents can choose either **A** or **R** to vote (deterministically) so that they can win in at least one world state. We find no equilibrium exists in such cases. Consequently, the informed majority decision cannot be aggregated.

Lemma 4. *For configuration $\{(\alpha_{\mathcal{L}}, \alpha_{\mathcal{H}}), (\mu, P_h^H, P_h^L)\}$ where the majority proportion $\alpha \leq \frac{1}{2M}$ and the number of agents $n \to \infty$, no strong Bayes Nash equilibrium exists in the majority vote mechanism.*

The following example applies our results to parameters in Table 1.

Example 2 (Threshold for the Existence of Strong Equilibrium). Within signal distributions given in Table 1, the maximum vote share of the informed majority decision cast by the majority type agents is $M = \frac{0.75}{0.5+0.75} = \frac{3}{5}$.

When the majority proportion, α, is greater than $\frac{1}{2M} = \frac{5}{6}$ (and $n \to \infty$), it forms a strong Bayes Nash equilibrium and aggregates the informed majority decision as long as the majority type-$\mathcal{H}$ agents adopt the optimal strategy.

When the majority proportion α is less than or equal to $\frac{5}{6}$, there is no strong Bayes Nash equilibrium. □

4 Results of Truthful Mechanism Design

In this section, we study the threshold for the majority proportion to ensure the informed majority decision in general mechanisms.

We design an easy-to-employ mechanism and get a more attainable threshold θ^*. Our mechanism has the lowest requirement for the majority proportion: when $\alpha \leq \theta^*$, the informed majority decision cannot always win in any anonymous mechanism, even as $n \to \infty$.

Theorem 2. *There exists an anonymous mechanism $\mathcal{M}$ with a critical threshold $\theta^* = \frac{1}{\Delta+1}$ for the majority proportion such that*

- *If majority proportion α exceeds threshold θ^*, i.e., $\alpha > \theta^*$, then truthful reporting becomes an $\epsilon(n)$-strong Bayes Nash equilibrium and this equilibrium ensures the informed majority decision with probability at least $p(n)$, where $\epsilon(n) \to 0$ and $p(n) \to 1$ as $n \to \infty$.*

Remark 1. *Our proposed mechanism provides a solution that aggregates the informed majority decision when agents truthfully report. However, there is no guarantee of agents' truthful reporting when $\alpha \leq \theta^*$. Our impossibility results show that, under some mild assumptions, any anonymous mechanism cannot lead to an ϵ-strong Bayes Nash equilibrium that aggregates the informed majority decision when $\alpha \leq \theta^*$. This complements the positive result in Theorem 2.*

The threshold of our mechanism θ^* is generally more attainable than the threshold of the majority vote mechanism $\theta_{\texttt{maj}}$.

Proposition 2. *For any configuration $\{(\alpha_{\mathcal{L}}, \alpha_{\mathcal{H}}), (\mu, P_h^H, P_h^L)\}$, it holds that $\theta_{maj} \geq \theta^*$, i.e., $\frac{1}{2M} \geq \frac{1}{\Delta+1}$. Particularly, the equality holds when $P_h^L + P_h^H = 1$.*

In the remaining of this section, we introduce our mechanism and intuitions behind Theorem 2 assuming $n \to \infty$. The complete proof and our impossibility results can be found in the full version of our paper.

4.1 Our Mechanism

Our mechanism draws inspiration from Schoenebeck and Tao's mechanism [46]. This work extends Prelec et al.'s surprisingly popular algorithm [43] to a truthful mechanism, where agents' posterior estimations of the signal frequencies serve as a threshold to distinguish two states. We follow this idea while a nuanced threshold value is elicited.

Our Mechanism

---

1. **Information Collection.**
   Each agent i reports his/her private type ($\mathcal{L}$ or $\mathcal{H}$), the received signal (l or h), and a threshold value $\hat{\delta}_i$ defined as $\frac{P_l^L}{P_l^L + P_h^H}$.
2. **Majority Type Identification.**
   Determine the majority-type, $\hat{M}$, from the agents' reports.
3. **Threshold Calculation.**
   Calculate the threshold, $\hat{\delta}$, as the median of the reported $\hat{\delta}_i$ values.
4. **State Assessment.**
   Set the true state $\hat{\omega}$ as L if the reporting frequency of l exceeds the threshold $\hat{\delta}$ and as H otherwise.
5. **Outcome Selection.**
   Choose the preferred alternative of majority-type $\hat{M}$ in state $\hat{\omega}$.

This mechanism elicits a threshold value $\hat{\delta}$ (expected to be $\delta = \frac{P_l^L}{P_l^L+P_h^H}$). Such a threshold value may seem artificial and challenging to estimate, as it demands that agents have a precise understanding of the parameters of the information structure. However, we can obtain this value in a user-friendly way by asking agents for their posterior forecasts on the world state (e.g., $\Pr[\omega = L \mid S_i = l]]$) and the estimations of signals received by peers (e.g., $\Pr[S_j = l \mid s_i = h]$). The details can be fond in the full version of our paper.

Below is a running example of our mechanism and Schoenebeck and Tao's mechanism [46]. As is shown, ours provides a more attainable threshold on the majority proportion to guarantee the informed majority decision under minority agents' group strategic behavior.

Example 3 (Running Example of Mechanisms). Given the signal distribution in Table 1 and a prior probability of state $\mu = \frac{1}{2}$, the thresholds for our mechanism and Schoenebeck and Tao's mechanism are calculated as follows.

If all the majority agents report truthfully, the threshold δ of ours is $\frac{P_l^L}{P_l^L+P_h^H} = \frac{0.5}{0.5+0.75} = \frac{2}{5}$. In contrast, in Schoenebeck and Tao's mechanism, the threshold values reported by agents are their posterior estimations about the frequency of signal l. These estimates come from agents' beliefs about the true world state, which is either $\Pr[\omega = H \mid S_i = l] = \frac{1}{3}$ or $\Pr[\omega = H \mid S_i = h] = \frac{3}{5}$. So the threshold values are either $\Pr[s_j = l \mid S_i = l] = \frac{1}{3} \times 0.25 + \frac{2}{3} \times 0.5 = \frac{5}{12}$ or $\Pr[s_j = l \mid S_i = h] = \frac{3}{5} \times 0.25 + \frac{2}{5} \times 0.5 = \frac{7}{20}$. The threshold δ' is the median of all the reported thresholds, so it must lie in the range between $\frac{7}{20}$ and $\frac{5}{12}$.

When the true world state is L, the frequency of signal l, which is 0.5, lies above both the threshold δ and δ'. On the other hand, the frequency of signal l, which is 0.25, is below the thresholds in state H.

The reporting frequency of l has the same properties as long as the proportion of the majority type, α, is large enough. However, when the majority proportion is smaller, for example, when $\alpha = 0.84$ and $n \to \infty$, the reporting frequency of l in state H may exceed the threshold in Schoenebeck and Tao's mechanism. That is, when the minority type agents uninformatively claim they receive signal l, the reporting frequency of l, i.e., $\alpha P_l^H + 1 - \alpha = 0.84 \times 0.25 + 0.16 = 0.37$, may exceed the threshold δ', which is in the range between $\frac{7}{20} = 0.35$ and $\frac{5}{12} \approx 0.42$.

On the other hand, our mechanism correctly infers the true world state if $\alpha > \frac{4}{5}$. In the above case, the reporting rate of signal l in state L and state H are $0.84 \times 0.5 + 0.16 = 0.58$ and 0.37 respectively. Our threshold $\delta = \frac{2}{5}$ succeeds in distinguishing two worlds. □

The following properties play key roles in our mechanism.

Proposition 3.

$$P_l^L > \frac{P_l^L}{P_l^L + P_h^H} > P_l^H \text{ and } P_h^L < \frac{P_h^H}{P_l^L + P_h^H} < P_h^H. \tag{5}$$

Furthermore, for $\alpha > \frac{1}{\Delta+1}$, we have the following inequalities:

$$\alpha P_l^L > \frac{P_l^L}{P_l^L + P_h^H} > \alpha P_l^H + (1-\alpha) \tag{6a}$$

$$\alpha P_h^L + (1-\alpha) < \frac{P_h^H}{P_l^L + P_h^H} < \alpha P_h^H \tag{6b}$$

The inequality chains in (5) ensure the aggregation of the informed majority decision when all the agents play truthfully. It says that the expected fraction of agents receiving signal l in state L, which is P_l^L, is above our threshold δ; that fraction in state H, which is P_l^H, falls below threshold δ. These properties ensure the correct state assessment in our mechanism.

The inequality chains in (6) are the key for the truthful-telling profile to be an approximate strong Bayes Nash equilibrium. Intuitively, it says that, when α is sufficiently large, the minority agents cannot invalidate (5) by voting for one alternative deterministically.

4.2 Analysis of Our Mechanism

In this section, we assume $n \to \infty$ and demonstrate properties of our mechanism.

Lemma 5. *If all agents report truthfully, our mechanism will output the informed majority decision with probability* 1 *when* $n \to \infty$.

When all agents report truthfully, our mechanism will correctly identify the majority type. By Proposition 3, the state assessment can find the true world state. Therefore, the informed majority decision prevails.

Truthful reporting is a "good" strategy profile, in the sense that it makes the informed majority decision win in our mechanism. Moreover, when the majority proportion is large enough, we find this profile forms a strong equilibrium.

Lemma 6. *For configuration* $\{(\alpha_{\mathcal{L}}, \alpha_{\mathcal{H}}), (\mu, P_h^H, P_h^L)\}$ *where the majority proportion* $\alpha > \frac{1}{\Delta+1}$ *and the number of agents* $n \to \infty$, *truthful reporting forms a strong Bayes Nash equilibrium in our mechanism.*

According to Lemma 5, when $n \to \infty$ and all agents report truthfully, our mechanism outputs the informed majority decision with probability 1. The majority-type agents already achieve their maximum utilities and have no incentive to deviate. So possible coalition D contains only minority agents.

The minority agents cannot alter the outcome of our mechanism: Since the majority type fraction is more than one half ($\alpha > \frac{1}{2}$), minority agents are unable to change the identified majority type $\hat{M}$. Similarly, these agents cannot change the median $\hat{\delta}$ calculated in Step 3. The only thing they can influence is the fraction of agents reporting signal l. However, by Proposition 3, the deviations in signal fraction cannot cross $\hat{\delta}$ in both states. So the minority coalition cannot alter the mechanism's output and truthful reporting is a strong Nash equilibrium.

Remark 2. *When $\alpha > \frac{1}{\Delta+1}$, our mechanism can lead to a "good" strong Bayes Nash equilibrium that aggregates the informed majority decision. Not only that, using the same idea in the majority vote mechanism, with $\alpha > \frac{1}{\Delta+1}$, we can show all the equilibria of our mechanism are "good".*

5 Comparing Two Mechanisms

In this section, we compare our mechanism with the majority vote mechanism. Both the majority vote and our truthful mechanism utilize the difference in signal frequencies across states, Δ, to distinguish the true state or the informed majority decision. However, they do this in different ways and result in different thresholds, $\theta_{\mathtt{maj}}$ and θ^*, for the majority proportion.

In our mechanism, the ratio $\frac{P_l^L}{P_l^L+P_h^H}$ serves as a threshold and the reporting signal frequencies of l in the two states naturally lie at both sides of the threshold. In particular, we have shown in Proposition 3 that the fraction of agents receiving signal l is below $\frac{P_l^L}{P_l^L+P_h^H}$ in state H, and above $\frac{P_l^L}{P_l^L+P_h^H}$ in state L. Our mechanism uses this observation to identify the true state.

In the majority vote mechanism, the midpoint, $\frac{1}{2}$, is the threshold of the vote share of alternative **A**. When the signals are unbiased, i.e., with $P_l^H < \frac{1}{2} < P_l^L$, the informed majority decision wins in both states as long as the majority agents vote according to their received signal. When the signals are biased, with one signal occurring more often than the other in both states, the majority-type agents need to strategically "shift" the signal frequencies P_l^H and P_l^L by adjusting their voting strategies. By making the adjusted frequencies on the opposite sides of $\frac{1}{2}$, the majority-type agents ensure their informed decision stands out.

Two mechanisms induce different majority proportion thresholds for the strong Bayes Nash equilibrium. The threshold is $\frac{1}{2M}$ in the majority vote mechanism and $\frac{1}{\Delta+1}$ in our mechanism. Notably, as we verify in Proposition 2, the majority vote has a stronger requirement to attain equilibrium because $\frac{1}{2M} \geq \frac{1}{\Delta+1}$ always holds.

For the existence of strong Bayes Nash equilibrium, which is equivalent to the perfect elicitation of the informed majority decision, the proportion of minority agents should be small so they cannot confuse the mechanism's judgment about the world state. Intuitively, the larger the difference between the two states, the harder the minority can confuse the mechanism. This difference corresponds to the difference in signal frequencies, Δ, in our mechanism. On the other hand, in the majority vote, it becomes the difference in the expected vote share of **A** under the majority type's optimal strategy, whose value is $p_{\mathbf{A}}^H(\delta_l^*, \delta_h^*) - p_{\mathbf{A}}^L(\delta_l^*, \delta_h^*) = \Delta_{\mathbf{A}}^H(\delta_l^*, \delta_h^*) - \Delta_{\mathbf{A}}^L(\delta_l^*, \delta_h^*) = \Delta \cdot (\delta_h^* + \delta_l^*)$. Given that the optimal strategy satisfies $\delta_h^*, \delta_l^* \in (0, \frac{1}{2}]$, the differential in majority voting, $\Delta \cdot (\delta_h^* + \delta_l^*)$, cannot surpass Δ. This leads to a higher threshold on the majority proportion in the majority vote mechanism.

In conclusion, our mechanism "places" a threshold $\frac{P_l^L}{P_l^L+P_h^H}$ between P_l^H and P_l^L, whereas, in the majority vote mechanisms, agents strategically "shift" the values of P_l^H and P_l^L to make them lie on different side of $\frac{1}{2}$. The shifting process reduces the margin Δ by a factor of $(\delta_h^* + \delta_l^*)$, which leaves more room for the minority type agents to flip the outcome of the election.

6 Future Work

One natural future work is to extend the characterization to the setting with general voters' utility functions $v_i(\cdot,\cdot)$ (i.e., beyond the two antagonistic types), as well as the setting with more than two states or more than two alternatives. More discussions about these are available in the full version of our paper.

Acknowledgments. This study was supported by the National Natural Science Foundation of China (No. 62172012 and No. 62102252).

Disclosure of Interests. The authors have no competing interests to declare that are relevant to the content of this article.

References

1. Acharya, A.: Information aggregation failure in a model of social mobility. Games Econom. Behav. **100**, 257–272 (2016)
2. Ali, S.N., Mihm, M., Siga, L.: Adverse selection in distributive politics. Available at SSRN 3579095 (2018)
3. Alonso, R., Câmara, O.: Persuading voters. Am. Econ. Rev. **106**(11), 3590–3605 (2016)
4. Austen-Smith, D., Banks, J.S.: Information aggregation, rationality, and the condorcet jury theorem. Am. Polit. Sci. Rev. **90**(1), 34–45 (1996)
5. Barelli, P., Bhattacharya, S., Siga, L.: Full information equivalence in large elections. Econometrica **90**(5), 2161–2185 (2022)
6. Berg, S.: Condorcet's jury theorem, dependency among jurors. Soc. Choice Welfare **10**(1), 87–95 (1993)
7. Bhattacharya, S.: Preference monotonicity and information aggregation in elections. Econometrica **81**(3), 1229–1247 (2013)
8. Bhattacharya, S.: Condorcet jury theorem in a spatial model of elections. Ph.D. thesis, Indian Institute of Management Calcutta (2023)
9. Boland, P.J., Proschan, F., Tong, Y.L.: Modelling dependence in simple and indirect majority systems. J. Appl. Probab. **26**(1), 81–88 (1989)
10. Callander, S.: Majority rule when voters like to win. Games Econom. Behav. **64**(2), 393–420 (2008)
11. Chakraborty, A., Ghosh, P.: Efficient equilibria and information aggregation in common interest voting games. Tech. rep, Mimeo (2003)
12. Chan, J., Gupta, S., Li, F., Wang, Y.: Pivotal persuasion. J. Econ. Theory **180**, 178–202 (2019)
13. Condorcet, M.d.: Essai sur l'application de l'analyse à la probabilité des décisions rendues à la pluralité des voix. Paris: L'Imprimerie Royale (1785)
14. Coughlan, P.J.: In defense of unanimous jury verdicts: mistrials, communication, and strategic voting. Am. Polit. Sci. Rev. **94**(2), 375–393 (2000)

15. De Condorcet, N.: Essai sur l'application de l'analyse à la probabilité des décisions rendues à la pluralité des voix. Cambridge University Press (2014)
16. Duggan, J., Martinelli, C.: A Bayesian model of voting in juries. Games Econom. Behav. **37**(2), 259–294 (2001)
17. Ekmekci, M., Lauermann, S.: Manipulated electorates and information aggregation. Rev. Econ. Stud. **87**(2), 997–1033 (2020)
18. Feddersen, T., Pesendorfer, W.: Voting behavior and information aggregation in elections with private information. Econometrica: J. Econometric Soc., 1029–1058 (1997)
19. Feddersen, T., Pesendorfer, W.: Convicting the innocent: the inferiority of unanimous jury verdicts under strategic voting. Am. Polit. Sci. Rev. **92**(1), 23–35 (1998)
20. Gerardi, D.: Jury verdicts and preference diversity. Am. Polit. Sci. Rev. **94**(2), 395–406 (2000)
21. Gibbard, A.: Manipulation of voting schemes: a general result. Econometrica: J. Econometric Soc., 587–601 (1973)
22. Grofman, B., Owen, G., Feld, S.L.: Thirteen theorems in search of the truth. Theor. Decis. **15**(3), 261–278 (1983)
23. Han, Q., Schoenebeck, G., Tao, B., Xia, L.: The wisdom of strategic voting. In: Proceedings of the 24th ACM Conference on Economics and Computation, pp. 885–905. ACM (2023)
24. Hanson, R.D.: Decision markets (1999)
25. Hosseini, H., Mandal, D., Shah, N., Shi, K.: Surprisingly popular voting recovers rankings, surprisingly! In: IJCAI, pp. 245–251. ijcai.org (2021)
26. Kim, J., Fey, M.: The swing voter's curse with adversarial preferences. J. Econ. Theory **135**(1), 236–252 (2007)
27. Ladha, K.K.: The condorcet jury theorem, free speech, and correlated votes. Am. J. Polit. Sci., 617–634 (1992)
28. Ladha, K.K.: Condorcet's jury theorem in light of de Finetti's theorem: majority-rule voting with correlated votes. Soc. Choice Welfare **10**(1), 69–85 (1993)
29. Martinelli, C.: Convergence results for unanimous voting. J. Econ. Theory **105**(2), 278–297 (2002)
30. McLennan, A.: Consequences of the condorcet jury theorem for beneficial information aggregation by rational agents. Am. Polit. Sci. Rev. **92**(2), 413–418 (1998)
31. McMurray, J.: Ideology as opinion: a spatial model of common-value elections. Am. Econ. J.: Microecon. **9**(4), 108–140 (2017)
32. McMurray, J.C.: Aggregating information by voting: the wisdom of the experts versus the wisdom of the masses. Rev. Econ. Stud. **80**(1), 277–312 (2013)
33. Meirowitz, A.: Informative voting and condorcet jury theorems with a continuum of types. Soc. Choice Welfare **19**, 219–236 (2002)
34. Miller, N.R.: Information, electorates, and democracy: some extensions and interpretations of the condorcet jury theorem. Inf. Pooling Group Decis. Making **2**, 173–192 (1986)
35. Monroe, M., Waggoner, B.: Public projects with preferences and predictions. arXiv preprint: arXiv:2403.01042 (2024)
36. Myerson, R.B.: Extended poisson games and the condorcet jury theorem. Games Econom. Behav. **25**(1), 111–131 (1998)
37. Nitzan, S., Paroush, J.: Collective decision making: an economic outlook. CUP Archive (1985)
38. Othman, A., Sandholm, T.: Decision rules and decision markets. In: AAMAS, pp. 625–632 (2010)

39. Owen, G., Grofman, B., Feld, S.L.: Proving a distribution-free generalization of the condorcet jury theorem. Math. Soc. Sci. **17**(1), 1–16 (1989)
40. Persico, N.: Committee design with endogenous information. Rev. Econ. Stud. **71**(1), 165–191 (2004)
41. Pesendorfer, W., Feddersen, T.: The swing voter's curse. Am. Econ. Rev. **86**(3), 408–424 (1996)
42. Prelec, D.: A Bayesian truth serum for subjective data. Science **306**(5695), 462–466 (2004)
43. Prelec, D., Seung, H.S., McCoy, J.: A solution to the single-question crowd wisdom problem. Nature **541**(7638), 532–535 (2017)
44. Razin, R.: Signaling and election motivations in a voting model with common values and responsive candidates. Econometrica **71**(4), 1083–1119 (2003)
45. Satterthwaite, M.A.: Strategy-proofness and arrow's conditions: Existence and correspondence theorems for voting procedures and social welfare functions. J. Econ. Theory **10**(2), 187–217 (1975)
46. Schoenebeck, G., Tao, B.: Wisdom of the crowd voting: truthful aggregation of voter information and preferences. NeurIPS **34**, 1872–1883 (2021)
47. Wit, J.: Rational choice and the condorcet jury theorem. Games Econom. Behav. **22**(2), 364–376 (1998)

Full Papers

Price of Non-discrimination in Public Combinatorial Contracts

Yiding Feng[1], Mengfan Ma[2(✉)], and Mingyu Xiao[3]

[1] Hong Kong University of Science and Technology, Hong Kong SAR, China
ydfeng@ust.hk

[2] Central China Normal University, Wuhan, China
mengfanma1@gmail.com

[3] University of Electronic Science and Technology of China, Chengdu, China
myxiao@uestc.edu.cn

Abstract. This paper addresses the challenge of contract design within modern social platforms such as YouTube, Medium, and TikTok, where the primary goal is to develop monetization incentives for content creators. The inherent complexities of these platforms, influenced by factors such as regulation and scalability, often hinder the provision of personalized contracts, despite the diversity among content creators.

We model this challenge as a multi-agent combinatorial contract design problem in which the principal (e.g., a digital platform) delegates an identical task (e.g., video production) to agents (e.g., content creators) and motivates them through contracts. Each agent's payment depends solely on the success of his assigned task, and the principal gains utility upon each successful task. Consequently, an optimal contract may vary among agents due to their heterogeneity, while a public contract offers the same terms to all agents. The *price of non-discrimination (PoN)* captures the difference in the principal's utility when applying non-discriminatory (public) contracts versus customized (optimal) contracts for individual content creators. We analyze how the price of non-discrimination interacts with the combinatorial structure of agents' technologies, establishing tight and nearly tight bounds on the price of non-discrimination across various scenarios.

Keywords: Algorithmic Game Theory · Price of Non-discrimination · Combinatorial Contracts · Public Contracts

1 Introduction

Contract theory, a cornerstone of microeconomics, is pivotal in service markets and was recognized by the 2016 Nobel Prize in Economics, awarded to

The full version of this paper including all missing proofs can be found at https://papers.ssrn.com/sol3/papers.cfm?abstract_id=4983784.

M. Mavronicolas et al. (Eds.): WINE 2024, LNCS 15534, pp. 25–41, 2026.
https://doi.org/10.1007/978-3-032-08560-3_2

Hart and Holmström. Recently, contract theory has gained attention in the CS-Econ community for its natural connections to optimization theory and applications in economic markets, including crowdsourcing platforms [29], social media, blockchain-based smart contracts [13], and incentivizing quality healthcare [6].

In the classical hidden-action principal-agent contract model [24,30], the principal aims to incentivize a single agent to undertake a costly action that determines the distribution of potential rewards to the principal. The principal provides a monetary transfer, or "contract", based on the observable reward outcome rather than the unobservable action taken by the agent. Given the contract, the agent chooses the action that maximizes his expected utility. The principal's objective is to maximize the expected reward, minus the transfer paid to the agent.

Large organizations frequently engage numerous agents through standardized agreements that are uniformly applied across groups of individuals or entities. For instance, retail giants like Walmart and McDonald's employ standardized contracts for their hourly employees. These contracts typically encompass consistent terms regarding wages, working hours, benefits, and company policies, ensuring fairness and uniformity across employees at similar levels [36]. Likewise, digital platforms such as YouTube adopt similar practices by providing standardized monetization terms to content creators. This ensures that all creators, irrespective of their content type or technical skills, operate under the same policies and compensation structures, streamlining management and fostering a level playing field [37]. Furthermore, non-standardized contracts may pose challenges in terms of regulatory compliance and legal obligations. For example, the Equal Employment Opportunity Commission (EEOC) in the United States prohibits employment discrimination [34]. Utilizing standardized contracts helps reduce the risk of inadvertent discrimination. We refer to such contracts, applied consistently across all agents, as *public contracts.*

The recent research on combinatorial contracts [14,16,18,19] explores the hidden-action principal-agent model within a combinatorial framework, capturing the complex interdependencies among the agent's actions. The key innovation in this model is that the agent can select any subset of actions as his strategy, rather than being limited to a single action.[1] Combinatorial contracts play a central role in the combinatorial markets of services, much like how combinatorial auctions are crucial in combinatorial markets of goods [33].

In the single-agent combinatorial contract model, introduced by [18], the principal delegates a task to an agent, which results in one of two outcomes: success, where the principal receives a reward of 1, or failure, where no reward is received. This task can be accomplished by taking a set of underlying costly actions, $\mathcal{A}$, consisting of m actions. Each subset of actions $S \subseteq \mathcal{A}$, a.k.a., strategy,

[1] To illustrate how the strategy of the agent can be combinatorial, consider the example of video creators on YouTube. Their actions can include hiring actors, adding special effects and graphics, using high-quality equipment, researching content ideas and trends, promoting content on social media, and more. Each action incurs some cost, and creators choose a combination of these actions as their strategies.

induces a probability distribution over outcomes. When an agent selects a set S, it yields an expected reward $f(S)$, where $f(\cdot)$ represents a reward function defined over subsets of actions. In other words, if the agent performs a set of actions $S \subseteq \mathcal{A}$, then the probability of success for the task is given by $f(S)$, and the total cost incurred by the agent is $\sum_{a \in S} c_a$, where c_a denotes the cost of action a. Typically, the tuple $(\mathcal{A}, f, c)$ is referred to as the technology of the agents.

In this work, we focus on the problem of public contracts in multi-agent combinatorial contract design. Our model involves a principal contracting with each of n agents for an identical task. Each agent i has $(\mathcal{A}, f_i, c_i)$ as his own technology[2]. The payment for each agent depends solely on the success of his assigned task, and the principal receives an additive reward of 1 for each successful completion. The utility that the principal derives from a single agent is referred to as the *contribution* of that agent. Thus, a contract in our model essentially consists of n sub-contracts, each tailored to an individual agent. An optimal contract is one that maximizes the principal's overall utility. Given the heterogeneity in agents' technologies, the sub-contracts in an optimal contract may differ across agents. We define the *price of non-discrimination (PoN)* as the worst-case ratio between the utility that the principal can achieve under an optimal contract and the utility under the best public contract, where all agents are offered the same terms. We explore how the PoN is influenced by the combinatorial nature of agents' technologies, providing tight and nearly tight bounds on the PoN for a variety of different settings.

1.1 Our Contributions

Our results are summarized in Table 1. First, we show that in the general environment, the PoN can be as large as the number of agents n (see the second row in Table 1). The bound n is shown to be tight by a single action instance that demonstrates pathological structures in agents' technologies: the expected reward among agents spans from $\frac{1}{L}$, $\frac{1}{L^2}$ to $\frac{1}{L^n}$ and the cost among agents varies from 0, $\frac{L-1}{L^n}$ to $\frac{L^n-1}{L^n}$, where L is a large positive number. Considering the tasks assigned to agents are identical in our model, such significant disparities in technology are rare and impractical.

We then focus on settings where the technologies of agents demonstrate some degree of homogeneity. Specifically, we consider the following three natural and orthogonal constraints on agents' technologies: (1) identical f_i's and single-parameter c_i's; (2) identical and subadditive f_i's; (3) bounded welfare (resp. marginal welfare) and supermodular (resp. gross substitutes) f_i's. Agents' cost functions are *single-parameter* if each action $a \in \mathcal{A}$ requires an intrinsic amount of effort, denoted by $c_0(a)$, and each agent $i \in \mathcal{N}$ has a cost per unit-of-effort parameter, denoted by μ_i. Agent i's *welfare* for taking action a is defined

[2] We assume that all agents share the same action set, which is justified by the identical nature of the tasks assigned to them. However, our results still hold even if each agent $i \in \mathcal{N}$ has a distinct action set $\mathcal{A}_i$.

as[3] $f_i(a) - c_i(a)$ and the corresponding *marginal welfare* is $f_i(a|\mathcal{A}) - c_i(a)$, where $f_i(a|\mathcal{A}) = f_i(\mathcal{A}) - f_i(\mathcal{A} \setminus \{a\})$.

In environments where agents possess identical success probability functions f and single-parameter cost functions, we demonstrate a significant reduction in the PoN from n to $\ln n$ (refer to the third row in Table 1). This logarithmic ratio is proven to be tight and surprisingly robust concerning the combinatorial structure of f and the range of the cost parameters $\{\mu_i\}_{i=1}^n$. Notably, this result is achieved without imposing constraints on f and $\{\mu_i\}_{i=1}^n$.

Table 1. Summary of our results for the PoN in different environments, where n is the number of agents and m is the number of actions. The notation "N/A" means no constraints, and "Identical" means the success probability functions are the same for all agents.

Success Probability Functions	Cost Functions	PoN	References
N/A	N/A	n	Theorem 2
Identical	Single-Parameter	$\ln n$	Theorem 3
Identical & Subadditive	N/A	$\tilde{\Theta}(m) \cdot \ln n$	Theorem 4
Supermodular	$\forall a \in \mathcal{A}, \forall i \in \mathcal{N}, f_i(a) - c_i(a) \geq \frac{1}{W}$	$\Theta(\log W)$	Theorem 5
	$\exists a \in \mathcal{A}, \forall i \in \mathcal{N}, f_i(a) - c_i(a) \geq \frac{1}{W}$	$O(W \log W)$	Theorem 5
Gross Substitutes	$\forall a \in \mathcal{A}, \forall i \in \mathcal{N}, f_i(a|\mathcal{A}) - c_i(a) \geq \frac{1}{W}$	$\Theta(\log W)$	Theorem 5
	$\exists a \in \mathcal{A}, \forall i \in \mathcal{N}, f_i(a|\mathcal{A}) - c_i(a) \geq \frac{1}{W}$	$O(W \log W)$	Theorem 5

For settings with identical and subadditive f_i's, the PoN scales as $\tilde{\Theta}(m) \cdot \ln n$, an interesting outcome particularly when the action set is small. In this analysis, we do not assume single-parameter cost functions, revealing that even a relaxed assumption introduces only a multiplicative factor linear in m to the PoN for a broad class of complement-free success probability functions. This aspect is underscored by the exponential number of strategies, 2^m, available to agents. We derive these results by bucketing agents by their success probability functions and demonstrating that subadditivity ensures only $\tilde{O}(m)$ non-empty buckets exist, each reducible to the $\ln n$-approximation instance in previous single-parameter environment.

At last, we consider the case that the welfare of actions is bounded below and show that the PoN is capped by a logarithmic factor of the welfare lower bound if the reward function is supermodular or gross substitutes. These findings stem from a structural condition on agents' optimal personalized contracts, which ensures a logarithmic PoN. Using the characterization of agents' critical values in [18,20,35], we show that bounded welfare (resp. marginal welfare) alongside supermodular (resp. gross substitutes) success probability functions are sufficient to meet this condition.

We consider the above results to be rather positive. They show that the PoN can be bounded by a logarithmic factor of the number of agents or the welfare

[3] With slight abuse of notations, we simplify notation $f_i(\{a\})$ as $f_i(a)$.

lower bound, or a linear factor of the number of actions[4], with natural and practical assumptions on agent's technologies. This is a significant improvement over the worst-case scenario where the PoN is as large as the number of agents.

1.2 Related Work

There has been a long line of research on the contract design problem. In this subsection, we mainly discuss the four directions most relevant to our work.

Combinatorial Contracts. The exploration of algorithmic aspects of contract theory began with the works of [4,22], which considered models where agents take explicitly defined actions. Particularly relevant to this paper is the research conducted by [18,20,35], which investigates algorithmic contract design for a single agent in a combinatorial setting. They demonstrate the inherent difficulty associated with submodular rewards and provide polynomial-time algorithms for gross substitutes and supermodular rewards. Additionally, [16] showed that for subadditive success probability functions, the gap between the optimal welfare and the principal's utility scales logarithmically with the size of the strategy space.

Multi-agent Contract Design. This paper is also related to algorithmic contract design for multiple agents, which can be categorized into two general models. The first model, examined by [10], is particularly relevant: agents complete individual tasks, with each agent's payment based solely on their individual performance. Under mild regularity conditions, they show that an optimal contract can be computed for supermodular rewards, with an approximation achievable for submodular rewards. The second model of multi-agent contract design, discussed by [4,16,19], significantly differs in that each agent's payment depends on the outcomes of other agents, indicating the presence of externalities. This model applies to scenarios where the principal seeks to incentivize a team of agents to collaboratively complete a single task, such as projects with clearly defined divisions of labor. In this context, the principal's payment to each agent is tied to the collective outcomes of their teamwork and agent types. Thus, the variation in payments among agents in their model is well-understood and may not be perceived as discriminatory. In our setting, however, we assume no externalities among agents, justified by the identical and independent nature of the tasks assigned. This scenario is commonly encountered in practice, especially in crowdsourcing and social media platforms.

Menus of Contracts. The notion of PoN is also closely related to research on menus of contracts [9,26,27,32], where an agent may belong to one of τ private types, drawn from a known prior distribution, with technologies that may differ by type. The considered benchmarks are the single contract, the menu of contracts, and the "first-best" benchmark—the optimal welfare. The single contract, as usual, offers a fixed contract to the agent, while the menu of contracts specifies a contract for each type. Although the motivations differ,

[4] In practice, the number of agents is usually much larger than the number of actions.

there are some technical parallels between our model and the menu-of-contracts model: we can interpret the τ types as τ agents in our model and also view the single contract as the public contract and the menu of contracts as the personalized contract. Thus, a lower bound on the gap between single contracts and menus of contracts is a lower bound for the PoN, while an upper bound on the gap between single contracts and optimal welfare bounds the PoN. In this direction, [26] demonstrated that the approximation ratio between the best single contract and the best menu of contracts is $\Omega(\max\{M, \log \tau\})$, while [27] showed that the ratio between the best single contract and the optimal welfare is $\Theta(M \log \tau)$, where M represents the number of strategies available to the agent. The key difference in our work is that previous research on menus of contracts does not consider the combinatorial structure of agents' technologies, which is central to our study. Notably, the number of strategies M in our model is 2^m, yet we still achieve a linear dependence on m in the PoN for identical and subadditive success probability functions. For single-parameter agents with identical success probability functions, we can achieve a tight bound of $\ln n$.

Linear Contracts. Another related line of research concerns the comparison between linear contracts and optimal contracts, as explored in [2,8,21,27]. Specifically, [21] studied linear contracts from a worst-case approximation perspective, establishing several tight parameterized approximation bounds, including linear dependence on the number of strategies and logarithmic dependence on the max-min reward ratio and cost ratio.

Lastly, and somewhat less related, single-parameter agents are analyzed in [2,3,7], while smaller action spaces are considered in [5] within an online learning framework. Furthermore, learning algorithms for contract design have also been studied in, e.g., [12,25,28,29,38].

2 Preliminaries

In this section, we first present the combinatorial contract design model. We then introduce the multi-agent setting and the concept of the price of non-discrimination. Finally, we discuss the structural results of the optimal personalized (resp. public) contracts, which are used in the later sections.

2.1 Single-Agent Combinatorial Contract Design

Our problem is built on the single-agent combinatorial contract design model introduced in [17]. In this model, a principal incentivizes an agent to complete a task, whose outcome is either success or failure.

Combinatorial Strategy, Reward Function, and Cost Function. There are m possible actions@ to accomplish the task, represented by the action set $\mathcal{A} = \{1, 2, \ldots, m\}$. The agent can adopt any subset of $\mathcal{A}$ as his *strategy* S, which forms the strategy set $2^{\mathcal{A}}$. After selecting a strategy $S \subseteq \mathcal{A}$, the task stochastically results in either success or failure. Success provides the principal with an additive reward of 1, while failure yields a reward of 0. We denote by set functions $f : 2^{\mathcal{A}} \to [0,1]$ and $c : 2^{\mathcal{A}} \to \mathbb{R}_{\geq 0}$ as the *reward function* and *cost function*, respectively. Specifically, for each strategy $S \subseteq \mathcal{A}$, $f(S)$ is the success probability of the task and $c(S)$ is the monetary cost spent by the agent. We refer to $(\mathcal{A}, f, c)$ as the *technology* of the agent.

Following the literature, we assume that the cost function c is *additive*, i.e., $c(S) = \sum_{a \in S} c(\{a\})$.[5] We assume that the reward function f is weakly increasing[6] and satisfies $f(\emptyset) = 0$. Moreover, we consider the following four important classes of reward functions.

- **(Gross substitutes)** We say reward function f is *gross substitutes* (GS) if for any two price vectors $\boldsymbol{p}, \boldsymbol{q} \in \mathbb{R}^m_{\geq 0}$ with $\boldsymbol{q} \geq \boldsymbol{p}$ (point-wise) and any $S \subseteq \mathcal{A}$ such that $(f(S) - \sum_{a \in S} p_a) \in \arg\max_{S^\dagger \subseteq \mathcal{A}} (f(S^\dagger) - \sum_{a \in S^\dagger} p_a)$, there is a $T \subseteq \mathcal{A}$ such that $(f(T) - \sum_{a \in S} q_a) \in \arg\max_{T' \subseteq \mathcal{A}} (f(T') - \sum_{a \in T'} q_a)$ and $T \supseteq \{a \in S : q_a = p_a\}$.
- **(Submodular)** We say reward function f is *submodular* if for any $S, T \subseteq \mathcal{A}$ such that $S \subseteq T$ and any $a \in \mathcal{A}$, $f(a|S) \geq f(a|T)$. Here $f(a|S) = f(S \cup \{a\}) - f(S \backslash \{a\})$ is the marginal value of action $a \in \mathcal{A}$ given strategy $S \subseteq \mathcal{A}$. An equivalent definition is that for any $S, T \subseteq \mathcal{A}$, $f(S) + f(T) \geq f(S \cup T) + f(S \cap T)$.
- **(Subadditive)** We say reward function f is *subadditive* if for any strategy $S, T \subseteq \mathcal{A}$, $f(S) + f(T) \geq f(S \cup T)$.
- **(Supermodular)** We say reward function f is *supermodular* if for any strategy $S, T \subseteq \mathcal{A}$ such that $S \subseteq T$ and any $a \in \mathcal{A}$, $f(a|S) \leq f(a|T)$. An equivalent definition is that for any $S, T \subseteq \mathcal{A}$, $f(S) + f(T) \leq f(S \cup T) + f(S \cap T)$.

The first three classes are considered as complement-free classes in the literature [15,31], where the containment is strict. The last class is used to characterize complementary actions [1,11,23].

Contract. To incentivize the agent to take costly strategies, the principal offers an outcome-based contract specified by $\alpha \in [0,1]$. In particular, the agent receives no payment if the task fails and receives α if the task succeeds. The constraint $\alpha \geq 0$ is consistent with the standard assumption of *limited liability*, meaning payments flow only from the principal to the agent. Furthermore, since the maximum reward for a task is 1, we impose $\alpha \leq 1$ as well.

Both the principal and agent are risk-neutral and have quasi-linear utilities. Given any contract $\alpha \in [0,1]$ and strategy $S \subseteq \mathcal{A}$, the expected utility of the principal (resp. agent) is $u(\alpha, S) = (1-\alpha)f(S)$ (resp. $\pi(\alpha, S) = \alpha f(S) - c(S)$). Facing a given contract α, the agent always chooses a best-responding strategy $S \in \arg\max_{S \subseteq \mathcal{A}} \pi(\alpha, S)$ that maximizes his expected utility and breaks tie in favor of the principal. We denote by $\mathcal{D}(\alpha)$ as the set of aforementioned best-responding strategies for contract α. With slight abuse of notations, we also

[5] When it is clear from the context, with slight abuse of notations, we simplify $c(\{a\})$ and $f(\{a\})$ as $c(a)$ and $f(a)$, respectively.

[6] A function f is weakly increasing if for any $S \subseteq T \subseteq \mathcal{A}$, $f(S) \leq f(T)$.

use $u(\alpha) \triangleq u(\alpha, S^*)$ and $\pi(\alpha) \triangleq \pi(\alpha, S^*)$, where $S^* \in \mathcal{D}(\alpha)$, to denote the principal and agent's expected utility under contract α. Finally, we denote by $V(\alpha) \triangleq f(S^*)$, where $S^* \in \mathcal{D}(\alpha)$, as the success probability (a.k.a., the expected reward) under contract α.

2.2 Multi-agent Setting and Price of Non-discrimination

Inspired by the applications in modern marketplaces, we consider the following generalization of the combinatorial contract design problem from the single-agent environments to the multi-agent environments. In this multi-agent model, a principal incentivizes each of n agents, denoted by $\mathcal{N}$, independently to complete an identical task.

For each agent $i \in \mathcal{N}$, we denote by f_i and c_i as his reward and cost function. We assume the success of agent i's task only depends on the strategy S_i selected by agent i and is independent of all other agents' strategies or task outcomes. Similar to f_i and c_i, we also use agent index i for notations $\pi_i, u_i, \mathcal{D}_i, V_i$ as well.

Due to this independence, the principal offers contracts α_i to each agent i separately. The goal of the principal is to design a profile contract $\boldsymbol{\alpha} = \{\alpha_i\}_{i \in \mathcal{N}}$ to maximize her total expected utility $U(\boldsymbol{\alpha}) \triangleq \sum_{i=1}^{n} u_i(\alpha_i)$. In this work, we are interested in the comparison between the following two types of contracts:

- **(Optimal personalized contract)** We say contract profile $\boldsymbol{\alpha}^* = \{\alpha_i^*\}_{i \in \mathcal{N}}$ is the *optimal personalized contract* if for each agent $i \in \mathcal{N}$, contract $\alpha_i^* \triangleq \arg\max_{\alpha_i \in [0,1]} u_i(\alpha_i)$ maximizes the principal's utility from this agent's task. That is, $\boldsymbol{\alpha}^* = (\alpha_1^*, \ldots, \alpha_n^*)$. We denote by $U^* \triangleq U(\boldsymbol{\alpha}^*)$ as the principal's total utility under the optimal personalized contract, and $u_i^* \triangleq u_i(\alpha_i^*)$ as the optimal utility of the principal from agent i.
- **(Optimal public contract)** We say contract profile $\boldsymbol{\alpha} = (\alpha^*)^{\otimes n}$ is the *optimal public contract* if contract $\alpha^* = \arg\max_{\alpha \in [0,1]} \sum_{i \in \mathcal{N}} u(\alpha)$ is a (single-valued) public contract that maximizes the principal's total utility. We denote by $\Pi(\alpha) \triangleq \sum_{i \in \mathcal{N}} u_i(\alpha)$ as the total utility of the principal under public contract α, and denote by $\Pi^* \triangleq \Pi(\alpha^*)$ as the principal's total utility under the optimal public contracts.

Since agents have heterogeneous reward and cost functions, the principal usually gains a strictly higher utility U^* via the optimal personalized contract than the utility Π^* via the optimal public contract.

Price of Non-discrimination. Due to various factors such as regulation and scalability, a principal in many real-world marketplaces may only allow the use of public contracts. For example, personalized contracts may be viewed as discrimination against agents. Motivated by this, we introduce the *price of non-discrimination (PoN)*, which is defined as the worst-case ratio between the utility the principal can gain under the optimal personalized contract and under the optimal public contract. Specifically, for a given instance $I = \{\mathcal{N}, \mathcal{A}, \{f_i\}_{i=1}^{n}, \{c_i\}_{i=1}^{n}\}$, we define the price of non-discrimination $\rho(I)$ for this instance as

$$\rho(I) \triangleq \frac{U^*(I)}{\Pi^*(I)}.$$

where we use $U^*(I)$ and $\Pi^*(I)$ to make explicit the dependence of the principal's utility on the model parameters: number of agents, number of actions, types of reward functions, and cost functions. The main objective of this paper is to study how this ratio performs across a wide range of scenarios (denoted by $\mathcal{I}$):

$$\rho \triangleq \sup_{I \in \mathcal{I}} \rho(I)$$

2.3 Structural Results of the Optimal Contract

In this subsection, we introduce some preliminary structural results about the optimal personalized contract and the optimal public contract, both of which are used in the later analysis.

Optimal personalized contract. We first introduce some insights regarding the structure of the optimal personalized contract of an arbitrary agent $i \in \mathcal{N}$. Since here we focus on the single-agent problem, we drop agent index i to simplify the notation.

Given contract $\alpha \in [0,1]$ and strategy $S \subseteq \mathcal{A}$, the agent's utility is $\pi(\alpha, S) = \alpha f(S) - c(S)$, which is an affine function of α. There are a total of 2^m such functions, one for each strategy $S \subseteq \mathcal{A}$, as shown in Fig. 1(a). Let $S_\alpha \in \mathcal{D}(\alpha)$ be the agent best-responding strategy under contract α_i. Then the optimal utility of the agent is $\pi(\alpha) = \alpha f(S_\alpha) - c(S_\alpha)$, which is the upper envelope of these 2^m affine functions. As such, it is continuous, monotonically non-decreasing, convex, continuous, and piecewise linear, as illustrated by pink line segments in Fig. 1(b).

Now we consider the utility of the principal (aka., the contribution from the agent) given contract α. It is $u(\alpha) = (1-\alpha)f(S_\alpha) = (1-\alpha)V(\alpha)$, which is illustrated by blue line segments in Fig. 1(b). As we can observe, it is a piecewise linear function that can be non-continuous and non-monotonic as a whole but is piecewise monotonically non-increasing. The x-intercepts of these linear functions are all 1. The y-intercepts and the slopes of the linear functions of u are $V(\alpha)$ and $-V(\alpha)$, respectively.

Observation 1. *For a single agent and any two contract $\alpha, \alpha^\dagger \in (0,1]$, it satisfies that (1) $\mathcal{D}(\alpha) = \mathcal{D}(\alpha^\dagger)$ if $V(\alpha) = V(\alpha^\dagger)$; and (2) $\mathcal{D}(\alpha) \cap \mathcal{D}(\alpha^\dagger) = \emptyset$ if $V(\alpha) \neq V(\alpha^\dagger)$.*

Next lemma establishes a monotonicity property of the reward as a function of contract α.

Lemma 1 (Proposition III.1 and Corollary III.2 in [17]). *For a single agent and any two contract $\alpha, \alpha^\dagger \in [0,1]$ with $\alpha \leq \alpha^\dagger$, it satisfies that $V(\alpha) \leq V(\alpha^\dagger)$. Moreover, for every strategy $S \in \mathcal{D}(\alpha)$ and strategy $S^\dagger \in \mathcal{D}(\alpha^\dagger)$, $f(S) \leq f(S^\dagger)$.*

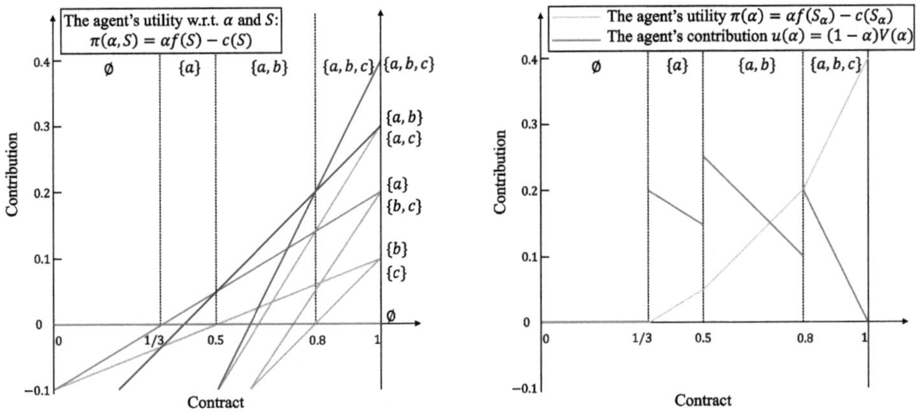

(a) The agent's utility as a function of contract and strategy.

(b) The agent's utility and contribution as a function of contract.

Fig. 1. In this example of a single agent, $\mathcal{A} = \{a, b, c\}$, f is additive with $f(a) = 0.3$, $f(b) = 0.2$, $f(c) = 0.5$. The cost function is $c(a) = 0.1$, $c(b) = 0.1$, $c(c) = 0.4$. The critical set are $\{\frac{1}{3}, 0.5, 0.8\}$. Agent i's best-responding set is $\emptyset$, $\{a\}$, $\{a, b\}$, $\{a, b, c\}$ for contracts in $[0, \frac{1}{3})$, $[\frac{1}{3}, 0.5)$, $[0.5, 0.8)$ and $[0.8, 1]$, respectively.

Based on the fact that agents break ties in favor of the principal, we next observe that V_i is right-continuous.

Lemma 2 (Observation III.3 in [17]). *For a single agent and any contract* $\alpha \in [0, 1]$, $\lim_{\epsilon \to 0^+} V(\alpha + \epsilon)$ *and* $\lim_{\epsilon \to 0^+} V(\alpha - \epsilon)$ *are well-defined. Moreover,* $V(\alpha) = \lim_{\epsilon \to 0^+} V(\alpha + \epsilon)$ *and thus the principal's utility* $u(\alpha) = (1 - \alpha)V(\alpha)$ *is also right-continuous*

Next, we capture the breakpoints of V or u. There exists $w \in [2^m - 1]$ and a series of α values $0 = \alpha^{(0)} < \alpha^{(1)} < \cdots < \alpha^{(w)} \le 1$ such that for every $\alpha \in [0, 1]$, $V(\alpha) = V(\alpha^{(k)})$, where $k = \max\{k \in [w] : \alpha^{(k)} \le \alpha\}$. Moreover, for every two $k, k^\dagger \in [w]$ with $k < k^\dagger$, $V(\alpha^{(k)}) < V_i(\alpha^{(k^\dagger)})$. Every $\alpha^{(k)}$ (except for $\alpha^{(0)}$) is said to be a *critical* value of the agent. We denote by $\mathcal{C} = \{\alpha^{(k)}\}_{k \in [w]}$ as the *critical set* which contains all critical values. As a sanity check, critical set $\mathcal{C}$ can be formally defined as

$$\mathcal{C} \triangleq \{\alpha \in (0, 1] \, : \, V(\alpha) \neq V(\alpha - \epsilon) \text{ for any small } \epsilon > 0\}.$$

At each critical value, the principal's utility u only jumps to a higher value, and the slope of u becomes steeper, which is observable in Fig. 1(b).

Optimal Public Contract. Now we discuss the structural insights about the optimal public contract. Recall that u_i is the principal's utility from agent i's task (aka., agent i's contribution), which is piecewise linear with jump discontinuities at agent i's critical values. Therefore, the principal's utility function under a public contract, which is the sum of all agents' contributions, is also piecewise linear, with jump discontinuities at the critical values of all agents. Then the optimal public contract must be one of the critical values of all agents, or 0.

Lemma 3. *The optimal public contract* $\alpha^* \in \{0\} \cup \bigcup_{i \in \mathcal{N}} \mathcal{C}_i$.

Proof. Let $\mathcal{C} = \bigcup_{i=1}^{n} \mathcal{C}_i$ be the set of all critical values of all agents. Let $\alpha^\dagger, \alpha^\ddagger \in \mathcal{C}$ be two neighboring critical values such that open interval $(\alpha^\dagger, \alpha^\ddagger) \cap \mathcal{C} = \emptyset$. By construction, we also have $(\alpha^\dagger, \alpha^\ddagger) \cap \mathcal{C}_i = \emptyset$ for each $i \in \mathcal{N}$. Thus, for any $\alpha \in [\alpha^\dagger, \alpha^\ddagger)$, any agent i's best-responding set $\mathcal{D}_i(\alpha)$ is an invariant. Take any $S \in \mathcal{D}_i(\alpha)$. Then agent i's contribution, $u_i(\alpha) = (1-\alpha) f_i(S)$, is increasing when α is decreasing in the interval $[\alpha^\dagger, \alpha^\ddagger)$. Therefore, $u_i(\alpha^\dagger) > u_i(\alpha)$ and public contract $\alpha^\dagger$ is weakly better than contract α. □

Though it is not the main focus of this paper, we also consider the computational complexity of finding an optimal public contract. The detailed results can be found in the full version.

Proposition 1. *For instances where all agents have gross substitutes or supermodular reward functions, the optimal public contract* α^* *can be computed in polynomial time. For instances where at least one agent has a submodular reward function, it is NP-hard to compute the optimal public contract.*

3 Price of Non-discrimination Characterization

In this section, we first consider the PoN in the general environment and show that the PoN can be as large as the number of agents. We then investigate how the PoN degrades when agents' technologies are subject to the constraints of single-parameter cost functions, small action sets, and bounded welfare. All missing proofs can be founded in the online full version of this paper.

3.1 General Environment

We show in this subsection that without any constraint on agents' technologies, the PoN can be as large as the number of agents n, even for instances with a singleton action set $\mathcal{A} = \{a\}$.

Given that the principal's utility is the summation of n agents' contributions, it is easy to see that $\rho \leq n$: the principal can simply offer the public contract that is the optimal personalized contract of the agent with the largest optimal personalized contribution. Next, we show this upper bound is tight by a specific instance I_{bad}, illustrated in Fig. 2.

Theorem 2. *For* n *agents with arbitrary reward functions* $\{f_i\}_{i=1}^{n}$, *and any additive cost functions* $\{c_i\}_{i=1}^{n}$, *the PoN satisfies* $\rho = n$. *Moreover, for every* n *and* $\epsilon > 0$, *there exists an instance* I_{bad} *with singleton action set (i.e.,* $m = 1$*) such that* $\rho(I_{\text{bad}}) \geq n - \epsilon$.

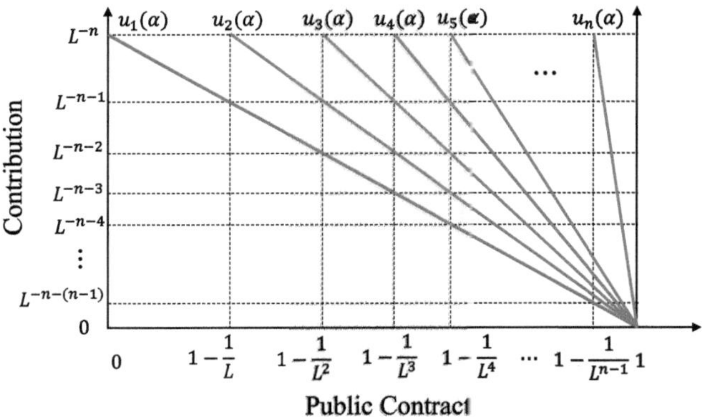

Fig. 2. The instance I_{bad} with $\rho = n$ in Theorem 2. Each blue line denotes the contribution curve of each agent.

The intuition behind the problematic instance I_{bad} is that for the optimal public contract, which is the optimal personalized contract of an agent, the principal achieves satisfactory utility from this agent but receives only a minuscule utility from the others. This large disparity arises from the pathological structure in agents' technologies: the expected reward of a (the slope of lines in Fig. 2) among agents spans from $\frac{1}{L}, \frac{1}{L^2}$ to $\frac{1}{L^n}$ and the cost of a among agents varies from $0, \frac{L-1}{L^n}$ to $\frac{L^n-1}{L^n}$. Besides, the welfare of agents is $\frac{1}{L^n}$, which can be arbitrarily small. Considering the tasks assigned to agents are identical in our model, such significant disparities in technology for the same task are rare and impractical. Disparities among content creators or freelancers undertaking the same task should not exhibit such dramatic variations.

In the subsequent subsections, we focus on settings where the technologies of agents demonstrate some degree of homogeneity. Specifically, we consider the following three natural and orthogonal constraints on agents' technologies: (1) identical f_i's and single-parameter c_i's; (2) identical and subadditive f_i's; (3) bounded welfare (resp. marginal welfare) and supermodular (resp. gross substitutes) f_i's.

3.2 Single-parameter Cost Function Environment

In this subsection, our focus is on the scenario where agents have identical reward function f and single-parameter cost functions $\{c_i\}_{i \in \mathcal{N}}$. Specifically, the agents' cost functions are *single-parameter* if each action $a \in \mathcal{A}$ requires a fixed intrinsic effort cost, denoted by $c_0(a)$, and each agent $i \in \mathcal{N}$ has a personal cost-per-unit-of-effort parameter (hereafter, cost parameter), denoted by μ_i. Formally, the cost function for each agent i is given by $c_i(a) = \mu_i \cdot c_0(a)$ for every action $a \in \mathcal{A}$.

We demonstrate that the PoN experiences a degradation from n to $\ln n$. This logarithmic ratio is proven to be tight and surprisingly robust concerning the combinatorial structure of reward function f and the range of the cost parameters $\{\mu_i\}_{i=1}^n$. The main result in this subsection is the following theorem:

Theorem 3. *For n agents with identical reward function f and single-parameter cost functions $\{c_i\}_{i\in\mathcal{N}}$, the PoN $\rho = \mathcal{H}_n$, where $\mathcal{H}_n = \sum_{i\in\mathcal{N}} \frac{1}{i}$ is the n-th harmonic number. Moreover, setting public contract as a certain agent's optimal personalized contract suffices to achieve the $\mathcal{H}_n$-approximation to the optimal utility.*

We first provide some intuition for the proof. Recall that the reason the PoN in the bad instance is n is that the optimal public contract only guarantees a high contribution from one agent, while the rest of the agents have low contributions, as shown in Fig. 2.

However, for agents with single-parameter cost functions, the optimal personalized contract α_i^* of agent i is guaranteed to achieve a contribution of at least u_i^* from any agent with a lower cost parameter, where u_i^* is the optimal personalized contribution of agent i (formalized in Lemma 4). Assuming the cost parameters are ordered such that $\mu_1 \leq \ldots \leq \mu_n$, the total contribution from agents in $[i]$ is at least $i \cdot u_i^*$. Let the public contract be the α_i^* such that $i \cdot u_i^*$ is the largest one among $\{u_1^*, 2\cdot u_2^*, \ldots, n\cdot u_n^*\}$ and we get $(\ln n)$-approximation to the optimal utility $U^* = \sum_{i\in\mathcal{N}} u_i^*$. See Fig. 3 for the lower bound instance.

Lemma 4. *Let $i, j \in \mathcal{N}$ be two agents with single-parameter cost functions and $\mu_i < \mu_j$. For i's any critical value $\alpha \in \mathcal{C}_i$, we have $u_i(\alpha) \geq u_j(\alpha)$.*

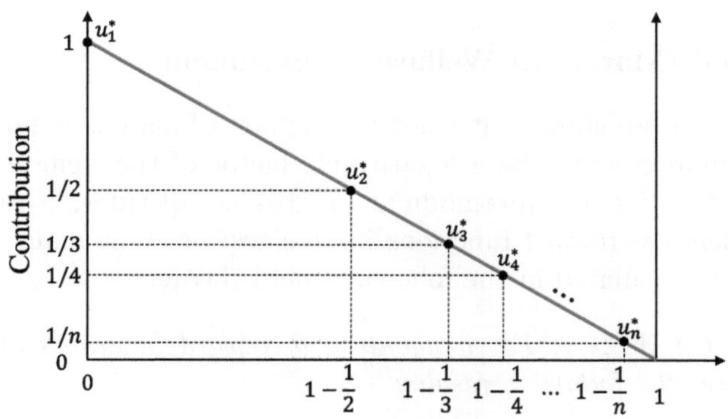

Fig. 3. The instance with $\rho = \mathcal{H}_n$ of Theorem 3. The black dots denote the point where each agent reach its optimal personalized contribution. Note that all agents' contribution function $u_i(\alpha)$'s share the same slope and the non-zero part of $u_i(\alpha)$ is the line segment between $(1-\frac{1}{i}, u_i^*)$ and $(1, 0)$. Thus by taking public contract $1-\frac{1}{i}$, the principal can achieve a utility of iu_i^*.

3.3 Small Action Set Environment

In this subsection, we demonstrate a result that is particularly interesting when the number of actions is small: for identical subadditive reward function f, the PoN is $\tilde{\Theta}(m) \cdot \ln n$.

Note that we do not make the assumption that agents have single-parameter cost functions in this subsection. Our result indicates that, for a large class of functions characterized as complement-free, the relaxation of the single-parameter cost function assumption only introduces a multiplicative factor linear in m to the PoN. This is particularly noteworthy considering the exponential number of strategies 2^m available to the agents.

Theorem 4. *For n agents with identical subadditive reward function f and m actions, the PoN ρ is $O(\frac{\log^2 m}{\log\log m} m) \cdot \ln n$. Moreover, exists an instance I with n agents, identical subadditive reward function f and m actions such that the PoN $\rho(I) = \tilde{\Omega}(m) \cdot \ln n$.*

Here we sketch the intuition behind the result. First, to deal with the exponential number of strategies, we bucket the strategies based on their reward. The sizes of these buckets follow a geometric progression with common ratio $(\log m)^{-1}$. By utilizing the subadditivity and monotonicity of the reward function, we can show that the number of non-empty buckets is $\tilde{O}(m)$. Next, considering the restricted instance in each bucket, we demonstrate that it can be reduced to the $\ln n$-approximation in Theorem 3, while only incurring a multiplicative loss of $\tilde{O}(m)$. As a result, we obtain an upper bound of $\tilde{O}(m) \cdot \ln n$ on the PoN.

3.4 Bounded (Marginal) Welfare Environment

In this subsection, we show that when the welfare of actions is bounded below, the PoN is bounded above by a logarithmic factor of the welfare lower bound if the reward function is supermodular or gross substitutes. Note that we do not require identical reward functions in this section. Our main result in this subsection is encapsulated in the following meta-theorem:

Theorem 5. *Let $W > 1$. For n agents with reward functions $\{f_i\}_{i=1}^n$ and m actions, we have the following results:*

- *If all reward functions f_i are supermodular and for any agent $i \in \mathcal{N}$ and any action $a \in \mathcal{A}$, the welfare $f(a) - c_i(a) \geq \frac{1}{W}$, then the PoN $\rho = O(\log W)$;*
- *If all reward functions f_i are supermodular and there exists an action $a \in \mathcal{A}$ such that the welfare $f(a) - c_i(a) \geq \frac{1}{W}$ for any agent $i \in \mathcal{N}$, then the PoN $\rho = O(W \log W)$;*
- *If all reward functions f_i are gross substitutes and for any agent $i \in \mathcal{N}$ and any action $a \in \mathcal{A}$, the marginal welfare $f(a|\mathcal{A}) - c_i(a) \geq \frac{1}{W}$, then the PoN $\rho = O(\log W)$;*

- *If all reward functions f_i are gross substitutes and there exists an action $a \in \mathcal{A}$ such that the welfare $f(a|\mathcal{A}) - c_i(a) \geq \frac{1}{W}$ for any agent $i \in \mathcal{N}$, then the PoN $\rho = O(W \log W)$.*

To prove the above results, we first identify a structural condition on agents' optimal personalized contracts that guarantees a logarithmic PoN. Then, we show that for supermodular (resp. gross substitutes) reward functions, bounded welfare (resp. marginal welfare) implies that such a condition is satisfied.

References

1. Abraham, I., Babaioff, M., Dughmi, S., Roughgarden, T.: Combinatorial auctions with restricted complements. In: Proceedings of the 13th ACM Conference on Electronic Commerce, EC '12, pp. 3–16. Association for Computing Machinery, New York (2012). https://doi.org/10.1145/2229012.2229016
2. Alon, T., Dütting, P., Li, Y., Talgam-Cohen, I.: Bayesian analysis of linear contracts. In: Proceedings of the 24th ACM Conference on Economics and Computation, EC '23, p. 6. Association for Computing Machinery, New York (2023). https://doi.org/10.1145/3580507.3597795
3. Alon, T., Dütting, P., Talgam-Cohen, I.: Contracts with private cost per unit-of-effort. In: Proceedings of the 22nd ACM Conference on Economics and Computation, EC '21, pp. 52–69. Association for Computing Machinery, New York (2021). https://doi.org/10.1145/3465456.3467651
4. Babaioff, M., Feldman, M., Nisan, N.: Combinatorial agency. In: Proceedings of the 7th ACM Conference on Electronic Commerce, pp. 18–28 (2006)
5. Bacchiocchi, F., Castiglioni, M., Marchesi, A., Gatti, N.: Learning optimal contracts: how to exploit small action spaces. In: The Twelfth International Conference on Learning Representations (2024). https://openreview.net/forum?id=WKuimaBj4I
6. Bastani, H., Bayati, M., Braverman, M., Gummadi, R., Johari, R.: Analysis of medicare pay-for-performance contracts. Available at SSRN 2839143 (2016)
7. Castiglioni, M., Chen, J., Li, M., Xu, H., Zuo, S.: A reduction from multi-parameter to single-parameter Bayesian contract design. arXiv preprint: arXiv:2404.03476 (2024)
8. Castiglioni, M., Marchesi, A., Gatti, N.: Bayesian agency: linear versus tractable contracts. In: Proceedings of the 22nd ACM Conference on Economics and Computation, EC '21, pp. 285–286. Association for Computing Machinery, New York (2021). https://doi.org/10.1145/3465456.3467602
9. Castiglioni, M., Marchesi, A., Gatti, N.: Designing menus of contracts efficiently: the power of randomization. In: Proceedings of the 23rd ACM Conference on Economics and Computation, EC '22, pp. 705–735 (2022)
10. Castiglioni, M., Marchesi, A., Gatti, N.: Multi-agent contract design: how to commission multiple agents with individual outcomes. In: Proceedings of the 24th ACM Conference on Economics and Computation, EC '23, pp. 412–448. Association for Computing Machinery, New York (2023). https://doi.org/10.1145/3580507.3597793

11. Chen, W., Teng, S., Zhang, H.: Capturing complementarity in set functions by going beyond submodularity/subadditivity. In: Blum, A. (ed.) 10th Innovations in Theoretical Computer Science Conference, ITCS 2019, 10–12 January 2019, San Diego, California, USA. LIPIcs, vol. 124, pp. 1–20. Schloss Dagstuhl - Leibniz-Zentrum für Informatik (2019). https://doi.org/10.4230/LIPICS.ITCS.2019.24
12. Chen, Y., Chen, Z., Deng, X., Huang, Z.: Are bounded contracts learnable and approximately optimal? arXiv preprint: arXiv:2402.14486 (2024)
13. Cong, L.W., He, Z.: Blockchain disruption and smart contracts. Rev. Finan. Stud. **32**(5), 1754–1797 (2019)
14. Deo-Campo Vuong, R., Dughmi, S., Patel, N., Prasad, A.: On supermodular contracts and dense subgraphs. In: Proceedings of the 2024 Annual ACM-SIAM Symposium on Discrete Algorithms (SODA), pp. 109–132. SIAM (2024)
15. Dobzinski, S., Nisan, N., Schapira, M.: Approximation algorithms for combinatorial auctions with complement-free bidders. Math. Oper. Res. **35**(1), 1–13 (2010) https://doi.org/10.1287/moor.1090.0436
16. Duetting, P., Ezra, T., Feldman, M., Kesselheim, T.: Multi-agent combinatorial contracts. arXiv preprint: arXiv:2405.08260 (2024)
17. Dütting, P., Ezra, T., Feldman, M., Kesselheim, T.: Combinatorial contracts. In: 2021 IEEE 62nd Annual Symposium on Foundations of Computer Science, FOCS '21, pp. 815–826. IEEE Computer Society, Los Alamitos (2022). https://doi.org/10.1109/FOCS52979.2021.00084
18. Dütting, P., Ezra, T., Feldman, M., Kesselheim, T.: Combinatorial contracts. In: 2021 IEEE 62nd Annual Symposium on Foundations of Computer Science (FOCS), pp. 815–826. IEEE (2022)
19. Dütting, P., Ezra, T., Feldman, M., Kesselheim, T.: Multi-agent contracts. In: Proceedings of the 55th Annual ACM Symposium on Theory of Computing, STOC '23, pp. 1311–1324. Association for Computing Machinery, New York (2023). https://doi.org/10.1145/3564246.3585193
20. Dütting, P., Feldman, M., Tzur, Y.G.: Combinatorial contracts beyond gross substitutes. In: Proceedings of the 2024 Annual ACM-SIAM Symposium on Discrete Algorithms (SODA), pp. 92–108. Society for Industrial and Applied Mathematics (2024). https://doi.org/10.1137/1.9781611977912.5
21. Dütting, P., Roughgarden, T., Talgam-Cohen, I.: Simple versus optimal contracts. In: Proceedings of the 20th ACM Conference on Economics and Computation, EC '19, pp. 369–387. Association for Computing Machinery, New York (2019). https://doi.org/10.1145/3328526.3329591
22. Feldman, M., Chuang, J., Stoica, I., Shenker, S.: Hidden-action in network routing. IEEE J. Sel. Areas Commun. **25**(6), 1161–1172 (2007)
23. Feldman, M., Friedler, O., Morgenstern, J., Reiner, G.: Simple mechanisms for agents with complements. In: Proceedings of the 2016 ACM Conference on Economics and Computation, EC '16, pp. 251–267. Association for Computing Machinery, New York (2016). https://doi.org/10.1145/2940716.2940735
24. Grossman, S.J., Hart, O.D.: An analysis of the principal-agent problem. In: Foundations of Insurance Economics: Readings in Economics and Finance, pp. 302–340. Springer (1992)
25. Guruganesh, G., et al.: Contracting with a learning agent. arXiv preprint: arXiv:2401.16198 (2024)
26. Guruganesh, G., Schneider, J., Wang, J., Zhao, J.: The power of menus in contract design. In: Proceedings of the 24th ACM Conference on Economics and Computation, EC '23, pp. 818–848. Association for Computing Machinery, New York (2023). https://doi.org/10.1145/3580507.3597735

27. Guruganesh, G., Schneider, J., Wang, J.R.: Contracts under moral hazard and adverse selection. In: Proceedings of the 22nd ACM Conference on Economics and Computation, pp. 563–582 (2021)
28. Han, M., Albert, M., Xu, H.: Learning in online principal-agent interactions: the power of menus. In: Proceedings of the AAAI Conference on Artificial Intelligence, vol. 38, no. 16, pp. 17426–17434 (2024). https://doi.org/10.1609/aaai.v38i16.29691
29. Ho, C.J., Slivkins, A., Vaughan, J.W.: Adaptive contract design for crowdsourcing markets: Bandit algorithms for repeated principal-agent problems. In: Proceedings of the Fifteenth ACM Conference on Economics and Computation, pp. 359–376 (2014)
30. Holmström, B.: Moral hazard and observability. Bell J. Econ., 74–91 (1979)
31. Lehmann, B., Lehmann, D., Nisan, N.: Combinatorial auctions with decreasing marginal utilities. In: Proceedings of the 3rd ACM Conference on Electronic Commerce, EC '01, pp. 18–28. Association for Computing Machinery, New York (2001). https://doi.org/10.1145/501158.501161
32. Papireddygari, M., Waggoner, B.: Contracts with information acquisition, via scoring rules. In: Proceedings of the 23rd ACM Conference on Economics and Computation, EC '22, pp. 703–704. Association for Computing Machinery, New York (2022). https://doi.org/10.1145/3490486.3538261
33. Pekeč, A., Rothkopf, M.H.: Combinatorial auction design. Manage. Sci. **49**(11), 1485–1503 (2003)
34. US EEOC: equal employment opportunity commission (2024). https://www.eeoc.gov/
35. Vuong, R.D.C., Dughmi, S., Patel, N., Prasad, A.: On Supermodular Contracts and Dense Subgraphs. Soc. Ind. Appl. Math., 109–132 (2024). https://doi.org/10.1137/1.9781611977912.6
36. Walmart: Working at Walmart (2024). https://corporate.walmart.com/about/working-at-walmart
37. YouTube: YouTube channel monetization policies (2024). https://support.google.com/youtube/answer/1311392?hl=en
38. Zhu, B., Bates, S., Yang, Z., Wang, Y., Jiao, J., Jordan, M.I.: The sample complexity of online contract design. In: Proceedings of the 24th ACM Conference on Economics and Computation, EC '23, p. 1188. Association for Computing Machinery, New York (2023). https://doi.org/10.1145/3580507.3597673

Barter Exchange with Bounded Trading Cycles

Yuval Emek(✉) and Matan-El Shpiro

Technion – Israel Institute of Technology, 3200003 Haifa, Israel
yemek@technion.ac.il

Abstract. Consider a *barter exchange* problem over a finite set of agents, where each agent owns an item and is also associated with a (privately known) *wish list* of items belonging to the other agents. An outcome of the problem is a (re)allocation of the items to the agents such that each agent either keeps her own item or receives an item from her (reported) wish list, subject to the constraint that the length of the trading cycles induced by the allocation is up-bounded by a prespecified *length bound* k. The utility of an agent from an allocation is 1 if she receives an item from her (true) wish list and 0 if she keeps her own item (the agent incurs a large dis-utility if she receives an item that is neither hers nor belongs to her wish list).

In this paper, we investigate the aforementioned barter exchange problem from the perspective of mechanism design without money, aiming for truthful (and individually rational) mechanisms whose objective is to maximize the *social welfare*. As the construction of a social welfare maximizing allocation is computationally intractable for length bounds $k \geq 3$, this paper focuses on (computationally efficient) truthful mechanisms that approximate the (combinatorially) optimal social welfare. We also study a more general version of the barter exchange problem, where the utility of an agent from participating in a trading cycle of length $2 \leq \ell \leq k$ is $\lambda(\ell)$, where λ is a general (monotonically non-increasing) *length function*. Our results include upper and lower bounds on the guaranteed approximation ratio, expressed in terms of the length bound k and the length function λ. On the technical side, our main contribution is an algorithmic tool that can be viewed as a truthful version of the *local search* paradigm. As it turns out, this tool can be applied to more general (bounded size) coalition formation problems.

Keywords: Barter Exchange · Truthful Mechanisms · Length Functions

1 Introduction

Barter markets, where players exchange goods without involving any payments, are a fundamental form of trade that dates back to the origin of mankind. Modern

A full version of this extended abstract appears in [7].

M. Mavronicolas et al. (Eds.): WINE 2024, LNCS 15534, pp. 42–60, 2026.
https://doi.org/10.1007/978-3-032-08560-3_3

forms of barter markets include the extensively studied kidney exchange market [3,4,6,8,12–15] as well as the emerging electronic markets for book exchange [11] and for recreational home exchange [9,10] (cf. [1]).

In the current paper, we study an abstract game theoretic model for barter exchange that was introduced by Roth, Sönmez, and Ünver [13] and studied further by Hatfield [8] and in a follow up paper of Roth et al. [14], as well as serving as the cornerstone for various other game theoretic models studied subsequently (refer to [7] for a detailed discussion). This model is defined over a set of agents, each holding an item that she wishes to exchange for one item from a certain subset of items held by the other agents.[1] The agents report their item wish lists to a mechanism that determines the outcome in the form of a collection of trading cycles, effectively reallocating the items among the agents.

A fundamental characteristic of the model of [13] is the $\{0,1\}$-preferences assumption, stating that the agents are indifferent about which item they receive as long as this item belongs to their wish list. Specifically, an agent that ends up with an item from her wish list enjoys a unit utility, whereas her utility is zero if she retains her original item (Roth et al. assume that the agent cannot receive a non-wished item). Among other aspects of this barter exchange model, Roth et al. [13] investigate the design of truthful (strategy proof) mechanisms with the objective of maximizing the social welfare.

Motivated by real-life difficulties in implementing long trading cycles, the authors of [8,13,14] advocate for mechanisms that generate trading cycles of bounded length. The extreme case of trading cycles that consist of (exactly) two agents reduces the mechanism designer's task to the classic combinatorial optimization problem of constructing a maximum matching in an undirected graph [13]. As demonstrated in [14] though, there is a significant social welfare gain in allowing slightly longer trading cycles, thus raising the challenge of designing truthful mechanisms for the *k-barter exchange* problem, namely, the problem of constructing a collection of trading cycles that maximizes the number of participating agents (and thus also the social welfare) under the constraint that each trading cycle is of length at most k, for small constants $k \geq 3$.

Example 1. For a concrete application, consider a home swap platform where users register their home for specific vacation dates and report on their wish list over other homes registered for the same dates. The platform then determines an exchange that specifies an assignment of the users to vacation homes.[2] A downside of this application is that a single user that defects and stays in her own home leads to a chain of cancellations affecting all participants of the corresponding trading cycle. Since users are likely to be highly upset by such forced

[1] Roth et al. use kidney exchange terminology in their model description. We prefer the more "neutral" agent-item terminology since we do not regard kidney exchange as this paper's main application.

[2] Although a more general abstraction of the home swap application may include ordinal utilities or heterogeneous cardinal utilities, we feel that the $\{0,1\}$-preferences assumption is reasonable also for this application, where users will probably refrain from including a home in their wish list unless it perfectly fits their needs.

cancellations, especially if they happen at the last minute, it makes sense for the platform to bound the maximum length k of the trading cycles, thus hedging the reputation loss resulting from unsatisfied users.

Of particular interest is the $k = 3$ case that comes with the desirable feature that each user is acquainted with all other users in her trading cycle (they are assigned to her home or she is assigned to theirs). In this case, a last minute cancellation is attributed to a person the user is actually aware of (rather than some obscure defector up the cancellation chain), hopefully preventing resentment towards the platform.

Unfortunately, as proved by Abraham et al. [2], solving the aforementioned k-barter exchange problem optimally is computationally intractable for any $k \geq 3$, irrespectively of truthfulness considerations. Consequently, the goal of this paper is to design efficient truthful mechanisms for the k-barter exchange problem, $k \geq 3$, that approximate the (combinatorially) optimal solution. Our main technical contribution is a novel framework for the design of truthful versions of *local search* algorithms for a related graph theoretic problem. Employing this framework, we develop the first efficient truthful k-barter exchange mechanism with a non-trivial approximation ratio for every constant $k \geq 3$.

While the $\{0, 1\}$-preferences assumption of [13] states that an agent has no priorities over the items in her wish list, it is often natural to assume that she prefers to receive a wished item through shorter trading cycles over longer ones (up to the length bound k). Indeed, the longer is the trading cycle, the higher is the risk for the exchange to break down due to defecting participants (cf. [6]). Moreover, a trading cycle of length 2 has the unique "accountability advantage" (in the aforementioned home swap application, this means that you stay in the home of the person who stays in your home) which is lost with longer trading cycles, so it is reasonable to assume that agents would prioritize trading cycles of length 2 even if longer trading cycles are allowed.

Motivated by these insights, we wish to extend the scope of our research to a more general barter exchange setting where each agent is associated with a non-increasing *length function* $\lambda : \{2, 3, \ldots, k\} \to (0, 1]$ so that her utility from receiving a wished item via a trading cycle of length $2 \leq \ell \leq k$ is $\lambda(\ell)$. As a first step in this direction, we assume that the same length function λ is shared by all agents, introducing the *(k, λ)-barter exchange* problem. It turns out that the combinatorial aspect of this problem remains intractable for any length function λ, thus we still aim for approximate solutions. Relying on the aforementioned local search framework, we develop efficient truthful (k, λ)-barter exchange mechanisms with a non-trivial approximation ratio for any choice of function λ. On the negative side, we establish approximability lower bounds, showing that the dependency of our mechanisms' approximation ratio on λ is optimal.

1.1 Model and Problem

Consider a collection of $n \in \mathbb{Z}_{>0}$ agents indexed by the integers in $[n]$. A *wish list* of agent $i \in [n]$ is a subset $W_i \subseteq [n] - \{i\}$ of the other agents. Let $\mathcal{W}_n =$

$2^{[n]-\{1\}} \times \cdots \times 2^{[n]-\{n\}}$ denote the set of all wish list vectors, indexed by the agents in $[n]$.

An *exchange* over $[n]$ is a bijection $\pi : [n] \to [n]$; we say that an agent $i \in [n]$ *partakes* in the exchange π if $\pi(i) \neq i$, denoting the set of agents that partake in π by $\mathcal{P}_\pi$. Observe that from a graph theoretic perspective, the exchange π induces a collection $\mathcal{C}_\pi$ of disjoint directed cycles on $\mathcal{P}_\pi$, referred to as the *trading cycles* of π, so that an arc $(i, j) \in \mathcal{P}_\pi \times \mathcal{P}_\pi$ is included in a trading cycle if and only if $\pi(i) = j$.[3] Given a wish list vector $W \in \mathcal{W}_n$, we say that the exchange π, as well as the trading cycles in $\mathcal{C}_\pi$, *respect* W if $\pi(i) \in W_i$ for each $i \in \mathcal{P}_\pi$.

Given a trading cycle c, let $\alpha(c) \subseteq [n]$ denote the set of agents that form c. The *length* of c is defined as $\mathrm{len}(c) = |\alpha(c)|$. The *length* of an exchange π, denoted by $\mathrm{len}(\pi)$, is defined to be the maximum length of any of its trading cycles, i.e., $\mathrm{len}(\pi) = \max\{\mathrm{len}(c) : c \in \mathcal{C}_\pi\}$. For an integer $k \geq 2$, let Π_n^k denote the set of all exchanges over $[n]$ whose length at most k. Given a wish list vector $W \in \mathcal{W}_n$, let $\Pi_W^k \subseteq \Pi_n^k$ denote the set of all exchanges over $[n]$ that respect W whose length is at most k. Likewise, let $\mathcal{C}_W^k = \bigcup_{\pi \in \Pi_W^k} \mathcal{C}_\pi$ denote the set of all trading cycles over $[n]$ that respect W whose length is at most k.

Given an integral *length bound* $k \geq 2$, an instance of the *k-barter exchange (k-BE)* problem over $n \in \mathbb{Z}_{>0}$ agents is characterized by a wish list $W_i^* \subseteq [n] - \{i\}$ for each agent $i \in [n]$, regarded as the agent's private information. The outcome space of the k-BE problem is the collection Π_n^k of exchanges over $[n]$ of length at most k. The utility of an agent $i \in [n]$ from an exchange $\pi \in \Pi_n^k$ is determined by the utility function $u : [n] \times \Pi_n^k \to \{-\infty\} \cup [0, 1]$ defined so that

$$u(i, \pi) = \begin{cases} 0, & i \notin \mathcal{P}_\pi \\ 1, & i \in \mathcal{P}_\pi \text{ and } \pi(i) \in W_i^* \\ -\infty, & i \in \mathcal{P}_\pi \text{ and } \pi(i) \notin W_i^* \end{cases} \tag{1}$$

(cf. [1,13]). This allows us to define the *social welfare* of π, denoted by $\mathrm{SW}(\pi)$, as the sum of all agent utilities, observing that if $\pi \in \Pi_{W^*}^k$, then $\mathrm{SW}(\pi) = |\mathcal{P}_\pi|$.

A (deterministic) *mechanism* for the k-BE problem is a function f that for any $n \in \mathbb{Z}_{>0}$, maps a given wish list vector $W \in \mathcal{W}_n$ to an exchange $f(W) \in \Pi_n^k$, regarding W_i as the wish list reported to the mechanism by agent $i \in [n]$. Following [13] (see also [1]), we make the simplifying *subset wish list assumption* stating that an agent's reported wish list is always a subset of her true wish list, i.e., $W_i \subseteq W_i^*$ for every $i \in [n]$.[4] The mechanism f is said to be *truthful* (a.k.a. *strategy proof*) if for every wish list vector $W \in \mathcal{W}_n$ (that satisfies $W_j \subseteq W_j^*$ for all $j \in [n]$) and for every agent $i \in [n]$, it holds that

$$u(i, f(W)) \leq u\left(i, f\left(W_{-i}, W_i^*\right)\right) ;$$

[3] Unless stated otherwise, the graph theoretic term cycle refers in this paper to a simple cycle, i.e., a cycle in which each vertex appears at most once.

[4] In [7], we present a simple method that lifts this assumption using an arbitrarily small amount of randomness.

in other words, it is a dominant strategy for agent i to report her (full) true wish list W_i^* to the mechanism.[5]

To ensure that mechanism f is *individually rational*, we further require that $f(W)$ respects W (i.e., $f(W) \in \Pi_W^k$), which means that the utility of the agents from f is never negative. Consequently, we can express the social welfare of the mechanism's outcome as

$$\mathrm{SW}(f(W)) = \sum_{i \in [n]} u(i, f(W)) = \sum_{c \in \mathcal{C}_{f(W)}} \mathrm{len}(c) = |\mathcal{P}_{f(W)}| . \tag{2}$$

We aim for truthful (and individually rational) mechanisms that maximize the social welfare of the exchanges they produce. To this end, we say that a truthful k-BE mechanism f admits *approximation ratio* $r \geq 1$ if for every $n \in \mathbb{Z}_{>0}$ and wish list vector $W^* \in \mathcal{W}_n$, it is guaranteed that

$$\mathrm{SW}(f(W^*)) \geq \frac{1}{r} \cdot \max \left\{ \mathrm{SW}(\pi) \,:\, \pi \in \Pi_{W^*}^k \right\} ;$$

that is, the benchmark of the approximation is the optimal exchange of length at most k that respects W^*. We typically regard the length bound k as a small constant (recall our particular interest in $k = 3$), whereas the number n of agents grows asymptotically. Our ultimate goal is to design truthful mechanisms whose approximation ratio is expressed as a function of k (i.e., constant approximation) and whose running time is polynomial in n (i.e., they are *computationally efficient*).

General Length Functions. As discussed earlier, we also consider a generalization of the k-BE problem, referred to as (k, λ)-*BE*, where $\lambda : \{2, 3, \ldots, k\} \to (0, 1]$ is a non-increasing *length function*. This generalization is obtained by redefining the utility function from (1) as

$$u(i, \pi) = \begin{cases} 0, & i \notin \mathcal{P}_\pi \\ \lambda(\mathrm{len}(c)), & i \in \mathcal{P}_\pi,\ \pi(i) \in W_i^*, \text{ and } c \text{ is the trading cycle in } \mathcal{C}_\pi \text{ s.t. } i \in \alpha(c) \\ -\infty, & i \in \mathcal{P}_\pi \text{ and } \pi(i) \notin W_i^* \end{cases} . \tag{3}$$

Consequently, if $\pi \in \Pi_{W^*}^k$, then the social welfare $\mathrm{SW}(\pi)$ of π becomes $\sum_{c \in \mathcal{C}_\pi} \mathrm{len}(c) \cdot \lambda(\mathrm{len}(c))$ and the social welfare expression in (2) is now replaced by

$$\mathrm{SW}(f(W)) = \sum_{i \in [n]} u(i, f(W)) = \sum_{c \in \mathcal{C}_{f(W)}} \mathrm{len}(c) \cdot \lambda(\mathrm{len}(c)) . \tag{4}$$

---

[5] Given a wish list vector $W \in \mathcal{W}_n$ and an agent $i \in [n]$, we adhere to the convention that W_{-i} denotes the vector obtained from W by omitting its ith coordinate, that is, $W_{-i} = (W_1, \ldots, W_{i-1}, W_{i+1}, \ldots, W_n)$. This notation is naturally extended to agent subsets $S \subseteq [n]$, using W_{-S} to denote the vector obtained from W by omitting its ith coordinate for every $i \in S$.

Similarly to the length bound k, the images $\lambda(2), \lambda(3), \ldots, \lambda(k)$ of the length function are also regarded as constants (part of the problem's specification).

To see that the (k, λ)-BE problem indeed generalizes the k-BE problem, simply take the *uniform* length function λ^u defined by setting $\lambda^u(\ell) = 1$ for every $2 \leq \ell \leq k$. In this regard, a length function λ is *non-uniform* if $\lambda(\ell) > \lambda(\ell')$ for some $2 \leq \ell < \ell' \leq k$.[6] Unless stated otherwise, we subsequently address the (k, λ)-BE problem in its general form that includes both uniform and non-uniform length functions λ. Although the k-BE problem is a special case of this more general problem, it is undoubtedly the most important special case and therefore, receives its own focus.

1.2 Our Contribution

As observed by Roth et al. [13], the 2-BE problem admits an efficient optimal truthful mechanism based on constructing a maximum size matching (breaking ties consistently) in the undirected graph $G = ([n], E_{\leftrightarrow})$ obtained from the reported wish list vector $W \in \mathcal{W}_n$ by setting $E_{\leftrightarrow} = \{\{i, j\} \mid i \in W_j \text{ and } j \in W_i\}$.[7] Consequently, our attention in the remainder of this paper is devoted to length bounds $k \geq 3$.

Arguably the simplest efficient (k, λ)-BE mechanism is the one we denote by `Greedy`. This naive mechanism iterates over the possible trading cycles c in increasing order of length $\text{len}(c) = 2, 3, \ldots, k$, greedily adding trading cycle c to $\mathcal{C}_\pi$ if c does not intersect with any trading cycle already included in $\mathcal{C}_\pi$.

Theorem 1. `Greedy` *is truthful and approximates the (k, λ)-BE problem within ratio k for every length bound $k \geq 3$ and (uniform or non-uniform) length function λ.*

Using the approximation bound of Theorem 1 as the starting point of our journey, this work is guided by the following two research questions: Does there exist an efficient truthful mechanism that approximates the (k, λ)-BE problem within ratio $r < k$? If so, how does the approximation ratio r depend on the length function λ? To answer these questions, we first investigate the uniform length function $\lambda = \lambda^u$.

---

[6] Note that scaling the images of λ by a constant factor has no effect on the approximation ratio.

[7] Notice that when $k = 2$, the length functions λ have no role as all trading cycles are of the same length.

Theorem 2. *The k-BE problem admits an efficient truthful mechanism with approximation ratio* $k-1+\epsilon$ *for every length bound* $k \geq 3$ *and constant* $\epsilon > 0$.[8] [9]

To the best of our knowledge, Theorem 2 provides the first non-trivial upper bound on the approximation ratio of efficient truthful k-BE mechanisms for any $k \geq 3$. Next, we turn our attention to non-uniform length functions $\lambda : \{2, 3, \ldots, k\} \to (0, 1]$ and define the parameter

$$\rho(\lambda) = \max_{2 \leq \ell < \ell' \leq k \,:\, \lambda(\ell) > \lambda(\ell')} \left\{ \max \left\{ \frac{\ell' \cdot \lambda(\ell')}{\lambda(\ell)}, \frac{\ell - 1}{\ell} \cdot \frac{\ell' \cdot \lambda(\ell')}{\lambda(\ell)} + 1 \right\} \right\} .$$

Theorem 3. *The* (k, λ)*-BE problem does not admit a truthful mechanism with approximation ratio strictly smaller than* $\rho(\lambda)$ *for every length bound* $k \geq 3$ *and non-uniform length function* λ.

An elaborate discussion of the properties of $\rho(\lambda)$ is provided in [7]. We emphasize that this lower bound is universal in the sense that it applies to *any* non-uniform length function λ. Moreover, it does not contradict Theorem 1 as $\rho(\lambda) < k$ for every non-uniform length function λ (see [7]). Surprisingly, $\rho(\lambda)$ plays a central role in our approximation ratio upper bound as well.

Theorem 4. *The* (k, λ)*-BE problem admits an efficient truthful mechanism with approximation ratio* $\max\{k - 1 + \epsilon, \rho(\lambda)\}$ *for every length bound* $k \geq 3$, *non-uniform length function* λ, *and constant* $\epsilon > 0$.

Refer to [7] for a discussion of the conditions under which the lower bound of Theorem 3 matches the upper bound of Theorem 4. The following theorem shows that the lower bound of Theorem 3 is tight if we disregard computational efficiency considerations.

Theorem 5. *The* (k, λ)*-BE problem admits a computationally inefficient truthful mechanism with approximation ratio* $\rho(\lambda)$ *for every length bound* $k \geq 3$ *and non-uniform length function* λ.

---

[8] The readers familiar with the work of Ashlagi and Roth [4] may wonder whether our Theorem 2 is implied by [4, Theorem 1], stating that any solution to the k-BE problem, which is maximal in terms of the set of partaking agents, is guaranteed to be a $(k-1)$-approximation. It is certainly not! The exchanges constructed by the (truthful) algorithm that lies at the heart of Theorem 2's proof are not necessarily maximal (see Sect. 5 for an example that demonstrates this fact), which forces us to use a considerably more involved argument when bounding the approximation ratio (see Sect. 4.1) in comparison to the 1-paragraph proof of [4, Theorem 1]'s upper bound. The existence of efficient constructions of (agent-)maximal exchanges, let alone truthful ones, remains an open question.

[9] The runtime of the mechanism promised in this theorem, as well as that of the one promised in Theorem 4, is up-bounded by $n^{O(k^2/\epsilon)}$ which is polynomial in n (only) as long as k and ϵ are constants. In the current paper, we do not focus on improving the dependency of the runtime bound on k and ϵ

Back to uniform length functions, as proved by Hatfield [8], the k-BE problem admits a computationally inefficient optimal truthful mechanism for every length bound $k \geq 3$, which means that Theorem 3 separates between the k-BE problem and the (k, λ)-BE problem with non-uniform λ. This leaves us with the following interesting open question: What is the best possible approximation ratio achievable by efficient truthful mechanisms for the k-BE problem? The existence of an optimal computationally inefficient truthful mechanism for this problem implies that any attempt to devise a non-trivial lower bound on the approximation ratio will have to rely on computational complexity hardness assumptions. Approximability lower bounds that employ such hardness assumptions are discussed in [7] for the combinatorial optimization aspect of the k-BE problem, irrespective of truthfulness considerations. Whether they can be improved by exploiting both the truthfulness requirement and the efficiency requirement remains an intriguing, and notoriously challenging, open question (we are unaware of any lower bound result that exploits both requirements concurrently in the context of mechanism design without money).

The mechanisms that lie at the heart of Theorem 2 and 4 belong to a family that we refer to as *standard local search mechanisms* (see Sect. 3 for a definition). It turns out that among the mechanisms in this family, the ones promised in Theorem 2 and 4 are optimal in terms of the dependency of their approximation ratio on k.

Theorem 6. *The (k, λ)-BE problem does not admit a truthful standard local search mechanism with approximation ratio strictly smaller than $k - 1$ for every length bound $k \geq 3$ and (uniform or non-uniform) length function λ.*

Forming Small Coalitions. The (k, λ)-BE problem is related to a family of problems whose outcome is a partition of the agent set $[n]$ into clusters, a.k.a. *coalitions*, of size at most k. Each agent $i \in [n]$ is associated with a wish list over the possible coalitions she is willing to partake in; her utility is 0, if she is assigned to a singleton coalition, and it is $\lambda(\ell)$ if she is assigned to a coalition of size $2 \leq \ell \leq k$ from her wish list. It turns out that the techniques used to establish our positive results apply also to this coalition formation setup (see the max-CGIS problem formulated in Sect. 2). Having said that, we subsequently restrict our attention to the BE problem family as it seems more natural, with more convincing applications, and given that our negative results are specific for this problem family. Refer to [7] for a further discussion.

1.3 Paper's Organization

The remainder of this paper is organized as follows. In Sect. 2, we present additional notation and terminology as well as some fundamental observations. In Sect. 3, we introduce the class of local search mechanisms and establish a sufficient condition for their truthfulness under the uniform length function. The (efficient) mechanism promised in Theorem 2 is then developed in Sect. 4. While we are unaware of any general approximability lower bounds for the uniform case (as discussed in Sect. 1.2, this is an important open question), in Sect. 5, we show that the approximation ratio analysis conducted in Sect. 4 is (nearly) tight for the specific mechanism of that section. Due to lack of space, the proofs of the remaining theorems in Sect. 1.2 (all theorems other than Theorem 2) are excluded from this extended abstract and deferred to the full version [7]. Nevertheless, a brief overview of Theorem 4's proof, that we view as slightly more important than the other missing proofs, is presented in Sect. 6.

2 Preliminaries

Graph Theoretic Definitions. Throughout, the term graph is reserved for undirected graphs, whereas directed graphs are referred to as digraphs. To avoid confusion, the terms vertex and arc refer to the basic elements of a digraph, whereas the terms node and edge refer to the basic elements of an (undirected) graph.

Consider an (undirected) graph $G = (V, E)$. For a node $v \in V$, let $N_G(v) = \{u \in V \mid \{u, v\} \in E\}$ denote the set of v's neighbors in G and let $\deg_G(v) = |N_G(v)|$ denote v's degree in G. The former notation is extended to node subsets $U \subseteq V$, denoting $N_G(U) = \bigcup_{v \in U} N_G(v)$.

Given a node subset $U \subseteq V$, the subgraph *induced* by U on G, denoted by $G[U]$, is the graph whose node set is U and whose edge set is $\{e \in E \mid e \subseteq U\}$. A d-*claw* is a complete bipartite graph with a single node on one side and d nodes on the other size. A graph is said to be d-*claw free* if it does not admit a d-claw as a node induced subgraph.

A node subset $I \subseteq V$ is an *independent set (IS)* in G if $\{u, v\} \notin E$ for any $u, v \in I$. An IS $I \subseteq V$ is *maximal* if U is not an IS for any $I \subset U \subseteq V$. Alternatively, I is a maximal IS if and only if it is both an IS and a dominating set in G.

Cycle Graphs. Consider a wish list vector $W \in \mathcal{W}_n$, $n \in \mathbb{Z}_{>0}$, and recall that $\mathcal{C}_W^k = \bigcup_{\pi \in \Pi_W^k} \mathcal{C}_\pi$ is the set of trading cycles over $[n]$ of length at most k that respect W. The k-*cycle graph* of W is an (undirected) graph $G = (V, E)$ defined over the trading cycle set $V = \mathcal{C}_W^k$ so that nodes $u, v \in V$ are adjacent under E if and only if their corresponding trading cycles share some (at least one) agents; that is, $\{u, v\} \in E$ if and only if $\alpha(u) \cap \alpha(v) \neq \emptyset$, recalling that $\alpha(u)$ and $\alpha(v)$ are the sets of agents that form the trading cycles u and v, respectively (see Fig. 1 for an illustration).

We extend the operator α from single nodes to node subsets $U \subseteq V$, defining $\alpha(U) = \bigcup_{v \in U} \alpha(v)$. In the converse direction, for an agent $i \in [n]$, let $\alpha^{-1}(i) = \{v \in V \mid i \in \alpha(v)\}$ be the set of nodes in which i *partakes*; notice that by definition, the nodes in $\alpha^{-1}(i)$ form a clique in G and every edge $e \in E$ is internal to the clique corresponding to (at least) one of the agents. We define the *length* of a node $v \in V$ to be the length of the corresponding trading cycle, i.e., $|\alpha(v)|$. Given a (uniform or non-uniform) length function $\lambda : \{2, 3, \ldots, k\} \to (0, 1]$, we define the *weight* $w(v) \in \mathbb{R}_{>0}$ of v to be $w(v) = |\alpha(v)| \cdot \lambda(|\alpha(v)|)$; the weight operator is also extended to node subsets $U \subseteq V$, defining $w(U) = \sum_{v \in U} w(v)$.

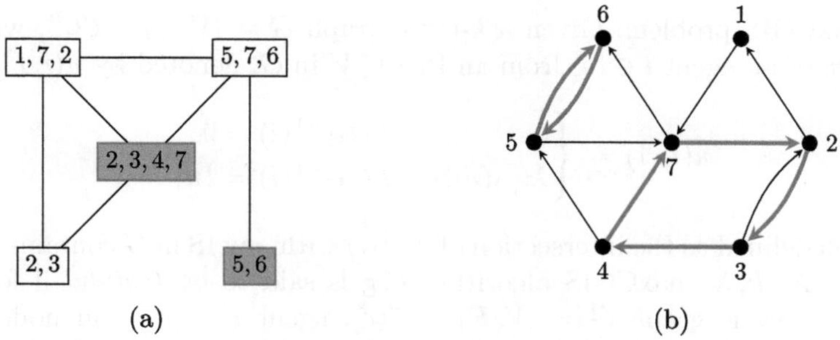

Fig. 1. The k-cycle graph depicted in Fig. 1a (where the agents in each node are listed according to their cyclic order) corresponds to the wish list vector represented by the digraph depicted in Fig. 1b under the length bound $k = 4$. The nodes marked in Fig. 1a form an IS that corresponds to the exchange (i.e., set of disjoint trading cycles) marked in Fig. 1b. This IS turns out to be an optimal solution for the k-maxCGIS problem, hence the corresponding exchange is an optimal solution for the k-BE problem.

Observation 7. *Consider a wish list vector $W \in \mathcal{W}_n$ and let $G = (V, E)$ be its corresponding k-cycle graph. There is a one-to-one correspondence between the exchanges in Π_W^k and the ISs in G. Moreover, if $I \subseteq V$ is the IS in G that corresponds to an exchange $\pi \in \Pi_W^k$, then $w(I) = \mathrm{SW}(\pi)$.*

Owing to Observation 7, we shall implement our (k, λ)-BE mechanisms by means of algorithms that construct an IS for a given k-cycle graph with the objective of maximizing its weight. It will be convenient to extend the input domain of these algorithms, allowing them to be invoked on k-cycle graphs that do not necessarily correspond to any particular wish list vector. Formally, we say that a graph $G = (V, E)$, equipped with an operator $\alpha : V \to 2^{[n]}$, is a *k-cycle graph* if (1) each node $v \in V$ is identified with a unique cyclic order over the agents in $\alpha(v)$ of length $|\alpha(v)| \leq k$;[10] and (2) $E = \{\{u, v\} \subseteq \mathrm{V} \mid \alpha(u) \cap \alpha(v) \neq \emptyset\}$. Let $\mathcal{CG}_n^k$ denote the family of k-cycle graphs over n agents. Notice that in general, the cyclic orders identified with the nodes of a k-cycle graph may not be compatible with the cycles in $\mathcal{C}_W^k$ for any wish list vector $W \in \mathcal{W}_n$; refer to [7] for an example of such a k-cycle graph.

The (k, λ)-maxCGIS Problem. We say that an algorithm `Alg` admits *approximation ratio* $r \geq 1$ for the maximum weight IS problem in k-cycle graphs under the length function λ, abbreviated as the *(k, λ)-maxCGIS* problem, if for every $n \in \mathbb{Z}_{>0}$ and for every k-cycle graph $G = (V, E) \in \mathcal{CG}_n^k$, it is guaranteed that `Alg`(G) is an IS in G and that $w(\texttt{Alg}(G)) \geq \frac{1}{r} w(I)$ for every IS I in G. The *k-maxCGIS* problem refers to the special case of (k, λ)-maxCGIS where $\lambda = \lambda^u$ is the uniform length function.

To account for the game theoretic aspects of the (k, λ)-BE problem, we introduce the notions of agent utility and truthfulness also in the context of the

[10] The cyclic orders are introduced only for the sake of distinguishing between the nodes (and ultimately, bounding their total number).

(k, λ)-maxCGIS problem. Given a k-cycle graph $G = (V, E) \in \mathcal{CG}_n^k$, we define the *utility* of an agent $i \in [n]$ from an IS $I \subseteq V$ in G, denoted by $u(i, I)$, to be

$$u(i,\ \mathrm{I}) = \begin{cases} 0, & I \cap \alpha^{-1}(i) = \emptyset \\ \lambda(|\alpha(v)|), & I \cap \alpha^{-1}(i) = \{v\} \end{cases};$$

this is well defined as the intersection of $\alpha^{-1}(i)$ with any IS in G contains at most one node. A (k, λ)-maxCGIS algorithm $\texttt{Alg}$ is said to be *truthful* if for every $n \in \mathbb{Z}_{>0}$, k-cycle graph $G = (V, E) \in \mathcal{CG}_n^k$, agent $i \in [n]$, and node subset $S \subseteq \alpha^{-1}(i)$, it is guaranteed that $u(i, \texttt{Alg}(G[V - S])) \leq u(i, \texttt{Alg}(G))$.

Observation 8. *An (efficient) truthful (k, λ)-maxCGIS algorithm implies an (efficient) truthful (k, λ)-BE mechanism with the same approximation ratio.*

It is important to point out that the truthfulness requirement from a (k, λ)-BE mechanism invoked on a wish list vector W is weaker than the truthfulness requirement from a (k, λ)-maxCGIS algorithm invoked on the corresponding k-cycle graph $G = (V, E)$ as the strategy space of the agents in the latter is more extensive than that of the former. Indeed, in the context of the (k, λ)-maxCGIS problem, an agent $i \in [n]$ may choose to "hide" from the algorithm a node subset $S \subseteq \alpha^{-1}(i)$ that does not correspond to omitting any agent subset from i's wish list W_i. We emphasize that while the stronger requirement from (k, λ)-maxCGIS algorithms makes our upper bounds "harder", we do not impose this stronger requirement in our impossibility results that merely exploit the requirements from a truthful mechanism as presented in Sect. 1.1.

The following structural observation regarding the family of k-cycle graphs plays a key role in the design of our approximation algorithms.

Observation 9. *Consider a k-cycle graph $G = (V, E) \in \mathcal{CG}_n^k$. The number of nodes in G satisfies $|V| \leq \sum_{h=2}^{k} \binom{n}{h}(h-1)! \leq \mathrm{poly}(n)$. Moreover, the number of nodes in any IS I in G satisfies $|I| \leq n/2$ and $|I \cap N_G(v)| \leq |\alpha(v)|$ for every node $v \in V$. In particular, the graph G is $(k + 1)$-claw free.*

We note that the reduction from the (combinatorial aspect of the) BE problem with length bound $k \geq 3$ to the problem of constructing a maximum weight IS in $(k + 1)$-claw free graphs, particularly line graphs of hypergraphs, is used also by Biro et al. [5]; refer to [7] for further discussion.

INPA Algorithms. An algorithm $\texttt{Alg}$ for the (k, λ)-maxCGIS problem is said to be *independent of non-partaking agents (INPA)* if for every $n \in \mathbb{Z}_{>0}$, k-cycle graph $G = (V, E) \in \mathcal{CG}_n^k$, agent $i \in [n] - \alpha(\texttt{Alg}(G))$, and node subset $S \subseteq \alpha^{-1}(i)$, it is guaranteed that $\texttt{Alg}(G[V - S]) = \texttt{Alg}(G)$. (Refer to [7] for a discussion of related game theoretic notions.) The importance of the INPA property of maxCGIS algorithms, specifically in the context of the uniform length function, is demonstrated by the following observation.

Observation 10. *For every length bound $k \geq 3$, any INPA algorithm for the k-maxCGIS problem is truthful.*

Proof. Let $\texttt{Alg}$ be an INPA algorithm for the k-maxCGIS problem. Fix some $n \in \mathbb{Z}_{>0}$ and consider a k-cycle graph $G = (V, E) \in \mathcal{CG}_n^k$ and an agent $i \in [n]$. If $i \in \alpha(\texttt{Alg}(G))$, then $u(i, \texttt{Alg}(G)) = 1$ and agent i cannot increase her utility by omitting any nodes from $\alpha^{-1}(i)$. So, assume that $i \notin \alpha(\texttt{Alg}(G))$ and consider a node subset $S \subseteq \alpha^{-1}(i)$. Since $\texttt{Alg}$ is INPA, it follows that $\texttt{Alg}(G[V - S]) = \texttt{Alg}(G)$, hence $i \notin \alpha(\texttt{Alg}(G[V - S]))$. Therefore, $u(i, \texttt{Alg}(G[V - S])) = 0 = u(i, \texttt{Alg}(G))$ which means that agent i does not improve her utility by omitting S from $\alpha^{-1}(i)$.

3 Local Search Algorithms

In this section, we introduce the class of local search (k, λ)-maxCGIS algorithms that is pivotal to our work. We then establish a sufficient condition for the truthfulness of algorithms in this class in the context of the k-maxCGIS problem (i.e., the special case of $\lambda = \lambda^u$).

Definition 1 (improvement rule, loyal, INPA, efficient). *An* improvement rule *r for the (k, λ)-maxCGIS problem is a function that maps a pair (G, I), consisting of a k-cycle graph $G = (V, E) \in \mathcal{CG}_n^k$, $n \in \mathbb{Z}_{>0}$, and an IS I in G, either to an IS I' in G satisfying $w(I') > w(I)$ or to $\perp$. The improvement rule r is said to be* loyal *if $r(G, I) = I' \neq \perp$ implies that $\alpha(I) \subseteq \alpha(I')$. The improvement rule r is said to be* independent of non-partaking agents (INPA) *if (1) $r(G, I) = I' \neq \perp$ implies that $r(G[V - S], I) = I'$ for every agent $i \in [n] - (\alpha(I) \cup \alpha(I'))$ and node subset $S \subseteq \alpha^{-1}(i)$; and (2) $r(G, I) = \perp$ implies that $r(G[V - S], I) = \perp$ for every agent $i \in [n] - \alpha(I)$ and node subset $S \subseteq \alpha^{-1}(i)$. Finally, the improvement rule r is said to be* efficient *if it admits a computationally efficient implementation.*

In words, the loyal property means that any agent that partakes in the original IS I must also partake in the new IS I'. The INPA property means that agents that do not partake in I nor in I' have no impact on the improvement rule's outcome. We can now define a local search algorithm over a list of improvement rules.

Definition 2 (local search algorithm). *A* local search *algorithm $\texttt{Alg}$ for the (k, λ)-maxCGIS problem is characterized by an (ordered) list $R = (r_1, \ldots, r_{|R|})$ of improvement rules. Given a k-cycle graph $G = (V, E) \in \mathcal{CG}_n^k$, $n \in \mathbb{Z}_{>0}$, the (deterministic) algorithm returns the IS $I = \texttt{Alg}(G)$ constructed via the following iterative process:*
(1) set $t \leftarrow 0$ and $I^t \leftarrow \emptyset$;
(2) if $r_j(G, I^t) = \perp$ for all $1 \leq j \leq |R|$, then return the set $I \leftarrow I^t$ and halt;
(3) let j^t be the smallest $1 \leq j \leq |R|$ such that $r_j(G, I^t) \neq \perp$;
(4) set $I^{t+1} \leftarrow r_{j^t}(G, I^t)$, set $t \leftarrow t + 1$, and go to step (2).
Let $\texttt{Alg}^t(G)$ denote the IS I^t associated with iteration t of this process.

It turns out that improvement rules which are loyal and INPA lead to an INPA (local search) algorithm.

Lemma 1. *Consider a local search algorithm* `Alg` *for the* (k, λ)*-maxCGIS problem characterized by a list* $R = (r_1, \ldots, r_{|R|})$ *of improvement rules. If all improvement rules in* R *are loyal and INPA, then* `Alg` *is INPA. Moreover, if* $|R|$ *is fixed and all improvement rules in* R *are efficient, then* `Alg` *is efficient.*

Proof (Proof Sketch (full proof is deferred to [7])). Consider an agent i who does not partake in `Alg`'s output. Since all improvement rules in R are loyal, it follows that i does not partake in any of the maintained ISs throughout `Alg`'s execution. That, combined with the fact that all improvement rules are INPA, implies that agent i does not affect the IS maintained at each step of the algorithm. Therefore, agent i cannot manipulate `Alg`'s output which means that `Alg` is INPA. By Observation 9, we know that the number of nodes in `Alg`'s output is at most $n/2$, which allows us to deduce that `Alg` is efficient.

Note that if we omit the requirement that all improvement rules in R are loyal, then there may exist an agent that partakes in one of the intermediate ISs maintained by `Alg`, but does not partake in `Alg`'s output. By deleting some of the nodes this agent partakes in, one can steer the execution of `Alg`, leading to a different output. Recalling that INPA implies truthfulness in the context of the k-maxCGIS problem, we obtain the following corollary from Observation 10 combined with Lemma 1.

Corollary 1. *Consider a local search algorithm* `Alg` *for the* k*-maxCGIS problem characterized by a fixed length list* R *of improvement rules. If all improvement rules in* R *are loyal, INPA, and efficient, then* `Alg` *is truthful and efficient.*

4 Upper Bound for the Uniform Length Function

We now turn to develop an efficient truthful algorithm for the k-maxCGIS problem, $k \geq 3$. Our algorithm is presented as a local search algorithm and its truthfulness is established based on the infrastructure developed in Sect. 3. Following that, in Sect. 4.1, we analyze the guaranteed approximation ratio of this algorithm and prove Theorem 2 (by virtue of Observation 8). We start with presenting the improvement rules based on which our algorithm is designed.

Expansion Improvement Rule. Given a k-cycle graph $G = (V, E) \in \mathcal{CG}_n^k$ and an IS I in G, the *expansion* improvement rule r_E returns the lexicographically first IS $I' = r_E(G, I)$ that satisfies (1) $I' \supset I$; and (2) $|I'| = |I| + 1$. If no such IS I' exists, then r_E returns $r_E(G, I) = \perp$. We refer to any IS that satisfies conditions (1) and (2) as a *candidate* of r_E. Refer to Figs. 2a–2c for an illustration of r_E.

Observation 11. *The expansion improvement rule* r_E *is loyal, INPA, and efficient.*

Proof. The fact that r_E is loyal follows from the requirement that $I' \supset I$. To see that r_E is INPA, observe that if $i \notin \alpha(I)$, then reducing $\alpha^{-1}(i)$ to $\alpha^{-1}(i) - S$ for some node subset $S \subseteq \alpha^{-1}(i)$ only decrease the collection of candidates from

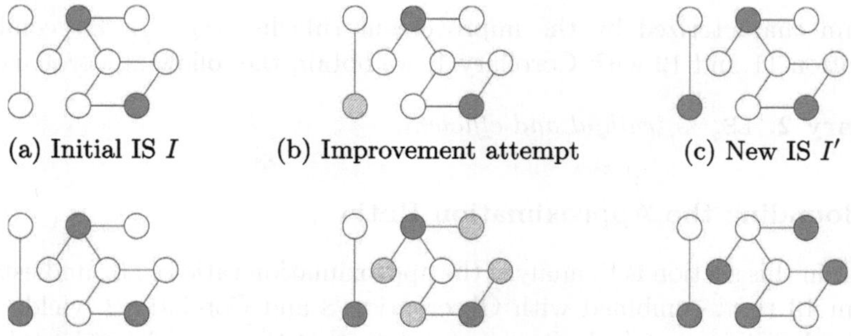

(a) Initial IS I (b) Improvement attempt (c) New IS I'

(d) Initial IS I (e) Improvement attempt (f) New IS I'

Fig. 2. The expansion improvement rule r_E (2a–2c) and the all-for-q improvement rule r_q (2d–2f).

which $r_E(G, I)$ is selected. Therefore, (1) if $r_E(G, I) = I' \neq \perp$ and $i \notin \alpha(I')$, then $r_E(G[V - S], I) = I'$; and (2) if $r_E(G, I) = \perp$, then $r_E(G[V - S], I) = \perp$. Finally, to see that r_E is efficient, notice that it can be implemented simply by inspecting all nodes in $V - I$. The assertion follows from the poly(n) size of G as promised in Observation 9.

All-For-q Improvement Rule. Given a k-cycle graph $G = (V, E) \in \mathcal{CG}_n^k$ and an IS I in G, the *all-for-q* improvement rule r_q for a constant $q \in \mathbb{Z}_{>0}$ returns the lexicographically first IS $I' = r_q(G, I)$ that satisfies (1) $\alpha(I') \supset \alpha(I)$; and (2) $I' = (I \cup X) - (N_G(X) \cap I)$, where $X \subseteq N_G(I)$ is an IS in G and $|N_G(X) \cap I| \leq q$. If no such IS I' exists, then r_q returns $r_q(G, I) = \perp$. We refer to any IS I' that satisfies conditions (1) and (2) as a *candidate* of r_q. Refer to Figs. 2d–2f for an illustration of r_q.

Observation 12. *The all-for-q improvement rule is loyal, INPA, and efficient for any constant* $q \in \mathbb{Z}_{>0}$.

Proof. The fact that r_q is loyal follows from requirement (1). To see that r_q is INPA, observe that if $i \notin \alpha(I)$, then reducing $\alpha^{-1}(i)$ to $\alpha^{-1}(i) - S$ for some node subset $S \subseteq \alpha^{-1}(i)$ can only decrease the collection of candidates from which $r_q(G, I)$ is selected. Therefore, (1) if $r_q(G, I) = I' \neq \perp$ and $i \notin \alpha(I')$, then $r_q(G[V - S], I) = I'$; and (2) if $r_q(G, I) = \perp$, then $r_q(G[V - S], I) = \perp$. Finally, to see that r_q is efficient, recall that Observation 9 guarantees that the k-cycle graph G is $(k + 1)$-claw free. Therefore, if $X \subseteq N_G(I)$ is an IS in G and $|N_G(X) \cap I| \leq q$, then $|X| \leq qk$. This means that one can implement the all-for-q improvement rule by inspecting all subsets of $V - I$ of size at most qk which is a constant by the assumption that k and q are constants. The assertion follows from the poly(n) size of G as promised in Observation 9.

The $\texttt{LS}_q$ *Algorithm.* Given a constant parameter $q \in \mathbb{Z}_{>0}$, our main algorithm for the k-maxCGIS problem, denoted by $\texttt{LS}_q$, is defined to be the local search algorithm characterized by the improvement rule list (r_E, r_q). By combining Observation 11 and 12 with Corollary 1, we obtain the following corollary.

Corollary 2. $\mathtt{LS}_q$ *is truthful and efficient.*

4.1 Bounding the Approximation Ratio

Our goal in this section is to analyze the approximation ratio of $\mathtt{LS}_q$ and establish Theorem 13 that, combined with Observation 8 and Corollary 2, yields Theorem 2 by choosing $q \geq 1/\epsilon$. In Sect. 5, we show that the upper bound established here on the approximation ratio of $\mathtt{LS}_q$ is nearly tight.

Theorem 13. $\mathtt{LS}_q$ *approximates the k-maxCGIS problem within ratio* $k - 1 + 1/q$.

Consider a k-cycle graph $G = (V, E) \in \mathcal{CG}_n^k$ input to $\mathtt{LS}_q$. Let $I = \mathtt{LS}_q(G)$ be the IS output by $\mathtt{LS}_q$ and let O be an optimal (maximum weight) IS in G. To establish Theorem 13, it suffices to show that if $H = (V_H, E_H)$ is a connected component of $G[I \cup O]$ with $I_H = V_H \cap I$ and $O_H = V_H \cap O$, then

$$w(O_H) \leq \left(k - 1 + \frac{1}{q}\right) w(I_H); \quad (5)$$

indeed, summing this inequality over all connected components H of $G[I \cup O]$ yields the desired bound $w(O) \leq \left(k - 1 + \frac{1}{q}\right) w(I)$.

Consider a connected component $H = (V_H, E_H)$ of $G[I \cup O]$ with $I_H = V_H \cap I$ and $O_H = V_H \cap O$. If $V_H = \{v\}$ is a singleton set, then the inclusion of the expansion improvement rule r_E in $\mathtt{LS}_q$ ensures that $I_H = \{v\}$, hence inequality (5) clearly holds. Assume hereafter that $|V_H| > 1$ and notice that in this case, H must be a bipartite graph with I_H on one side and O_H on the other.

The construction of H implies that $(O \cup I_H) - O_H$ is an independent set in G, thus the optimality of O ensures that $w(I_H) \leq w(O_H)$. If $w(I_H) = w(O_H)$, then inequality (5) holds trivially, so assume hereafter that $w(I_H) < w(O_H)$. By Observation 9, we know that $\deg_H(v) \leq |\alpha(v)| = w(v)$ for every $v \in I_H$. We shall establish inequality (5) by proving Lemma 2 and 3 for which we need the following combinatorial proposition (whose proof is deferred to [7]).

Proposition 1. *Let* $P = (A \uplus B, E)$ *be a connected bipartite graph and let* $w : A \cup B \to \{2, 3, \ldots, k\}$ *be a weight function that satisfies* $w(a) \geq \deg_P(a)$ *for each node* $a \in A$. *Then,* $w(B) \leq w(A)\left(k - 1 + \frac{1}{|A|}\right)$. *Moreover, if there exists a node* $a \in A$ *with* $w(a) > \deg_P(a)$, *then* $w(B) \leq w(A)(k-1)$.

Lemma 2. *If* $\deg_H(v) = w(v)$ *for every* $v \in I_H$, *then* $w(O_H) < \left(k - 1 + \frac{1}{q}\right) w(I_H)$.

Proof. If $|I_H| > q$, then by plugging $A = I_H$ and $B = O_H$ into Proposition 1, we conclude that $w(O_H) \leq \left(k - 1 + \frac{1}{|I_H|}\right) w(I_H) < \left(k - 1 + \frac{1}{q}\right) w(I_H)$, thus proving the assertion. So, assume towards contradiction that $|I_H| \leq q$, and recall that $I = \mathtt{LS}_q(G)$ is the IS returned by $\mathtt{LS}_q$ on G. By the construction of H, we know that $I' = (I \cup O_H) - I_H$ is an IS in G. We shall reach a contradiction by showing that I' is a candidate of r_q when applied to G and I, which means that the application of $\mathtt{LS}_q$ to G should not have returned I.

To this end, consider a node $v \in I_H$. The assumption that $\deg_H(v) = w(v)$ implies that $|N_H(v)| = |\alpha(v)|$. As $N_H(v)$ is an IS (recall that $N_H(v) \subseteq O_H$), we know that $\alpha(u) \cap \alpha(u') = \emptyset$ for every $u, u' \in N_H(v)$. This means that for every agent $i \in \alpha(v)$, there exists a unique node $u = u(i) \in N_H(v)$ such that $i \in \alpha(u)$. In particular, we conclude that $\alpha(v) \subseteq \alpha(N_H(v)) \subseteq \alpha(O_H)$, hence $\alpha(I_H) \subseteq \alpha(O_H)$. Therefore, $\alpha(I) \subseteq \alpha(I')$ and the assumption that $w(I_H) < w(O_H)$ guarantees that $\alpha(I) \subset \alpha(I')$. Since $|I_H| \leq q$ and since $O_H \subseteq N_G(I)$, it follows that I' is indeed a candidate of r_q, realized by taking $X = O_H$.

Lemma 3. *If there exists a node $v \in I_H$ such that $\deg_H(v) < w(v)$, then $w(O_H) \leq (k-1)w(I_H)$.*

Proof. Follows directly from Proposition 1 by plugging $A = I_H$ and $B = O_H$.

5 The Analysis of $\mathtt{LS}_q$ is Nearly Tight

Fix $k = 3$ and any parameter $q \in \mathbb{Z}_{>0}$. In this section, we present an example for a k-cycle graph $G_{\text{bad}} = (V, E) \in \mathcal{CG}_n^k$ with an optimal IS $O \subset V$ such that

$$\frac{w(O)}{w(\mathtt{LS}_q(G_{\text{bad}}))} = k - 1 + \frac{1}{q+1},$$

thus demonstrating that the analysis in Sect. 4.1 is nearly tight. The k-cycle graph G_{bad} is illustrated in Fig. 3. (Alternatively, the graph illustrated in Fig. 3 can be a connected component in a larger k-cycle graph, allowing n to become arbitrarily large with respect to q.) The IS $I \subset V$ that consists of the blue (solid) rectangles admits a weight of $w(I) = 3(q+1)$. A quick inspection reveals that $r_E(G_{\text{bad}}, I) = r_q(G_{\text{bad}}, I) = \perp$, where r_E and r_q are the expansion and all-for-q improvement rules, respectively. Therefore, with the right assignment of agent identities, we can ensure that $I = \mathtt{LS}_q(G_{\text{bad}})$. On the other hand, the IS $O \subset V$ that consists of the red (dashed) rectangles admits a weight of $w(O) = 3(2q+3)$, hence $\frac{w(O)}{w(\mathtt{LS}_q(G_{\text{bad}}))} = k - 1 + \frac{1}{q+1}$.

The k-cycle graph G_{bad} also demonstrates that the IS constructed by algorithm $\mathtt{LS}_q$ may be far from maximal in terms of the set of partaking agents (see footnote 8).

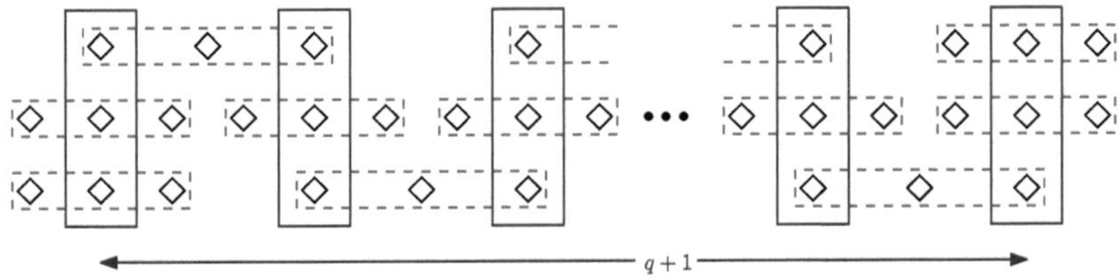

Fig. 3. The k-cycle graph $G_{\mathrm{bad}} = (V, E)$ whose nodes are depicted by the $q + 1$ blue (solid) rectangles and the $2q+3$ red (dashed) rectangles, defined over a set of $n = 6q+9$ agents depicted by the black diamond shapes, so that the diamond shape corresponding to agent $i \in [n]$ is contained in the rectangle corresponding to node $v \in V$ if (and only if) $i \in \alpha(v)$.

6 Brief Overview of the Non-uniform Case

To establish Theorem 4, we prove that for every length bound $k \geq 3$ and non-uniform length function λ, there exists an efficient (k, λ)-maxCGIS algorithm that given a k-cycle graph G, constructs an IS I in G whose weight $w(I)$ approximates that of an optimal IS in G within ratio at most $\max\{k-1+\epsilon, \rho(\lambda)\}$, where $\epsilon > 0$ is an arbitrarily small constant (recall the definition of $\rho(\lambda)$ in Sect. 1.2). in the current section, we provide an informal overview of this proof; due to lack of space, the full proof is deferred to [7].

The promised algorithm, denoted by $\mathtt{NU}_q$, is composed from "simpler" algorithms by means of *algorithm concatenation*, namely, the process of applying maxCGIS algorithms consecutively so that each algorithm is invoked on the subgraph of G that remains after removing the nodes in the ISs constructed by the previous algorithms and their neighbors. Owing to the agents' incentives, we focus on the concatenation of algorithms that consider nodes in non-decreasing order of length. In [7], we develop a sufficient condition for the truthfulness of maxCGIS algorithm resulting from such a concatenation.

For $2 \leq \ell \leq k$, let $\mathtt{Greedy}^{\ell}$ be the local search algorithm whose sole improvement rule is the expansion improvement rule r_E (defined in Sect. 4), restricted to nodes v of length $|\alpha(v)| = \ell$. Let ℓ^* be the largest $2 \leq \ell \leq k - 1$ that satisfies $\lambda(\ell) > \lambda(k)$ (this is well defined as λ is non-uniform). Algorithm $\mathtt{NU}_q$ is constructed as the concatenation

$$\mathtt{NU}_q = \mathtt{Greedy}^2 \cdot \mathtt{Greedy}^3 \cdots \mathtt{Greedy}^{\ell^*} \cdot \mathtt{LS}_q\,,$$

where $\mathtt{LS}_q$ is the local search algorithm introduced in Sect. 4. In other words, algorithm $\mathtt{NU}_q$ constructs the promised IS by greedily adding nodes of non-decreasing length, removing the added nodes and their neighbors from G, until all nodes in the remaining cycle graph G_{ls} are of length larger than ℓ^*. By the choice of ℓ^*, we know that $\lambda(|\alpha(v)|) = \lambda(k)$ for all nodes v in G_{ls}, so the invocation of $\mathtt{LS}_q$ on G_{ls} actually behaves like the uniform case. The truthfulness of algorithm $\mathtt{NU}_q$ is established by showing that the corresponding concatenation satisfies the aforementioned sufficient truthfulness condition.

Regarding the approximation ratio, we first claim that for every $2 \leq \ell \leq k$, if the concatenated algorithm $\texttt{Greedy}^2 \cdot \texttt{Greedy}^3 \cdots \texttt{Greedy}^\ell$ is invoked on a cycle graph all of whose nodes are of length at most ℓ, then the constructed IS is guaranteed to be an ℓ-approximation of the optimal IS (in the same cycle graph).[11] Building on this claim and some additional combinatorial arguments, we prove that

$$w(O_{gr}) \leq w(I_{gr}) \cdot \max\{k-1, \rho(\lambda)\},$$

where I_{gr} is the IS constructed by the invocation of $\texttt{Greedy}^2 \cdot \texttt{Greedy}^3 \cdots \texttt{Greedy}^{\ell^*}$ on G and O_{gr} is an optimal IS in the cycle graph $G_{gr} = G - G_{ls}$. On the other hand, by employing Theorem 13, we prove that

$$w(O_{ls}) \leq w(I_{ls}) \cdot (k - 1 + 1/\,\mathrm{q}),$$

where I_{ls} is IS constructed by the invocation of $\texttt{LS}_q$ on the cycle graph G_{ls} (recalling that this maxCGIS instance is de facto uniform) and O_{ls} is an optimal IS in G_{ls}. The promised approximation upper bound is then established by combining these two inequalities.

References

1. Abbassi, Z., Haghpanah, N., Mirrokni, V.: Exchange market mechanisms without money. In: Web and Internet Economics: 11th International Conference, WINE 2015, Amsterdam, The Netherlands, 9–12 December 2015, Proceedings, vol. 9470, p. 429 (2015)
2. Abraham, D.J., Blum, A., Sandholm, T.: Clearing algorithms for barter exchange markets: enabling nationwide kidney exchanges. In: MacKie-Mason, J.K., Parkes, D.C., Resnick, P. (eds.) Proceedings 8th ACM Conference on Electronic Commerce (EC-2007), San Diego, California, USA, 11–15 June 2007, pp. 295–304. ACM (2007)
3. Ashlagi, I., Fischer, F., Kash, I.A., Procaccia, A.D.: Mix and match: a strategyproof mechanism for multi-hospital kidney exchange. Games Econom. Behav. **91**, 284–296 (2015)
4. Ashlagi, I., Roth, A.E.: Free riding and participation in large scale, multi-hospital kidney exchange. Theor. Econ. **9**(3), 817–863 (2014)
5. Biro, P., Manlove, D.F., Rizzi, R.: Maximum weight cycle packing in directed graphs, with application to kidney exchange programs. Discrete Math., Algorithms Appl. **1**(04), 499–517 (2009)
6. Dickerson, J.P., Procaccia, A.D., Sandholm, T.: Failure-aware kidney exchange. Manage. Sci. **65**(4), 1768–1791 (2019)
7. Emek, Y., Shpiro, M.E.: Barter exchange with bounded trading cycles (2024). https://arxiv.org/abs/2410.06683
8. Hatfield, J.W.: Pairwise kidney exchange: comment. J. Econ. Theory **125**(2), 189–193 (2005)
9. Home Exchange. https://www.homeexchange.com/
10. Home Exchange with Love Home Swap. https://www.lovehomeswap.com/. House Swap Holidays

[11] In passing, this claim establishes Theorem 1 as algorithm `Greedy` is, in fact, the concatenation $\texttt{Greedy}^2 \cdot \texttt{Greedy}^3 \cdots \texttt{Greedy}^k$; refer to [7] for further details.

11. Trade Used Books with PaperBack Swap. https://www.paperbackswap.com/. The world's largest book club
12. Roth, A.E., Sönmez, T., Ünver, M.U.: Kidney exchange. Q. J. Econ. **119**(2), 457–488 (2004)
13. Roth, A.E., Sönmez, T., Ünver, M.U.: Pairwise kidney exchange. J. Econ. theory **125**(2), 151–188 (2005)
14. Roth, A.E., Sönmez, T., Ünver, M.U.: Efficient kidney exchange: coincidence of wants in markets with compatibility-based preferences. Am. Econ. Rev. **97**(3), 828–851 (2007)
15. Ünver, M.U.: Dynamic kidney exchange. Rev. Econ. Stud. **77**(1), 372–414 (2010)

Matching with Nested and Bundled Pandora Boxes

Robin Bowers[(✉)] and Bo Waggoner

University of Colorado Boulder, 1111 Engineering Drive, Boulder, CO 80309, USA
robin.bowers@colorado.edu

Abstract. We consider max-weighted matching with costs for learning the weights, modeled as a "Pandora's Box" on each endpoint of an edge. Each vertex has an initially-unknown value for being matched to a neighbor, and an algorithm must pay some cost to observe this value. The goal is to maximize the total matched value minus costs. Our model is inspired by two-sided settings, such as matching employees to employers. Importantly for such settings, we allow for negative values, causing existing approaches to fail. We first prove impossibility results for algorithms in two natural classes. Any algorithm that "bundles" the two Pandora boxes incident to an edge is an $o(1)$-approximation. Likewise, any "vertex-based" algorithm, which uses properties of the separate Pandora's boxes but does not consider the interaction of their value distributions, is an $o(1)$-approximation. Instead, we utilize *Pandora's Nested-Box Problem*, i.e. multiple stages of inspection. We give a self-contained, fully constructive optimal solution to the nested-boxes problem, which may have structural observations of interest compared to prior work. By interpreting each edge as a nested box, we leverage this solution to obtain a constant-factor approximation algorithm. Finally, we show any "edge-based" algorithm, which considers the interactions of values along an edge but *not* with the rest of the graph, is also an $o(1)$-approximation.

1 Introduction

We consider a problem of weighted matching under uncertainty about weights, motivated by two-sided matching settings such as job search websites and school admissions. In many of these matching applications, the values of each side for a potential match are initially unknown until some effort or cost is expended to inspect and discover the value. Investigating the quality of a school or job may be time consuming, and sorting through applicants may be costly for a company. The challenge in these settings is balancing the costly acquisition of information against the need to find high-value matches. We ask whether it is possible for an efficient algorithm to navigate this tradeoff.

Recent work on market design has studied the problem of matching with information acquisition, typically from the perspective of a platform designer in a strategic environment (see Sect. 1.2). This paper takes a step back to consider a simpler, purely algorithmic problem. However, all of our results have practical

M. Mavronicolas et al. (Eds.): WINE 2024, LNCS 15534, pp. 61–83, 2026.
https://doi.org/10.1007/978-3-032-08560-3_4

implications for design of two-sided matching platforms, particularly a number of impossibilities for practically-motivated classes of algorithms. Both these negative results and our eventual positive results depend on new insights into the *Pandora's Box* model.

Pandora's Box. We use Weitzman Weitzman, M.L. [1979]'s Pandora's Box model of costly information acquisition. In this model, each Pandora box is parameterized by an "inspection cost" to open the box and a distribution over the value inside (drawn independently). In Weitzman's original problem, an algorithm may choose any sequence of inspections, then stop at any time and claim the value inside some opened box. The goal is to maximize the final value minus all inspection costs paid.

Our Model. In *Pandora's Matching Problem*, we are given a graph with initially uncertain values on edges. Each edge of the graph has two Pandora boxes, one at each endpoint, representing the values of each of the two parties involved in a potential match. The weight of an edge is the sum of the values on its endpoints, which may each be inspected separately at a cost. An algorithm may conduct any sequence of inspections, then stop and output a matching consisting of fully-inspected edges. The performance, or "welfare", of an algorithm is the sum of edge weights in the matching minus the sum of inspection costs paid. We consider the approximation factor, i.e. worst-case ratio of expected welfare to that of the optimal algorithm (whatever it may be).

Our model captures two-sided preferences in matching markets with costly inspection on both sides. For example, a company seeking to hire a new worker must conduct time-intensive interviews to determine the fit of a candidate, while the candidate must do the same. Similarly, a student applying to colleges must invest the time to determine if each school she considers would be a good fit, as applying to every possible school is infeasible. A university spends time and money processing applications, and cannot give every application a full evaluation. We are interested in the consequences of this two-sided inspection dynamic for algorithm design.

We will relate our problem to *Pandora's Nested-Box Problem*, in which opening a box uncovers another costly box within, along with some information about the contents. We give a new treatment of the nested-boxes problem and apply it to address Pandora's Matching Problem.

Negative Values. Prior work of Waggoner Bowers and Waggoner [2024] implies an approximation algorithm for the *positive-values* setting where all values are positive and exceed inspection costs in expectation. However, many two-sided matching applications involve negative values. For example, a candidate generally has a negative value for being assigned a job and needs to be paid a wage in order to do it. [1] Because of this, we are particularly motivated by a

[1] Simple adjustments to reduce to the positive setting do not work. E.g., adding the wage of the job to the job's value distorts the social welfare unless we also subtract it from the employers' value for hiring the employee, and in general there will be pairs where the sum is negative. Some of our negative results formalize the impossibility of such approaches.

general-values setting. Waggoner Bowers and Waggoner [2024] state obtaining an approximation algorithm in the general-values setting as a main open problem.

1.1 Results

We expect the problem to be NP-hard, as even the special case of one Pandora box per edge is suspected to be NP-hard (e.g. Singla Singla, S. [2018]). However, a (1/2)-approximation to that special case is given by Kleinberg et al. Kleinberg, R., et al. [2016] and Singla Singla, S. [2018], utilizing a generalization of Weitzman's descending procedure to mimic a greedy matching. Motivated by this approach, we first consider simple classes of approximation algorithms.

Bundled Box & Vertex-Based Algorithms. A first natural approach is to "bundle" the two boxes on each edge into one larger Pandora box: if inspecting the edge, always inspect both boxes simultaneously. Another possible approach is a generalization of the approach taken by Waggoner Bowers and Waggoner [2024] and Singla Singla, S. [2018], which we call *vertex-based* algorithms.

Unfortunately, Sect. 3 finds that both bundled box and vertex-based approaches fail in general when applied to the two-sided matching setting.

Theorem 1. *No bundled-box or vertex-based algorithm for Pandora's Matching Problem can guarantee a positive approximation factor.*

These impossibilities also have implications for practical platform design. The bundled setting corresponds to traditional interviewing, where both parties must agree to "inspect" together. This result suggests that welfare may be lost from markets that require participants to coordinate their information-gathering efforts. Similarly, a vertex-based approach is one that allows for decentralized decisionmaking. For example, the matching mechanism in Bowers and Waggoner Bowers and Waggoner [2024] does not share information about the participants' preferences with each other. The failure of vertex-based algorithms implies that some information-sharing is necessary to coordinate good matches.

Nested Boxes. Section 4 turns to the *Pandora's Nested-Box Problem*, also called Pandora's problem with multiple stages of inspection, to account for interactions between the boxes along an edge. Kleinberg, R., et al. [2016] provide a generalization of Weitzman's indices and our descending procedure to the multi-stage inspection setting. We give a self-contained, constructive treatment with intuition for the algorithm that we believe is of independent value (see Sect. 1.2).

Fixed-Orientation Algorithms. In Sect. 5, we utilize our nested-boxes machinery to create two approximation algorithms for Pandora's Matching Problem.

Theorem 2. (Main Result). *There exists an algorithm which obtains a* $1/4$*-approximation for Pandora's Matching Problem. The guarantee holds even if each edge's pair of values may be arbitrarily correlated with each other.*

Both of these algorithms work by first fixing an order of inspection (an "orientation") for each edge, and then running the $1/2$-approximation descending procedure on the fixed-orientation setting. Finally, we consider simpler deterministic algorithms, and demonstrate that a very natural class of *edge-based fixed-orientation* algorithms also cannot achieve a constant approximation factor.

Theorem 3. *No edge-based fixed-orientation algorithm for Pandora's Matching Problem can guarantee a positive approximation factor.*

1.2 Related Work

Pandora's Box and Its Variants The Pandora's Box Problem was first introduced by Weitzman Weitzman, M.L. [1979]. Our terminology and machinery for the problem follow Kleinberg et al. Kleinberg, R., et al. [2016], who used the problem in an auction setting, as well as a one-sided matching setting (i.e. matching with a single Pandora box on each edge).

This approach was extended to many combinatorial optimization problems with information acquisition by Singla Singla, S. [2018]. Our problem is not addressed by his results in particular because greedy max-weight matching, if there is a weight on each endpoint of an edge, is not a *frugal algorithm*: it needs to consider the contributions of pairs of items (both endpoints of an edge), while a frugal algorithm considers one at a time. In a sense, our impossibility result for vertex-based algorithms in Sect. 3 implies that no variant of the frugal framework can solve our problem.

For what we call *Pandora's Nested-Box Problem*, also referred to as multiple stages of inspection, there are a number of recent works adjacent to ours. Most important, the Weitzman index approach and optimal descending algorithm was shown to extend to this case by Kleinberg et al. Kleinberg, R., et al. [2016]. The treatment there is quite general (in particular, handling countably infinite stages of inspection), but it is also quite complex and nonconstructive. In Sect. 4, we include a concrete and direct extension of the Weitzman index machinery to the nested-boxes setting. We present the key features in the main body of the paper because, in addition to enabling our final results, they do provide some novel points of utility and interest:

- A simple and direct construction with a much shorter, self-contained treatment. Given an example (e.g. a Bernoulli value whose bias is selected over the course of the inspections), it is not necessarily clear from Kleinberg et al. Kleinberg, R., et al. [2016] how exactly to compute the Weitzman indices; here we give a direct formula.
- Illustration of several subtle points that arise in the analysis, such as a precise and constructive formulation of the *non-exposed* condition (also called exercise in the money, claim above the cap, etc.).
- Simple and concrete machinery – Weitzman indices (also called strike prices or reservation values) capped values (also called covered call values or deferred values), and a general "amortization lemma" – that is more amenable to adaptation to different environments.

Gupta et al. Gupta et al. [2019] provides a generalization of Singla Singla, S. [2018]'s frugal algorithms to a multistage inspection model for combinatorial optimization. They consider optimization over boxes represented by Markov chains with costs for advancement, where a chain must be advanced to a terminal state in order for the box to be added to the solution. Gupta et al. Gupta et al. [2019] derives a parallel to the Weitzman index machinery of Kleinberg et al. Kleinberg, R., et al. [2016] in the discrete Markov decision process setting. Their approach is essentially a generalization of our Sect. 4, except with a focus on compatibility with frugal algorithms. As mentioned above, our two-sided matching problem is incompatible with frugal algorithms, so the results of Gupta et al. (2019) Gupta et al. [2019] do not apply to our main problem.

Several other works study variants of the nested-boxes problem, but they are much farther from our setting. Ke and Villas-Boas (2019) Tony Ke, T. and Miguel Villas-Boas, J [2019] study a model in which a decisionmaker gathers information about each of a set of alternatives in continuous time for continuous cost, and stops at any time to choose an alternative. (In particular, full inspection is not obligatory.) Aouad et al. Aouad, A., et al. [2020] consider a model with two stages of inspection, partial and full. However, the algorithm does not have to undertake partial inspection of an option prior to a full inspection. (Full inspection is still obligatory in order to claim.) In this setting, Aouad et al. Aouad, A., et al. [2020] show that a descending index-based policy is no longer optimal, studying features of the optimal solution and approximation algorithms. Boodaghians et al. Boodaghians, S. et al. [2020] study a somewhat-related setting that can be viewed as Pandora's box problem where boxes must be opened in a particular order, particularly when the ordering follows a linear or DAG structure.

There has been a variety of other recent work on variants of the Pandora's Box Problem, including when inspection is non-obligatory to claim the item (Doval Doval, L. [2018], Beyhaghi and Kleinberg Beyhaghi, H. and Kleinberg, R. [2019] , Beyhaghi and Cai Beyhaghi, H and Cai, L. [2023], Fu et al., Fu, H., et al. [2023]) and when the values in the boxes are correlated Beyhaghi and Cai Chawla, S., et al. [2020]. All of the above settings and a number of others are discussed in the recent survey of Beyhaghi and Cai Beyhaghi, H and Cai, L. [2023].

Market Design. A number of recent works in market design or mechanism design have utilized Pandora's box as a model for initially-unknown values. Closest to our work is owers and Waggoner Bowers and Waggoner [2024]. The authors consider essentially the same model as we do, but with each endpoint representing a strategic agent. They study a descending-price mechanism, an extension of Kleinberg et al. Kleinberg, R., et al. [2016] proposed by Waggoner and Weyl Waggoner, B. and Weyl, E.G. [2019], in which agents maintain bids on each of their neighbors in the graph and are matched and pay their bids from the highest-total edge downward. With the assumption of nonnegative values, Bowers and Waggoner Bowers and Waggoner [2024] prove a constant-factor Price of Anarchy guarantee, implying that a constant-factor approximation is achievable in that setting efficiently (up to computation of an equilibrium; we extend their result into an efficient algorithm in the online appendix for completeness).

They give partial results for the general-values case, but leave the existence of a constant-factor approximation as a main open question.

Other works in the market design literature generally focus more on modeling of matching marketplaces than we do here. Immorlica et al. Immorlica, N., et al. [2021] studies design of a platform to facilitate two-sided matching with inspection costs. There is a continuous population of agents and matching occurs in continuous time, with the platform directing agents of a given type to "meet" and inspect counterparts of some other type. The authors give a 1/4-approximation to the problem of finding an optimal flow process, despite adhering to incentive-compatibility constraints. Immorlica et al. Immorlica, N., et al. [2020] and Hakimov et al. Hakimov, R., et al. [2023] both study the college application and admissions process with costly information acquisition through a theoretical lens; Immorlica et al. Immorlica, N., et al. [2020] include a case where each student is modeled as a Pandora box (two-sided matching with one-sided information acquisition). Other works such as Ashlagi et al. Ashlagi I., et al. [2020] study the cost of learning preferences through a lens of communication complexity. Chade et al. Chade, H., et al. [2017] surveys prior approaches in the economics literature to matching markets with information acquisition, but the models and techniques are quite different from our problem.

2 Preliminaries

In this section, we introduce notation and define the main problem studied in this paper, Pandora's Matching Problem. We also introduce tools from prior work for Pandora's box analysis.

2.1 Model and Notation

Recall that in the traditional Pandora's Box Problem, Pandora is presented with n boxes. Formally, a *Pandora box* is a pair (D, c) consisting of the distribution D of the value $v \in \mathbb{R}$ inside the box along with the cost $c \in \mathbb{R}_{\geq 0}$ to open the box and observe the value. Pandora can pay the costs and open as many boxes as she wishes, but can only claim the value from one box.

To adapt the problem to a two-sided matching setting, we let each edge $\{i, j\}$ in a graph have a different Pandora box associated with each endpoint: vertex i has some initially-unknown value for matching to j, and vice versa. Formally, the input to the *Pandora's Matching Problem* consists of:

- A graph $G = (V, E)$. We do not require the graph to be bipartite.
- For each edge $\{i, j\} \in E$: a pair of Pandora boxes (D_{ij}, c_{ij}) and (D_{ji}, c_{ji}).

An algorithm iteratively chooses an unopened box (i, j), i.e. an endpoint of some edge of the graph, and "opens" it by paying the cost c_{ij}, learning v_{ij}. Each v_{ij} is drawn from its respective distribution D_{ij} independently. At any point, the algorithm may choose to stop inspecting and output a matching of G consisting of fully-inspected edges. Note that we distinguish the edge $\{i,j\}$ from the two boxes on its endpoints, (i,j) and (j,i).

Given an algorithm and an input instance $\mathcal{I} = (G, \{D_{ij}\}, \{c_{ij}\})$, we let $\mathbb{I}_{ij}$ be the indicator variable that box (i,j) is opened (or "inspected"), and let $\mathbb{A}_{ij}$ be the indicator variable that box (i,j) is included in the matching (or "claimed"). Formally, the matching requirement is $\mathbb{A}_{ij} = \mathbb{A}_{ji} \quad (\forall \{i,j\} \in E)$ along with $\sum_{j \in V} \mathbb{A}_{ij} \leq 1 \quad (\forall i \in V)$.

We impose the classic "obligatory inspection" requirement, that both endpoints of an edge must be inspected before the edge can be included in the matching. Formally, $\mathbb{A}_{ij} \leq \mathbb{I}_{ij}$ for each box (i,j). The performance of algorithm ALG on the instance $\mathcal{I}$ is

$$\text{Welfare}(\textsc{Alg}) := \mathbb{E}\left[\sum_{(i,j)} \mathbb{A}_{ij} v_{ij} - \mathbb{I}_{ij} c_{ij}\right],$$

which we refer to as the *welfare* of the algorithm.[2] We define the *welfare contribution* of box (i,j) to be $\mathbb{A}_{ij} v_{ij} - \mathbb{I}_{ij} c_{ij}$. We denote the optimal algorithm by OPT. We say an algorithm ALG achieves an approximation factor α or is an α-approximation if, for all instances $\mathcal{I}$,

$$\frac{\text{Welfare}(\textsc{Alg})}{\text{Welfare}(\textsc{Opt})} \geq \alpha.$$

2.2 Pandora's Toolbox

We recall several tools from the existing Pandora's Box literature. The optimal algorithm for the original problem of Weitzman Weitzman, M.L. [1979] relies on indices that we will call σ_{ij} for each box (i,j).

Definition 1 (Weitzman Index & Capped Value). *The* Weitzman index *of a Pandora box (D_{ij}, c_{ij}) is the unique σ_{ij} such that*

$$\mathop{\mathbb{E}}_{v_{ij} \sim D_{ij}} \left[(v_{ij} - \sigma_{ij})^+\right] = c_{ij},$$

and the capped value *of the box is the random variable $\kappa_{ij} = \min(v_{ij}, \sigma_{ij})$.*

Here $(\cdot)^+ := \max\{\cdot, 0\}$. Discussions, such as a financial-options interpretation, can be found in other works such as Kleinberg et al. Kleinberg, R., et al. [2016], Beyhaghi and Cai Beyhaghi, H and Cai, L. [2023].

Definition 2. *An algorithm is* exposed *on box (i,j) if there is a nonzero probability that $\mathbb{I}_{ij} = 1$, $v_{ij} > \sigma_{ij}$, and $\mathbb{A}_{ij} = 0$. Otherwise, the algorithm is* non-exposed *on box (i,j).*

[2] The "welfare" terminology (i) references our motivating economic applications and (ii) differentiates from the term "value" which refers to the contents of Pandora boxes.

In other words, an algorithm is exposed on a box if it opens the box, finds that the value is larger than the index, and ultimately fails to claim that box. We call an algorithm *non-exposed* if it is non-exposed on all boxes.

Lemma 1. (Kleinberg et al. *Kleinberg, R., et al. [2016]***).** *For any algorithm, for any Pandora box* (i, j),

$$\mathbb{E}\left[\mathbb{A}_{ij} v_{ij} - \mathbb{I}_{ij} c_{ij}\right] \leq \mathbb{E}\left[\mathbb{A}_{ij} \kappa_{ij}\right],$$

with equality if and only if the policy is non-exposed on box (i, j).

In the classic Pandora's Box Problem, recall that there are n Pandora boxes and Pandora must ultimately select just one. As discussed in Kleinberg et al. Kleinberg, R., et al. [2016], Lemma 1 implies optimality of a simple *descending* procedure, which inspires approaches in this work. In the descending procedure, boxes are opened in order of their index from largest to smallest. If at any point the largest value observed so far is larger than all remaining indices, the search is stopped and that value is claimed. This algorithm is non-exposed. Therefore, summing the equality in Lemma 1 over all boxes, its welfare is equal to its total expected capped value. But it always claims the box with the highest capped value. Lemma 1 implies that this maximizes the possible welfare of any policy that claims at most one box, so the descending procedure is optimal.

3 Initial Attempts

We first consider two natural classes of algorithms with intuitive interpretations, and demonstrate that neither of them can achieve welfare guarantees in the general-values setting.

3.1 Bundled-Box Algorithms

Singla Singla, S. [2018] provides a general welfare approximation guarantee for the one-sided matching setting, in which each edge has only one Pandora box on it. Can our two-box setting be reduced to this setting, by simply forcing boxes on an edge to always be opened simultaneously? In many real-world settings such as interviewing, where a company and employee must both invest simultaneously to reveal their values, it is natural to require the algorithm to "bundle" the two boxes incident to an edge. Formally, a *bundled-box algorithm* is one that, if it opens one box on an edge, always opens the other next. In particular, it satisfies $\mathbb{I}_{ij} = \mathbb{I}_{ji} \;\; (\forall \{i, j\} \in E)$.

A bundled-box algorithm functionally reduces each edge to a single Pandora box, with value the sum of individual values and cost the sum of costs. We model the bundled box on edge $\{i, j\}$ as having a combined value $v'_{\{i,j\}} = v_{ij} + v_{ji}$, and we define the new value distribution $v'_{\{i,j\}} \sim D'_{\{i,j\}}$, the convolution of D_{ij} and D_{ji}.

If we let OPT' denote the optimal algorithm for this setting, then Kleinberg et al. Kleinberg, R., et al. [2016] and Singla Singla, S. [2018] both give 1/2-approximations to OPT', extending Weitzman's descending procedure: inspect the edges from largest index downward, matching any edge once it has been inspected and its value is larger than all remaining indices, and removing edges from consideration once they are not legal to add to the matching.

The Power of Bundled Boxes. The key question is then whether OPT' can approximate OPT, i.e. whether bundling comes at a high cost. On one hand, OPT clearly has significant additional power over simultaneous inspection in choosing inspections. On the other, edges are ultimately bundled when claimed by a matching. Here, we show that bundling the edges can perform arbitrarily poorly relative to algorithms that inspect asynchronously.

Proposition 1. *No bundled-box algorithm can guarantee better than an $O(\frac{1}{n})$-approximation factor for Pandora's Matching Problem on n vertices.*

The proof, given in full in the online version of this paper, provides an instance on which no bundled-box algorithm can approximate OPT to a constant factor.

The key feature of the proof is the observation that a bundled-box algorithm is forced to pay to open boxes with unhelpful information, which can be avoided by an algorithm inspecting endpoints independently. This result may have implications for the design of matching platforms such as labor markets, where it suggests enabling at least some asynchronous inspection.

3.2 Vertex-Based Algorithms

We next consider algorithms which treat the boxes independently of one another, making decisions to open or permanently accept boxes independently of any other boxes. A natural approach is to directly extend Weitzman Weitzman, M.L. [1979]'s optimal index-based approach to this setting. For example, one could inspect the boxes from largest Weitzman index downward, matching an edge once both of its endpoints are inspected and the sum of the values is large enough. Approaches like this match the *frugal algorithm* framework of Singla Singla, S. [2018]. We find that, when values must be positive, the mechanism of Bowers and Waggoner Bowers and Waggoner [2024] yields a 1/4-approximation algorithm, but that no vertex-based algorithm can achieve a constant approximation for general values.

Definition 3. *An algorithm is* vertex-based *if, for each ordered pair (i, j), the algorithm uses only Weitzman's index σ_{ij}, the inspection cost of a vertex c_{ij}, and (when revealed) the value v_{ij}, but does not have access to the distribution D_{ij}.*

To enable analysis, Definition 3 formalizes vertex-based algorithms by limiting the information available to the algorithm. A vertex-based algorithm knows everything about the setting except for the distribution of values D_{ij} at each endpoint. We replace the knowledge of D_{ij} with direct access to the Weitzman index σ_{ij} instead.

Positive Values. In the *positive-values* setting of Pandora's Matching Problem, value distributions D_{ij} are supported on the positive reals and $\mathbb{E}[v_{ij}] \geq c_{ij}$ for all (i, j). This implies $\sigma_{ij} \geq 0$ and $\kappa_{ij} \geq 0$ as well.

This restricted setting is already nontrivial. For example, our impossibility result of Proposition 1 holds even in the positive-valued setting, showing that no bundled-box algorithm can obtain a better-than-$O(1/n)$ approximation ratio. However, there is a vertex-based algorithm which achieves a 1/4-approximation to the optimal expected welfare in the positive-values setting. This is a corollary of the results of Bowers and Waggoner Bowers and Waggoner [2024], which gives a descending-price style mechanism that achieves a constant Price of Anarchy by focusing on the higher-capped-value endpoint of each edge. We extract a simple approximation algorithm for completeness, with details and proof available in the online appendix.

Proposition 2. *There exists a vertex-based algorithm which achieves a* 1/4*-approximation in the positive-values setting of Pandora's Matching Problem.*

General Values and Failure of Vertex-Based Algorithms. However, this result hinges on the nonnegativity of values: it will match an edge when one endpoint has a positive value, even if the other endpoint's value is arbitrarily negative. We find that this problem is not fixable for any vertex-based algorithm when values may be negative.

Proposition 3. *No vertex-based algorithm can guarantee any positive approximation factor for Pandora's Matching Problem.*

The proof of Proposition 3 is available in the online version of the paper, and relies on a single-edge counterexample. In this example, a vertex-based algorithm cannot distinguish the endpoints to pick an inspection ordering but an optimal policy must inspect them in a specific order.

This negative result has implications for platform or mechanism design: an algorithm must coordinate a nontrivial amount of information-sharing in order to guarantee high welfare in general. We must turn to algorithms which consider edges as a whole, rather than the individual indices of the endpoints.

The Correct Way to Combine Endpoints' Information If we must combine endpoints' information to make good decisions, how should we do so? An edge in the Pandora's Matching Problem involves *multiple stages* of information gathering: first we inspect one endpoint and learn something about the edge's total value, and later we inspect the other. Therefore, we must put the matching problem on hold and investigate *Pandora's Nested-Box Problem.* Once we have a solution to that problem, we will come back and apply it in the matching setting.

4 Pandora's Nested-Box Problem

To solve the general problem of matching with vertex-based Pandora boxes, we will need a new approach. We turn to the model of *Pandora's Nested-Box Problem.* In this model, when Pandora first pays the cost to open a box, there is a second box inside, with its own associated cost; and so on, with the final reward inside the last inner box.

While we are coining the term "Nested Boxes" to refer to this problem, it has previously been studied and an optimal solution is known (Kleinberg et al., Kleinberg, R., et al. [2016]). In fact, just as with the standard Pandora's Box Problem, this follows directly from the Gittins Index Theorem as a special case (Gittins, Gittins, J.C. [1979]). However, just as with the standard Pandora's Box Problem, a constructive solution can give insights about the structure of the problem.

4.1 Model

For clarity, we will introduce terminology that distinguishes between *baskets* and *boxes* (illustrated in Fig. 1). Using this terminology, in the traditional Pandora's Box Problem, Pandora is presented with n baskets, and each basket initially contains a single box. When Pandora chooses to inspect the box in basket i, she pays the cost c_i to open its box, finds a reward v_i, throws away the box packaging, and places v_i in the basket. In the end, Pandora can pick one basket with a reward in it to take home. We modify this problem into the *Pandora's Nested-Box Problem.*

Pandora's Nested-Boxes. A *Pandora basket* i, defined by a pair (D_i, c_i), contains a nested set of $d \geq 1$ boxes. For $\ell = 1, \ldots, d-1$, opening the ℓth box in basket i provides the algorithm with a signal $s_i^{(\ell)}$ regarding the value v_i of the basket. Opening the dth box reveals v_i. The signals and final value are drawn jointly from a known distribution, i.e. $(s_i^{(1)}, \ldots, s_i^{(d-1)}, v_i) \sim D_i$. The sequence of costs $c_i = (c_i^{(1)}, \ldots, c_i^{(d)})$ to perform each inspection is also known.

We refer to the most-recently opened box ℓ in a basket as the current *stage* of a basket: When we have opened no boxes, we are in stage zero. After opening all d boxes, we are on the dth stage, and after claiming the item, the $d+1$st. We modify the indicator variables $\mathbb{A}$ and $\mathbb{I}$ to accommodate nested boxes. $\mathbb{A}_i$ continues to denote the claiming of basket i, but now the indicator variable for opening the ℓth box in basket i is $\mathbb{I}_i^{(\ell)}$. We will let $\mathbb{I}_i^{(d+1)} = \mathbb{A}_i$ for notational convenience in inductive statements. We refer to opening a nested box or claiming the final item as *advancing* the stage of the basket.

All previous boxes $k = 1, \ldots, \ell - 1$ in the basket must be opened before opening nested box ℓ, and all d boxes in a basket must have been opened to claim the contents of the basket. Formally, we require for any algorithm that $\mathbb{I}_i^{(1)} \geq \cdots \geq \mathbb{I}_i^{(d)} \geq \mathbb{I}_i^{(d+1)} = \mathbb{A}_i$.

Optimization Problems. In the spirit of Singla Singla, S. [2018], we allow for general optimization problems. In such a problem we are given a set of

Fig. 1. $n = 3$ Pandora baskets with $d = 2$ nested boxes, advanced to various stages with signals and/or values revealed. All distributions and costs are known ahead of time.

Pandora baskets, each specified by (D_i, c_i), along with a constraint on which subsets may be claimed. The algorithm performs any sequence of steps in which it selects a basket and advances it, and eventually stops and claims a subset of fully-advanced baskets that satisfies the constraint. In particular, in the Pandora's Nested-Box Problem, the algorithm may claim at most one basket. The *welfare contribution* of basket i is $\mathbb{A}_i v_i - \sum_{\ell=1}^{d} \mathbb{I}_i^{(\ell)} c_i^{(\ell)}$. The expected welfare of an algorithm is the sum of the expected welfare contributions of the baskets.

Remarks. The model and results in this section can be generalized so that the "depth" d of each basket may be a different integer, and so that the costs of inner boxes are random variables, with each $c_i^{(\ell)}$ a function of the signals $s_i^{(1)}, \dots, s_i^{(\ell-1)}$. Essentially no change to the below analysis is required. The case of possibly-infinite depth is treated in Kleinberg et al. Kleinberg, R., et al. [2016].

4.2 Pandora's Toolbox for Nested Boxes

Classic Pandora's Box analyses depend on the proper definitions of the index and capped value, as well as on the key lemma (Lemma 1). We need to generalize these definitions correctly and prove the analogous lemma. Kleinberg et al. Kleinberg, R., et al. [2016] give more general but nonconstructive definitions, and fleshing out these details reveals a few interesting subtleties and useful intuitions.

Conceptually, when defining an index $\sigma_i^{(\ell)}$ for the nested box ℓ, we would like to capture the usefulness of its contents, box $\ell + 1$, by some "proxy value" $\hat{v}_i^{(\ell+1)}$, such that the index could be assigned in the classical way as the solution to $\mathbb{E}[(\hat{v}_i^{(\ell+1)} - \sigma_i^{(\ell)})^+] = c_i^{(\ell)}$. In fact, this turns out to be possible, where the "proxy value" is none other than the capped value of box $\ell + 1$. This allows us to define $\kappa_i^{(\ell)}$ and $\sigma_i^{(\ell)}$ together via backward induction.

Definition 4. Nested Weitzman Index & Capped Value). *The* nested Weitzman index *for box i at the ℓth stage of inspection is the random variable $\sigma_i^{(\ell)}$, a function of $s_i^{(1)}, \dots, s_i^{(\ell-1)}$, that uniquely solves*

$$\mathbb{E}\left[\left(\kappa_i^{(\ell+1)} - \sigma_i^{(\ell)}\right)^+ \,\middle|\, s_i^{(1)}, \dots, s_i^{(\ell-1)}\right] = c_i^{(\ell)},$$

where $\kappa_i^{(\ell)}$ *is the* capped value $\kappa_i^{(\ell)} = \min\{\sigma_i^{(\ell)}, \kappa_i^{(\ell+1)}\}$ *for* $1 \le \ell \le d$, *and we define* $\kappa_i^{(d+1)} = v_i$. *We also define* $\sigma_i^{(d+1)} = v_i$.

Nested Non-exposure. We also adapt the notion of non-exposure to fit the nested box setting. Intuitively, potential for exposure occurs when the algorithm opens a box ℓ in basket i and receives a "good news" signal, resulting in a high Weitzman index for the next box. To avoid exposure, the algorithm should continue by opening the next box in this basket. Given our claim that $\kappa_i^{(\ell)}$ represents the value of the ℓth box, we might like to define exposure as a case where $\kappa_i^{(\ell+1)} > \sigma_i^{(\ell)}$, and the algorithm has advanced basket i to stage ℓ, yet the algorithm does not advance it to stage $\ell + 1$. Consider a case, however, where $\sigma_i^{(1)} < \kappa_i^{(3)} < \sigma_i^{(2)}$ and we have advanced to stage 2. In this case, not advancing to stage 3 would cause an exposure *relative to the decision to advance to stage* 2, because $\kappa_i^{(3)} > \sigma_i^{(1)}$, even if there is no exposure relative to $\sigma_i^{(2)}$. This intuition motivates our definition of non-exposure.

Let $\gamma_i^{(\ell)} = \min_{1 \le \ell' \le \ell} \sigma_i^{(\ell')}$, the minimum of the first ℓ Weitzman indices for basket i. We note that $\gamma_i^{(d+1)} = \kappa_i^{(1)} = \min_\ell \sigma_i^{(\ell)}$. See Figs. 2 and 3 for a visual depiction.

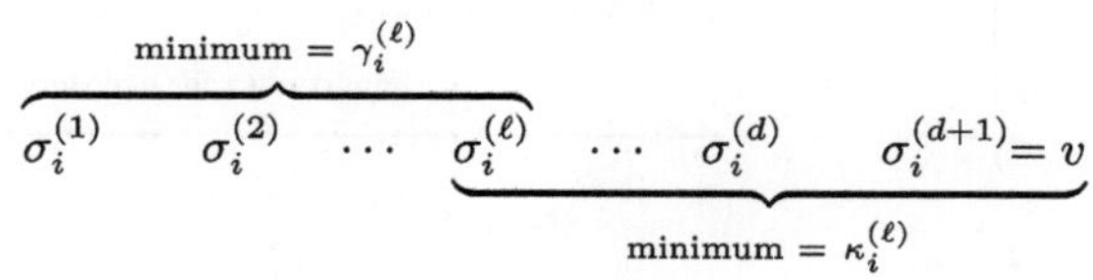

Fig. 2. The definitions of $\kappa_i^{(\ell)}$ and $\gamma_i^{(\ell)}$ in terms of the nested indices $\sigma_i^{(\ell)}$ of a Pandora basket i. Note that $\gamma_i^{(1)}, \gamma_i^{(2)}, \dots$ is a weakly decreasing sequence, while $\kappa_i^{(1)}, \kappa_i^{(2)}, \dots$ is weakly increasing (see Fig. 3).

Definition 5. *An algorithm is* exposed *on basket i if there is nonzero probability that there exists* $\ell \in \{1, \dots, d\}$ *such that* $\mathbb{I}_i^{(\ell)} > \mathbb{I}_i^{(\ell+1)}$ *but* $\gamma_i^{(\ell)} < \kappa_i^{(\ell+1)}$. *Otherwise, it is* non-exposed *on basket i. An algorithm is non-exposed if it is non-exposed on all baskets.*

Definition 5 raises an apparent obstacle. In the classic Pandora's Box problem, it is obvious how to keep an algorithm non-exposed: if the value is higher than the index, claim it. But here, at an arbitrary stage ℓ, $\kappa_i^{(\ell+1)}$ is not visible to the algorithm, as its realization depends on what would happen if the basket continued to advance. Therefore, it is not obvious how to guarantee non-exposure. The resolution comes from observing that $\kappa_i^{(\ell+1)} \le \sigma_i^{(\ell+1)}$. Therefore, we can prevent exposure by always continuing to advance if the next box's Weitzman index is larger than $\gamma_i^{(\ell)}$, the minimum of the indices so far. It is necessary to compare to $\gamma_i^{(\ell)}$ rather than $\sigma_i^{(\ell)}$ by the previous discussion.

The Key Lemma. In the classic setting, the key lemma (Lemma 1) upper-bounds the expected welfare contribution of a Pandora box, with equality iff non-exposure. It can also be referred to as an amortization lemma, as it relates the average welfare to the average capped value. In our nested setting, the analogue is the following.

Lemma 2. *For any algorithm, the expected welfare contribution of a Pandora basket i satisfies*

$$\mathbb{E}\left[\mathbb{A}_i v_i - \sum_{\ell=1}^{d} \mathbb{I}_i^{(\ell)} c_i^{(\ell)}\right] \leq \mathbb{E}\left[\mathbb{A}_i \kappa_i^{(1)}\right],$$

with equality if and only if the algorithm is non-exposed on basket i.

We prove Lemma 2 in the online appendix.

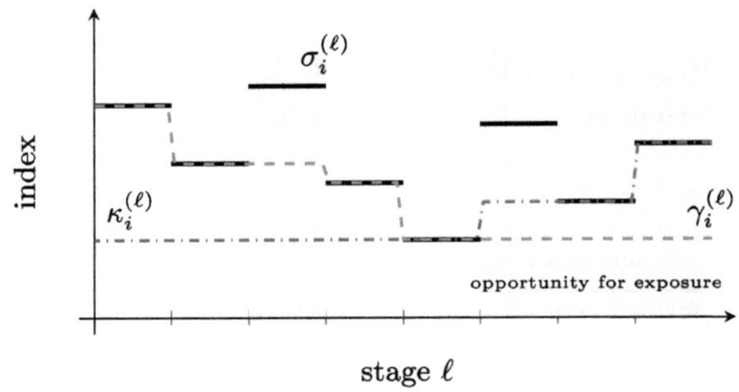

Fig. 3. One realization of the Weitzman indices $\sigma_i^{(1)}, \ldots$ when opening nested boxes in a Pandora basket i. The minimum index so far is $\gamma_i^{(\ell)}$ and the minimum yet to come (including the final value) is the capped value $\kappa_i^{(\ell)}$. Yellow regions represent possible opportunities for exposure: if the index $\sigma_i^{(\ell)}$ of the next box under consideration is higher than the current $\gamma_i^{(\ell-1)}$, it is possible that $\kappa_i^{(\ell)} > \gamma_i^{(\ell-1)}$, in which case failing to continue opening would cause exposure.

It is not immediately clear that non-exposed algorithms exist, as the conditions in Definition 5 rely on information unavailable to a decision-maker at the time of each choice.

4.3 Descending Procedures

Recall that *advancing* a basket refers to opening the next box or claiming the item if all boxes are opened.

Definition 6. *Call an algorithm a* descending procedure *if:* (a) *it maintains a non-increasing subset of baskets, the* eligible *set,* (b) *at each step it advances the eligible basket with the largest Weitzman index, and* (c) *it never removes a basket from the eligible set in the same step in which the basket was advanced.*

The idea of the eligible subset is that in optimization problems such as matching, the constraints often cause an algorithm to permanently rule out some basket from further consideration.

Lemma 3. *Any descending procedure is non-exposed.*

The proof of Lemma 3 is provided in the online version of this paper. We will crucially use Lemma 3 to show our main positive result for Pandora's Matching Problem.

As a warm-up to matching, we sketch a proof that the natural descending algorithm is optimal for Pandora's Nested-Box Problem, i.e. the case where Pandora may select any one box. The descending algorithm always advances the basket i with maximum current Weitzman index σ_i^t, until one has been claimed or all indices are negative. Its capped value is pointwise maximum, i.e. $\sum_i \mathbb{A}_i \kappa_i^{(1)} = \max_i (\kappa_i^{(1)})^+$ in all realizations. The key point in the proof is that at each time basket i was advanced a stage, its Weitzman index was larger than all other baskets', so its minimum index exceeds the minimum of the others, i.e. $\kappa_i^{(1)} \geq \kappa_j^{(1)}$ for all j. Now, the algorithm is indeed a *descending procedure*, so it is non-exposed by Lemma 3. By Lemma 2 then, *(a)* its welfare is equal to its expected capped value $\mathbb{E}[\max_i (\kappa_i^{(1)})^+]$, and *(b)* no algorithm can do better.

5 Approximation for Pandora's Matching Problem

We now return to the matching setting, ready to utilize nested boxes. In order to obtain an approximation algorithm for Pandora's Matching Problem, we consider *oriented graphs* where inspections of each edge must proceed in a certain order. We use this algorithm to create approximation algorithms to the general problem, even with some correlation between values. Finally, we consider a tempting simplification to "edge-based" algorithms.

5.1 Matching with Oriented Graphs

We first suppose that each edge is restricted to a particular order of inspection, e.g. (i, j) must be inspected prior to (j, i). This is a natural setting for some applications. A job-search platform may require candidates to submit resumes before companies can screen them: inspections must begin on the candidate side before moving to the employer side of the match. Formally, we capture the predetermined order of inspection with an orientation of the graph.

Definition 7. *An* orientation $\mathcal{O}$ *of an undirected graph* $G = (V, E)$ *is a set of ordered pairs of vertices such that, for every edge* $\{i, j\} \in E$, $\mathcal{O}$ *contains exactly one of* (i, j) *or* (j, i).

We refer to an ordered pair $(i, j) \in \mathcal{O}$ as an *oriented edge*. Given an instance and an orientation $\mathcal{O}$ of the edges, we say an algorithm is $\mathcal{O}$*-oriented* if for all realizations its sequence of inspections respects the orientation, i.e. if $(i, j) \in \mathcal{O}$, then endpoint (i, j) must be inspected prior to (j, i), if (j, i) is inspected. If an algorithm is $\mathcal{O}$-oriented for some $\mathcal{O}$, we call it a *fixed-orientation algorithm.*

Comparison to Committing Policies. Fixed-orientation algorithms may be reminiscent of *committing policies* appearing in related work on Pandora's Box with Non-obligatory Inspection, e.g. Beyhaghi and Kleinberg Beyhaghi, H. and Kleinberg, R. [2019]. Committing policies fix the sequence of box inspections in advance, though they may choose their stopping time dynamically. Although a fixed-orientation algorithm "commits" to an orientation $\mathcal{O}$, it will in general adaptively decide which edge or "basket" to advance next, depending on information revealed at previous stages. In other words, it only commits to order of inspection *within* each edge. Another difference is that, in the nonobligatory setting, we can reduce WLOG to a set of $n + 1$ commitments with n boxes; it is not obvious how to reduce the $2^{|E|}$ orientations of a Pandora's Matching Problem instance.

Orientations and Nested Pandora Boxes. For fixed-orientation algorithms, edges can be represented as Pandora baskets with two nested boxes, i.e. two stages of inspection. Opening the first box reveals the value of one endpoint, and the opening second reveals the full value of the edge. For example, if $(i, j) \in \mathcal{O}$, then edge $\{i, j\}$ is a basket where the first box costs c_{ij} to open and reveals signal v_{ij}, and opening the second box for cost c_{ji} reveals the combined value $v_{ij} + v_{ji}$. To advance to the third stage, or "claim the basket", is to add the edge to our matching.

We now consider optimizing expected welfare when restricted to a fixed orientation $\mathcal{O}$. Even under the fixed-orientation restriction, the optimal matching algorithm is unclear. The problem is more general than matching with a single Pandora box on each edge, a problem suspected to be NP-hard. But for a given orientation $\mathcal{O}$, the natural descending procedure achieves a 1/2-approximation to the optimal $\mathcal{O}$-oriented algorithm.

Algorithm 1. $\mathcal{O}$-Oriented Descending Procedure. Given a fixed orientation $\mathcal{O}$, for steps $t = 1, \ldots$:

- For each $(i, j) \in \mathcal{O}$, let σ_{ij}^t denote the current Weitzman index of basket (i, j).
- Advance the basket (i, j) with largest σ_{ij}^t.
  - If advancing results in "claiming" the basket, then add edge $\{i, j\}$ to the matching. Delete any edges incident to i or j from $\mathcal{O}$.
- If at any point the largest σ_{ij}^t is negative or $\mathcal{O}$ is empty, stop.

This is indeed a *descending procedure* (Definition 6), as it maintains a non-increasing eligible set (technically, $\mathcal{O}$ union the matched edges) and always advances and retains the largest-index basket in that set. Lemma 3 therefore gives the following.

Corollary 1. *For any orientation $\mathcal{O}$, the $\mathcal{O}$-Oriented Descending Procedure is non-exposed.*

We can continue with an analysis analogous to that of Kleinberg et al. Kleinberg, R., et al. [2016] for the case of matching with a Pandora box on each edge. Given $\mathcal{O}$, let *$\mathcal{O}$-κ-Greedy* refer to the following matching (used only for analysis): for each oriented edge $(i,j) \in \mathcal{O}$, let $x_{\{i,j\}} = \kappa_{ij}^{(1)}$, the capped value for basket (i,j); take the matching produced by the weighted greedy algorithm with edge weights $x_{\{i,j\}}$.

Lemma 4. *For any orientation $\mathcal{O}$, with probability one, the $\mathcal{O}$-Oriented Descending Procedure 1 produces the same matching as $\mathcal{O}$-κ-Greedy.*

We leave the proof of 4 to the appendix, available online.

Proposition 4. *The $\mathcal{O}$-Oriented Descending Procedure (Algorithm 1) achieves at least 1/2 the expected welfare of the optimal $\mathcal{O}$-oriented algorithm.*

Proof. Let $\mathbb{A}_{ij}^{\text{OPT}}$ be the indicator that edge $\{i,j\}$ is claimed by the optimal $\mathcal{O}$-oriented algorithm. For any realization of all values, consider a graph with edge weights $\{\kappa_{ij}^{(1)}\}$. In this graph, let $\mathbb{A}_{ij}^{G}$ be the indicator for $\{i,j\}$ being included in the greedy matching, i.e. in $\mathcal{O}$-κ-Greedy; and let $\mathbb{A}_{ij}^{*}$ be the corresponding indicator for the maximum weighted matching. The welfare of the $\mathcal{O}$-Oriented Descending Procedure satisfies

$$
\begin{aligned}
\text{Welfare}(\text{ALG}) &= \mathbb{E} \sum_{(i,j)\in\mathcal{O}} \mathbb{A}_{ij}\kappa_{ij}^{(1)} && \text{Corollary1} \\
&= \mathbb{E} \sum_{(i,j)\in\mathcal{O}} \mathbb{A}_{ij}^{G}\kappa_{ij}^{(1)} && \text{Lemma 4} \\
&\geq \frac{1}{2}\mathbb{E} \sum_{(i,j)\in\mathcal{O}} \mathbb{A}_{ij}^{*}\kappa_{ij}^{(1)} && \text{weighted matching fact} \\
&\geq \frac{1}{2}\mathbb{E} \sum_{(i,j)\in\mathcal{O}} \mathbb{A}_{ij}^{\text{OPT}}\kappa_{ij}^{(1)} && \text{optimality of } \{\mathbb{A}_{ij}^{*}\} \\
&\geq \frac{1}{2}\text{Welfare}(\text{OPT}) && \text{Lemma 2.}
\end{aligned}
$$

Remark. Even with a fixed orientation, we do not know how to improve on a 1/2-approximation factor in polynomial time. We are limited by reliance on greedy matchings, which can only achieve 1/2 of the optimal weighted matching even without inspections.

5.2 Main Positive Result

We now finally provide two approximation algorithms for Pandora's Matching Problem, both of which first pick an orientation $\mathcal{O}$ and then run the $\mathcal{O}$-Oriented Descending Procedure.

The first algorithm we consider is randomized, and orients each edge independently. We will first present and analyze the randomized algorithm, before presenting a deterministic algorithm using concepts developed in the randomized setting.

Algorithm 2. Randomized Pandora Matching Algorithm. Construct an orientation $\mathcal{O}$ by orienting each edge $\{i, j\}$ as either (i, j) or (j, i) uniformly and independently at random. Run the $\mathcal{O}$-Oriented Descending procedure.

Note that by construction, $\mathcal{O}$ is drawn from the uniform distribution over all orientations for the graph.

Theorem 4. *In Pandora's Matching Problem, Algorithm 2 achieves at least a* $1/4$*-approximation of the optimal welfare.*

Before proving the approximation guarantee, we first define some helpful notation and an upper-bound on the value of an optimal algorithm. We will also use these tools in defining our deterministic algorithm.

Definition 8. *The* reverse *of an orientation* $\mathcal{O}$ *is* $\hat{\mathcal{O}} = \{(j, i) \mid (i, j) \in \mathcal{O}\}$.

The key step in the analysis of the randomized algorithm is an upper bound on the welfare of the optimal algorithm relative to an orientation and its reverse. The crux of this bound is that any realization of an algorithm induces an orientation on edges it inspects, based on the order of inspection. The welfare an edge generates in an algorithm, then, can be "charged" across any orientation $\mathcal{O}$ and its reverse, according to which orientation matches the direction in which the edge is inspected.

Lemma 5. *For any orientation* $\mathcal{O}$ *and it s reverse* $\hat{\mathcal{O}}$*, the welfare of* OPT *on Pandora's Matching Problem is upper-bounded by*

$$\text{Welfare}(\text{OPT}) \leq 2 \cdot \text{Welfare}(\mathcal{O}\text{-}desc.) \quad + \quad 2 \cdot \text{Welfare}(\hat{\mathcal{O}}\text{-}desc.).$$

The proof of this lemma is given in the online appendix. This bound on the optimal welfare of any algorithm for Pandora's Matching Problem allows a quick proof of Theorem 4, also available online.

Given the heavy reliance of Algorithm 2 on randomization of orientation, we also present a deterministic algorithm based on determining an orientation and then running the associated descending procedure. A similar âĂIJbest-of-two-worldsâĂİ technique is used in Beyhaghi and Kleinberg Beyhaghi, H. and Kleinberg, R. [2019] in a different context, explored in the online appendix.

This algorithm differs from the randomized algorithm in that the final orientation is determined globally. In the randomized algorithm we select the orientation of each edge independently, but the Best-of-Two-Worlds algorithm compares two orientations over the entire graph and selects one.

Theorem 5. *In Pandora's Matching Problem, the Best-of-Two-Worlds algorithm achieves at least a* $1/4$*-approximation of the optimal welfare.*

Extension to Correlated-Within-Edges Model. We can extend these positive results to a model where each pair of values v_{ij}, v_{ji} on an edge are arbitrarily correlated with each other and independent of the other edges. In fact, we can extend this a bit further: in the *correlated-within-edges model*, we suppose that each Pandora box on an endpoint (i, j) reveals for cost c_{ij} an arbitrary signal s_{ij}, the total value on an edge is an arbitrary measurable function $f_{\{i,j\}}(s_{ij}, s_{ji})$, and for each edge, the pair (s_{ij}, s_{ji}) are drawn jointly from a distribution $D_{\{i,j\}}$, independently of all other pairs.

In this setting, the definition of an algorithm is unchanged, i.e. it adaptively pays to open boxes and eventually outputs a matching of fully-opened edges. The proofs of Proposition 4 and Lemma 5 go through unchanged, and the proofs of both algorithms hold with values correlated along edges. In the proof of Lemma 5, an edge is modeled as a pair of arbitrary Pandora baskets (i, j) and (j, i), where an algorithm must pick one of the baskets at the time of first inspection of the edge. For any orientation $\mathcal{O}$, i.e. a commitment on each edge to one of its two possible baskets, the proof then cites Proposition 4. And Proposition 4 addresses the problem of matching with a single, arbitrary Pandora basket per edge, utilizing our generic results for Nested Boxes in Sect. 4.

Observation 1. *The* 1/4*-approximations of the Randomized Pandora Matching and Best-of-Two-Worlds algorithms continue to hold in the correlated-within-edges model.*

5.3 Dessert?

We have finally presented two constant-factor approximation algorithms for Pandora's Matching Problem, one randomized and one deterministic. The extreme simplicity of our randomized algorithm's edge orientation choice hints that a simpler deterministic approach should be possible, potentially following the structure of the randomized algorithm: for each edge, pick an orientation independently of any other edge orientations. This would greatly simplify our deterministic approximation algorithm, and such an algorithm would be a nice "dessert". In particular, a natural approach might be to orient each edge according to which "basket" has a higher initial Weitzman index, then run the Oriented Descending Procedure. We consider the following generalization of this approach:

Definition 9. *An* edge-based fixed-orientation algorithm *is one that, given an instance of Pandora's Matching Problem, constructs an orientation $\mathcal{O}$ by applying a deterministic rule to each edge $\{i, j\}$ that depends only on the parameters of that edge, $D_{ij}, c_{ij}, D_{ji}, c_{ji}$, and then runs some $\mathcal{O}$-oriented algorithm.*

Unfortunately, we have the following result.

Theorem 6. ("No dessert"). *No edge-based fixed-orientation algorithm guarantees any positive approximation for Pandora's Matching Problem.*

Intuition. The proof relies on a particular edge, shown in Fig. 4. Every edge-based fixed orientation algorithm either orients that edge in one direction or the other. If there is no outside option, then the (i, j) orientation is arbitrarily better, because we want to match the edge if and only if $v_{ij} > 0$, and we discover this

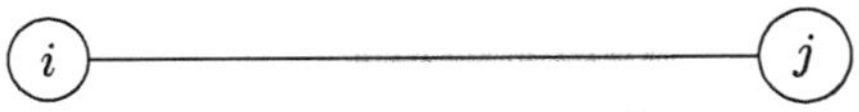

$$v_{ij} = \begin{cases} \frac{1}{\alpha^3} & \text{w.prob. } \alpha \\ 0 & \text{else} \end{cases} \qquad v_{ij} = \begin{cases} 0 & \text{w.prob. } \alpha^2 \\ -\frac{1}{\alpha^3} + \frac{1}{\alpha} & \text{else} \end{cases}$$

$$c_{ij} = 1 \qquad c_{ji} = 1 - \alpha$$

Fig. 4. An edge used to prove Theorem 6. Without an outside option, it is optimal to inspect i before j (Instance 1), but with an outside option of $\frac{1}{\alpha}$, it is optimal to inspect j before i (Instance 2).

immediately, saving the cost of inspecting (j,i) when $v_{ij} = 0$. But if there is an outside option of $\frac{1}{\alpha}$, then the basket is worthwhile if and only if both boxes contain their high value (a "success"). Here the (j,i) orientation is arbitrarily better, as the overall probability of a success is the same but we pay a second inspection cost with only α^2 probability instead of α.

For the skeptical reader, the online appendix contains an alternative proof of Theorem 6 that does not use any Pandora box machinery.

Proof. Recall that $\sigma_{ij}^{(1)}$ is the initial Weitzman index for the basket corresponding to orientation (i,j), and $\kappa_{ij}^{(1)}$ is the capped value, a random variable. By Lemma 2, the expected welfare contribution of a basket (i,j) is upper-bounded $\mathbb{E}[\mathbb{A}_{ij}\kappa_{ij}^{(1)}]$, with equality if the algorithm is non-exposed. In particular, since $\kappa_{ij}^{(1)} \le \sigma_{ij}^{(1)}$, we can upper-bound the welfare contribution of a box under any algorithm by the maximum $\sigma_{ij}^{(1)}$ over any basket available to it.

We record the following (calculations appear in the online appendix):

- $\sigma_{ij}^{(1)} = \frac{1}{\alpha} - 1$.
- $\kappa_{ij}^{(1)} = \frac{1}{\alpha} - 1$ with probability α, otherwise negative.
- $\sigma_{ji}^{(1)} = \frac{1}{\alpha^2} - \frac{1}{\alpha}$.
- $\kappa_{ji}^{(1)} = \frac{1}{\alpha^2} - \frac{1}{\alpha}$ with probability α^3, otherwise nonpositive.

Instance 1. We consider a graph consisting of just the edge in Fig. 4. An algorithm can orient the edge as a basket (i,j) or as a basket (j,i). In each case, the descending procedure is optimal and achieves welfare equal to expected capped value (Sect. 4). With a single (i,j) basket, the descending procedure obtains welfare $\mathbb{E}[(\kappa_{ij})^+] = \alpha\left(\frac{1}{\alpha} - 1\right) = 1 - \alpha$. With (j,i), the welfare is $\mathbb{E}[(\kappa_{ji}^{(1)})^+] = \alpha^3\left(\frac{1}{\alpha^2} - \frac{1}{\alpha}\right) = \alpha - \alpha^2$. So the orientation (j,i) first is an approximation of $\frac{\alpha-\alpha^2}{1-\alpha} = \alpha$. This holds for arbitrarily small $\alpha > 0$, so no algorithm that orients the edge as (j,i) can guarantee a positive approximation.

Instance 2. We will consider a star graph with i at the center. There is a special edge $\{i,k\}$ with $v_{ik} = \frac{1}{\alpha}$, $c_{ik} = 0$, $v_{ki} = 0$, $c_{ki} = 0$. In other words, i has an "outside option" $\frac{1}{\alpha}$ it can match to at any time. Every other edge in the graph $\{i,j_1\},\ldots,\{i,j_m\}$ is an independent copy of the edge $\{i,j\}$ in Fig. 4. The number of copies will be $m = \omega(\frac{1}{\alpha^3})$. A edge-based fixed-orientation algorithm must orient each copy in the same direction, as it applies the same deterministic rule to each edge.

After orienting every edge in the (j, i) direction, we have an instance of Nested Pandora's Boxes with $m + 1$ baskets. The optimal welfare, achieved by the descending procedure, is equal to the expected maximum capped value of a basket. With $\omega(\frac{1}{\alpha^3})$ baskets, some basket has capped value $\frac{1}{\alpha^2} - \frac{1}{\alpha}$ with probability $1 - o(1)$; so the algorithm's expected welfare is $(\frac{1}{\alpha^2} - \frac{1}{\alpha})(1 - o(1))$.

After orienting every edge in the (i, j) direction, the same argument holds, except that the capped value of every basket is always at most $\frac{1}{\alpha} - 1$. These are actually less than the capped value of the outside option, which is deterministically $\frac{1}{\alpha}$, which is the maximum capped value and therefore the best any algorithm can do.

The ratio of the optimal (i, j) orientation welfare to (j, i) is, asymptotically, $\frac{1/\alpha}{1/\alpha^2} = \alpha$. This holds as $\alpha \to 0$, so no algorithm that orients $\{i, j\}$ as (i, j) has a positive approximation guarantee.

The analysis of instance 1 provides an interesting corollary.

Corollary 2. *Even with just a single edge, inspecting according to the orientation (i, j) with the larger Weitzman index $\sigma_{ij}^{(1)}$ can be arbitrarily suboptimal.*

6 Discussion

Summary. Prior work had considered matching with a single Pandora box on each edge. A satisfactory solution exists for that problem with by-now standard techniques for information-acquisition settings. In this paper, we considered Pandora's Matching Problem, a two-sided matching setting in which each participant i and j in a potential match $\{i, j\}$ have an initially-hidden value for the match, modeled as one Pandora box on each endpoint of an edge. We allowed for negative values, an important case for settings such as labor markets.

We found that this problem poses significant challenges for standard approaches in ways that likely have implications for the design of matching platforms. In particular, the following approaches all fail to guarantee any positive approximation factor.

1. A *bundled-box* approach, where inspections of both boxes incident to an edge must occur simultaneously. This suggests that interview stages on matching platforms can be costly in terms of welfare, while asynchronous inspections should be enabled.
2. A *vertex-based* approach, where decisions are made based on values and Weitzman indices at the endpoints, but the algorithm does not consider the interaction of the value distributions of the endpoints. (This approach gives an approximation factor in the positive-values setting.) This suggests that platforms should treat an edge not as two separate boxes, but as one basket of *nested boxes* that undergoes a multi-stage information-gathering process.

3. An *edge-based* approach, where the order of inspection within an edge is prescribed by some deterministic rule independent of the rest of the graph. This suggests that platforms should coordinate the order of inspections within edges (i.e. i first or j first) in a global fashion.

To evade all of these pitfalls, we proposed a Best-of-Two-Worlds algorithm that: (1) dynamically and partially inspects edges, only coming back to complete an edge's inspection if prudent; (2) treats each edge as a multi-stage Pandora basket with nested boxes; (3) coordinates the decision of how to orient each edge globally by considering the "two worlds". We showed that this algorithm gives an efficient 1/4-approximation for Pandora's Matching Problem.

6.1 Future Directions

A number of questions remain for two-sided matching with inspection costs. One interesting question is in the difficulty of the Pandora's Matching Problem itself. Is there an algorithm for the Pandora's Matching Problem that beats the 1/2-approximation factor from greedy matching?

Open Problem 1. *Is Pandora's Matching Problem APX-hard, or can a PTAS be found?*

It is also open to give a substantially different approximation algorithm than Best-of-Two-Worlds.

Extensions. There are other extensions of the Pandora's Box model which may also be interesting for modeling two-sided matching settings. The *nonobligatory* setting allows boxes to be claimed uninspected. If, for example, a student knows she likely has a high value for attending a certain school, she may forgo the cost of visiting the school and directly accept an offer. The simpler *Pandora's Box with Nonobligatory Inspection* problem without the matching constraint is already known to be NP-hard (Fu et al., Fu, H., et al. [2023]), but existing approximation schemes (Fu et al., Fu, H., et al. [2023], Beyhaghi and Cai Beyhaghi, H and Cai, L. [2023]) give hope that approximation may be tractable in the matching setting as well.

Section 5 showed that edge-based fixed-orientation algorithms are, in general, arbitrarily suboptimal. However, it is possible that under natural restrictions on the value distributions, such algorithms do achieve constant-factor approximations. Understanding settings in which such algorithms can safely be applied could have implications for design of platforms which fix an orientation for practicality.

Acknowledgements. This work was supported by the National Science Foundation Award #2329431.

References

Aouad, A., Ji, J., Shaposhnik, Y.: The pandora's box problem with sequential inspections. Journal SSRN 3726167 (2020)

Ashlagi, I., Braverman, M., Kanoria, Y., Shi, P.: Clearing matching markets efficiently: informative signals and match recommendations. Manag. Sci. **665**, 2163–2193 (2020)
Beyhaghi, H., Cai, L.: Pandora's problem with nonobligatory inspection: optimal structure and a PTAS. In: Proceedings of the 55th Annual ACM Symposium on Theory of Computing (STOC 2023). Association for Computing Machinery, pp. 803–816 (2023)
Beyhaghi, H and Cai, L.: Recent developments in the pandora's box problem: variants and applications. SIGecom Exchanges **21**, 1 (2023)
Beyhaghi, H., Kleinberg, R.: Pandora's problem with nonobligatory inspection. In: Proceedings of the 2019 ACM Conference on Economics and Computation (EC 2019). Association for Computing Machinery, 131–132 (2019)
Boodaghians, S., Fusco, F., Lazos, P., Leonardi, S.: Pandora's box problem with order constraints (EC 2020). Assoc. Comput. Mach., 439–458 (2020)
Bowers, R., Waggoner, B.: High-welfare matching markets via descending price. In: Nature, S. (ed.) Web and Internet Economics, Jugal Garg, Max Klimm, and Yuqing Kong, pp. 59–76. Switzerland, Cham (2024)
Chade, H., Eeckhout, J., Smith, L.: Sorting through search and matching models in economics. J. Econ. Literature **55**(2), 493–544 (2017)
Chawla, S., Gergatsouli, E., Teng, Y., Tzamos, C., Zhang, R.: Pandora's box with correlations: Learning and approximation. In: IEEE 61st Annual Symposium on Foundations of Computer Science (FOCS), pp. 1214–1225. IEEE (2020)
Doval, L.: Whether or not to open Pandora's box. J. Econ. Theory **175**, 127–158 (2018)
Fu, H., Li, J., Liu, D.: Pandora box problem with nonobligatory inspection: hardness and approximation scheme. In Proceedings of the 55th Annual ACM Symposium on Theory of Computing, pp. 789–802 (2023)
Gittins, J.C.: Bandit processes and dynamic allocation indices. J. Royal Stat. Soc. Seri. B (Methodol.) **41**(2), 148–177 (1979)
Gupta, A., Jiang, H., Scully, Z., Singla, S.: The Markovian Price of Information. In: Lodi, A., Nagarajan, V. (eds.) IPCO 2019. LNCS, vol. 11480, pp. 233–246. Springer, Cham (2019). https://doi.org/10.1007/978-3-030-17953-3_18
Hakimov, R., Kübler, D., Pan, S.: Costly information acquisition in centralized matching markets. Quan. Econ. **14**(4), 1447–1490 (2023)
Immorlica, N., Leshno, J., Lo, I., Lucier, B.: Information acquisition in matching markets: the role of price discovery. Available at SSRN 3705049 (2020)
Immorlica, N., Lucier, B., Manshadi, V.H., Wei, L.: Designing approximately optimal search on matching platforms. In: Proceedings of the 2021 ACM Conference on Economics and Computation (EC 2021), Péter Biró, Shuchi Chawla, and Federico Echenique (Eds.). Association for Computing Machinery, pp. 632–633 (2021)
Tony Ke, T., Miguel Villas-Boas, J.: Optimal learning before choice. J. Econ. Theory **180**, 383–437 (2019)
Kleinberg, R., Waggoner, B., Weyl, E.G.: Descending price optimally coordinates search. In: Proceedings of the 2016 ACM Conference on Economics and Computation (EC 2016). Association for Computing Machinery, pp. 23–24 (2016)
Singla, S.: The price of information in combinatorial optimization. In: Proceedings of the Twenty-Ninth Annual ACM-SIAM Symposium on Discrete Algorithms (SODA). SIAM, pp. 2523–2532 (2018)
Waggoner, B., Weyl, E.G.: Matching markets via descending price. Available at SSRN 3373934 (2019)
Weitzman, M.L.: Optimal search for the best alternative. Econometrica **47**(3), 641–654

Online Matching Meets Sampling Without Replacement

Zhiyi Huang(✉), Chui Shan Lee, Jianqiao Lu, and Xinkai Shu

The University of Hong Kong, Pokfulam Road, Hong Kong, Hong Kong
{zhiyi,cslee,jqlu,xkshu}@cs.hku.hk

Abstract. Sampling without replacement is a natural online rounding strategy for converting fractional bipartite matching into an integral one. In Online Bipartite Matching, we can use the Balance algorithm to fractionally match each online vertex, and then sample an unmatched offline neighbor with probability proportional to the fractional matching. In Online Stochastic Matching, we can take the solution to a linear program relaxation as a reference, and then match each online vertex to an unmatched offline neighbor with probability proportional to the fractional matching of the online vertex's type. On the one hand, we find empirical evidence that online matching algorithms based on sampling without replacement outperform existing algorithms. On the other hand, the literature offers little theoretical understanding of the power of sampling without replacement in online matching problems.

This paper fills the gap in the literature by giving the first non-trivial competitive analyses of sampling without replacement for online matching problems. In Online Stochastic Matching, we develop a potential function analysis framework to show that sampling without replacement is at least 0.707-competitive. The new analysis framework further allows us to derandomize the algorithm to obtain the first polynomial-time deterministic algorithm that breaks the $1 - \frac{1}{e}$ barrier. In Online Bipartite Matching, we show that sampling without replacement provides provable online correlated selection guarantees when the selection probabilities correspond to the fractional matching chosen by the Balance algorithm. As a result, we prove that sampling without replacement is at least 0.513-competitive for Online Bipartite Matching.

Keywords: Online Matching · Randomized Algorithms

1 Introduction

Matching problems have been a central topic in theoretical computer science for decades. They arise in a wide range of applications. Online advertising platforms match users and advertisers. Ride-hailing services assign drivers to riders. Kidney exchange programs pair patients and donors. In many of these scenarios, the goal

This work is supported by an NSFC grant (No. 6212290003).

M. Mavronicolas et al. (Eds.): WINE 2024, LNCS 15534, pp. 84–101, 2026.
https://doi.org/10.1007/978-3-032-08560-3_5

is to match a set of agents or objects to another set of agents or objects, subject to some constraints or preferences. The success of these applications heavily relies on the efficiency and effectiveness of the algorithms used to solve matching problems.

With the rise of online platforms and their need for real-time decision-making, online matching algorithms have gained much attention in recent years. A prominent model is the Online Bipartite Matching problem proposed by Karp et al. [28]. It considers a bipartite graph with online vertices on the left and offline vertices on the right. The online vertices arrive one at a time, and the online algorithms need to decide how to match each online vertex as it arrives without information about future online vertices and their edges.

Researchers have developed two approaches to design algorithms for Online Bipartite Matching. On the one hand, Karp et al. [28] proposed the RANKING algorithm and proved that it achieves the optimal $1 - \frac{1}{e}$ competitive ratio. It randomly ranks the offline vertices at the beginning, and then greedily matches each online vertex to the highest-ranked unmatched neighbor.

On the other hand, Kalyanasundaram and Pruhs [26] considered the BALANCE algorithm (a.k.a., WATER-FILLING or WATER-LEVEL). It may be viewed as a fractional online matching algorithm that splits one unit of each online vertex among its neighbors while balancing the total amounts of online vertices allocated to different offline vertices. Moreover, there have been recent studies on online rounding algorithms that convert BALANCE and its variants into integral algorithms for Online Bipartite Matching. For instance, Gao et al. [16] designed the BALANCE OCS algorithm building on the online correlated selection technique, which was first proposed by Fahrbach et al. [13], then further developed in a series of works [5,16,23,36]. Buchbinder et al. [9] and Naor et al. [35] also studied online rounding algorithms for matching and how to use negative correlation to improve rounding efficiency.

Despite these efforts, there has been no theoretical understanding of arguably the most natural rounding algorithm, which samples an unmatched neighbor for each online vertex with probability proportional to the fractional allocation of BALANCE. Since this approach effectively picks an offline vertex via sampling without replacement (SWOR) with sampling masses given by BALANCE, we call the resulting online matching algorithm BALANCE SWOR.

We next consider a related problem known as Online Stochastic Matching, which was first introduced by Feldman et al. [14]. It is also about online matching on bipartite graphs but the online vertices therein are generated by a random process that is known to the algorithms from the beginning.

There is a rich literature on Online Stochastic Matching, in which all algorithms follow a two-step framework. First, algorithms find a fractional bipartite matching between the offline vertices and the online vertex types, which represents the likelihood that they are matched in the optimal matching of the randomly generated bipartite graph. This fractional matching is often computed by solving a linear program (LP) constructed based on the stochastic process that generates online vertices. Then, when an online vertex arrives, algorithms use different online rounding strategies to make a matching decision, based on the current set of unmatched offline vertices and the fractional matching of the online vertex's type.

For example, the SUGGESTED MATCHING algorithm [14] first samples an offline neighbor (regardless of whether it is matched) with probabilities equal to the fractional matching of the online vertex's type. It then matches the online vertex to this neighbor if it is still unmatched. This simple approach is shown to be $(1 - \frac{1}{e})$-competitive for Online Stochastic Matching.

Since then, many efforts [4,7,14,18,20,24,33,38] have been devoted to improving the online rounding strategy by sampling a second offline neighbor, usually correlated to the sampling result of the first neighbor, so that algorithms could try the second neighbor when the first is already matched. This is known as the power of two choices in Online Stochastic Matching, resulting in online algorithms with competitive ratios strictly better than $1 - \frac{1}{e}$. Huang et al. [21] and Tang et al. [37] explored the power of multiple choices in Online Stochastic Matching by adopting online correlated selection algorithms as their rounding strategies. The POISSON OCS algorithm by Huang et al. [21] achieves the state-of-the-art competitive ratio of 0.716.

Once again, despite the long line of research, we still have little understanding on arguably the most natural rounding algorithm, which samples an unmatched neighbor with probability proportional to the fractional matching of the online vertex's type. We will refer to this algorithm as STOCHASTIC SWOR, where SWOR stands for sampling without replacement.

1.1 Our Contributions

We first examine the mentioned BALANCE SWOR and STOCHASTIC SWOR algorithms through empirical experiments to determine whether it is worthwhile to further study them theoretically. Our experiments compare them with existing online matching algorithms on six real-world datasets that were used in a recent empirical study by Borodin et al. [6].

To our surprise, BALANCE SWOR and STOCHASTIC SWOR outperform existing algorithms that admit non-trivial theoretical guarantees. Among algorithms for Online Bipartite Matching, BALANCE SWOR achieves slightly better results compared to both BALANCE OCS and RANKING. Among algorithms for Online Stochastic Matching, STOCHASTIC SWOR and a derandomized version called REGULARIZED GREEDY, which we will explain shortly, perform slightly better than POISSON OCS, which in turn shows substantially better results than other algorithms from the Online Stochastic Matching literature. See Table 1 for some of our empirical results, and our full paper [19] for more details.

Conceptually, the empirical experiments suggest that a simple rounding strategy like sampling without replacement may be sufficient in the context of online matching problems, and may even be slightly superior compared to more involved techniques like online correlated selection. The rest of the paper is therefore devoted to providing non-trivial competitive analyses of BALANCE SWOR and STOCHASTIC SWOR.

Table 1. Empirical results for the algorithms considered in this paper (in bold) and top-performing existing algorithms with known theoretical guarantees. See [19] for the full results.

Algorithm	Caltech36	Reed98	CE-GN	CE-PG	beause	mbeaflw
Regularized Greedy	0.928	0.929	0.984	0.990	0.962	0.966
Stochastic SWOR	0.929	0.927	0.958	0.962	0.959	0.975
Poisson OCS	0.929	0.926	0.957	0.960	0.958	0.974
Balance SWOR	0.874	0.873	0.943	0.950	0.943	0.971
Balance OCS	0.871	0.870	0.942	0.949	0.942	0.970
Ranking	0.859	0.859	0.934	0.944	0.936	0.966

Competitive Analysis of Stochastic SWOR. This paper gives the first non-trivial analysis of STOCHASTIC SWOR, showing that it is at least 0.707-competitive for Online Stochastic Matching. Although it is worse than the state-of-the-art 0.716 ratio [21], it surpasses a natural barrier 0.706 suffered by all algorithms prior to the recent introduction of the Natural LP by [20,21].

Our analysis develops a *potential function analysis framework* that is novel in the context of Online Stochastic Matching. Up to normalization, we may assume without loss of generality that online vertices arrive within the time horizon 0 to 1. We design a potential function $\Phi(t)$ for any time $0 \le t \le 1$, which depends on the LP-based fractional matching and the current set of unmatched offline vertices. Intuitively, $\Phi(t)$ is a lower bound for the expected number of matches the online algorithm would get after time t. To substantiate this intuition, $\Phi(t)$ shall satisfy three conditions. The first one is a boundary condition at the beginning such that $\Phi(0) \ge \Gamma \cdot \text{OPT}$, where Γ is the competitive ratio and OPT is the optimal objective. The second one is a boundary condition at the end, namely, $\Phi(1) = 0$. Finally, we will show that the online algorithm's expected match rate weakly dominates the potential function's expected decrease rate, i.e., $\mathbf{E}\left[A(t) + \Phi(t)\right]$ is non-decreasing in $0 \le t \le 1$ where $A(t)$ denotes the size of algorithm's matching up to time t.

As natural as it may seem, this potential function analysis framework has never been employed in the literature of Online Stochastic Matching. We argue that it might be the right approach. In fact, the folklore yet computationally intractable optimal algorithm for Online Stochastic Matching fits into this framework. Here we let $\Phi(t)$ be the best expected matching size achievable by online algorithms after time t, which can be computed in exponential time by backward induction. When an online vertex arrives, the optimal online algorithm matches it to an unmatched offline neighbor such that the potential function decreases the least.

The challenge is therefore to design a computationally tractable potential function for the stated competitive ratio. Our potential computes the contribution of each online vertex type separately and then sums them up. Consider for simplicity an online type i with unit arrival rate. Suppose that the fractional matching assigns an x_i fraction of i to *its currently unmatched offline neighbors.* If this is at the beginning, type i's contribution would be mainly proportional to x_i, because the LP-based fractional matching serves as a reference that implicitly accounts for the likelihood that all i's neighbors are already matched when a type-i vertex comes. As time proceeds, type i's contribution to the potential would smoothly transition to be proportional to $\min\{\frac{x_i}{\theta}, 1\}$ for some parameter $0 \le \theta \le \frac{1}{2}$. This is because as we get closer to the end, it is less likely that all i's neighbors would get matched within the remaining time. In the extreme case when it is already at the end, type i's contribution should simply be proportional to its unit arrival rate. The mentioned functional form interpolates between the two extremes, and is inspired by the analysis of the TOP HALF SAMPLING algorithm by Huang et al. [21]. Finally, we optimize θ to get the 0.707 competitive ratio.

Deterministic Online Stochastic Matching Algorithm. Due to the potential function analysis, we can further derandomize the algorithm by matching each online vertex to an unmatched offline neighbor for which the potential function decreases the least. The algorithm-dependent part of the potential's decreased amount may be viewed as a regularization term, which measures how likely an offline neighbor could get matched later if we do not match it now. We therefore call this algorithm REGULARIZED GREEDY. By design, it is at least 0.707-competitive for Online Stochastic Matching. This is the first polynomial-time deterministic algorithm that breaks the $1 - \frac{1}{e}$ barrier.Empirical experiments further suggest that the derandomized REGULARIZED GREEDY is slightly better than the randomized STOCHASTIC SWOR on real-world graphs.

Competitive Analysis of Balance SWOR. Last but not least, we offer the first non-trivial analysis of the BALANCE SWOR algorithm, proving that it is at least 0.513-competitive for Online Bipartite Matching. Our analysis is based on the following observation which is simple in hindsight. Although sampling without replacement fails to give any worst-case guarantee with respect to (w.r.t.) the online correlated selection problem [16], in BALANCE SWOR it only needs to handle sampling probabilities chosen by the BALANCE algorithm, which by design balances the total sampling probabilities assigned to different offline vertices. This observation circumvents the worst-case instance for sampling without replacement. As a result, we derive a restricted form of online correlated selection guarantees, and the stated competitive ratio for Online Bipartite Matching.

1.2 Other Related Work

After Karp et al. [28], Aggarwal et al. [1] extended the RANKING algorithm and the $(1 - \frac{1}{e})$ competitive ratio to vertex-weighted matching. Feldman et al. [14] proved that no online algorithm can achieve a non-trivial worst-case competitive ratio for edge-weighted matching. They further proposed the *free disposal model* and an optimal $(1 - \frac{1}{e})$-competitive algorithm in this model under a large-market assumption. AdWords is another noteworthy variant of Online Bipartite

Matching, which was proposed by Mehta et al. [34]. They gave an optimal $(1-\frac{1}{e})$-competitive algorithm for AdWords also under a large-market assumption. The algorithms by Feldman et al. [14] and Mehta et al. [34] may be viewed as generalizations of the BALANCE algorithm. Subsequently, these results have been streamlined and integrated within the online primal-dual framework [8,11,12]. The aforementioned online correlated selection technique contributed to breaking the $\frac{1}{2}$ barrier in edge-weighted matching with free disposal [5,13,16,36] and AdWords [23], without the large-market assumptions.

These extensions of Online Bipartite Matching have also been studied in the stochastic model. Huang et al. [21] proposed POISSON OCS and TOP HALF SAMPLING which get the state-of-the-art 0.716 and 0.706 competitive ratios for vertex-weighted matching and edge-weighted matching with free disposal respectively. For edge-weighted matching without free disposal, Huang et al. [21] proved a 0.703 upper bound, separating it from the other settings. Yan [38] recently proposed the MULTISTAGE SUGGESTED MATCHING algorithm which breaks the $1-\frac{1}{e}$ barrier, and this result was further improved by Feng et al. [15]. Aouad and Ma [3] studied edge-weighted online stochastic matching with correlated online vertex arrivals. Ma et al. [31] studied individual fairness maximization and group fairness maximization in online stochastic matching with integral arrival rate.

The literature has also studied the random order model of online matching which lies between the worst-case and stochastic models. For (unweighted) Online Bipartite Matching with random arrival order, Karande et al. [27] and Mahdian and Yan [32] showed that RANKING's competitive ratio is between 0.696 and 0.727. For vertex-weighted matching, Huang et al. [22] and Jin and Williamson [25] proposed variants of RANKING that break the $1-\frac{1}{e}$ barrier. For edge-weighted matching without free disposal, Kesselheim et al. [29] generalized the secretary algorithm to get an optimal $\frac{1}{e}$ competitive ratio. Finally, Devanur and Hayes [10], Kesselheim et al. [30], and Agrawal and Devanur [2] proposed $1-\epsilon$ competitive algorithms for AdWords under a large-market assumption.

2 Preliminaries

2.1 Online Bipartite Matching

We first define the Online Bipartite Matching problem, which was proposed by Karp et al. [28]. Consider an undirected bipartite graph $G = (I, J, E)$, where I and J are the sets of left-hand-side and right-hand-side vertices respectively, and E is the set of edges. Define I_j to be the set of online neighbors of an offline vertex $j \in J$, and define J_i similarly for $i \in I$. In the beginning, an online algorithm knows only the right-hand-side vertices, but not the left-hand-side vertices and edges. Then, the left-hand-side vertices arrive one by one. When a left-hand-side vertex $i \in I$ arrives, the algorithm observes its incident edges, and must immediately and irrevocably decide how to match i. Based on the distinct nature of left-hand-side and right-hand-side vertices, we will refer to them as online and offline vertices respectively.

We consider maximizing the size of the matching. Following the standard competitive analysis, we compare the size of an online algorithm's matching, taking expectation over the algorithm's randomness, with the size of the optimal matching in hindsight, i.e., the best achievable objective if we had complete information about the bipartite graph $G = (I, J, E)$. The competitive ratio of an online algorithm is the minimum of this ratio among all bipartite graphs and all arrival orders of online vertices. An algorithm is Γ-competitive if its competitive ratio is at least Γ.

2.2 Online Stochastic Matching

Next we present the Online Stochastic Matching problem introduced by Feldman et al. [14], and the Natural LP recently developed by Huang and Shu [20].

Consider an undirected bipartite graph $G = (I, J, E)$, where I is the set of *online vertex types*, J the set of offline vertices, and E the set of edges. Let I_j and J_i denote the neighborhoods of $j \in J$ and $i \in I$. We will refer to G as the *type graph.* Online vertices of each type $i \in I$ arrives from time 0 to 1 by a Poisson process with arrival rate $\lambda_i > 0$.An online algorithm knows the type graph and the arrival rates from the beginning, but not the realization of online vertex arrivals. When an online vertex arrives, the algorithm sees its type i and thus the incident edges, and must immediately and irrevocably decide how to match it.

We still consider maximizing the cardinality of the matching. On the one hand, we have the expected size of an online algorithm's matching over the algorithm's randomness and the random online vertex arrivals. On the other hand, we consider the expected size of the optimal matching in hindsight over the random online vertex arrivals, i.e., the best achievable objective in expectation if we always choose the optimal matching w.r.t. the realized bipartite graph. The competitive ratio of an online algorithm is the minimum of this ratio among all possible type graphs and arrival rates. An algorithm is Γ-competitive if its competitive ratio is at least Γ.

Natural Linear Program. Consider $x = (x_{ij})_{(i,j)\in E}$ where x_{ij} corresponds to the probability that offline vertex j is matched to an online vertex of type i. We will use the following Natural LP:

$$\begin{array}{lll} \text{maximize} & \sum_{(i,j)\in E} x_{ij} & \\ \text{subject to} & \sum_{j\in J_i} x_{ij} \le \lambda_i & \forall i \in I \\ & \sum_{i\in S} x_{ij} \le 1 - \exp\big(-\sum_{i\in S}\lambda_i\big) & \forall j \in J, \forall S \subseteq I_j \\ & x_{ij} \ge 0 & \forall (i,j) \in E \end{array} \quad (1)$$

The Natural LP plays two roles in Online Stochastic Matching. First, it is a computationally efficient upper bound of the optimal objective, because it can be solved in polynomial time using the ellipsoid method and a polynomial time separation oracle for its second constraint [20]. Further, online algorithms can use the optimal solution of the Natural LP, which we will abuse notation and denote also $x = (x_{ij})_{(i,j)\in E}$, as a reference for making online decisions.

Lemma 1 (Converse Jensen Inequality, c.f., Huang and Shu [20]). *For any convex $f : [0,1] \to \mathbb{R}$ such that $f(0) = 0$, and any offline vertex $j \in J$:*

$$\sum_{i\in I} \lambda_i f\Big(\frac{x_{ij}}{\lambda_i}\Big) \le \int_0^\infty f(e^{-\lambda})d\lambda \ .$$

Denote function $\max\{x, 0\}$ as x^+. For any $0 \le \theta \le \frac{1}{2}$, consider $f(x) = \frac{1}{\theta}\big(x - (1-\theta)\big)^+$. We have the next corollary of the Converse Jensen Inequality.

Corollary 1. *For any $0 \le \theta \le \frac{1}{2}$ and any offline vertex $j \in J$:*

$$\frac{1}{\theta}\sum_{i\in I} \big(x_{ij} - (1-\theta)\lambda_i\big)^+ \le 1 + \frac{1-\theta}{\theta}\ln(1-\theta) \ .$$

3 Stochastic Sampling Without Replacement

This section presents the first nontrivial competitive analysis of the STOCHASTIC SWOR algorithm. We further derandomize it and get a deterministic algorithm called REGULARIZED GREEDY which achieves the same competitive ratio. We first formally define the STOCHASTIC SWOR algorithm.

Stochastic Sampling Without Replacement

Input at the beginning:

- Type graph $G = (I, J, E)$;
- Arrival rates $(\lambda_i)_{i\in I}$;
- Solution of the Natural LP $(x_{ij})_{(i,j)\in E}$.

When an online vertex of type $i \in I$ arrives at time $0 \le t \le 1$:

- Match it to an unmatched offline neighbor j with probability proportional to x_{ij}.

STOCHASTIC SWOR is at least weakly better than SUGGESTED MATCHING, which samples an offline vertex j regardless of whether it is already matched. Hence, STOCHASTIC SWOR is at least $1 - \frac{1}{e} \approx 0.632$-competitive. Our main result is a substantially better ratio.

Theorem 1. STOCHASTIC SWOR *is 0.707-competitive for Online Stochastic Matching.*

3.1 Potential Function Analysis of Online Stochastic Matching Algorithms

Potential Function Analysis Framework. For any time $0 \le t \le 1$, let $A(t)$ denote the number of edges matched by the online algorithm. We keep track of the unmatched portion of fractional matching $(x_{ij})_{(i,j)\in E}$ using the following notations:

$$x_{ij}(t) = \begin{cases} x_{ij} & if j is unmatched \\ 0 & otherwise \end{cases}, \; x_j(t) = \sum_{i\in I_j} x_{ij}(t) \;,\; x(t) = \{x_{ij}(t)\}_{(i,j)\in E} \,. \tag{2}$$

Further, we define auxiliary variables for match rates, i.e., the values of $x_{ij}(t)$ divided by the arrival rate λ_i of online vertex type i:

$$\rho_{ij}(t) = \frac{x_{ij}(t)}{\lambda_i}\,, \quad \rho_i(t) = \sum_{j\in J_i} \rho_{ij}(t) \,. \tag{3}$$

We will design a potential function $\Phi(t)$, which depends on $x(t)$, i.e., the unmatched portion of the fractional matching, and the current time $0 \le t \le 1$. To prove that the competitive ratio is at least Γ, our potential function needs to satisfy that:

(A) $\Phi(0) \ge \Gamma \cdot \text{OPT}$;
(B) $\Phi(1) = 0$; and
(C) $\mathbf{E}\left[A(t) + \Phi(t)\right]$ is non-decreasing in $0 \le t \le 1$.

Since $A(0) = 0$ by definition, combining these conditions gives $\mathbf{E}\,A(1) \ge \Phi(0) \ge \Gamma \cdot \text{OPT}$.

Remark 1. The value of this analysis framework lies beyond the online unweighted stochastic matching problem. It can also be extended to online edge-weighted stochastic matching with free disposal. Using this framework, we propose a deterministic online algorithm for this problem that achieves the same competitive ratio as the state-of-the-art Top Half Sampling by Huang et al. [21].Readers may refer to the appendix of our full paper [19] for details.

Design of Potential Function. Next, for some parameter $0 \le \theta \le \frac{1}{2}$ to be determined, define function $p : [0,1] \to [0,1]$ as:

$$p(\rho) = \min\left\{\frac{\rho}{\theta}, 1\right\} .$$

Initially, each online type i's contribution to our potential function is mainly proportional to $\lambda_i \rho_i(t)$. In other words, we directly take the LP-based fractional matching as our reference, since that implicitly takes into account the possibility that all offline neighbors are already matched when a type-i online vertex arrives. As time proceeds, an online type i's contribution will smoothly transition to be proportional to $\lambda_i p(\rho_i(t))$. This is because with less time remaining, it is less likely that all offline neighbors would get matched after time t. We can therefore expect an online type i's contribution to be closer to its arrival rate λ_i even if $\rho_i(t)$ is small.

To formalize this transition, we further define functions $\alpha, \beta : [0,1] \to [0,1]$ as follows:

$$\alpha(t) = 1 - \frac{\frac{1}{\theta}e^{-(1-\ln(1-\theta))(1-t)} - (1-\ln(1-\theta))e^{-\frac{1}{\theta}(1-t)}}{\frac{1}{\theta} - 1 + \ln(1-\theta)} ,$$

$$\beta(t) = \frac{e^{-(1-\ln(1-\theta))(1-t)} - e^{-\frac{1}{\theta}(1-t)}}{\frac{1}{\theta} - 1 + \ln(1-\theta)} ,$$

derived by solving the second-order differential equations in the next lemma. Our potential function is:

$$\Phi(t) = \alpha(t) \sum_{(i,j)\in E} x_{ij}(t) + \beta(t) \sum_{i\in I} \lambda_i p\big(\rho_i(t)\big) .$$

We next summarize the properties that characterize functions $\alpha(t)$ and $\beta(t)$ as a lemma, which we will use in our analysis. The proofs are basic calculus and hence omitted. Interested reader may also plot both functions for $0 \le t \le 1$, e.g., with $\theta = \frac{1}{2}$, to verify that $\alpha(t)$ is larger than $\beta(t)$ for small t, but $\beta(t)$ becomes the dominant term when t tends to 1.

Lemma 2. *Functions $\alpha(t)$ and $\beta(t)$ satisfy the following properties:*

1. $\alpha(0) + \beta(0) = 1 - \dfrac{\left(\frac{1}{\theta} - 1\right) e^{-(1-\ln(1-\theta))} + \ln(1-\theta)e^{-\frac{1}{\theta}}}{\frac{1}{\theta} - 1 + \ln(1-\theta)}$.
2. $\alpha(1) = \beta(1) = 0$.
3. $\alpha(t) + \dfrac{1}{\theta}\beta(t) \le 1$ *for any* $0 \le t \le 1$.
4. $\dfrac{d}{dt}\alpha(t) = -\dfrac{1-\ln(1-\theta)}{\theta}\beta(t)$.
5. $\dfrac{d}{dt}\beta(t) = \alpha(t) + \left(\frac{1}{\theta} + 1 - \ln(1-\theta)\right)\beta(t) - 1$.

3.2 Proof of Theorem 1

We next verify the aforementioned sufficient conditions for competitive ratio:

$$\Gamma = 1 - \frac{\left(\frac{1}{\theta} - 1\right) e^{-(1-\ln(1-\theta))} + \ln(1-\theta)e^{-\frac{1}{\theta}}}{\frac{1}{\theta} - 1 + \ln(1-\theta)} . \tag{4}$$

Theorem 1 then follows by choosing $\theta = 0.4254$ which gives $\Gamma \approx 0.7078 > 0.707$.

Condition 3.1. Since $p(\rho) \geq \rho$ for any $0 \leq \rho \leq 1$, we have:

$$\sum_{i \in I} \lambda_i p\big(\rho_i(0)\big) \geq \sum_{i \in I} \lambda_i \rho_i(0) = \sum_{(i,j) \in E} x_{ij}(0) = \text{OPT} .$$

Therefore, we get that:

$$\Phi(0) = \alpha(0) \sum_{(i,j) \in E} x_{ij}(0) + \beta(0) \sum_{i \in I} \lambda_i p\big(\rho_i(0)\big) \geq \big(\alpha(0) + \beta(0)\big) \cdot \text{OPT} .$$

Further by Property 1 of Lemma 2 and by the definition of Γ in Eqn. (4), we have $\alpha(0) + \beta(0) = \Gamma$.

Condition 3.1. It follows by Property 2 of Lemma 2.

Condition 3.1. For any time $0 \leq t \leq 1$, and conditioned on any set of matched offline vertices at time t and the corresponding $x(t)$, we next show that the derivative of $\mathbf{E}\left[A(t) + \Phi(t) \mid x(t)\right]$ is nonnegative. The derivative can be expressed as:

$$\begin{aligned}\frac{d}{dt}\mathbf{E}\big[A(t)|x(t)\big] + \alpha(t)\frac{d}{dt}\mathbf{E}\Big[\sum_{(i,j) \in E} x_{ij}(t)|x(t)\Big] + \beta(t)\frac{d}{dt}\mathbf{E}\Big[\sum_{i \in I} \lambda_i p\big(\rho_i(t)\big)|x(t)\Big] \\ + \Big(\frac{d}{dt}\alpha(t)\Big) \sum_{(i,j) \in E} x_{ij}(t) + \Big(\frac{d}{dt}\beta(t)\Big) \sum_{i \in I} \lambda_i p\big(\rho_i(t)\big)\end{aligned} \tag{5}$$

Define the match rate of edge (i, j) at time t as:

$$y_{ij}(t) = \lambda_i \cdot \frac{\rho_{ij}(t)}{\rho_i(t)} .$$

where λ_i is the arrival rate of i, and $\frac{\rho_{ij}(t)}{\rho_i(t)}$ is the probability of sampling j conditioned on i's arrival by the definition of sampling without replacement.

Further let $y_j(t) = \sum_{i \in I_j} y_{ij}(t)$ be the total match rate of offline vertex j at time t. The first line of Eqn. (5) equals:

$$\sum_{j \in J} y_j(t) \Big(1 - \alpha(t) x_j(t) - \beta(t) \sum_{i \in I_j} \lambda_i \big(p(\rho_i(t)) - p(\rho_i(t) - \rho_{ij}(t))\big)\Big) . \tag{6}$$

The three terms inside the big parentheses correspond to the respective changes in $\mathbf{E}\big[A(t) \mid x(t)\big]$, $\alpha(t)\mathbf{E}\big[\sum_{(i,j) \in E} x_{ij}(t) \mid x(t)\big]$ and $\beta(t)\mathbf{E}\big[\sum_{i \in I} \lambda_i p(\rho_i(t)) \mid x(t)\big]$ when j gets matched.

The coefficient of $y_j(t)$ in Eqn. (6) is at least $1 - \alpha(t) - \frac{1}{\theta}\beta(t)$, because the derivative of function p is at most $\frac{1}{\theta}$ and $\sum_{i \in I_j} \lambda_i \rho_{ij}(t) = x_j(t) \leq 1$. It is therefore nonnegative by Property 3 of Lemma 2. Hence, replacing $y_{ij}(t)$ with

$y_{ij}(t)p(\rho_i(t))$ makes Eqn. (5) weakly smaller. Further, since $x_j(t) \leq 1$, replacing $\alpha(t)x_j(t)$ with $\alpha(t)$ also makes Eqn. (5) weakly smaller. Hence, we conclude that:

$$\begin{aligned}(5) \geq &\big(1-\alpha(t)\big)\sum_{i\in I}\lambda_i p(\rho_i(t)) \\ &-\beta(t)\sum_{j\in J}\Big(\sum_{i\in I_j} y_{ij}(t)p(\rho_i(t))\Big)\Big(\sum_{i\in I_j}\lambda_i\big(p(\rho_i(t))-p(\rho_i(t)-\rho_{ij}(t))\big)\Big) \\ &+\Big(\frac{d}{dt}\alpha(t)\Big)\sum_{(i,j)\in E} x_{ij}(t)+\Big(\frac{d}{dt}\beta(t)\Big)\sum_{i\in I}\lambda_i p(\rho_i(t)) \,. \end{aligned} \tag{7}$$

If we treat functions α, β, and their derivatives as coefficients, then the second term on the right-hand-side of Eqn. (7) is quadratic, while the other terms are linear. The next lemma bounds this term by a linear combination of two linear terms $\sum_{i\in I}\lambda_i p(\rho_i(t))$ and $\sum_{(i,j)\in E} x_{ij}(t)$ in this inequality. Its proof is deferred to the full paper [19].

Lemma 3. *For any* $0 \leq t \leq 1$, *the following inequality holds:*

$$\begin{aligned}&\sum_{j\in J}\Big(\sum_{i\in I_j} y_{ij}(t)p(\rho_i(t))\Big)\Big(\sum_{i\in I_j}\lambda_i\big(p(\rho_i(t))-p(\rho_i(t)-\rho_{ij}(t))\big)\Big) \\ &\leq \Big(\frac{1}{\theta}+1-\ln(1-\theta)\Big)\sum_{i\in I}\lambda_i p(\rho_i(t)) - \frac{1-\ln(1-\theta)}{\theta}\sum_{(i,j)\in E} x_{ij}(t). \end{aligned} \tag{8}$$

We remark that a similar lemma holds if on the left we use the original match rates $y_{ij}(t)$ instead of the scaled $y_{ij}(t)p(\rho_i(t))$, but the first term on the right would be $\sum_{i\in I}\lambda_i$. This is not an existing term in Eqn. (7), and therefore we would not have the cancellation in our subsequent analysis. This is why we artificially scale the match rates in the previous step of our analysis.

Lemma 3 is partly inspired by Lemma 5 of Huang et al. [21]. The main differences are threefold. Firstly and most importantly, our approach decouples the (lower bounds of) match rates of online vertices in the lemma with the actual match rates of the online algorithm. Following the approach of Huang et al. [21], the *algorithm* would need to artificially decrease the match rate of each online vertex i by a $p(\rho_i(t))$ factor, and we would not be able to analyze Stochastic SWOR. Second, we distribute the match rate of each online vertex i to its offline neighbors proportional to the fractional matching $x(t)$, while their approach considered a distorted distribution of the match rate, which once again disagrees with Stochastic SWOR. Last but not least, our approach extends to any $0 \leq \theta \leq \frac{1}{2}$ while their approach only allows $\theta = \frac{1}{2}$.

Applying Lemma 3 to Eqn. (7), we have:

$$\begin{aligned}(5) \geq &\Big(1-\alpha(t)-\beta(t)\Big(\frac{1}{\theta}+1-\ln(1-\theta)\Big)+\frac{d}{dt}\beta(t)\Big)\sum_{i\in I}\lambda_i p(\rho_i(t)) \\ &+\Big(\beta(t)\frac{1-\ln(1-\theta)}{\theta}+\frac{d}{dt}\alpha(t)\Big)\sum_{(i,j)\in E} x_{ij}(t)\,. \end{aligned}$$

This equals zero by Properties 4 and 5 of Lemma 2. Hence, we complete the verification of Condition 3.1 and thus the proof of Theorem 1.

3.3 Regularized Greedy

This subsection derandomizes Stochastic SWOR and introduces a deterministic algorithm called Regularized Greedy which is also 0.707-competitive.

Observe that Conditions 3.1 and 3.1 always hold, independent of the online algorithm's decisions. Therefore, the competitive analysis from the last subsection applies to any algorithm that satisfies Condition 3.1, i.e., the monotonicity of $\mathbf{E}\left[A(t)+\Phi(t)\right]$ for $0 \le t \le 1$. Conditioned on the set of matched offline vertices and the corresponding $x(t)$ at time t, it suffices to ensure that the derivative $\frac{d}{dt}\mathbf{E}\left[A(t)+\Phi(t) \mid x(t)\right]$ is nonnegative, which can be written as follows:

$$\overbrace{\frac{d}{dt}\mathbf{E}\big[A(t)|x(t)\big] + \alpha(t)\frac{d}{dt}\mathbf{E}\Big[\sum_{(i,j)\in E} x_{ij}(t)|x(t)\Big] + \beta(t)\frac{d}{dt}\mathbf{E}\Big[\sum_{i\in I}\lambda_i p(\rho_i(t))|x(t)\Big]}^{depends on algorithm's decisions}$$
$$+ \underbrace{\Big(\frac{d}{dt}\alpha(t)\Big)\sum_{(i,j)\in E} x_{ij}(t) + \Big(\frac{d}{dt}\beta(t)\Big)\sum_{i\in I}\lambda_i p(\rho_i(t))}_{independent of algorithm's decisions}. \tag{9}$$

Given any online algorithm A, we can write the first line above, i.e., the part that depends on the algorithm's decisions, as follows:

$$\sum_{i\in I}\lambda_i\sum_{j\in J_i}\mathbf{Pr}\big[A \text{ matches } i \text{ to } j \mid x(t)\big]$$
$$\Big(1 - \underbrace{\alpha(t)x_j(t) - \beta(t)\sum_{i'\in I_j}\lambda_{i'}\Big(p\big(\rho_{i'}(t)\big) - p\big(\rho_{i'}(t)-\rho_{i'j}(t)\big)\Big)}_{\text{regularization term}}\Big)$$

Intuitively, this regularization term is our estimated likelihood that offline vertex j would be matched in the remaining time, and thus the cost of matching it right now.

The analysis of Stochastic SWOR may be viewed as a constructive proof of the *existence of online algorithms* which ensure nonnegativity of the derivative in Eq. (9). Instead, we may directly maximize the first line of Eq. (9): conditioned on the arrival of an online vertex of type i at time t, match it to an unmatched offline neighbor j with the smallest regularization term, i.e.:

$$\alpha(t)x_j(t) + \beta(t)\sum_{i'\in I_j}\lambda_{i'}\Big(p\big(\rho_{i'}(t)\big) - p\big(\rho_{i'}(t)-\rho_{i'j}(t)\big)\Big). \tag{10}$$

Regularized Greedy

Input at the beginning:

- Type graph $G = (I, J, E)$;
- Arrival rates $(\lambda_i)_{i \in I}$;
- Solution $(x_{ij})_{(i,j) \in E}$ of the Natural LP.

When an online vertex of type $i \in I$ arrives at time $0 \le t \le 1$:

- Match it to an unmatched offline neighbor j that minimizes Eqn. (10).

As a corollary of the definition of REGULARIZED GREEDY and the analysis of STOCHASTIC SWOR, we have the next theorem which makes REGULARIZED GREEDY the first polynomial-time deterministic algorithm that breaks the $1 - \frac{1}{e}$ barrier in Online Stochastic Matching.

Theorem 2. REGULARIZED GREEDY *is* 0.707*-competitive for Online Stochastic Matching.*

4 Balance Sampling Without Replacement

This section gives the first non-trivial competitive analysis of the BALANCE SWOR algorithm. We assume without loss of generality that $I = \{1, 2, \ldots, n\}$ for some positive integer n, where $i \in I$ is the i-th online vertex by their arrival order.

BALANCE SWOR consists of two parts. First, it runs an unbounded version of the BALANCE algorithm to maintain a fractional allocation of online vertices to offline vertices. When each online vertex $i \in I$ arrives, the unbounded BALANCE algorithm allocates an x_{ij} amount of vertex i to each neighbor $j \in J_i$ such that (1) in total one unit of vertex i is allocated to its neighbors, i.e., $\sum_{j \in J_i} x_{ij} = 1$, (2) every neighbor with $x_{ij} > 0$ is allocated with the same total amount of online vertices (including the allocation of i), i.e., has the same $\sum_{i' \le i} x_{i'j}$, and (3) every neighbor with $x_{ij} = 0$ must have already been allocated with a weakly larger total amount of online vertices.

Next, it resorts to sampling without replacement to find an integral matching, using the fractional allocation given by unbounded BALANCE as a reference. That is, it matches online vertex i to an unmatched offline neighbor j with probability proportional to x_{ij}.

BALANCE SWOR

State variable: (for every offline vertex $j \in J$)

- $y_j^i = \sum_{i' \leq i} x_{ij}$, total amount of online vertices 1 to i that is allocated to j; let $y_j^0 = 0$.

When an online vertex of $i \in I$ arrives:

- Find threshold $\bar{y} > 0$ such that $\sum_{(i,j) \in E} (\bar{y} - y_j^{i-1})^+ = 1$.
- Let $x_{ij} = (\bar{y} - y_j^{i-1})^+$ for every neighbor j.
- Match i to an unmatched neighbor j with probability proportional to x_{ij}.

Trivially, BALANCE SWOR is at least 0.5-competitive because it is a greedy algorithm that always matches an online vertex whenever there is an unmatched neighbor. The main result of this section is the first competitive analysis of BALANCE SWOR that is strictly better than 0.5. Its proof is deferred to the full paper [19].

Theorem 3. BALANCE SWOR *is at least* 0.513*-competitive for Online Bipartite Matching.*

Comparison with Balance OCS. Gao et al. [16] proposed a similar algorithm called BALANCE OCS, based on their improved online correlated selection (OCS) technique. Their algorithm also runs the unbounded BALANCE algorithm to obtain a fractional allocation $x = (x_{ij})_{(i,j) \in E}$. Instead of sampling an unmatched neighbor j with probability proportional to x_{ij} like BALANCE SWOR, their algorithm samples with probability proportional to $w(y_j^{i-1}) \cdot x_{ij}$ with function:

$$w(y) = \exp\left(y + \frac{1}{2}y^2 + \frac{4 - 2\sqrt{3}}{3}y^3\right).$$

Although the adjustment to the sampling probabilities made by Gao et al. [16] is necessary for online correlated selection, our empirical study suggests that it is redundant, if not harmful, in the context of Online Bipartite Matching. Theorem 3 is the first step towards a theoretical explanation to this empirical observation.

5 Discussion

We conclude the paper with a brief discussion on our intuition of why rounding by sampling without replacement may be better than online correlated selection for online matching problems.

Online correlated selection, by design, leads to a stronger notion of competitiveness: for every offline vertex, the probability that it is matched at the end is at least Γ times how much it is fractionally matched. Sampling without replacement is only $1 - \frac{1}{e}$ competitive w.r.t. this stronger notion. To see this, consider the following example from Gao et al. [16]. Fix an offline vertex j. Suppose that in each round $i = 1, 2, \ldots, n$, vertex j is paired with a fresh online vertex j_i such that the fractional matching matches j by $\frac{1}{n}$ and j_i by $1 - \frac{1}{n}$. Hence, vertex j is matched at the end with probability only $1 - (1 - \frac{1}{n})^n \approx 1 - \frac{1}{e}$. Online correlated selection algorithms adjust the weights to slighter favor vertex j in later rounds if it is still unmatched, in order to be competitive for every vertex. We observe that, however, the overall matching is in fact extremely good in this example. Since we have a fresh offline vertex in every round, every online vertex gets matched with certainty. Therefore, we believe that the adjustments to sampling weights may be redundant; we just need an appropriate analysis framework that considers the overall matching quality like the potential analysis introduced in this paper, instead of considering each offline vertex separately.

We hope that the results and techniques from this paper could serve as building blocks for further research on sampling without replacement as an online rounding strategy, and on the design and analysis of deterministic online algorithms for online problems under stochastic models.

References

1. Aggarwal, G., Goel, G., Karande, C., Mehta, A.: Online vertex-weighted bipartite matching and single-bid budgeted allocations. In: Proceedings of the 22nd Annual ACM-SIAM Symposium on Discrete Algorithms, pp. 1253–1264. SIAM (2011)
2. Agrawal, S., Devanur, N.R.: Fast algorithms for online stochastic convex programming. In: Proceedings of the 26th Annual ACM-SIAM Symposium on Discrete Algorithms, pp. 1405–1424. SIAM (2015)
3. Aouad, A., Ma, W.: A nonparametric framework for online stochastic matching with correlated arrivals. In: Proceedings of the 24th ACM Conference on Economics and Computation, p. 114. ACM, New York (2023)
4. Bahmani, B., Kapralov, M.: Improved bounds for online stochastic matching. In: Proceedings of the 18th Annual European Symposium on Algorithms, pp. 170–181. Springer (2010)
5. Blanc, G., Charikar, M.: Multiway online correlated selection. In: Proceedings of the 62nd Annual IEEE Symposium on Foundations of Computer Science, pp. 1277–1284. IEEE (2021)
6. Borodin, A., Karavasilis, C., Pankratov, D.: An experimental study of algorithms for online bipartite matching. J. Exper. Algorithmics **25**, 1–37 (2020)
7. Brubach, B., Sankararaman, K.A., Srinivasan, A., Xu, P.: New algorithms, better bounds, and a novel model for online stochastic matching. In: Proceedings of the 24th Annual European Symposium on Algorithms. Schloss Dagstuhl-Leibniz-Zentrum fuer Informatik (2016)
8. Buchbinder, N., Jain, K., Naor, J.: Online primal-dual algorithms for maximizing ad-auctions revenue. In: Proceedings of the 15th Annual European Symposium on Algorithms, pp. 253–264. Springer (2007)
9. Buchbinder, N., Naor, J., Wajc, D.: Lossless online rounding for online bipartite matching (despite its impossibility). In: Proceedings of the 34th Annual ACM-SIAM Symposium on Discrete Algorithms, pp. 2030–2068. SIAM (2023)

10. Devanur, N.R., Hayes, T.P.: The AdWords problem: online keyword matching with budgeted bidders under random permutations. In: Proceedings of the 10th ACM Conference on Electronic Commerce, pp. 71–78 (2009)
11. Devanur, N.R., Huang, Z., Korula, N., Mirrokni, V.S., Yan, Q.: Whole-page optimization and submodular welfare maximization with online bidders. ACM Trans. Econ. Comput. **4**(3), 1–20 (2016)
12. Devanur, N.R., Jain, K., Kleinberg, R.D.: Randomized primal-dual analysis of ranking for online bipartite matching. In: Proceedings of the 24th Annual ACM-SIAM Symposium on Discrete Algorithms, pp. 101–107. SIAM (2013)
13. Fahrbach, M., Huang, Z., Tao, R., Zadimoghaddam, M.: Edge-weighted online bipartite matching. J. ACM **69**(6), 1–35 (2022)
14. Feldman, J., Mehta, A., Mirrokni, V., Muthukrishnan, S.: Online stochastic matching: beating $1 - \frac{1}{e}$. In: Proceedings of the 50th Annual IEEE Symposium on Foundations of Computer Science, pp. 117–126. IEEE (2009)
15. Feng, Y., Qiu, G., Wu, X., Zhou, S.: Improved competitive ratio for edge-weighted online stochastic matching. arXiv preprint arXiv:2302.05633 (2023)
16. Gao, R., He, Z., Huang, Z., Nie, Z., Yuan, B., Zhong, Y.: Improved online correlated selection. In: Proceedings of the 62nd Annual IEEE Symposium on Foundations of Computer Science, pp. 1265–1276. IEEE (2021)
17. Goel, G., Mehta, A.: Online budgeted matching in random input models with applications to AdWords. In: Proceedings of the 19th Annual ACM-SIAM Symposium on Discrete Algorithms, pp. 982–991 (2008)
18. Haeupler, B., Mirrokni, V.S., Zadimoghaddam, M.: Online stochastic weighted matching: improved approximation algorithms. In: Proceedings of the 7th International Workshop on Internet and Network Economics, pp., pp. 170–181. Springer (2011)
19. Huang, Z., Lee, C.S., Lu, J., Shu, X.: Online matching meets sampling without replacement. arXiv preprint arXiv:2410.06868 (2024)
20. Huang, Z., Shu, X.: Online stochastic matching, Poisson arrivals, and the natural linear program. In: Proceedings of the 53rd Annual ACM SIGACT Symposium on Theory of Computing, pp. 682–693 (2021)
21. Huang, Z., Shu, X., Yan, S.: The power of multiple choices in online stochastic matching. In: Proceedings of the 54th Annual ACM SIGACT Symposium on Theory of Computing, pp. 91–103 (2022)
22. Huang, Z., Tang, Z.G., Wu, X., Zhang, Y.: Online vertex-weighted bipartite matching: Beating $1-\frac{1}{e}$ with random arrivals. ACM Trans. Algorithms **15**(3), 1–15 (2019)
23. Huang, Z., Zhang, Q., Zhang, Y.: Adwords in a panorama. In: Proceedings of the 61st Annual IEEE Symposium on Foundations of Computer Science, pp. 1416–1426. IEEE (2020)
24. Jaillet, P., Lu, X.: Online stochastic matching: new algorithms with better bounds. Math. Oper. Res. **39**(3), 624–646 (2014)
25. Jin, B., Williamson, D.P.: Improved analysis of RANKING for online vertex-weighted bipartite matching in the random order model. In: Proceedings of the 17th International Conference on Web and Internet Economics, pp. 207–225. Springer (2022)
26. Kalyanasundaram, B., Pruhs, K.R.: An optimal deterministic algorithm for online b-matching. Theoret. Comput. Sci. **233**(1–2), 319–325 (2000)
27. Karande, C., Mehta, A., Tripathi, P.: Online bipartite matching with unknown distributions. In: Proceedings of the 43rd Annual ACM Symposium on Theory of Computing, pp. 587–596 (2011)

28. Karp, R.M., Vazirani, U.V., Vazirani, V.V.: An optimal algorithm for on-line bipartite matching. In: Proceedings of the 22nd Annual ACM Symposium on Theory of Computing, pp. 352–358 (1990)
29. Kesselheim, T., Radke, K., Tönnis, A., Vöcking, B.: An optimal online algorithm for weighted bipartite matching and extensions to combinatorial auctions. In: Proceedings of the 21st Annual European Symposium on Algorithms, pp. 589–600. Springer (2013)
30. Kesselheim, T., Radke, K., Tönnis, A., Vöcking, B.: Primal beats dual on online packing LPs in the random-order model. SIAM J. Comput. **47**(5), 1939–1964 (2018)
31. Ma, W., Xu, P., Xu, Y.: Fairness maximization among offline agents in online-matching markets. In: Proceedings of the 17th International Workshop on Internet and Network Economics, pp. 16–27. Springer (2023)
32. Mahdian, M., Yan, Q.: Online bipartite matching with random arrivals: an approach based on strongly factor-revealing LPs. In: Proceedings of the 43rd Annual ACM Symposium on Theory of Computing, pp. 597–606 (2011)
33. Manshadi, V.H., Oveis Gharan, S., Saberi, A.: Online stochastic matching: online actions based on offline statistics. Math. Oper. Res. **37**(4), 559–573 (2012)
34. Mehta, A., Saberi, A., Vazirani, U., Vazirani, V.: Adwords and generalized online matching. J. ACM **54**(5), 22–es (2007)
35. Naor, J.S., Srinivasan, A., Wajc, D.: Online dependent rounding schemes. arXiv preprint arXiv:2301.08680 (2023)
36. Shin, Y., An, H.C.: Making three out of two: Three-way online correlated selection. In: Proceedings of the 32nd International Symposium on Algorithms and Computation. Schloss Dagstuhl-Leibniz-Zentrum für Informatik (2021)
37. Tang, Z.G., Wu, J., Wu, H.: (Fractional) online stochastic matching via fine-grained offline statistics. In: Proceedings of the 54th Annual ACM SIGACT Symposium on Theory of Computing, pp. 77–90 (2022)
38. Yan, S.: Edge-weighted online stochastic matching: beating $1 - \frac{1}{e}$. arXiv preprint arXiv:2210.12543 (2022)

Stochastic Online Metric Matching: Adversarial Is No Harder Than Stochastic

Amin Saberi[1], Mingwei Yang[1(✉)], and Sophie H. Yu[2]

[1] Department of Management Science and Engineering, Stanford University, Stanford, CA, USA
{saberi,mwyang}@stanford.edu

[2] Wharton School of Business, University of Pennsylvania, Philadelphia, PA, USA
hysophie@wharton.upenn.edu

Abstract. We study the stochastic online metric matching problem. In this problem, m servers and n requests are located in a metric space, where all servers are available upfront and requests arrive one at a time. In particular, servers are adversarially chosen, and requests are independently drawn from a known distribution. Upon the arrival of a new request, it needs to be immediately and irrevocably matched to a free server, resulting in a cost of their distance. The objective is to minimize the total matching cost.

In this paper, we show that the problem can be reduced to a more accessible setting where both servers and requests are drawn from the same distribution by incurring a moderate cost. Combining our reduction with previous techniques, for $[0,1]^d$ with various choices of distributions, we achieve improved competitive ratios and nearly optimal regrets in both balanced and unbalanced markets. In particular, we give $O(1)$-competitive algorithms for $d \geq 3$ in both balanced and unbalanced markets with smooth distributions. Our algorithms improve on the $O((\log\log\log n)^2)$ competitive ratio of Gupta et al. (ICALP'19) for balanced markets in various regimes, and provide the first positive results for unbalanced markets.

Keywords: Online algorithm · Stochastic matching · Metric space

1 Introduction

Online matching forms a central topic of online algorithms and has received extensive attention due to the surge of online matching markets in recent years. Some of its prominent applications include online advertising, job search, and dating, where the constant growth of online advertising platforms has particularly spurred numerous research from both algorithmic and economic perspectives. A seminal paper [15] proposes the online bipartite matching problem and presents optimal deterministic and randomized algorithms. Since then, elegant

The full version of our paper is available on arXiv: https://arxiv.org/abs/2407.14785.

M. Mavronicolas et al. (Eds.): WINE 2024, LNCS 15534, pp. 102–118, 2026.
https://doi.org/10.1007/978-3-032-08560-3_6

algorithmic results have been devised to accommodate more diverse environments. See the surveys by [11] and by [17] for a more comprehensive introduction.

In recent years, another online matching paradigm, online (bipartite) minimum cost matching, has emerged. Motivated by the application of ride-hailing, in this problem, m nodes from one side of the market are called *servers* and are fully known upfront. On the other side of the market, which is often referred to as the *request* side, n nodes arrive sequentially. Upon the arrival of a request i, it has to be immediately and irrevocably assigned to an available server j, incurring a cost $\delta(i, j)$, and both request i and server j are removed.[1] The objective is to minimize the total matching cost.

Due to the importance of application to ride-hailing, it is reasonable to assume that both servers and requests are located in the same metric space, and the cost incurred by matching a server and a request is measured by the distance between them. A majority of literature on online metric matching focuses on the *adversarial* model [18,20,22], where the locations of all servers and requests are adversarially chosen, or the weaker *random order* model [21], which assumes that the order of adversarially chosen requests is randomly permuted. See Sect. 1.2 for a more comprehensive literature review in these settings.

The online matching model with adversarial online arrivals is usually deemed too pessimistic since practical instances generally exhibit a more stochastic nature. To model a practical and hence more tractable scenario, the stochastic online metric matching problem, where the requests are independently drawn from a known probability distribution, has attracted considerable attention in the past few years [4,6,9]. The most basic model assumes *balanced markets*, which contain an equal number of servers and requests. For the Euclidean space $[0, 1]^d$, where d is a constant not scaling with n, [14] proposes the *Hierarchical Greedy* algorithm with a constant competitive ratio when all servers and requests are uniformly distributed, and [6] achieves nearly optimal regret guarantees when servers and requests follow different distributions. [4] analyzes the performance of the *Greedy* algorithm, whose prominence is empirically validated, and shows that Greedy is $O(1)$-competitive when servers and requests are uniformly distributed in $[0, 1]$.

[2] notices that the potential value of excess servers has been overlooked by the literature and considers *unbalanced markets*, which possess at most a linear amount of excess servers. When servers and requests are uniformly distributed in $[0, 1]$, they show that the total matching cost of Greedy can be substantially improved by transitioning from balanced markets to unbalanced markets, and their results imply that Greedy is $O(\log^3 n)$-competitive. [4] subsequently pinpoints the competitive ratio of Greedy to be $O(1)$ under the same setting. Moreover, Hierarchical Greedy also admits a constant competitive ratio in unbalanced markets when servers and requests are uniformly distributed in $[0, 1]^d$ for all $d \geq 1$ [14].

---

[1] We always assume $m \geq n$ to guarantee that every request can be served.

In contrast to the fruitful achievements when both servers and requests are stochastic, it appears to be more challenging to establish positive results for adversarial servers and stochastic requests. Specifically, in balanced markets, [9] gives an $O((\log\log\log n)^2)$-competitive algorithm for general metrics and distributions, which further achieves a constant competitive ratio for tree metrics. Moreover, in balanced markets with requests uniformly distributed in $[0,1]$, [4] shows that Greedy is $\Theta(\log n)$-competitive and Hierarchical Greedy is $\Omega(n^{1/4})$-competitive.

1.1 Our Results

As our main contribution, assuming requests are stochastic, we present a reduction from the setting with adversarial servers to that with stochastic servers: If an algorithm $\mathcal{A}$ exists for balanced markets with both servers and requests randomly drawn from the same distribution $\mathbb{D}$, we can use $\mathcal{A}$ in a black-box manner to derive another algorithm $\mathcal{A}'$ for unbalanced markets with servers adversarially chosen and requests randomly drawn from $\mathbb{D}$ (Theorem 1). Moreover, if the markets are balanced, which can be viewed as a special case of unbalanced markets, $\mathcal{A}'$ admits the same competitive ratio as $\mathcal{A}$ up to a constant factor. Hence, by utilizing the recent progress on the setting with stochastic servers and requests [6,14], and applying our reduction techniques, we then describe the following results.

Table 1. Competitive ratios for the Euclidean space $[0,1]^d$ with adversarial servers and stochastic requests. All previous state-of-the-art results are given in black fonts, where we use $\omega(1)$ to denote $O((\log\log\log n)^2)$, and all new results achieved in this paper are marked in red. In particular, for balanced markets with general distributions, the competitive ratio of $O(1)$ for $d=1$ is obtained by [9] and this paper (see the full version of our paper), which implies the two bounds on its left, and the competitive ratio of $O((\log\log\log n)^2)$ for $d\geq 2$ is achieved by [9], which implies all $O((\log\log\log n)^2)$ bounds on its left.

	Balanced Markets			Unbalanced Markets	
	Uniform Dis.	Smooth Dis.	General Dis.	Smooth Dis.	General Dis.
$d=1$	$O(1)$	$O(1)$	$O(1)$	$O(\log n)$	$O(\log n)$
$d=2$	$O(1)$ $(\omega(1))$	$\omega(1)$	$\omega(1)$	$O(\sqrt{\log n})$	$O(\log n)$
$d\geq 3$	$O(1)$ $(\omega(1))$	$O(1)$ $(\omega(1))$	$\omega(1)$	$O(1)$	$O(\log n)$

Multiplicative Bound (Competitive Ratio). We first introduce the competitive ratio results for the Euclidean space $[0,1]^d$ achieved by combining our reduction with algorithms of prior work, which are summarized in Table 1. Recall that the competitive ratio of an algorithm is defined as the worst ratio between the cost incurred by the algorithm and the optimal offline cost, and can be viewed as a multiplicative guarantee. We start by showing that combining our reduction with the Hierarchical Greedy algorithm of [14] leads to an $O(1)$-competitive algorithm in balanced markets for all $d \geq 1$ with adversarial servers and uniformly distributed requests (Theorem 2).

Next, we describe our results in balanced markets with servers adversarially chosen and requests randomly drawn from a smooth distribution, where a distribution is smooth if it admits a density function upper bounded by a constant. We adopt the Simulate-Optimize-Assign-Repeat (SOAR) algorithm of [6] as the algorithm $\mathcal{A}$ required by the reduction, which is originally used to give nearly optimal regrets in balanced markets when servers and requests follow different distributions. We show that this leads to an $O(1)$-competitive algorithm for $d \geq 3$ (Theorem 3).

Finally, we present our results in unbalanced markets with adversarial servers and stochastic requests. When requests are randomly drawn from a smooth distribution, by applying the same approach as for balanced markets, the resulting algorithm admits a competitive ratio of $O(\sqrt{n})$ for $d = 1$, $O(\sqrt{\log n})$ for $d = 2$, and $O(1)$ for $d \geq 3$ (Theorem 4). To improve the undesirable competitive ratio for $d = 1$, we show that the $O(\log n)$ competitive ratio of the algorithm proposed by [10], which holds in balanced markets when both servers and requests are adversarially located in a doubling metric[2], is preserved in unbalanced markets (Corollary 2).

Additive Bound (Regret). We turn to our regret results for the Euclidean space $[0,1]^d$. Since the regret of an algorithm is defined as the worst difference between the average cost incurred by the algorithm and the average optimal offline cost, it can be viewed as an additive guarantee. Recall that in balanced markets with servers and requests following different but arbitrary distributions, [6] shows that the SOAR algorithm admits a regret of $O(n^{-\frac{1}{2}})$ for $d = 1$, $O(n^{-\frac{1}{2}} \log n)$ for $d = 2$, and $O(n^{-\frac{1}{d}})$ for $d \geq 3$, which is optimal for $d \neq 2$ and optimal up to a logarithmic factor for $d = 2$. By combining our reduction with the SOAR algorithm, we show that the same regrets are achievable even in unbalanced markets with servers adversarially chosen and requests randomly drawn from an arbitrary distribution (Theorem 5). Since balanced markets form a special case of unbalanced markets and our setting with adversarial servers is more general, the lower bounds by [6] continue to hold in our setting, and hence, our regret guarantees are nearly optimal.

Additional Results. As alluded to previously, we generalize the $O(\log n)$ competitive ratio of the algorithm by [10] for doubling metrics with adversarial servers and requests from balanced markets to unbalanced markets. This is achieved based on the observation that every greedy algorithm with a tie-breaking rule

[2] A metric space is said to be *doubling* if a doubling constant $M > 0$ exists such that for all $r > 0$, every ball of radius r can be covered by at most M balls of radius $r/2$. In particular, the Euclidean space $[0,1]^d$ is a doubling metric for constant d.

satisfying certain properties admits the same competitive ratio in balanced and unbalanced markets (Theorem 6). As a by-product, we similarly show that the $O(\log^3 n)$ competitive ratio of the algorithm by [18] in balanced markets with servers and requests adversarially located in an arbitrary metric is retained in unbalanced markets (Corollary 1). We remark here that all prior results for adversarial servers and requests are established in balanced markets, and it is not immediate whether they can be carried over to unbalanced markets.

Moreover, in the full version of our paper, we extend our reduction to the case where cost functions only *approximately* satisfy the triangle inequality. Specifically, we say that a cost function δ satisfies the *η-approximate triangle inequality* for $\eta > 0$ if for all x, y, z, we have $\delta(x, z) \leq \eta \cdot (\delta(x, y) + \delta(y, z))$. To showcase the generality of our approach, we show that combining our generalized reduction with the SOAR algorithm leads to positive results for $[0, 1]^d$ with cost function $\|\cdot\|_2^p$ for some $p \geq 1$.

Finally, in the full version of our paper, we show that combining our reduction with the SOAR algorithm gives an alternative $O(1)$-competitive algorithm in balanced markets with tree metrics and general distributions, recovering the result by [9].

1.2 Related Work

For general metrics with adversarial servers and requests, [13,16], and [21] respectively give $(2n-1)$-competitive deterministic algorithms, which is also shown to be tight for deterministic algorithms. For randomized algorithms, an $O(\log^3 n)$-competitive algorithm is proposed by [18]; it is subsequently improved by [5] to $O(\log^2 n)$, which remains the best upper bound. In the random-order model, the deterministic algorithm of [21] admits a competitive ratio of $O(\log n)$, which is optimal even for randomized algorithms, and [8] shows that the competitive ratio of Greedy lies between $\Omega(n^{0.22})$ and $O(n)$.

For the line metric with adversarial servers and requests, [13] shows that the competitive ratio of Greedy is at least $2^n - 1$, and a deterministic algorithm with a sublinear competitive ratio is known [3]. Furthermore, [19] presents an $O(\log^2 n)$-competitive deterministic algorithm, which is later shown to be $O(\log n)$-competitive [22]. Moving to randomized algorithms, [10] proposes an $O(\log n)$-competitive algorithm for the more general doubling metrics. For lower bounds, [7] shows that no deterministic algorithm can achieve a competitive ratio strictly better than 9.001; it is then improved to $\Omega(\sqrt{\log n})$, which holds even for randomized algorithms [20].

Various papers study other aspects of stochastic online metric matching. [24] shows that Greedy admits a competitive ratio of $O(\sqrt{n})$ on the unit disk. In addition, [23] characterizes the cost of Greedy in unbalanced markets with servers and requests uniformly distributed in $[0, 1]$ using a partial differential equation approach.

2 Preliminaries

In our model, n requests and m servers are located in a metric space (X, δ), where $n \leq m$ and $\delta : X \times X \to \mathbb{R}_{\geq 0}$ is a cost function satisfying the triangle inequality. Denote the locations of servers as $S = \{s_1, \ldots, s_m\}$ and the locations of requests as $R = \{r_1, \ldots, r_n\}$. The locations of servers are adversarially chosen and all available at the beginning. The requests arrive one by one, and their locations are independently drawn from a known distribution $\mathbb{D}$ over X. Upon the arrival of a new request, we need to match it to an available server immediately and irrevocably, and then both the request and the matched server leave the system. The cost of matching a server s and a request r is $\delta(s, r)$, and we aim to minimize the total matching cost.

For all algorithm $\mathcal{A}$ and set of servers S with $|S| \geq n$, denote $\text{cost}_{\mathcal{A}}(S, n)$ as the expected cost of $\mathcal{A}$ when the initial set of servers is S and n requests are randomly drawn from $\mathbb{D}$, and denote $\text{cost}_{\mathcal{A}}(n)$ as the expected cost of $\mathcal{A}$ when $|S| = |R| = n$ and all servers and requests are randomly drawn from $\mathbb{D}$. Given a matching M between two sets S and R, for each matched element $s \in S$ (resp. $r \in R$), we use $M(s)$ (resp. $M(r)$) to denote the element in R (resp. S) that s (resp. r) is matched to.

We define the smoothness property of distributions as follows, which requires distributions to admit an upper bounded density function.

Definition 1. β-Smoothness. *We say that a distribution $\mathbb{D}$ over X is β-*smooth *if $\mathbb{D}$ admits a density function $\mu_{\mathbb{D}}$ such that $\mu_{\mathbb{D}}(x) \leq \beta$ for all $x \in X$.*

2.1 Benchmark

We adopt the optimal offline solution, defined as the optimal matching cost without any future uncertainty, as our benchmark. For all set of servers S and set of requests R such that $|S| \geq |R|$, denote $\text{OPT}(S, R)$ as the minimum cost of all matchings between S and R that match all requests. Given a set of servers S with $|S| \geq n$ and a distribution for request arrival $\mathbb{D}$, the optimal offline solution on S with n requests randomly drawn from $\mathbb{D}$ is defined as $\text{OPT}(S, n) = \mathbb{E}_{R \sim \mathbb{D}^n} [\text{OPT}(S, R)]$, where $R \sim \mathbb{D}^n$ denotes that the requests in R are independently drawn from $\mathbb{D}$. Furthermore, define $\text{OPT}(n) = \mathbb{E}_{S,R \sim \mathbb{D}^n} [\text{OPT}(S, R)]$.

We use *competitive ratio* and *regret* to measure the performance of online algorithms. For given metric space (X, δ) and distribution $\mathbb{D}$, we say that the competitive ratio of an algorithm $\mathcal{A}$ is α if for all concerned instances[3] (S, n),

$$\text{cost}_{\mathcal{A}}(S, n) \leq \alpha \cdot \text{OPT}(S, n) .$$

Moreover, define the regret of an algorithm $\mathcal{A}$ for an instance (S, n) as

$$\text{Regret}(\mathcal{A}; S, n) \triangleq \frac{1}{n} \left(\text{cost}_{\mathcal{A}}(S, n) - \text{OPT}(S, n) \right) .$$

[3] We require $|S| = n$ when considering balanced markets and $n \leq |S| \leq O(n)$ for unbalanced markets.

2.2 Optimal Offline Solutions for Euclidean Space

A considerable portion of this paper focuses on Euclidean space. In this subsection, we assume the metric space to be $[0,1]^d$ with the Euclidean distance and bound the optimal offline solutions for smooth distributions and general distributions. All the omitted proofs in this subsection can be found in the full version of our paper.

For an arbitrary distribution $\mathbb{D}$, the following upper bounds for $\mathrm{OPT}(n)$ are known from the optimal transport literature.

Lemma 1 (Theorem 3.11 in [12]). *For all $d \geq 1$ and distribution $\mathbb{D}$ on $[0,1]^d$,*

$$\mathrm{OPT}(n) \leq \begin{cases} O(\sqrt{n}) & d = 1 \\ O(\sqrt{n}\log n) & d = 2 \\ O(n^{1-\frac{1}{d}}) & d \geq 3 \end{cases}.$$

When restricted to smooth distributions, we improve the upper bound for $d = 2$ in Lemma 1 by generalizing the proof of the upper bound in the Ajtai-Komlós-Tusnády Matching Theorem [1], which asserts that $\mathrm{OPT}(n) = \Theta(\sqrt{n \log n})$ for uniform distribution $\mathbb{D}$ and $d = 2$.

Lemma 2. *Assume $\mathbb{D}$ to be β-smooth for some constant $\beta > 0$. For all $d \geq 1$,*

$$\mathrm{OPT}(n) \leq \begin{cases} O(\sqrt{n}) & d = 1 \\ O(\sqrt{n \log n}) & d = 2 \\ O(n^{1-\frac{1}{d}}) & d \geq 3 \end{cases}.$$

Next, we establish the lower bounds for smooth distributions in balanced markets.

Lemma 3. *Assume $\mathbb{D}$ to be β-smooth for some constant $\beta > 0$. For all $d \geq 1$,*

$$\mathrm{OPT}(n) \geq \begin{cases} \Omega(\sqrt{n}) & d = 1 \\ \Omega(n^{1-\frac{1}{d}}) & d \geq 2 \end{cases}.$$

The lower bound in Lemma 3 for $d = 1$ follows from the quantile mapping and the well-known $\Omega(\sqrt{n})$ lower bound for the optimal offline solution with uniform distribution, and the lower bounds for $d \geq 2$ rely on nearest-neighbor-distance.

Finally, we present the lower bounds for smooth distributions in unbalanced markets, which are proved by employing nearest-neighbor-distance as well.

Lemma 4. *Assume $\mathbb{D}$ to be β-smooth for some constant $\beta > 0$. For all $d \geq 1$ and $S \subseteq [0,1]^d$ with $n \leq |S| \leq O(n)$, $\mathrm{OPT}(S, n) \geq \Omega(n^{1-\frac{1}{d}})$.*

2.3 The Simulate-Optimize-Assign-Repeat Algorithm

The Simulate-Optimize-Assign-Repeat (SOAR) algorithm is proposed by [6] under the setting in which both servers and requests are stochastic. The SOAR algorithm, whose guarantee is stated below, will serve as an important building block of our algorithms and be invoked in a black-box manner.

Lemma 5. *[6] Suppose that $m = n$ and $\mathbb{D}$ is a distribution on X, where the cost function δ is* not *required to satisfy the triangle inequality. When servers and requests are randomly drawn from $\mathbb{D}$, an algorithm exists with an expected cost of $\sum_{t=1}^{n} \frac{1}{t} \cdot \mathrm{OPT}(t)$.*

3 Reducing Adversarial to Stochastic

In this section, we present an algorithmic framework to transform an algorithm for stochastic servers and requests into an algorithm for adversarial servers and stochastic requests at a moderate cost.

Theorem 1. *Given a distribution $\mathbb{D}$ on a metric space (X, δ), assume the existence of an algorithm $\mathcal{A}$ such that $\mathrm{cost}_{\mathcal{A}}(n) \leq \alpha \cdot \mathrm{OPT}(n)$. Then, an algorithm $\mathcal{A}'$ exists such that $\mathrm{cost}_{\mathcal{A}'}(S, n) \leq \mathrm{OPT}(S, n) + \mathrm{cost}_{\mathcal{A}}(n)$ for every set of servers S with $|S| \geq n$. If we further have $|S| = n$, then $\mathrm{cost}_{\mathcal{A}'}(S, n) \leq (2\alpha+1) \cdot \mathrm{OPT}(S, n)$.*

3.1 Proof of Theorem 1

Assuming that requests are stochastic, we start by lower bounding the optimal offline solution in balanced markets with adversarial servers. In particular, we show that it is at least the optimal offline solution in balanced markets with stochastic servers and requests up to a constant factor.

Lemma 6. *For all metric space (X, δ), for all distribution $\mathbb{D}$ over X and set of servers S' with $|S'| = n$, $\mathrm{OPT}(S', n) \geq \frac{1}{2} \cdot \mathrm{OPT}(n)$.*

Proof. Given a set of servers S' with $|S'| = n$, we claim that $\mathrm{OPT}(S, R) \leq \mathrm{OPT}(S, S') + \mathrm{OPT}(S', R)$ for all set of servers S and set of requests R with $|S| = |R| = n$. To see this, let M_1 be the perfect matching between S and S' with the minimum cost and M_2 be the perfect matching between S' and R with the minimum cost. Note that

$$\begin{aligned}\mathrm{OPT}(S, S') + \mathrm{OPT}(S', R) &= \sum_{s' \in S'} \delta(M_1(s'), s') + \sum_{s' \in S'} \delta(s', M_2(s')) \\ &\geq \sum_{s' \in S'} \delta(M_1(s'), M_2(s')) \\ &\geq \mathrm{OPT}(S, R),\end{aligned}$$

where the first inequality holds by the triangle inequality. Taking expectations on both sides over $S, R \sim \mathbb{D}^n$, we obtain

$$\mathrm{OPT}(n) \leq \mathbb{E}_{S,R \sim \mathbb{D}^n} [\mathrm{OPT}(S, S') + \mathrm{OPT}(S', R)] = 2 \cdot \mathrm{OPT}(S', n),$$

which concludes the proof.

Now, we are ready to prove Theorem 1. Given a distribution $\mathbb{D}$ over a metric space (X, δ), let $\mathcal{A}$ be an algorithm such that $\text{cost}_{\mathcal{A}}(n) \le \alpha \cdot \text{OPT}(n)$, and we construct an algorithm $\mathcal{A}'$ for adversarial servers and stochastic requests as follows. Given a set of servers S, we randomly generate a set of points $I \sim \mathbb{D}^n$, and let $M^{S,I}$ be the matching between S and I with the minimum cost that matches all points in I. Then, we run $\mathcal{A}$ with the set of initial servers being I. For each new request r, suppose that $\mathcal{A}$ matches it to $i \in I$, and we match r to $M^{S,I}(i) \in S$ in $\mathcal{A}'$.

Next, we analyze the expected cost of $\mathcal{A}'$. By the triangle inequality, the expected cost of $\mathcal{A}'$ is upper bounded by

$$\begin{aligned} \text{cost}_{\mathcal{A}'}(S, n) &\le \mathbb{E}_{I \sim \mathbb{D}^n}\left[\sum_{i \in I} \delta(M^{S,I}(i), i) + \text{cost}_{\mathcal{A}}(I, n)\right] \\ &= \text{OPT}(S, n) + \text{cost}_{\mathcal{A}}(n)\,, \end{aligned}$$

which concludes the first part in Theorem 1. Furthermore, when $|S| = n$, by the guarantee of $\mathcal{A}$ and Lemma 6, we obtain

$$\text{cost}_{\mathcal{A}'}(S, n) \le \text{OPT}(S, n) + \alpha \cdot \text{OPT}(n) \le (2\alpha + 1) \cdot \text{OPT}(S, n)\,,$$

concluding the second part in Theorem 1.

4 Euclidean Space with Smooth Distributions

In this section, we assume the metric space to be $[0, 1]^d$ with the Euclidean distance and $\mathbb{D}$ to be β-smooth[4] for some constant $\beta > 0$, and we will apply Theorem 1 to derive algorithms for both balanced and unbalanced markets. In Sect. 4.1, we focus on uniform distribution, which is a special case of smooth distributions, and give an $O(1)$-competitive algorithm for all $d \ge 1$ in balanced markets.

Next, we consider balanced markets with smooth distributions in Sect. 4.2. Recall that when $d = 1$, an $O(1)$-competitive algorithm in balanced markets for general distributions[5] is given by [9]. Moreover, the above algorithm admits an incentive ratio of $O((\log\log\log n)^2)$ for general metrics and distributions. However, for $d = 2$ and smooth distributions, we are only able to estimate the optimal offline solution up to a $\Theta(\sqrt{\log n})$ factor, which prevents our approach from achieving an $o(\sqrt{\log n})$ competitive ratio. Hence, we only focus on the case where $d \ge 3$. In particular, we present an $O(1)$-competitive algorithm for all $d \ge 3$.

[4] The smoothness of distributions is a mild technical assumption allowing us to lower bound $\text{OPT}(n)$. To see the necessity, consider $\mathbb{D}$ a non-smooth distribution supported on a singleton, in which case $\text{OPT}(n) = 0$.

[5] In the full version of our paper, we also give an alternative $O(1)$-competitive algorithm for $d = 1$ in balanced markets with general distributions.

Finally, in Sect. 4.3, we extend our results for smooth distributions to unbalanced markets that possess at most linearly many excess servers, i.e., $n \leq m \leq O(n)$. We give an algorithm with a competitive ratio of $O(\sqrt{n})$ for $d = 1$, $O(\sqrt{\log n})$ for $d = 2$, and $O(1)$ for $d \geq 3$.

4.1 Warm-Up: Balanced Markets with Uniform Distribution

In this subsection, we consider balanced markets and assume $\mathbb{D}$ to be uniform distribution. We show that combining Theorem 1 with the Hierarchical Greedy algorithm of [14] results in an $O(1)$-competitive algorithm for all $d \geq 1$.

We present our algorithm in the following theorem.

Theorem 2. *Suppose that $m = n$. For all $d \geq 1$, an $O(1)$-competitive algorithm exists for adversarial servers and uniformly distributed requests.*

Proof. Recall that [14] gives an algorithm with a competitive ratio of $O(1)$ for all $d \geq 1$ when both servers and requests follow uniform distribution. Combining the above algorithm with Theorem 1 concludes the proof.

4.2 Balanced Markets

In this subsection, we assume $d \geq 3$ and consider balanced markets with smooth distributions. We will prove that the SOAR algorithm admits an $O(1)$ competitive ratio for $d \geq 3$ when servers and requests are randomly drawn from $\mathbb{D}$. Together with Theorem 1, this leads to an $O(1)$-competitive algorithm for $d \geq 3$ with adversarial servers and stochastic requests.

We first present our main theorem in this subsection.

Theorem 3. *Suppose that $m = n$ and $\mathbb{D}$ is β-smooth for some constant $\beta > 0$. When servers are adversarially chosen and requests are randomly drawn from $\mathbb{D}$, an $O(1)$-competitive algorithm exists for $d \geq 3$.*

The rest of this section is devoted to proving Theorem 3. By Theorem 1, it suffices to show that the SOAR algorithm is $O(1)$-competitive for $d \geq 3$ when both servers and requests are randomly drawn from $\mathbb{D}$. We compute the expected cost of the SOAR algorithm in the following lemma. Even though we only consider $d \geq 3$ in this subsection, the costs of the SOAR algorithm for $d \leq 2$ computed by Lemma 7 will be useful afterward.

Lemma 7. *Suppose that $m = n$ and $\mathbb{D}$ is β-smooth for some constant $\beta > 0$. An algorithm $\mathcal{A}$ exists such that*

$$\mathrm{cost}_{\mathcal{A}}(n) \leq \begin{cases} O(\sqrt{n}) & d = 1 \\ O(\sqrt{n \log n}) & d = 2 \\ O(n^{1-\frac{1}{d}}) & d \geq 3 \end{cases} .$$

Proof. Let $\mathcal{A}$ be the algorithm as per Lemma 5, and we estimate the cost of $\mathcal{A}$ for $d = 1$, $d = 2$, and $d \geq 3$ separately. Recall that an upper bound for $\text{OPT}(n)$ is given by Lemma 2. For $d = 1$,

$$\text{cost}_{\mathcal{A}}(n) = \sum_{t=1}^{n} \frac{1}{t} \cdot \text{OPT}(t) = O\left(\sum_{t=1}^{n} \frac{1}{\sqrt{t}}\right) = O\left(\int_{1}^{n} \frac{1}{\sqrt{x}} \mathrm{d}x\right) = O\left(\sqrt{n}\right).$$

For $d = 2$,

$$\begin{aligned}\text{cost}_{\mathcal{A}}(n) &= \sum_{t=1}^{n} \frac{1}{t} \cdot \text{OPT}(t) = O\left(\sum_{t=1}^{n} \sqrt{\frac{\log t}{t}}\right) \\ &= O\left(\sqrt{\log n} \cdot \int_{1}^{n} \frac{1}{\sqrt{x}} \mathrm{d}x\right) = O\left(\sqrt{n \log n}\right).\end{aligned}$$

For $d \geq 3$,

$$\text{cost}_{\mathcal{A}}(n) = \sum_{t=1}^{n} \frac{1}{t} \cdot \text{OPT}(t) = O\left(\sum_{t=1}^{n} \frac{1}{t^{1/d}}\right) = O\left(\int_{1}^{n} \frac{1}{x^{1/d}} \mathrm{d}x\right) = O\left(n^{1-\frac{1}{d}}\right).$$

Combining all the above concludes the proof.

Finally, combining Lemma 7 with the lower bounds for $\text{OPT}(n)$ given by Lemma 3 concludes the proof of Theorem 3.

4.3 Unbalanced Markets

We extend our results for smooth distributions to unbalanced markets satisfying $n \leq m \leq O(n)$.

Theorem 4. *Suppose that $n \leq m \leq O(n)$ and $\mathbb{D}$ is β-smooth for some constant $\beta > 0$. When servers are adversarially chosen and requests are randomly drawn from $\mathbb{D}$, an algorithm exists with a competitive ratio of $O(\sqrt{n})$ for $d = 1$, $O(\sqrt{\log n})$ for $d = 2$, and $O(1)$ for $d \geq 3$.*

Proof. By Lemma 7, an algorithm $\mathcal{A}$ exists such that

$$\text{cost}_{\mathcal{A}}(n) \leq \begin{cases} O(\sqrt{n}) & d = 1 \\ O(\sqrt{n \log n}) & d = 2 \\ O(n^{1-\frac{1}{d}}) & d \geq 3 \end{cases}.$$

Moreover, by Theorem 1, an algorithm $\mathcal{A}'$ exists satisfying

$$\text{cost}_{\mathcal{A}'}(S, n) \leq \text{OPT}(S, n) + \text{cost}_{\mathcal{A}}(n)$$

for every set of servers S with $n \leq |S| \leq O(n)$. Combining the lower bounds for $\text{OPT}(S, n)$ given by Lemma 4 concludes the proof.

5 Euclidean Space with General Distributions

In this section, we assume the metric space to be $[0,1]^d$ with the Euclidean distance. We do not make any assumptions on the distribution and derive algorithms with nearly optimal regret bounds for all $d \geq 1$ in unbalanced markets with at most linearly many excess servers. In particular, our results recover the nearly optimal regret bounds by [6], which is established in balanced markets with servers and requests following different distributions. In the weaker setting studied by [6], the regret bound is optimal for $d \neq 2$ and optimal up to a logarithmic factor for $d = 2$, implying that our regret bounds are also nearly optimal.

Theorem 5. *Suppose that $n \leq m \leq O(n)$ and $\mathbb{D}$ is an arbitrary distribution on $[0,1]^d$. An algorithm $\mathcal{A}'$ exists that, for every set of servers S with $|S| = m$, admits a regret of*

$$\text{Regret}(\mathcal{A}'; S, n) \leq \begin{cases} O(n^{-\frac{1}{2}}) & d = 1 \\ O(n^{-\frac{1}{2}} \log n) & d = 2 \\ O(n^{-\frac{1}{d}}) & d \geq 3 \end{cases}.$$

Proof. For an arbitrary distribution $\mathbb{D}$, by applying the same arguments as in the proof of Lemma 7 except that the upper bound for $\text{OPT}(n)$ is given by Lemma 1 instead of Lemma 2, we obtain an algorithm $\mathcal{A}$ such that

$$\text{cost}_{\mathcal{A}}(n) \leq \begin{cases} O(\sqrt{n}) & d = 1 \\ O(\sqrt{n} \log n) & d = 2 \\ O(n^{1-\frac{1}{d}}) & d \geq 3 \end{cases}.$$

By Theorem 1, an algorithm $\mathcal{A}'$ exists satisfying $\text{cost}_{\mathcal{A}'}(S, n) \leq \text{OPT}(S, n) + \text{cost}_{\mathcal{A}}(n)$ for every set of servers S with $n \leq |S| \leq O(n)$. As a result, the regret of $\mathcal{A}'$ equals

$$\text{Regret}(\mathcal{A}'; S, n) = \frac{1}{n} \left(\text{cost}_{\mathcal{A}'}(S, n) - \text{OPT}(S, n) \right) \leq \frac{\text{cost}_{\mathcal{A}}(n)}{n},$$

which concludes the proof.

We briefly explain the reason behind the contrast that in unbalanced markets with $d \leq 2$, our approach admits a nearly optimal regret (Theorem 5) with a suboptimal competitive ratio (Theorem 4). In fact, our regret guarantee indicates that $\text{cost}_{\mathcal{A}}(S, n) - \text{OPT}(S, n)$ is small, yet we have no control over the lower bound for $\text{OPT}(S, n)$ even with smooth distributions, which prevents us from achieving a nearly optimal competitive ratio.

6 Adversarial Setting

In this section, we focus on a more general setting where both servers and requests are adversarially chosen. This setting has been extensively studied in the past [5,22], whereas all known guarantees are established in balanced markets. We show that the competitive ratio guarantees in balanced markets of algorithms satisfying certain natural conditions can be carried over to unbalanced markets, and the resulting competitive ratios in unbalanced markets only depend on the number of requests. Combined with the algorithms of prior work, we obtain competitive ratio guarantees in unbalanced markets with adversarial servers and requests, which in turn leads to improved guarantees in unbalanced markets with adversarial servers and stochastic requests. In particular, we show that both the $O(\log^3 n)$-competitive algorithm of [18] for general metrics and the $O(\log n)$-competitive algorithm of [10] for doubling metrics admit the same competitive ratios in unbalanced markets. The latter result also improves the $O(\sqrt{n})$ competitive ratio for $d = 1$ in unbalanced markets with servers adversarially chosen and requests drawn from a smooth distribution in Theorem 4. All the omitted proofs in this section can be found in the full version of our paper.

Given the execution of an online algorithm $\mathcal{A}$ on an instance with initial servers S, we use $N_{\mathcal{A}}(r \mid S)$ to denote the set of available servers closest to the request r upon its arrival. We say that an online algorithm $\mathcal{A}$ is *greedy* if it always (randomly) matches a new request to one of the closest available servers, i.e., it matches r to a server in $N_{\mathcal{A}}(r \mid S)$. In particular, every greedy algorithm is completely determined by a (randomized) tie-breaking rule such that when facing multiple closest available servers upon the arrival of a new request, the algorithm applies the tie-breaking rule to (randomly) decide who to match with. We say that the tie-breaking rule of a greedy algorithm $\mathcal{A}$ is *local* if it only depends on the set of closest available servers $N_{\mathcal{A}}(r \mid S)$. Moreover, the local tie-breaking rule of a greedy algorithm $\mathcal{A}$ is called *non-increasing* if the probability of selecting each server in $N_{\mathcal{A}}(r \mid S)$ does not increase by adding additional servers into $N_{\mathcal{A}}(r \mid S)$, i.e., for all sets of servers S', S'' with $S' \subseteq S''$, for every $s \in S'$,

$$\begin{aligned} &\mathbb{P}\left\{\mathcal{A} \text{ matches } r \text{ to } s \mid N_{\mathcal{A}}(r \mid S) = S'\right\} \\ &\geq \mathbb{P}\left\{\mathcal{A} \text{ matches } r \text{ to } s \mid N_{\mathcal{A}}(r \mid S) = S''\right\} . \end{aligned}$$

We show that the competitive ratio guarantee for every greedy algorithm with a local and non-increasing tie-breaking rule in balanced markets can be carried over to unbalanced markets.

Theorem 6. *For an arbitrary metric, if a greedy algorithm $\mathcal{A}$ with a local and non-increasing tie-breaking rule is α-competitive in balanced markets with adversarial servers and requests, then $\mathcal{A}$ is α-competitive in unbalanced markets with adversarial servers and requests.*

We remark here that the resulting competitive ratio α for unbalanced markets can depend on the number of requests but is independent of the number of excess servers.

Due to the space limit, we only outline the proof of Theorem 6 here. Firstly, we observe that it suffices to show that adding new servers does not increase the cost of $\mathcal{A}$ with a fixed set of requests. Then, we compare the costs of $\mathcal{A}$ with and without new servers by carefully coupling the (randomized) decisions made by $\mathcal{A}$ with and without new servers for the first arrived request. Finally, the proof is completed by reduction on the number of requests.

6.1 Applications on Hierarchically Separated Tree Metrics

Now, we apply Theorem 6 to algorithms of prior work. Recall that for balanced markets with adversarial servers and requests, the algorithm of [18], abbreviated as the *MNP* algorithm, is $O(\log^3 n)$-competitive for general metrics, and the *Random-Subtree* algorithm of [10] is $O(\log n)$-competitive for doubling metrics. We show that the competitive ratios of both algorithms are retained in unbalanced markets.

Both the MNP and Random Subtree algorithms are designed based on the same framework. They first randomly embed the original metric into a *Hierarchically Separated Tree (HST)* metric, which preserves the distance between every pair of nodes up to an $O(\log n)$ distortion. Next, they show that a randomized greedy subroutine with a specific tie-breaking rule admits a competitive ratio of β for the HST metric. Overall, this leads to an $O(\beta \log n)$-competitive algorithm for the original metric. Hence, for our purpose, it suffices to show that the tie-breaking rule of the greedy subroutine for the HST metric is local and non-increasing.

Recall that an HST is a rooted tree satisfying the following property: For each subtree, the distances from the subtree's root to all leaf nodes in the subtree are identical. Moreover, the metric embedding phase of the algorithms will map each location in the original metric to a leaf node of the HST. From now on, we will restrict our attention to the HST metric with all servers and requests locating at the HST's leaf nodes.

We start by analyzing the greedy subroutine adopted by the MNP algorithm, which works as follows. Upon the arrival of a new request r, assuming the set of initial servers to be S, denote $A(r \mid S)$ as the lowest ancestor of r the subtree rooted at whom contains at least one available server, and the subroutine matches r to a uniformly random server among all available servers in the subtree rooted at $A(r \mid S)$. In the following lemma, we show that this tie-breaking rule is local and non-increasing, which implies that the greedy subroutine in the MNP algorithm satisfies the conditions in Theorem 6.

Lemma 8. *The greedy subroutine in the MNP algorithm admits a local and non-increasing tie-breaking rule.*

The following corollary is a direct consequence of combining Theorem 6 and Lemma 8 with the competitive ratio guarantee of the MNP algorithm in balanced markets [18].

Corollary 1. *Suppose that m servers and n requests are adversarially chosen with $n \leq m$. For an arbitrary metric, an $O(\log^3 n)$-competitive algorithm exists.*

We remark here that for general metrics with adversarial servers and requests, a randomized algorithm with an improved competitive ratio of $O(\log^2 n)$ exists [5]. However, this algorithm does not belong to the family of greedy algorithms, and hence, Theorem 6 does not apply.

Next, we analyze the greedy subroutine adopted by the Random-Subtree algorithm, which works as follows. Upon the arrival of a new request r, recall that $A(r \mid S)$ denotes the lowest ancestor of r the subtree rooted at whom contains at least one available server, assuming the set of initial servers to be S. Starting off at $A(r \mid S)$, the subroutine repeatedly moves us to a uniformly random subtree among all subtrees of the current node containing at least one available server until we reach a leaf node, and r is matched to a uniformly random available server at the leaf node. In the following lemma, we show that this tie-breaking rule is local and non-increasing, which implies that the greedy subroutine in the Random Subtree algorithm satisfies the conditions in Theorem 6.

Lemma 9. *The greedy subroutine in the Random Subtree algorithm admits a local and non-increasing tie-breaking rule.*

Combining Theorem 6 and Lemma 9 with the competitive ratio guarantee of the Random Subtree algorithm in balanced markets [10], we derive the competitive ratio guarantee of the Random Subtree algorithm in unbalanced markets in the following corollary.

Corollary 2. *Suppose that m servers and n requests are adversarially chosen with $n \leq m$. For doubling metrics, an $O(\log n)$-competitive algorithm exists.*

7 Discussion and Future Directions

In this paper, for balanced markets, we achieve an $O(1)$ competitive ratio for the Euclidean space with smooth distributions and $d \neq 2$. Nevertheless, it still remains open to give an $O(1)$-competitive algorithm for general metrics and distributions. The potential approaches for achieving this goal comprise (1) improving the analysis of the algorithm of [9], which is proved to be $O((\log\log\log n)^2)$-competitive without any super-constant lower bound known, and (2) applying Theorem 1 and the SOAR algorithm to the simplified model given by [9]. The main technical barrier underlying the latter approach lies in the difficulty of characterizing $\mathrm{OPT}(n)$ for general metrics and distributions.

We also derive an algorithm in unbalanced markets for the Euclidean space and smooth distributions with improved competitive ratios. Our algorithm admits an optimal constant competitive ratio for $d \geq 3$, leaving the improvement for $d \leq 2$ an intriguing future direction. Given that the line metric, i.e., $d = 1$, is typically the most tractable case [4,9], it is surprising that it becomes the most intractable one with the presence of excess servers, rendering its resolution more appealing.

Moreover, for adversarial servers and requests, we show that the competitive ratio of every greedy algorithm with a local and non-increasing tie-breaking rule in balanced markets is preserved in unbalanced markets. However, our characterization fails to capture all the existing algorithms. In this most extensively studied setting, our understanding of the achievable guarantees in unbalanced markets remains very limited, and nothing beyond our results is known. An exciting future direction is to generalize the performance of the existing algorithms with provable guarantees in balanced markets to unbalanced markets.

Finally, in the full version of our paper, we showcase the generality of our reduction by establishing positive results for cost functions only approximately satisfying the triangle inequality. Conversely, a strong impossibility result for achieving a bounded competitive ratio without any assumption on cost functions exists [9]. Our results can be viewed as pushing the boundary within which a bounded competitive ratio is achievable, and it would be interesting to further characterize the boundary between possibility and impossibility.

Acknowledgments. We thank Yash Kanoria, Akshit Kumar, Tristan Pollner, Shayan Talaei, Kiana Ashgri, and Yilun Chen for their helpful discussions, and the anonymous reviewers for their valuable comments. Amin Saberi is supported in part by NSF Awards CCF2209520, CCF2312156, and a gift from CISCO.

References

1. Ajtai, M., Komlós, J., Tusnády, G.E.: On optimal matchings. Comb. **4**(4), 259–264 (1984)
2. Akbarpour, M., Alimohammadi, Y., Li, S., Saberi, A.: The value of excess supply in spatial matching markets. In: EC, p. 62. ACM (2022)
3. Antoniadis, A., Barcelo, N., Nugent, M., Pruhs, K., Scquizzato, M.: A o(n)-competitive deterministic algorithm for online matching on a line. Algorithmica **81**(7), 2917–2933 (2019)
4. Balkanski, E., Faenza, Y., Périvier, N.: The power of greedy for online minimum cost matching on the line. In: EC, pp. 185–205. ACM (2023)
5. Bansal, N., Buchbinder, N., Gupta, A., Naor, J.: A randomized o(log2 k)-competitive algorithm for metric bipartite matching. Algorithmica **68**(2), 390–403 (2014)
6. Chen, Y., Kanoria, Y., Kumar, A., Zhang, W.: Feature based dynamic matching. In: EC, p. 451. ACM (2023)
7. Fuchs, B., Hochstättler, W., Kern, W.: Online matching on a line. Theor. Comput. Sci. **332**(1–3), 251–264 (2005)
8. Gairing, M., Klimm, M.: Greedy metric minimum online matchings with random arrivals. Oper. Res. Lett. **47**(2), 88–91 (2019)
9. Gupta, A., Guruganesh, G., Peng, B., Wajc, D.: Stochastic online metric matching. In: ICALP. LIPIcs, vol. 132, pp. 67:1–67:14. Schloss Dagstuhl - Leibniz-Zentrum für Informatik (2019)
10. Gupta, A., Lewi, K.: The online metric matching problem for doubling metrics. In: Czumaj, A., Mehlhorn, K., Pitts, A., Wattenhofer, R. (eds.) ICALP 2012. LNCS, vol. 7391, pp. 424–435. Springer, Heidelberg (2012). https://doi.org/10.1007/978-3-642-31594-7_36

11. Huang, Z., Tang, Z.G., Wajc, D.: Online matching: a brief survey. CoRR abs/2407.05381 (2024)
12. Hundrieser, S., Staudt, T., Munk, A.: Empirical optimal transport between different measures adapts to lower complexity (2022)
13. Kalyanasundaram, B., Pruhs, K.: Online weighted matching. J. Algorithms **14**(3), 478–488 (1993)
14. Kanoria, Y.: Dynamic spatial matching. In: EC, pp. 63–64. ACM (2022)
15. Karp, R.M., Vazirani, U.V., Vazirani, V.V.: An optimal algorithm for on-line bipartite matching. In: STOC, pp. 352–358. ACM (1990)
16. Khuller, S., Mitchell, S.G., Vazirani, V.V.: On-line algorithms for weighted bipartite matching and stable marriages. Theor. Comput. Sci. **127**(2), 255–267 (1994)
17. Mehta, A.: Online matching and ad allocation. Found. Trends Theor. Comput. Sci. **8**(4), 265–368 (2013)
18. Meyerson, A., Nanavati, A., Poplawski, L.J.: Randomized online algorithms for minimum metric bipartite matching. In: SODA, pp. 954–959. ACM Press (2006)
19. Nayyar, K., Raghvendra, S.: An input sensitive online algorithm for the metric bipartite matching problem. In: FOCS, pp. 505–515. IEEE Computer Society (2017)
20. Peserico, E., Scquizzato, M.: Matching on the line admits no $o(\sqrt{\log n})$-competitive algorithm. ACM Trans. Algorithms **19**(3), 28:1–28:4 (2023)
21. Raghvendra, S.: A robust and optimal online algorithm for minimum metric bipartite matching. In: APPROX-RANDOM. LIPIcs, vol. 60, pp. 18:1–18:16. Schloss Dagstuhl - Leibniz-Zentrum für Informatik (2016)
22. Raghvendra, S.: Optimal analysis of an online algorithm for the bipartite matching problem on a line. In: SoCG. LIPIcs, vol. 99, pp. 67:1–67:14. Schloss Dagstuhl - Leibniz-Zentrum für Informatik (2018)
23. Sentenac, F., Noiry, N., Lerasle, M., Ménard, L., Perchet, V.: Online matching in geometric random graphs. CoRR abs/2306.07891 (2023)
24. Tsai, Y.T., Tang, C.Y., Chen, Y.Y.: Average performance of a greedy algorithm for the on-line minimum matching problem on euclidean space. Inf. Process. Lett. **51**(6), 275–282 (1994)

Time-Efficient Algorithms for Nash-Bargaining-Based Matching Market Models

Ioannis Panageas[1,2], Thorben Tröbst[1(✉)], and Vijay V. Vazirani[1]

[1] Department of Computer Science, University of California, Irvine, Donald Bren Hall, 92697 Irvine, CA, USA
{ipanagea,vazirani}@ics.uci.edu, t.troebst@uci.edu
[2] Archimedes Unit, Athena Research Center, Artemidos 1, 151 25 Marousi, Greece

Abstract. In the area of matching-based market design, existing models using cardinal utilities suffer from two deficiencies: First, the Hylland-Zeckhauser (HZ) mechanism [17], which has remained a classic in economics for one-sided matching markets, is intractable; computation of even an approximate equilibrium is PPAD-complete [7,27]. Second, there is an extreme paucity of such models. This led [16] to define a rich collection of Nash-bargaining-based models for one-sided and two-sided matching markets, in both Fisher and Arrow-Debreu settings, together with very fast implementations using available solvers and very encouraging experimental results.

In this paper, we give fast algorithms with proven running times for the models of [16] using the techniques of multiplicative weights update (MWU) and conditional gradient descent (CGD). Additionally, we make the following contributions:

1. By [24], a linear one-sided Nash-bargaining-based matching market satisfies envy-freeness within factor two. We show that the other models satisfy approximate equal-share fairness, where the exact factor depends on the utility function being used in the particular model.
2. We define a Nash-bargaining-based model for non-bipartite matching markets and give fast algorithms for it using conditional gradient descent.

Keywords: Matching markets · Nash bargaining · Multiplicative weights update · Gradient descent

I. Panageas—Supported by start-up grant and ICS research award. This work has been partially supported by project MIS 5154714 of the National Recovery and Resilience Plan Greece 2.0 funded by the European Union under the NextGenerationEU Program.
T. Tröbst—Supported in part by NSF grant CCF-1815901.
V. V. Vazirani—Supported in part by NSF grants CCF-1815901 and CCF-2230414.

M. Mavronicolas et al. (Eds.): WINE 2024, LNCS 15534, pp. 119–136, 2026.
https://doi.org/10.1007/978-3-032-08560-3_7

1 Introduction

For a mechanism to be highly impactful, it must have both good game-theoretic properties and computational efficiency. Our paper deals with cardinal-utility matching markets, for which the most prominent mechanism, for the key case of a linear one-sided matching market, was due to Hylland and Zeckhauser [17]. This pricing-based mechanism has all the game-theoretic properties one could ask for; however, it turns out to be computationally intractable in theory and practice. Fortunately a viable alternative has emerged, namely a Nash-bargaining-based model [16]. Very recent work [24] has shown that it possesses good game-theoretic properties, and since it involves solving a convex program, its solution can be found in polynomial time using the ellipsoid algorithm; however, the latter is far from practical. The purpose of this paper is to rectify the last deficiency by giving time-efficient algorithms with proven running times; for this purpose we use the techniques of multiplicative weights update (MWU) and conditional gradient descent (CGD).

The Hylland-Zeckhauser (HZ) mechanism [17] uses the power of the pricing mechanism to produce allocations that are Pareto optimal and envy-free [17], and the mechanism is incentive compatible in the large [15]. However, the recent works [7,27] show that computation of even an approximate HZ equilibrium is PPAD-complete. HZ is also intractable in practice: The only known method for computing it uses algebraic cell decomposition [2,4], which requires exploration of $n^{5n^2} > 10^{80}$ cells for a problem of size $n = 5$. Additionally, the area of matching-based market design suffers from a second deficiency: There is an extreme paucity of models using cardinal utilities. This stands in sharp contrast with general equilibrium theory, which defined and extensively studied several fundamental market models to address a number of specialized and realistic situations, e.g., see [19]. Lastly, cardinal utility functions produce higher quality allocations as compared to ordinal utilities, leading to significant gain in efficiency, e.g., [18] give a striking example in which an allocation under cardinal utilities improves each of the n agents by a factor of $\theta(\log(n))$ over the allocation made under ordinal utilities.

For all the reasons stated above, designing an alternative mechanism for cardinal utility matching markets became a pressing issue. Fortunately, for traditional market models, an alternative to the pricing-based mechanism had been studied in the past: [25] gave a Nash-bargaining-based mechanism for the linear Arrow-Debreu model; see Sect. 1.1 for the point of view that led to this work. Building on this idea, [16] defined a rich collection of Nash-bargaining-based models for one-sided and two-sided matching markets, in both Fisher and Arrow-Debreu settings. Since the Nash bargaining solution is captured by a convex program, it can be efficiently computed to any degree of accuracy using the ellipsoid algorithm [14,28]. However, since the ellipsoid algorithm is very slow, [16] gave implementations using available solvers, and very favorable experimental results. As stated above, our paper gives time-efficient algorithms with proven running times.

By definition, the Nash bargaining solution is Pareto optimal; however it is neither envy-free nor incentive compatible. Recently [24] have shown that the Nash-bargaining-based mechanism for the linear one-sided matching market satisfies envy-freeness within factor two and incentive compatibility within factor two; moreover, both bounds are tight. Furthermore they showed that finding an envy-free and Pareto optimal solution is PPAD-hard. Hence the game-theoretic properties of the one-sided model are almost as good as is feasible.

[24] also gave examples of two-sided matching markets for which Pareto optimal and envy-free allocations don't exist, hence precluding pricing-based mechanisms. In contrast, the Nash-bargaining-based approach [16] easily yields models for these and more general matching market settings. Furthermore, in lieu of envyfreeness, in the current paper we show that these models satisfy an approximate notion of equal-share fairness where the exact factor depends on the utility function being used in the particular model, see Lemmas 2, 8 and 9. We also define, for the first time, a model for non-bipartite matching markets – it is also Nash-bargaining-based – and give fast algorithms for it using conditional gradient descent. The roommates problem under cardinal utilities is an application of this model.

Our fast, combinatorial[1] algorithms are based on the techniques of multiplicative weights update (MWU) and conditional gradient descent (CGD); we solve our non-bipartite matching market model using CGD only. In every case, we study not only the Fisher but also the Arrow-Debreu version; the latter is also called the exchange version. Here is a summary of the models and the algorithmic techniques used by us: One-sided linear Fisher (MWU and CGD); One-sided linear Arrow-Debreu (MWU and CGD); One-sided separable, piecewise-linear concave (SPLC) Fisher and Arrow-Debreu (MWU and CGD); Two-sided linear Fisher and Arrow-Debreu (CGD); Two-sided SPLC Fisher and Arrow-Debreu (CGD); and Non-bipartite linear Fisher and Arrow-Debreu (CGD).

The recent computer science revolutions of the Internet and mobile computing led to the launching of highly impactful and innovative one-sided and two-sided matching markets, such as online advertisement platforms (Google Ads), ride-hailing services (Uber, Lyft), food delivery services (Doordash, Uber Eats), vacation rentals (Airbnb, VRBO), freelancing (Taskrabbit, Upwork) and online dating services (Match.com, OkCupid). In turn it led to the formation of the new area of online and matching-based market design, e.g., see [10]. Our paper is motivated by this challenge and opportunity.

Technical Contributions: The main difference between market models in general equilibrium theory and matching markets is that in the latter, each agent needs to be allocated exactly one unit of goods, the latter being probability shares over goods. Once a fractional allocation is computed, it is rounded using the Birkhoff-von Neumann Theorem. This restriction of one unit is also the reason that both our algorithmic techniques call for additional ideas, over and above previous works. Our contributions fall into three categories:

[1] We use the term "combinatorial" in the same sense as Schrijver [23], i.e., the algorithm does not use an LP or convex program solver.

1). Modeling: Our model for non-bipartite matching markets, given in Sect. 3, is the first such model – it is also Nash-bargaining-based.

2). Multiplicative Weights Update (MWU): We provide an MWU algorithm which gives an ϵ-approximate Nash bargaining solution in time $\frac{n^3 \log n}{\epsilon^2}$. The algorithm uses the primal-dual scheme and therefore yields both allocations and prices. Our techniques deviate from the literature due to additional constraints, namely matching, and the presence of endowments. The main idea underlying our approach is to create a feasibility program, see (F-LiAD) , the feasible solutions of which are exactly the optimal primal and dual values of the variables of Program (1). We then find an approximate feasible solution to this program using MWU, yielding an approximate Nash bargaining solution. Although other works in the past have also followed this approach of first finding a feasibility program, e.g., to solve linear programs and find equilibria in markets [3,11], in our case this step was much more challenging as expounded below.

In (F-LiAD) , each agent aims to maximize her utility subject to budget constraints, i.e., for each agent i the following holds: $\sum_j x_{ij}(p_j + q_i) \leq 1 + c_i \min_j \frac{u_{ij}}{p_j+q_i}$. Observe that both the left and the right hand sides of the inequality involve prices; in particular, in the case of Linear Fisher Program, the right hand side is just 1. To come up with the right budget constraints, we used the KKT conditions. The KKT conditions would yield that each agent i maximizes her utility u_i subject to budget constraints that would involve the utility u_i itself. This self-reference had to be removed.

Note that the aforementioned challenge becomes even harder for the case of SPLC utilities where the correct constraints would involve prices in addition to allocations for all segments in the utility functions. The rest of the analysis of MWU algorithm follows a potential function argument adapted appropriately to our setting.

3). Conditional Gradient Descent (CGD): First order methods, in particular projected gradient descent, have proven successful for solving large scale market equilibrium problems. The challenges to extending this approach to our models are two-fold. We get around these hurdles by resorting to a conditional gradient method instead.

a). The objective function of each of the convex programs we study is logarithmic. Therefore, its gradient is neither Lipschitz nor smooth. It must therefore be modified in order to guarantee these properties. [12] circumvent this problem for the Fisher market by replacing the objective by its quadratic extension below a certain point. This is obtained through the well-known equal-share fairness property, i.e. the fact that in an equilibrium we have $u_i \geq \frac{1}{n}\sum_{j\in G} u_{ij}$ for all agents i. Since our models are more involved, we need to extend this property by introducing approximations. We show that our models satisfy approximate notions of equal-share fairness, which allows us to bound the equilibrium solution away from the boundary. In particular, this holds even in the case of Arrow-Debreu extension of models, i.e., with initial endowments. We also show that these bounds are sharp up to a constant factor, which may be of independent interest.

b). In the Fisher setting, the efficiency of the projected gradient descent method relies on projections onto the simplex which are not time-consuming. On the other hand, in our models, the feasible region is generally given by matching or flow polytopes, for which projections are significantly more expensive. Therefore, we employ a conditional gradient method instead which relies on combinatorial matching and flow algorithms that are very efficient in practice. In the case of SPLC utilities we show how an additional "shifting step" can be used to decrease the dependence on the (rather large) diameter of the feasible region.

1.1 Related Results

For a detailed overview of the related literature, we refer to the full version of the paper due to space constraints.

2 Basic Solution Concepts

Nash Bargaining Game: An *n-person Nash bargaining game* consists of a pair $(\mathcal{N}, c)$, where $\mathcal{N} \subseteq \mathsf{R}_+^n$ is a compact, convex set and $c \in \mathcal{N}$. The set $\mathcal{N}$ is called the *feasible set* – its elements are vectors whose components are the utilities that the n players can simultaneously accrue. Point c is the *disagreement point* representing the utilities of the players if they fail to come to an agreement.

The set of n agents will be denoted by A and the agents will be numbered $1, 2, \ldots n$. Instance $(\mathcal{N}, c)$ is said to be *feasible* if there is a point in $\mathcal{N}$ at which each agent does strictly better than in the disagreement point. In game theory it is customary to assume that the given Nash bargaining problem $(\mathcal{N}, c)$ is feasible; we will make this assumption as well.

Nash [21] asked the question which point in $\mathcal{N}$ players would reach through bargaining. He identified four axioms for such a point: Pareto optimality, symmetry, invariance under affine transformations of utilities, and independence of irrelevant alternatives. Via an elegant proof, Nash showed:

Theorem 1 (Nash *[21]***).** *If the game $(\mathcal{N}, c)$ is feasible then there is a unique point in $\mathcal{N}$ satisfying the axioms stated above. Moreover, this point is obtained by maximizing $\Pi_{i \in A}(v_i - c_i)$ over $\boldsymbol{v} \in \mathcal{N}$.*

Nash bargaining can be used to solve problems of fair division, in which case the resulting allocations turn out to be remarkably fair, see for example [1,6,8,9,20]. Note that the objective function, i.e. the product of agents' utilities, is often referred to as Nash social welfare (NSW). Since this objective function is log-concave, the solution to the Nash bargaining problem can be captured via a convex program and thus can be approximated arbitrarily well in polynomial time [14,28] if separation over $\mathcal{N}$ can be implemented in polynomial time.

One-Sided Matching Market: Let $A = \{1, 2, \dots n\}$ be a set of n agents and $G = \{1, 2, \dots, n\}$ be a set of n indivisible goods. Each agent i has a utility function u_{ij} over goods $j \in G$. The goal is to design a mechanism to allocate exactly one good to each agent, so that properties such as Pareto optimality are satisfied. Goods are rendered divisible by assuming that there is one unit of probability share of each good. Let x_{ij} be the allocation of probability share that agent i receives of good j. Then, $\sum_j u_{ij} x_{ij}$ is the *expected utility* accrued by agent i.

To make it a matching market, an additional constraint is that the total probability share allocated to each agent is one unit, i.e., the entire allocation must form a *fractional perfect matching* in the complete bipartite graph over vertex sets A and G. Now, a solution can be viewed as a doubly stochastic matrix. The Birkhoff-von Neumann algorithm [5,29] then extracts a random underlying perfect matching in such a way that the expected utility accrued to each agent from the integral perfect matching is the same as from the fractional perfect matching. Since *ex ante* Pareto optimality implies *ex post* Pareto optimality, the integral allocation will also be Pareto optimal.

Two-Sided Matching Market: Our two-sided matching market model consist of a set $A = \{1, 2, \dots n\}$ of agents and a set $J = \{1, 2, \dots, n\}$ of jobs. Each agent i has a utility function u_{ij} over goods $j \in G$ and each job j has a utility function w_{ij} over agents $i \in A$. Let x_{ij} be the allocation of probability share that agent i receives of good j. Then, $\sum_j u_{ij} x_{ij}$ is the expected utility accrued by agent i and $\sum_i w_{ij} x_{ij}$ is the expected utility accrued by job j. As in the one-sided case, the Birkhoff-von Neumann procedure [5,29] then extracts a random underlying perfect matching from x.

Non-bipartite Matching Market: Our non-bipartite matching market model consist of a set $A = \{1, 2, \dots, n\}$ of agents and utilities for each pair $u_{ij} \in \boldsymbol{R}_+$ which i and j accrue on getting matched to each other. Let x_{ij} be the fractional extent to which i and j are matched. Then, $\sum_j u_{ij} x_{ij}$ is the utility accrued by agent i. The allocation vector $x \in \mathsf{R}_+^E$ is said to be a *fractional matching in* G if it is a convex combination of (integral) matchings in G. As in the bipartite case, it is possible to take a fractional matching and, in combinatorial, polynomial time, decompose it into a convex combination of most n^2 integral matchings [22,26].

3 Nash-Bargaining-Based Models

We will define four one-sided matching market models based on our Nash bargaining approach. For the case of linear utilities, we have defined both Fisher and Arrow-Debreu versions, namely *LiF* and *LiAD*. For more general utility functions we have defined only the Arrow-Debreu version; the Fisher version is obtained by setting all disagreement utilities to zero. For each model, we give the convex program whose optimal solution captures the Nash bargaining solution.

Our one-sided matching market models consist of a set $A = \{1, 2, \dots n\}$ of agents and a set $G = \{1, 2, \dots, n\}$ of infinitely divisible goods; observe that there is an equal number of agents and goods. There is one unit of each good and each agent needs to be allocated a total of one unit of goods. Hence the allocation needs to be a fractional perfect matching, as defined next.

Definition 1. *Let us name the coordinates of a vector* $x \in \mathsf{R}_+^{n^2}$ *by pairs* i, j *for* $i \in A$ *and* $j \in G$. *Then* x *is said to be a* fractional perfect matching *if* $\sum_j x_{ij} = 1$ $\forall i \in A$ *and* $\sum_i x_{ij} = 1$ $\forall j \in G$.

An equilibrium allocation can be viewed as a doubly stochastic matrix, and the Birkhoff-von Neumann procedure [5,29] can be used to extract a random underlying perfect matching in such a way that the expected utility accrued to each agent from the integral perfect matching is the same as from the fractional perfect matching.

1). Under the *linear Fisher Nash bargaining one-sided matching market*, abbreviated *LiF*, each agent $i \in A$ has a linear utility function. Corresponding to each fractional perfect matching x, there is a vector v_x in the feasible set $\mathcal{N}$; its components are the utilities derived by the agents under the allocation given by x. The disagreement point is the origin. The setup of *LiF* is identical to that of the HZ mechanism; the difference lies in the definition of the solution to an instance.

2). Under the *linear Arrow-Debreu Nash bargaining one-sided matching market*, abbreviated *LiAD*, each agent $i \in A$ has a linear utility function, as above. Additionally, we are specified an initial fractional perfect matching x_I which gives the initial endowments of the agents. These two pieces of information define the utility accrued by each agent from her initial endowment; this is her disagreement point c_i. We will assume that the problem is feasible in the sense that there is a fractional perfect matching, defining a redistribution of the goods, under which each agent i derives strictly more utility than c_i. Each vector $v \in \mathcal{N}$ is as defined in *LiF*. Henceforth, we will consider the slightly more general problem in which are specified only a (feasible) disagreement point c and not the initial endowments x_I.

The convex program for *LiAD* is given in (1).

$$\max \quad \sum_{i \in A} \log(u_i(x) - c_i) \tag{1a}$$

$$\text{s.t.} \quad \sum_{i \in A} x_{ij} \leq 1 \quad \forall j \in G, \tag{1b}$$

$$\sum_{j \in G} x_{ij} \leq 1 \quad \forall i \in A, \tag{1c}$$

$$x \geq 0. \tag{1d}$$

For $c_i = 0$, we can prove the following approximate equal-share fairness property which guarantees that every agent achieves at least $1/2$ of the utility that they would get under the equal share matching that assigns $\frac{1}{n}$ to all edges.

Lemma 1. *Let x be an optimal solution to* (1) *with all $c_i = 0$, then for all agents i we have $u_i(x) \geq \frac{1}{2n} \sum_{j \in G} u_{ij}$.*

In the case of non-zero c_i, the equal share matching may no longer be feasible and so this notion loses some meaning. However, we will give an analogous weaker bound in Sect. 5, specifically in Lemma 9.

3). Economists like to model *diminishing marginal utilities for goods* by considering concave utility functions. Since we are in a fixed-precision model of computing, we will consider separable, piecewise-linear concave (SPLC) utility functions. The *separable, piecewise-linear concave Arrow-Debreu Nash bargaining one-sided matching market*, abbreviated *SAD*, is analogous to LiAD, with the difference that each agent has a separable, piecewise-linear concave utility function, hence generalizing the linear utility functions specified in LiAD. When there are no disagreement utilities c_i, we call this model *SF*.

For each agent i and good j, function $f_i^j : \mathbb{R}_+ \to \mathbb{R}_+$ gives the utility derived by i as a function of the amount of good j she receives. Each f_i^j is a non-negative, non-decreasing, piecewise-linear, concave function. The overall utility of buyer i, $u_i(x)$, for bundle $x = (x_1, \ldots, x_n)$ of goods, is additively separable over the goods, i.e., $u_i(x) = \sum_{j \in G} f_i^j(x_j)$. We will assume that each f_i is given as list S_{ij} of *segments* which have a length of l_{ijk} and a slope of u_{ijk}. Due to the concavity assumption, we can then formulate SAD as convex program (2).

$$
\begin{aligned}
\max \quad & \sum_{i \in A} \log(u_i(x) - c_i) \\
\text{s.t.} \quad & \sum_{i \in A} \sum_{k \in S_{ij}} x_{ijk} \leq 1 \qquad \forall j \in G, \\
& \sum_{j \in G} \sum_{k \in S_{ij}} x_{ijk} \leq 1 \qquad \forall i \in A, \\
& \qquad\qquad x_{ijk} \leq l_{ijk} \quad \forall i \in A, \forall j \in G, \forall k \in S_{ij}, \\
& \qquad\qquad\quad x \geq 0.
\end{aligned} \tag{2}
$$

As was the case for linear utilities, a $\frac{1}{2}$-approximate notion of equal-share fairness holds if $c_i = 0$ for all i. For simplicity assume that there is some ℓ such that for all i, j, we have $S_{ij} = \{1, \ldots, \ell+1\}$ and $\sum_{k=1}^{\ell} l_{ijk} = 1$. Note that due to the concavity of the utilities, the utility of player i in the equal-share allocation is at most $\frac{1}{n} \sum_{j \in G} \sum_{k=1}^{\ell} u_{ijk} l_{ijk}$.

Lemma 2. *Let x be an optimal solution to* (1) *with $c_i = 0$ then for all i, we have $u_i(x) \geq \frac{1}{2n} \sum_{j \in G} \sum_{k=1}^{\ell} u_{ijk} l_{ijk}$.*

For the remaining models we will not get such strong equal-share fairness properties. However, weakened versions are shown in in the full version of this paper.

3.1 Two-Sided Matching Markets

1.) Our two-sided matching market model consist of a set $A = \{1, 2, \ldots n\}$ of agents and a set $J = \{1, 2, \ldots, n\}$ of jobs. For uniformity, we have assumed that there is an equal number of agents and jobs, though the model can be easily enhanced and made more general. Our goal is to find an integral perfect matching between agents and jobs; however, we will relax this to finding a fractional perfect matching, x, followed by rounding as described above. We will explicitly define only the simplest case of two-sided markets; more general models follow along the same lines as one-sided markets.

Under the *linear Arrow-Debreu bargaining two-sided matching market*, abbreviated *2AD*, the utility accrued by agent $i \in A$ under allocation x, $u_i(x) = \sum_{j \in J} u_{ij} x_{ij}$, where u_{ij} is the utility accrued by i if she were assigned job j integrally. Analogously, the utility accrued by job $j \in J$ under allocation x, $w_j(x) = \sum_{i \in A} w_{ij} x_{ij}$, where w_{ij} is the utility accrued by j if it were assigned to i integrally.

In keeping with the axiom of symmetry under Nash bargaining, we will posit that the desires of agents and jobs are equally important and we are led to defining the feasible set in a $2n$ dimensional space, i.e., $\mathcal{N} \subseteq \mathsf{R}^{2n}_{+}$. The first n components of feasible point $v \in \mathcal{N}$ represent the utilities derived by the n agents, i.e., $u_i(x)$, and the last n components the utilities derived by the n jobs, i.e., $w_j(x)$, under a fractional perfect matching x. We seek the Nash bargaining point with respect to disagreement utilities c_i for all agents $i \in G$ and d_j for jobs $j \in J$. If all c_i and d_j are 0, we call this model reduces to the *linear Fisher bargaining two-sided matching market* or *2LF*. A convex program of *2AD* is given in (3).

$$\begin{aligned}
\max \quad & \sum_{i \in A} \log(u_i(x) - c_i) \; + \sum_{j \in J} \log(w_j(x) - d_j) \\
\text{s.t.} \quad & \sum_{i \in A} x_{ij} \leq 1 \quad \forall j \in J, \\
& \sum_{j \in J} x_{ij} \leq 1 \quad \forall i \in A \;\text{ and } x \geq 0.
\end{aligned} \tag{3}$$

2.) We may also extend this model even further by allowing SPLC utilities on both sides just like in the one-sided case. This defines the *SPLC Arrow-Debreu bargaining two-sided matching market* or *2SAD*.

3.2 Non-bipartite Matching Market

Our non-bipartite matching market model consist of a set $A = \{1, 2, \ldots, n\}$ of agents and utilities for each pair $u_{ij} \in \boldsymbol{R}_{+}$ which i and j accrue on getting matched to each other. Let $G = (A, E)$ be the complete graph on n vertices, i.e., all edges are present. The vertices correspond to agents. A vector $x \in \mathsf{R}^{E}_{+}$ is said to be a *fractional matching in G* if it is a convex combination of (integral) matchings in G. The goal is to find a Nash-bargaining solution with respect to disagreement utilities c_i for all agents i. This is a *linear Arrow-Debreu bargaining non-bipartite matching market* or *NBAD*. If $c_i = 0$ for all agents, then we call this the *linear Fisher bargaining non-bipartite matching market* or *NBLF*.

Due to a classic result by Edmonds we have that x is a fractional matching if and only if $x(\delta(i)) \leq 1$ for all i and $x(E(B)) \leq \frac{|B|-1}{2}$ for all sets $B \subseteq A$ of odd cardinality. For ease of notation let us denote the collection of all such odd subsets B of A by $\mathcal{O}$. This motivates the convex program:

$$\begin{aligned} \max \quad & \sum_{i \in A} \log(u_i(x) - c_i) \\ \text{s.t.} \quad & x(\delta(i)) \leq 1 && \forall i \in A, \\ & x(E(B)) \leq \frac{|B|-1}{2} && \forall B \in \mathcal{O} \text{ and } x \geq 0. \end{aligned} \tag{4}$$

We remark that as in the bipartite case, it is possible to take a fractional matching and (in a combinatorial, polynomial time manner), decompose it into a convex combination of most n^2 integral matchings. This was shown by Padberg and Wolsey [22]; for a modern proof see [26]. This allows a similar rounding strategy if one desires integral allocations.

4 Multiplicative Weights Update

In this section we prove that Multiplicative Weights Update (MWU) converges to an ϵ-approximate Nash bargaining solution for model LiAD. The main result is given in the following theorem:

Theorem 2. *The following hold:*

- *MWUA (Algorithm 1) computes an ϵ-approximate Nash bargaining solution (see Definition 2) of Program (1), i.e., model LiAD, after $O\left(\frac{n \log n}{\epsilon^2}\right)$ iterations and each iterate can be implemented in $O(n^2)$ time.*
- *Moreover, in the full version of the paper we show that MWUA (Algorithm 2 in full version) computes an ϵ-approximate Nash bargaining solution of model SAD, after $O\left(\frac{n \log n}{\epsilon^2}\right)$ iterations. Each iteration can be implemented in $O(l \cdot n^2)$ time, where l is the number of segments of the piecewise linear utilities.*

We provide the analysis of the case of LiAD, i.e., linear utilities with endowments. The proofs and the modified Lemmas so that the analysis carries over for piecewise linear utilities can be found in the full version of the paper.

4.1 From Optimization to Feasibility

Multiplicative Weights Update Algorithm (MWUA) [3] has found numerous applications in Game Theory, e.g., has been used for computing Nash Equilibrium in zero-sum games and potential games or Correlated Equilibrium in general games. Another surprising application, is that MWUA can be used to find approximately feasible points for linear programs. Inspired by the latter, we use MWUA to find an (approximately) feasible solution to the feasibility program below (see Definition 2 for approximate feasibility):

$$
\begin{aligned}
\mathrm{CP}_i(p,q) &\le u_i(x) \quad \forall i \in A,\\
\sum_{i\in A} x_{ij} &\le 1 \quad \forall j \in G,\\
\sum_{j\in G} x_{ij} &\le 1 \quad \forall i \in A, \qquad\qquad \text{(F-LiAD)}\\
\sum_{j\in G} p_j + \sum_{i\in A} q_i &= n + \sum_{i\in A} c_i \min_{j\in G} \frac{p_j+q_i}{u_{ij}},\\
x,p,q &\ge 0,
\end{aligned}
$$

$$
\text{where } \mathrm{CP}_i(p,q) = \max\left\{ \sum_{j\in G} u_{ij}y_j \,\middle|\, \sum_{j\in G}(p_j+q_i)y_j \le 1 + c_i \min_{j\in G}\frac{p_j+q_i}{u_{ij}}, y \ge 0 \right\}.
$$

We note that $\mathrm{CP}_i(p,q)$ stands for"convex program" that player i needs to solve to maximize her utility subject to budget constraints. These budgets constraints are chosen carefully so that Lemma 3 below works and can be derived through KKT conditions.

We are able to show that the Feasibility Program (F-LiAD) actually contains *only* the optimal primal-dual solution of the convex program (1), namely any feasible point of (F-LiAD) satisfies the KKT of (1). This implies that as long as MWUA finds a solution that is approximately feasible for (F-LiAD) , the solution will satisfy approximately the KKT conditions and hence will be approximately optimal. The first step towards the proof is to argue that if (F-LiAD) is feasible, that is there exist (x^*, p^*, q^*) that satisfy the constraints in (F-LiAD) , then and only then (x^*, p^*, q^*) are the primal-dual variables that satisfy the KKT conditions. This is given in the following Lemma:

Lemma 3 (Optimization to Feasibility). *Let x^* be an optimal solution of Program (1), then there exist p^*, q^* so that (x^*, p^*, q^*) is feasible for (F-LiAD) . Conversely, assume that (x^*, p^*, q^*) is a feasible solution for (F-LiAD) , then x^* is an optimal solution for Program (1).*

A corollary that can be derived from Lemma 3 is that an approximate feasible solution of (F-LiAD) is an approximately optimal Nash bargaining solution of Program (1). Before we state the corollary, we provide the formal definition of an

ϵ-approximate feasible solution to (F-LiAD) and ϵ-approximate Nash bargaining solution.

Definition 2 (Approximately feasible). *A point $(\tilde{x}, \tilde{p}, \tilde{q})$ is called ϵ-approximate feasible of (F-LiAD) , if it satisfies non-negativity, price and max-utility constaints but might violate allocation constraints by an additive ϵ; that is we have $\sum_{i\in A} \tilde{x}_{ij} \leq 1+\epsilon$, for all $j \in G$ and $\sum_{j\in G} \tilde{x}_{ij} \leq 1+\epsilon$, for all $i \in A$. Moreover, we define an ϵ-approximate Nash bargaining solution for program (1) to be a feasible allocation x such that for each agent i, if i changes his allocation from x_i to some feasible allocation x'_i, he cannot gain more than an additive ϵ in his utility.*

Corollary 1. *Let $(\tilde{x}, \tilde{p}, \tilde{q})$ be a ϵ-approximate feasible point of (F-LiAD) . Then $\frac{1}{1+\epsilon} \cdot \tilde{x}$ is a $O(\epsilon)$-approximate Nash bargaining solution of Program (1).*

Algorithm 1 Multiplicative Weight Update

$\tilde{p}^{(1)} \leftarrow \mathbf{1}, \tilde{q}^{(1)} \leftarrow \mathbf{1}$ initialization
for $t = 1$ to T **do**
  Rescale $\tilde{p}^{(t)}, \tilde{q}^{(t)}$ so that $\sum_{j\in G} p_j^{(t)} + \sum_{i\in A} q_i^{(t)} = n + \sum_{i\in A} c_i \min_{j\in G} \frac{p_j^{(t)}+q_i^{(t)}}{u_{ij}}$.
  for $i = 1$ to n **do**
    $x_i^{(t)} \leftarrow \arg\max_y \left\{ \sum_{j\in G} u_{ij} y_j : \sum_{j\in G} \left(p_j^{(t)} + q_i^{(t)}\right) y_j \leq 1 + c_i \min_{j\in G} \frac{p_j^{(t)}+q_i^{(t)}}{u_{ij}} \right\}$.
  for $j = 1$ to n **do**
    $d_j^{(t)} \leftarrow \sum_{i\in A} x_{ij}^{(t)}$
    $h_j^{(t)} \leftarrow \sum_{i\in G} x_{ji}^{(t)}$
  $\sigma^{(t)} \leftarrow \min\left(\frac{1}{\max_{j\in G} d_j^{(t)}}, \frac{1}{\max_{i\in A} h_i^{(t)}}\right)$
  for $j = 1$ to n **do**
    $\tilde{p}_j^{(t+1)} \leftarrow \tilde{p}_j^{(t)}(1 + \epsilon\sigma^{(t)} d_j^{(t)})$
    $\tilde{q}_j^{(t+1)} \leftarrow \tilde{q}_j^{(t)}(1 + \epsilon\sigma^{(t)} h_j^{(t)})$
return $\overline{x} \leftarrow \frac{\sum_{t=1}^T \sigma^{(t)} x^{(t)}}{\sum_{t=1}^T \sigma^{(t)}}$ (matrix), $\overline{p} \leftarrow \frac{\sum_{t=1}^T \sigma^{(t)} p^{(t)}}{\sum_{t=1}^T \sigma^{(t)}}$, $\overline{q} \leftarrow \frac{\sum_{t=1}^T \sigma^{(t)} q^{(t)}}{\sum_{t=1}^T \sigma^{(t)}}$.

4.2 MWUA Analysis for LiAD via (F-LiAD)

Since we showed equivalence (Lemma 3) between Program (1) and (F-LiAD) , we will apply MWUA on the latter (see Algorithm 1). The rest of the section will be focusing on proving convergence and bounding the rate of convergence of MWUA. We shall show that after $O\left(\frac{n \log n}{\epsilon^2}\right)$ iterations, MWUA will reach a point $(\tilde{x}, \tilde{p}, \tilde{q})$ that is an ϵ-approximate feasible solution for (F-LiAD) and hence $\frac{1}{1+\epsilon} \cdot \tilde{x}$ will be an $O(\epsilon)$-approximate optimal solution for each agent in Program (1), as stated in the aforementioned Corollary 1.

Before we proceed with the analysis of Algorithm 1, we need to argue that the rescaling step in the For-loop is well-defined, i.e., we can always rescale $\tilde{p}^{(t)}, \tilde{q}^{(t)}$ to $p^{(t)}, q^{(t)}$ so that

$$\sum_{j\in G} p_j^{(t)} + \sum_{i\in A} q_i^{(t)} = n + \sum_{i\in A} c_i \min_{j\in G} \frac{p_j^{(t)} + q_i^{(t)}}{u_{ij}}.$$

This is captured by the following lemma:

Lemma 4 (Rescaling). *Assume (F-LiAD) is feasible. Given* $\tilde{p}, \tilde{q} \geq 1$*, we can always rescale them to* p, q *so that* $\sum_{j\in G} p_j + \sum_{i\in A} q_i = n + \sum_{i\in A} c_i \min_{j\in G} \frac{p_j+q_i}{u_{ij}}$.

Potential function. To bound the rate of convergence of MWUA and show that it actually reaches an $O(\epsilon)$-approximate feasible point of (F-LiAD) , we use a potential function argument, an idea that is quite common in online learning literature [3]. The potential function at iterate t is defined to be (sum of prices at iterate t)

$$\Phi(t) = \sum_{j\in G} \tilde{p}_j^{(t)} + \sum_{i\in A} \tilde{q}_i^{(t)}. \tag{5}$$

We need to find upper and lower bounds on $\Phi(T)$ which will enable us to bound the convergence rate of MWUA (T denotes the number of iterations of MWUA). The first lemma gives an upper bound on the potential function Φ.

Lemma 5 (Bounds on $\Phi(T)$). *The following holds:*

$$\Phi(T) \leq 2n \cdot exp\left(\epsilon \sum_{t=1}^{T-1} \sigma^{(t)}\right) \quad and \tag{6}$$

$$\Phi(T) \geq exp\left(\epsilon(1-\epsilon) \max\left(\max_{j\in G} \sum_{t=1}^{T} \sigma^{(t)} d_j^{(t)}, \max_{i\in A} \sum_{t=1}^{T} \sigma^{(t)} h_i^{(t)}\right)\right), \tag{7}$$

where $\sigma^{(t)} := \min\left(\frac{1}{\max_{j\in G}\sum_{i\in A} x_{ij}^{(t)}}, \frac{1}{\max_{i\in A}\sum_{j\in G} x_{ij}^{(t)}}\right)$, $d_j^{(t)} := \sum_{i\in A} x_{ij}^{(t)}$ *and* $h_i^{(t)} := \sum_{j\in G} x_{ij}^{(t)2}$.

Proving Theorem 2. We provide the important steps behind the proof of Theorem 2. For a complete proof, see full version. The idea is to consider the weighted time average of the allocations $x^{(t)}$ and prices $p^{(t)}, q^{(t)}$ that is $\overline{x} = \frac{\sum_{t=1}^{T} \sigma^{(t)} x^{(t)}}{\sum_{t=1}^{T} \sigma^{(t)}}$

[2] See Algorithm 1. $\sigma^{(t)}$ is essentially the renormalization parameter so that the "utility" of each constraint/expert in MWU is bounded by one.

and $\overline{p} = \frac{\sum_{t=1}^{T} \sigma^{(t)} p^{(t)}}{\sum_{t=1}^{T} \sigma^{(t)}}$ and $\overline{q} = \frac{\sum_{t=1}^{T} \sigma^{(t)} q^{(t)}}{\sum_{t=1}^{T} \sigma^{(t)}}$ in a window $[1, T]$ in which T is chosen to be $\Theta\left(\frac{2n \log 2n}{\epsilon^2}\right)$. Exploiting the fact that CP_i is a convex function, Lemma 5 and for the proper choice of T one can show that $\mathrm{CP}_i(\overline{p}, \overline{q}) \leq u_i(\overline{x})$ for all i as well as $\sum_j \overline{x}_{ij} \leq 1 + O(\epsilon)$ and $\sum_i \overline{x}_{ij} \leq 1 + O(\epsilon)$ for all i and j respectively. Thus, $(\overline{x}, \overline{p}, \overline{q})$ is an $O(\epsilon)$-approximate feasible solution of (F-LiAD) .

5 Conditional Gradient Algorithms

Since finding Nash-bargaining points can be captured by a convex program, we can leverage techniques from convex optimization such as gradient descent to compute rapidly converging approximations. We will now describe a conditional gradient type algorithm which is relatively simple and projection free while converging at a rate of $O(1/\epsilon)$.

5.1 Linear Utilities (LiF)

Let us first consider the simplest setting of a one-sided market with linear utilities, i.e. the model LiF. We are given a set A of agents and G of goods with $|A| = |G| = n$ and utilities u_{ij} for all $i \in A$ and $j \in G$. Our goal is to solve the convex program (1) for $c_i = 0$.

In the following we will assume that the utilities have been rescaled so that $\max\{u_{ij} \mid j \in G\} = 1$ for all i and $\sum_{j \in G} u_{ij} \geq \frac{1}{\kappa}$ for some value κ. In particular, the objective function is 1-strongly concave with respect to $u_i(x)$ but not x_{ij}. Note that since $u_i(x)$ is not bounded from below, the objective function is neither Lipschitz nor smooth.

By Lemma 1, we can restrict ourselves to the region where $u_i(x) \geq \frac{1}{2\kappa n}$ without changing the optimum solution. However, while general purpose projected gradient descent algorithms can achieve $O(\log(1/\epsilon))$ convergence rates on this modified problem, we wish to exploit the combinatorial structure of P. Therefore we will modify the objective function instead.

Define the quadratic extension of the logarithm at $x_0 > 0$ by

$$\eta(x; x_0) := \begin{cases} \log(x_0) + (x - x_0)\frac{1}{x_0} - \frac{1}{2}(x - x_0)^2 \frac{1}{x_0^2} & \text{if } x \leq x_0, \\ \log(x) & \text{otherwise.} \end{cases}$$

Then $\eta(x; x_0)$ is $\frac{1}{x_0^2}$-smooth everywhere, i.e. its gradient is $\frac{1}{x_0^2}$-Lipschitz. Consider now the modified convex program

$$\begin{aligned} \max \quad & \sum_{i \in A} \eta\left(u_i(x); \frac{1}{2\kappa n}\right) \\ \text{s.t.} \quad & \sum_{i \in A} x_{ij} \leq 1 \quad \forall j \in G, \\ & \sum_{j \in G} x_{ij} \leq 1 \quad \forall i \in A \text{ and } x \geq 0. \end{aligned} \tag{8}$$

Lemma 6. *Every optimum solution to* (8) *is also optimum for* (1) *with* $c_i = 0$.

In the following let $\phi(x) := \sum_{i \in A} \log u_i(x)$ and $\psi(x) := \sum_{i \in A} \eta\left(u_i(x); \frac{1}{2\kappa n}\right)$ be the objective functions of the original CP and (8) respectively.

Lemma 7. *Let x be an arbitrary allocation and x^* an optimal one. Let $\psi(x^*) - \psi(x) \leq \epsilon$ for some $\epsilon > 0$. Then $||u(x) - u(x^*)||^2 \leq 2\epsilon$. Moreover, if $\delta \in (0, 1/2)$, and $\epsilon = \min\left\{\delta, \frac{\delta^{2/3}}{8n^2\kappa^2}\right\}$, then $\phi(x^*) - \phi(x) \leq O(1)\delta$.*

These two lemmas together imply that it suffices to solve (8) . Approximate solutions to this modified program are also approximate solutions for the original CP, both wrt. the Euclidean norm on the utility vectors and the original objective function. Finding an approximate solution in a combinatorial way can be achieved using the conditional gradient method over the matching polytope; see Algorithm 2. Note that the gradient of ψ is easily computable.

Algorithm 2 Conditional Gradient for program (8)

```
1: x^(0) ≡ 0
2: for t ← 1,...,T do
3:     for i ∈ A, j ∈ G do
4:         w_ij^(t) ← ∂_ij ψ(x^(t−1))
5:     y^(t) ← max-weight matching with weights w^(t)
6:     x^(t) ← (1 − 2/(t+1)) x^(t−1) + 2/(t+1) y^(t)
7: return x^(T)
```

Theorem 3. *Algorithm 2 returns some x with $\psi(x^*) - \psi(x) \leq \epsilon$ in $O\left(\frac{n^3\kappa^2}{\epsilon}\right)$ many iterations. Each iteration can be implemented in $O(n^3)$ time.*

We remark that it is in principle possible to achieve faster convergence rates for conditional gradient type algorithms by leveraging strong concavity in addition to smoothness [13]. Note that while ψ is strongly concave in the utilities, it is unfortunately not strongly concave in the allocation x.

5.2 Non-bipartite Matching Markets (NBLF)

The result from the previous section can be extended to the non-bipartite setting, i.e. model NBLF. Recall that we have a set of n agents A with non-negative utilities u_{ij} for all $i, j \in A$. The goal is then to solve the convex program (4) with $c_i = 0$.

Assume that the utilities have been rescaled so that $\max\{u_{ij} \mid j \in A\} = 1$ for all i and $\sum_{j \in A} u_{ij} \geq \frac{1}{\kappa}$ for some value κ. Since it is possible to optimize over the matching polytope using combinatorial methods, the primary ingredient is to once again bound the utilities away from 0.

Lemma 8. *Let x be an optimal solution to (4) with $c_i = 0$, then for all agents i we have*

$$u_i(x) \geq \frac{1}{2n^2} \sum_{j \in G} u_{ij} \geq \frac{1}{2\kappa n^2}.$$

This bound may initially seem weak when compared to the $\frac{1}{2n} \sum_{j \in G} u_{ij}$ bound from Lemma 1 of the previous section but one can show that this bound is tight up to a constant factor. However, this utility bound is still adequate in order to show convergence of the conditional gradient method with the modified objective function $\psi(x) := \sum_{i \in A} \eta\left(u_i(x); \frac{1}{2n^2\kappa}\right)$.

Theorem 4. *Algorithm 2 returns some x with $\psi(x^*) - \psi(x) \leq \epsilon$ in $O\left(\frac{n^5\kappa^2}{\epsilon}\right)$ many iterations. Each iteration can be implemented in $O(n^3)$ time.*

5.3 Extension to Endowments (LiAD)

Let us now return to the case of linear utilities but in which agents come with preexisting endowments, i.e. each agent i has some disagreement utility c_i and the goal is then to solve the convex program (1). The added difficulty of this setting is that, in general, a feasible solution may not exist. We thus assume that we are given a feasible $\hat{x}$ such that $u_i(\hat{x}) \geq (1+\delta)c_i$ for all i and some $\delta > 0$. We call this δ the *feasibility gap* of the instance.

In practice there are two likely ways to obtain a δ. Either one solves the above linear program to find the optimal δ or the disagreement utilities c_i are defined as some constant fraction of the agents' utilities over their initial endowments. More precisely, assume that each agent i comes to the market with endowments e_{ij} over the goods j. Then one may simply define $c_i := \frac{1}{1+\delta} \sum_{j \in G} u_{ij} e_{ij}$. The market will then guarantee that no agent gets worse by a factor of more than $(1+\delta)$ which is an approximate notion of individual rationality.

Regardless of where δ comes from, we may use it to derive a similar lower bound as in Lemma 1. Once again, assume that the utilities have been rescaled so that $\max\{u_{ij} \mid j \in G\} = 1$ for all i and $\sum_{j \in G} u_{ij} \geq \frac{1}{\kappa}$ for some $\kappa > 0$.

Lemma 9. *Let x be an optimal solution to (1), then for all agents i we have*

$$u_i(x) \geq c_i + \frac{1}{2n^2(1+1/\delta)} \sum_{j \in G} u_{ij} \geq c_i + \frac{1}{2n^2(1+1/\delta)\kappa}.$$

The modified convex program will thus use the objective function $\psi(x) := \sum_{i \in A} \eta\left(u_i(x) - c_i; \frac{1}{2n^2(1+1/\delta)\kappa}\right)$ and we can once again apply Algorithm 2.

Theorem 5. *Algorithm 2 returns x with $\psi(x^*) - \psi(x) \leq \epsilon$ in $O\left(\frac{n^5(1+1/\delta)^2\kappa^2}{\epsilon}\right)$ many iterations. Each iteration can be implemented in $O(n^3)$ time.*

We remark that an analogous result to Lemma 7 holds also in this setting, i.e. convergence in the modified objective ψ implies convergence of the agents' utilities in the Euclidean norm. Thus, if one chooses ϵ on the order of $O\left(\frac{1}{n\sqrt{(1+1/\delta)\kappa}}\right)$, then one is guaranteed that $u_i(x) \geq c_i$ for all agents i.

Finally, in the full version of the paper we show that this can be extended to a much more general setting of two-sided SPLC utilities with endowments (i.e. model 2SAD).

References

1. Abebe, R., Cole, R., Gkatzelis, V., Hartline, J.D.: A truthful cardinal mechanism for one-sided matching. In: Proceedings of the Fourteenth Annual ACM-SIAM Symposium on Discrete Algorithms, pp. 2096–2113. SIAM (2020)
2. Alaei, S., Jalaly Khalilabadi, P., Tardos, E.: Computing equilibrium in matching markets. In: Proceedings of the 2017 ACM Conference on Economics and Computation, pp. 245–261 (2017)
3. Arora, S., Hazan, E., Kale, S.: The multiplicative weights update method: a meta-algorithm and applications. Theory Comput. **8**(1), 121–164 (2012)
4. Basu, S., Pollack, R., Roy, M.: A new algorithm to find a point in every cell defined by a family of polynomials. Quantifier Elimination and Cylindrical Algebraic Decomposition, B. Caviness and J. Johnson eds., Springer-Verlag, to appear (1995)
5. Birkhoff, G.: Tres observaciones sobre el algebra lineal. Univ. Nac. Tucuman, Ser. A **5**, 147–154 (1946)
6. Caragiannis, I., Kurokawa, D., Moulin, H., Procaccia, A.D., Shah, N., Wang, J.: The unreasonable fairness of maximum Nash welfare. ACM Trans. Econ. Comput. (TEAC) **7**(3), 1–32 (2019)
7. Chen, T., Chen, X., Peng, B., Yannakakis, M.: Computational hardness of the hylland-Zeckhauser scheme. In: Proceedings of the 2022 ACM-SIAM Symposium on Discrete Algorithms (2022)
8. Cole, R., et al.: Convex program duality, Fisher markets, and Nash social welfare. In: Proceedings of the 2017 ACM Conference on Economics and Computation, pp. 459–460 (2017)
9. Cole, R., Gkatzelis, V.: Approximating the Nash social welfare with indivisible items. SIAM J. Comput. **47**(3), 1211–1236 (2018)
10. Echenique, F., Immorlica, N., (Eds.), V.V.V.: Online and Matching-Based Market Design. Cambridge University Press (2022). To Appear
11. Fleischer, L., Garg, R., Kapoor, S., Khandekar, R., Saberi, A.: A fast and simple algorithm for computing market equilibria. In: International Workshop on Internet and Network Economics, pp. 19–30. Springer (2008)
12. Gao, Y., Kroer, C.: First-order methods for large-scale market equilibrium computation. Adv. Neural Inf. Process. Syst. **33** (2020)
13. Garber, D., Hazan, E.: A linearly convergent variant of the conditional gradient algorithm under strong convexity, with applications to online and stochastic optimization. SIAM J. Optim. **26**(3), 1493–1528 (2016)
14. Grötschel, M., Lovász, L., Schrijver, A.: Geometric algorithms and combinatorial optimization, vol. 2. Springer Science & Business Media (2012)

15. He, Y., Miralles, A., Pycia, M., Yan, J.: A pseudo-market approach to allocation with priorities. American Econ. J. Microecon. **10**(3), 272–314 (2018)
16. Hosseini, M., Vazirani, V.V.: Nash-bargaining-based models for matching markets: One-sided and two-sided; fisher and arrow-debreu. In: Proceedings of the 2022 Conference on Innovations in Theoretical Computer Science, ITCS (2022)
17. Hylland, A., Zeckhauser, R.: The efficient allocation of individuals to positions. J. Polit. Econ. **87**(2), 293–314 (1979)
18. Immorlica, N., Lucier, B., Weyl, G., Mollner, J.: Approximate efficiency in matching markets. In: International Conference on Web and Internet Economics, pp. 252–265. Springer (2017)
19. Mas-Colell, A., Whinston, M.D., Green, J.R., et al.: Microeconomic Theory, vol. 1. Oxford University Press, New York (1995)
20. Moulin, H.: Fair division in the age of internet. Annual Review of Economics (2018)
21. Nash, J.: Two-person cooperative games. Econometrica: J. Econ. Soc. pp. 128–140 (1953)
22. Padberg, M.W., Wolsey, L.A.: Fractional covers for forests and matchings. Math. Program. **29**(1), 1–14 (1984)
23. Schrijver, A.: Theory of Linear and Integer Programming. John Wiley & Sons, New York, NY (1986)
24. Tröbst, T., Vazirani, V.V.: Cardinal-utility matching markets: The quest for envy-freeness, pareto-optimality, and efficient computability. In: Proceedings of the 2024 ACM Conference on Economics and Computation (2024)
25. Vazirani, V.V.: The notion of a rational convex program, and an algorithm for the Arrow-Debreu Nash bargaining game. J. ACM **59(2)** (2012)
26. Vazirani, V.V.: An extension of the Birkhoff-von Neumann Theorem to non-bipartite graphs. arXiv preprint arXiv:2010.05984 (2020)
27. Vazirani, V.V., Yannakakis, M.: Computational complexity of the Hylland-Zeckhauser scheme for one-sided matching markets. In: 12th Innovations in Theoretical Computer Science Conference, ITCS 2021, January 6-8, 2021, Virtual Conference (2021)
28. Vishnoi, N.: Algorithms for Convex Optimization. Cambridge University Press (2021). To Appear
29. Von Neumann, J.: A certain zero-sum two-person game equivalent to the optimal assignment problem. Contrib. Theor. Games **2**, 5–12 (1953)

Price Competition in Linear Fisher Markets: Stability, Equilibrium and Personalization

Juncheng Li(✉) and Pingzhong Tang

Tsinghua University, Beijing, China
lijuncheng13@gmail.com

Abstract. Linear Fisher market is one of the most fundamental economic models. The market is traditionally examined on the basis of individual's price-taking behavior. However, this assumption breaks in markets such as online advertising and e-commerce, where several oligopolists dominate the market and are able to compete with each other via strategic actions. Motivated by this, we study the price competition among sellers in linear Fisher markets. From an algorithmic game-theoretic perspective, we establish a model to analyze behaviors of buyers and sellers that are driven by utility-maximizing purposes and also constrained by computational tractability. The main economic observation is the role played by personalization: the classic benchmark market outcome, namely competitive equilibrium, remains to be a steady-state if every buyer must be treated "equally"; however, sellers have the incentive to personalize, and as a result the market would become more unpredictable and less efficient. In addition, we build a series of algorithmic and complexity results along the road to justify our modeling choices and reveal market structures. We find interesting connections between our model and other computational problems such as stable matching, network flow, etc. We believe these results and techniques are of independent interest.

1 Introduction

Fisher market (and the more general Arrow-Debreu market [2,23]) is one of the most foundational economic models. It serves as an excellent first approximation to those real-world markets that consist of a *large* number of buyers and sellers such that no individual has the power to effectively influence prices. In addition, due to the many nice properties of the market solution, named *competitive equilibrium*, it is also widely implemented in markets where there is only a *single* benevolent seller/owner who wants to allocate items to buyers/recipients in a fair and envy-free way [16,29]. The model and its idea keeps receiving attention in the literature as it is applied and adapted to various situations, such as fair division of indivisible items [3,6,7,9] or a continuum of items [18], recommender systems [19–21] and online advertising auction markets [10,11].

In spite of the broad applicability of the model and the appeal of competitive equilibrium, the price-taking assumption fails to hold in markets like oligopolies,

M. Mavronicolas et al. (Eds.): WINE 2024, LNCS 15534, pp. 137–154, 2026.
https://doi.org/10.1007/978-3-032-08560-3_8

where the items are controlled by a group of sellers who are powerful enough to set prices arbitrarily for items of its own. In these industries, the market outcome is not only driven by the omnipotent power of overall supply and demand, but also the revenue-seeking behaviors of individual sellers. Seller competition is an ongoing topic in the literature of game theory, from the classic models of competition on the basis of prices and production quantities (see, e.g., Chap. 6 of the book by Dixon [13]), to the more complex ones like the model by Selcuk [27] and Paes Leme et al. [26] where sellers compete for buyers' money through selecting suitable trading mechanisms. We cannot review the vast literature exhaustively here, but to the best of our knowledge, no existing work puts seller competition in the context of (or close to) the fundamental Fisher market model. In this paper, we initiate the study of this problem and approach it from both *game-theoretic* and *computational* perspectives.

Starting with the work by Megiddo and Papadimitriou [24], theoretical computer scientists began to notice an important yet previously overlooked aspect of economic models – the computation of solution concepts. If an equilibrium cannot be computed efficiently, it would be hard, if not impossible, to implement for the market owner and less likely to emerge as the outcome of social learning. In a linear Fisher market, the competitive equilibrium is not only poly-time solvable [15,25,28], but could also be reached through well-motivated fast-converging dynamics [8,32]. It is therefore of great computational soundness. On the other hand, if some equilibrium concept is shown to be hard to achieve, those economically weaker but computationally tractable solutions would become more favorable.[1]

In this paper, we propose a new price competition model over linear Fisher markets, and characterize the set of market outcomes that are both economically stable and computationally feasible. Our model incorporates several features absent in traditional formulations, and in particular the role of buyer *personalization* is examined. The market is more well-behaved without personalization, but sellers have the incentive to do so, and the resulting market might become more unpredictable and less efficient. Besides revealing market structures, our computational results also demonstrate some interesting relationships between our model and problems such as stable matching and network flow. These new connections and techniques may be of independent interest as well.

1.1 Markets Driven by Price Competition

A linear Fisher market consists of a set of divisible items and a set of budget-constrained buyers with linear utilities. There is a long line of research on Fisher markets that sits at the intersection of theoretical computer science and economics. Our research question is of great importance to the understanding of the price competition among sellers in linear Fisher markets. One of the real-world examples is online advertising: for instance, YouTube Shorts, Instagram

[1] One such example is *coarse correlated equilibrium*, which can be reached through decentralized no-regret learning in a large variety of games where Nash equilibrium is hard to compute.

Reels and TikTok all feature short videos and compete for very similar audiences. In theory, researchers also show that online advertising markets can be well captured by the linear Fisher market model [10,11] (but they study a single advertising platform rather than multiple competing ones). Below we will introduce the key features brought by the price-setting power of sellers, and the rationale behind them.

In Fisher markets, buyers are traditionally assumed to be *supply-unaware* in the sense that they demand and only demand items with the maximum *bang-per-bucks* (the value acquired for each unit of money spent on the item) regardless of supply. In real-world markets, however, people seldom let their money sit idle. So in our model we let trade happen whenever possible, even if the bang-per-buck is not optimal.

In a price-taking market, the role played by sellers is negligible, since a change of item ownership never affects the equilibrium outcome. In our model, besides the power to set prices, sellers could make decisions based on the aggregated behaviors of buyers on all the items in their possession. Consequently, item-ownership becomes influential.

Traditionally there is no buyer personalization in Fisher markets, which is a natural result of information symmetry and perfect competition. With the advent of big data, however, sellers are now able to literally personalize for every individual. E.g., in online advertising, sellers (such as Google and Meta) make the definitive decisions on pricing and allocation. Trade happens in real time and no arbitrage is possible. The items sold (opportunities to place an ad) are themselves personalized. This motivates us to observe the implication of personalization and it turns out to be of great consequence.

As the main focus of this paper is the competition among sellers, the complex strategic behaviors of buyers are not considered, except that each buyer must be served a bundle of items optimal in hindsight.[2] Previously researchers [1,5] studied a game model on the opposite side, where buyers can *misreport* their valuations and the market owner will allocate items via the competitive equilibrium given the reported valuations. This is hard to achieve in markets where it is the seller who has the information advantage. For instance, in online advertising, valuations are predicted based on the data and machine learning model in the control of the seller, who would typically hide most of the detailed information from buyers for both privacy and profit considerations.

1.2 Our Results

We propose a model over linear Fisher market with additionally a set of sellers, each owning a subset of items of the underlying market. Each seller's strategy consists of two parts: *prices* and *priorities*. Prices are allowed to be personalized,

[2] In real-world markets such as online advertising, there are indeed few opportunities for buyers to be strategic. Alternative buyer models might include, e.g., the allowance of arbitrage or buyers forming coalitions, which constitute interesting directions to explore but are beyond the scope of this paper.

and the priority is a total preorder dictating who the seller prefers to sell an item to (independent of prices).[3] A myopic seller may use a *natural* priority that fully agrees with the pricing, which represents a greedy revenue-seeking behavior. Possibly surprisingly, we will show that a natural priority may be suboptimal, so arbitrary priorities constitute a necessary strategic component in general. We do not incorporate complex buyer behaviors in our model; the only requirement is that they should be served an optimal bundle *in hindsight.*

Given sellers' prices and priorities, we impose that the trade outcome (also referred to as the allocation) should satisfy *stability*, which dictates that no pair of a buyer and a seller has the incentive to deviate and form a new trade such that both are made strictly happier (w.r.t. valuation for buyer and priorities for seller). Besides its intuitive economic implication, we show that a stable allocation always exists and is strongly poly-time computable. It is therefore justified as a requisite of the trade outcome. A stable allocation is generally not unique, which can be interpreted as an inherent uncertainty of the market and is interesting in its own right.

We proceed to analyze different categories of markets characterized by the degree to which the market is personalized.

Non-personalized Pricing with Natural Priorities. We first consider the case where no personalization is allowed, and find an interesting "one-sided" tractability result: it is strongly poly-time solvable to find the (market-wise) revenue-maximizing stable allocation, while it is APX-hard to minimize it. Moreover, the set of revenue-optimal[4] stable allocations coincides perfectly with the set of seller-wise Pareto-optimal ones. Therefore we further impose that the trade outcome should satisfy revenue-optimality. Nonetheless, the issue of multiplicity remains. As a result, we adopt a maximin approach and study the *maximin equilibrium* where no seller could *guarantee* itself a higher revenue after deviation regardless of the allocation selection. We show that the competitive equilibrium always forms a maximin equilibrium, but is generally not the unique one. We also explore properties of those non-competitive equilibria via case study, which include market non-clearance, ownership awareness, and what kind of sellers may gain or lose in the competition.

Non-personalized Pricing with Arbitrary Priorities. Things become complicated even when just the priorities are personalized.[5] It is now APX-hard to

[3] As an illustration, consider e-commerce platforms. Though each item comes with a posted price, it can be personalized through coupons, and priorities can be implemented (implicitly) by the recommendation algorithm.

[4] In this paper, "revenue-optimal" is used interchangeably with "revenue-maximizing".

[5] In reality, sometimes the pricing is expected to be non-personalized, while a personalization of priority is possibly easier to emerge. For instance, in ride-hailing, the pricing depends only on buyer-oblivious features like the driver/vehicle and the route, but the platform (such as Uber and Didi) controls the assignment of drivers to passengers, which creates the kind of personalized priorities we study here. Note that, though ride-hailing serves as a good example for non-personalized pricing with arbitrary priorities, it does not perfectly fit into the Fisher market model. Nonethe-

find either the revenue-maximizing or the revenue-minimizing stable allocation, which indicates that the market outcome is relatively unpredictable. We also give an example where multiplicity is not an issue but there is a profitable deviation for a seller at the competitive equilibrium. This shows the incentive of a seller to personalize even just the priorities, and the inefficiency it might bring.

Personalized Pricing. Interestingly, with personalized pricing, we identify a natural subset of prices and priorities that could lead to a *unique* stable allocation, as long as there is no tie involved (and priorities should be natural). However, the APX-hardness remains in general, which implies that different tie-breaking choices may lead to significant market changes. Furthermore, we give an example where, given any prices and priorities, there always exists a profitable deviation for some seller. These evidence shows the market is highly unpredictable.

Competitive Pricing for Duopoly. Delving deeper into the aforementioned example, we find that it is partly due to the static nature of our model that no equilibrium exists. Hence we deviate a little from the main model to incorporate some dynamic feature, i.e., sellers are now allowed to retaliate the opponent by matching the deviated prices (more precisely the bang-per-bucks). The model is defined only for duopoly as the price-matching behavior is more meaningful with two sellers. In general, the market outcome in this new model is still unpredictable, so we focus on a specific class of markets that share the same underlying Fisher market with the no-equilibrium example while having different item ownership. We show that the competitive equilibrium will form a stationary structure when the market is relatively *balanced*, while less efficient outcomes will appear as sellers become more different from each other.

1.3 Computational Techniques and Structures

In the main text we will focus on the analysis of the market model. Here we give an overview of the techniques used in and the insights gained from our computational results, and refer readers to the appendix of the full version of our paper [22] for the details.

The backbone of the algorithm computing a stable allocation (Theorem 1 and Appendix A [22]) is adapted from the classic Gale-Shapley algorithm [17]: here buyers propose to items in decreasing order of preferences, and items reject buyers with lower priorities. However, a naive adaption to the fractional case will result in excessively long or even infinite loops. Much efforts are thus dedicated to find these loops and resolve them efficiently. The definition of stable allocation actually bears some resemblance to stable matching, but neither of these two concepts contains the other as a special case: in our model the preference is connected with money, while in stable matching there is no concept of budgets.

---

less, we believe our model could shed some light into the understanding of these markets.

An interesting new connection is established between our non-personalized market and network flow. A revenue-optimal stable allocation (with arbitrary pricing) can be encoded in a flow network, and computed with a particular implementation of the Edmonds-Karp algorithm [14] such that certain economic property (a partial form of stability) is maintained for every intermediate flow throughout the execution (Theorem 2 and Appendix B.1 [22]). Some proofs (including those for personalized markets) can be viewed intuitively as examining how the flow/allocation is "dragged" towards a certain direction by the "gravity" of stability. Prior to our work, network flow has been used in Fisher markets, albeit in ways quite different from ours. Végh [30] and Végh [31] encode the computation of competitive equilibrium in linear Fisher markets as a single-shot generalized minimum-cost flow and generalized maximum flow problem, respectively. Devanur et al. [12] compute the competitive equilibrium by solving a series of maximum flows corresponding to different price vectors. In their model, buyers only demand items they prefer the most and items will never become under-demanded during the execution of the algorithm.

The "one-sided" tractability of revenue optimization in non-personalized pricing markets resembles the case for maximal/minimal matching, and we show that their computational structures are indeed closely related (see Appendix B.1 [22]).

Most of our APX-hardness results on revenue maximization/minimization also imply APX-completeness, since the revenues of two stable allocations differ by at most a factor of 2 (which is tight; see Appendix C.3 [22]) and a stable allocation is poly-time computable.

2 Preliminaries

A **Fisher market** consists of: (1) a set J of m divisible items, each with a supply (or quantity) normalized to 1; (2) a set I of n buyers; (3) a budget B_i for each buyer i; (4) a utility function $u_i : [0,1]^{n\times m} \to \mathbb{R}$ for each buyer i, which takes the allocation of the whole market $x \in [0,1]^{n\times m}$ as the input (budgets have no value for buyers). In this paper, there is no externality and u_i depends only on the allocation received by buyer i itself. Note that traditionally there is no explicit mention of seller or item-ownership.

A Fisher market is **linear**, if u_i has the form $u_i(x) = \sum_j v_{i,j}x_{i,j}$, where $x_{i,j}$ is the fraction of item j allocated to buyer i, and $v_{i,j} \geq 0$ is the value of item j to buyer i. For each item, there is at least one buyer i with $v_{i,j} > 0$. Each item will come with a price p_j, and buyers will try to acquire as much utility as possible within their budget constraints.

A **competitive equilibrium** (p^*, x^*) of a linear Fisher market consists of a pricing vector (or simply pricing or prices) $p^* \in [0,+\infty)^m$ and an allocation $x^* \in [0,1]^{n\times m}$ satisfying:

- Buyer optimization or envy-freeness: $x_i^* \in \operatorname*{argmax}_{x_i\in[0,+\infty)^m:\sum_j p_j^* x_{i,j}\leq B_i} \sum_j v_{i,j}x_{i,j}$;
- Market clearing: $\sum_i x_{i,j}^* = 1, \forall j$.

Later we will study equilibrium concepts other than the competitive equilibrium. For convenience, we will refer to the allocation/pricing at the corresponding equilibrium as the *equilibrium allocation/pricing* if its meaning is unambiguous from the context. Otherwise we will explicitly associate the quantity with the equilibrium name, e.g., competitive equilibrium allocation/pricing.

Below we state some well-known results that will be frequently used:

- An equilibrium always exists. Equilibrium pricing p^* is unique. There may be multiple equilibrium allocations, but buyers' utilities are identical across them.
- Buyers always deplete their budgets at equilibrium.
- Given prices p, the **bang-per-buck** of item j for buyer i is $\frac{v_{i,j}}{p_j}$, the value acquired for each unit of money spent on buying item j. At equilibrium, every buyer only gets items with the *highest* bang-per-buck, i.e., if $x_{i,j} > 0$, then $\frac{v_{i,j}}{p_j} \geq \frac{v_{i,j'}}{p_{j'}}, \forall j'$. Note that this requirement will be relaxed in our price competition model.

3 Fixed Pricing Markets and Stability

Definition 1. *A* ***fixed pricing market*** *is defined over a linear Fisher market with a set of sellers, of which each seller k owns a disjoint subset J_k of items. Trade proceeds with two stages:*

(1, pricing and prioritizing) Each seller k sets a ***personalized price*** *$p_{i,j}$ for each buyer i and each item $j \in J_k$. Besides, each seller also chooses a total preorder $\geq_j$, called* ***priority****, over buyers for each item $j \in J_k$. A priority is* ***natural w.r.t. prices*** *p, if $i \geq_j i'$ whenever $p_{i,j} \geq p_{i',j}$. When we focus on* ***non-personalized*** *pricing, we mean the price (not priority) of an item should be equal across buyers. Given the pricing, call the total preorder $\geq_i$ that satisfies $j \geq_i j'$ whenever $\frac{v_{i,j}}{p_{i,j}} \geq \frac{v_{i,j'}}{p_{i,j'}}$ the* ***preference*** *of buyer i over items. We will also say an item* prioritizes *some buyer or a buyer* prefers *some item (over the other).*

(2, trading) Trade happens according to some allocation x, where $x_{i,j} \in [0,1]$ is the fraction of item j allocated to buyer i, and $\sum_i x_{i,j} \leq 1, \forall j$. The utility of each buyer is the total valuation it acquires, and the revenue of each seller is the total payment it receives.

Fixed pricing markets set the stage for the analysis of seller competition in this paper. The definition is mostly straightforward: sellers post prices and trade happens.[6] A natural priority represents the seller's *greedy* revenue-optimizing

[6] Our model does not only capture markets with posted prices. E.g., in online advertising markets with second price auction and auto-bidders, prices (and priorities) are determined in real time by both buyer competition (via second price) and the platform (via instruments such as boosts and reserve prices; see, e.g., [4]). In addition, auto-bidders using the multiplicative pacing strategy behave in precisely the same way as in our model, i.e., they buy items in decreasing order of bang-per-bucks until depleting budgets [11].

behavior to sell an item with a higher price. Preference represents a similar rationale on the buyer side. Later we will show that a natural priority may be myopic and sub-optimal at length. Therefore it is economically significant to allow arbitrary priority.

In Definition 1, we do not impose any condition on the trade outcome x. Below we will introduce a minimal requirement on x (given pricing p), and show that it is both economically and computationally reasonable.

Definition 2. *Given an allocation x, let $r_i = \min_{j:x_{i,j}>0} \frac{v_{i,j}}{p_{i,j}}$ be the minimum bang-per-buck among items bought by buyer i. x is* ***stable*** *for buyer i, if it satisfies: (1) budget-feasible: $\sum_j x_{i,j}p_{i,j} \leq B_i$; (2) if $x_{i,j} > 0$, then $v_{i,j} > 0$; (3) if $x_{i,j} < 1$, then either (3.a) the bang-per-buck of j is no larger than r_i and i's budget is already depleted, or (3.b) item j is fully allocated and $i' \geq_j i$ for every other buyer i' with $x_{i',j} > 0$. x is stable if it is stable for all buyers.*

Given prices, priorities and an allocation, a fraction of item j is said to be ***available*** *to buyer i, if it is either unsold, bought by buyer i itself, or bought by some buyer i' with $i >_j i'$ (i.e., $i \geq_j i'$ and $i' \not\geq_j i$). An item is available to buyer i, if a positive fraction of it is available to buyer i.*

From a single buyer's perspective, given its available items, the optimal bundle is the solution of a fractional knapsack problem, i.e., buying available items in decreasing order of bang-per-bucks until depleting its budget or buying all available items. Stability ensures that each buyer get its optimal bundle in hindsight. As for sellers, if an item is available to buyer i, its owner will happily sell (a fraction of) it to buyer i (by withholding the item from buyers with strictly lower priorities, if necessary). If an allocation is not stable, there would be a pair of buyer and seller who would like to deviate from it and make themselves strictly happier with respect to preference and priority, respectively. Therefore stability is a necessary economic requirement. Furthermore, a stable allocation always exists and is poly-time computable.

Theorem 1 (Existence and computation of stable allocation). *A stable allocation always exists and can be computed in strongly poly-time. If all inputs are rational numbers, there exists a poly-time computable stable allocation where all entries are also rational numbers.*

The algorithm and proof is given in Appendix A of the full version of our paper [22]. Stability does not always lead to a unique allocation. Nonetheless, the multiplicity captures an uncertain aspect of the market, into which we will probe further in later sections.

Definition 3. *Given prices p, an allocation x is called* ***compatible*** *for buyer i, if x is budget-feasible and each item j with $x_{i,j} > 0$ has a bang-per-buck that is weakly larger than the bang-per-buck of any item that is currently available to it. In other words, buyer i gets its optimal bundle over available items with its currently spent money. Allocation x is compatible, if it is compatible for all buyers.*

Lemma 1. *A compatible allocation is stable, if and only if, for each buyer i, either $B_i = \sum_j x_{i,j}p_{i,j}$, or it gets all items available to it.*

Compatibility characterizes a transient "stable" state by focusing only on the budgets already spent. It will be useful in the market analysis.

4 Non-personalized Pricing Markets with Natural Priorities

In this section, we study markets that are *fully non-personalized*: every buyer is treated equally with respect to both pricing and priority. A fraction of item j is available to buyer i if and only if it is either not allocated, or owned by buyer i itself. Instability only occurs when there exist a buyer i_0 and an item j_0 such that $v_{i_0,j_0} > 0, \sum_i x_{i,j_0} < 1$ and it holds that either (1) buyer i_0's remaining budget is not zero or (2) buyer i_0 strictly prefers j_0 to some other item j_1 with $x_{i_0,j_1} > 0$. Missing proofs in this section are given in Appendix B of the full version of our paper [22].

valuation	item 1 (seller 1)	item 2 (seller 2)
buyer 1 (budget 1)	1	1
buyer 2 (budget 1)	0	1

4.1 Seller-Side Optimality and Equilibrium Concept

We begin with the example depicted in the above table. If $p_1 = p_2 = 1$, it is easy to check that $x_{1,2} = 1, x_{1,1} = x_{2,1} = x_{2,2} = 0$ forms a stable allocation. However, it is seller-wise (weakly) Pareto-dominated in revenue by another stable allocation: $y_{1,1} = y_{2,2} = 1, y_{1,2} = y_{2,1} = 0$. This motivates us to wonder whether allocations such as x are less likely to occur in general.

To formalize and justify the above intuition, we have the following complexity result and Pareto-efficiency guarantee.[7]

Theorem 2 (One-sided tractability of revenue optimization). *A stable allocation with maximum revenue can be computed in strongly poly-time. On the other side, it is APX-hard to compute the minimum revenue over all stable allocations.*

Corollary 1 (Seller-side Pareto-optimality). *For any compatible allocation x, there is a revenue-optimal stable allocation y item-wise (thus seller-wise) covering x, i.e., $\sum_i x_{i,j} \leq \sum_i y_{i,j}, \forall j \in J$. Since stability implies compatibility, any revenue-optimal stable allocation is seller-wise Pareto-optimal (in allocation and revenue).*

[7] Corollary 1 is categorized as a corollary since it is a direct consequence of the *proof* of Theorem 2.

Corollary 1 shows that the set of market-wise revenue-optimal stable allocations is exactly the set of seller-side Pareto-optimal ones. The contrast between the hardness of minimizing revenue and the tractability of maximizing revenue further supports imposing revenue-optimality as a market requisite. Therefore it should be reasonable to restrict our attention to revenue-optimal stable allocation in fully non-personalized markets. Nonetheless, revenue-optimality is still unable to produce a unique allocation for any pricing, which hinders us from formulating a normal form game of price competition. To circumvent it, we adopt a maximin approach and define the following equilibrium concept to capture those states that we believe are the most likely to emerge as the market outcome.

Definition 4. *Given pricing p, let $R_k(p)$ be the minimum revenue received by seller k over all market-wise revenue-optimal stable allocations with pricing p. In a non-personalized pricing market with natural priorities, a* ***maximin equilibrium*** *(p, x) consists of a pricing p and an allocation x satisfying: (1) x is a revenue-optimal stable allocation with p; (2) for any seller k and any pricing p' that differs from p only on items owned by k, $R_k(p') \leq \sum_{i \in I, j \in J_k} p_j x_{i,j}$.*

At a non-maximin-equilibrium state, there will be a deviation for some seller that *guarantees* itself a strictly higher revenue. Our equilibrium concept eliminates these clearly unsteady outcomes. Though in reality a seller might attempt a possibly profitable (but non-guaranteed) deviation, the maximin equilibrium is economically meaningful and it includes many interesting market states as will be detailed below.

4.2 Equilibrium Analysis

By Corollary 1, any compatible allocation has a revenue-optimal stable allocation that item-wise covers it. To show a deviation cannot improve the deviator's worst-case revenue, one sufficient condition is to find a compatible allocation where buyers already spend too much on other sellers, ensuring the deviator cannot be better off even if all the unspent money goes to it. The above idea will be repeatedly used in this section to establish maximin equilibrium.

Theorem 3. *Any competitive equilibrium (p^*, x^*) is also a maximin equilibrium.*

Proof. Consider any unilateral deviated price vector p that differs from p^* for a single seller k. Let x^* be a competitive equilibrium allocation. Let H be the set of items with $p_j > p_j^*$. Construct x as follows:

$$x_{i,j} = \begin{cases} x_{i,j}^*, & \text{if } j \notin H; \\ 0, & \text{if } j \in H. \end{cases}$$

By the property of competitive equilibrium, items outside H are fully allocated in x, and if $x_{i,j}^* > 0$, then $\frac{p_{j'}^*}{v_{i,j'}} \geq \frac{p_j^*}{v_{i,j}}, \forall j'$. As a result, if $x_{i,j} > 0$, then $j \notin H$

and $\frac{p_{j'}}{v_{i,j'}} > \frac{p^*_{j'}}{v_{i,j'}} \geq \frac{p^*_j}{v_{i,j}} \geq \frac{p'_j}{v_{i,j}}, \forall j' \in H$. Therefore x is compatible, and by corollary 1, there exists a revenue-optimal stable allocation y such that $\sum_i x_{i,j} \leq \sum_i y_{i,j}, \forall j \in J$. This implies that items are fully sold in y for all sellers other than k, so k's worst-case revenue at p cannot exceed its competitive equilibrium revenue. □

Theorem 3 shows that, even if sellers have the price-setting power, as long as no personalization is allowed, the competitive equilibrium remains to be a steady-state. This strengthens the foundation of this classic equilibrium concept from a game-theoretic perspective of seller competition. Also note that the competitive equilibrium is ownership-oblivious, while in general the set of maximin equilibria is not a singleton and it changes as the ownership changes.

To explore further properties of the equilibria, consider the following representative example.

Example 1. The market consists of 2 buyers and 2 sellers. Each seller has a budget of 2, and the valuations and item-ownership are given in the following table.

	seller 1	seller 2	
	item 1	item 2	item 3
buyer 1 (budget 2)	2	1	0
buyer 2 (budget 2)	1/3	1	1

At the unique competitive equilibrium, $p_1^* = 2, p_2^* = p_3^* = 1$, buyer 1 gets item 1, and buyer 2 gets item 2 and 3.

Multiplicity of maximin equilibria. There is a maximin equilibrium with pricing: $p_1 = \frac{4}{3}, p_2 = \frac{2}{3}, p_3 = 2$. With this pricing, there is a unique revenue-optimal stable allocation, where buyer 1 gets item 1 and 2, while buyer 2 gets item 3. To see it is a maximin equilibrium, consider all the possible deviations as follows:

- Seller 2 strictly increases p_2. Then allocation x with $x_{1,1} = 1$ and $x_{i,j} = 0, \forall (i,j) \neq (1,1)$ is compatible, so seller 2 cannot increase worst-case revenue.
- p_2 does not increase. Allocation x with $x_{1,1} = x_{1,2} = 1$ and 0 otherwise is compatible. Note that we do not need to consider deviation of p_3.
- Seller 1 increases p_1. Allocation x with $x_{1,2} = x_{2,3} = 1$ and 0 otherwise is compatible.
- Seller 1 at most earns the price of item 1, so decreasing p_1 cannot increase revenue.

Market Non-clearance. From the above analysis we can see the price of item 3 will not bring extra incentive issues for both sellers, as long as it remains more attractive than item 1 for buyer 2. Therefore raising p_3 to 3 (with buyer 2 still depleting its budget on item 3) also leads to a maximin equilibrium. In this case, item 3 is not fully sold.

Importance of Overall Competitiveness. Intuitively, the above market non-clearing maximin equilibrium exists due to the fact that seller 1 is too weak to be competitive against seller 2 with respect to buyer 2 (even if $v_{2,1}$ is not very small compared to $v_{2,2} + v_{2,3}$). Actually 4/3 is the minimum revenue seller 1 would get from any maximin equilibrium. It can only be improved if $v_{2,1}$ is raised to at least 2/3 (with other valuations unchanged). In contrast, seller 2 is competitive enough with respect to buyer 1, so it can earn from the price competition. $v_{1,2}$ is actually *tight* in terms of the best revenue seller 2 could get at maximin equilibrium. In comparison, the competitive equilibrium remains the same if we raised $v_{2,1}$ to 1 or decreased $v_{1,2}$ to 0.

Ownership Awareness. Suppose item 3 is now owned by seller 1. Then the pricing $p_1 = \frac{4}{3}, p_2 = \frac{2}{3}, p_3 = 2$ ceases to be a maximin equilibrium. In this case, seller 2 could raise p_2 to 1 such that item 2 could remain as the most preferred item of buyer 2 and would always be fully sold at maximin equilibrium.

In summary, within a fully non-personalized market, there is generally some inherent uncertainty that cannot be resolved by just pricing and competition (e.g., due to user habits, brand awareness, etc.). The competitive equilibrium still forms a stable structure from a risk-averse point of view, though other (possibly less efficient) outcomes are also likely to emerge as a steady-state.

5 Non-personalized Pricing Markets with Arbitrary Priorities

In this section, we consider the personalization of just priorities. We will first show that, though it seems to be a minor modification, the market outcome immediately becomes more unpredictable as arbitrary priorities are introduced.

Theorem 4. *In non-personalized pricing markets with arbitrary priorities, for fixed pricing and priorities, the problem of finding the maximum revenue over stable allocations is APX-hard. Furthermore, for fixed pricing, it is APX-hard to find the maximum revenue over all stable allocations and all priorities.*

The proof of Theorem 4 shares the same key construction of the proof of Theorem 6. See the proof and remarks in Appendix C.2 of the full version of our paper [22] on how the construction therein can be adapted to prove Theorem 4.

Theorem 4 demonstrates two layers of intractability. First, combing with Theorem 2, we have that both maximizing and minimizing the market-wise revenue is intractable for fixed strategies of sellers. This makes it evident that the market outcome is relatively unpredictable. In addition, in markets like ride-hailing where prices change less frequently while priorities are more flexible, the second half of Theorem 4 shows that it is also hard to find the optimal priorities for both the whole market and individual sellers.

Moreover, even if we take the conservative maximin approach, the competitive equilibrium may cease to be a stationary state. To see this, consider Example 1 again. Recall that, at competitive equilibrium, $p_1^* = 2$, $p_2^* = p_3^* = 1$, buyer 1 gets item 1 and buyer 2 gets item 2 and 3. Priorities do not affect allocation here. Suppose seller 2 deviates to $p_2 = 2/3$, $p_3 = 2$ and strictly prioritizes buyer 1 over buyer 2 for item 2. Since buyer 1 now strictly prefers item 2 over item 1, it will buy item 2 in whole and deplete the rest of budget on item 1. Buyer 2 also prefers item 2 the most, but it is not prioritized to buy. So it could only buy item 3 at a higher price (which is still more attractive then the now unsold item 1). As a result, seller 2 can guarantee itself a higher revenue by deviating from the competitive equilibrium pricing. Note that in this example no allocation selection is involved.

6 Personalized Pricing Markets

The situation becomes even more complex with price personalization. We start with an interesting result that identifies a natural situation where the allocation is unique and thus exempted from the selection problem. Missing proofs in this section are given in Appendix C of the full version of our paper [22].

Theorem 5. *If items' priorities $\geq_j$ and buyers' preferences over items $\geq_i$ are all strict, and $\geq_j$ is natural, then there is a unique stable allocation.*

However, even with natural priorities, the market becomes unpredictable when ties exist.

Theorem 6. *The problem of finding the stable allocation with the maximum revenue is APX-hard, even if the priorities are natural. The APX-hardness of minimizing revenue is implied by Theorem 2.*

Note that ties can be easily broken by small changes, e.g., subsidies/coupons in e-commerce, boosts in auto-bidding, or an update of the machine learning model that matches buyers and items. Hence the hardness result indicates that these perturbations might bring a large overhaul of the trading outcome. The proof also shows that a natural priority may not be optimal, even if the personalized prices are all distinct from one another.

To make things worse, a market may admit no stationary state, regardless of the selection criterion, as shown by the following result.

Theorem 7. *Consider a market consisting of 2 buyers and 2 sellers. The valuations and item-ownership are given in Table 1.*

With any pricing of seller 2, seller 1 always has a response guaranteeing a revenue strictly larger than 2, regardless of allocation selection.

Proof. We analyze seller 1's best responses for different combinations of $p_{1,2}$ and $p_{2,2}$.

Table 1. A fixed pricing market without stationary state.

valuation	item 1 (seller 1)	item 2 (seller 2)
buyer 1 (budget 2)	1	2
buyer 2 (budget 2)	2	1

- $p_{1,2} > 0, p_{2,2} > 1$. Consider $p_{1,1} = \frac{1}{4}p_{1,2}$, and $p_{2,1} = 1 + p_{2,2}$. Since $p_{1,1} < \frac{1}{2}p_{1,2}$ and $p_{2,1} < 2p_{1,2}$, both buyers strictly prefer item 1, and thus for any stable allocation, either both buyers spend all budgets on item 1, or item 1 is fully sold. For the former case, seller 1's revenue is 4. Below assume item 1 is fully sold.
  - If $p_{1,1} \geq p_{2,1} > 2$, seller 1's revenue is at least $\min(p_{1,1}, p_{2,1}) > 2$.
  - If $p_{1,1} < p_{2,1}$, then seller 1 prefers buyer 2 and buyer 2 will spend all its budget on item 1. But buyer 2 cannot buy item 1 in whole, so a fraction of item 1 will be sold to buyer 1, and seller 1's total revenue is also strictly larger than 2.
- $p_{1,2} < 2, p_{2,2} \leq 1$. Then seller 2's revenue is strictly less than 2. Seller 1 can accommodate all the rest budgets simply by choosing $p_{1,1} = p_{2,1} = 4$.
- $p_{1,2} > 4, p_{2,2} \leq 1$. Choose $p_{1,1} = 2 + \epsilon$ for some small $\epsilon > 0$ such that $p_{1,1} < \frac{p_{1,2}}{2}$ and buyer 1 strictly prefers item 1 to item 2. Then buyer 1 will deplete its budget on item 1 but cannot buy it entirely, and buyer 2 will spend a positive amount on the remainder of item 1 since item 2 costs at most 1 dollar.
- $p_{1,2} \in (2, 4], p_{2,2} \leq 1$. Consider $p_{1,1} = \frac{1-\epsilon}{2} \cdot p_{1,2}$ and $p_{2,1} = \frac{2}{\epsilon}$. By theorem 5, the stable allocation is unique. Buyer 2 spends at most 1 dollar on item 2, so it spends at least 1 dollar on item 1 (it has higher priority) but buys at most a fraction of ϵ. Buyer 1 strictly prefers item 1, so the rest part of item 1 (excluding the fraction sold to buyer 2) must be sold to buyer 1, which generates a revenue of at least $\frac{(1-\epsilon)^2}{2} \cdot p_{1,2}$. With sufficiently small ϵ, we have $\frac{(1-\epsilon)^2}{2} \cdot p_{1,2} + 1 > 2$.
- $p_{1,2} = 2, p_{2,2} \leq 1$. Choose $p_{1,1} = \epsilon$ and $p_{2,1} = \frac{1}{\epsilon}$ for sufficiently small ϵ. By theorem 5, the stable allocation is unique. Buyer 1 strictly prefers item 1, buyer 2 strictly prefers item 2, but $p_{1,1}$ and $p_{2,2}$ are both smaller than 2, so both items will be fully sold in stable allocation. On the other hand, buyer 1 has top priority on item 2, buyer 2 has top priority on item 1, and $p_{1,2}$ and $p_{2,1}$ are both no less than 2, so both buyers will deplete their budgets in stable allocation. Furthermore, buyer 1 must spend some money on item 1 since item 1 cannot be fully sold to buyer 2. Therefore buyer 1 spends strictly less than 2 on item 2, and the revenue of seller 2 is also strictly less than 2. Since both buyers will deplete budgets, seller 1's revenue will be strictly larger than 2.

□

By symmetry, seller 2 always has a response guaranteeing a revenue strictly larger than 2, as well. Clearly the revenues of both sellers cannot exceed 2 simultaneously. Therefore there exists no equilibrium no matter how it is defined. From the proof we can also see how the prices can be personalized to benefit the seller.

6.1 Competitive Duopoly Pricing Markets

One contributor to the success of the deviation in Theorem 7 is that the opponent's pricing is always fixed. Consider the seemingly promising strategy for buyer 2: $p_{1,2} = 2, p_{2,2} \leq 1$ (which is similar to the competitive equilibrium pricing). After seller 1 pushes the bang-per-buck for buyer 1 high enough, some money of buyer 2 will be attracted from item 2 to item 1, and thus seller 2's revenue will decrease. However, seller 1 can always deplete buyer 2's remaining budget (after buying item 2) by pushing the bang-per-buck for buyer 2 low enough.

Suppose, however, that seller 1 does carry out the deviation. Seller 2 might not sit still, and there is a straightforward (not necessarily optimal) countermeasure: lowering the bang-per-buck for buyer 2 to a level that is slightly higher than seller 1. This will not change buyer preference, and seller 2 can at least earn some more money from buyer 1.

Motivated by the above observation, we attempt to get some glimpse into the possible steady states of a personalized market by incorporating a "competitive" feature into the model, such that buyer's preference is still based on each seller's reported prices, but trading is settled with the *lowest* bang-per-buck among all sellers. Note that this can be approximated arbitrarily closely through the process described above. To make it amenable, we also introduce *bang-per-buck pricing*[8] and only define this modified market for two sellers.

Definition 5. *A* ***competitive duopoly pricing market*** *is defined over a linear Fisher market with two sellers. The trade proceeds with three stages:*

(1, reporting bang-per-buck) Seller 1 and 2 each sets a bang-per-buck $a_i \geq 0$ *and* $b_i \geq 0$ *for each buyer i.*

(2, pricing and prioritizing) Buyer i's preference depends on which one of a_i *and* b_i *is smaller. Buyers have no preference over items owned by the same seller since the bang-per-bucks are equalized. Item j's price for buyer i is instead given by* $p_{i,j} = v_{i,j} \cdot \max(a_i, b_i)$*. Priorities should be natural w.r.t. p.*

(3, trading) Trade happens according to an allocation x that is stable w.r.t. the priorities and preferences defined above (instead of the clearing prices, which give the same bang-per-buck for all items).

Note that the market is still unpredictable in general, since the construction used in the proof of Theorem 6 implies that the problem is APX-hard even for the restricted case where (1) item's priorities are natural and strict among buyers with non-zero valuations; (2) items can be grouped into two sets, and every buyer is indifferent among items in the same group.

[8] It relates to auto-bidding in first price auctions markets [10], and it provides a weak form of optimality for the seller. See Appendix C.4 [22] for detailed discussion.

Below we do not bother with the allocation selection problem. Instead we focus only on a generalized version of the market in Table 1 where the selection rule will not be an issue. Then we could directly analyze the Nash equilibrium of the normal form game, i.e., the bang-per-buck profile (a, b) such that each seller is best responding to its opponent.

Table 2. The underlying market is identical to the one in Table 1, but the item ownership is different.

	seller 1		seller 2	
	item 1	item 2	item 3	item 4
buyer 1	s	$2t$	$2(1-t)$	$1-s$
buyer 2	$2s$	t	$1-t$	$2(1-s)$

Example 2. Consider the market depicted in Table 2 where each buyer has a budget of 2. The underlying Fisher market is essentially identical to the one in Table 1: one item, denoted as j_1, that buyer 1 values 2 dollars and buyer 2 values 1 dollar, and the other item j_2 that buyer 1 values 1 dollar and buyer 2 values 2 dollars. The difference lies in the ownership: seller 1 now owns a fraction t of j_1 and a fraction s of item j_2, while seller 2 owns the rest.

Proposition 1. *The competitive equilibrium bang-per-buck profile* $a_1 = a_2 = b_1 = b_2 = 1$ *forms a Nash equilibrium if and only if* $s, t \in [1/3, 2/3]$.

Intuitively, with $s, t \in [1/3, 2/3]$, the sellers are kind of *similar* to each other in both the volume and the quality of the items they provide. This makes the market competitive enough to reach an efficient state.

Proposition 2. *With* $s = 1$ *and* $t = 0$ *(the ownership in Table 1),* $a_2 = b_1 = \frac{4}{3}, a_1 = b_2 = 1$ *forms a Nash equilibrium. Both buyers will get half of both items at the Nash equilibrium.*

In this case, though with similar market sizes, each seller provides a distinctive product that attracts a different audience. Though the market clears, the allocation is not Pareto-optimal for buyers. In this particular case, seller competition hurts buyers. Sellers seem to get nothing from personalization, but if one of them did not do so, the opponent would gain the advantage.

7 Conclusion and Future Works

In this paper, we examine the price competition among sellers within the foundational model of linear Fisher market. We show that, given sellers' pricing (and priorities), the set of allocations satisfying the "stability" property constitutes the steady-states of the market. Stable allocations are both economically reasonable and computationally practical, but not unique, which represents an uncertain aspect of the market. We move on to studying markets with different degrees of personalization, which turn out to have quite distinct properties. If no buyer personalization is allowed, the competitive equilibrium remains to form a steady-state (the maximin equilibrium), albeit not the unique one. On the other hand, personalization could bring sellers more revenue, but the resulting market becomes more unpredictable, as evidenced by the two-sided hardness of maximizing/minimizing revenue and the non-existence of (static) equilibrium.

There are several interesting directions for future research. Most of our theorems (except for Theorem 7) hold regardless of the number of sellers, but the properties of various solution concepts are mainly examined via small instances containing two sellers. Though they provide some meaningful insights, it is important to explore these properties in a more general sense. In Sect. 6 we give much evidence to show that personalized markets are unpredictable. It is of great importance to explore what kinds of (possibly dynamic) states these markets might end up with (Sect. 6.1 serves as a preliminary attempt in this direction).

References

1. Adsul, B., Babu, C.S., Garg, J., Mehta, R., Sohoni, M.: Nash Equilibria in fisher market. In: Kontogiannis, S., Koutsoupias, E., Spirakis, P.G. (eds.) SAGT 2010. LNCS, vol. 6386, pp. 30–41. Springer, Heidelberg (2010). https://doi.org/10.1007/978-3-642-16170-4_4
2. Arrow, K.J., Debreu, G.: Existence of an equilibrium for a competitive economy. Econometrica: J. Econ. Soc. pp. 265–290 (1954)
3. Babaioff, M., Nisan, N., Talgam-Cohen, I.: Fair allocation through competitive equilibrium from generic incomes. In: FAT, p. 180 (2019)
4. Balseiro, S., Deng, Y., Mao, J., Mirrokni, V., Zuo, S.: Robust auction design in the auto-bidding world. Adv. Neural. Inf. Process. Syst. **34**, 17777–17788 (2021)
5. Brânzei, S., Chen, Y., Deng, X., Filos-Ratsikas, A., Frederiksen, S., Zhang, J.: The Fisher market game: Equilibrium and welfare. In: Proceedings of the AAAI Conference on Artificial Intelligence, vol. 28 (2014)
6. Budish, E.: The combinatorial assignment problem: approximate competitive equilibrium from equal incomes. J. Polit. Econ. **119**(6), 1061–1103 (2011)
7. Cole, R., et al.: Convex program duality, Fisher markets, and Nash social welfare. In: Proceedings of the 2017 ACM Conference on Economics and Computation, pp. 459–460 (2017)
8. Cole, R., Fleischer, L.: Fast-converging tatonnement algorithms for one-time and ongoing market problems. In: Proceedings of the fortieth Annual ACM Symposium on Theory of Computing, pp. 315–324 (2008)
9. Cole, R., Gkatzelis, V.: Approximating the Nash social welfare with indivisible items. In: Proceedings of the Forty-seventh Annual ACM Symposium on Theory of Computing, pp. 371–380 (2015)
10. Conitzer, V., et al.: Pacing equilibrium in first price auction markets. Manage. Sci. **68**(12), 8515–8535 (2022)
11. Conitzer, V., Kroer, C., Sodomka, E., Stier-Moses, N.E.: Multiplicative pacing equilibria in auction markets. Oper. Res. **70**(2), 963–989 (2022)

12. Devanur, N.R., Papadimitriou, C.H., Saberi, A., Vazirani, V.V.: Market equilibrium via a primal-dual algorithm for a convex program. J. ACM (JACM) **55**(5), 1–18 (2008)
13. Dixon, H.D.: Surfing economics. Surfing Econ. pp. 1–222 (2017)
14. Edmonds, J., Karp, R.M.: Theoretical improvements in algorithmic efficiency for network flow problems. J. ACM (JACM) **19**(2), 248–264 (1972)
15. Eisenberg, E., Gale, D.: Consensus of subjective probabilities: The pari-mutuel method. Ann. Math. Stat. **30**(1), 165–168 (1959)
16. Foley, D.: Resource allocation and the public sector. Yale Econ. Essays **7**, 45 (1967)
17. Gale, D., Shapley, L.S.: College admissions and the stability of marriage. Am. Math. Mon. **69**(1), 9–15 (1962)
18. Gao, Y., Kroer, C.: Infinite-dimensional Fisher markets: Equilibrium, duality and optimization. In: Proceedings of the AAAI Conference on Artificial Intelligence. vol. 35, pp. 5432–5439 (2021)
19. Kroer, C., Peysakhovich, A.: Scalable fair division for at most one preferences. arXiv preprint arXiv:1909.10925 (2019)
20. Kroer, C., Peysakhovich, A., Sodomka, E., Stier-Moses, N.E.: Computing large market equilibria using abstractions. In: Proceedings of the 2019 ACM Conference on Economics and Computation, pp. 745–746 (2019)
21. Kroer, C., Sinha, D., Zhang, X., Cheng, S., Zhou, Z.: Fair notification optimization: An auction approach. arXiv preprint arXiv:2302.04835 (2023)
22. Li, J., Tang, P.: Price competition in linear fisher markets: Stability, equilibrium and personalization. arXiv preprint arXiv:2407.11869 (2024)
23. McKenzie, L.: On equilibrium in Graham's model of world trade and other competitive systems. Econometrica: J. Econ. Soc. pp. 147–161 (1954)
24. Megiddo, N., Papadimitriou, C.H.: On total functions, existence theorems and computational complexity. Theoret. Comput. Sci. **81**(2), 317–324 (1991)
25. Orlin, J.B.: Improved algorithms for computing Fisher's market clearing prices: Computing Fisher's market clearing prices. In: Proceedings of the Forty-second ACM Symposium on Theory of Computing, pp. 291–300 (2010)
26. Paes Leme, R., Sivan, B., Teng, Y.: Why do competitive markets converge to first-price auctions? In: Proceedings of The Web Conference 2020, pp. 596–605 (2020)
27. Selcuk, C.: Auctions vs. fixed pricing: Competing for budget constrained buyers. Games Econ. Behav. **103**, 262–285 (2017)
28. Shmyrev, V.I.: An algorithm for finding equilibrium in the linear exchange model with fixed budgets. J. Appl. Ind. Math. **3**, 505–518 (2009)
29. Varian, H.R.: Equity, envy, and efficiency. J. Econ. Theor. **9**(1), 63–91 (1974)
30. Végh, L.A.: Strongly polynomial algorithm for a class of minimum-cost flow problems with separable convex objectives. In: Proceedings of the Forty-fourth Annual ACM Symposium on Theory of Computing, pp. 27–40 (2012)
31. Végh, L.A.: Concave generalized flows with applications to market equilibria. Math. Oper. Res. **39**(2), 573–596 (2014)
32. Zhang, L.: Proportional response dynamics in the Fisher market. Theoret. Comput. Sci. **412**(24), 2691–2698 (2011)

Proportionally Representative Clustering

Haris Aziz[1], Barton E. Lee[2], Sean Morota Chu[1], and Jeremy Vollen[1(✉)]

[1] UNSW Sydney, Sydney, Australia
{haris.aziz,j.vollen}@unsw.edu.au
[2] ETH Zurich, Zurich, Switzerland
bartonlee@ethz.ch

Abstract. In recent years, there has been a surge in effort to formalize notions of fairness in machine learning. We focus on *centroid clustering*—one of the fundamental tasks in unsupervised machine learning. We propose a new axiom "proportionally representative fairness" (PRF) that is designed for clustering problems where the selection of centroids reflects the distribution of data points and how tightly they are clustered together. Our fairness concept is not satisfied by existing fair clustering algorithms. We design efficient algorithms to achieve PRF both for unconstrained and discrete clustering problems. Our algorithm for the unconstrained setting is also the first known polynomial-time approximation algorithm for the well-studied Proportional Fairness (PF) axiom [13]. Our algorithm for the discrete setting also matches the best known approximation factor for PF.

Keywords: clustering · algorithmic fairness · computational social choice

1 Introduction

As artificial intelligence and machine learning serve as the 'new electricity' of systems, it is becoming increasingly important to ensure that AI systems embody societal values such as privacy, transparency, and fairness. Fairness is a growing concern, as AI systems are used to make decisions that can critically affect our daily lives, finances, and careers [16]. The impulse to tackle the issue of fairness in AI is prominently reflected in several governmental policy documents [3,17,22], the emergence of dedicated conferences on Ethical AI, and the formation of AI ethics boards by various big tech companies. Over the past years, several developments have taken place in the theory and application of fairness in supervised learning. One prominent approach is to use information derived from labelled data in supervised learning to formalize some notions of equity within and across (pre-specified and) protected attributes [19]. In contrast, in several models of unsupervised learning, protected attributes or their labelling are unknown. Fairness in unsupervised machine learning is becoming important with the growth of data collection both in public (e.g., via IoT devices) and private sectors (e.g.,

M. Mavronicolas et al. (Eds.): WINE 2024, LNCS 15534, pp. 155–171, 2026.
https://doi.org/10.1007/978-3-032-08560-3_9

map services and social media). Unfettered and irresponsible use of such data has the danger of increasing inequity [35].

We focus on fairness in clustering, which is one of the most widely applied tasks in unsupervised machine learning. A (centroid selection) clustering problem has a metric space $\mathcal{X}$ with a distance measure $d : \mathcal{X} \times \mathcal{X} \rightarrow \mathbb{R}^+ \cup \{0\}$, a multiset $\mathcal{N} \subseteq \mathcal{X}$ of n data points (agents), a set $\mathcal{M}$ of candidate centers, and positive integer $k \leq n$. In many clustering problems, each data point corresponds to an individual's attribute profile; thus, we refer to the data points as *agents*. The goal is to find a size-k subset of centers, $Y \subseteq \mathcal{M} : |Y| = k$. Such *centroid selection* problems can be widely applied to various applications including recommender systems and data-compression. The problem also captures social planning scenarios where k facility locations need to be chosen to serve a set of agents.

In such a (centroid selection) clustering problem, there are two versions of the problem. In the *discrete* version, the set of candidate centers $\mathcal{M}$ is a finite subset of $\mathcal{X}$. In the *unconstrained* or *continuous* setting, $\mathcal{M} = \mathcal{X}$. We first focus on the unconstrained setting as done by [34]. Later we discuss how our concepts and ideas extend to the discrete setting. For a location i and set of locations $S \subseteq \mathcal{M}$, we will denote $\min_{c \in S} d(i, c)$ as $d(i, S)$. Standard clustering solutions include k-center, k-medians, and k-means.[1] These standard objectives can be viewed as achieving some form of global welfare but are not well-suited for proportional representational and fairness concerns. The problems of computing outcomes optimising these objectives are also NP-hard [4,32].

We seek a suitable concept for clustering that captures general fairness principles such as non-discrimination, equality of opportunity, and equality of outcome. In many applications, the data points correspond to real individuals or their positions or opinions. Such individuals may expect the clustering outcome to capture natural fairness requirements. Previous axiomatic studies of clustering have measured fair representation by the distance from each data point to the nearest centroid. We initiate a proportional representation perspective on fairness in clustering. In capturing this desiderata axiomatically, we require a property in which representation guarantees depend on the number (or proportion) of data points and how "tightly" they are clustered together.

Our perspective is motivated by applications in which groups of individuals (to which the data points correspond) may expect to receive centroid representation proportional to their size. For instance, the centroids may be citizens selected to make representative decisions on behalf of the public, as is done in a jury or a sortition panel. Similarly, computing a proportionally fair clustering is also meaningful in facility location, where the location of the facility should depend on the density and the number of people it is serving. More generally, our axioms and algorithms are well-motivated in any setting in which one may want to take a representative sample from datapoints between which a distance

[1] The k-center solution outputs a subset $Y \in \arg\min_{W \subseteq \mathcal{M}, |W|=k} \max_{i \in \mathcal{N}} d(i, W)$. The k-medians solution outputs a subset $Y \in \arg\min_{W \subseteq \mathcal{M}, |W|=k} \sum_{i \in \mathcal{N}} d(i, W)$. The k-means solution outputs a subset $Y \in \arg\min_{W \subseteq \mathcal{M}, |W|=k} \sum_{i \in \mathcal{N}} d(i, W)^2$.

metric is well-defined. In this paper, we consider the following fundamental research question. *What is a meaningful concept of proportional fairness in centroid clustering? Is such a concept guaranteed to be achievable?*

Contributions. In contrast to traditional optimization objectives in clustering and centroid selection, we take an axiomatic fairness perspective. We first identify a potential shortcoming of some recently introduced concepts that aim to capture proportional fairness for clustering; in particular, we show how the concepts do not capture an intuitive and minimal requirement of proportional representation called *unanimous proportionality.* We then propose a new axiom called *proportional representative fairness (PRF)*, which is our central conceptual contribution.

We propose PRF for both the unconstrained and discrete clustering setting. We discuss how PRF overcomes some of the drawbacks of the previously introduced concepts. In particular, it implies the minimal requirement of unanimous proportionality. PRF has several other desirable features: a PRF outcome is guaranteed to exist, the axiom does not require a preset specification of protected groups and it is robust to outliers and multiplicative scaling. We show that none of the traditional clustering algorithms or previously introduced algorithms for fair clustering satisfy PRF.

We design polynomial-time algorithms to achieve PRF both in the unconstrained and discrete settings.[2] We call our algorithms *SEAR (Spatial Expanding Approval Rule)* as they are inspired by using core ideas from the ordinal multiwinner voting Expanding Approvals Rule (EAR) of [6] but carefully use spatial information to make synchronized decisions. We prove that our algorithm for the unconstrained setting also gives 3-approximate *Proportional Fairness (PF)* [13], and hence is the first known polynomial-time approximation algorithm for PF on general metric spaces. We also prove that our algorithm for the discrete setting matches the best known approximation of *Proportional Fairness (PF)*, thereby providing no compromise on PF approximation but additionally satisfying PRF. We summarize our theoretical contributions in Table 1 via a comparison of our algorithms with the best guarantees previously known. All missing proofs can be found in the full version of our paper [7].

2 Related Work

2.1 Fair Clustering

Fairness in machine learning and computer systems is a vast topic [25]—for an overview, see [16] or the survey by [33]. We focus on the fundamental learning problem of clustering, which is a well-studied problem in computer science,

[2] The results for unconstrained and discrete settings are incomparable in the following sense. The guarantee of existence of a fair outcome in the discrete setting guarantees the existence of a fair outcome in the unconstrained setting; however, a polynomial-time algorithm for computing a fair outcome in the discrete setting does not imply the same for the unconstrained setting.

Table 1. Theoretical comparison of our algorithms (Algorithm 1 for the unconstrained setting and Algorithm 2 for the discrete setting) with Greedy Capture (GC), which gives the best known approximate PF guarantees. Unconstrained and discrete GC results come from [34] and [13], respectively.

	Unconstrained			Discrete		
	ρ-PF	PRF	Complex.	ρ-PF	PRF	Complex.
Our Algorithms (SEAR)	**3**	✓	**in P**	$\approx$ **2.4**	✓	**in P**
Greedy Capture	≈ 2.4	-	NP-hard	≈ 2.4	-	in P

statistics, and operations research [23]. Our approach is axiomatic, focused on the concept of proportional fairness (or representation), and does not rely on the use of a pre-specified or protected set of attributes. Thus, the most closely related papers are those of [13,31,34], and [26], which we review below.

[13] proposed the concept of *proportional fairness* (PF) for the discrete setting, which is based on the entitlement of groups of agents of large enough size. They reasoned that PF is a desirable fairness concept but it is not guaranteed to be achievable for every instance. On the other hand, they showed that outcomes satisfying reasonable approximations of PF are guaranteed to exist and can be computed in polynomial time via the 'Greedy Capture' algorithm.

[34] analyzed the proportional fairness axiom in the unconstrained setting and presented similar results to those obtained by [13]. They showed that the natural adaptation of the Greedy Capture algorithm finds an approximately PF outcome in Euclidean space. On the other hand, they showed that checking whether a PF outcome exists is NP-hard and that the clustering returned by the 'greedy capture' algorithm cannot be computed in polynomial time unless P = NP. [31] consider the same discrete clustering setting as [13] and propose a stronger version of proportional fairness called *core fairness*. They present results on lower and upper approximation bounds for various restrictions of the discrete setting.

[26] consider the discrete clustering problem in which the set of agents and candidate centers coincide. For this setting, they propose a fairness concept that has a similar goal as that of [13]. In subsequent work, it has been referred to as *individual fairness* [11,37]. Individual fairness requires that, for each agent i, there is a selected center that is at most $r(i)$ from i, where $r(i)$ is the smallest radius around i that includes n/k agents.

In a paper inspired by our work, [28] analyze the key algorithms and axioms in our paper in further detail and show that the axioms that we propose also imply approximations of various fairness concepts introduced in the literature. Drawing an analogy to multiwinner voting axioms, they refer to our proposed axiom PRF for the discrete setting as mPJR and also define a weaker axiom which they call mJR. They take a deep dive into the logical relations between the axioms and approximate notions of proportional fairness, individual fairness, core fairness, and another core-inspired generalization of proportional fairness [20]. They prove that any approximation to PF is also an approximation to individual fairness and vice versa. In addition to the algorithms we formalize in this paper, they also consider Greedy Capture [13] and a randomized version of this algorithm known as Fair Greedy Capture [20]. A key takeaway from our paper is reaffirmed by their work, namely that our algorithms are the best-known algorithms for satisfying proportional representation axioms while also giving non-trivial approximation guarantees with respect to other fairness notions.

There are also several concepts and algorithms in clustering where the protected groups are pre-specified based on protected attributes [1,2,8,9,14,15,21,29]. In contrast, we are seeking fairness properties that apply to arbitrarily defined groups of all possible sizes. In our proposed axiom, these groups are *endogenously* determined from data points; in some real-world settings, this is a legal requirement since it can be forbidden by law to have decision-making processes use certain attributes as inputs. Clustering can also be viewed as a facility location problem [24], where multiple facilities (centers) are to be located in a metric space; for a recent overview, see the survey by [12].

2.2 Comparison with Multi-winner Voting

The clustering problem especially has natural connections with multi-winner voting in which agents and centers in the clustering context correspond to agents and candidates in multi-winner voting.

The literature on proportionally representative multi-winner voting includes axioms such as JR, PJR, EJR, and FJR for dichotomous preferences [5,30], axioms such as PSC for rankings [18], and axioms such as Generalised PSC [5,6] or Rank-PJR+ [10] for weak orders. Naively applying these axioms to our spatial problem results in a reduction from our unconstrained problem to a multiwinner voting problem with an infinite number of candidates. Conversely, our fairness (or proportional representation) axioms take into account the additional distance-based cardinal information available when the candidates belong to a general metric space. Hence, our axioms formalize a notion of proportional representation for m-dimensional Euclidean space for $m > 2$. In contrast to ordinal algorithms for multi-winner voting (such as the Expanding Approvals Rule (EAR) [6] or Single Transferable Vote), the algorithms in this paper carefully account for additional spatial information to make synchronized decisions.

[27] consider a similar setting to ours but explore to what extent algorithms that are agnostic to cardinal spatial information and only use ordinal ranking of pairwise distances can achieve approximate fairness. They define (α, γ)-proportional representation, a powerful strengthening of approximate core fairness [31]. Their notion captures (and strengthens) the ideal of core stability in this setting: no coalition of agents S large enough to "deserve" t centroids should be able to propose an alternative set of t centroids which they all prefer to their individually best size-t subset of the selected centroids. They then parameterize this idea by α, the required size of a coalition which deserves t centroids, and γ, the degree to which the alternative centroid selection better represents the

coalition. Similar to ours, their property effectively addresses the shortfalls of existing concepts with respect to proportional representation (see Definition 2). In contrast to theirs, the properties we study do not require parameterization and are guaranteed to exist. [27] show that the (ordinal multi-winner) Expanding Approvals Rule (EAR) of [6] is (α, γ)-representative with $\gamma \leq 1 + 6.71\alpha/(\alpha - 1)$ and also satisfies 5.71-approximation to proportional fairness. They are able to improve on these bounds with a modified version of Greedy Capture when cardinal information is known.

Proportionally representation in multi-winner voting (and more generally participatory budgeting) has also been studied when agents have cardinal utilities. Based on prior work for approvals [5,36] study desirable axioms such as FJR and EJR for positive, *additive* utilities, and rules such as the *Method of Equal Shares* (MES). We make no assumption of additive utilities, nor do we ever consider total or average distances. Under additive cardinal utilities, a group of agents of proportion large enough to "deserve" multiple candidates can (in some instances) be satisfied with respect to EJR by a single candidate which gives some agent very high utility. Because our model and axioms take an agnostic approach to cardinal utility values, we instead place requirements on the selection of a proportional number of candidates. One possible approach to our problem is to take distances as proxy for utilities and to apply a rule like MES. However, translating to cardinal utilities necessitates the loss of important spatial information as different cardinal utilities consistent with the distances give different MES outcomes. One possibility is to set utility to be negation of the distance. But the setting and axioms of Peters et al. assume *positive*, additive utilities.

3 Towards an Axiom Capturing Proportional Representation

Proportional fairness for clustering has been considered in a series of recent papers. [13] first proposed a proportional fairness axiom that requires that there is no set of agents of size at least n/k that can find an unselected center that is "better" (i.e., located closer than the closest selected center) for each of them.

Definition 1 Proportional Fairness *[13]). $X \subseteq \mathcal{M}$ with $|X| = k$ satisfies proportional fairness if $\forall S \subseteq \mathcal{N}$ with $|S| \geq \lceil \frac{n}{k} \rceil$ and for all $c \in \mathcal{M}$, there exists $i \in S$ with $d(i,c) \geq d(i,X)$.*

The idea of proportional fairness was further strengthened to core fairness that requires that there is no set of agents of size at least n/k that can demand a more preferred location for a center [31]. One rationale for proportional fairness is explained in the following example, which was first presented by [34] and then reused by [31]. We reuse the same example. Consider a set of data points/agents. The agents are divided into 11 subsets of clusters each of which is densely clustered. One cluster of agents has size 10,000. The other 10 clusters have sizes 100 each. The big cluster is very far from the smaller clusters. The small clusters are relatively near to each other. [34] and [31] are of the view that any set of centers that satisfies a reasonable notion of proportional fairness would place 10 centers for the big cluster and 1 center serving the small clusters. Next, we point out that proportional fairness and core fairness do not necessarily satisfy this minimal requirement, which we formalize via the *unanimous proportionality* axiom below.

Definition 2 (Unanimous Proportionality (UP)). *$X \subseteq \mathcal{M}$ with $|X| = k$ satisfies* unanimous proportionality *if $\forall S \subseteq \mathcal{N}$ with $|S| \geq \ell\lceil \frac{n}{k} \rceil$ and each agent in S has the same location $x \in \mathcal{X}$, then ℓ possible locations closest to x are selected.*

UP captures a principle which [13] highlighted as desirable—in their words, one which requires that *"any subset of at least $r\lceil n/k \rceil$ individuals is entitled to choose r centers."*[3] The following example shows that proportional fairness [13] does not imply Unanimous Proportionality. Similarly, it can be shown that the fairness concept proposed by [26] does not imply unanimous fairness.

Example 1 (Proportional fairness and core fairness). Suppose agents are located in the $[0, 1]$ interval and $k = 11$. 10,000 agents are at point 0 and 1000 agents are at point 1. UP requires that 10 centers are at or just around point 0 and 1 center is at point 1. However, placing 1 center at point 0 and 10 centers at point 1 satisfies the proportional fairness concept of [13] and also the core fairness concept of [31].

The example above shows that there is a need to formalize new concepts in order to capture proportional representative fairness. The property should imply unanimous proportionality but should also be meaningful if no two agents/data points completely coincide in their location. One reason why the existing fairness concepts do not capture UP is that they make an implicit assumption that an agent *only* cares about the distance from the nearest center and not on how many centers are nearby. As discussed in the introduction, this assumption is not necessarily accurate in contexts such as data abstraction, sortition, and political representation.

In addition to capturing proportional representation in a robust sense, we would also like our axiom (and algorithms) to guarantee the existing proportional fairness concepts, at least in an approximate fashion. Since a PF outcome may not exist, [13] and [34] studied an approximate notion of proportional fairness in the discrete and unconstrained settings, respectively.

Definition 3 ρ(-approximate Proportional Fairness). *$X \subseteq \mathcal{M}$ with $|X| = k$ satisfies ρ-approximate PF (or ρ-PF) if $\forall S \subseteq N$ with $|S| \geq \lceil \frac{n}{k} \rceil$ and for all $c \in \mathcal{M}$, there exists $i \in S$ with $\rho \cdot d(i, c) \geq d(i, X)$.*

[3] In the term unanimous proportionality, unanimous refers to the condition where a set of agents have the same location, and proportionality refers to the requirements that such agents get an appropriate number of centers in proportion to the number of agentss.

We seek to devise an axiom which is compatible with approximate PF, and subsequently to design algorithms which satisfy our axiom while also providing good approximate guarantees with respect to PF. We close the section with an example presented by [34] that shows that a PF outcome may not exist. We will use the same example to illustrate our new fairness concept in the next section.

Example 2 Consider Fig. 1. For the 6 agents and $k = 3$, it has been shown by [34] that a PF outcome may not exist.

Fig. 1. An example instance with 6 agents and $k = 3$ for which a PF outcome does not exist.

4 Fairness for the Unconstrained Setting

We propose a new concept *Proportionally Representative Fairness (PRF)* that captures proportional representation fairness concerns. The intuition behind the concept is based on two natural principles: (1) a set of agents that is large enough deserves to have a proportional number of centers 'near' it, and (2) the requirement of proximity of the nearby centers to a subset of agents should also depend on how densely together the particular subset of agents are. We use these two principles to propose a new fairness concept.

Definition 4 (Proportionally Representative Fairness (PRF) for Unconstrained Clustering). *An outcome $X \subseteq \mathcal{M}$ with $|X| = k$ satisfies* Proportionally Representative Fairness (PRF) *if the following holds. For any set of agents $S \subseteq \mathcal{N}$ of size at least $|S| \geq \ell n/k$ such that the maximum distance between pairs of agents in S is y, the following holds: $|c \in X : \exists i \in S \text{ s.t } d(i, c) \leq y| \geq \ell$.*

PRF has the following features: it implies unanimous proportionality; it is oblivious to and does not depend on the specification of protected groups or sensitive attributes; it is robust to outliers in the data, since the fairness requirements are for groups of points that are sizable enough; and it is scale invariant to multiplication of distances (i.e., PRF outcomes are preserved if all the distances are multiplied by a constant).

Next, let us illustrate an example where proportional fairness is not achievable but one of the most natural clustering outcomes satisfies PRF.

Example 3 (Requirements of PRF). We revisit Example 2 to illustrate the requirements that PRF imposes. For each set S of size at least $\ell\lceil \frac{n}{k} \rceil = 2\ell$, PRF requires ℓ centers in the relevant neighborhood of agents in S (see Fig. 2).

Therefore, for agent 1 and 2, PRF requires that one selected candidate center should be in one of the circles around 1 or 2. For the set of agents $\{1, 2, 3\}$, PRF also requires that one candidate center should be in one of the circles around 1, 2 or 3. A natural solution is to have one center located between agents 1, 2, and 3, another center located between agents 4, 5, and 6, and the final center located somewhere not too far from all agents (e.g., between the two groups of agents). Such solutions are intuitively reasonable and fair and, furthermore, satisfy PRF; however, as mentioned in Example 2, this instance does not admit a proportionally fair solution.

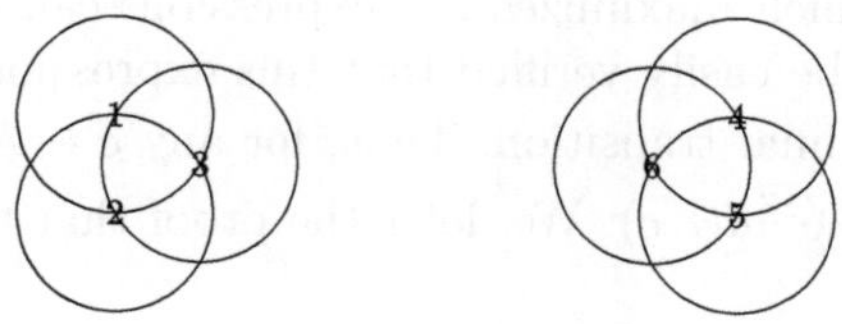

Fig. 2. Some of the requirements of PRF for instance in Example 2.

In addition to capturing natural outcomes when PF outcomes may not exist, PRF actually guarantees a ρ-PF outcome for $\rho \approx 3.6$ in the unconstrained setting on general metric spaces.

Proposition 1. *For metric spaces, if an outcome $X \subseteq \mathcal{M}$ with $|X| = k$ satisfies PRF for Unconstrained Clustering, then X is $\frac{3+\sqrt{17}}{2}$-approximate PF, and there exists an instance for which this bound is tight.*

Proof. Let X be an outcome which satisfies PRF for Unconstrained Clustering. Fix $S \subseteq N, c \in \mathcal{M}$ with $|S| \geq \lceil \frac{n}{k} \rceil$.

Let y be the maximum distance between pairs of agents in S, i.e., $y = \max_{i,j \in S} d(i,j)$, and let i_1 and i_2 be two agents in S that maximize this distance. We assume that $y > 0$; otherwise, if $y = 0$, then every agent in S is located at the same point, and there exists $x \in X$ such that $d(i,x) = 0 \leq d(i,c)$ for all $i \in S$. Denote by i^* the agent in S for which the distance to c is maximized. We have that $d(i^*,c) \geq \max(d(i_1,c), d(i_2,c)) \geq \frac{1}{2}(d(i_1,c) + d(i_2,c)) \geq \frac{y}{2}$.

Next, since $|S| \geq \frac{n}{k}$ and every pair of agents in S are a distance of at most y from each other, it follows from X satisfying Unconstrained PRF that there exist $x \in X, i \in S$ such that $d(i,x) \leq y$. Using this along with the triangle inequality, we can establish the following upper bound: $d(i^*,x) \leq d(i^*,c) + d(c,x) \leq d(i^*,c) + d(i,c) + y$. From these observations, the minimum PF approximation ratio obtained by i or i^* becomes

$$\begin{aligned}
&\min\left(\frac{d(i,x)}{d(i,c)}, \frac{d(i^*,x)}{d(i^*,c)}\right)\\
&\leq \min\left(\frac{y}{d(i,c)}, \frac{d(i^*,c)+d(i,c)+y}{d(i^*,c)}\right)\\
&\leq \min\left(\frac{y}{d(i,c)}, 3+\frac{2\cdot d(i,c)}{y}\right)\\
&\leq \max_{z\geq 0}(\min(z, 3+2/z)) = \frac{3+\sqrt{17}}{2}.
\end{aligned}$$

Note that the final inequality above holds since replacing $\frac{y}{d(i,c)}$ by the non-negative number z, which maximizes the expression, can only weakly increase the expression. It can be easily verified that this expression is maximized when $z = \frac{3+\sqrt{17}}{2}$, hence the final transition. Thus, for any $c \in \mathcal{M}$, there exists $i \in S$ such that $d(i, X) \leq \frac{3+\sqrt{17}}{2} d(i, c)$. We defer the proof that this bound is tight to the full version.

Next, we highlight that the well-known k-means or k-median solutions do not necessarily satisfy PRF. Example 4 is adapted from an example by [13].

Example 4. k-means does not satisfy PRF). Consider Fig. 3 in which there are $n/3$ agents uniformly distributed on the perimeter of each of the three circles. If $k = 3$, a k-means or k-median solution will place one center by the small circles and two centers in the big circle. In contrast, PRF requires that each of the circles gets its own centroid. It can also be shown that k-center does not satisfy PRF.

Fig. 3. The k-means solution may not satisfy PRF.

On face value, it is not clear whether an outcome satisfying PRF exists as similar concepts such as PF cannot always be guaranteed [34]. Secondly, even if a PRF outcome is guaranteed to exist, the complex constraints seem computationally challenging. PRF is a property that requires conditions on an exponential number of subsets of agents. For each subset of agents, it enforces representational constraints pertaining to an infinite number of neighborhood distances. Perhaps surprisingly, we show that not only is a PRF outcome guaranteed to exist, it can be computed in polynomial time.

The algorithm intuitively works as follows. Firstly, instead of considering infinite possible centers, we restrict our attention to the n candidates centers that coincide with each agent's location. Each agent is given an initial weight of 1. This weight dynamically decreases over the course of the algorithm. The algorithm can be viewed as gradually expanding the neighborhoods of each of the agents.[4] Instead of continuously expanding the neighborhood, we only consider at most n^2 possible values of neighborhood radii. If there exists some location in the candidate set such that it is in the intersection of the neighborhoods of a group of agents with total weight of at least n/k, we select one such candidate, and reduce the agents' collective weight by n/k.[5] The process is continued until k centers are chosen. The algorithm is formally specified as Algorithm 1. The following is the main result of the present section and summarizes the theoretical guarantees of Algorithm 1.

Theorem 1. *Algorithm 1 terminates in polynomial time $O(n^4k^2)$ and returns a set of k centers which satisfy PRF and 3-approximate PF.*

We will now formally prove two key lemmas used to prove Theorem 1.

Lemma 1. *Algorithm 1 returns an outcome that satisfies PRF.*

Proof. Suppose for contradiction that the algorithm finds a set of k centers W that violates PRF. In that case, there exist a set of agents $S \subseteq \mathcal{N}$ of size at least $|S| \geq \ell n/k$ such that the maximum distance between pairs of agents in S is y but the number of locations that are within y of some agent in S is at most $\ell - 1$. In that case, consider the snapshot of the algorithm when neighborhood distance $y = \max_{i,j \in S} d(i, j)$ is considered. At this point, $|c \in X : \exists i \in S \text{ s.t } d(i, c) \leq y| \leq \ell - 1$. Since agents in $\mathcal{N}$ have only used their weight towards selecting locations within y up to this point, it follows that $\sum_{i \in S} w_i \geq \ell\frac{n}{k} - (\ell-1)\frac{n}{k} = \frac{n}{k}$. Therefore, the agents in S still have total weight at least $\frac{n}{k}$ to select one more location that is a distance of at most y from some agent in S. Hence, at this stage, when the neighborhood distance is y, the algorithm would have selected at least one more center within distance y than in the outcome W.

Lemma 1 implies guaranteed existence of an outcome satisfying our axiom, which was one of our goals in devising PRF. In conjunction with Proposition 1,

[4] The idea of expanding neighborhoods is used in other algorithms in the literature, such as the Greedy Capture algorithm for the unconstrained setting studied by [34]. Unlike our algorithm, it does not involve reweighting of agents. More importantly, Greedy Capture does not satisfy PRF and is not polynomial time. Our algorithm uses reweighting of agents as is done by EAR for multi-winner voting [6]. Our algorithm can be viewed as using key ideas from the EAR for committee voting while taking into account additional spatial information to make synchronized decisions.

[5] We point out that the reweighting process can be done in any manner that results in the group's total weight reducing by n/k. Two natural possibilities would be (i) reducing weights as equally as possible, or (ii) reducing weights proportional to agent distances from the selected candidate.

Algorithm 1 SEAR (Spatial Expanding Approval Rule) for Unconstrained Clustering

Input: metric space $\mathcal{X}$ with a distance measure $d : \mathcal{X} \times \mathcal{X} \to \mathbb{R}^+ \cup \{0\}$, a finite multiset $\mathcal{N} \subseteq \mathcal{X}$ of n data points (agents), and positive integer k.
Output: A multiset of k centers
$w_i \longleftarrow 1$ for each $i \in \mathcal{N}$
Consider the set $D = \{d(i, i') \mid i, i' \in N\}$. Order the entries in D as $d_1 \leq d_2 \leq \cdots d_{|D|}$.
$\mathcal{M} \longleftarrow \mathcal{N}$
$j \longleftarrow 1$
$W \longleftarrow \emptyset$
while $|W| < k$ **do**
  $C^* = \{c \in \mathcal{M} \mid \sum_{i \in \mathcal{N} | d(i,c) \leq d_j} w_i \geq n/k\}$
  if $C^* = \emptyset$ **then**
    $j \longleftarrow j + 1$
  else
    Select some candidate c^* from C^* such that $c^* = \arg\max_{c' \in C^*} \sum_{i \in \mathcal{N} | d(i,c') \leq d_j} w_i$:
    $W \longleftarrow W \cup \{c^*\}$; $\mathcal{M} \longleftarrow \mathcal{M} \setminus \{c^*\}$
    $N' \longleftarrow \{i \in \mathcal{N} : d(i, c^*) \leq d_j\}$
    Modify the weights of agents in N' so the total weight of agents in N', i.e., $\sum_{i \in N'} w_i$, decreases by exactly n/k.
Return W

the lemma also gives us that the outcome returned by Algorithm 1 is $\frac{1}{2}(3 + \sqrt{17})$-PF. This is noteworthy as it establishes Algorithm 1 as the first known polynomial-time approximation algorithm for PF on general metric spaces in the unconstrained setting. As we will show next, Algorithm 1 actually provides the stronger guarantee of 3-PF.

Lemma 2. *Algorithm 1 finds an outcome that satisfies* 3*-approximate PF.*

Proof. For an instance of unconstrained clustering, let X be the outcome returned by Algorithm 1. Fix $S \subseteq N, c \in \mathcal{M}$ with $|S| \geq \lceil \frac{n}{k} \rceil$. We will show that there exists an agent $i \in S$ and center $x \in X$ such that $d(i, x) \leq 3d(i, c)$. We assume that $S \cap X = \emptyset$, since otherwise $d(i, x) = 0$ for some agent and center, and we are done.

At the start of the algorithm, the collective weight of agents in S is $|S|$. When the algorithm terminates, the collective weight of S is zero. Thus, in some iteration, the collective weight of S decreases for the first time. We denote the centroid selected in this iteration by x^* and denote an arbitrary agent whose weight is decreased in this iteration by i^*. Let y be the maximum distance between i^* and any other agent in S.

We first point out that $y \geq d(i^*, x^*)$. To see this, suppose for contradiction $y < d(i^*, x^*)$. Since y and $d(i^*, x^*)$ are both distances between pairs of agents in N, $y = d_j$ and $d(i^*, x^*) = d_{j'}$ for some j and j'. Furthermore, $j < j'$, since

D is in increasing order and by assumption that $y < d(i^*, x^*)$. Thus, when the algorithm considers neighborhoods of $d_j = y$, the collective weight of S has not yet decreased, and we have that

$$\sum_{i \in \mathcal{N} | d(i, i^*) \leq y} w_i \geq |S| \geq \lceil \frac{n}{k} \rceil \geq \frac{n}{k}.$$

Therefore, before transitioning to the next element in D, the algorithm will either add i^* to X or the weight of some agent in S will decrease. The first possibility contradicts $S \cap X = \emptyset$. For the second possibility, by assumption, it must be that x^* was added to X. But, since $d(i^*, x^*) > y$, $i^* \notin N'$ and w_{i^*} does not decrease, which gives us a contradiction.

We divide the remainder of the argument into cases. First, consider the case in which $d(i^*, c) \geq \frac{y}{3}$. It follows that $d(i^*, x^*) \leq y \leq 3d(i^*, c)$. If, instead, $d(i^*, c) < \frac{y}{3}$, then consider the agent $j \in S \setminus \{i\}$ such that $d(i^*, j) = y$. By the triangle inequality, we have that

$$\begin{aligned} d(j, c) &\geq y - d(i^*, c) \geq \frac{2y}{3} \\ &\geq \frac{1}{3}(d(j, i^*) + d(i^*, x^*)) \geq \frac{1}{3}d(j, x^*). \end{aligned}$$

□

5 Fairness for the Discrete Setting

We presented PRF as a desirable concept for the unconstrained setting. It is not immediate how to adapt the concept to the discrete setting. Applying the unconstrained PRF definition to the discrete setting leads to the issues of non-existence. On the other hand, some natural attempts to account for the coarseness of the candidate space, $\mathcal{M}$, can lead to a concept that is too weak. For example, one alternative is to require that an outcome X be such that if, for any set of agents $S \subseteq \mathcal{N}$ of size at least $|S| \geq \ell n/k$ such that the maximum distance between pairs of agents in S is y, the following holds: $|c \in X : \exists i \in S \text{ s.t } d(i, c) \leq y| \geq \min\{\ell, |\cup_{j \in S} B_y(j) \cap \mathcal{M}|\}$, where $B_r(i)$ denote a ball of radius r around agent i. Alternatively, we could replace the right-hand side of the final inequality with $\min\{\ell, |\cap_{i \in S} B_y(i) \cap \mathcal{M}|\}$. Both of these versions are too weak. To see this, suppose $k = 1$, all agents are located at 0, and $\mathcal{M} = (\ldots, -2, -1, 1, 2, 3, \ldots)$, then neither version places any restriction on the facility location and do not even imply UP.

To resolve these issues, we need to take into account that the nearest candidate locations may be very far from a subset of agents. We adapt our PRF concept for the discrete setting via a careful adaptation of the PRF concept for the unconstrained setting.

Definition 5. (Proportionally Representative Fairness (PRF) for Discrete Clustering). *For any set of agents $S \subseteq \mathcal{N}$ of size at least $|S| \geq \ell n/k$, if there are $\ell' \leq \ell$ candidates from $\mathcal{M}$ that are at distance at most y from all agents in S, the following holds: $|c \in X : \exists i \in S \text{ s.t } d(i, c) \leq y| \geq \ell'$.*

Algorithm 2 SEAR (Spatial Expanding Approval Rule) for Discrete Clustering

Input: metric space $\mathcal{X}$ with a distance measure $d : \mathcal{X} \times \mathcal{X} \rightarrow \mathbb{R}^+ \cup \{0\}$, a finite multiset $\mathcal{N} \subseteq \mathcal{X}$ of n data points (agents), a finite set of candidate locations $\mathcal{M}$, and positive integer k.
Output: A multiset of k centers
$w_i \longleftarrow 1$ for each $i \in \mathcal{N}$
Consider the set $D = \{d(i, c) \mid i \in \mathcal{N}, c \in \mathcal{M}\}$. Order the entries in D as $d_1 \leq d_2 \leq \cdots \leq d_{|D|}$.
$j \longleftarrow 1$
$W \longleftarrow \emptyset$
while $|W| < k$ **do**
  $C^* = \{c \in \mathcal{M} \mid \sum_{i \in \mathcal{N} | d(i,c) \leq d_j} w_i \geq n/k\}$
  if $C^* = \emptyset$ **then**
    $j \longleftarrow j + 1$
  else
    Select some candidate c^* from C^* such that $c^* = \arg\max_{c' \in C^*} \sum_{i \in \mathcal{N} | d(i,c') \leq d_j} w_i$:
    $W \longleftarrow W \cup \{c^*\}$; $\mathcal{M} \longleftarrow \mathcal{M} \setminus \{c^*\}$
    $N' \longleftarrow \{i \in \mathcal{N} : d(i, c^*) \leq d_j\}$
    Modify the weights of agents in N' so the total weight of agents in N', i.e., $\sum_{i \in N'} w_i$, decreases by exactly n/k.
Return W

We begin by highlighting the fact that, in the discrete setting, PRF implies $(1 + \sqrt{2})$-approximate PF, the closest PF approximation known to exist in the discrete setting [13]. Our proof follows similarly to Theorem 1 from [13].

Proposition 2. *For metric spaces, if an outcome $X \subseteq \mathcal{M}$ with $|X| = k$ satisfies PRF for Discrete Clustering, then X is $(1 + \sqrt{2})$-approximate PF.*

For the discrete setting, we propose a new algorithm (Algorithm 2). Algorithm 2 is similar to Algorithm 1, which was designed for the unconstrained setting. In fact, Algorithm 1 can be viewed as first setting $\mathcal{M}$ to the multiset of candidate locations corresponding to agents in $\mathcal{N}$ and then running Algorithm 2. Similar to Algorithm 1, Algorithm 2 terminates in polynomial time and returns k centers and its output satisfies PRF.

Theorem 2. *Algorithm 1 terminates in polynomial time $O(|\mathcal{M}|^2 n^2)$ and returns a set of k centers which satisfy PRF and $(1 + \sqrt{2})$-approximate PF.*

Notably, by satisfying PRF, Algorithm 2 matches the best known PF approximation factor in the discrete setting. As the following remark highlights, however, the existing algorithms do not capture PRF.

Remark 1. Chen et al. [13] present Greedy Capture. An equivalent algorithm, called ALG_g, is also presented by Li et al. [31]. However, these algorithms do not satisfy PRF. In fact, for given k, these algorithms may fail to output k candidate locations.[6]

In the full version of our paper, we experimentally compare Algorithm 2 to ALG_g and a variant of k-means. We show that it performs especially well when the mean squared distance of the agents is calculated based on their distance from not only the closest center but also the next j-closest centers (for large j). This provides another perspective indicating that, in situations where a "fair" clustering should account for an agent's distance from more than just her closest center, our algorithm is better suited than the existing alternatives. We also consider two natural and stronger notions of PRF in the discrete setting (that also apply to the continuous setting). We show that the versions that we consider do not guarantee the existence of a fair outcome.

6 Discussion

We revisited a classical AI problem with a novel solution concept that builds on ideas from social choice and fair division. We proposed a natural fairness concept called PRF for clustering. PRF is guaranteed to exist and has desirable features: it implies unanimous proportionality, does not require the specification of protected groups, and is robust to outliers and multiplicative scaling. Even if our only focus is on the PF axiom, we make significant progress on efficient algorithms for approximate PF for general metric spaces. We did not focus on strategic aspects. Unfortunately, PRF and PF are incompatible in general with the incentive of an agent to report their location truthfully (see full version). A natural next step is to identify stronger versions of PRF that still guarantee existence. It will be interesting to consider refinements of our algorithms which make more clever choices when selecting candidates with sufficient weighted support. Another direction for future research is to understand how much impact the constraint of PRF imposes on standard optimization objectives such as for k-means, k-median, or k-center.

Acknowledgments. Haris Aziz is supported by the NSF-CSIRO project on"Fair Sequential Collective Decision-Making". We thank John Dickerson and Bo Li for helpful comments.

[6] To see this, let $k = 3$ and let $\boldsymbol{x} = (0, 0, 1)$. Chen et al.'s Greedy Capture and Li et al.'s ALG_g will select candidate locations 0 and 1, but will not output a third location. Of course, this issue could be rectified by arbitrarily choosing a third candidate location—but then there is no guarantee that the set of locations would satisfy PRF.

References

1. Abbasi, M., Bhaskara, A., Venkatasubramanian, S.: Fair clustering via equitable group representations. In: Proceedings of the 2021 ACM Conference on Fairness, Accountability, and Transparency, pp. 504–514, FAccT '21, Association for Computing Machinery, New York, NY, USA (2021)
2. Ahmadian, S., Epasto, A., Kumar, R., Mahdian, M.: Clustering without over-representation. In: Proceedings of the 25th ACM SIGKDD International Conference on Knowledge Discovery & Data Mining, pp. 267–275, ACM (2019)
3. AI-Committee: The national artificial intelligence research and development strategic plan: 2019 update. Technical Report, National Science and Technology Council, Washington, D.C. (2019)
4. Aloise, D., Deshpande, A., Hansen, P., Popat, P.: Np-hardness of Euclidean sum-of-squares clustering. Mach. Learn. **75**(2), 245–248 (2009)
5. Aziz, H., Brill, M., Conitzer, V., Elkind, E., Freeman, R., Walsh, T.: Justified representation in approval-based committee voting. Soc. Choice Welfare **48**(2), 461–485 (2017). https://doi.org/10.1007/s00355-016-1019-3
6. Aziz, H., Lee, B.: The expanding approvals rule: improving proportional representation and monotonicity. Soc. Choice Welfare **54**(1), 1–45 (2020)
7. Aziz, H., Lee, B.E., Chu, S.M., Vollen, J.: Proportionally representative clustering. arXiv preprint arXiv:2304.13917 (2023)
8. Backurs, A., Indyk, P., Onak, K., Schieber, B., Vakilian, A., Wagner, T.: Scalable fair clustering. In: Proceedings of the 36th International Conference on Machine Learning, ICML, Proceedings of Machine Learning Research, vol. 97, pp. 405–413, PMLR (2019)
9. Bera, S.K., Chakrabarty, D., Flores, N., Negahbani, M.: Fair algorithms for clustering. In: Advances in Neural Information Processing Systems 32: Annual Conference on Neural Information Processing Systems 2019, NeurIPS 2019, pp. 4955–4966 (2019)
10. Brill, M., Peters, J.: Robust and verifiable proportionality axioms for multiwinner voting. In: EC '23: The 24th ACM Conference on Economics and Computation, Boulder, CO, USA, July 11 - 15, 2022, ACM (2023)
11. Chakrabarty, D., Negahbani, M.: Better algorithms for individually fair k-clustering. CoRR **abs/2106.12150** (2021)
12. Chan, H., Filos-Ratsikas, A., Li, B., Li, M., Wang, C.: Mechanism design for facility location problems: A survey. In: Proceedings of the 30th International Joint Conference on Artificial Intelligence (IJCAI) (2021)
13. Chen, X., Fain, B., Lyu, L., Munagala, K.: Proportionally fair clustering. In: Proceedings of the 36th International Conference on Machine Learning, ICML, Proceedings of Machine Learning Research, vol. 97, pp. 1032–1041, PMLR (2019)
14. Chierichetti, F., Kumar, R., Lattanzi, S., Vassilvitskii, S.: Fair clustering through fairlets. In: Proceedings of the 31st Annual Conference on Neural Information Processing Systems (NeurIPS), pp. 5029–5037 (2017)
15. Chiplunkar, A., Kale, S., Ramamoorthy, S.N.: How to solve fair k-center in massive data models. In: Proceedings of the 37th International Conference on Machine Learning, ICML 2020, vol. 119, pp. 1877–1886, PMLR (2020)
16. Chouldechova, A., Roth, A.: A snapshot of the frontiers of fairness in machine learning. Commun. ACM **63**(5), 82–89 (2020)
17. Congressional-Research-Service: Artificial intelligence: Background, selected issues, and policy considerations. Technical Report, Congressional Research Service Reports (2021)

18. Dummett, M.: Voting Procedures. Oxford University Press (1984)
19. Dwork, C., Hardt, M., Pitassi, T., Reingold, O., Zemel, R.S.: Fairness through awareness. In: Innovations in Theoretical Computer Science 2012, Cambridge, MA, USA, January 8-10, 2012, pp. 214–226, ACM (2012)
20. Ebadian, S., Micha, E.: Boosting sortition via proportional representation. arXiv preprint arXiv:2406.00913 (2024)
21. Esmaeili, S.A., Brubach, B., Srinivasan, A., Dickerson, J.P.: Fair clustering under a bounded cost. In: Proceedings of the 35th Annual Conference on Neural Information Processing Systems (NeurIPS) (2021)
22. European Commission: White paper on artificial intelligence: a european approach to excellence and trust. White Paper COM(2020) 65 final, European Commission, Brussels (2020)
23. Everitt, B.S., Landau, S., Leese, M.: Cluster Analysis. Wiley Publishing, 4th edn. (2009)
24. Farahani, R.Z., Hekmatfar, M.: Facility location: concepts, models, algorithms and case studies. Springer Science & Business Media (2009)
25. Friedman, B., Nissenbaum, H.: Bias in computer systems. ACM Trans. Inf. Syst. **14**(3), 330–347 (1996)
26. Jung, C., Kannan, S., Lutz, N.: Service in your neighborhood: Fairness in center location. In: 1st Symposium on Foundations of Responsible Computing, FORC 2020, June 1-3, 2020, Harvard University, Cambridge, MA, USA (virtual conference), LIPIcs, vol. 156, pp. 5:1–5:15 (2020)
27. Kalayci, Y.H., Kempe, D., Kher, V.: Proportional representation in metric spaces and low-distortion committee selection. In: Thirty-Eighth AAAI Conference on Artificial Intelligence, AAAI 2024, pp. 9815–9823, AAAI Press (2024)
28. Kellerhals, L., Peters, J.: Proportional fairness in clustering: A social choice perspective. CoRR **abs/2310.18162** (2023)
29. Kleindessner, M., Samadi, S., Awasthi, P., Morgenstern, J.: Guarantees for spectral clustering with fairness constraints. In: Proceedings of the 36th International Conference on Machine Learning (ICML), vol. 97, pp. 3458–3467, PMLR (2019)
30. Lackner, M., Skowron, P.: Multi-Winner Voting with Approval Preferences. Springer Briefs in Intelligent Systems, Springer (2023)
31. Li, B., Li, L., Sun, A., Wang, C., Wang, Y.: Approximate group fairness for clustering. In: Proceedings of the 38th International Conference on Machine Learning, Proceedings of Machine Learning Research, vol. 139, pp. 6381–6391, PMLR (2021)
32. Megiddo, N., Supowit, K.J.: On the complexity of some common geometric location problems. SIAM J. Comput. **13**(1), 182–196 (1984)
33. Mehrabi, N., Morstatter, F., Saxena, N., Lerman, K., Galstyan, A.: A survey on bias and fairness in machine learning. ACM Comput. Surv. **54**(6) (2021)
34. Micha, E., Shah, N.: Proportionally fair clustering revisited. In: 47th International Colloquium on Automata, Languages, and Programming, ICALP 2020, vol. 168, pp. 85:1–85:16 (2020)
35. O'Neil, C.: Weapons of Math Destruction: How Big Data Increases Inequality and Threatens Democracy. Crown Publishing Group, USA (2016)
36. Peters, D., Pierczyński, G., Skowron, P.: Proportional participatory budgeting with additive utilities. In: Proceedings of the 35th Conference on Neural Information Processing Systems (NeurIPS), pp. 12726–12737 (2021)
37. Vakilian, A., Yalçiner, M.: Improved approximation algorithms for individually fair clustering. CoRR **abs/2106.14043** (2021)

The Fairness of Maximum Nash Social Welfare Under Matroid Constraints and Beyond

Yuanyuan Wang, Xin Chen(✉), and Qingqin Nong

Ocean University of China, Qingdao 266100, China
wyy8088@stu.ouc.edu.cn, {chenxin1403,qqnong}@ouc.edu.cn

Abstract. We study the problem of fair allocation of a set of indivisible items among agents with additive valuations, under matroid constraints and two generalizations: p-extendible system and independence system constraints. The objective is to find fair and efficient allocations in which the subset of items assigned to every agent satisfies the given constraint. We focus on a common fairness notion of envy-freeness up to one item (EF1) and a well-known efficient (and fair) notion of the maximum Nash social welfare (Max-NSW). By using properties of matroids, we demonstrate that the Max-NSW allocation, implying Pareto optimality (PO), achieves a tight 1/2-EF1 under matroid constraints. This result resolves an open question proposed in prior literature [26]. In particular, if agents have 2-valued ($\{1, a\}$) valuations, we prove that the Max-NSW allocation admits $\max\{1/a^2, 1/2\}$-EF1 and PO. Under strongly p-extendible system constraints, we show that the Max-NSW allocation guarantees $\max\{1/p, 1/4\}$-EF1 and PO for identical binary valuations. Indeed, the approximation of 1/4 is the ratio for independence system constraints and additive valuations. Additionally, for lexicographic preferences, we study possibly feasible allocations other than Max-NSW admitting exactly EF1 and PO under the above constraints.

Keywords: Nash social welfare · EF1 · PO · Matroids · Independence systems

1 Introduction

Fair allocation of indivisible items among agents is a highly active problem in computational economics and artificial intelligence, due to its growing applications, e.g., Public Housing [10], Courses Allocation [6], the website of Spliddit (spliddit.org) [14], and the company Fair Outcomes (fairoutcomes.com). Central to this problem falls into two categories: fairness and efficiency. To capture

This work is supported in part by the National Natural Science Foundation of China (Nos. 12171444, 12301418, 12471306) and Natural Science Foundation of Shandong (No. ZR2022QA014).

M. Mavronicolas et al. (Eds.): WINE 2024, LNCS 15534, pp. 172–189, 2026.
https://doi.org/10.1007/978-3-032-08560-3_10

fairness, envy-freeness (EF, [13]) is a compelling criterion: every agent prefers her own bundle to another one's allocated bundle. Unfortunately, an EF allocation does not always exist even for allocating one indivisible item between two agents. This motivates a series of less stringent notions of envy-freeness, e.g., envy-freeness up to one item (EF1). For general monotone valuations, an EF1 allocation can be computed in polynomial time [21]. For finding efficient EF1 allocations, we are interested in a well-known criterion of Nash social welfare (NSW) that calculates the product of agents' utilities. An allocation maximizing the Nash social welfare (Max-NSW) guarantees EF1 [8] and PO.

While allocating indivisible items, we are more interested in relevant constraints, such as, cardinality constraints [4], budget constraints [27], scheduling constraints [20]) and conflicting constraints [17]. These categories of constraints can be formulated as special cases of independence system constraints. An independence system is a system $(E, \mathcal{F})$, where E is a finite item set and $\mathcal{F}$ is a collection of (independent) subsets of E that has hereditary property: if $D \in \mathcal{F}$ and $C \subseteq D$ then $C \in \mathcal{F}$. As a powerful abstraction of independence, matroid structures [11,15,16,22] have played a prominent role in combinatorial optimization. An independence system $(M, \mathcal{F})$ is a matroid if $\mathcal{F}$ also has augmentation property: for $C, D \in \mathcal{F}$ with $|C| < |D|$, there is an item $x \in D \setminus C$ such that $C + x \in \mathcal{F}$[1]. In particular, cardinality constraints are equivalent to partition constraints [4].

Due to the flexible properties of matroid (or independence system) structures, it is challenging to explore fair and efficient allocations under matroid constraints (or more general independence system constraints). Indeed, Suksompong [26] illustrated that the existence problem of EF1 and PO allocation remained open under matroids constraints (in Sect. 4). In this paper, we are interested in making efforts towards this direction by Max-NSW allocations, approximate EF1 allocations, properties of matroids and other effective techniques.

1.1 Our Contribution

We have strived to examine the existence of Max-NSW and approximate EF1 allocations under three prominent constraints: matroids, p-extendible systems, and independence systems (the relationships between the three can be seen in Fig. 1). Throughout, we select a specific Max-NSW allocation guaranteeing Pareto Optimality (PO) under all three classes of constraints. The main contributions in this paper are summarized in Table 1.

Matroid Constraints (Sect. 3). A system $(M, \mathcal{F})$ is a matroid if M is a finite item set, and $\mathcal{F}$ satisfies that i) hereditary property: if $D \in \mathcal{F}$ and $C \subseteq D$, then $C \in \mathcal{F}$; ii) augmentation property: if $C, D \in \mathcal{F}$ and $|D| > |C|$, then there is an item $x \in D \setminus C$ such that $C + x \in \mathcal{F}$. For general additive valuations, we illustrate that every Max-NSW allocation achieves 1/2-EF1 under matroid constraints. This approximation ratio cannot be improved since we exemplify an

[1] The notation $C + x$ means $C \cup \{x\}$, likewise $C - x$ means $C \setminus \{x\}$.

instance that for arbitrary $\varepsilon > 0$, there is no feasible Max-NSW and $(1/2+\varepsilon)$-EF1 allocation under partition matroids. In particular, for identical valuations, Biswas et al. [4] have shown that an exact EF1 and PO allocation was guaranteed to exist by the Max-NSW allocation under matroid constraints. In this settings, we find an exact EF1 and PO allocation by leximin ordering. Furthermore, if all agents have binary valuations, Dror et al. [11] computed an exact EF1 and PO allocation under partition matroids. We explore the extended settings of 2-valued valuations that each item's value falls into $\{1, a\}$ with $a > 1$ for each agent, and strikingly demonstrate that every Max-NSW allocation is $\max\{1/a^2, 1/2\}$-EF1 and PO under partition matroids.

Beyond Matroid Constraints (Sect. 4). The results focus on two general classes p-extendible systems and independence systems. We restrict the p-extendible system [22] to a strongly p-extendible system $(M, \mathcal{F})$ (Definition 7): if $C \in \mathcal{F}, D \in \mathcal{F}$ with $C \subset D$ and if $H \cap C = \emptyset$ such that $C \cup H \in \mathcal{F}$, then there exists a subset $Y \subseteq D \setminus C$ with $|Y| \leq p|H|$ such that $D \setminus Y \cup H \in \mathcal{F}$. Furthermore, we focus on strongly p-extendible systems, which still include all matroids. For identical binary valuations, we show that every Max-NSW allocation achieves $\max\{1/p, 1/4\}$-EF1 and PO under strongly p-extendible systems. Independence systems are systems that satisfy only hereditary property. For additive valuations, every Max-NSW allocation is 1/4-EF1 and the approximation is already tight. Interestingly, we further consider lexicographic preferences [24], under independence system constraints, and compute an exact EF1 and PO allocation by a greedy-method algorithm.

Table 1. Summary of our main results. Identical, bi. and identical-bi refer to identical additive, binary additive, and identical binary additive valuations respectively. LB and UB refer to lower bounds and upper bounds respectively. "⇒ PO" refers to outcomes guarantee Pareto optimality.

Constraints	Valuations	Max-NSWs (⇒ PO)		Others (⇒ PO)
		LB	UB	
Partition Matroids	additive	$\frac{1}{2}$-EF1	$(\frac{1}{2}+\varepsilon)$-EF1	1-EF1 (bi.) [11]
Matroids	additive	$\frac{1}{2}$-EF1	$(\frac{1}{2}+\varepsilon)$-EF1	
	identical	1-EF1 [4]		1-EF1(by leximin)
	$\{1, a\}$-valued	$\max\{\frac{1}{a^2}, \frac{1}{2}\}$-EF1		
Strongly p-Extendible Systems	identical-bi	$\max\{\frac{1}{p}, \frac{1}{4}\}$-EF1	$(\frac{2}{3}+\varepsilon)$-EF1	
Independence Systems	additive	$\frac{1}{4}$-EF1	$(\frac{1}{4}+\varepsilon)$-EF1	
	lexicographic			1-EF1 (by RR alg.)

1.2 Related Work

There is a vast literature on fairly and efficiently allocating indivisible items without constraints. In particular, Caragiannis et al. [7,8] originated that a Max-NSW allocation is both strikingly EF1 and approximately Maximin share for additive valuations. Barman et al. [2] provided an efficient greedy algorithm to compute a Max-NSW allocation for binary valuations while finding Max-NSW allocations is APX-hard in general [19]. Benabbou et al. [3] showed that Max-NSW and leximin allocations both processed the EF1 property for matroid rank valuations. Amanatidis et al. [1] established that a Max-NSW allocation was always envy-freeness up to any item (EFX, [9]) for 2-valued valuations. Plaut and Roughgarden [23] EFX allocations and PO allocations were not compatible with general valuations. Recently, Feldman et al. [12] demonstrated an optimal tradeoff that for any $0 \leq \alpha \leq 1$, an α-EFX allocation can guarantee $1/(\alpha+1)$-Max-NSW for additive valuations.

Our work is related to fair and efficient allocations in constrained settings. Wu et al. [27] showed that under budget constraints, the Max-NSW allocation achieved 1/4-EF1 with tight approximation ratio. We extend this positive result to more general constraints (independence systems). Biswas and Barman [4] firstly considered special matroid constraints. They developed efficient algorithms to compute EF1 allocations under laminar matroids for identical additive valuations. If items are labeled by goods or chores, Shoshan et al. [25] provided a polynomial-time algorithm to compute feasible allocations under partition constraints guaranteed PO and EF up to one good and one chore. Dror et al. [11] demonstrated the existence of EF1 allocations by devising algorithms on the settings of heterogeneous partition matroids, and n agents with binary additive valuations (or two agents with general additive valuations). Besides, more literature studied other fairness allocations (e.g., Maximin share) under matroid constraints, which can be traced back to references [15,17,18]. Beyond the above special settings, we will focus on central constraints of general matroids, p-extendible systems and even independent systems in this paper, and we further explore the efficiency of feasible allocations.

2 Preliminaries

Given an instance $\mathcal{I} = (N, M, (v_i)_{i \in N})$, where $M = \{g_1, \cdots, g_m\}$ is a finite set of items (or goods), $N = \{1, \ldots, n\}$ is set of agents. Each agent i has an individual valuation function $v_i(\cdot) : 2^M \to \mathbb{R}_{\geq 0}$. Throughout, we assume that $v_i(\cdot)$ is normalized: $v_i(\emptyset) = 0$, and monotone: $v_i(S) \leq v_i(T)$ for all $S \subseteq T \subseteq M$. Our goal is to find fair and efficient allocations under broad constraints: matroids, p-extendible systems, and independence systems. The relationship between these three classes of constraints is in Fig. 1.

2.1 Independence Systems and Matroids

Let $\mathcal{M} = (E, \mathcal{F})$ be a set system including a finite set E and a collection $\mathcal{F} \subseteq 2^E$ over a set E. An independence system satisfies hereditary property.

Definition 1 (Independence System). *A set system $\mathcal{M} = (E, \mathcal{F})$ is an independence system if it satisfies the following properties:*

(1) $\emptyset \in \mathcal{F}$;
(2) if $D \in \mathcal{F}$ and $C \subseteq D$, then $C \in \mathcal{F}$;

The elements of $\mathcal{F}$ are called independent set.

Based on Definition 1 and augmentation property, we define the central structure in combinatorial optimization, called matroids.

Definition 2 (Matroid). *An independence system $\mathcal{M} = (E, \mathcal{F})$ is a matroid if it satisfies the following property:*

(3) if $C, D \in \mathcal{F}$ and $|D| > |C|$, then there exists $x \in D \setminus C$ such that $C + x \in \mathcal{F}$.

In addition, an independent set $B \in \mathcal{F}$ is a base if for all $x \in E \setminus B$, $B + x \notin \mathcal{F}$.

Indeed, a base of $\mathcal{M}$ is a maximal independent subset of E. Matroid structures has been extensively studied in combinatorial optimization. The following lemmas illustrate key conclusions on matroids.

Lemma 1 ([5]). Let $\mathcal{M} = (E, \mathcal{F})$ be a matroid. If $B \in \mathcal{F}$ and $B' \in \mathcal{F}$ are two bases, then there exists a bijection $\sigma : B \to B'$ such that $B + \sigma(z) - z$ and $B' + z - \sigma(z)$ are also bases in $\mathcal{F}$ for all $z \in B$.

Corollary 1. *Let $\mathcal{M} = (E, \mathcal{F})$ be a matroid. If $I, J \in \mathcal{F}$ are two independent sets with same cardinality, i.e., $|I| = |J| = k$, then there exists a bijection $\sigma : I \to J$ such that $I + \sigma(z) - z$ and $J + z - \sigma(z)$ are also in $\mathcal{F}$ for all $z \in I$.*

In fact, Corollary 1 is a direct result of applying Lemma 1 to a special case of matroid, called the *truncated matroid.* To be detailed, given a matroid $\mathcal{M} = (E, \mathcal{F})$, $\mathcal{M}_k = (E, \mathcal{F}_k)$ is a truncated matroid of $\mathcal{M}$ if $\mathcal{F}_k = \{S \in \mathcal{F} : |S| \leq k\}$. Furthermore, combining capacities and categories, we can define another special case of matroid, a *partition matroid* $\mathcal{M} = (E, \mathcal{F})$ if $E = \{E_1, \ldots, E_\ell\}$ can be partitioned into $\ell \leq |E|$ categories with corresponding capacities $k_1, \ldots, k_\ell$, such that $\mathcal{F} = \{S \subseteq E : |S \cap E_i| \leq k_i, \forall i = 1, \ldots, \ell\}$. Specially, if there is a single category, a partition matroid is referred as a *uniform matorid*.

If we relax the augmentation property on matroids, a generalized structure can be derived, called p-extendible system.

Definition 3 (p-extendible System [22]). An independence system $\mathcal{M} = (E, \mathcal{F})$ is a p-extendible system if it satisfies the following property:

(4) if $C \in \mathcal{F}, D \in \mathcal{F}$ with $C \subset D$ and if $x \notin C$ such that $C + x \in \mathcal{F}$, then there exists a subset $Y \subseteq D \setminus C$ with $|Y| \leq p$ such that $D \setminus Y + x \in \mathcal{F}$.

Lemma 2 ([22].) *The system $(E, \mathcal{F})$ is a matroid if and only if is 1-extendible.*

Based on the above lemma, every matroid is also a p-extendible system for any $p \in N^+$. We remark that many natural problems fall in p-extendible framework, such as maximum weight b-matching, maximum profit scheduling, and matroid intersection.

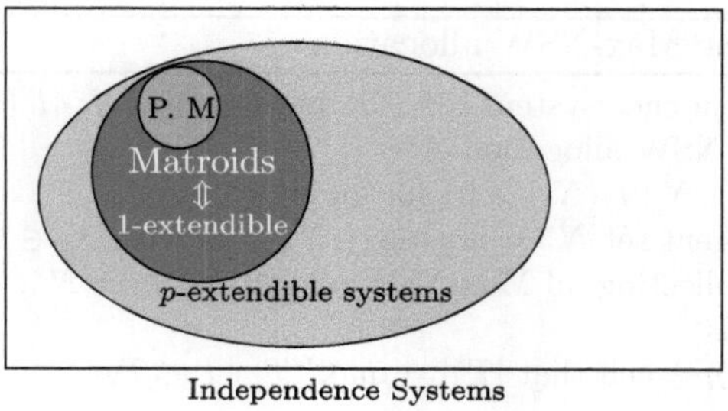

Fig. 1. Relations between several classes of constraints on independence systems. P. M refers to partition matroids.

2.2 Fairness and Efficiency Criteria

Given an independence system $\mathcal{M} = (M, \mathcal{F})$ and an instance $(M, N, (v_i)_{i\in N})$. A *feasible allocation* X refers to a collection of n disjoint bundles $X_1, \ldots, X_n$ such that $\cup_{i\in N} X_i \subseteq M$ and every bundle $X_i \subseteq M$ is an independent set in $\mathcal{F}$. Notice that all agents obey one independence system $(M, \mathcal{F})$. Due to independence constraints, a *complete allocation* X (s.t. $\cup_{i\in N} X_i = M$) may not exist, and thus we admit incomplete allocations with unallocated items, i.e., $M \setminus (\cup_{i\in N} X_i) \neq \emptyset$. Unless otherwise specified, we will write "an allocation" instead of "a feasible allocation" in the remainder.

In the independence settings (matroids, p-extendible or independence systems), our work is to explore "fair" and "efficient" allocations. The fairness notions we are interested in are based on envy-freeness. An allocation X is *envy-free* (EF) if $X_i \in \mathcal{F}$ for every agent $i \in N$ and $v_i(X_i) \geq v_i(X_j)$ for every pair of agents i and j. Due to independence constraints, a trivial but inefficient $A = (\emptyset, \ldots, \emptyset)$ is EF under $(M, \mathcal{F})$. We are also interested in a relaxed notion of classical EF property.

Definition 4 (EF1). *Let $\alpha \in [0, 1]$ be any constant. An allocation X is α-approximate envy-free up to one item (α-EF1) if every bundle X_i is in $\mathcal{F}$ and*

$$\forall i, j \in N, \exists Z \subseteq X_j, |Z| \leq 1, v_i(X_i) \geq \alpha \cdot v_i(X_j \setminus Z).$$

In particular, a 1-EF1 allocation is also called an exact EF1 allocation.

Besides the notion of EF1, we are also interested in the maximum Nash social welfare, which is a prominent notion of both fairness and efficiency.

Definition 5 (Max-NSW). *An allocation A^* maximizes the Nash social welfare (NSW) if*

$$NSW(A^*) = \max\{NSW(X) : X_i \in \mathcal{F},\ \forall i\},$$

where $NSW(X) = \prod_{i\in N} v_i(X)$ is the Nash social welfare of X.

In particular, we select the Max-NSW allocation returned by Algorithm 1.

Algorithm 1: The Max-NSW allocation

Input: an independence system $(M, \mathcal{F})$, an instance $(N, M, (v_i)_{i \in N})$
Output: A Max-NSW allocation A^*
Denote $N_X = \{i \in N : v_i(X_i) > 0\}$ for an allocation X;
Select a largest agent set $N^* = \arg\max\{|N_X| : X \text{ with } X_i \in \mathcal{F},\ \forall i\}$;
Denote by $\mathcal{T}_{N^*}$ collecting all Max-NSW allocations over N^*;
if $|\mathcal{T}_{N^*}| > 1$ **then**
  Select $T^* \in \mathcal{T}_{N^*}$ such that $|T^*| = \max\{|T| : T \in \mathcal{T}_{N^*}\}$;
else
  $\mathcal{T}_{N^*} = \{T^*\}$;
$A_i^* \leftarrow T_i^*$ for all $i \in N^*$ and $A_i^* \leftarrow \emptyset$ otherwise;
Return $A^* = (A_1^*, \ldots, A_n^*)$

Specifically, Algorithm 1 firstly select a largest agent set N^* that simultaneously received positive utilities. Then select a Max-NSW allocation T^* that maximizes the number of agents in N^*. Finally, construct the output allocation by giving bundle T_i^* to agents in N^* and giving nothing to the remaining agents.

Based on Algorithm 1, we show a crucial observation to capture that every Max-NSW allocation is PO.

Definition 6 (PO). *An allocation A^* is Pareto optimal (PO) if no alternative allocation Pareto dominates A^*. Specifically, for every alternative X,*

$$\exists i \in N, v_i(X) > v_i(A_i^*) \Rightarrow \exists j \in N, v_j(X_i) < v_j(A_j^*).$$

Observation 1 *For general additive valuations, every Max-NSW allocation is also a PO allocation under independence system (including matroid or p-extendible system) constraints.*

3 Matroid Constraints

In this section, we explore the existence of EF1 and PO allocations under matroid constraints, which capture important properties, such as heredity and augmentation. Recall that for binary valuations, Dror et al. [11] computed an exact EF1 and PO allocation under partition matroid constraints. We will explore the results for general additive, 2-valued additive and identical additive valuations.

3.1 General Additive Valuations

For general additive valuations, we firstly state an negative result by constructing a worst instance under partition matroids.

Lemma 3. *For any $\varepsilon > 0$, there exists an instance that no Max-NSW allocation can guarantee $(\frac{1}{2} + \varepsilon)$-EF1 under partition matroid constraints.*

Table 2. A worst instance under matroid constraints.

g_j	g_1	$\cdots$	g_{k+1}	g_{k+2}	$\cdots$	g_{2k+2}
$v_1(g_j)$	$2k+1-\delta$	$\cdots$	$2k+1-\delta$	k	$\cdots$	k
$v_2(g_j)$	k	$\cdots$	k	0	$\cdots$	0

Proof. Given an instance $\mathcal{I} = (N, M, (v_i)_{i \in N})$ with a matroid $\mathcal{M} = (M, \mathcal{F})$. $M = \{g_1, \ldots, g_{2k+2}\}$ is an item set, and $\mathcal{F} = \{S \mid |S| \leq k+1, S \subset M\}$ is a collection of independent sets. There are two agents $N = \{1, 2\}$ with valuations $v_i(\cdot)$ as Table 2, where $k > 0$ is sufficiently large, and $\delta > 0$ is sufficiently small.

Next, we will calculate Max-NSW allocations in this instance. Let X be any feasible allocation under matroid $\mathcal{M}$, where $X = (X_1, X_2)$ and $|X_i| \leq k+1$ for any agent $i = 1, 2$. Assume w.l.o.g that X_1 contain $x \in [0, k+1]$ items with the value of $2k+1-\delta$. Due to $|X_i| \leq k+1$ for $i = 1, 2$, we can get that $v_1(X_1) \leq (k+1-\delta)x + k(k+1)$, $v_2(X_2) \leq -kx + k(k+1)$, and $NSW(X) \leq \varphi(x)$, where $\varphi(x) = -(k^2+k-k\delta)x^2 + (k^2+k-(k^2+k)\delta)x + k^2(k+1)^2$. Since that $\varphi(x)$ is decreasing over $[1, k+1]$ and $X \neq A^*$, we have that $NSW(X) < \max\{\varphi(0), \varphi(1)\} = k^2(k+1)^2$. Thus, the allocation, that agent 1 gets $\{g_{k+2}, \ldots, g_{2k+2}\}$ and agent 2 gets $\{g_1, \ldots, g_{k+1}\}$, is the unique Max-NSW allocation under matroid $\mathcal{M}$. Let $A^* = (A_1^*, A_2^*)$ be an allocation with $A_1^* = \{g_{k+2}, \ldots, g_{2k+2}\}$ and $A_2^* = \{g_1, \ldots, g_{k+1}\}$. It holds that

$$v_1(A_1^*) = k(k+1) < \left(\frac{1}{2} + \frac{1}{2k}\right) \cdot v_1(A_2 - g).$$

Define $\varepsilon = \frac{1}{2k}$, we can obtain the conclusion that no Max-NSW allocation can guarantee $(\frac{1}{2} + \varepsilon)$-EF1 under partition matroid constraints. □

Now, we prove an important theorem of the section, which establishes an $\frac{1}{2}$-EF1 and PO allocation.

Theorem 1. *For general additive valuations, every Max-NSW allocation guarantees $\frac{1}{2}$-EF1 and PO under matroid constraints.*

Proof. Given an instance $\mathcal{I} = (N, M, (v_i)_{i \in N})$ with a matroid $\mathcal{M} = (M, \mathcal{F})$. $M = \{g_1, \ldots, g_m\}$ is item set and $N = \{1, \ldots, n\}$ is agent set. Let $A = (A_1, \ldots, A_n)$ be a Max-NSW allocation. By Corollary 2, A is a PO allocation. The following will show that A is 1/2-EF1. Suppose that there exist $i, j \in N$ such that

$$v_i(A_i) < \frac{1}{2} v_i(A_j - g), \forall g \in A_j. \tag{1}$$

We will discuss from two cases depending on whether $NSW(A) > 0$ or not.
Case 1. $NSW(A) > 0$. Due to A is feasible under matroid $\mathcal{M}$, we have that $A_i \in \mathcal{F}$, $A_j \in \mathcal{F}$. If $|A_i| = |A_j|$, based on Corollary 1, there exists a bijection $\phi : A_i \to A_j$ such that $A_i - g + \phi(g) \in \mathcal{F}$ and $A_j - \phi(g) + g \in \mathcal{F}$. Otherwise if $|A_i| \neq |A_j|$, we can assume w.l.o.g. that $|A_i| < |A_j|$. Based on the definition

of matroid, there must exist an item set $C = \{g_j^1, \dots, g_j^l\} \subseteq A_j$ which suffices that $A_i \cup \{g_j^1, \dots, g_j^l\} \in \mathcal{F}$ and $|A_i \cup \{g_j^1, \dots, g_j^l\}| = |A_j|$. We can also build a bijection from $A_i' = A_i \cup C$ to A_j, and every $g_0^k \in C$ is a copy item of g_j^k with $v_i(g_0^k) = v_j(g_0^k) = 0$, $v_i(A_i) = v_i(A_i')$. Hence, we define $\phi : A_i' \to A_j$ as a bijection from A_i' to A_j.

For any item pair $(g, \phi(g)) \in \phi$, we define the value $\rho(g, \phi(g))$ as following:

$$\rho(g, \phi(g)) = \frac{v_i(\phi(g)) - v_i(g)}{v_j(\phi(g)) - v_j(g)}.$$

Observe that $\rho(g, \phi(g)) = 0$ if $v_i(\phi(g)) - v_i(g) = 0$ and $v_j(\phi(g)) - v_j(g) \neq 0$, $\rho(g, \phi(g)) = \infty$ if $v_j(\phi(g)) - v_j(g) = 0$ and $v_i(\phi(g)) - v_i(g) \neq 0$. We denote by $P = \{(g, \phi(g)), v_i(\phi(g)) - v_i(g) > 0\}$ and $Q = \{(g, \phi(g)), v_i(\phi(g)) - v_i(g) \leq 0\}$ the two types of sets of item pairs. For agent i, denote by $A_{iP}' = \{g \in A_i', \exists \phi(g) \in A_j, s.t. (g, \phi(g)) \in P\}$ and $A_{iQ}' = \{g \in A_i', \exists \phi(g) \in A_j, s.t. (g, \phi(g)) \in Q\}$ the item sets over A_i'. Similarly, for agent j, we denote $A_{jP} = \{\phi(g) \in A_j, \exists g \in A_i, s.t. (g, \phi(g)) \in P\}$, $A_{jQ} = \{\phi(g) \in A_j, \exists g \in A_i', s.t. (g, \phi(g)) \in Q\}$. Let $(g^*, \phi(g^*)) = \arg\max_{(g,\phi(g)) \in P} \rho(g, \phi(g))$, we construct another allocation $B = (B_1, \dots, B_n)$ where $B_i = A_i' + \phi(g^*) - g^*, B_j = A_j + g^* - \phi(g^*)$, and $B_k = A_k$ for any $k \neq i, j$. We will prove that

$$v_i(B_i)v_j(B_j) - v_i(A_i)v_j(A_j) > 0. \tag{2}$$

If the above Inequality (2) holds, then $NSW(B) > NSW(A)$, which is contradicted with $NSW(A) = \max\{NSW(X) \mid X \text{ is feasible}\}$. Thus, the following will focus on the proof of Inequality (2).

Based on the definition of A_{iQ}' and A_{jQ}, $v_i(A_{jQ}) - v_i(A_{iQ}') \leq 0$, it holds that

$$\begin{aligned} v_i(A_j) - v_i(A_i') &= [v_i(A_{jP}) - v_i(A_{iP}')] + [v_i(A_{jQ}) - v_i(A_{iQ}')] \\ &\leq [v_i(A_{jP}) - v_i(A_{iP}')]. \end{aligned} \tag{3}$$

Due to Inequality (1), there must exist $(g, \phi(g)) \in P$ such that $v_i(\phi(g)) > v_i(g)$. It means that $(g, \phi(g)) \in P$ and $\phi(g) \in A_{jP}$. Therefore, $(g^*, \phi(g^*)) \in P$.

If $v_j(\phi(g^*)) - v_j(g^*) \leq 0$, it is easy to verify that $NSW(B) > NSW(A)$, which is contradicted with that A is a Max-NSW allocation.

If $v_j(\phi(g^*)) - v_j(g^*) > 0$, we can obtain that

$$\frac{v_i(A_{jP}) - v_i(A_{iP}')}{v_j(A_{jP})} \geq \frac{v_i(A_{jP}) - v_i(A_{iP}')}{v_j(A_j)} \geq \frac{v_i(A_j) - v_i(A_i')}{v_j(A_j)} > \frac{v_i(A_i') + v_i(\phi(g^*))}{v_j(A_j)}.$$

The first inequality holds because $v_j(A_j) \geq v_j(A_{jP})$, the second inequality holds because of Inequality (3), and the third inequality holds because of Inequality (1). Furthermore, by the definition of $(g^*, \phi(g^*))$, it holds that

$$\frac{v_i(\phi(g^*)) - v_i(g^*)}{v_j(\phi(g^*)) - v_j(g^*)} \geq \frac{v_i(A_{jP}) - v_i(A_{iP}')}{v_j(A_{jP}) - v_j(A_{iP}')} \geq \frac{v_i(A_{jP}) - v_i(A_{iP}')}{v_j(A_{jP})} > \frac{v_i(A_i') + v_i(\phi(g^*))}{v_j(A_j)}.$$

Notice that $v_i(\phi(g^*)) - v_i(g^*) > 0$ and $v_i(A_i') + v_i(\phi(g^*)) > 0$, we have

$$[v_i(\phi(g^*)) - v_i(g^*)] \cdot v_j(A_j) > [v_i(A_i') + v_i(\phi(g^*))] \cdot [v_j(\phi(g^*)) - v_j(g^*)].$$

Thus,

$$\begin{aligned} &v_i(B_i)v_j(B_j) - v_i(A_i)v_j(A_j) \\ &\quad = [v_i(\phi(g^*)) - v_i(g^*)] \cdot v_j(A_j) - [v_i(A_i') + v_i(\phi(g^*))] \cdot [v_j(\phi(g^*)) - v_j(g^*)] \\ &\qquad - v_i(g^*) \cdot [v_j(g^*) - v_j(\phi(g^*))] > 0. \end{aligned}$$

Case 2. $NSW(A) = 0$. If $v_i(A_i) \geq 0$ and $v_j(A_j) > 0$, then the proof process is similar as Case 1. Otherwise if $v_i(A_i) \geq 0$ and $v_j(A_j) = 0$, we can construct another allocation $B = (B_1, \ldots, B_n)$ where $B_i = A_j$, $B_j = A_i$, $B_k = A_k$ for all $k \neq i, j$. It can be verified that $|\{i \in N : v_i(B_i) > 0\}| > |\{i \in N : v_i(A_i) > 0\}|$, which is contradicted with the definition of A.

To be concluded, $A = (A_1, \ldots, A_n)$ is a 1/2-EF1 and PO allocation. □

Corollary 2. *For general additive valuations, every Max-NSW allocation guarantees $\frac{1}{2}$-EF1 and PO under partition matroid constraints.*

Combining with Lemma 3, we remark that the approximation of 1/2 is a tight ratio under partition matroid constraints.

3.2 Special Cases of Additive Valuations

This section considers two special cases of additive valuations: identical additive and 2-valued additive valuations. For identical additive valuations, Biswas et al. [4] showed that the Max-NSW allocation admitted exact EF1 and PO under matroid constraints. We provide an exact EF1 and PO allocation in this settings by leximin ordering.

Theorem 2. *For identical valuations, every leximin allocation guarantees exactly EF1 and PO under matroid constraints.*

Proof. Given an instance $\mathcal{I} = (N, M, v)$ with a matroid $\mathcal{M} = (M, \mathcal{F})$. $M = \{g_1, \ldots, g_m\}$ is item set and $N = \{1, \ldots, n\}$ is agent set. Notice that all agents share an identical valuation function $v(\cdot)$. Let $A = (A_1, \ldots, A_n)$ be a leximin allocation such that $A_i \in \mathcal{F}$ for all i. By the definition of leximin, $A \succeq X$ for any allocation X and "$\succeq$" is referred to a comparison operator. It is clear that A is a PO allocation. Thus, the following will show that A is exactly EF1.

Suppose that A is not EF1, there exists two agents $i, j \in N$, $v(A_i) < v(A_j - g)$ for any $g \in A_j$. If $|A_j| < |A_i|$, by the definition of $\mathcal{M}$, there exists $\{g_i^1, \ldots, g_i^l\} \subseteq A_i$ such that $A_j \cup \{g_i^1, \ldots, g_i^l\} \in \mathcal{F}$ and $|A_j \cup \{g_i^1, \ldots, g_i^l\}| = |A_i|$. Denote by $C = \{g_0^1, \ldots, g_0^l\}$ an item set where every $g_0^k \in C$ is a copy item of $g_i^k \in A_i$ with $v_i(g_0^k) = v_j(g_0^k) = 0$. Thus, based on Corollary 1, there exists a bijection $\phi : A_i \to A_j \cup C$ such that $A_k - g + \phi(g) \in \mathcal{F}$ for any $k \in \{i, j\}$ and any $g \in A_k$. Due to the assumption of $v(A_i) < v(A_j)$, there must exist one pair $(g_i, \phi(g_i))$ such that $v(g_i) < v(\phi(g_i))$. Thus, we can construct another feasible allocation $B = (B_1, \ldots, B_n)$ where $B_i = A_i - g_i + \phi(g_i)$, $B_j = A_j - \phi(g_i) + g_i$, $B_k = A_k$ for all $k \neq i, j$. Notice that $v(B_i) > v(A_i)$ and $v(B_j) \geq v(A_j - \phi(g_i)) > v(A_i)$. Thus, we have $A \prec B$ which is contradicted with $A \succeq B$.

For the remaining case of $|A_j| \geq |A_i|$, we can build similar contradictions by the above analysis. Hence the leximin allocation A is exactly EF1 and PO under matroid constraints. □

The following theorem will focus on 2-valued valuations. Let $a > 1$ be a constant. For any agent $i \in N$, her valuation $v_i(\cdot)$ is 2-valued if $v_i(g) \in \{1, a\}$ for any $g \in M$.

Theorem 3. *For 2-valued valuations with* $\{1, a\}$, *every Max-NSW allocation guarantees* $\max\{\frac{1}{a^2}, \frac{1}{2}\}$*-EF1 and PO under matroid constraints.*

Proof. Given an instance $\mathcal{I} = (N, M, (v_i)_{i \in N})$ with a matroid $\mathcal{M} = (M, \mathcal{F})$. $M = \{g_1, \ldots, g_m\}$ is a item set, $N = \{1, \ldots, n\}$ is a agent set and $v_i(g) \in \{1, a\}$, $a > 1$. Let $A = (A_1, \ldots, A_n)$ be a Max-NSW allocation. Due to Theorem 1, A is 1/2-EF1 and PO. The following will illustrate that A is also $\frac{1}{p}$-EF1.

Suppose that A is not $\frac{1}{a^2}$-EF1, there exists two agents $i, j \in N$, $v(A_i) < \frac{1}{a^2} v(A_j - g)$ for any $g \in A_j$. We will discuss from two cases depending on whether $|A_j| \leq |A_i|$ or not.

Case 1. If $|A_j| \leq |A_i|$, there exists one pair (g_i, g_j) such that $A_i + g_j - g_i \in \mathcal{F}$, $A_j + g_i - g_j \in \mathcal{F}$, and $v_i(g_i) = 1, v_i(g_j) = a$. Construct another feasible allocation $B = (B_1, \ldots, B_n)$ as follows: $B_i = A_i - g_i + g_j$, $B_j = A_j - g_j + g_i$, $B_k = A_k$ for all $k \neq i, j$. For simplicity, denote $\Delta = NSW(B) - NSW(A)$. The following illustrates that $NSW(B) > NSW(A)$, which contradicts the definition of A. If $v_j(g_i) \leq v_j(g_j)$, it holds that $\Delta = (a-1)v_j(A_j) - (v_j(g_j) - v_j(g_i))v_i(A_i) - (a-1)(v_j(g_j) - v_j(g_i)) \geq (a-1)v_j(A_j) > 0$. Otherwise, $\Delta = (a-1)(v_j(A_j) - v_i(A_i) - (a-1)) > (a-1)(v_j(A_j - g_j) - \frac{1}{a^2} v_i(A_j - g_j)) > 0$.

Case 2. If $|A_j| > |A_i|$, there exists $g_j \in A_j$ such that $A_i + g_j \in \mathcal{F}$. Construct another feasible allocation $B = (B_1, \ldots, B_n)$ as follows: $B_i = A_i + g_j$, $B_j = A_j - g_j$, $B_k = A_k$ for all $k \neq i, j$. The following illustrates that $NSW(B) > NSW(A)$, which contradicts the definition of A. In fact, if $v_i(g_j) = v_j(g_j)$, $\Delta = v_i(g_j)(v_j(A_j - g_j) - \frac{1}{a^2} v_i(A_j - g_j)) > 0$. If $v_i(g_j) > v_j(g_j)$, $\Delta = a v_j(A_j) - v_i(A_i) - a > 0$. For the remaining case of $v_i(g_j) < v_j(g_j)$, $v_i(B_i)v_j(B_j) - v_i(A_i)v_j(A_j) = v_j(A_j) - a v_i(A_i) - a > v_j(A_j - g_i) - \frac{1}{a} v_i(A_j - g_j) > 0$. □

4 Beyond Matroid Constraints

In this section, we proceed to study EF1 and PO allocations under constraints that go beyond matroids. The classes of constraints we mainly focus on are p-extendible systems and independent systems.

4.1 p-Extendible System Constraints

As mentioned previously, a p-extendible system $(M, \mathcal{F})$ satisfies a key property that if $C \in \mathcal{F}, D \in \mathcal{F}$ with $C \subset D$ and if $x \notin C$ such that $C + x \in \mathcal{F}$, then there exists an item set $Y \subseteq D \setminus C$ with $|Y| \leq p$ such that $D \setminus Y + x \in \mathcal{F}$. Notice that the swap-out set Y is independent in $\mathcal{F}$ while adding a single item x. However, if adding several items H to D, Y would not always admit independence.

Building on the above observation, we will state a negative result for identical and binary valuations.

Lemma 4. *For identical binary valuations, there exists an instance, in which a Max-NSW allocation is exactly $\frac{2}{5}$-EF1 under 2-extendible system constraints.*

Proof. Given a 2-extendible system $(M, \mathcal{F})$, where $M = \{g_1, \ldots, g_8\}$, $\mathcal{F} = P(\{g_1, g_2\}) \cup P(\{g_1, g_3, g_4, g_7, g_8\}) \cup P(\{g_2, g_3, g_4, g_7, g_8\}) \cup P(\{g_3, g_4, g_5, g_6, g_7, g_8\})$. Specifically, $P(S) = \{T : T \subseteq S\}$ is denoted by the power set of an item set S. There are two agents with identical additive valuations $v(\cdot)$ and $v(g) = 1$ for every single item $g \in M$. It is easy to verify that $\max_X \prod_{i=1}^n v_i(X_i) = 12$, and $A_1 = \{g_1, g_2\}$, $A_2 = \{g_3, g_4, g_5, g_6, g_7, g_8\}$ is a Max-NSW allocation. Thus, $v(A_1) = 2 = \frac{2}{5} \cdot v(A_2 - g)$ for all $g \in A_2$. □

In the 2-extendible system in Lemma 4, when we consider adding an independent set to another independent set, let $H = \{g_1, g_2\}$, $C = \emptyset$, $D = \{g_3, g_4, g_5, g_6, g_7, g_8\}$, there exists the unique subset $Y \subseteq D \setminus C$ with $Y = \{g_3, g_4, g_5, g_6, g_7, g_8\}$ such that $D \setminus Y \cup H \in \mathcal{F}$, but $|Y| > 2 \cdot 2$. Thus, we next introduce strongly p-extendible systems satisfying that the swapped-out Y always satisfies $|Y| \leq p|H|$.

Definition 7 (Strongly p-extendible System). *A strongly p-extendible system is a pair $(E, \mathcal{F})$ where $\mathcal{F} \subseteq 2^E$ is a collection of subsets of E which satisfies the following properties: (1) $\emptyset \in \mathcal{F}$; (2) if $D \in \mathcal{F}$ and $C \subseteq D$, then $C \in \mathcal{F}$; (3) if $C \in \mathcal{F}, D \in \mathcal{F}$ with $C \subset D$ and if $H \cap C = \emptyset$ such that $C \cup H \in \mathcal{F}$, then there exists a subset $Y \subseteq D \setminus C$ with $|Y| \leq p|H|$ such that $D \setminus Y \cup H \in \mathcal{F}$.*

Proposition 1. *The system $(E, \mathcal{F})$ is a matroid if and only if is strongly 1-extendible.*

Now, we establish main results about strongly p-extendible systems.

Lemma 5. *For identical binary valuations, there exists an instance, in which, Max-NSW and EF1 are incompatible and all Max-NSW allocations are exactly $\frac{2}{3}$-EF1 under strongly 2-extendible system constraints.*

Proof. Given a 2-extendible system $(M, \mathcal{F})$, where $M = \{g_1, \ldots, g_8\}$, $\mathcal{F} = P(\{g_1, g_2\}) \cup P(\{g_1, g_3, g_4\}) \cup P(\{g_2, g_3, g_4\}) \cup P(\{g_3, g_4, g_5, g_6\})$. Specifically, $P(S) = \{T : T \subseteq S\}$ is denoted by the power set of an item set S. There are two agents with identical additive valuations $v(\cdot)$ and $v(g) = 1$ for every single item $g \in M$. It is easy to verify that $\max_X \prod_{i=1}^n v_i(X_i) = 8$, and $A_1 = \{g_1, g_2\}$, $A_2 = \{g_3, g_4, g_5, g_6\}$ is the unique Max-NSW allocation. Thus, $v(A_1) = 2 = \frac{2}{3} \cdot v(A_2 - g)$ for all $g \in A_2$. □

Theorem 4. *For identical binary valuations, every Max-NSW allocation guarantees $\frac{1}{p}$-EF1 under strongly p-extendible system constraints.*

Proof. Given an instance $\mathcal{I} = (N, M, v)$ with a p-extendible system $\mathcal{M} = (M, \mathcal{F})$. $M = \{g_1, \ldots, g_m\}$ is a item set, $N = \{1, \ldots, n\}$ is a agent set and $v(\cdot)$ is a binary valuation for all agents. Let A^* be a Max-NSW allocation. Suppose, for contradiction, that A^* is not $1/p$-EF1, and that there exist two agents $i, j \in N$ and any single item $g \in A_j^*$,

$$v(A_i^*) < \frac{1}{p} \cdot (v(A_j^*) - v(g)), \quad \forall g \in A_j^*. \tag{4}$$

Case 1. If there exists $g^* \in A_j^*$ such that $A_i^* + g^* \in \mathcal{F}$. Let B be an alternative allocation such that $B_i = A_i^* + g^*$, $B_j = A_j^* - g^*$, and $B_k = A_k$ for all $k \neq i, j$. Due to $A_i^* + g^* \in \mathcal{F}$ and $A_k^* \in \mathcal{F}$ for all k, we get that B is feasible under matroid $(M, \mathcal{F})$. Furthermore, $NSW(B) = v(B_i)v(B_j)\prod_{k \neq i,j} v(B_k) = NSW(A^*) + v(g)[v(A_j^*) - v(A_i^*) - v(g)] > NSW(A^*)$, where the inequality holds by Inequality (5). It contradicts the optimality of $NSW(A^*)$.
Case 2. If $A_i^* + g \notin \mathcal{F}$ for all $g \in A_j^*$. By the definition of p-extendible system, for every item $g \in A_i^*$, we can construct a feasible bundle $(A_j^* + g) \setminus Y_g \in \mathcal{F}$ by adding g to A_j^* and deleting some subset $Y_g \subset A_j^*$ with $|Y_g| \leq p$. Iteratively moving all items from A_i^* to A_j^*, we can construct another feasible bundle by further deleting at most $p|A_i^*|$ items from A_j^*. Combining non-wastefulness of A^*, $|A_j^*| > p|A_i^*| + 1$. Thus, there exists $Y \subset A_j^*$ with $|Y| \leq p|A_i^*|$ and $A_i^* \cup A_j^* \setminus Y \in \mathcal{F}$.

(i) If $|Y| > |A_i|$, then assume that $|Y| = |A_i^*| + d$. Combining with $|Y| \leq p|A_i^*|$, $0 < d \leq (p-1)|A_i^*|$. Let B be an alternative allocation such that $B_i = Y$, $B_j = A_i^* + A_j^* - Y$, and $B_k = A_k^*$ for all k. Thus, $NSW(B) = NSW(A^*) + d(|A_j^*| - |A_i^*| - d) \geq NSW(A^*) + d(|A_j^*| - p|A_j^*|) > NSW(A^*)$, contradicting the optimality of $NSW(A^*)$.

(ii) If $|Y| \leq |A_i|$, then we can construct an extension C such that $Y \subset C \subset A_j^*$ and $|C| = |A_i^*| + 1$. Notice that $C \in \mathcal{F}$. Let B be an alternative allocation where $B_i = C$, $B_j = A_i^* + A_j^* - C$, and $B_k = A_k^*$ for all k. Thus, we obtain that $NSW(B) = NSW(A^*) + (|A_j^*| - |C|) > NSW(A^*)$ which contradicts the optimality of $NSW(A^*)$. □

4.2 Independent System Constraints

This section focus on independence system constraints with two cases of valuations: general additive valuations and lexicographic preferences. We start with two negative results in additive settings.

Lemma 6. *There exists an instance that no $(\frac{1}{4} + \varepsilon)$-EF1 and Max-NSW allocation under independence system constraints.*

Proof. Given an independence system $(M, \mathcal{F})$, where $M = \{g_1, \ldots, g_m\}$ with $m = 1 + 1/\delta$, $\mathcal{F} = P(\{g_1\}) \cup P(\{g_2, \ldots, g_m\})$, and $\delta > 0$ be a small enough constant. There are two agents $\{1, 2\}$ that have identical valuation function $v(\cdot)$, and the values are as follows: $v(g_1) = 1 + \delta$, $v(g_k) = 4\delta$ for all single item $g_k \neq g_1$.

Consider an arbitrary feasible allocation X under $(M, \mathcal{F})$. Due to identical valuation $v(\cdot)$, assume w.l.o.g. that $v(X_1) \leq v(X_2)$. If $g_1 \in X_1 \cup X_2$, then

$NSW(X) = v(X_1)v(X_2) \leq (1+\delta)(4\delta \cdot 1/\delta) = 4(1+\delta)$ since one agent has the only one large item g_1 and the other one have at most the remaining small items with value of 4δ. Otherwise if $X_1 \cup X_2 \subseteq M - g_1$, then $NSW(X) \leq 2 \cdot (4\delta \cdot 1/(2\delta)) = 4$, and the equality holds when both agents have $1/(2\delta)$ small items. Thus, $NSW(X) \leq 4(1+\delta)$ for all X. Let A^* be an allocation that $A_1^* = \{g_1\}$ and $A_2^* = \{g_2, \ldots, g_m\}$. By the definition of $(M, \mathcal{F})$, $A_1^*, A_2^* \in \mathcal{F}$ and thus A^* is a feasible allocation. Therefore, A^* is the unique Max-NSW allocation with $NSW(A^*) = 4(1+\delta)$. However, $v(A_1^*) = (1-\delta^2)/4 \cdot v(A_2^* - g) < 1/4 \cdot v_1(A_2^* - g)$ for all $g \in A_2^*$. Thus, A^* is not $(1/4+\varepsilon)$-EF1 for any $\varepsilon > 0$. □

Next, we establish the main result about general additive valuations.

Theorem 5. *For general additive valuations, every Max-NSW allocation guarantees $\frac{1}{4}$-EF1 and PO under independence system constraints.*

Proof. Given an independence system $(M, \mathcal{F})$ and any corresponding instance $\mathcal{I} = (N, M, (v_i)_{i \in N})$. Let A^* be a Max-NSW allocation returned by Algorithm 1. We will analyze it in two situations as follows:

Case 1. $NSW(A^*) > 0$. In this case, $v_k(A_k^*) > 0$ for every agent k. We assume that A^* is not 1/4-EF1, that is, there exist two agents $i, j \in N$ and any item $g \in A_j^*$ such that $v_j(A_j^* - g) > 4 \cdot v_i(A_i^*)$. The following will construct an alternative allocation.

Consider agent j's bundle $A_j^* = \{g_1, \ldots, g_t\}$, where $v_j(h_1) \leq \cdots \leq v_j(g_t)$. Notice that the item g_t has the maximum value in A_j^*. Define $H_1 = \{g_j \in A_j^* - g_t : j \equiv 1 (mod\ 2)\}$ and similarly define $H_2 = \{g_j \in A_j^* - g_t : j \equiv 0 (mod\ 2)\}$. Due to $A_j^* \in \mathcal{F}$, $H_1, H_2 \in \mathcal{F}$. Furthermore, at least one of the following inequalities holds: i) $v_j(H_1) \leq v_j(H_2) \leq v_j(H_1 + g_t)$; ii) $v_j(H_2) \leq v_j(H_1) \leq v_j(H_2 + g_t)$. Let $H_{max} = \arg\max\{v_i(H_1), v_i(H_2)\}$ be the preferable bundle for agent i. Then, we construct an alternative allocation B, where $B_i = H_{max}$, $B_j = A_j^* \setminus H_{max}$, and $B_k = A_k^*$ for all $k \neq i, j$. By the definition of independence system, B is a feasible allocation under $(M, \mathcal{F})$. Therefore, $v_i(B_i) = v_i(H_{max}) \geq 1/2 \cdot v_i(A_j^* - h_t) > 2 \cdot v_i(A_i^*)$, $v_j(B_j) = v_j(A_j^*) - v_j(H_{max}) \geq 1/2 \cdot v_j(A_j^*)$, and thus $NSW(B) = v_i(B_i)v_j(B_j)\prod_{k \neq i,j} v_k(B_k) > v_i(A_i)v_j(A_j)\prod_{k \neq i,j} v_k(A_k) = NSW(A^*)$, which gives us the desire contradiction.

Case 2. $NSW(A^*) = 0$. Define $N_X^+ = \{k \in N, v_k(X_k) > 0\}$ as the agents that obtain positive values from allocation X. Due to $v_i(\emptyset) = 0$, $v_i(A_j^* - g) > 0$, and thus $|A_j^*| \geq 2$. Based on Algorithm 1, $v_j(A_j^*) > 0$ and $|N_{A^*}^+| \geq |N_X^+|$ for any Max-NSW allocation X.

If agent i also has positive valuation, i.e., $v_i(A_i^*) > 0$, we can construct an alternative allocation by similar method in Case 1, which towards a contradiction with the definition of A^*. Otherwise if $i \notin N_{A^*}^+$, then $v_i(A_i^*) = 0$. Let B be an alternative allocation where $B_i = A_j^* - g$, $B_j = \{g\}$, $B_k = A_k^*$ for all k, and item g is an arbitrary item of A_j^*. By the definition of independence system, B is a feasible allocation under $(M, \mathcal{F})$. Moreover, $v_i(B_i) = v_i(A_j^* - g) > 0$, $v_j(B_j) = v_j(\{g\}) > 0$, and $v_k(B_k) = v_k(A_k^*)$ for all k. Thus, $|N_B^+| > |N_{A^*}^+|$, contradicting the definition of A^*. □

Corollary 3. *For general additive valuations, every Max-NSW allocation guarantees* $\max\{\frac{1}{p}, \frac{1}{4}\}$*-EF1 and PO under strongly p-extendible system constraints.*

4.3 Lexicographic Preferences

Heretofore, we have studied additive valuations and several special cases. Although these valuations crucial in practice, they are not worked on the scenes with ordering preferences. This section introduces lexicographic preferences.

Algorithm 2: Round-Robin under Independence System Constraints

---

Input: An IS $(M, \mathcal{F})$, $v_i(\cdot)_{i\in N}$, and an independence oracle.
Output: An allocation $A = (A_1, \ldots, A_{|N|})$.
Initialize: $(A_1, \ldots, A_{|N|}) = (\emptyset, \ldots, \emptyset)$;
for $r = 1$ *to* $\lceil \frac{|M|}{n} \rceil$ **do**
  for $i = 1$ *to* $|N|$ **do**
    if $\exists g \in M$, *s.t.* $A_i + g \in \mathcal{F}$ **then**
      Let $g^* \in \arg\max_{g\in M}\{v_i(g) : A_i + g \in \mathcal{F}\}$;
      $A_i = A_i + g^*$;
      $M = M - g^*$;
    else
      $N = N - i$;
Return $A = (A_1, \ldots, A_{|N|})$.

Definition 8 (Lexicographic Preference). *Given an instance* $(M, N, (v_i)_{i\in N})$. *Every agent i has a linear ordering* $\pi_i = (g_{i1}, \ldots, g_{im})$ *such that* $v_i(g_{i1}) > \cdots > v_i(g_{im})$. *Let* $(o_S(g_{i1}), \ldots, o_S(g_{im}))$ *be a vector of bundle S allocated to agent i:* $o_S(g_{ik}) = v(g_{ik})$ *for all* $g_{ik} \in S$ *and* $o_S(g_{ik}) = 0$ *otherwise. For any two bundles* $X, Y \subseteq M$, *agent i's preference is induced by lexicographic order as follows:*

$$v_i(X) > v_i(Y) \Leftrightarrow \exists \ell, o_X(g_{i\ell}) > o_Y(g_{i\ell}) \text{ and } \forall k < \ell, o_X(g_{ik}) = o_Y(g_{ik}).$$

To compute an exact EF1 and Max-NSW allocation, we present Algorithm 2 by Round-Robin method. In this algorithm, we make a slight modification of the classic Round-Robin that if any item is added to someone's bundle would break the independence requirement, he is being skipped over.

Next, we provide a positive result about lexicographic preferences.

Theorem 6. *For lexicographic preferences, the allocation computed by Algorithm 2 guarantees exactly EF1 and PO under independence system constrains.*

Proof. Given an instance $\mathcal{I} = (N, M, (v_i)_{i\in N})$ with an independence system $\mathcal{M} = (M, \mathcal{F})$, where $M = \{g_1, \ldots, g_m\}$ is a item set, $N = \{1, \ldots, n\}$ is a agent set and $v_i(g) \in \{1, a\}$, $a > 1$. Let A be the allocation returned by Algorithm 2. For each t-th iteration of outer for-loop, denote by h_i^t the item allocated to agent i in the t-th iteration.

To prove that A is EF1, we consider arbitrary two agents $i, j \in N$. By the procedure of Algorithm 2, $v_i(h_i^t) \geq v_i(h_j^{t+1})$ for every iteration $1 \leq t \leq \lceil \frac{m}{n} \rceil$. If $i < j$, then agent i select items before agent j in the first iteration. Thus, $v_i(h_i^1) \geq v_i(g)$ for all $g \in A_j$. By lexicographic preference, $v_i(A_i) \geq v_i(A_j)$. Otherwise if agent j select her item before agent i in each iteration. Due to $v_i(h_i^t) \geq v_i(h_j^{t+1})$ for each iteration t, $v_i(A_i) \geq v_i(A_j - h_j^1)$. Hence, the output A is exactly EF1.

To prove that A is PO, we suppose that B is an alternative allocation such that $B_k \in \mathcal{F}$ for all k, where $v_d(B_d) > v_d(A_d)$ for some agent d, and $v_k(B_k) \geq v_k(A_k)$ for any other $k \neq d$. Since $v_d(B_d) > v_d(A_d)$, there must exist an item $o \in B_d \setminus A_d$ such that $v_d(o) > v_d(h_i^1)$. By the procedure of Algorithm 2, the item o is allocated to an agent $c < d$, which selects her item before d in the first iteration. Thus, $o = h_c^1 \in A_c \setminus B_c$. Due to $v_c(B_c) \geq v_c(A_c)$, there exists another item $o' \in B_c \setminus A_c$ satisfying $v_c(o') > v_c(o)$. Since that agent 1 obtain her most preferable item h_1^1, $v_1(h_1^1) \geq v_1(g)$ for all $g \in B_1$, we have $v_1(A_1) \geq v_1(B_1)$. By repeatedly finding o', there must exist an agent $1 \leq a < d$ satisfying $v_a(A_a) > v_a(B_a)$. It contradicts the assumption of A. Thus, the output A is PO. □

Although Algorithm 2 performs perfect in lexicographic preferences, this algorithm performs unsatisfactory in additive valuations.

Lemma 7. *For additive valuations, there exists an instance that the allocation returned by Algorithm 2 may not guarantee* $(\frac{1}{2}+\varepsilon)$*-EF1 and PO under partition matroid constraints.*

Proof. Given an instance $\mathcal{I} = (N, M, (v_i)_{i\in N})$ with an independence system $\mathcal{M} = (M, \mathcal{F})$. $M = \{g_1, \ldots, g_{3\eta}\}$ divided by η categories $C_1 = \{g_1, \ldots, g_{2\eta}\}$, $C_k = \{g_{2\eta+k}\}$ for any other $k \neq 1$, $\mathcal{F} = \{S \subseteq M, |S \cap C_1| \leq \eta, |S \cap C_k| \leq 1 \text{ for } k \neq 1\}$. There are two agents having identical additive valuation functions $v(\cdot)$: $v(g) = 1$ for all $g \in M$.

Let $\eta > 1$ be a sufficiently large constant. Execute the procedure of Algorithm 2, agent 1 gets $\{g_{2\eta+1}, \ldots g_{3\eta}\}$, agent 2 gets $\{g_1, \ldots, g_\eta\}$ in the first η iterations. By the definition of $\mathcal{F}$, in each iteration $\eta < t \leq 2\eta$, agent 1 gets one more item $g_{\eta+t}$ but agent 2 cannot get any new item. Thus, $A_1 = \{g_{\eta+1}, \ldots, g_{3\eta}\}$ and $A_2 = \{g_1, \ldots, g_\eta\}$. It implies that $v(A_2) < (\frac{1}{2} + \frac{1}{2\eta-1}) \cdot v(A_1 - g)$. Hence, the lemma holds by setting $\varepsilon = \frac{1}{2\eta-1}$. □

5 Conclusions and Future Directions

The tension between fairness and efficiency is a focal point of allocation problems; this tension appears even without computational considerations. When every allocation is feasible, Max-NSW allocations ensure EF1 and PO for additive valuations. However, with constraints, there is an insurmountable gap between EF1 and Max-NSW, so the approximation analysis comes in handy. In this paper, we study the relatively broad constrained settings that agents have additive valuations and obtain several tight results for approximate EF1 and Max-NSW allocations. We show that, under matroid constraints, every Max-NSW allocation is $\frac{1}{2}$-EF1 and PO. Under beyond matroid constraints, the approximation ratio is relaxed to $\frac{1}{4}$ in other independence systems. When we set the constraints on strongly p-extendible systems, in which the swapped-out set Y is independent, every Max-NSW allocation is $\max\{\frac{1}{p}, \frac{1}{4}\}$-EF1 and PO for identical binary valuations. As a matter of course, the most obvious open question is whether for additive valuations, $\frac{1}{p}$-EF1 and Max-NSW are compatible.

Other promising future directions are: (i) generalizing our results beyond additive valuations, and (ii) designing efficient algorithms to compute approximate EF1 and PO allocations under matroid or independence system constraints.

References

1. Amanatidis, G., Birmpas, G., Filos-Ratsikas, A., Hollender, A., Voudouris, A.A.: Maximum Nash welfare and other stories about EFX. Theoret. Comput. Sci. **863**, 69–85 (2021)
2. Barman, S., Krishnamurthy, S.K., Vaish, R.: Greedy algorithms for maximizing Nash social welfare. In: Proceedings of the 17th International Conference on Autonomous Agents and MultiAgent Systems, pp. 7–13 (2018)
3. Benabbou, N., Chakraborty, M., Igarashi, A., Zick, Y.: Finding fair and efficient allocations for matroid rank valuations. ACM Trans. Econ. Comput. **9**(4), 1–41 (2021)
4. Biswas, A., Barman, S.: Fair division under cardinality constraints. In: Proceedings of the 27th International Joint Conference on Artificial Intelligence, pp. 91–97 (2018)
5. Brualdi, R.A.: Comments on bases in dependence structures. Bull. Aust. Math. Soc. **1**(2), 161–167 (1969)
6. Budish, E., Cachon, G.P., Kessler, J.B., Othman, A.: Course match: a large-scale implementation of approximate competitive equilibrium from equal incomes for combinatorial allocation. Oper. Res. **65**(2), 314–336 (2017)
7. Caragiannis, I., Kurokawa, D., Moulin, H., Procaccia, A.D., Shah, N., Wang, J.: The unreasonable fairness of maximum Nash welfare. In: Proceedings of the 2016 ACM Conference on Economics and Computation, pp. 305–322 (2016)
8. Caragiannis, I., Kurokawa, D., Moulin, H., Procaccia, A.D., Shah, N., Wang, J.: The unreasonable fairness of maximum Nash welfare. ACM Trans. Econ. Comput. (TEAC) **7**(3), 1–32 (2019)
9. Chaudhury, B.R., Garg, J., Mehlhorn, K.: EFX exists for three agents. In: Proceedings of the 21st ACM Conference on Economics and Computation, pp. 1–19 (2020)
10. Deng, Y., Sing, T.F., Ren, C.: The story of Singapore's public housing: from a nation of home-seekers to a nation of homeowners. In: The Future of Public Housing, pp. 103–121 (2013)

11. Dror, A., Feldman, M., Segal-Halevi, E.: On fair division under heterogeneous matroid constraints. In: Proceedings of the AAAI Conference on Artificial Intelligence, vol. 35, pp. 5312–5320 (2021)
12. Feldman, M., Mauras, S., Ponitka, T.: On optimal tradeoffs between EFX and Nash welfare. In: Proceedings of the AAAI Conference on Artificial Intelligence, vol. 38, pp. 9688–9695 (2024)
13. Foley, D.: Resource allocation and the public sector. Yale Econ. Essays **7**, 45–98 (1967)
14. Goldman, J., Procaccia, A.D.: Spliddit: unleashing fair division algorithms. ACM SIGecom Exchanges **13**(2), 41–46 (2015)
15. Gourvès, L., Monnot, J.: On maximin share allocations in matroids. Theoret. Comput. Sci. **754**, 50–64 (2019)
16. Gourvès, L., Monnot, J., Tlilane, L.: Near fairness in matroids. In: Proceedings of the Twenty-first European Conference on Artificial Intelligence, pp. 393–398 (2014)
17. Hummel, H., Hetland, M.L.: Guaranteeing half-maximin shares under cardinality constraints. In: Proceedings of the 21st International Conference on Autonomous Agents and Multiagent Systems, pp. 1633–1635 (2022)
18. Hummel, H., Hetland, M.L.: Maximin shares under cardinality constraints. In: European Conference on Multi-Agent Systems, pp. 188–206 (2022)
19. Lee, E.: APX-hardness of maximizing Nash social welfare with indivisible items. Inf. Process. Lett. **122**, 17–20 (2017)
20. Li, Z., Vetta, A.: The fair division of hereditary set systems. ACM Trans. Econ. Comput. (TEAC) **9**(2), 1–19 (2021)
21. Lipton, R.J., Markakis, E., Mossel, E., Saberi, A.: On approximately fair allocations of indivisible goods. In: Proceedings of the 5th ACM Conference on Electronic Commerce, pp. 125–131 (2004)
22. Mestre, J.: Greedy in approximation algorithms. In: European Symposium on Algorithms, pp. 528–539. Springer (2006)
23. Plaut, B., Roughgarden, T.: Almost envy-freeness with general valuations. SIAM J. Discret. Math. **34**(2), 1039–1068 (2020)
24. Saban, D., Sethuraman, J.: A note on object allocation under lexicographic preferences. J. Math. Econ. **50**, 283–289 (2014)
25. Shoshan, H., Hazon, N., Segal-Halevi, E.: Efficient nearly-fair division with capacity constraints. In: Proceedings of the 2023 International Conference on Autonomous Agents and Multiagent Systems, pp. 206–214 (2023)
26. Suksompong, W.: Constraints in fair division. ACM SIGecom Exchanges **19**(2), 46–61 (2021)
27. Wu, X., Li, B., Gan, J.: Budget-feasible maximum Nash social welfare is almost envy-free. In: Proceedings of the 30th International Joint Conference on Artificial Intelligence, IJCAI-21, pp. 465–471 (2021)

Fair Interventions in Weighted Congestion Games

Miriam Fischer[1](✉), Martin Gairing[2], and Dario Paccagnan[1]

[1] Department of Computing, Imperial College London, Huxley Building, 180 Queen's Gate, London SW7 2AZ, UK
{m.fischer21,d.paccagnan}@imperial.ac.uk

[2] Department of Computer Science, University of Liverpool, Ashton Street, Liverpool L69 3BX, UK
m.gairing@liverpool.ac.uk

Abstract. In this work we study the power and limitations of fair interventions in weighted congestion games. Specifically, we focus on interventions that aim at improving the equilibrium quality (price of anarchy) and are fair in a suitably defined sense. Within this setting, we provide three key contributions.

First, we show that no fair intervention can reduce the price of anarchy below a given factor depending solely on the class of latencies considered. Interestingly, this lower bound is unconditional, i.e., it applies regardless of how much computation interventions are allowed to use.

Second, we design a taxation mechanism that is fair and achieves a price of anarchy matching this unconditional lower bound, all the while being polynomial-time computable.

Third, we show that no intervention (fair or not) can achieve a better approximation if polynomial computability is required. We do so by proving that the minimum social cost is NP-hard to minimize below a factor identical to the one previously introduced. In doing so, our work shows that the algorithm proposed by Makarychev and Sviridenko [29] to tackle optimization problems with a "diseconomy of scale" is optimal, and provide a novel way to derandomize its solution via equilibrium computation.

Keywords: Weighted Congestion Games · Price of Anarchy · Taxation Mechanism · Hardness of Approximation · Polynomial Time Algorithms

1 Introduction

The performance of many distributed systems often depends on the choices made by myopic users, who take decision based on their individual interests and without regard to the overall system performance. In this context, the presence of selfish users often leads to suboptimal outcomes and the celebrated notion of

M. Mavronicolas et al. (Eds.): WINE 2024, LNCS 15534, pp. 190–206, 2026.
https://doi.org/10.1007/978-3-032-08560-3_11

price of anarchy [28] constitutes a commonly employed metric to quantify the resulting performance degradation.

Against this backdrop, (weighted) congestion games [34] have played a starring role in the development of a rich body of literature primarily focusing on quantifying the price of anarchy, thanks to which we now have a complete characterization of this metric. However, within this setting, much less is known regarding the ability of *interventions* – a means for the system designer to modify the users' perceived costs – to improve the equilibrium quality. Indeed, although a multitude of approaches have been proposed, including coordination mechanisms [13], Stackelberg strategies [17], taxation mechanisms [7,9,33], signaling [5,23], network design techniques [19,36], cost-sharing methods [20,21], it is still unclear how powerful interventions are in reducing the price of anarchy, especially when polynomial computability is required.

Concurrently, there has also been a surge of interest in developing algorithms that are *fair* in the sense that they do not discriminate between players with similar or identical attributes, with examples ranging from classification [15] to online decision-making [24] and allocation of indivisible items [2]. It is therefore natural to study how effective *fair interventions* (see Sect. 1.4 for a definition) are in reducing the price of anarchy. In particular, it is interesting to study whether the fairness constraint imposes any additional limitation on the achievable price of anarchy, and how such additional requirement intertwines with the computational complexity of the resulting interventions.

These questions can be summarized as follows:

- *How effective are fair interventions in reducing the price of anarchy?*
- *Do fairness requirements limit the ability of interventions to improve the price of anarchy?*

In this paper we settle these questions for the class of weighted congestion games. We do so both in the setting where no limitation is imposed on the intervention's computational requirement, and in the case where polynomial computability is desired. Perhaps interestingly, our work shows that when efficient computation is needed, fair interventions are as a powerful mean to reduce the price of anarchy as any other.

1.1 Our Contributions and Their Implications

Three major technical contributions substantiate and expand upon the previous claims:

C1) **Unconditional Lower Bound for Fair Interventions.** We give a lower bound on the achievable price of anarchy of fair interventions that is independent of their computational complexity, but depends solely on the class of latency functions employed (Theorem 1). Stated differently, we show that no fair intervention can achieve a price of anarchy below this bound. To the best of our knowledge, our work is the first to analyze unconditional lower bounds for fair interventions.

C2) **Optimal and Efficiently-computable Fair Interventions.** We provide a framework to design fair interventions, in the form of taxation mechanisms, whose price of anarchy matches the lower bound previously introduced, and show that such interventions are polynomially-computable (Theorem 2). In the full version of the manuscript [16], we provide explicit expressions of the optimal mechanism for the widely studied class of polynomial latency functions. Even without fairness constraints, the problem of designing polynomially-computable interventions to reduce the price of anarchy has not been previously studied beyond linear latency functions [9].

C3) **Limitations of Unfair Interventions.** We complement the previous results by showing that no intervention, regardless of whether it is fair or not, can reduce the price of anarchy below the factor previously derived, if polynomial computability is required. We prove this result by showing that the problem of minimizing the social cost is NP-hard to approximate below a factor identical to that already obtained (Theorem 3). No tight hardness of approximation was known for weighted congestion games.

Taken together these contributions show that: i) The proposed taxation mechanisms are optimal amongst all fair intervention in the sense that no fair intervention can induce a better price of anarchy, regardless of its computational complexity. ii) The proposed taxation mechanisms are also optimal within all polynomially-computable interventions, fair or not, and more generally also against any polynomially-computable algorithm. This is interesting as, prior to this work, it was unclear whether (fair) interventions were a sufficiently powerful tool to achieve optimal approximation in weighted congestion games. Our paper answers this question in the positive.

Finally, although the following observations are direct consequences of the previous contributions, we highlight them separately, as we believe they may be of independent interest.

C4) **Optimal Approximation, and Matching Hardness.** Makarychev and Sviridenko [29] study approximation algorithms for optimization problems with a "diseconomy of scale", to which weighted congestion games belong. Interestingly, the hardness factor we obtain in C3) matches the approximation factor derived by [29] and thus fully settles the approximability of this important class of problems. Finally, we note that by combining our results in C2) with existing algorithms for the computation of correlated equilibria (e.g., no-regret dynamics), one obtains a polynomial-time algorithm to minimize the social cost that achieves the best possible approximation (Corollary 1). This approach provides an alternative to the randomized rounding techniques proposed by Makarychev and Sviridenko [29].

Three final comments are in order. First, note that our unconditional lower bound only requires a very mild notion of fairness (see Sect. 1.4, Interventions & Fairness for details on its definition). This makes the unconditional lower bound very strong, as the impossibility result we derive holds over a large class of interventions, and strengthening the notion of fairness can only worsen the lower bound. At the same time, we note that the taxation mechanism we present satisfies a much stronger fairness notion, as players using the same resource are treated equally even if they have different weights. Second, we observe that the price of anarchy obtained by our taxation mechanism holds both for mixed Nash equilibria and correlated/coarse correlated equilibria. This extension is significant as it gives performance bounds that apply also when players revise their action and achieve low regret [38]. Third, our work settles the question of whether approximating the minimum social cost is strictly harder in weighted congestion games compared to their unweighted variant. Perhaps interestingly, our work shows that the two problems are equally difficult to approximate only when the class of latency functions considered is closed under *abscissa scaling*[1], which includes the important class of polynomial latency functions. We refer to [16] for details and proof.

1.2 Techniques

Three separate sets of techniques are needed to prove our results.

The lower bound for fair interventions is shown using, at its core, a symmetry argument. Specifically, the lower bound is obtained by exhibiting an instance of weighted congestion games where any fair intervention induces high cost at the worst mixed Nash equilibrium, relative to the optimal allocation. Interestingly, the instance utilized consists of a network with parallel links and identical players. In these settings, fair interventions maintain the symmetry of the game owing to their definition, so that a symmetric mixed Nash equilibrium must exist [31]. One then concludes by showing that any symmetric mixed profile, whether or not an equilibrium, has the desired high social cost.

Taxation mechanisms are a natural type of intervention often considered to reduce the price of anarchy, and their *fairness* follows readily from their definition. Contrary to that, providing a recipe to design taxation mechanisms that are both efficiently computable *and* achieve the desired approximation is significantly more challenging. We achieve this goal following a two-pronged approach. First, we compute a fractional solution to a Linear Programming (LP) relaxation for the problem of minimizing the social cost, and then design taxes which, using the solution as input, make sure that the resulting equilibria achieve the desired approximation.

In more detail, we leverage the LP relaxation of [29] to minimizing social cost in combinatorial optimization problems with a diseconomy of scale. While this relaxation has an exponential number of decision variables (for each resource there is a variable for each subset of players), the resulting LP can be solved in polynomial time to arbitrary precision using the ellipsoid method. We adapt this relaxation to weighted congestion games, and show how its (non-integral) solution can be used as input to our taxation mechanism, which defines taxes for each resource. Using critical inequalities based on stochastic orders of sums

[1] A class of latency functions $\mathcal{L}$ is closed under abscissa scaling if, for any function $\ell \in \mathcal{L}$ and $\alpha \geq 0$, the function h with $h(x) = \ell(\alpha x)$ is in $\mathcal{L}$.

of Poisson random variables, we prove that this taxation mechanism enforces equilibria with the desired price of anarchy. We note that this is challenging, as the LP relaxation of [29] was designed to compute an allocation with low social cost in a *centralized* setting, and not in the game-theoretic setting we consider here. Our result thus shows that, whether agents are strategic or not, it is possible to achieve the same optimal approximation.

For polynomial latency functions, we obtain closed-form expressions of the taxation mechanism, which we present in the full version of the manuscript [16]. We do so by showing that the moments of a weighted sum of independent Poisson random variables can be computed through a binomial-based recursion, which might be of independent interest. The proof heavily relies on moment generating functions and generalizes a result by Gould and Quaintance [22, Eq 4.2].

The result regarding hardness of approximation is obtained through an approximation preserving reduction [3] from the corresponding problem for unweighted congestion games [33]. Although weighted congestion games are significantly more expressive than their unweighted variant, we are able to apply such approximation preserving reduction owing to the following crucial observation: weighted congestion games where players have *identical* weights are sufficient to achieve the largest possible inapproximability factor, once again highlighting the importance of symmetry.

Finally, we note that this paper differs significantly from recent work on the design of taxes for *unweighted* congestion games [33]. First, Paccagnan and Gairing [33] do not consider fairness constraints and therefore do not study how such requirement impacts the achievable performance, e.g., the unconditional lower bound in Theorem 1. Second, their work is limited to the unweighted setting, and, perhaps unsurprisingly, their techniques do not carry over to the setting considered here. For example, in the unweighted case, they obtain taxation mechanisms through the solution of a minimization problem where the cost on each resource is replaced by a Poisson approximant. While this problem can be solved in polynomial time for the unweighted case, it can not be used here as it gives rise to a non-convex optimization problem for the weighted case. Third, obtaining explicit expression for optimal taxation mechanisms requires new ideas, even for the widely studied class of polynomial latency functions, as players' individual weights have to be accounted for.

Lastly, as we will show, the NP-hardness factor for minimizing the social cost is also different in the weighted and unweighted settings.

1.3 Related Work

Price of Anarchy. Since its introduction [28], the *price of anarchy* has been extensively studied in various setting, and for both unweighted and weighted congestion games we now have a rich theory. This includes exact bounds for congestion games with linear [4,12] and polynomial [1] latency functions. These results led to the development of Roughgarden's *smoothness framework* [38], which elegantly distilled and formulated previous ideas for bounding the price of anarchy in a unified framework. Roughgarden [38] also showed that upper bounds achieved through the smoothness framework extend all the way to *coarse correlated equilibria*, which can be computed by *no-regret* algorithms. Bhawalkar et al. [6] extended those results to weighted congestion games.

Fairness. While fairness notions have been widely studied, one of the most commonly employed is that of *individual fairness* [15], which originated in the design of classification algorithms. Such notion requires that individuals with similar attributes receive similar outcomes. In this respect, our notion of fair interventions (see Sect. 1.4 for a definition) is closely aligned to the former, with the crucial difference that we do not define similarity as a continuous metric, but rather only with regards to identical players. In this respect, our notion of fairness imposes weaker constraints than that of individual fairness, resulting in a very strong unconditional lower bound. Indeed, many of the interventions studied in the literature implicitly satisfy our fairness definition and, therefore, must obey our unconditional lower bound. These include cost-sharing mechanisms [20,21], public signaling in singleton congestion games [41], cost-balancing tolls in atomic symmetric network congestion games [18] and taxation mechanisms by [7,9,32]. At the same time, we observe that the interventions we propose satisfy a much stronger fairness notion as players selecting the same resource are treated equally, even if they have different weights.

Finally, we observe that, within the realm of congestion games, the fairness of *allocations* has been studied for both the non-atomic [14,35] and the atomic [10,26] variants of congestion games, though with significant differences. Most notably, these works focus on the fairness of a given allocation, and not on how that allocation can be induced by fair interventions.

Interventions & Taxation Mechanisms. Different approaches, such as coordination mechanisms [13], Stackelberg strategies [17], taxation mechanisms [9], signaling [5], cost-sharing strategies [20,21], have been proposed to cope with the performance degradation associated to selfish decision making. Amongst them, taxation mechanisms have attracted significant attention thanks to their ability to indirectly influence the resulting system performance. Their design was initiated by Caragiannis et al. [9] for linear (weighted) congestion games, and further studied by [7,32] in the polynomial setting. Limitedly to linear congestion games, Caragiannis et al. [9] provide efficiently-computable taxes whose price of anarchy matches the one we achieve. For polynomial congestion games, Bilò and Vinci [7] also propose taxation mechanisms with a price of anarchy matching ours. However, there is a fundamental difference: the taxes [7] proposes require knowledge of an *optimal* allocation, and therefore can not be efficiently computed, as minimizing the social cost is NP-hard. In contrast, our taxation mechanism achieves the same price of anarchy whilst also being *efficiently* computable.

Computational Lower Bounds. Meyers and Schulz [30] show that minimizing the social cost in unweighted congestion games is NP-hard to approximate for general convex latency functions, while Roughgarden [37] showed that computational lower bounds on minimizing the social cost translate to lower bounds on the price of anarchy. Specifically, for polynomial latency functions of maximum degree D and non-negative coefficients, he provided the first lower bound, parametrized by D, on approximating social cost. For unweighted congestion games, tight hardness factors for approximating the minimum social cost have recently been given by Paccagnan and Gairing [33] for general latency functions. No tight hardness factor was known for weighted congestion games.

Unconditional Lower Bounds. The aforementioned lower bounds are conditional on polynomial-time computability and do not consider fairness constraints. Unconditional lower bounds have been obtained for Stackelberg strategies in unweighted congestion games with linear latencies [8,17] and, in weighted congestion games, for cost-sharing mechanisms [20,27] and taxes for linear latency functions [9]. However, these works restrict to specific classes of interventions, and *do not* consider fairness. Our work significantly departs from these as it provides - for general latency functions - unconditional lower bounds for *any* intervention that is *fair*. For linear latency functions, we recover the unconditional lower bound on taxes by Caragiannis et al. [9], and thus provide a fairness interpretation of their work.

1.4 Weighted Congestion Games and Taxation Mechanisms

Notation. Let $\mathbb{N}, \mathbb{N}_0, \mathbb{R}, \mathbb{R}_{\geq 0}, \mathbb{R}_{>0}$ denote the sets of natural numbers, natural numbers including zero, real numbers, non-negative real numbers, and positive real numbers. Given $m \in \mathbb{N}$, let $[m]$ denote the set $\{1, \ldots, m\}$. Throughout we will denote with $\mathrm{Poi}(x)$ a Poisson distribution with parameter x and with $\mathrm{Bin}(n, p)$ a Binomial distribution with n trials and success probability p for each trial. We will also use $P \leq_{cx} Q$ to denote that distribution P is smaller-equal than distribution Q in the convex order sense, that is, $\mathbb{E}_P[f(P)] \leq \mathbb{E}_Q[f(Q)]$ for all convex functions $f : \mathbb{R} \to \mathbb{R}$ [40].

Weighted Congestion Games. In a *weighted congestion game* we are given a set of players $[N] = \{1, \ldots, N\}$, and a set of resources $\mathcal{R}$. Each player $i \in [N]$ has a *weight* $w_i \in \mathbb{R}_{>0}$ and a *strategy set* $\mathcal{A}_i \subseteq 2^{\mathcal{R}}$ from which player i choses an *action* $a_i \in \mathcal{A}_i$. In an *allocation* $a = (a_1, \ldots, a_N) \in \mathcal{A}$, each player chooses an action from their strategy profile, where we denote by $\mathcal{A} = \mathcal{A}_1 \times \ldots \times \mathcal{A}_N$ the set of feasible allocations.

Given an allocation a, denote $x_r(a)$ the *load* of players currently selecting resource $r \in \mathcal{R}$, i.e., $x_r(a) = \sum_{i \in [N]: r \in a_i} w_i$. The cost for using each resource $r \in \mathcal{R}$ depends only on the load of players concurrently selecting that resource, and is defined by a *latency function* $\ell_r : \mathbb{R}_{>0} \to \mathbb{R}_{\geq 0}$. Throughout the paper we assume that we can evaluate $\ell_r(x)$ at any given load x in polynomial time with respect to the length of the input representation of ℓ_r and x. The cost of a player i is obtained by summing the costs of all resources selected, that is

$$\mathrm{C}_i(a) = \sum_{r \in a_i} \ell_r(x_r(a)).$$

Throughout this paper, we define the shorthand $c_r(x) = x\ell_r(x)$ for all x and $r \in \mathcal{R}$.

In a *mixed Nash equilibrium* every player i chooses a probability distribution σ_i over their strategy set $\mathcal{A}_i$, such that no player can unilaterally improve by changing to another action; formally, for the product distribution $\sigma = \sigma_1 \times \cdots \times \sigma_N$, $\mathbb{E}_{a\sim\sigma}[C_i(a)] \leq \mathbb{E}_{a_{-i}\sim\sigma_{-i}}[C_i(a_i', a_{-i})]$, for every $i \in [N]$ and $a_i' \in \mathcal{A}_i$. Similarly a joint probability distribution σ over allocations is a *coarse correlated equilibrium* if for every $i \in [N]$ and $a_i' \in \mathcal{A}_i$, $\mathbb{E}_{a\sim\sigma}[C_i(a)] \leq \mathbb{E}_{a\sim\sigma}[C_i(a_i', a_{-i})]$. Finally, the *social cost* represents the weighted sum of the players' costs

$$\mathrm{SC}(a) = \sum_{i\in[N]} w_i C_i(a) = \sum_{r\in\mathcal{R}} x_r(a)\ell_r(x_r(a)). \tag{1}$$

For a mixed or coarse correlated profile σ the social cost is defined accordingly as $\mathbb{E}_{a\sim\sigma}[SC(a)]$.

Throughout the manuscript, we make the following standard assumption on each latency function of the underlying congestion game.

Assumption 1. *The function $\ell : \mathbb{R}_{\geq 0} \to \mathbb{R}_{\geq 0}$ is non-decreasing and semi-convex.*[2]

Further, we denote with $\mathcal{G}$ the set of all congestion games where all latency functions $\{\ell_r\}_{r\in\mathcal{R}}$ belong to a given set of functions $\mathcal{L}$. Given an instance $G = (N, \{w_i\}_{i=1}^N, \mathcal{R}, \{\mathcal{A}_i\}_{i=1}^N, \{\ell_r\}_{r\in\mathcal{R}})$ of congestion game, we denote with `MinSC` the problem of minimizing the social cost in (1).

Interventions & Fairness. As self-interested decision making often deteriorates the system performance, interventions aim to ameliorate this issue. Formally, an intervention $I : G \mapsto \bar{G}$ is a map that associates a congestion game G to a new game $\bar{G}$ (not necessarily a congestion game), where the players are the same, but their perceived cost has been redefined to be $\bar{C}_i(a)$, and their action sets have been (potentially) reduced to $\bar{\mathcal{A}}_i \subseteq \mathcal{A}_i$. We emphasize that although I changes players' *perceived* cost from $C_i(a)$ to $\bar{C}_i(a)$, the social cost (1) remains unchanged.

A *fair* intervention gives identical players identical treatment. That is, if i and j have identical action sets and weights, then $\bar{\mathcal{A}}_i = \bar{\mathcal{A}}_j$ and $\bar{C}_i(a) = \bar{C}_j(\pi_{ij}(a))$, with $\pi_{ij}(a)$ permuting a_i and a_j in allocation a. This implies that if a fair intervention is applied to a symmetric congestion game G, the resulting game $\bar{G}$ is also symmetric.[3] Note that this is a very mild request, and corresponds to considering interventions that do not distinguish the players identity in $\bar{G}$ if these where not distinguishable in G. When defined this way, fair interventions include

[2] The function $\ell(x)$ is semi-convex in its domain if the function $c(x) = x\ell(x)$ is convex, in the discrete sense, in the same domain.

[3] In symmetric games, all players have identical action sets, and the cost a player incurs only depends on its action, and the actions played by all players, but not on who plays them, see [11,39].

a wide range of commonly employed mechanisms to improve the equilibrium performance, most notably taxation mechanisms, cost-sharing mechanisms, public signaling, and many others.

Taxation Mechanisms. In Sect. 3, we will study *taxation mechanisms*, a specific type of intervention that only modifies the players' costs. Formally, a taxation mechanism $T : G \times r \to \tau_r$ associates an instance G, and a resource $r \in \mathcal{R}$ to a taxation function $\tau_r : \mathbb{R} \to \mathbb{R}_{\geq 0}$. Note that each taxation function τ_r is (possibly) congestion-dependent, that is, it associates the load on resource r to the corresponding tax. As a consequence, each player i experiences a cost factoring both the cost associated to the chosen resources, and the tax, i.e.,

$$\bar{C}_i(a) = \sum_{r \in a_i} [\ell_r(x_r(a)) + \tau_r(x_r(a))].$$

We note that defined this way, taxation mechanisms are fair interventions as all players using a specific resource r experience the same tax τ_r.

Price of Anarchy. As typical in the literature, we measure the performance of a given intervention I using the ratio between the social cost incurred at the worst-performing outcome and the minimum social cost. Given the self-interested nature of the players, an outcome is commonly described by any of the following classical equilibrium notions: pure or mixed Nash equilibria, correlated or coarse correlated equilibria. However, we recall that within the class of weighted congestion games, existence is guaranteed for mixed Nash equilibria and correlated or coarse correlated equilibria, but not necessarily for pure Nash equilibria. When considering mixed Nash equilibria, the performance of an intervention I is gauged using the notion of *price of anarchy* [28], i.e.,

$$\text{PoA}(I) = \sup_{G \in \mathcal{G}} \frac{\texttt{NECost}(G, I)}{\texttt{MinCost}(G)}, \tag{2}$$

where $\texttt{MinCost}(G) = \min_{a \in \mathcal{A}} \text{SC}(a)$ is the minimum social cost for instance G, and $\texttt{NECost}(G, I)$ denotes the highest expected social cost at a mixed Nash equilibrium obtained when employing the intervention I on the game G. By definition, $\text{PoA}(I) \geq 1$ and the lower the price of anarchy, the better performance I guarantees. While it is possible to define the notion of price of anarchy for each and every equilibrium class mentioned, we do not pursue this direction, as we will show that all these metrics coincide within our setting. Thus, we will simply use $\text{PoA}(I)$ to refer to the efficiency values of *any* and *all* these equilibrium classes. Finally, we observe that, while interventions influence the players' perceived cost, they do not impact the expression of the social cost, which is still of the form in (1).

1.5 Roadmap

The remainder of the manuscript is organized as follows. In Sect. 2, we provide the unconditional lower bound for fair interventions. In Sect. 3, we provide fair, polynomially computable taxes that achieve a price of anarchy matching the lower bound previously derived. In Sect. 4 we show hardness of approximating the minimum social cost and limitations of unfair interventions. Due to space limitations, we refer to [16] for all proofs.

2 Lower Bound on Price of Anarchy of Fair Interventions

Our first result shows that fair interventions incur a fundamental limitation in their ability to improve the equilibrium performance, and that such limitation is independent of how much computational power the intervention is given, i.e., $I(G)$ might even require exponential computation in the size of G. This is the case already in the setting where all resources have the same latency ℓ and all players are symmetric, i.e., they have the same action sets and weights.

Theorem 1. *In weighted congestion games where all resources have the same latency function ℓ, there exists no fair intervention that induces a price of anarchy lower than $\rho_\ell - \varepsilon$, for any $\varepsilon > 0$, where*

$$\rho_\ell = \sup_{t>0} \frac{\mathbb{E}_{P\sim Poi(1)}[tP\ell(tP)]}{t\ell(t)}, \qquad (3)$$

and we define $\ell(0) = 0$. If $\rho_\ell = \infty$, the price of anarchy of any fair intervention is unbounded.

We further note that for class $\mathcal{L}$ of non-negative linear combinations of $\{1, x, \dots, x^D\}$, the highest degree monomial x^D determines $\rho = \max_{\ell\in\mathcal{L}} \rho_\ell$, which reduces to the $(D+1)$'st Bell number. We present the proof in [16].

3 Fair, Optimal, and Efficiently-Computable Taxes

In this section we provide a general recipe to design efficiently-computable taxation mechanisms with a price of anarchy matching the lower bound on fair interventions previously derived. We will then specialize this approach to the case of polynomial latencies, for which we will derive explicitly given taxation mechanisms.

We note that the taxation mechanisms we study are not only fair in the sense of Sect. 1.4, but satisfy a *much stronger* fairness notion in that players using the same resources are treated equally, regardless of their weight.

Our approach will make use of two crucial ingredients which we now introduce: i) a *linear programming relaxation of* `MinSC`, and ii) a *parametrized taxation mechanism.* At its core, our approach will derive a set of parameters from an optimal solution of such linear program, and use them for the aforementioned taxation mechanism. Towards this goal, we make the following minimal assumption, which we require in this section.

Assumption 2. *Players' weights $\{w_i\}_{i=1}^N$ are multiples of some factor $\delta \geq 0$ and w_i/δ are polynomially bounded in N and $1/\varepsilon$, where ε is the desired accuracy.*

We note that this assumption, also used in [29], is purely technical as it if often possible to round the weights of a given game so that the resulting cost changes by a small quantity (see [29, Theorem A.1]). Recall the shorthand $c_r(x) = x\ell_r(x)$ for all x and $r \in \mathcal{R}$.

3.1 LP Relaxation of MinSC

The first ingredient our approach employs is a suitably defined relaxation of the original (integral) MinSC problem. While it is straightforward to formulate MinSC as the following integer program (where y_{i,a_i} equals one if player i selects action $a_i \in \mathcal{A}_i$ and zero else),

$$\begin{aligned}
\min \quad & \sum_{r\in\mathcal{R}} c_r\left(\sum_{i\in[N]} w_i v_{r,i}\right) & \\
\text{subject to} \quad & v_{r,i} = \sum_{a_i\in\mathcal{A}_i : r\in a_i} y_{i,a_i} & \text{for all } r\in\mathcal{R}, i\in[N], \\
& \sum_{a_i\in\mathcal{A}_i} y_{i,a_i} = 1 & \text{for all } i\in[N], \\
& y_{i,a_i}\in\{0,1\} & \text{for all } a_i\in\mathcal{A}_i, i\in[N],
\end{aligned} \tag{4}$$

we do not employ its natural relaxation obtained removing the integral constraints, but instead make use of a more refined one following an approach similar to that in [29].

Towards presenting the relaxation, it is convenient to introduce variables $z_{S,r} \in \{0,1\}$ for $S \subseteq [N]$, $r \in \mathcal{R}$, encoding what subset of players utilize resource r, where $\sum_{S\subseteq[N]} z_{S,r} = 1$, i.e., only one subset is active. The previous program can then be equivalently written as the following linear program in the variables $z_{S,r}$, $v_{r,i}$, y_{i,a_i}, upon additionally restricting the variables $z_{S,r}$ to be binary, i.e., $z_{S,r} \in \{0,1\}$,

$$\begin{aligned}
\min \quad & \sum_{r\in\mathcal{R}}\sum_{S\subseteq[N]} c_r\left(\sum_{i\in S} w_i\right)\cdot z_{S,r} & \\
\text{subject to} \quad & \sum_{S\subseteq[N]} z_{S,r} = 1 & \text{for all } r\in\mathcal{R}, \\
& \sum_{S\subseteq[N]: i\in S} z_{S,r} = v_{r,i} & \text{for all } r\in\mathcal{R}, i\in[N], \\
& z_{S,r}\geq 0 & \text{for all } r\in\mathcal{R}, S\subseteq[N], \\
& v_{r,i} = \sum_{a_i\in\mathcal{A}_i : r\in a_i} y_{i,a_i} & \text{for all } r\in\mathcal{R}, i\in[N], \\
& \sum_{a_i\in\mathcal{A}_i} y_{i,a_i} = 1 & \text{for all } i\in[N], \\
& y_{i,a_i}\geq 0 & \text{for all } a_i\in\mathcal{A}_i, i\in[N].
\end{aligned} \tag{5}$$

Program (5) constitutes the linear programming relaxation of (4) we will use in the remainder of this manuscript. Due to the presence of $z_{S,r}$, $S \subseteq [N]$, this linear program has exponentially many decision variables. Yet, following a similar approach to that of [29], one can show that (5) can be solved to arbitrary precision in polynomial time. We provide the proof in the full version [16] of this manuscript.

3.2 Parametrized Taxation Mechanisms

Towards introducing the taxation mechanism, it is convenient to define the following functions $p_r : \mathbb{R}^N \to \mathbb{R}$ (one for each resource)

$$p_r(v_r) = \mathbb{E}\left[c_r\left(\sum_{i\in[N]} w_i P_i\right)\right], \tag{6}$$

where $P_i \sim \mathrm{Poi}\,(v_{r,i})$ are independent Poisson random variables, $c_r : \mathbb{R} \to \mathbb{R}$, defined by $c_r(x) = x\ell_r(x)$, is convex, and $v_{r,i}$ denotes the i-th entry of the vector $v_r \in \mathbb{R}^N$.[4]

Throughout, we assume we can evaluate (6) for given v_r in polynomial time with respect to the input representation of the game. We show that this is the case for the widely studied class weighted congestion games with polynomial latencies by obtaining an explicit expression. We refer the interested reader to the full version of the paper [16] for a proof.

The parametrized taxation mechanism we employ $\tau_r : \mathbb{R}\times\mathbb{R}^N \to \mathbb{R}$ associates each resource r to a corresponding tax $\tau_r(x; v_r)$ that depends on the total load x on that resource and on the previously introduced parameter v_r. Letting

$$\bar{\ell}_r(x; v_r) = \tau_r(x; v_r) + \ell_r(x)$$

represent the modified latency function accounting for both the original latency and the imposed tax, we define such taxation mechanism through the following recursion

$$x_r\bar{\ell}_r(x_r; v_r) - \sum_i w_i v_{r,i}\bar{\ell}_r(x_r + w_i; v_r) = c_r(x_r) - p_r(v_r) \qquad \forall x_r, v_r \geq 0, \tag{7}$$

where we additionally set $\bar{\ell}_r(x_r, 0) = \ell_r(x_r)$ for all $x_r \geq 0$, and $\bar{\ell}_r(0, v_r) = 0$ for all $v_r \geq 0$.

3.3 An Optimal Taxation Mechanism

Equipped with the parametrized taxation mechanism defined in (7), and with the linear programming relaxation of (5), we are now ready to provide a general

---

[4] In the ensuing proof, $v_{r,i}$ will represent the proportion of player i's weight allocated to resource r.

methodology to derive taxation mechanisms with optimal approximation. We begin with the general case, and then specialize this to polynomial latency functions. In the following, denote $P(\lambda) \sim \text{Poi}(\lambda)$ a Poisson random variable with mean $\lambda \geq 0$.

Theorem 2. *Consider a weighted congestion game G where each latency belongs to a common set of functions $\mathcal{L}$ and satisfies Assumption 1. Let $(\bar{z}, \bar{v}, \bar{y})$ be an optimal solution of* (5). *Let $\bar{\ell}_r(x; \bar{v}_r)$ satisfy recursion* (7) *and be non-decreasing in the first argument, $r \in \mathcal{R}$. Consider the taxation mechanism T that uses the perceived latency functions $\bar{\ell}_r(x; \bar{v}_r)$.*

- *Then, the price of anarchy of T is no higher than $\rho = \sup_{\ell \in \mathcal{L}} \rho_\ell$.*
- *Further, for any choice of $\varepsilon > 0$, one can compute* in polynomial time *taxes whose price of anarchy is no-higher than $\rho + \varepsilon$. The result holds for mixed Nash and correlated/coarse correlated equilibria.*

We analyze the widely studied class of polynomial latency functions with non-negative coefficients, i.e.:

$$\ell_r(x) = \sum_{d=0}^{D} b_d^r x^d \quad \text{, where } b_d^r \geq 0 \text{ for all } d \text{ and } r.$$

Importantly, we obtain explicit expressions of the taxation mechanism for this class. As an example, for the monomial $\ell(x) = x$, we obtain

$$\bar{\ell}(x; \bar{v}) = \sum_{i \in [N]} \bar{v}_i w_i + x,$$

while for the monomial $\ell(x) = x^2$, we obtain

$$\bar{\ell}_r(x; \bar{\text{v}}) = \left(\sum_{i \in [N]} \bar{v}_i w_i \right)^2 + 2 \left(\sum_{i \in [N]} \bar{v}_i w_i^2 \right) + \left(\sum_{i \in [N]} \bar{v}_i w_i \right) x + x^2.$$

Due to space limitations, we refer to [16] for a complete analysis and explicit expressions for higher degrees.

3.4 Tight Polynomial Algorithms

Since the result in Theorem 2 holds also for correlated/coarse correlated equilibria, we can leverage existing polynomial time algorithms to compute such equilibria and achieve an approximation matching the corresponding price of anarchy. This is a direct consequence of Theorem 2 and polynomial computability of correlated equilibria [25].

Corollary 1. *Consider a weighted congestion game G where each latency belongs to a common set of functions $\mathcal{L}$ and satisfies Assumption 1 and Assumption 2. Let $(\bar{z}, \bar{v}, \bar{y})$ be an optimal solution of* (5), *let $\bar{\ell}_r(x; \bar{v}_r)$ satisfy recursion* (7) *and be non-decreasing in the first argument, for $r \in \mathcal{R}$. Let $\rho = \sup_{\ell \in \mathcal{L}} \rho_\ell$. For any $\varepsilon > 0$, there exists a polynomial time algorithm to compute an allocation with a cost lower than $(\rho + \varepsilon) \cdot \min_{a \in \mathcal{A}} \mathrm{SC}(a)$.*

The approximation ratio presented in Corollary 1 can be achieved, for example, as follows. Given a desired tolerance $\varepsilon > 0$, we design a taxation mechanism ensuring a price of anarchy of $\sup_{\ell \in \mathcal{L}} \rho_\ell + \varepsilon/2$, which can be done in polynomial time thanks to Theorem 2. We use such taxation mechanisms, and compute an exact correlated equilibrium in polynomial time leveraging the result of Jiang and Leyton-Brown [25], who guarantees that the resulting correlated equilibrium has polynomial-size support. We then compute a correlated equilibrium and enumerate all pure strategy profiles in its support, identifying with a^* that with lowest cost. Since the price of anarchy bounds of Theorem 2 hold for correlated equilibria, the pure strategy profile a^* inherits a matching or better approximation ratio.

4 NP-Hardness of Approximation and Limitations of Unfair Interventions

Having shown a fair, efficiently-computable taxation mechanism with best-possible price of anarchy amongst all fair interventions, we investigate whether unfair interventions can achieve strictly better price of anarchy than their fair counterpart. Perhaps surprisingly, the answer is in the negative when polynomial computability is required for the intervention and the equilibrium concept. To obtain this result, we first show that minimizing the social cost is NP-hard to approximate below a factor identical to the lower bound on the price of anarchy of fair interventions (Theorem 3). Combined with the polynomial computability of correlated equilibria [25], and using [37], this immediately gives the following result:

Corollary 2. *In weighted congestion game G where each latency belongs to a common set of functions $\mathcal{L}$ and satisfies Assumption 1, no polynomially computable intervention, fair or unfair, achieves a price of anarchy less than $\rho_\ell = \sup_{\ell \in \mathcal{L}}$ for correlated equilibria, assuming* P$\neq$ NP.

Note that the polynomial time requirement is necessary as otherwise the intervention could just compute the optimum and prescribe the users to play accordingly.

We further point out that minimizing the social cost in weighted congestion games belongs to the more general class of *Optimization Problems with Diseconomies of Scale* studied by Makarychev and Sviridenko [29]. Hence, the ensuing theorem not only establishes latency function dependent hardness results for weighted congestion games, but also is the first to show tightness for the broader class of optimization problems with diseconomies of scale. Specifically, our result shows that the approximation algorithm in [29] is the best possible already for very simple settings, i.e., optimization problems with diseconomies of scale where the constraint set is a simplex, all resources have the same latency function, and all weights are equal. This was not known, and we believe may be of independent interest.

Theorem 3. *In weighted congestion games where all resources feature the same latency function $\ell : \mathbb{R}_{>0} \to \mathbb{R}_{>0}$ and ℓ is non-decreasing semi-convex,* `MinSC` *is* NP*-hard to approximate within any factor smaller than ρ_ℓ as defined in (3), i.e.*

$$\rho_\ell = \sup_{t>0} \frac{\mathbb{E}_{P \sim Poi(1)}[tP\ell(tP)]}{t\ell(t)}.$$

If $\rho_\ell = \infty$, then `MinSC` *is* NP*-hard to approximate within any finite factor. These NP-hardness results hold even if all players have the same weight w.*

Naturally, Theorem 3 applies directly to richer classes of congestion games, whereby the players' weights or the resources' latency functions can differ. For example, if the latency function of each resource can be constructed by non-negative linear combination of given functions $\{\ell_1, \ldots, \ell_m\}$, then `MinSC` is NP-hard to approximate within any factor smaller than $\max_{j \in \{1,\ldots,m\}} \rho_{\ell_j}$, i.e., smaller than that produced by the worst function.

Acknowledgements. This research was supported in part by a Google DeepMind Scholarship awarded to Miriam Fischer, and in part by EPSRC grant EP/Y001001/1, funded by the International Science Partnerships Fund (ISPF) and UKRI. We thank these institutions for their financial support.

References

1. Aland, S., Dumrauf, D., Gairing, M., Monien, B., Schoppmann, F.: Exact price of anarchy for polynomial congestion games. SIAM J. Comput. **40**(5), 1211–1233 (2011)
2. Amanatidis, G., et al.: Fair division of indivisible goods: Recent progress and open questions. Artif. Intell. **322**, 103965 (2023)
3. Ausiello, G., Paschos, V.T.: Approximation preserving reductions, chap. 12, pp. 351–380. Wiley (2014)
4. Awerbuch, B., Azar, Y., Epstein, A.: The price of routing unsplittable flow. SIAM J. Comput. **42**(1), 160–177 (2013)
5. Bhaskar, U., Cheng, Y., Ko, Y.K., Swamy, C.: Hardness results for signaling in bayesian zero-sum and network routing games. In: Conitzer, V., Bergemann, D., Chen, Y. (eds.) Proceedings of the 2016 ACM Conference on Economics and Computation, EC 2016, Maastricht, The Netherlands, 24–28 July 2016, pp. 479–496. ACM (2016)
6. Bhawalkar, K., Gairing, M., Roughgarden, T.: Weighted congestion games: the price of anarchy, universal worst-case examples, and tightness. ACM Trans. Econ. Comput. (TEAC) **2**(4), 1–23 (2014)

7. Bilò, V., Vinci, C.: Dynamic taxes for polynomial congestion games. ACM Trans. Econ. Comput. **7**(3), 15:1–15:36 (2019)
8. Bilò, V., Vinci, C.: On stackelberg strategies in affine congestion games. Theory Comput. Syst. **63**, 1228–1249 (2019)
9. Caragiannis, I., Kaklamanis, C., Kanellopoulos, P.: Taxes for linear atomic congestion games. ACM Trans. Algorithms **7**(1), 13:1–13:31 (2010)
10. Chakrabarty, D., Mehta, A., Nagarajan, V.: Fairness and optimality in congestion games. In: Riedl, J., Kearns, M.J., Reiter, M.K. (eds.) Proceedings 6th ACM Conference on Electronic Commerce (EC-2005), Vancouver, BC, Canada, 5–8 June 2005, pp. 52–57. ACM (2005)
11. Cheng, S.F., Reeves, D.M., Vorobeychik, Y., Wellman, M.P.: Notes on equilibria in symmetric games. In: Proceedings of the 6th International Workshop on Game Theoretic and Decision Theoretic Agents GTDT 2004, pp. 71–78 (2004)
12. Christodoulou, G., Koutsoupias, E.: The price of anarchy of finite congestion games. In: Gabow, H.N., Fagin, R. (eds.) Proceedings of the 37th Annual ACM Symposium on Theory of Computing, Baltimore, MD, USA, 22–24 May 2005, pp. 67–73. ACM (2005)
13. Christodoulou, G., Koutsoupias, E., Nanavati, A.: Coordination mechanisms. Theoret. Comput. Sci. **410**(36), 3327–3336 (2009)
14. Correa, J.R., Schulz, A.S., Stier-Moses, N.E.: Fast, fair, and efficient flows in networks. Oper. Res. **55**(2), 215–225 (2007)
15. Dwork, C., Hardt, M., Pitassi, T., Reingold, O., Zemel, R.: Fairness through awareness. In: Proceedings of the 3rd Innovations in Theoretical Computer Science Conference, pp. 214–226 (2012)
16. Fischer, M., Gairing, M., Paccagnan, D.: Fair interventions in weighted congestion games. arXiv preprint arXiv:2311.16760 (2023)
17. Fotakis, D.: Stackelberg strategies for atomic congestion games. Theory Comput. Syst. **47**(1), 218–249 (2010)
18. Fotakis, D., Spirakis, P.G.: Cost-balancing tolls for atomic network congestion games. Internet Math. **5**(4), 343–363 (2008)
19. Gairing, M., Harks, T., Klimm, M.: Complexity and approximation of the continuous network design problem. SIAM J. Optim. **27**(3), 1554–1582 (2017)
20. Gairing, M., Kollias, K., Kotsialou, G.: Existence and efficiency of equilibria for cost-sharing in generalized weighted congestion games. ACM Trans. Econ. Comput. **8**(2), 11:1–11:28 (2020)
21. Gkatzelis, V., Kollias, K., Roughgarden, T.: Optimal cost-sharing in weighted congestion games. In: Liu, T.-Y., Qi, Q., Ye, Y. (eds.) WINE 2014. LNCS, vol. 8877, pp. 72–88. Springer, Cham (2014). https://doi.org/10.1007/978-3-319-13129-0_6
22. Gould, H.W., Quaintance, J.: A linear binomial recurrence and the bell numbers and polynomials. Appl. Anal. Discret. Math. 371–385 (2007)
23. Griesbach, S.M., Hoefer, M., Klimm, M., Koglin, T.: Public signals in network congestion games. In: Proceedings of the 23rd ACM Conference on Economics and Computation, EC 2022, p. 736 (2022)
24. Gupta, S., Kamble, V.: Individual fairness in hindsight. J. Mach. Learn. Res. **22**(1), 6386–6420 (2021)
25. Jiang, A.X., Leyton-Brown, K.: Polynomial-time computation of exact correlated equilibrium in compact games. Games Econom. Behav. **91**, 347–359 (2015)
26. Kleinberg, J., Rabani, Y., Tardos, É.: Fairness in routing and load balancing. In: 40th Annual Symposium on Foundations of Computer Science (Cat. No. 99CB37039), pp. 568–578. IEEE (1999)

27. Kollias, K., Roughgarden, T.: Restoring pure equilibria to weighted congestion games. ACM Trans. Econ. Comput. (TEAC) **3**(4), 1–24 (2015)
28. Koutsoupias, E., Papadimitriou, C.: Worst-case equilibria. In: Annual Symposium on Theoretical Aspects of Computer Science, pp. 404–413. Springer (1999)
29. Makarychev, K., Sviridenko, M.: Solving optimization problems with diseconomies of scale via decoupling. J. ACM **65**(6), 42:1–42:27 (2018)
30. Meyers, C.A., Schulz, A.S.: The complexity of welfare maximization in congestion games. Networks **59**(2), 252–260 (2012)
31. Nash, J.: Non-cooperative games. Ann. Math. **2**, 286–295 (1951)
32. Paccagnan, D., Chandan, R., Ferguson, B.L., Marden, J.R.: Optimal taxes in atomic congestion games. ACM Trans. Econ. Comput. (TEAC) **9**(3), 1–33 (2021)
33. Paccagnan, D., Gairing, M.: In congestion games, taxes achieve optimal approximation. Oper. Res. **72**(3), 966–982 (2023)
34. Rosenthal, R.W.: A class of games possessing pure-strategy nash equilibria. Internat. J. Game Theory **2**(1), 65–67 (1973)
35. Roughgarden, T.: How unfair is optimal routing? In: Symposium on Discrete Algorithms: Proceedings of the Thirteenth Annual ACM-SIAM Symposium on Discrete Algorithms, pp. 203–204 (2002)
36. Roughgarden, T.: On the severity of braess's paradox: designing networks for selfish users is hard. J. Comput. Syst. Sci. **72**(5), 922–953 (2006)
37. Roughgarden, T.: Barriers to near-optimal equilibria. In: 55th IEEE Annual Symposium on Foundations of Computer Science, FOCS 2014, Philadelphia, PA, USA, 18–21 October 2014, pp. 71–80. IEEE Computer Society (2014)
38. Roughgarden, T.: Intrinsic robustness of the price of anarchy. J. ACM **62**(5), 32:1–32:42 (2015)
39. Ryan, C., Jiang, A.X., Leyton-Brown, K.: Symmetric games with piecewise linear utilities. In: Proceedings of the Behavioral and Quantitative Game Theory: Conference on Future Directions, p. 1 (2010)
40. Shaked, M., Shanthikumar, J.G.: Stochastic Orders. Springer (2007)
41. Zhou, C., Nguyen, T.H., Xu, H.: Algorithmic information design in multi-player games: possibilities and limits in singleton congestion. In: Proceedings of the 23rd ACM Conference on Economics and Computation, p. 869 (2022)

Fair and Almost Truthful Mechanisms for Additive Valuations and Beyond

Biaoshuai Tao[1(✉)] and Mingwei Yang[2]

[1] Shanghai Jiao Tong University, Shanghai 200240, China
bstao@sjtu.edu.cn
[2] Stanford University, Stanford, CA 94305, USA
mwyang@stanford.edu

Abstract. We study the problem of fairly allocating indivisible goods among n strategic agents. It is well-known that truthfulness is incompatible with any meaningful fairness notions. We bypass the strong negative result by considering the concept of *incentive ratio*, a relaxation of truthfulness quantifying agents' incentive to misreport. Previous studies show that Round-Robin, which satisfies *envy-freeness up to one good (EF1)*, achieves an incentive ratio of 2 for additive valuations.

In this paper, we explore the incentive ratio achievable by fair mechanisms for various classes of valuations besides additive ones. We first show that, for arbitrary $\epsilon > 0$, every $(\frac{1}{2} + \epsilon)$-EF1 mechanism for additive valuations admits an incentive ratio of at least 1.5. Then, using the above lower bound for additive valuations in a black-box manner, we show that for arbitrary $\epsilon > 0$, every ϵ-EF1 mechanism for *cancelable* valuations admits an infinite incentive ratio. Moreover, for *subadditive cancelable* valuations, we show that Round-Robin, which satisfies EF1, achieves an incentive ratio of 2, and every $(\varphi - 1)$-EF1 mechanism admits an incentive ratio of at least φ with $\varphi = (1 + \sqrt{5})/2 \approx 1.618$. Finally, for *submodular* valuations, we show that Round-Robin, which satisfies $\frac{1}{2}$-EF1, admits an incentive ratio of n.

Keywords: Fair division · Mechanism design · Incentive ratio

1 Introduction

In recent years, the field of discrete fair division has received extensive attention. In the canonical model, m indivisible goods, which are positively valued, are divided among a group of n competing agents in a fair manner without disposal. Arguably, the most appealing fairness notion is *envy-freeness (EF)*, which is defined as each agent weakly preferring his own bundle to any other agent's bundle. However, in the indivisible regime, EF allocations may not exist, even approximately: Consider the instance with two agents competing for one good. Given the intractability of EF, some of its natural relaxations are *envy-freeness*

The full version of our paper is available on arXiv: https://arxiv.org/abs/2306.15920.

M. Mavronicolas et al. (Eds.): WINE 2024, LNCS 15534, pp. 207–224, 2026.
https://doi.org/10.1007/978-3-032-08560-3_12

up to one good (EF1) proposed by [41] and [26], and *envy-freeness up to any good (EFX)* introduced by [29] and [37]. In an EF1 (EFX) allocation, agent i may envy agent j, but the envy can be eliminated by removing one (any) good from agent j's bundle. It is known that EF1 allocations always exist and can be computed in polynomial time for general valuations [41], whereas the existence of EFX allocations remains open even for additive valuations. See the surveys by [5] and by [13] for more details of the non-strategic setting.[1]

Nevertheless, in most practical situations, agents have no incentive to faithfully report their preferences if misreporting leads to a better outcome from their perspective. This results in a game-theoretic perspective of the fair division problem, under which the quest for fairness becomes significantly less tractable. Given that malicious behaviors of agents can result in severe fairness and welfare losses, a large body of research papers focus on designing mechanisms that are both truthful and fair [8,28,41,42]. Unfortunately, [4] shows that truthfulness is incompatible with any reasonable fairness concept if monetary transfers are prohibited, and this even holds for two agents with additive valuations.

Given the impossibility of combining fairness and truthfulness, the next direction to pursue is devising sufficiently fair mechanisms, which serves as the primary objective in our setting, while remaining close to truthfulness. Even though agents are still incentivized to manipulate under the relaxed requirement for truthfulness, there still exist reasons to settle for slight relaxation. Firstly, we measure agents' incentive to manipulate, which will be specified later, in a worst-case sense, and thus, the incentive for misreporting is very likely to be even smaller in the concerned applications. Moreover, performing effective manipulations is costly since it requires knowing other agents' preferences, which are usually difficult to acquire, and the best response turns out to be computationally intractable even for several elementary mechanisms including Round-Robin and cut-and-choose [6,11].

As a natural relaxation of truthfulness, the notion of *incentive ratio* has been widely studied under the contexts of Fisher markets [31], resource sharing [32–34], housing markets [47], and resource allocation [20,39,49]. Informally, the incentive ratio of a mechanism is defined as the worst-case ratio between the utility that an agent gains by manipulation and his utility under truthful telling. The definition of incentive ratio is also closely related to the popular notion of *approximate Nash equilibrium* [27,46].

In this paper, we explore the possibility of simultaneously achieving fairness and a small incentive ratio.[2] In the most well-studied setting of additive valuations, [49] shows that Round-Robin, which satisfies EF1, admits an incentive ratio of 2. Inspired by the recent focus on fair division for more general valuations in both the non-strategic setting [17,21,30] and the strategic setting [7], we also consider valuation classes that largely generalize additivity. In particular, one of our main focuses is the class of *cancelable* valuations, which generalizes

[1] These two surveys are later combined into one [3].

[2] This resembles the concept of *price of fairness* [16,19], which captures the efficiency loss in fair allocations as opposed to the incentive loss.

budget-additive, unit-demand, and multiplicative valuations [21] and has found its applications in various fair division results [1,7,21]. In addition, we also look into *subadditive* and *submodular* valuations, which constitute fundamental properties in combinatorial optimization and have been of recent interest in the fair division literature [7,17,30].

1.1 Our Contributions

In this paper, we study the incentive ratio achievable by fair mechanisms and give several positive and negative results for various classes of valuations, which are summarized in Table 1. In particular, we provide the first incentive ratio lower bound strictly larger than 1 by a constant as well as the first bounded incentive ratio upper bound beyond additive valuations. In more detail, we describe our main contributions as follows:

- For additive valuations, we show that every $(\frac{1}{2}+\epsilon)$-EF1 mechanism admits an incentive ratio of at least 1.5, where $\epsilon > 0$ can arbitrarily depend on n and m (Theorem 1). This result largely improves the incentive ratio lower bound of strictly larger than 1 by [7] and rules out the possibility of achieving $(\frac{1}{2}+\epsilon)$-EF1 together with an incentive ratio of arbitrarily close to 1.
- For cancelable valuations, we show that every ϵ-EF1 mechanism admits an infinite incentive ratio, where $\epsilon > 0$ can arbitrarily depend on n and m (Theorem 2). In particular, our proof utilizes our incentive ratio lower bound for approximately EF1 mechanisms with additive valuations in a *black-box* manner, and our result holds even for *multiplicative* valuations, which constitute a subclass of cancelable valuations.
- We show that the impossibility result for cancelable valuations can be bypassed by the additional property of subadditivity. Specifically, for subadditive cancelable valuations, we show that Round-Robin, which is known to satisfy EF1 [7], admits an incentive ratio of 2 (Theorem 3), thereby proving a separation between the cancelable and subadditive cancelable cases. On the negative side, we show that every $(\varphi-1)$-EF1 mechanism for subadditive cancelable valuations admits an incentive ratio of at least φ with $\varphi=(1+\sqrt{5})/2\approx 1.618$ (Theorem 4), improving the lower bound of 1.5 given under the additive case.
- For submodular valuations, we show that a generalization of Round-Robin, which is proved to satisfy $\frac{1}{2}$-EF1 [7], admits an incentive ratio of n (Theorem 5).

Finally, although Round-Robin is known to be prominent with additive valuations in both the non-strategic and strategic settings (see Sect. 1.2), its properties for more general valuations are less explored [7,22]. As a by-product, our positive results, which are all established via Round-Robin, further characterize the incentive guarantees of Round-Robin beyond additive valuations.

Table 1. Results for the incentive ratio achievable by mechanisms satisfying a specified fairness criterion for a certain class of valuations, where $\varphi = (1+\sqrt{5})/2 \approx 1.618$ and $\epsilon > 0$ denotes a small real number arbitrarily depending on n and m. The upper bound of 2 for additive valuations is shown by [49], and the lower bound of strictly larger than 1 for submodular valuations is implied by the negative result of [4] for additive valuations. The remaining bounds in the table are proved in this paper. In particular, the lower bound of 1.5 for additive valuations holds for $(\frac{1}{2}+\epsilon)$-EF1, and the lower bound of φ for subadditive cancelable valuations holds for $(\varphi-1)$-EF1.

Valuations	Fairness	Incentive Ratio
Additive	EF1	$[1.5, 2]$
Subadditive Cancelable	EF1	$[\varphi, 2]$
Cancelable	ϵ-EF1	∞
Submodular	1/2-EF1	$(1, n]$

1.2 Related Work

The mechanism design aspect of fair division is also extensively studied when resources are *divisible*, which is out of the scope of this paper, and we refer to the recent paper [24] and the references therein for a more comprehensive overview. It is worth mentioning that in divisible resource allocations, the *Maximum Nash Welfare (MNW)* and *Probabilistic Serial (PS)* rules, which satisfy multiple fairness and efficiency properties, are shown to admit an incentive ratio of 2 [20,31,39,40]. Hence, by *implementing* the fractional allocations induced by the MNW or PS rules over certain integral fair allocations, randomized fair mechanisms for indivisible resources are obtained, which satisfy desirable *ex-ante* incentive ratio guarantees promised by the fractional allocations [10,35].

Besides incentive ratio, various paradigms for bypassing the strong impossibility of combining truthfulness and fairness are also proposed. Several recent papers [6,7] initiate the study of *equilibrium fairness*, which explores the fairness guarantees of the allocations induced by *pure Nash equilibria (PNE)* with respect to the underlying true valuations. [6] shows that Round-Robin achieves desirable equilibrium fairness properties for additive valuations. Later on, [7] generalizes the equilibrium fairness guarantees of Round-Robin to cancelable and submodular valuations. In addition, other relaxed notions of truthfulness are proposed, including the *ex-ante truthfulness* [25,43], *maximin strategy-proofness* [23], *non-obvious manipulability* [44,45,48], and *risk-averse truthfulness* [24]. Finally, another series of research considers the restricted category of dichotomous valuations and aims to design truthful mechanisms accompanied by desirable fairness and efficiency properties [15,18,38].

Apart from its desirable incentive ratio and equilibrium fairness guarantees mentioned previously, Round-Robin appears as an essential tool for various fair division problems with additive valuations. Without strategic agents, its variants are applied to produce approximate *maximin share fair* allocations [9,17], EF1 allocations for *mixed goods and chores* (i.e., items with negative values) [12], and more. In the strategic setting, [45] shows that Round-Robin is not obviously manipulable, and [36] establishes that a variant of Round-Robin is *Bayesian incentive compatible* when agents' priors satisfy a neutrality condition.

2 Preliminaries

As conventions, given a mapping $f : X \to Y$, let $f^{-1}(y) = \{x \in X \mid f(x) = y\}$ for every $y \in Y$ and $f(X') = \{f(x) \mid x \in X'\}$ for every $X' \subseteq X$.

Let $G = \{g_1, \dots, g_m\}$ denote the set of m goods and $N = [n]$ be the set of n agents. A *bundle* is a subset of G. An *allocation* $A = (A_1, \dots, A_n)$ is defined as a partition of G satisfying $A_i \cap A_j = \emptyset$ for all $i \neq j$ and $\bigcup_{i=1}^{n} A_i = G$, where A_i denotes the bundle received by agent i.

We assume that each agent i is associated with a non-negative valuation $v_i(G')$ for each set of goods $G' \subseteq G$; for convenience, we write $v_i(g)$, $v_i(S - g)$ and $v_i(S + g)$ instead of $v_i(\{g\})$, $v_i(S \setminus \{g\})$ and $v_i(S \cup \{g\})$. We assume that every v_i is *normalized*, i.e., $v_i(\emptyset) = 0$, and *monotone*, i.e., $v_i(S) \leq v_i(T)$ for all $S \subseteq T \subseteq G$. We also adopt the shortcut $v_i(T \mid S)$ for the *marginal value* of a set of goods T with respect to a set of goods S, i.e., $v_i(T \mid S) = v_i(T \cup S) - v_i(S)$. For each agent i, we say that v_i is

- *subadditive*, if $v_i(S \cup T) \leq v_i(S) + v_i(T)$ for all $S, T \subseteq G$.
- *submodular*, if $v_i(g \mid S) \geq v_i(g \mid T)$ for all $S \subseteq T \subseteq G$ and $g \in G \setminus T$.
- *cancelable*, if $v_i(S + g) > v_i(T + g) \implies v(S) > v(T)$ for all $S, T \subseteq G$ and $g \in G \setminus (S \cup T)$.
- *additive*, if $v_i(S \cup T) = v_i(S) + v_i(T)$ for all $S, T \subseteq G$ with $S \cap T = \emptyset$.

Note that although both submodular and (subadditive) cancelable valuations are strict superclasses of additive valuations, neither one is a superclass of the other [7]. Given an allocation A, define the *utility* of agent i as $v_i(A_i)$.

We define the fairness notion considered in this paper as follows.

Definition 1 (α-EF1). *An allocation $A = (A_1, \dots, A_n)$ is said to satisfy* α-envy-freeness up to one good (α-EF1) *for $\alpha \in [0, 1]$ if for all $i, j \in N$, either $A_j = \emptyset$ or there exists $g \in A_j$ such that $v_i(A_i) \geq \alpha \cdot v_i(A_j - g)$. If A satisfies 1-EF1, we simply say that A satisfies EF1.*

A *mechanism* $\mathcal{M}$ takes a valuation profile $\boldsymbol{v} = (v_1, \dots, v_n)$ as input, and outputs an allocation $\mathcal{M}(\boldsymbol{v}) = (\mathcal{M}_1(\boldsymbol{v}), \dots, \mathcal{M}_n(\boldsymbol{v}))$, where $\mathcal{M}_i(\boldsymbol{v})$ denotes the bundle received by agent i. Each agent has an underlying true valuation and is required to report a (possibly fake) valuation to the mechanism. We adopt the notion of *incentive ratio* to quantify the degree of untruthfulness of a mechanism.

Definition 2 (Incentive Ratio). *The* incentive ratio *of a mechanism $\mathcal{M}$ is defined as*

$$\sup_{n,m} \sup_{v_1,\dots,v_n} \sup_{i \in [n]} \sup_{v_i'} \frac{v_i(\mathcal{M}_i(v_1, \dots, v_i', \dots, v_n))}{v_i(\mathcal{M}_i(v_1, \dots, v_i, \dots, v_n))}.$$

Observe that the incentive ratio of every mechanism is at least 1 by setting $v'_i = v_i$. If the incentive ratio of a mechanism is 1, then we say that the mechanism is *truthful.*

2.1 Strongly Desire and Control

Recall that [4] proposes the notions of *strongly desire* and *control* in the context of truthfulness with two agents. We generalize these notions and accommodate them to the concept of incentive ratio. In this subsection, we assume that there are $n = 2$ agents and do not make any restrictions on valuations except for the default that they are normalized and monotone.

For $\alpha \geq 1$, we say that an agent i *α-strongly desires* a good g if he values g strictly more than all goods in $G \setminus \{g\}$ combined multiplying by α, i.e., $v_i(g) > \alpha \cdot v_i(G - g)$. Next, we define the notion of α-control.

Definition 3 (α-Control). *Given a mechanism $\mathcal{M}$ and $\alpha \geq 1$, we say that an agent i α-*controls *a good g with respect to $\mathcal{M}$ if for every profile $\boldsymbol{v}$ where agent i α-strongly desires g, $g \in \mathcal{M}_i(\boldsymbol{v})$.*

Given a mechanism $\mathcal{M}$ and $\alpha \geq 1$, every good g is α-controlled by at most one agent with respect to $\mathcal{M}$ since when both agents α-strongly desire g, only one agent can receive it. Moreover, assuming that $\mathcal{M}$ admits an incentive ratio of α, we show in the following lemma that every good is α-controlled by exactly one agent with respect to $\mathcal{M}$.

Lemma 1. *Given a mechanism $\mathcal{M}$ with an incentive ratio of $\alpha \geq 1$, every $g \in G$ is α-controlled by exactly one agent with respect to $\mathcal{M}$.*

Proof. Let $\boldsymbol{v} = (v_1, v_2)$ be a profile where both agents α-strongly desire g. Assume without loss of generality that $g \in \mathcal{M}_1(\boldsymbol{v})$, and we show that g is α-controlled by agent 1 with respect to $\mathcal{M}$. Let $\boldsymbol{v}' = (v'_1, v'_2)$ be an arbitrary profile in which agent 1 α-strongly desires g, and we aim to show that $g \in \mathcal{M}_1(\boldsymbol{v}')$. Initially, consider the intermediate profile $\boldsymbol{v}^* = (v_1, v'_2)$. If $g \in \mathcal{M}_2(\boldsymbol{v}^*)$, then agent 2 would deviate from $\boldsymbol{v}$ to $\boldsymbol{v}^*$ to improve his utility in $\boldsymbol{v}$ by strictly more than α times. Hence, by the incentive ratio α of $\mathcal{M}$, $g \in \mathcal{M}_1(v^*)$. Similarly, if $g \in \mathcal{M}_2(\boldsymbol{v}')$, then agent 1 would deviate from $\boldsymbol{v}'$ to $\boldsymbol{v}^*$ to improve his utility in $\boldsymbol{v}'$ by strictly more than α times. Hence, by the incentive ratio α of $\mathcal{M}$, $g \in \mathcal{M}_1(\boldsymbol{v}')$, concluding that agent 1 α-controls g with respect to $\mathcal{M}$.

3 Additive Valuations

In this section, we consider additive valuations and show an incentive ratio lower bound of 1.5 for $(\frac{1}{2} + \epsilon)$-EF1 mechanisms, where $\epsilon > 0$ can arbitrarily depend on n and m.

Theorem 1. *Every $(\frac{1}{2} + \epsilon)$-EF1 mechanism for additive valuations admits an incentive ratio of at least* 1.5, *where $\epsilon > 0$ can arbitrarily depend on n and m.*

The rest of this section is devoted to proving Theorem 1, for which we construct a series of profiles and show that $(\frac{1}{2}+\epsilon)$-EF1 and an incentive ratio of strictly smaller than 1.5 cannot be simultaneously guaranteed in all these profiles. Assume for contradiction that a $(\frac{1}{2}+\epsilon)$-EF1 mechanism $\mathcal{M}$ for additive valuations exists with an incentive ratio of α satisfying $1 \le \alpha < 1.5$. Suppose that there are $n = 2$ additive agents and $m = 7$ goods. For every $i \in [2]$, denote N_i as the set of goods α-controlled by agent i with respect to $\mathcal{M}$. By Lemma 1, (N_1, N_2) forms a partition of G. Without loss of generality, assume that $|N_1| \ge 4$ and $\{g_1, g_2, g_3, g_4\} \subseteq N_1$. Denote $G' = \{g_1, g_2, g_3, g_4\}$, and every constructed additive valuation v in the proof will satisfy $v(G \setminus G') = 0$. For simplicity, we assume goods in $G \setminus G'$ always to be assigned to agent 1, and we omit them when describing valuations and allocations.

We start with the profile $\boldsymbol{v}^{(0)} = (v_1, v_2)$ where

		g_1	g_2	g_3	g_4
$\boldsymbol{v}^{(0)}$	v_1	1	1	1	1
	v_2	1	1	1	1

By the $(\frac{1}{2}+\epsilon)$-EF1 property of $\mathcal{M}$, $|\mathcal{M}_1(\boldsymbol{v}^{(0)})| = |\mathcal{M}_2(\boldsymbol{v}^{(0)})| = 2$. Without loss of generality, assume that $\mathcal{M}_1(\boldsymbol{v}^{(0)}) = \{g_1, g_2\}$ and $\mathcal{M}_2(\boldsymbol{v}^{(0)}) = \{g_3, g_4\}$.

Let δ be an arbitrary real number with $\delta \ge 5$, and we consider the next profile $\boldsymbol{v}^{(1)} = (v_1', v_2)$ where

		g_1	g_2	g_3	g_4
$\boldsymbol{v}^{(1)}$	v_1'	1	1.5	0	0
	v_2	1	1	1	1

We claim that $\mathcal{M}_1(\boldsymbol{v}^{(1)}) = \{g_1, g_2\}$ and $\mathcal{M}_2(\boldsymbol{v}^{(1)}) = \{g_3, g_4\}$. Firstly, by the $(\frac{1}{2}+\epsilon)$-EF1 property of $\mathcal{M}$, $|\mathcal{M}_2(\boldsymbol{v}^{(1)})| \ge 2$. Moreover, if $|\{g_1, g_2\} \cap \mathcal{M}_1(\boldsymbol{v}^{(1)})| < 2$, by deviating from $\boldsymbol{v}^{(1)}$ to $\boldsymbol{v}^{(0)}$, agent 1 can increase his utility in $\boldsymbol{v}^{(1)}$ by a factor of

$$\frac{v_1'(\mathcal{M}_1(\boldsymbol{v}^{(0)}))}{v_1'(\mathcal{M}_1(\boldsymbol{v}^{(1)}))} \ge \frac{v_1'(\{g_1, g_2\})}{v_1'(g_2)} = \frac{2.5}{1.5} > \alpha,$$

violating the incentive ratio α of $\mathcal{M}$. Hence, $\{g_1, g_2\} \subseteq \mathcal{M}_1(\boldsymbol{v}^{(1)})$, and it follows that $\mathcal{M}_1(\boldsymbol{v}^{(1)}) = \{g_1, g_2\}$ and $\mathcal{M}_2(\boldsymbol{v}^{(1)}) = \{g_3, g_4\}$.

We proceed to the next profile $\boldsymbol{v}^{(2)} = (v_1'', v_2)$ where
Analogous to $\boldsymbol{v}^{(1)}$, we can show that $\mathcal{M}_1(\boldsymbol{v}^{(2)}) = \{g_1, g_2\}$ and $\mathcal{M}_2(\boldsymbol{v}^{(2)}) = \{g_3, g_4\}$.

In the next profile, we modify the valuation of agent 2. Define $\boldsymbol{v}^{(3)} = (v_1', v_2')$ where

		g_1	g_2	g_3	g_4
$\boldsymbol{v}^{(2)}$	v_1''	1.5	1	0	0
	v_2	1	1	1	1

		g_1	g_2	g_3	g_4
$\boldsymbol{v}^{(3)}$	v_1'	1	1.5	0	0
	v_2'	0	δ	1	1

We claim that $\mathcal{M}_1(\boldsymbol{v}^{(3)}) = \{g_1, g_2\}$ and $\mathcal{M}_2(\boldsymbol{v}^{(3)}) = \{g_3, g_4\}$. Firstly, since agent 1 α-strongly desires g_2 in $\boldsymbol{v}^{(3)}$, $g_2 \in \mathcal{M}_1(\boldsymbol{v}^{(3)})$ by the assumption that agent 1 α-controls g_2 with respect to $\mathcal{M}$. Moreover, if $|\{g_3, g_4\} \cap \mathcal{M}_2(\boldsymbol{v}^{(3)})| < 2$, by deviating from $\boldsymbol{v}^{(3)}$ to $\boldsymbol{v}^{(1)}$, agent 2 can increase his utility in $\boldsymbol{v}^{(3)}$ by a factor of

$$\frac{v_2'(\mathcal{M}_2(\boldsymbol{v}^{(1)}))}{v_2'(\mathcal{M}_2(\boldsymbol{v}^{(3)}))} \geq 2 > \alpha,$$

violating the incentive ratio α of $\mathcal{M}$. Hence, $\{g_3, g_4\} \subseteq \mathcal{M}_2(\boldsymbol{v}^{(3)})$. Finally, if $|\mathcal{M}_2(\boldsymbol{v}^{(3)})| > 2$, by deviating from $\boldsymbol{v}^{(1)}$ to $\boldsymbol{v}^{(3)}$, agent 2 can increase his utility in $\boldsymbol{v}^{(1)}$ by a factor of

$$\frac{v_2(\mathcal{M}_2(\boldsymbol{v}^{(3)}))}{v_2(\mathcal{M}_2(\boldsymbol{v}^{(1)}))} \geq \frac{3}{2} > \alpha,$$

violating the incentive ratio α of $\mathcal{M}$. As a result, $|\mathcal{M}_2(\boldsymbol{v}^{(3)})| \leq 2$, and it follows that $\mathcal{M}_1(\boldsymbol{v}^{(3)}) = \{g_1, g_2\}$ and $\mathcal{M}_2(\boldsymbol{v}^{(3)}) = \{g_3, g_4\}$.

In the next profile, we manage to allocate g_2 to agent 2. Define $\boldsymbol{v}^{(4)} = (v_1'', v_2'')$ where

		g_1	g_2	g_3	g_4
$\boldsymbol{v}^{(4)}$	v_1''	1.5	1	0	0
	v_2''	δ	δ	1	1

We claim that $g_1 \in \mathcal{M}_1(\boldsymbol{v}^{(4)})$ and $g_2 \in \mathcal{M}_2(\boldsymbol{v}^{(4)})$. Firstly, since agent 1 α-strongly desires g_1 in $\boldsymbol{v}^{(4)}$, $g_1 \in \mathcal{M}_1(\boldsymbol{v}^{(4)})$ by the assumption that agent 1 α-controls g_1 with respect to $\mathcal{M}$. Moreover, if $\{g_1, g_2\} \subseteq \mathcal{M}_1(\boldsymbol{v}^{(4)})$, then $\mathcal{M}(\boldsymbol{v}^{(4)})$ is not $(\frac{1}{2} + \epsilon)$-EF1 for agent 2, violating the $(\frac{1}{2} + \epsilon)$-EF1 property of $\mathcal{M}$. Hence, $|\{g_1, g_2\} \cap \mathcal{M}_1(\boldsymbol{v}^{(4)})| \leq 1$, and it follows that $g_1 \in \mathcal{M}_1(\boldsymbol{v}^{(4)})$ and $g_2 \in \mathcal{M}_2(\boldsymbol{v}^{(4)})$.

We present our final profile to derive a contradiction. Define $\boldsymbol{v}^{(5)} = (v_1'', v_2')$ where
Firstly, if $|\{g_1, g_2\} \cap \mathcal{M}_1(\boldsymbol{v}^{(5)})| < 2$, by deviating from $\boldsymbol{v}^{(5)}$ to $\boldsymbol{v}^{(3)}$, agent 1 can increase his utility in $\boldsymbol{v}^{(5)}$ by a factor of

$$\frac{v_1''(\mathcal{M}_1(\boldsymbol{v}^{(3)}))}{v_1''(\mathcal{M}_1(\boldsymbol{v}^{(5)}))} \geq \frac{v_1''(\{g_1, g_2\})}{v_1''(\{g_1\})} = \frac{2.5}{1.5} > \alpha,$$

		g_1	g_2	g_3	g_4
$\boldsymbol{v}^{(5)}$	v_1''	1.5	1	0	0
	v_2'	0	δ	1	1

violating the incentive ratio α of $\mathcal{M}$. Hence, $\{g_1, g_2\} \subseteq \mathcal{M}_1(\boldsymbol{v}^{(5)})$. However, by deviating from $\boldsymbol{v}^{(5)}$ to $\boldsymbol{v}^{(4)}$, agent 2 can increase his utility in $\boldsymbol{v}^{(5)}$ by a factor of

$$\frac{v_2'(\mathcal{M}_2(\boldsymbol{v}^{(4)}))}{v_2'(\mathcal{M}_2(\boldsymbol{v}^{(5)}))} \geq \frac{v_2'(g_2)}{v_2'(\{g_3, g_4\})} = \frac{\delta}{2} > \alpha,$$

violating the incentive ratio α of $\mathcal{M}$. This concludes the proof of Theorem 1.

4 Cancelable Valuations

In this section, we consider cancelable valuations and show that every ϵ-EF1 mechanism admits an infinite incentive ratio, where $\epsilon > 0$ can arbitrarily depend on n and m.

Theorem 2. *Every ϵ-EF1 mechanism for cancelable valuations admits an infinite incentive ratio, where $\epsilon > 0$ can arbitrarily depend on n and m.*

The rest of this section is devoted to proving Theorem 2. In particular, we establish Theorem 2 by showing a stronger statement that every ϵ-EF1 mechanism for *multiplicative* valuations, which constitute a subset of cancelable valuations [21], admits an infinite incentive ratio. Recall that a valuation v is multiplicative if $v(S) = \prod_{g \in S} v(g)$ for every $S \subseteq G$ with $|S| \geq 1$. Moreover, since we assume valuations to be normalized and monotone, a multiplicative valuation v should also satisfy $v(\emptyset) = 0$ and $v(g) \geq 1$ for every $g \in G$.

Assume for contradiction that there exists an ϵ-EF1 mechanism $\mathcal{M}^c$ for multiplicative valuations with an incentive ratio of $\alpha < \infty$, where α can arbitrarily depend on n and m. We construct a mechanism $\mathcal{M}$ for additive valuations as follows, which we will show that satisfies $\frac{9}{10}$-EF1 with an incentive ratio of at most 1.1, violating Theorem 1. Given $\delta > 0$ and an additive valuation v, define v^δ as the multiplicative valuation satisfying $v^\delta(\emptyset) = 0$ and $v^\delta(S) = \exp(\delta \cdot v(S))$ for every $S \subseteq G$ with $|S| \geq 1$, and it is easy to verify that v^δ is normalized and monotone. Given as input an additive valuation profile $(v_1, \ldots, v_n)$, let $\delta > 0$ be a sufficiently large real number such that for all $i \in N$ and $S \subseteq G$ with $v_i(S) > 0$,

$$\frac{\max\{\ln(1/\epsilon), \ln \alpha\}}{\delta \cdot v_i(S)} \leq \frac{1}{10}, \tag{1}$$

and $\mathcal{M}$ outputs the allocation $\mathcal{M}^c(v_1^\delta, \ldots, v_n^\delta)$. Note that δ is a function of $(v_1, \ldots, v_n)$, ϵ, and α.

We first show that $\mathcal{M}$ satisfies $\frac{9}{10}$-EF1 for additive valuations.

Lemma 2. *$\mathcal{M}$ satisfies $\frac{9}{10}$-EF1 for additive valuations.*

Proof. Fix an additive valuation profile $(v_1, \dots, v_n)$, and let $A = \mathcal{M}(v_1, \dots, v_n) = \mathcal{M}^c(v_1^\delta, \dots, v_n^\delta)$. Fix $i, j \in N$, and we show that A satisfies $\frac{9}{10}$-EF1 for the pair of agents (i, j) with respect to $(v_1, \dots, v_n)$, i.e., if $A_j \neq \emptyset$, then there exists $g \in A_j$ such that $v_i(A_i) \geq \frac{9}{10} \cdot v_i(A_j - g)$. By the ϵ-EF1 property of $\mathcal{M}^c$, if $A_j \neq \emptyset$, then there exists $g \in A_j$ such that $v_i^\delta(A_i) \geq \epsilon \cdot v_i^\delta(A_j - g)$, which is equivalent to

$$\exp(\delta \cdot v_i(A_i)) \geq \epsilon \cdot \exp(\delta \cdot v_i(A_j - g)). \tag{2}$$

Assume that $v_i(A_j - g) > 0$, since otherwise, $\frac{9}{10}$-EF1 is straightforwardly satisfied for pair (i, j) with respect to $(v_1, \dots, v_n)$. Taking logarithm on both sides of (2) and rearranging the terms, we obtain

$$v_i(A_i) \geq \frac{\ln \epsilon}{\delta} + v_i(A_j - g) \geq \frac{9}{10} \cdot v_i(A_j - g),$$

where the second inequality holds by the assumption that $v_i(A_j - g) > 0$ and (1), implying that A satisfies $\frac{9}{10}$-EF1 for pair (i, j) with respect to $(v_1, \dots, v_n)$. Therefore, A satisfies $\frac{9}{10}$-EF1 with respect to $(v_1, \dots, v_n)$, concluding that $\mathcal{M}$ satisfies $\frac{9}{10}$-EF1 for additive valuations.

Next, we show an incentive ratio upper bound of 1.1 for $\mathcal{M}$ with additive valuations.

Lemma 3. *$\mathcal{M}$ admits an incentive ratio of at most* 1.1 *for additive valuations.*

Proof. Due to symmetry, it is sufficient to show that agent 1 cannot increase his utility under $\mathcal{M}$ by a factor of strictly larger than 1.1 via misreporting. Fix an additive valuation profile $(v_1, \dots, v_n)$, and suppose that agent 1 manipulates his valuation as $\hat{v}_1$. By the definition of $\mathcal{M}$, the utilities of agent 1 with and without manipulation are $v_1(\mathcal{M}_1(\hat{v}_1, v_2, \dots, v_n)) = v_1(\mathcal{M}_1^c(\hat{v}_1^\delta, v_2^\delta, \dots, v_n^\delta))$ and $v_1(\mathcal{M}_1(v_1, \dots, v_n)) = v_1(\mathcal{M}_1^c(v_1^\delta, \dots, v_n^\delta))$, respectively. By the incentive ratio α of $\mathcal{M}^c$,

$$v_1^\delta(\mathcal{M}_1^c(\hat{v}_1^\delta, v_2^\delta, \dots, v_n^\delta)) \leq \alpha \cdot v_1^\delta(\mathcal{M}_1^c(v_1^\delta, \dots, v_n^\delta)). \tag{3}$$

Taking logarithm on both sides and by the definition of v_1^δ, (3) is equivalent to

$$\delta \cdot v_1(\mathcal{M}_1^c(\hat{v}_1^\delta, v_2^\delta, \dots, v_n^\delta)) \leq \ln \alpha + \delta \cdot v_1(\mathcal{M}_1^c(v_1^\delta, \dots, v_n^\delta)). \tag{4}$$

If $v_1(\mathcal{M}_1^c(v_1^\delta, \dots, v_n^\delta)) = 0$, then $v_1(\mathcal{M}_1^c(\hat{v}_1^\delta, v_2^\delta, \dots, v_n^\delta)) = 0$ by (4) and (1), which implies that agent 1 cannot increase his utility under $\mathcal{M}$ by misreporting $\hat{v}_1$. From now on, we assume that $v_1(\mathcal{M}_1^c(v_1^\delta, \dots, v_n^\delta)) > 0$.

Note that by dividing $\delta \cdot v_1(\mathcal{M}_1^c(v_1^\delta, \dots, v_n^\delta)) > 0$ on both sides of (4) and rearranging the terms, we obtain

$$\frac{v_1(\mathcal{M}_1^c(\hat{v}_1^\delta, v_2^\delta, \dots, v_n^\delta))}{v_1(\mathcal{M}_1^c(v_1^\delta, v_2^\delta, \dots, v_n^\delta))} \leq \frac{\ln \alpha}{\delta \cdot v_1(\mathcal{M}_1^c(v_1^\delta, \dots, v_n^\delta))} + 1 \leq 1.1, \tag{5}$$

where the last inequality holds by the assumption that $v_1(\mathcal{M}_1^c(v_1^\delta, \ldots, v_n^\delta)) > 0$ and (1). Hence, by misreporting $\hat{v}_1$, agent 1 can increase his utility under $\mathcal{M}$ by a factor of

$$\frac{v_1(\mathcal{M}_1(\hat{v}_1, v_2, \ldots, v_n))}{v_1(\mathcal{M}_1(v_1, v_2, \ldots, v_n))} = \frac{v_1(\mathcal{M}_1^c(\hat{v}_1^\delta, v_2^\delta, \ldots, v_n^\delta))}{v_1(\mathcal{M}_1^c(v_1^\delta, v_2^\delta, \ldots, v_n^\delta))} \leq 1.1,$$

where the equality holds by the definition of $\mathcal{M}$ and the inequality holds by (5), concluding that the incentive ratio of $\mathcal{M}$ is upper bounded by 1.1 for additive valuations.

Finally, combining Lemma 2 and Lemma 3, it follows that $\mathcal{M}$ satisfies $\frac{9}{10}$-EF1 with an incentive ratio of at most 1.1 for additive valuations, which contradicts Theorem 1, concluding the proof of Theorem 2.

5 Subadditive Cancelable Valuations

We have shown in Theorem 2 that every EF1 mechanism for cancelable valuations admits an infinite incentive ratio. In this section, we show that this impossibility result can be bypassed with the additional property of subadditivity. In particular, for subadditive cancelable valuations, we show that Round-Robin, which satisfies EF1, admits an incentive ratio of 2. Then, we complement our positive result by providing an incentive ratio lower bound of $\varphi = (1+\sqrt{5})/2 \approx 1.618$ for $(\varphi - 1)$-EF1 mechanisms with subadditive cancelable valuations, improving the lower bound of 1.5 implied by Theorem 1.

5.1 Upper Bound

We first present our positive result. Recall that Round-Robin, which is formally presented in Mechanism 1, consists of multiple *rounds*, and at each round, agents alternately receive an available good with the highest value. When multiple goods have the same value, we assume agents always break ties lexicographically, i.e., breaking ties in favor of the choice with the smallest index. We call the process that an agent receives a good at a *stage*, and there are m stages in total.

Mechanism 1. Round-Robin

```
1: S = G; (A_1, ..., A_n) = (∅, ..., ∅); k = ⌈m/n⌉
2: for r = 1, ..., k do
3:     for i = 1, ..., n do
4:         g = arg max_{h∈S} v_i(h)          ▷ Break ties lexicographically.
5:         A_i = A_i ∪ {g}       ▷ The current agent receives his favorite available good.
6:         S = S \ {g}                        ▷ The good is no longer available.
7: return A = (A_1, ..., A_n)
```

For cancelable valuations, [7] shows that Mechanism 1 satisfies EF1, and hence, Mechanism 1 admits an infinite incentive ratio by Theorem 2. Nevertheless, we show that the incentive ratio of Mechanism 1 can be improved to 2 with the additional property of subadditivity.

Theorem 3. *Mechanism 1 admits an incentive ratio of* 2 *for subadditive cancelable valuations.*

The rest of this subsection is devoted to proving Theorem 3. Note that the lower bound in Theorem 3 is implied by the lower bound for the incentive ratio of Mechanism 1 for additive valuations [49], and it remains to prove the upper bound.

We first present a crucial property of cancelable valuations.

Lemma 4 ([7]). *Suppose that $v(\cdot)$ is cancelable. Let $X = \{x_1, \dots, x_k\} \subseteq G$ and $Y = \{y_1, \dots, y_k\} \subseteq G$. If $v(x_j) \geq v(y_j)$ for every $j \in [k]$, then $v(X) \geq v(Y)$.*

Since every agent i cannot alter the goods chosen in the first $i-1$ stages by manipulation, it is sufficient to prove the incentive ratio for agent 1. Assuming all agents report truthfully, we renumber the goods so that for every $i \in [m]$, g_i is the good received by some agent in stage i, i.e., g_i is the favorite good among all the remaining goods for the agent who is designated to receive a good in stage i. For $i \in \{0, \dots, m\}$, define $G_i = \{g_{i+1}, \dots, g_m\}$ as the set of remaining goods at the end of stage i and B_i as the set of goods received by agent 1 until the end of stage i.

Now, assume that agent 1 manipulates his valuation, and we run Mechanism 1 again on the manipulated valuation profile. For every $i \in \{0, \dots, m\}$, let g'_i be the good received by some agent in stage i, and define $G'_i = \{g'_{i+1}, \dots, g'_m\}$ and B'_i analogously. A crucial observation, which will be formalized later, is that $G'_i \setminus G_i$ includes all the goods in $\{g_1, \dots, g_i\}$ "left" to the subsequent stages by agent 1 via manipulation, and in the extreme case, they will all end up being received by agent 1.

For every $i \in \{0, \dots, m\}$, let $X_i = B'_i \cup (G'_i \setminus G_i)$ be the set of goods possibly obtained by agent 1 among goods in $\{g_1, \dots, g_i\}$. Our goal is to establish good-wise comparisons between goods in X_i and B_i. Specifically, we hope to assign each good in X_i to a good in B_i such that the value of the former with respect to agent 1's true valuation is upper bounded by that of the latter, and each good in B_i is assigned with at most two goods in X_i. In particular, we show that there exist mappings $M_i : X_i \to B_i$ for every $i \in \{0, \dots, m\}$ satisfying the following properties:

1. for every $g \in X_i$, $v_1(g) \leq v_1(M_i(g))$, and
2. for every $g \in B_i$, $|M_i^{-1}(g)| \leq 2$.

We demonstrate the implication of the existence of such mappings. With the mapping $M_m : X_m \to B_m$ satisfying both properties, by Property 2 , there exists a partition (R_1, R_2) of $X_m = B'_m$ such that for every $g \in B_m$, $|M_m^{-1}(g) \cap R_1| \leq 1$

and $|M_m^{-1}(g) \cap R_2| \leq 1$. Furthermore, by Property 1, we apply Lemma 4 twice with respectively $X = R_1$, $Y = M_m(R_1)$ and $X = R_2$, $Y = M_m(R_2)$ to obtain

$$v_1(R_1) \leq v_1(M_m(R_1)) \leq v_1(B_m)$$

where the second inequality holds by the monotonicity of v_1, and similarly,

$$v_1(R_2) \leq v_1(M_m(R_2)) \leq v_1(B_m).$$

As a result, by the subadditivity of v_1,

$$v_1(B'_m) \leq v_1(R_1) + v_1(R_2) \leq 2v_1(B_m).$$

This indicates that agent 1 cannot gain a utility of more than $2v_1(B_m)$ by manipulation, where $v_1(B_m)$ equals his utility when reporting truthfully, concluding the proof of Theorem 3.

It remains to prove the existence of such mappings, which is provided in the following lemma.

Lemma 5. *There exist mappings $M_i : X_i \to B_i$ for every $i \in \{0, \dots, m\}$ satisfying Properties 1 and 2 .*

Proof. We say that M_i is *valid* if it satisfies Properties 1 and 2 . For $i \in [m]$, let $b_i \in B_i$ be the good received by agent 1 the latest among all goods in B_i when he reports truthfully. Observe that $v_1(b_i) \geq \max_{g \in G_i} v_1(g)$ by the description of Mechanism 1. For $i = 0$, $X_0 = \emptyset$ since $B_0 = B'_0 = \emptyset$ and $G_0 = G'_0 = G$, and hence, a valid mapping M_0 straightforwardly exists. Assume for induction that M_{k-1} is a valid mapping with $k \in [m]$, and we show how to construct a valid mapping M_k based on M_{k-1}. For convenience, we call the goods in $X_k \setminus X_{k-1}$ as new goods and all other goods in X_k as old goods. We define $M_k(g') = M_{k-1}(g')$ for each old good g', and it remains to specify $M_k(g)$ for each new good g

If $k = cn + 1$ for some $c \in \mathbb{Z}_{\geq 0}$, i.e., it is agent 1's turn to receive his favorite good, then we have $B_k = B_{k-1} \cup \{g_k\}$, $B'_k = B'_{k-1} \cup \{g'_k\}$, and $b_k = g_k$. Note that the only possible new goods are g_k and g'_k. For each new good g, let $M_k(g) = g_k$. If $g_k = g'_k$, then it is easy to verify that M_k satisfies both Properties 1 and 2. Now, assume that $g_k \neq g'_k$, and we show that M_k constructed above is a valid mapping. Firstly, Property 2 is satisfied as only g_k and g'_k might be contained in $M_k^{-1}(b_k)$. Besides, if g_k is a new good, then Property 1 straightforwardly holds for g_k. Finally, if g'_k is a new good, then $g'_k \notin X_{k-1}$, which implies that $g'_k \in G_{k-1}$, and by the definition of g_k, we have $v_1(g'_k) \leq \max_{g \in G_{k-1}} v_1(g) \leq v_1(g_k) = v_1(M_k(g'_k))$. Thus, Property 1 also holds for g'_k. As a result, M_k is a valid mapping.

If $k = cn + j$ for some $c \in \mathbb{Z}_{\geq 0}$ and $j \in \{2, \dots, n\}$, then we have $B_k = B_{k-1}$ and $B'_k = B'_{k-1}$. Note that the only possible new good is g_k. Assume that $g_k \neq g'_k$, as otherwise, we have $X_k = X_{k-1}$ and we are done. If $g_k \notin G'_k$, then we are also done since $X_k \subseteq X_{k-1}$. Now, assume that $g_k \in G'_k \subseteq G'_{k-1}$. Since g_k is agent j's favorite good in G_{k-1} and agent j, who reports truthfully, receives $g'_k \neq g_k$ in stage k when agent 1 manipulates his valuation, we have

$g'_k \notin G_{k-1}$, which implies that $g'_k \in G'_{k-1} \setminus G_{k-1} \subseteq X_{k-1}$. Thus, $G'_k \setminus G_k = ((G'_{k-1} \setminus G_{k-1}) \cup \{g_k\}) \setminus \{g'_k\}$, which implies that g_k is a new good and $g'_k \notin X_k$. Let $M_k(g_k) = M_{k-1}(g'_k)$, and it is easy to verify that Property 2 are satisfied for M_k (Let $M_k(g_k) = M_{k-1}(g'_k) = h$. Then, g'_k was a preimage of h in M_{k-1}, and it is replaced by g_k in M_k. Therefore, the number of preimages of h is unchanged). To see that Property 1 holds, note that

$$v_1(g_k) \leq v_1(b_k) = \min_{g \in B_k} v_1(g) \leq v_1(M_{k-1}(g'_k)) = v_1(M_k(g_k)),$$

where the last inequality holds by $M_{k-1}(g'_k) \in B_k$. As a result, M_k is a valid mapping.

5.2 Lower Bound

In this subsection, we give our improved incentive ratio lower bound for $(\varphi - 1)$-EF1 mechanisms.

Theorem 4. *Let $\varphi = (1+\sqrt{5})/2$. Every $(\varphi-1)$-EF1 mechanism for subadditive cancelable valuations admits an incentive ratio of at least φ.*

The proof of Theorem 4, which can be found in the full version of our paper, is a modification of the proof of Theorem 1. Interestingly, we only need to replace one constructed additive valuation in the proof of Theorem 1 with a non-additive valuation, and all other valuations remain the same.

6 Submodular Valuations

In this section, we consider submodular valuations and show that a generalization of Round-Robin, which satisfies $\frac{1}{2}$-EF1, admits an incentive ratio of n.

Given that Mechanism 1 is not known to possess any fairness property for submodular valuations, we consider a generalization of Round-Robin presented in Mechanism 2. In particular, instead of receiving an available good with the highest value, agents alternately receive an available good with the highest marginal value with respect to the current bundle. [7] shows that Mechanism 2 satisfies $\frac{1}{2}$-EF1 for submodular valuations. We show in the following theorem that Mechanism 2 admits an incentive ratio of n for submodular valuations.

Theorem 5. *Mechanism 2 admits an incentive ratio of n for submodular valuations.*

Due to the space limit, we only outline the proof of Theorem 5 here and defer it to the full version of our paper. In particular, the upper bound stems from the observations that it suffices to prove the incentive ratio for agent 1, and the utility of agent 1 always constitutes at least a $1/n$ fraction of his value for G. For the lower bound, we construct an instance with valuations defined by *coverage functions* and show that an agent can increase his utility by a factor of n via misreporting.

Mechanism 2. Round-Robin for Submodular Valuations

$S = G$; $(A_1, \dots, A_n) = (\emptyset, \dots, \emptyset)$; $k = \lceil m/n \rceil$
for $r = 1, \dots, k$ **do**
  for $i = 1, \dots, n$ **do**
    $g = \arg\max_{h \in S} v_i(h \mid A_i)$ ▷ Break ties lexicographically.
    $A_i = A_i \cup \{g\}$ ▷ The current agent receives his favorite available good.
    $S = S \setminus \{g\}$ ▷ The good is no longer available.
return $A = (A_1, \dots, A_n)$

7 Discussion and Future Directions

In this paper, we provide both positive and negative results for the incentive ratio achievable by fair mechanisms for various categories of valuations and leave many open problems. The most interesting future direction is to close the gaps between the incentive ratio upper and lower bounds.

In addition, we only consider additive, (subadditive) cancelable, and submodular valuations in this paper, while the fair division problem with other valuation classes has also received extensive attention [2,30]. In particular, our understanding of the compatibility between fairness and incentive ratios remains quite limited for submodular and more general valuations. Hence, it would be intriguing to investigate broader valuation classes including subadditive valuations and *fractionally subadditive (XOS)* valuations, where the latter constitute a strict superset of submodular valuations and a strict subset of subadditive valuations.

For the efforts toward answering the above questions, besides Theorem 5, we further show that for XOS valuations, Mechanism 2 admits an incentive ratio of $\lceil m/n \rceil$ and does not provide any fairness guarantee. Moreover, we show that the *Envy-Graph Procedure* mechanism, which produces EF1 allocations for general valuations [41] and, to the best of our knowledge, remains the only known EF1 mechanism even for submodular valuations, admits an infinite incentive ratio for additive valuations. The proofs of all the aforementioned results can be found in the full version of our paper. Notice that all the these results are only for particular mechanisms, and it is unclear if there exist other mechanisms that can simultaneously achieve (approximate) EF1 and low incentive ratios.

Finally, the results on divisible resource allocations suggest that allowing randomization usually leads to substantial improvements in incentive guarantees [14,20,43]. Hence, it is also natural to study the incentive ratio of randomized fair mechanisms in the indivisible setting.

Acknowledgments. The research of Biaoshuai Tao was supported by the National Natural Science Foundation of China (No. 62102252).

Disclosure of Interests. The authors have no competing interests to declare that are relevant to the content of this article.

References

1. Akrami, H., Alon, N., Chaudhury, B.R., Garg, J., Mehlhorn, K., Mehta, R.: EFX: a simpler approach and an (almost) optimal guarantee via rainbow cycle number. In: EC, p. 61. ACM (2023)
2. Akrami, H., Rezvan, R., Seddighin, M.: An EF2X allocation protocol for restricted additive valuations. In: IJCAI, pp. 17–23. ijcai.org (2022)
3. Amanatidis, G., et al.: Fair division of indivisible goods: recent progress and open questions. Artif. Intell. **322**, 103965 (2023)
4. Amanatidis, G., Birmpas, G., Christodoulou, G., Markakis, E.: Truthful allocation mechanisms without payments: characterization and implications on fairness. In: EC, pp. 545–562. ACM (2017)
5. Amanatidis, G., Birmpas, G., Filos-Ratsikas, A., Voudouris, A.A.: Fair division of indivisible goods: a survey. In: IJCAI, pp. 5385–5393. ijcai.org (2022)
6. Amanatidis, G., Birmpas, G., Fusco, F., Lazos, P., Leonardi, S., Reiffenhäuser, R.: Allocating indivisible goods to strategic agents: pure nash equilibria and fairness. In: Feldman, M., Fu, H., Talgam-Cohen, I. (eds.) WINE 2021. LNCS, vol. 13112, pp. 149–166. Springer, Cham (2022). https://doi.org/10.1007/978-3-030-94676-0_9
7. Amanatidis, G., Birmpas, G., Lazos, P., Leonardi, S., Reiffenhäuser, R.: Round-robin beyond additive agents: existence and fairness of approximate equilibria. In: EC, pp. 67–87. ACM (2023)
8. Amanatidis, G., Birmpas, G., Markakis, E.: On truthful mechanisms for maximin share allocations. In: IJCAI, pp. 31–37. IJCAI/AAAI Press (2016)
9. Amanatidis, G., Markakis, E., Nikzad, A., Saberi, A.: Approximation algorithms for computing maximin share allocations. ACM Trans. Algor. **13**(4), 52:1–52:28 (2017)
10. Aziz, H.: Simultaneously achieving ex-ante and ex-post fairness. In: Chen, X., Gravin, N., Hoefer, M., Mehta, R. (eds.) WINE 2020. LNCS, vol. 12495, pp. 341–355. Springer, Cham (2020). https://doi.org/10.1007/978-3-030-64946-3_24
11. Aziz, H., Bouveret, S., Lang, J., Mackenzie, S.: Complexity of manipulating sequential allocation. In: AAAI, pp. 328–334. AAAI Press (2017)
12. Aziz, H., Caragiannis, I., Igarashi, A., Walsh, T.: Fair allocation of indivisible goods and chores. Auton. Agents Multi Agent Syst. **36**(1), 3 (2022)
13. Aziz, H., Li, B., Moulin, H., Wu, X.: Algorithmic fair allocation of indivisible items: a survey and new questions. SIGecom Exch. **20**(1), 24–40 (2022)
14. Aziz, H., Ye, C.: Cake cutting algorithms for piecewise constant and piecewise uniform valuations. In: Liu, T.-Y., Qi, Q., Ye, Y. (eds.) WINE 2014. LNCS, vol. 8877, pp. 1–14. Springer, Cham (2014). https://doi.org/10.1007/978-3-319-13129-0_1
15. Babaioff, M., Ezra, T., Feige, U.: Fair and truthful mechanisms for dichotomous valuations. In: AAAI, pp. 5119–5126. AAAI Press (2021)
16. Barman, S., Bhaskar, U., Shah, N.: Optimal bounds on the price of fairness for indivisible goods. In: Chen, X., Gravin, N., Hoefer, M., Mehta, R. (eds.) WINE 2020. LNCS, vol. 12495, pp. 356–369. Springer, Cham (2020). https://doi.org/10.1007/978-3-030-64946-3_25
17. Barman, S., Krishnamurthy, S.K.: Approximation algorithms for maximin fair division. ACM Trans. Econ. Comput. **8**(1), 5:1–5:28 (2020)
18. Barman, S., Verma, P.: Truthful and fair mechanisms for matroid-rank valuations. In: AAAI, pp. 4801–4808. AAAI Press (2022)

19. Bei, X., Lu, X., Manurangsi, P., Suksompong, W.: The price of fairness for indivisible goods. Theory Comput. Syst. **65**(7), 1069–1093 (2021)
20. Bei, X., Tao, B., Wu, J., Yang, M.: The incentive guarantees behind nash welfare in divisible resources allocation. In: WINE. Lecture Notes in Computer Science. Springer, Heidelberg (2023)
21. Berger, B., Cohen, A., Feldman, M., Fiat, A.: Almost full EFX exists for four agents. In: AAAI, pp. 4826–4833. AAAI Press (2022)
22. Bouveret, S., Lang, J.: Manipulating picking sequences. In: ECAI. Frontiers in Artificial Intelligence and Applications, vol. 263, pp. 141–146. IOS Press (2014)
23. Brams, S.J., Jones, M.A., Klamler, C., et al.: Better ways to cut a cake. Not. AMS **53**(11), 1314–1321 (2006)
24. Bu, X., Song, J., Tao, B.: On existence of truthful fair cake cutting mechanisms. Artif. Intell. **319**, 103904 (2023)
25. Bu, X., Tao, B.: Truthful and almost envy-free mechanism of allocating indivisible goods: the power of randomness. CoRR arxiv:2407.13634 (2024)
26. Budish, E.: The combinatorial assignment problem: approximate competitive equilibrium from equal incomes. In: BQGT, p. 74:1. ACM (2010)
27. Caragiannis, I., Fanelli, A., Gravin, N., Skopalik, A.: Approximate pure nash equilibria in weighted congestion games: existence, efficient computation, and structure. ACM Trans. Econ. Comput. **3**(1), 2:1–2:32 (2015)
28. Caragiannis, I., Kaklamanis, C., Kanellopoulos, P., Kyropoulou, M.: On low-envy truthful allocations. In: Rossi, F., Tsoukias, A. (eds.) ADT 2009. LNCS (LNAI), vol. 5783, pp. 111–119. Springer, Heidelberg (2009). https://doi.org/10.1007/978-3-642-04428-1_10
29. Caragiannis, I., Kurokawa, D., Moulin, H., Procaccia, A.D., Shah, N., Wang, J.: The unreasonable fairness of maximum nash welfare. ACM Trans. Econ. Comput. **7**(3), 12:1–12:32 (2019)
30. Chaudhury, B.R., Garg, J., Mehta, R.: Fair and efficient allocations under subadditive valuations. In: AAAI, pp. 5269–5276. AAAI Press (2021)
31. Chen, N., Deng, X., Tang, B., Zhang, H.R., Zhang, J.: Incentive ratio: a game theoretical analysis of market equilibria. Inf. Comput. **285**(Part), 104875 (2022)
32. Chen, Z., Cheng, Y., Deng, X., Qi, Q., Yan, X.: Agent incentives of strategic behavior in resource exchange. Disc. Appl. Math. **264**, 15–25 (2019)
33. Cheng, Y., Deng, X., Li, Y.: Tightening up the incentive ratio for resource sharing over the rings. In: IPDPS, pp. 127–136. IEEE (2020)
34. Cheng, Y., Deng, X., Li, Y., Yan, X.: Tight incentive analysis on sybil attacks to market equilibrium of resource exchange over general networks. In: EC, pp. 792–793. ACM (2022)
35. Freeman, R., Shah, N., Vaish, R.: Best of both worlds: ex-ante and ex-post fairness in resource allocation. In: EC, pp. 21–22. ACM (2020)
36. Gkatzelis, V., Psomas, A., Tan, X., Verma, P.: Getting more by knowing less: bayesian incentive compatible mechanisms for fair division. CoRR arxiv:2306.02040 (2023)
37. Gourvès, L., Monnot, J., Tlilane, L.: Near fairness in matroids. In: ECAI. Frontiers in Artificial Intelligence and Applications, vol. 263, pp. 393–398. IOS Press (2014)
38. Halpern, D., Procaccia, A.D., Psomas, A., Shah, N.: Fair division with binary valuations: one rule to rule them all. In: Chen, X., Gravin, N., Hoefer, M., Mehta, R. (eds.) WINE 2020. LNCS, vol. 12495, pp. 370–383. Springer, Cham (2020). https://doi.org/10.1007/978-3-030-64946-3_26
39. Huang, H., Wang, Z., Wei, Z., Zhang, J.: Bounded incentives in manipulating the probabilistic serial rule. J. Comput. Syst. Sci. **140**, 103491 (2024)

40. Li, B., Sun, A., Xing, S.: Bounding the incentive ratio of the probabilistic serial rule. In: AAMAS, pp. 1128–1136. International Foundation for Autonomous Agents and Multiagent Systems/ACM (2024)
41. Lipton, R.J., Markakis, E., Mossel, E., Saberi, A.: On approximately fair allocations of indivisible goods. In: EC, pp. 125–131. ACM (2004)
42. Markakis, E., Psomas, C.-A.: On worst-case allocations in the presence of indivisible goods. In: Chen, N., Elkind, E., Koutsoupias, E. (eds.) WINE 2011. LNCS, vol. 7090, pp. 278–289. Springer, Heidelberg (2011). https://doi.org/10.1007/978-3-642-25510-6_24
43. Mossel, E., Tamuz, O.: Truthful fair division. In: Kontogiannis, S., Koutsoupias, E., Spirakis, P.G. (eds.) SAGT 2010. LNCS, vol. 6386, pp. 288–299. Springer, Heidelberg (2010). https://doi.org/10.1007/978-3-642-16170-4_25
44. Ortega, J., Segal-Halevi, E.: Obvious manipulations in cake-cutting. In: Social Choice and Welfare, pp. 1–20 (2022)
45. Psomas, A., Verma, P.: Fair and efficient allocations without obvious manipulations. In: NeurIPS (2022)
46. Rubinstein, A.: Settling the complexity of computing approximate two-player nash equilibria. SIGecom Exch. **15**(2), 45–49 (2017)
47. Todo, T.: Analysis of incentive ratio in top-trading-cycles algorithms. In: Proceedings of the Annual Conference of JSAI 33rd (2019), pp. 2F1E304–2F1E304. The Japanese Society for Artificial Intelligence (2019)
48. Troyan, P., Morrill, T.: Obvious manipulations. J. Econ. Theory **185**, 104970 (2020)
49. Xiao, M., Ling, J.: Algorithms for manipulating sequential allocation. In: AAAI, pp. 2302–2309. AAAI Press (2020)

An Algorithmic Theory of Simplicity in Mechanism Design

Diodato Ferraioli[1(✉)] and Carmine Ventre[2]

[1] Università degli Studi di Salerno, 84016 Fisciano, SA, Italy
dferraioli@unisa.it
[2] King's College London, Strand, London WC2R 2LS, UK
carmine.ventre@kcl.ac.uk

Abstract. A growing body of work in economics and computation focuses on the trade-off between implementability and simplicity in mechanism design. The goal is to develop a theory that not only allows to design an incentive structure easy to grasp for imperfectly rational agents, but also understand the ensuing limitations on the class of mechanisms that enforce it. In this context, OSP mechanisms have assumed a prominent role since they provably account for the absence of contingent reasoning skills, a specific cognitive limitation. For single-dimensional agents, it is known that OSP mechanisms need to use certain greedy algorithms. In this work, we introduce a notion that interpolates between OSP and SOSP, a more stringent notion where agents only plan a subset of their own future moves. We provide an algorithmic characterization of this novel class of mechanisms for single-dimensional domains and binary allocation problems, that precisely measures the interplay between simplicity and implementability. We further show how mechanisms based on reverse greedy algorithms (a.k.a., deferred acceptance auctions) are algorithmically more robust to imperfectly rationality than those adopting greedy algorithms.

1 Introduction

A set of agents need to determine an outcome that will affect each of them. In these cases, a mechanism is designed to interact with the agents and compute an outcome based on the decisions made during the interaction. In its more general form, the mechanism can be modelled as an extensive-form game. To guarantee properties on the quality of the solution computed (e.g., optimality for an objective function of interest) mechanisms are required to provide incentives for the agents to behave in a predictable and desirable way. For example, in dominant-strategy incentive-compatible mechanisms, it is pointwise optimal for the agents to behave correctly and truthfully report their private information to the mechanism. This property can however be too weak for certain agents, as observed experimentally [4,14]. In particular, different extensive-form implementations of the mechanism can lead to different degrees of strategic confusion. Consider, for example, software agents that take decisions during the execution of the mechanism. These agents could take actions that are irrational from the economic point of view when they have been "badly" programmed, either because the programmer misunderstood the incentive structure in

M. Mavronicolas et al. (Eds.): WINE 2024, LNCS 15534, pp. 225–242, 2026.
https://doi.org/10.1007/978-3-032-08560-3_13

place or due to computational barriers preventing from comparing payoffs for each strategy adopted by the other agents [18]. A less complex decision process would be guaranteed by associating one outcome/payoff to each possible action taken by the agent, since she would simply need to rank her own actions independently of the behavior of the other agents. This is the idea behind obviously strategyproof (OSP) mechanisms [16]. OSP is a stronger notion of incentive-compatibility that takes a conservative view and requires honesty to be an obviously dominant strategy: the worst payoff that can be achieved with it is not worse than the best possible payoff obtained with a different strategy, where worst and best are chosen over the possible strategies of the remaining agents.

OSP is shown to capture the incentives of a specific form of imperfect rationality—absence of contingent reasoning skills. From the computational point of view, OSP is known to be intimately linked with greedy algorithms for single-dimensional agents: OSP is equivalent to certain well-defined combinations of greedy algorithms that are suitably monotone in the agent's private information [7,12]. This algorithmic lens allows to focus on the quality of the solutions output by OSP mechanisms, measured in terms of the approximation guarantee to a given objective function, and conclude that these monotone greedy algorithms can perform well for some binary allocation problems [7] but, in general, less well for richer solution spaces [12].

Does OSP encapsulate strategic simplicity in mechanism design? Let us consider agents with limited foresight, who can only devise a plan for their next k moves (their *planning horizon*) but cannot anticipate what they will do after that. It may not be simple enough for agents to go beyond that horizon in their reasoning; in chess, for example, if white can always win, any winning strategy is obviously dominant but requires to look at many moves in the future meaning that the strategic choices in chess are far from obvious [20]. Similarly, software agents may need more data to be "retrained" and get a more granular picture of the scenarios that go beyond their k-th future move and plan accordingly. What is simplicity in mechanism design for these agents? OSP would result simple for agents with infinite planning horizons (e.g., agents who need not retrain and have all the data available from the beginning). A solution concept called Strong OSP (SOSP) has been recently introduced in the literature to define simplicity for agents who cannot forecast any of their own future moves [20] (e.g., those who need to retrain after each move). Can we interpolate between these two extremes? Is, in case, the class of implementable mechanisms larger the longer the agents' planning horizons?

Our Contribution. We introduce the notion of k-step OSP mechanisms to capture the incentive compatibility of agents who decide their next move by only planning for the subsequent k steps and leave the remaining moves undecided. Thus, OSP corresponds to the case in which $k = \infty$ since, when determining their next move, agents reason about all their subsequent actions in the extensive-form mechanism. SOSP, on the contrary, corresponds to the case in which $k = 0$ in that no extra future action is planned for in addition to the one under consideration. We fully characterize k-step OSP mechanisms as follows:

Main Theorem 1 (informal). *A mechanism is k-step OSP for single-parameter agents and binary allocation problems if and only if each agent receives at most $k+2$ "OSP queries", the $(k+2)$-th (if any) being "payoff-neutral up to one type".*

In the informal statement above, "OSP query" means that the extensive-form mechanism must satisfy the ensuing OSP constraints, whereas "payoff-neutral up to one type query" means that the payoff for the queried agent i is essentially determined for all her moves but potentially one; this move corresponds to a single type of i for which her outcome can still be undecided at that point. Whilst it is expected that the queries must be OSP, it is surprising that there is clean connection between the number of possible queries that each agent can be asked and her planning horizon, with up to $k+1$ "unrestricted" queries to each agent i and one closing query that not only can affect the outcome that the other players will receive but also i's (albeit very limitedly)[1]. Thus, our notion fully captures the trade-off between simplicity and implementability. The more sophisticated the decision making of the agents is, the more mechanisms are incentive compatible. In particular, our notion gives rise to a fully nested class of mechanisms since every k-step OSP mechanism is also k'-step OSP for $k' > k$.

To prove this characterization, we develop the theoretical foundations to study k-step OSP mechanisms. We prove that it is w.l.o.g. to focus on a certain family of implementations and find a structural property of k-OSP constraints that limits the class of queries that the mechanism can make. We subsequently give a version of the Taxation Principle for k-OSP mechanisms by proving that from a certain point of the extensive-form game, outcomes (and payments) need to be constant for the agent being queried. Few additional structural properties are then proved to bound the number of queries to $k+2$ and determine the power of the last query.

This result complements the results by Pycia and Troyan [20] for binary allocation problems from two perspectives. Firstly, our characterization says that each agent has two payoff relevant queries in an SOSP mechanism. Under an assumption of rich domains, Pycia and Troyan [20] prove instead that agents only take one payoff-relevant move in SOSP mechanisms. Thus, our result highlights that for general single-dimensional type domains, there can be a second query which affects the outcome of the queried agent in a very specific and limited way. Secondly, Pycia and Troyan [20] consider a notion of k-step dominance, which is very related to ours, in that agents think about k future self moves when taking a decision. It turns out that their definition is technically different, with consequences on the boundaries between simplicity and implementability. In [20], although agents take into account their subsequent k moves they do not commit to them but rather follow the principle of "cross[ing] that bridge when you come to it" (i.e., make choices as they arise) [21]. It is thus assumed that the agents optimistically complete the "honest" path of play beyond their planning horizon when taking a decision. To some extent, our notion formalizes the "look before you leap" (i.e., create a complete contingent plan for the possible k future decisions one may have) model of decision making [21].

[1] This connection is neither an "innate" nor a straightforward property of k-step OSP mechanisms, since for larger outcome spaces there may be k-step OSP mechanisms with more than $k+2$ queries.

Agents take a pessimistic perspective by considering the worst possible way in which they can complete their "honest" path of play beyond their planning horizon. Consequently, our approach is more restrictive but more robust to decisions made outside the agents' planning horizons. (Please see Example 1 for an illustration of the differences between the two notions.) Importantly, within their modeling of decision making, Pycia and Troyan [20] prove that one future move is enough for 1-step simple mechanisms, for which honesty is one-step dominant, to collapse to OSP mechanisms. In fact, agents can at each step reconsider whether the plan that was 1-step dominant at the previous decision point is still the right way to go; intuitively, this freedom allows the agents to reconstruct an obviously dominant strategy step by step. Our notion instead draws a more fine grained boundary between the mechanisms that can be implemented and the behavioral model of the agents interacting with them.

Example 1 (Ascending price auctions). Consider an ascending auction for a single good, where at each non-terminal information set, an agent is called to play and has two actions, Stay In or Drop Out. The payoff of an agent i is equal to the agent's valuation v_i minus her payment if the agent is allocated the good and 0 otherwise. The price for the good weakly increases along each path of play. The auction ends when there is only one agent who has not dropped out; she wins the good and pays the price associated to the last time she moved. We remark that the ascending auction is OSP because the strategy of staying in as long as the current price is below the agent's valuation is obviously dominant, as shown in [16].

A one-step dominant strategy at information set I^* is the following. If $p(I^*)$, the price at I^*, is not higher than v_i, then Stay In and Drop Out at the subsequent information set I along a path of play. If the price at I^* is higher than v_i, then Drop Out at I^* and the subsequent information set I. This is one-step dominant because in both cases, the minimum from following the strategy is 0 (dropping out eventually) which is not worse than other strategies at I^* which can only guarantee a payoff of 0. Importantly, in her decision process, agent i is ignoring future paths of play wherein the agent will not drop out. However, player i must not necessarily drop out at I but can still stay in if the price at I is not higher than v_i. Agent i can thus eventually have inconsistent plans (cf. Remark 1 in [20]). Moreover, there are paths where i could, beyond her planning horizon, stay in even for prices higher than v_i.

A 1-step obviously dominant strategy instead mimics the aforementioned obviously dominant strategy but limits it to a planning horizon of two moves. So the agent would look at both $p(I^*)$ and $p(I)$, prices at I^* and I, and Stay In (Drop Out, resp.) at $I' \in \{I^*, I\}$ if $p(I') \leq v_i$ ($p(I') > v_i$, resp.). Incidentally, and differently from above, our notion leads to consistent strategies since the decision of agent i at I will follow the same recommendation dictated by the strategy at I^*. This strategy is 1-step obviously dominant only when the minimum payoff that can be reached by completing it in any possible way would be better than a deviation. So, for example, if the type domain of agent i were \$1, \$2, \$3, \$4 and \$5, and her type were \$4 then, by following the strategy, agent i would have stayed in for prices \$2 and \$3 (for her first two moves). To evaluate the 1-step obvious dominance of the strategy, agent i would then take the worst possible path of play that her future self could take, e.g., the decision to stay in for a price of $5 leading to a payoff of -1. The best possible payoff by deviating is 0 (e.g., dropping out).

To further explore the boundary between implementabilty and simplicity, we study the extent to which the approximation guarantee of OSP deteriorates when future self moves are not fully accounted for in decision making. We build on our first theorem to characterize what social choice functions can be implemented by k-step OSP mechanisms. Towards this end, we define a version of cycle monotonicity for k-step OSP; as in the cases of SP and OSP (amongst others) cycle monotonicity is a useful tool in that it allows to separate the allocation function (whose approximation guarantee we want to bound) from the payment function. The construction of this useful toolkit requires some further work to conveniently rewrite k-step OSP constraints. We are then able to algorithmically characterize k-step OSP mechanisms, by relating the cycles of the OSP graph [6] with those of the k-step OSP graph we define.

Main Theorem 2 (informal). *There is a k-step OSP mechanism implementing binary social choice function f if and only if f is two-way greedy with k-limitable priority functions.*

The algorithmic nature of OSP for binary allocation problems can be explained in terms of two-way greedy algorithms [7]. These are algorithms that build the eventual solution by either greedily including agents with low cost (high valuation, resp.) – termed forward greedy algorithm – or by greedily excluding agents that have a high cost (low valuation, resp.) – termed reverse greedy algorithm – for minimization (maximization, respectively) problems.[2] These algorithms adopt adaptive priority functions (that is, priority functions that depend on the past choices made by the algorithm) to define the greedy order in which agents have to be processed. The restriction from OSP to k-step OSP limits the class of priority functions that can adopted – it must be the case that each agent i is at the top of the priority list for no more than $k+2$ times excluding the instances in which i is consecutively at the top. These are k-limitable priority functions. Loosely speaking, this means that the greedy algorithm must be ready to decide whether an agent is included or excluded from the solution in no more than $k+2$ steps, thus clearly demarcating the algorithmic limitations that k-step OSP imposes.

We conclude this work by exploring the power of this class of algorithms. We focus here on p-systems, a class of problems for which it is known that greedy algorithms can compute a solution with approximation at most p. To what extent can this result be replicated for k-step OSP mechanisms? The size of the type domain of the agents and the format of the greedy algorithm play a crucial role. Let d denote half the size of the agents' type domain (for simplicity of exposition assume that d is an integer); we obtain the following tight result.

[2] In general, different agents can face different directions; in specific circumstances, it is possible to move from one direction to the other even for single agents, cf. Section 5.

Main Theorem 3 (informal). *There is a p-approximate k-limitable reverse greedy algorithm for p-systems whenever $k \geq d-2$. There is an instance of a 1-system (i.e., a weighted matroid) for which no k-limitable two-way (forward, respectively) greedy algorithm has bounded approximation for $k < d-2$ ($k < 2d-3$, respectively).*

We can then conclude that, in the worst case, there is no gradual degradation of the performances of k-step OSP mechanisms; as the value of k decreases, there is a stark dichotomy. Essentially, the two-way greedy algorithm needs to traverse the domain of each agent to compute good solutions when the types in the domain are sufficiently far apart, meaning that k must be large enough. It is conceivable to imagine that were it possible to cluster the domain of the agents around k centres without losing too much granularity, then better results would be possible. Interestingly, reverse greedy needs half the number of priority functions of forward greedy. It was already known that forward greedy is less robust than reverse greedy to economic desiderata (such as, individual rationality) of OSP mechanisms, due to the characterization of OSP payments in [11]. Our result further shows that reverse greedy is not only economically but also algorithmically more robust than forward greedy to imperfect rationality.

Related Work. OSP mechanisms have been extensively studied in various contexts, including stable matchings and single-peaked domains [2,3,22]; settings without monetary transfers [19]; binary allocation and single-dimensional problems [7,12]; and, combinatorial auctions and machine scheduling [5,15]. Extensions and variants of OSP include monitoring and verification in OSP mechanisms [8,9]; k-OSP, interpolating between OSP and classical strategyproofness [10]; and, Non-Obviously Manipulable (NOM) mechanisms [1,23]. These studies have provided characterizations, impossibility results, and approximation bounds for various OSP mechanism design problems.

2 Preliminaries

Let N be a set of n *selfish agents* and let $\mathcal{S}$ be a set of feasible *outcomes*. Each agent i has a *type* $t_i \in D_i$, where D_i is the *domain* of i. The type t_i is assumed to be *private knowledge* of agent i. We let $t_i(X) \in \mathbb{R}$ denote the *cost* of agent i with type t_i for the outcome $X \in \mathcal{S}$. When costs are negative, it means that the agent has a profit from the solution, called *valuation*.

We would like to run a *mechanism* in order to select an outcome and assign opportune payments. I.e., the mechanism implements a pair (f, p), where f (termed *social choice function* or, simply, algorithm) maps the actions taken by the agents to a feasible solution in $\mathcal{S}$, and p maps the actions taken by the agents to *payments*. Note that payments need not be positive. Each selfish agent i is equipped with a *quasi-linear utility function* $u_i\colon D_i \times \mathcal{S} \to \mathbb{R}$: for $t_i \in D_i$ and for an outcome $X \in \mathcal{S}$ returned by a mechanism $\mathcal{M}$, $u_i(t_i, X)$ is the utility that agent i has for the implementation of outcome X when her type is t_i, i.e., $u_i(t_i, X) = p_i - t_i(X)$. In this work we will focus on *single-parameter* settings, that is, the case in which the private information of each bidder i is a single real number t_i and $t_i(X)$ can be expressed as $t_i w_i(X)$ for some publicly known function w_i.

In order to implement (f, p), we design a game Γ for the agents to play. Specifically, Γ is an imperfect-information, extensive-form game with perfect recall, defined in the standard way: $\mathscr{H}$ is a finite collection of partially ordered *histories* (i.e., sequences of moves). At every non-terminal history $h \in \mathscr{H}$, one agent $i \in N$ is called to play and has a finite set of *actions* $A(h)$ from which to choose. At some history h it may be also possible that the mechanism M plays by taking a random choice: specifically, we denote with $\omega(h)$ the realization of the mechanism's random choice at history h, and with $\omega = (\omega(h))_{h \in \mathscr{H}:\, M \text{ plays at } h}$ the mechanism's random choices along the entire game. Each terminal history is associated with an outcome $X \in \mathscr{S}$, and agents receive utility $u_i(t_i, X)$. We let $h' \subseteq h$ denote that h' is a subhistory of h (equivalently, h is a continuation history of h'), and write $h' \subset h$ when $h' \subseteq h$ but $h \neq h'$. When useful, we sometimes write $h = (h', a)$ to denote the history h that is reached by starting at history h' and following the action $a \in A(h)$.

An *information set* I of agent i is a set of histories such that for any $h, h' \in I$ and any subhistories $\tilde{h} \subseteq h$ and $\tilde{h}' \subseteq h'$ at which i moves, at least one of the following conditions holds: (i) there is a history $\tilde{h}^* \subseteq \tilde{h}$ such that $\tilde{h}^*$ and $\tilde{h}'$ are in the same information set, $A(\tilde{h}^*) = A(\tilde{h}')$, and i makes the same move at $\tilde{h}^*$ and $\tilde{h}'$; (ii) there is a history $\tilde{h}^* \subseteq \tilde{h}'$ such that $\tilde{h}^*$ and $\tilde{h}$ are in the same information set, $A(\tilde{h}^*) = A(\tilde{h})$, and i makes the same move at $\tilde{h}^*$ and $\tilde{h}$. We denote by $I(h)$ the information set containing history h. Roughly speaking, an information set collects all the histories that an agent is unable to distinguish.

A *strategy* for a player i in game Γ is a function S_i that specifies an action at each one of her information sets. When we want to refer to the strategies of different types t_i of agent i, we write $S_i(t_i)$ for the strategy followed by agent i of type t_i; in particular, $S_i(t_i)(I)$ denotes the action chosen by agent i with type t_i at information set I. We use $\mathbf{S}(\mathbf{t}) = (S_i(t_i))_{i \in N}$ to denote the strategy profile for all of the agents when the type profile is $\mathbf{t} = (t_i)_{i \in N}$. An *extensive-form mechanism* is an extensive-form game together with a profile of strategies $\mathbf{S}$.

Given an agent i, an information set I is an *earliest point of departure* between strategy S_i and S_i' in game Γ if I is on the path of play under both S_i and S_i' and both strategies choose the same action at all earlier information sets, but choose a different action at I. I.e., I is the earliest information set at which these two strategies diverge. Note that for two strategies, there will in general be multiple earliest points of departure.

For an agent i with preference t_i, strategy S_i *k-step obviously dominates* strategy S_i' in game Γ if, starting at any earliest point of departure I between S_i and S_i', the outcome that maximize the utility of i among the ones reachable following S_i' at I is weakly worse than the outcome that achieves the minimum utility among the ones reachable by following S_i at I, where worst (best, resp.) cases are determined by considering any future play by other agents (including random choices of the mechanism) and any future play of agent i that coincides with strategy S_i (S_i', resp.) in all information sets I' following I such that i plays at most k times between I (excluded) and I' (included). In particular we simply say that S_i *obviously domi-*

nates S'_i when $k = \infty$, and that it *strongly obviously dominates* S'_i when $k = 0$. If a strategy S_i (k-step / strongly) obviously dominates all other S'_i, then we say that S_i is *(k-step / strongly) obviously dominant.* If a mechanism implements f by guaranteeing that each player has a (k-step / strongly) obviously dominant strategy, we say that it is *(k-step / strongly) obviously strategy-proof* ((k-step / S) OSP).

Round-Table Mechanisms. Mackenzie [17] proved that for OSP mechanisms, it is without loss of generality to consider mechanisms $\mathcal{M}$ with a specific format, named *round table mechanisms,* defined as follows. $\mathcal{M}$ is defined by a triple $(f, p, \mathcal{T})$ where, as above, the pair (f, p) determines the outcome of the mechanism, and $\mathcal{T}$ is a tree, called *implementation tree,* such that:

- Every leaf ℓ is labeled with a possible outcome of the mechanism $(X(\ell), p(\ell))$, where $X(\ell) \in \mathcal{S}$ and $p(\ell) \in \mathbb{R}^n$;
- Each internal node u in the implementation tree $\mathcal{T}$ defines the following:
  - An agent $i = i(u)$ to whom the mechanism makes some query. Each possible answer to this query leads to a different child of u.
  - A subdomain $D(u) = (D_i(u), D_{-i}(u))$ containing all types profiles that are *compatible* with u (or available at u), i.e., with all the answers to the queries from the root down to node u. Specifically, the query at node u defines a partition of the current domain of i, $D_i(u)$ into $k \geq 2$ subdomains, one for each of the k children of node u. Thus, the domain of each of these children will have, as for the domain of i, the subdomain of $D_i(u)$ corresponding to a different answer of i at u, and an unchanged domain for the other agents. We also say that action at u *signals* the associated subdomain.

Observe that, according to the definition above, for every type profile $\mathbf{b} = (b_i \in D_i)_{i \in N}$ there is only one leaf $\ell = \ell(\mathbf{b})$ such that $\mathbf{b}$ belongs to $D^{(\ell)}$. Similarly, to each leaf ℓ there is at least a profile $\mathbf{b}$ that belongs to $D^{(\ell)}$. This allows as to simplify the notation: indeed, we can define $\mathcal{M}(\mathbf{b}) = (X(\ell), p(\ell))$ for $\ell = \ell(\mathbf{b})$. Similarly, we can simply write $f(\mathbf{b}) = (f_1(\mathbf{b}), \ldots, f_n(\mathbf{b})) = X(\ell)$ and $p(\mathbf{b}) = (p_1(\mathbf{b}), \ldots, p_n(\mathbf{b})) \in \mathbb{R}^n = p(\ell)$, and $u_i(t_i, \mathcal{M}(b_i, \mathbf{b}_{-i})) = p_i(b_i, \mathbf{b}_{-i}) - t_i(f(b_i, \mathbf{b}_{-i}))$. For the single-parameter setting, considered in this work, we can further simplify the notation, by setting $t_i(X) = t_i f_i(\mathbf{b})$ when we want to express the cost of a single-parameter agent i of type t_i for the output of social choice function f when the actions taken by the agent lead to the leaf ℓ associated with bid vector $\mathbf{b}$.

Two type profiles $\mathbf{b}$, $\mathbf{b}'$ are said to *diverge* (or to be separated) at a node u of $\mathcal{T}$ if this node has two children v, v' such that $\mathbf{b} \in D(v)$, whereas $\mathbf{b}' \in D(v')$. For every such node u, we say that $i(u)$ is the *divergent agent* at u. We sometimes abuse notation and we say that two types t_i and b_i of the agent $i = i(u)$ diverge (are separated) at u.

A round-table mechanism $\mathcal{M}$ is OSP if for every agent i with real type t_i, for every vertex u such that $i = i(u)$, for every $\mathbf{b}_{-i}, \mathbf{b}'_{-i}$ (with $\mathbf{b}'_{-i}$ not necessarily different from $\mathbf{b}_{-i}$), and for every $b_i \in D_i$, with $b_i \neq t_i$, such that $(t_i, \mathbf{b}_{-i})$ and $(b_i, \mathbf{b}'_{-i})$ are compatible with u, but diverge at u, it holds that $u_i(t_i, \mathcal{M}(t_i, \mathbf{b}_{-i})) \geq u_i(t_i, \mathcal{M}(b_i, \mathbf{b}'_{-i}))$. Roughly speaking, OSP requires that each time agent i is asked to take a decision that depends on her type, the worst utility that she can get if she henceforth behaves according to her true type is at least the best utility she can get by deviating.

We will extend this definition as follows. Let us first start with some useful definitions. For each agent i, and for each vertex u in the implementation tree $\mathcal{T}$ such that $i = i(u)$, and every u' along the path from u to a leaf, we say that u' belongs to the *k-step limit* $\mathcal{L}_k(u)$ of u if u' is the k-th node along this path (u excluded) in which i is queried or, if i is queried in less than k nodes along this path, then u' is the leaf. Moreover, for each agent i, and for each vertex u in the implementation tree $\mathcal{T}$ such that $i = i(u)$, the *k-step neighborhood* $\mathcal{N}_k(u)$ of u consists of each node of $\mathcal{T}$ that appears in the path between u and some $u' \in \mathcal{L}_k(u)$. Then, for each agent i and for each vertex u in the implementation tree $\mathcal{T}$ such that $i = i(u)$, two (not necessarily different) profiles **a** and **b** are said to be *k-step unseparated* at u if **a** and **b** do not diverge either at u or in every node in the k-step neighborhood of u. Finally, for each agent i and profile **a**, and for each vertex u in the implementation tree $\mathcal{T}$ such that $i = i(u)$ and **a** is available at u, the *k-step equivalence class* of **a** at u is $\Gamma_u^k(\mathbf{a}) = \{\mathbf{b} \colon \mathbf{b}$ is k-step unseparated from **a** at $u\}$.

Then, a round-table mechanism $\mathcal{M}$ is *k-step OSP* if for every agent i with real type t_i, for every vertex u such that $i = i(u)$, for every $\mathbf{b}_{-i}, \mathbf{b}'_{-i}, \mathbf{b}''_{-i}$ (with $\mathbf{b}'_{-i}, \mathbf{b}''_{-i}$ not necessarily different from $\mathbf{b}_{-i}$ and different from each other), and for every $b_i, b'_i \in D_i$ such that $(t_i, \mathbf{b}_{-i}), (b_i, \mathbf{b}'_{-i}) \in \Gamma_u^k(t_i, \mathbf{b}_{-i})$, while $(t_i, \mathbf{b}_{-i})$ and $(b'_i, \mathbf{b}''_{-i})$ diverge at u, it holds $u_i(t_i, \mathcal{M}(b_i, \mathbf{b}'_{-i})) \geq u_i(t_i, \mathcal{M}(b'_i, \mathbf{b}''_{-i}))$. Roughly speaking, k-step OSP requires that, at each time step agent i is asked to take a decision, the worst utility that she can get if she now behaves according to her true type is at least the best utility she can get by behaving differently, *even if she later (i.e., after the next k queries received by the mechanism) changes her mind and plays as if her type was an untruthful but still available one.*

Mackenzie [17] proved that if there is an OSP mechanism implementing a social choice function f, then this can be also implemented by an OSP round-table mechanism. We show that this is true also for k-step OSP. We will thus focus on round-table mechanisms in this paper.

Theorem 1. *There is a k-step OSP mechanism implementing a social choice f if and only if f can be also implemented by a k-step OSP round-table mechanism.*

3 Characterizing the Implementation Tree

$\mathcal{M} = (f, \cdot, \mathcal{T})$ is said to be *almost-ordered*, if for every agent i, every node u of $\mathcal{T}$ such that $i = i(u)$, and every pair of profiles **a** and **b** separated at u such that $f_i(\mathbf{a}) > f_i(\mathbf{b})$ we have that $c_i < d_i$ where $c_i = \max\{x_i \mid \exists \mathbf{x}_{-i} \colon (x_i, \mathbf{x}_{-i}) \in \Gamma_u^k(\mathbf{a})\}$ and $d_i = \min\{x_i \mid \exists \mathbf{x}_{-i} \colon (x_i, \mathbf{x}_{-i}) \in \Gamma_u^k(\mathbf{b})\}$. Roughly speaking, $\mathcal{M}$ is almost ordered if for every pair L and R of the subsets in which $D_i(u)$ is partitioned at u, either $\mathcal{M}$ assigns the same outcome to all types in $L \cup R$ or these two sets are ordered (i.e., the types in one set are all smaller than the types in the other). Due to page limit, proofs are deferred to full version.

Lemma 1. *An extensive-form mechanism $\mathcal{M} = (f, p)$ with implementation tree $\mathcal{T}$ is k-step OSP if and only if for all i, all vertices u of $\mathcal{T}$ such that $i = i(u)$, and every pair* *of profiles* **a** *and* **b** *separated at u, the following k*-step OSP constraint *holds for all* $(c_i, \mathbf{c}_{-i}) \in \Gamma_u^k(\mathbf{a})$*:*

$$p_i(b_i, \mathbf{b}_{-i}) - p_i(a_i, \mathbf{a}_{-i}) \leq c_i \left(f_i(b_i, \mathbf{b}_{-i}) - f_i(a_i, \mathbf{a}_{-i})\right). \qquad (1)$$

Theorem 2. *A k-step OSP mechanism* $\mathcal{M}$ *implementing social function f with implementation tree* $\mathcal{T}$ *is almost-ordered.*

Let us focus on settings with binary outcomes, i.e., $\mathcal{S} = \{0,1\}^n$ and consider a mechanism $\mathcal{M}$ implementing (f, p) with implementation tree $\mathcal{T}$. We next provide some useful definitions.

Definition 1 (Suffix/Prefix of Type Domains). *Given a node* $u \in \mathcal{T}$ *and agent* $i = i(u)$ *we say that the domain* $D_i(u)$ *is* prefix *if* $\max\{t \in D_i(u)\} < \min\{t \in D_i \setminus D_i(u)\}$*, i.e., it only contains the smaller types in the domain of u (with larger types already removed in the queries preceding u). Similarly, we say that* $D_i(u)$ *is* suffix *if* $\min\{t \in D_i(u)\} > \max\{t \in D_i \setminus D_i(u)\}$.

Definition 2 (Revelation, Extremal and (In)Effective Queries). *Given a node* $u \in \mathcal{T}$ *and a type* $t \in D_i(u)$, $i = i(u)$*, we say that the query at this node is a(n)*

- **Revelation query** *if agent i is asked to reveal her type.*
- **Extremal query** *if agent i is asked to separate one extreme (i.e., the minimum or maximum type) from the rest of her current domain* $D_i(u)$.
- **Ineffective query** *if for every* $\mathbf{x}_{-i}$ *available at u, and every* $t', t'' \in D_i(u)$, $f_i(t', \mathbf{x}_{-i}) = f_i(t'', \mathbf{x}_{-i})$ *and* $p_i(t', \mathbf{x}_{-i}) = p_i(t'', \mathbf{x}_{-i})$*. In words, regardless of how types are partitioned by this query, i receives the same outcome and payment for every given profile* $\mathbf{x}_{-i}$.
- **Strongly Ineffective query** *if for all* $\mathbf{x}_{-i}, \mathbf{x}'_{-i} \in D_{-i}(u)$ *and all* $t', t'' \in D_i(u)$*, we have* $f_i(t', \mathbf{x}_{-i}) = f_i(t'', \mathbf{x}'_{-i})$ *and* $p_i(t', \mathbf{x}_{-i}) = p_i(t'', \mathbf{x}'_{-i})$*. In words, in such a query agent i receives the same outcome and payment for each possible profile available at that history.*
- **Only-t effective query** *if for each* $\mathbf{x}_{-i}$ *available at u and every* $t', t'' \in D_i(u)$ *with* $t', t'' \neq t$, $f_i(t', \mathbf{x}_{-i}) = f_i(t'', \mathbf{x}_{-i})$ *and* $p_i(t', \mathbf{x}_{-i}) = p_i(t'', \mathbf{x}_{-i})$*, and there is* $\mathbf{y}_{-i}$ *available at u such that* $f_i(t, \mathbf{y}_{-i}) \neq f_i(t', \mathbf{y}_{-i})$*. In words, for each profile of other agents' actions, the outcome and payment received by i is the same except for the type t.*
- **Strongly only-t effective query** *if for each* $\mathbf{x}_{-i}, \mathbf{x}'_{-i}$ *available at u and every* $t', t'' \in D_i(u)$ *such that* $t', t'' \neq t$, $f_i(t', \mathbf{x}_{-i}) = f_i(t'', \mathbf{x}'_{-i})$ *and* $p_i(t', \mathbf{x}_{-i}) = p_i(t'', \mathbf{x}'_{-i})$*, and there is* $\mathbf{y}_{-i}$ *available at u such that* $f_i(t, \mathbf{y}_{-i}) \neq f_i(t', \mathbf{y}_{-i})$*. That is, i receives exactly the same outcome and payment in each profile available at u, except for the ones in which she has type t.*

We concentrate on (strongly) only-maximum (minimum, resp.) effective queries, that is, the (strongly) only-t queries for which $t = \max\{t \in D_i(u)\}$ *(*$t = \min\{t \in D_i(u)\}$*, resp.).*

Definition 3 (*k*-limited mechanism). *A mechanism* $\mathcal{M}$ *implemented by a tree* $\mathcal{T}$ *is k*-limited *if for each agent i, and for every path P from the root of* $\mathcal{T}$ *to a leaf, one of the following holds:*

- *i is queried at most $k+1$ times along P;*
- *i is queried at most $k+2$ times along P* and *at the node u corresponding to the $(k+2)$-th query: (i) either $|D_i(u)| = 2$ or $D_i(u)$ is prefix; and (ii) the query at u is either a strongly ineffective revelation query, or a strongly only-maximum effective revelation query or an only-maximum effective extremal query that separates the maximum type in $D_i(u)$ from the rest of the domain;*
- *i is queried at most $k+2$ times along P* and *at the node u corresponding to the $(k+2)$-th query we have that: (i) $D_i(u)$ is suffix; and (ii) the query at u is either a strongly ineffective revelation query, a strongly only-minimum effective revelation query or it is an only-minimum effective extremal query that separates the minimum type in $D_i(u)$ from the rest of the domain.*

Theorem 3. *If there is a k-step OSP mechanism $\mathcal{M}$ that implements (f,p) with implementation tree $\mathcal{T}$, then there is a k-limited k-step OSP mechanism $\mathcal{M}'$ implementing (f,p).*

The rest of this section proves Theorem 3. We first prove some preliminary properties of a k-step OSP mechanism $\mathcal{M}$, which amount to a version of the taxation principle for k-step OSP.

Lemma 2 (Taxation Principle for k-step OSP). *Let $\mathcal{M}$ be a k-step OSP mechanism that implements (f,p) with implementation tree $\mathcal{T}$. For all i and for each vertex u in $\mathcal{T}$ such that $i = i(u)$, take any three profiles $\mathbf{a} = (a_i, \mathbf{a}_{-i})$, $\mathbf{c} = (c_i, \mathbf{a}_{-i})$, and $\mathbf{d} = (d_i, \mathbf{a}_{-i})$ such that (i) $a_i < c_i < d_i$, (ii) $\mathbf{a}, \mathbf{c}, \mathbf{d} \in \Gamma_u^k(\mathbf{a})$, and (iii) there is $u' \notin \mathcal{N}_k(u)$ such that $i = i(u')$, $\mathbf{a}$, $\mathbf{c}$, and $\mathbf{d}$ are available at u', and two among $\mathbf{a}$, $\mathbf{c}$, and $\mathbf{d}$ are separated at u'. We have that:*

1. *if there are $\mathbf{b}$ and $\mathbf{b}'$ separated from $\mathbf{a}, \mathbf{c}, \mathbf{d}$ by i along the path from u to u' such that $b_i > d_i$ and $b'_i < a_i$, then $f_i(\mathbf{a}) = f_i(\mathbf{c}) = f_i(\mathbf{d})$ and $p_i(\mathbf{a}) = p_i(\mathbf{c}) = p_i(\mathbf{d})$ (Outer-sandwich separations);*
2. *if there is $\mathbf{b}$ separated from $\mathbf{a}, \mathbf{c}, \mathbf{d}$ by i along the path from u to u' such that $a_i < b_i < d_i$, then $f_i(\mathbf{a}) = f_i(\mathbf{c}) = f_i(\mathbf{d})$ and $p_i(\mathbf{a}) = p_i(\mathbf{c}) = p_i(\mathbf{d})$ (Inner-sandwich separation);*
3. *if there is $\mathbf{b}$ separated from $\mathbf{a}, \mathbf{c}, \mathbf{d}$ by i along the path from u to u' such that $b_i > d_i$ then $f_i(\mathbf{a}) = f_i(\mathbf{c})$ and $p_i(\mathbf{a}) = p_i(\mathbf{c})$ (Top separation);*
4. *if there is $\mathbf{b}$ separated from $\mathbf{a}, \mathbf{c}, \mathbf{d}$ by i along the path from u to u' such that $b_i < a_i$ then $f_i(\mathbf{c}) = f_i(\mathbf{d})$ and $p_i(\mathbf{c}) = p_i(\mathbf{d})$ (Bottom separation).*

Mechanism $\mathcal{M}$ with implementation tree $\mathcal{T}$ is *almost k-limited* if $\mathcal{T}$ is such that for each agent i and every path P from the root of $\mathcal{T}$ to a leaf, the following conditions are satisfied:

- for $u \in P$ corresponding to the $(k+2)$-th query to i, if we denote with a_i and d_i the minimum and the maximum of the types of i available at u respectively, we must have that
  - if $D_i(u)$ contains at least three types and it is neither prefix nor suffix (i.e., in the previous $k+1$ queries to i, either i separated from $D_i(u)$ types b_i and b'_i such that $b_i > d_i$ and $b'_i < a_i$, or i separated from $D_i(u)$ type b_i such that $a_i < b_i < d_i$), then the $(k+2)$-th query is ineffective;

- if $D_i(u)$ contains at most two types or it is prefix (i.e., in the previous $k+1$ queries to i, i separated from $D_i(u)$ only types b_i such that $b_i > d_i$), then the $(k+2)$-th query is either ineffective or it is only-maximum effective;
- if $D_i(u)$ contains at least three types and it is suffix (i.e., in the previous $k+1$ queries to i, i separated from the domain only types b_i such that $b_i < a_i$), then the $(k+2)$-th query is either ineffective or it is only-minimum effective.

– every $u \in P$ corresponding to the q-th query to i with $q \geq k+3$ are ineffective.

Considering the first query along a path P, Lemma 2 restricts the $(k+2)$-th query on P as in the first bullet point (the three cases corresponding to sandwich, top, and bottom separations, respectively). Moreover, given these properties, the subsequent queries must be ineffective (as requested by the second condition of almost k-limited mechanisms). Thus, Lemma 2 can be restated as follows.

Corollary 1. *If a mechanism $\mathcal{M}$ that implements (f,p) with implementation tree $\mathcal{T}$ is k-step OSP, then $\mathcal{T}$ is almost k-limited.*

We further limit the effectiveness of the $(k+2)$-th and following queries: if a query is ineffective for a pair of separated types then it is strongly ineffective for those types.

Lemma 3 (Strong Ineffectiveness upon Separation). *If a mechanism $\mathcal{M}$ that implements (f,p) with implementation tree $\mathcal{T}$ is k-step OSP, then for every i, and every path P of $\mathcal{T}$, every node $u \in P$ such that $i(u)=i$, if there are $t,t' \in D_i(u)$ such that for every $\mathbf{x}_{-i}$ available at u we have that $f_i(t,\mathbf{x}_{-i}) = f_i(t',\mathbf{x}_{-i})$, and t,t' are separated at u, then for every $\mathbf{x}_{-i},\mathbf{x}'_{-i}$ available at u we have that $f_i(t,\mathbf{x}_{-i}) = f_i(t',\mathbf{x}'_{-i})$ and $p_i(t,\mathbf{x}_{-i}) = p_i(t',\mathbf{x}'_{-i})$.*

Lemma 3 states that if in path P of $\mathcal{T}$ the $(k+2)$-th query to i is ineffective, then it must be strongly ineffective. Similarly, if in some path P of $\mathcal{T}$ the $(k+2)$-th query to i is only-maximum (only-minimum, resp.) effective, and the types of i different from the maximum (minimum, resp.) are separated at the $(k+2)$-th or successive query, then the $(k+2)$-th query is strongly only-maximum (only-minimum, resp.) effective. In order to stress that in $\mathcal{T}$ this property must be satisfied, we say that the corresponding mechanism is *strongly almost k-limited*. Note that the $(k+2)$-th query of a strongly almost k-limited mechanism can still be extremal if this is the last query.

Lemmas 2 and 3 imply that a k-step OSP mechanism must be strongly almost k-limited. However, in such a mechanism we are still allowed to query agent i more than $k+2$ times, or to have a $(k+2)$-th query that is neither a revelation nor an extremal query. However we can show that it is possible to transform $\mathcal{T}$ in order to achieve the desired reduction to a k-limited mechanism. In particular, we will see that we can always guarantee that the $(k+2)$-th is a revelation query. Theorem 3 then follows from the repeated application of these transformations, since for each path, agent i is either queried at most $k+1$ times, or she is queried $k+2$ times, the last being a revelation or an extremal query.

4 Cycle-Monotonicity for k-Step OSP

For each i, for each path P from the root of $\mathcal{T}$ to the node u corresponding to the $(k+2)$-th query to i, if it exists, or to a leaf otherwise, we partition the domain D_i of types of i in classes $D^1_{i,P},\dots,D^{\ell}_{i,P}$, where $\ell = \min\{k+2, q+1\}$, q denoting the number of queries to i along this path; $D^j_{i,P}$, for $j < \ell$ contains all types of i available at the j-th query at i, but not at the $(j+1)$-th query to i along P, and $D^{\ell}_{i,P} = D_i \setminus \bigcup_{j=1}^{\ell-1} D^j_{i,P}$.

Suppose that, along the path P, i is queried at least $k+2$ times, and let u be the node corresponding to the $(k+2)$-th query at i. Let $D^{k+2,\overline{E}}_{i,P}$ be a maximal set of types of i available at u that receive the same outcome for each possible fixed profile of the remaining agents available at u, i.e., $D^{k+2,\overline{E}}_{i,P} \subseteq D^{k+2}_{i,P}$ such that for every $t, t' \in D^{k+2,\overline{E}}_{i,P}$ and every $\mathbf{x}_{-i}$ available at u we have that $f_i(t, \mathbf{x}_{-i}) = f(t', \mathbf{x}_{-i})$, and for every $t \in D^{k+2,\overline{E}}_{i,P}$ and every $t'' \in D^{k+2,E}_{i,P} = D^{k+2}_{i,P} \setminus D^{k+2,\overline{E}}_{i,P}$ there is $\mathbf{y}_{-i}$ available at u for which $f_i(t, \mathbf{y}_{-i}) \neq f_i(t'', \mathbf{y}_{-i})$. Note that if the $(k+2)$-th query is (strongly) ineffective, we have that $D^{k+2,\overline{E}}_{i,P} = D^{k+2}_{i,P}$. Moreover, if the $(k+2)$-th query is (strongly) only-maximum or only-minimum effective, then $D^{k+2,E}_{i,P}$ contains only the one extremal type for which the $(k+2)$-th query is effective.

Given this partition of the type domain of agent i, we then partition the profiles in equivalence classes as follows: for each agent i, for each path P from the root of $\mathcal{T}$ to a node u corresponding to the $(k+2)$-th query to i, if it exists, or to a leaf otherwise, we define the equivalence classes $\Lambda^j_{i,P}(b) = \{(x_i, \mathbf{x}_{-i}) \mid x_i \in D^j_{i,P}, \mathbf{x}_{-i}$ available at u, $f_i(x_i, \mathbf{x}_{-i}) = b\}$, for $b \in \{0,1\}$ and $j = q+1$ if in P agent i is queried at most $q \le k+1$ times, and $j \in \{(k+2, E), (k+2, \overline{E})\}$ otherwise. Let Λ_i be the set that contains all the equivalence classes defined for agent i. Moreover, we abuse notation and we set $f_i(\Lambda^j_{i,P}(0)) = 0$ and $f_i(\Lambda^j_{i,P}(1)) = 1$.

Definition 4. *(k-step OSP-graph) Let f be a social choice function and $\mathcal{T}$ be an implementation tree. We define for every agent i, the k-step OSP-graph $\mathcal{O}^k_{i,f,\mathcal{T}}$ with Λ_i as the set of vertices, and an edge e between $\Lambda, \Lambda' \in \Lambda_i$ exists if there are $\mathbf{x} \in \Lambda$ and $\mathbf{x}' \in \Lambda'$ such that $\mathbf{x}$ and $\mathbf{x}'$ have been separated in $\mathcal{T}$ by i. Moreover, we set* $w(e) = \min_{x_i \mid \exists \mathbf{x}_{-i}:\,(x_i,\mathbf{x}_{-i}) \in \Lambda} x_i (f_i(\Lambda') - f_i(\Lambda))$.

We say that the k-step OSP cycle monotonicity (k-step OSP CMON) property holds if, for all i, the graph $\mathcal{O}^k_{i,f,\mathcal{T}}$ does not have negative weight cycles.

Theorem 4. *A mechanism $\mathcal{M}$ with implementation tree $\mathcal{T}$ is k-step OSP for a social function f on finite domains if and only if it is k-limited and k-step OSP CMON holds.*

5 Algorithmic Characterization

We start by recalling the characterization for $k = \infty$ [7]: there is an OSP mechanism implementing f if and only f is *two-way greedy implementable.* Let us define this notion. Agent i is *revealable* at node u under social choice function f if either $f_i(\mathbf{x}) = 1$ for every $\mathbf{x}$ such that $x_i \in D_i(u)$ and $x_i < \max D_i(u)$, and $\mathbf{x}_{-i}$ is compatible with u, or $f_i(\mathbf{x}) = 0$ for every $\mathbf{x}$ such that $x_i \in D_i(u)$ and $x_i > \min D_i(u)$, and $\mathbf{x}_{-i}$ is compatible with u. Algorithm $f\colon \times_{i=1}^n D_i \to \mathcal{S}$ is *two-way greedy implementable* if it can be implemented by a round-table mechanism with implementation tree $\mathcal{T}$ such that:

- at each node u the agent $i = i(u)$ separates her domain $D_i(u)$ in two ordered subsets $L_i(u)$ and $R_i(u)$, i.e., $\max L_i(u) < \min R_i(u)$, such that at least one of the two is a singleton, i.e., it contains a single type, being the minimum in $D_i(u)$ (if $L_i(u)$ is a singleton) or the maximum (if $R_i(u)$ is a singleton). If $L_i(u)$ is a singleton, then the agent i receives outcome 1 in every profile compatible with $L_i(u) \times D_{-i}(u)$, and we say that i interacts with the mechanism in a greedy fashion. If $R_i(u)$ is a singleton, then i receives outcome 0 in every profile compatible with $R_i(u) \times D_{-i}(u)$, and we say that i interacts with the mechanism in a reverse greedy fashion;
- each agent i is not allowed to interleave interaction in a greedy fashion with interaction in a reverse greedy fashion until she is revealable. Specifically, agent i at node u is either revealable, or can interact with the mechanism in a greedy fashion if and only if she i interacted with the mechanism in a greedy fashion at every node u' along the path between the root of $\mathcal{T}$ and u such that $i = i(u')$.

While the definition of two-way greedy implementable mechanism appears to be quite involved, it establishes a very strong relationship between two-way greedy implementable algorithms and greedy algorithms [7]. Essentially every greedy algorithm and every reverse greedy algorithm (a.k.a., deferred-acceptance algorithm) is two-way greedy, and hence can be turned in an OSP mechanism. This continues to hold if the algorithm is allowed to greedily insert into the solutions some components (i.e., interact greedily with some agents) and reverse greedily remove from the solutions other components (i.e., interact reverse greedily with other agents). We will finally observe that whenever an agent is revealable at some node u, we can ask the agent to reveal her type without affecting the OSPness of the mechanism.

We can prove there is a useful relation between negative cycles in the k-step OSP-graph, and negative cycles in the ∞-step OSP-graph: there is no negative cycle in $\mathcal{O}^k_{i,f,\mathcal{T}}$ if and only if there is no negative cycle in $\mathcal{O}^\infty_{i,f,\mathcal{T}}$. However, two-way greedy implementation of f assumes that the implementation tree $\mathcal{T}$ is binary and makes only extremal queries. Clearly, each implementation tree $\mathcal{T}'$ for which there are no negative cycles in $\mathcal{O}^\infty_{i,f,\mathcal{T}'}$, regardless of the kind of queries that are performed in $\mathcal{T}'$, can be transformed in an implementation tree $\mathcal{T}$ with binary extremal queries that still has no negative cycle through a *serialization* procedure (see [7, Observation 3 and Theorem 4]). Similarly, a binary implementation tree $\mathcal{T}$ with extremal queries and no negative cycle in $\mathcal{O}^\infty_{i,f,\mathcal{T}}$ can be transformed in an implementation tree $\mathcal{T}'$ with generic queries but still no negative cycle through a simple *compression* procedure: every two consecutive nodes u and u' such that $i = i(u) = i(u')$ can be merged in a single node U with outgoing edges leading to $v_1, \ldots, v_x$, and to $v'_1, \ldots, v'_y$, where $v_1, \ldots, v_x$ are the children of u different from u', and $v'_1, \ldots, v'_y$ are the children of u'. However, these serialization/compression operations will change the number of queries done to each agent. While this does not matter for ∞-step OSPness, it turns out to be very relevant for k-step OSPness. Hence, we will introduce a way to keep the number of queries to a given agent unchanged after the operations of serialization/compression. Specifically, we say that a binary implementation tree $\mathcal{T}$ with extremal queries is *k-limitable* if and only if the implementation tree $\mathcal{T}'$ achieved through the compression procedure described above is *k-limited.* (For convenience we are abusing a bit our terminology by calling k-limited the implementation tree $\mathcal{T}'$ rather than the mechanism $\mathcal{M}$ using $\mathcal{T}'$.)

Theorem 5. *There is a k-step OSP mechanism implementing f iff f is two-way greedy implementable and the corresponding implementation tree $\mathcal{T}$ is k-limitable.*

Essentially, Theorem 5 states that the OSP characterization in term of greedy algorithms continues to hold even for k-step OSPness. However, we here further require a constraint to be satisfied, i.e., that the implementation tree is k-limited (in its compressed version) or k-limitable (in the serialized version). The effect of this limitation is very heavy in the case of SOSP mechanisms.

6 Approximation Guarantee of k-Step OSP Mechanisms

In this section we apply our characterization to quantify the restriction that k-step OSP imposes on the quality of the algorithmic solution that can be implemented. We will focus on maximization problems; agents' types will thus be a non-negative valuation (i.e., a non-positive cost) for each algorithmic allocation received.

To introduce our questions of interest, we start by discussing *single-item auctions*: n agents have a valuation $v_i \in D_i$ for the item, and we are willing to sell the item to the agent with the highest valuation. Note that in order to fully define the social function f for this problem, we need to specify how ties are broken. We will assume that ties are broken in favor of the agent with the smallest index. We will also assume for simplicity that every agent i has a domain $D_i = D$. The ascending price auction discussed in Example 1 is an OSP mechanism that solves this problem optimally. We can rephrase it in terms of two-way greedy implementation as follows: mark all agents as available; for t from the smallest type to the second largest type in D or until there is only one available agent, ask each available agent i in order of their index whether her type is t, and in case of positive answer, mark the agent as unavailable; assign the item to the agent with smallest index among the available ones. Actually, this is not the unique OSP auction implementing the social function of interest (e.g., we can query agents from the largest to the second smallest type and allocate the item upon a positive answer).

Our question is then: can we implement single-item auctions with a k-step OSP mechanism? If not, how large can be the *price of limited foresight*? To answer these questions, observe that the English auction described above queries agents $|D|-1$ times, the last query being a revelation strongly only-minimum effective query. Hence, this mechanism is k-limitable, and, by Theorem 5, k-step OSP for $k \geq |D|-3$.

However, the mechanism fails to be $(|D|-4)$-limited. Indeed, after the $(|D|-3)$-th query to agent i, her domain contains the three largest types, and for each action of other agents compatible with the $(|D|-3)$-th query, the outcome of this agent should be the same when agent i has the smallest and the second smallest type. But the ascending price auction described above does not provide such a guarantee. Can there be another k-step OSP mechanism implementing the social choice function f? Or some social function f' differing from f only in the tie-breaking rule (and hence, still returning an allocation that maximizes the social welfare)? We next address these questions in a more general setting than single-item auctions.

p-Systems. We will now focus on the class of problems that satisfy the *downward closed property*. In these problems we are given a set E of elements, a weight $w(e) \in \mathbb{R}$ for each $einE$, and a set $\mathcal{F}$, containing subsets $S \subseteq E$, named *feasible solutions*, that enjoy the following property: if $S \in \mathcal{F}$, then $T \in \mathcal{F}$ for every $T \subseteq S$, and we aim to select the feasible solution of maximum total weight, i.e. $S^* = \arg\max_{S \in \mathcal{F}} \sum_{e \in S} w(e)$.

Specifically, we focus on *p-systems*: these are special problems for which it is known that the *greedy* algorithm, that processes elements e in decreasing order of their weight, and includes them in the current solution S unless $S \cup \{e\} \notin \mathcal{F}$, and the *reverse greedy* algorithm, that processes elements e in increasing order of their weight, and remove any solution S containing e from the set $\mathcal{O}$ of maximal feasible solutions unless this empties $\mathcal{O}$, have approximation p. It is easy to check that the problem of social-welfare single-item auction is an example of 1-system (a.k.a., matroid). (For a formal definition of p-systems, we refer the reader to [13].) This, in turn, means that for all these problems, it is possible to design an OSP mechanism that is able to return a p-approximate feasible solution for the problem [7].

We defer to the full version the proof that it is always possible to implement the reverse greedy algorithm defined above as a k-limitable two-way greedy algorithm when $k \geq \left\lceil \frac{|D|}{2} \right\rceil - 2$, and thus, by Theorem 5, that a k-step OSP algorithm exists that returns a p-approximate solution for every p-system. We also show that, if no constraint is given on the values in D, then no k-step OSP mechanism exists that is able to return a bounded approximation of the optimal solution, whenever $k < \left\lceil \frac{|D|}{2} \right\rceil - 2$.

Theorem 6. *There is a two-way greedy algorithm, which (i) is k-limitable for $k \geq \left\lceil \frac{|D|}{2} \right\rceil - 2$; and, (ii) returns a p-approximation of the feasible set of maximum weight for every p-system. Conversely, for every $\rho > 0$, every $d = |D| \geq 5$, and every $k < \frac{d}{2} - 2$ ($k < d - 3$, resp.), there is a p-system $(E, \mathcal{F})$ such that no k-limitable two-way greedy (greedy, resp.) algorithm returns a feasible solution of total weight not larger than a $\frac{1}{\rho}$ fraction of the total weight of the optimal feasible solution.*

7 Conclusions

In this work, we have studied the algorithmic robustness of OSP to agents that are not able to perform contingent reasoning and think about their future actions. Specifically, we introduce a novel notion, termed k-step OSP, that smoothens the notions of OSP (where absence of contingent reasoning is the only cognitive limitation) and SOSP (where in addition agents are unable to think about any of their future actions) by maintaining the assumption that agents are not able to think contingently but allowing them a foresight of k self moves ahead. We provide an algorithmic characterization of these mechanisms for single-dimensional agents and binary outcomes, via the introduction of a new cycle monotonicity toolkit. We apply our characterization to downward-closed maximization problems and prove that the performance of

OSP can deteriorate when k is small in comparison to the type space of the agents. En route, we prove that reverse greedy algorithms are more robust than greedy algorithms to this worsening of the performances.

A natural open problems left by this work is to understand the extent to which the findings above hold for non-binary allocation problems, and quantify the limitations of limited planning horizons for other optimization problems, such as scheduling related machines that is now fully understood for OSP [12]. We highlight that our framework can also be adopted for non-binary allocation problems to prove that queries from the $(k+2)$-th onward must be limited, but this does not necessarily lead to mechanisms with only $k+2$ queries since the Taxation Principle for k-step OSP would be less clean as the relative weight of types and outcomes would matter.

Acknowledgements. Diodato Ferraioli is supported by the PNRR project FAIR – Future AI Research (PE00000013) under the NRRP MUR program funded by NextGenerationEU. Carmine Ventre acknowledges funding from the UKRI Trustworthy Autonomous Systems Hub (EP/V00784X/1).

References

1. Archbold, T., De Keijzer, B., Ventre, C.: Willy wonka mechanisms. In: Proceedings of the 23rd International Conference on Autonomous Agents and Multiagent Systems (AAMAS 2024) (2024)
2. Arribillaga, R.P., Massó, J., Neme, A.: All sequential allotment rules are obviously strategy-proof. Theor. Econ. **18**(3), 1023–1061 (2023)
3. Ashlagi, I., Gonczarowski, Y.A.: Stable matching mechanisms are not obviously strategy-proof. J. Econ. Theory **177**, 405–425 (2018)
4. Ausubel, L.M.: An efficient ascending-bid auction for multiple objects. Am. Econ. Rev. **94**(5), 1452–1475 (2004)
5. Ferraioli, D., Meier, A., Penna, P., Ventre, C.: Obviously strategyproof mechanisms for machine scheduling. In: Bender, M.A., Svensson, O., Herman, G. (eds.) 27th Annual European Symposium on Algorithms, ESA 2019, vol. 144, pp. 46:1–46:15 (2019)
6. Ferraioli, D., Meier, A., Penna, P., Ventre, C.: New constructions of obviously strategyproof mechanisms. Math. Oper. Res. **48**(1), 332–362 (2023)
7. Ferraioli, D., Penna, P., Ventre, C.: Two-way greedy: algorithms for imperfect rationality. In: Feldman, M., Fu, H., Talgam-Cohen, I. (eds.) WINE 2021. LNCS, vol. 13112, pp. 3–21. Springer, Cham (2022). https://doi.org/10.1007/978-3-030-94676-0_1
8. Ferraioli, D., Ventre, C.: Probabilistic verification for obviously strategyproof mechanisms. In: Proceedings of the Twenty-Seventh International Joint Conference on Artificial Intelligence, IJCAI-18, pp. 240–246 (2018)
9. Ferraioli, D., Ventre, C.: Approximation guarantee of OSP mechanisms: the case of machine scheduling and facility location. Algorithmica **83**(2), 695–725 (2021)
10. Ferraioli, D., Ventre, C.: Obvious strategyproofness, bounded rationality and approximation. Theory Comput. Syst. **66**(3), 696–720 (2022)
11. Ferraioli, D., Ventre, C.: Explicit payments for obviously strategyproof mechanisms. In: Proceedings of the 2023 International Conference on Autonomous Agents and Multiagent Systems, AAMAS 2023, London, United Kingdom, 29 May–2 June 2023, pp. 21257–21336. ACM (2023)

12. Ferraioli, D., Ventre, C.: On the connection between greedy algorithms and imperfect rationality. In: Leyton-Brown, K., Hartline, J.D., Samuelson, L. (eds.) Proceedings of the 24th ACM Conference on Economics and Computation, EC 2023, London, United Kingdom, 9–12 July 2023, pp. 657–677. ACM (2023)
13. Hausmann, D., Korte, B., Jenkyns, T.A.: Worst case analysis of greedy type algorithms for independence systems. In: Combinatorial Optimization, pp. 120–131 (1980)
14. Kagel, J.H., Harstad, R.M., Levin, D.: Information impact and allocation rules in auctions with affiliated private values: a laboratory study. Econometrica **55**(6), 1275–1304 (1987)
15. de Keijzer, B., Kyropoulou, M., Ventre, C.: Obviously strategyproof single-minded combinatorial auctions. In: ICALP, pp. 71:1–71:17 (2020)
16. Li, S.: Obviously strategy-proof mechanisms. Am. Econ. Rev. **107**(11), 3257–3287 (2017)
17. Mackenzie, A.: A revelation principle for obviously strategy-proof implementation. Games Econ. Behav. **124**, 512–533 (2020)
18. Nisan, N., Roughgarden, T., Tardos, E., Vazirani, V. (eds.): Algorithmic Game Theory (2017)
19. Pycia, M., Troyan, P.: Obvious dominance and random priority. In: Proceedings of the 2019 ACM Conference on Economics and Computation, EC '19, p. 1. Association for Computing Machinery, New York (2019)
20. Pycia, M., Troyan, P.: A theory of simplicity in games and mechanism design. Econometrica **91**(4), 1495–1526 (2023)
21. Savage, L.: Foundations of Statistics. John Wiley and Sons, Hoboken (1954)
22. Troyan, P.: Obviously strategy-proof implementation of top trading cycles. Int. Econ. Rev. **60**(3), 1249–1261 (2019)
23. Troyan, P., Morrill, T.: Obvious manipulations. J. Econ. Theory **185**, 104970 (2020)

Proportional Dynamics in Linear Fisher Markets with Auto-bidding: Convergence, Incentives and Fairness

Juncheng Li[(✉)] and Pingzhong Tang

Tsinghua University, Beijing, China
lijuncheng13@gmail.com

Abstract. Proportional dynamics, originated from peer-to-peer file sharing systems, models a decentralized price-learning process in Fisher markets. Previously, items in the dynamics operate independently of one another, and each is assumed to belong to a different seller. In this paper, we show how it can be generalized to the setting where each seller brings *multiple* items and buyers allocate budgets at the granularity of *sellers* rather than individual items. The generalized dynamics consistently converges to the competitive equilibrium, and interestingly relates to the auto-bidding paradigm currently popular in online advertising auction markets. In contrast to peer-to-peer networks, the proportional rule is not imposed as a protocol in auto-bidding markets. Regarding this incentive concern, we show that buyers have a strong tendency to follow the rule, but it is easy for sellers to profitably deviate (given buyers' commitment to the rule). Based on this observation, we further study the seller-side deviation game and show that it admits a unique pure Nash equilibrium. Though it is generally different from the competitive equilibrium, we show that it attains a good fairness guarantee as long as the market is competitive enough and not severely monopolized.

Keywords: Proportional dynamics · Auto-bidding · Linear Fisher market · Nash social welfare

1 Introduction

Market dynamics describes the interplay among market members in response to various signals. For a solution concept of a model, besides its economic properties, if there exists some natural and computationally efficient dynamics that leads to it, it is much more likely to emerge in real-world markets. The modern theory of general equilibrium is pioneered by Walras [28] with the well-known dynamics named *tâtonnement process.* It captures the intuitive phenomenon where the price increases if the demand is greater than supply, and lowers otherwise. However, tâtonnement does not specify a rule to determine intermediate off-equilibrium outcomes [4]. Buyers interact with items by reporting their demands given the current pricing, and it is when at least an approximate equilibrium is

M. Mavronicolas et al. (Eds.): WINE 2024, LNCS 15534, pp. 243–260, 2026.
https://doi.org/10.1007/978-3-032-08560-3_14

reached that trade really happens. The requirement of a centralized coordinator who reports prices and collects demands is also not always realistic [9,15]. Moreover, the discrete version of the process may not even converge to the competitive equilibrium in the foundational linear Fisher market [19], which models the classic scenario where a set of items is to be sold to a set of budget-constrained buyers with linear utilities.

Besides tâtonnement process, another dynamics that received a lot of attention in the last decade is the *proportional dynamics* (or proportional response dynamics). In this process, at each round, buyers are required to bid for every item and pay their bids in advance. After all the bids are collected, each item is allocated to buyers *in proportion to their bids*. Conversely, buyers' bids are generated by allocating budgets among items *in proportion to the utility* they receive from each item in the last round. In contrast to tâtonnement process, proportional dynamics operates in a fully decentralized way, requires no parameter configuration, and always converges to the competitive equilibrium in linear Fisher markets with a fast rate [30].

Despite its mathematical attractiveness, the two-sided proportional allocation rule is less common to be observed in the real world (especially compared to the natural price update rule of tâtonnement [30]), apart from the peer-to-peer (P2P) file sharing systems from which it is motivated [29]. In this paper, we generalize proportional dynamics to the case where each seller comes with *multiple* items and buyers allocate budgets on the basis of sellers rather than items. In particular, our new formulation bears a close resemblance to the *online advertising* markets with first price auction and auto-bidding, which puts it in a context of great practical significance.

In today's online advertising auction markets, instead of directly competing with each other in every auction, advertisers could only participate through proxy *auto-bidders* provided by the seller. Within an advertising platform, the seller has a strong control over all the components, while buyers are largely price-taking. Importantly, with budget-constrained value-maximizing auto-bidders, the online advertising auction market is exactly a linear Fisher market, and the steady-state named *pacing equilibrium* formed by auto-bidders also has a deep relationship to the competitive equilibrium of the linear Fisher market [11,12].

Motivated by this connection, we incorporate the auto-bidding auction model into proportional dynamics, and explore three research questions. First, to justify that our generalization is reasonable and meaningful, we need to establish that the dynamics converges consistently to the competitive equilibrium while maintaining a reasonable convergence rate. Second, from the perspective of algorithmic game theory, we study the incentives for buyers and sellers to adhere to or deviate from the proportional allocation rule. Finally, since the incentive analysis shows that buyers are more inclined to follow the rule while sellers are prone to deviate, we are interested in finding out which state the strategic behaviors of sellers might bring the market to, and whether a certain degree of fairness could still be guaranteed on the buyer side.

1.1 Our Results

In this paper, we extend proportional dynamics to the one-seller multi-item setting, which to the authors' knowledge has not been studied in the literature. The generalized dynamics works as follows. At the beginning of each round, each buyer reallocates its budget among sellers in proportion to the utility it receives from each seller in the last time step.[1] Within each sub-market that consists of all the items owned by a single seller, items are sold through simultaneous first price auctions (one for each item). There is an auto-bidder provided by the seller using the *multiplicative pacing* strategy to bid on behalf of each buyer, i.e., the bid for every item must have the form αv, where α is the *pacing multiplier* chosen by the auto-bidder to pace the rate at which the budget is spent and will be uniformly applied to *all* items, and v is the valuation of the item to the buyer. Given the amount of money paid in advance to the seller, the allocation is determined by presuming that auto-bidders reach the pacing equilibrium [11], at which the pacing multiplier of every buyer is simultaneously optimal in hindsight. The pacing equilibrium is unique with respect to the utility received by each buyer, therefore the dynamics is fully deterministic.

We first show that, if both sides of the market truthfully implement the rule specified above, the dynamics always converges to the competitive equilibrium. The one-seller multi-item setting brings new difficulties, and we will see in the proof how pacing equilibrium helps us overcome them and hence justifies itself as a meaningful generalization of the per-item proportional rule. For the rate of convergence, we show that the average *Eisenberg-Gale objective*[2] function converges to the optimal value at a rate of $O(T^{-1})$, where T denotes the number of rounds.

We proceed to explore the incentive of buyers and sellers to implement the (generalized) proportional rule. For buyers, though not always optimal, the proportional rule gives a 2-approximation to the optimal utility. Moreover, the exact utility maximization problem is both non-convex and non-smooth, and its optimization requires non-local information that is hard to acquire in reality.[3] Therefore buyers have a strong motivation to follow the proportional rule.

On the other hand, sellers have readily available instruments to deviate from the vanilla version of first price auction. If the mechanism of *additive boosts* is allowed (which exists almost everywhere in real-world auto-bidding markets and is widely studied in the literate; see, e.g., [2]), sellers can subsidy or penalize auto-bidders at each auction while a generalized pacing equilibrium can still be

---

[1] As in the original formulation, to make it work properly, every buyer should initially put a non-zero amount of money on *every* seller.

[2] This objective function is maximized at the competitive equilibrium and relates to the well-known fairness measure, the Nash social welfare. Therefore it serves as a reasonable measure of the distance from an off-equilibrium state to the competitive equilibrium. Its definition can be found in Sect. 2.1.

[3] Besides, in practice, even knowing its sub-optimality, equalizing bang-per-bucks (or return-on-investments) among sellers is widely accepted and easily interpretable, particularly for non-expert buyers.

reached. Moreover, given any budget profile submitted by buyers, it is easy for the seller to compute proper boosts to make *any* allocation as the generalized pacing equilibrium within its sub-market. Therefore sellers can directly manipulate the allocation to compete for buyers' budgets, anticipating their proportional allocating behavior. We show that the competitive equilibrium gives an incentive ratio of 5 to each seller, i.e., a unilateral deviation from the competitive equilibrium could bring the manipulator a revenue of at most 5 times the original. Nonetheless, the revenue optimization problem is convex and the information requirement is easy to meet. Therefore sellers are still prone to deviate.

Based on the incentive analysis, we formulate a seller-side game to study the consequence of seller deviation when buyers commit to the proportional rule. We show that the game always admits a unique pure Nash equilibrium (PNE). Though it generally does not agree with the competitive equilibrium, it guarantees a Nash social welfare at least a $(1-\Delta)$ fraction of the optimal, where $\Delta \in (0,1]$ is a parameter characterizing the degree of *monopolization* of the market. Therefore as long as the market is not severely monopolized, the fairness of the competitive equilibrium can be largely preserved at the PNE.

1.2 Related Work

Our work sits at the intersection of two important lines of literature, namely the study on proportional dynamics in Fisher markets and on auto-bidding in online advertising auction markets. Researchers in the former area typically stick to the setup where each seller brings only one item. To the authors' knowledge, our work is the first to put the dynamics in a context where each seller brings multiple items. In contrast, in auto-bidding markets each seller (advertising platform) naturally brings multiple items. Importantly, Conitzer et al. [11,12] relates the auto-bidding markets with either first or second price auction to linear Fisher markets. They find that, with first price auction, the pacing equilibrium formed by auto-bidders can be viewed as a slightly adapted version of the competitive equilibrium,[4] while with second price auction, it should additionally assume that each buyer is supply-aware. These works constitute the theoretical foundation for us to incorporate auto-bidding into proportional dynamics in Fisher markets. Nonetheless, Conitzer et al. [11,12] consider the market owned by a single seller without outside competitors, while we extend it to the setting where multiple sellers compete with each other. Paes Leme et al. [25] and Despotakis et al. [14] also consider the competition among multiple auto-bidding platforms, in an attempt to explain the recent trend in the industry that more and more platforms shift from second price to first price auction, but their models are not built upon Fisher markets. Below we will review related works from different lines of literature in more detail.

[4] The difference lies in that they study budget-constrained *quasi-linear* utility-maximizing advertisers, while in Fisher markets value-maximizing buyers are considered (as done in this paper). To the authors' knowledge, value-maximizers significantly outweigh quasi-linear utility-maximizers in trading volume nowadays.

Online Advertising Auction Markets with Auto-bidding. Today auto-bidders are not yet diverse and powerful enough to optimize any utility function. Linear-utility maximization with budget-constraint (the type of auto-bidder studied in this paper; see, e.g., [18]) is one of the most adopted options. Moreover, with limited feedback from the seller, typically buyers could only make decisions based on its payment and acquired value on each seller, even though they are not truly linear-utility maximizers. As a result, linear Fisher markets serve as a good first approximation to online advertising markets and a starting point for more complex models [11,12,22]. Another constraint also frequently used in practice is Return-On-Investment (ROI, or equivalently Return-On-Ad-Spend, ROAS) [2,20], and sometimes both budget and ROI constraints are imposed [1,13]. The multiplicative pacing strategy is one of the most implemented strategies in the industry. It is applicable for both budget and ROI constrained bidders and for both first and second price auction [11,12,22].

Both first and second price auction are usually augmented with the mechanism of additive boosts, which exists almost everywhere in the industry and is also widely studied in the literature. Previously researchers [2,13] mainly focus on the effect of boosts within the sub-market owned by a single seller. Our work complements this line of research by showing how it could be utilized in the competition among multiple platforms.

First price auction, the auction format studied in this paper, is gaining popularity in recent years in the industry. Researchers also show that, when auto-bidders apply the multiplicative pacing strategy, first price auction outperforms second price in many regards [11,22], which endows sellers with a strong motivation to adopt it. In particular, the pacing equilibrium can be computed efficiently in large-scale first price auction markets [18], which renders our dynamics computationally practical. Though second price auction cannot be fully captured in our framework, with additive boosts and reserve prices [2], sellers can manipulate the allocation in a similar way as dictated in Sect. 5. Therefore our results could also provide some insights into it.

Trading Post Game and Proportional Dynamics. The item-side proportional allocation rule is first studied by Shapley and Shubik [26]. They propose the trading post game to study markets involving non-price-taking buyers, where each item comes with a trading post to handle bid collection and item allocation. The rule also emerges in the mechanism design literature on various resource allocation settings. Kelly [21] shows that, in a computer network, if the capacity of each link is allocated in proportion to the bid of each demand, the social welfare can be maximized at equilibrium. Feldman et al. [17] apply the rule in allocating computing resources and show that the Nash equilibrium attains a good performance on both efficiency and fairness.

Wu and Zhang [29] first propose the proportional dynamics for a special case of Fisher markets modeling P2P file sharing systems such as BitTorrent. Within this particular context, each individual is simultaneously a buyer and a seller, and the proportional rule is indeed implemented as the protocol. Note that, though a P2P network seems to be an exchange (Arrow-Debreu) market, this first proportional dynamics does not directly generalize to the exchange model [4] since there is no money involved. After Zhang [30] generalizes the dynamics to general Fisher markets, researchers further extend it to various settings and establish its convergence [4,7,8].

Computation of Competitive Equilibrium. The computation of solution concepts is an ongoing topic in the study of economic models. For Fisher markets, the tractability of computing competitive equilibrium largely depends on the utility functions of buyers. For linear utilities, besides decentralized dynamics, the competitive equilibrium can also be computed in a centralized way via convex programs [16,27] or combinatorial algorithms (e.g., [24]). For other families of utilities, e.g., additively separable and PLC utilities, the problem becomes PPAD-hard [6]. These complexity results put a limit on what we could expect from a decentralized dynamics, particularly on the rate of convergence. In this paper, we show an *ergodic* rate[5] $O(T^{-1})$ of convergence for the Eisenberg-Gale objective, which is identical to the rate established by Brânzei et al. [4] for proportional dynamics in Arrow-Debreu markets. Previously a fast rate of convergence of proportional dynamics is mainly established via interpreting the dynamics as a decentralized mirror descent process over a market-wise convex program [3,7]. It is interesting to study whether such a connection exists for our one-seller multi-item setting or the Arrow-Debreu markets.

Nash Social Welfare and Fairness. Nash social welfare (NSW) is a well-established measure of fairness in budget-based allocation. In linear Fisher markets, the competitive equilibrium is characterized by the Eisenberg-Gale program [16] that optimizes the NSW. When valuations are unknown to the market owner, Cole et al. [10] propose a truthful mechanism for buyers to achieve a 2.718-approximation of the optimal NSW. The Shapley-Shubik trading post game, though untruthful, is shown to give a 2-approximation of NSW at any PNE, and it also guarantees proportionality for every individual buyer [5]. Note that the game considered in Sect. 6 captures the competition among *sellers*, while the aforementioned works focus on the buyer-side strategic behaviors. Our work is quite distinct from previous results, and it is interesting to see that buyers can provide themselves a certain degree of fairness by collectively committing to the proportional allocation rule.

2 Model Description

2.1 Linear Fisher Markets and Competitive Equilibrium

A **linear Fisher market** consists of a set J of m divisible items (each with a normalized 1 unit of supply) and a set I of n buyers. Each buyer i has a budget B_i. The utility function u_i of buyer i has the form $u_i(x) = \sum_j v_{i,j}x_{i,j}$, where $x_{i,j} \in [0, 1]$ is the fraction of item j allocated to buyer i, and $v_{i,j} \geq 0$ is the value

[5] I.e., the rate at which the average (instead of the last iterate) of some value converges to the limit point.

of item j to buyer i. For each item, there is at least one buyer i with $v_{i,j} > 0$. The total budgets of all buyers is denoted as $B = \sum_i B_i$.

A **competitive equilibrium** (p^*, x^*) consists of a pricing $p^* \in [0, +\infty)^m$ and an allocation $x^* \in [0, 1]^{n\times m}$ that satisfies:

- Each buyer gets allocated an optimal bundle it affords: $x_i^* \in \arg\max_{x_i \in [0,+\infty)^m : \sum_j p_j^* x_{i,j} \le B_i} \sum_j v_{i,j} x_{i,j}$;
- The market clears: $\sum_i x_{i,j}^* = 1, \forall j$.

The competitive equilibrium is also known to satisfy the following properties.

- x is an equilibrium allocation if and only if it solves the **Eisenberg-Gale (EG) program**:

$$
\begin{aligned}
\max \quad & \sum_i B_i \ln u_i \\
\text{s.t.} \quad & u_i \le \sum_j v_{i,j} x_{i,j}, \forall i; \\
& \sum_i x_{i,j} \le 1, \forall j; \\
& x_{i,j} \ge 0, \forall i, j.
\end{aligned}
$$

  $\sum_i B_i \ln u_i$ is called the **EG-objective**. Note its relationship with our fairness measure NSW, defined as follows:

$$
\text{NSW}(x) = \prod_i (u_i(x))^{\frac{B_i}{B}} .
$$

- An equilibrium always exists, and is unique with respect to prices and buyers' utilities. Buyers always deplete their budgets at equilibrium.
- Fixing p, the **bang-per-buck** of item j for buyer i is $\frac{v_{i,j}}{p_j}$, the value received by buyer i for each unit of money paid for item j. At equilibrium, every buyer gets its most wanted items with respect to the bang-per-buck, i.e., if $x_{i,j} > 0$, then $\frac{v_{i,j}}{p_j} \ge \frac{v_{i,j'}}{p_{j'}}, \forall j'$.

2.2 Auto-bidding Markets and Pacing Equilibrium

The traditional Fisher market model is *seller-oblivious*, where sellers do not have the power to affect the price and the item-ownership has no effect on the competitive equilibrium. In this paper, we assume that the items are partitioned into several subsets $\{J_k\}$, each owned by a single seller k. The set of all sellers is denoted as K. We will consider the incentive issues of sellers and how their strategic behaviors would influence the market outcome.

For each sub-market owned by a single seller, we assume that items are sold through simultaneous first price auctions (one auction for one item) with an auto-bidding mechanism. Each item is allocated to the buyer with the highest bid, and the winner should pay its bid.[6] In contrast to bidding for each item directly,

[6] Items are allowed to be allocated fractionally, as is commonly done in the auto-bidding literature.

in an auto-bidding market, each buyer only submits in advance the amount of money it is willing to spend on the seller. The bidding is instead handled by an auto-bidder (one for each buyer) using the *multiplicative pacing* strategy, i.e., there is a *pacing multiplier* $\alpha_i > 0$ for each buyer i such that buyer i's bid for item j is $\alpha_i v_{i,j}$. The steady-state reached by the auto-bidders is called a (first price) **pacing equilibrium** (α, x) [11], which consists of pacing multipliers α and allocation x such that: (1) item j is sold at price $p_j = \max_i \alpha_i v_{i,j}, \forall j$; (2) if $x_{i,j} > 0$, $\alpha_i v_{i,j} = p_j$; (3) $\sum_i x_{i,j} = 1, \forall j$; (4) $\sum_j p_j x_{i,j} = B_i, \forall i$. The rule of first price auction is enforced by conditions (1–3), while condition (4) shows that each auto-bidder best responds to its opponents. To see the latter, note that both the received utility and the payment of each auto-bidder are monotone with respect to the pacing multiplier α_i, therefore depleting the budget is always the optimal solution.

Note that, given buyers' submitted budgets, the sub-market owned by each seller is itself a linear Fisher market. The pacing equilibrium actually captures the same state (regarding allocation and payment) to the competitive equilibrium of this sub-market: here p_j is exactly the price at the competitive equilibrium, and α_i is the inverse of the maximum bang-per-buck over all items. Though technically equivalent, pacing equilibrium brings a new perspective to the classic Fisher market model, and even inspires new algorithms computing the competitive equilibrium at a large scale [18]. We refer readers to [11] and its related works for more information. In particular, though multiplicative pacing is not optimal in hindsight for each buyer at the level of individual auctions, the resulting overall allocation is in many regards better than the second price pacing equilibrium [11] that guarantees each buyer the ex-post optimality. Combined with the on-going trend in the industry that platforms keep moving from second to first price auction, it is practically significant to understand the price-forming dynamics among competing sellers using first price auction and auto-bidding.

2.3 Proportional Dynamics in Auto-Bidding Markets

With the introduction of auto-bidding, the generalized proportional dynamics is formally defined as follows.

Definition 1. *The* ***proportional dynamics*** *in auto-bidding markets is an iterative process where buyers repeatedly allocate their budgets among sellers and sellers allocate items via pacing equilibrium. Let $B_i(k,t)$ be the amount of money buyer i allocates to seller k at time t, and $u_i(k,t)$ be the utility received by buyer i at the pacing equilibrium of the sub-market owned by seller k. Let $u_i(K,t) = \sum_k u_i(k,t)$. Buyers adopt the following update rule: $B_i(k,t+1) = \frac{u_i(k,t)}{u_i(K,t)} \cdot B_i$.*

With our definition, the original proportional dynamics [30] is the special case where each seller owns only one item. The buyer-side proportional rule follows the same spirit of the original formulation (with only a change of granularity), and in Sect. 3 we will see why allocation via pacing equilibrium is a meaningful generalization of the proportional rule for sellers. Note that for each buyer

i, at time t, the only information required to update its budget allocation is $\{u_i(k,t)\}_{k\in K}$, which will always be known to it in the real world.

Below we will denote by $B_i(k,*)$ the money of buyer i spent on seller k at an arbitrary (but fixed) competitive equilibrium, and $u_i(K,*)$ the (unique) utility of buyer i at the competitive equilibrium. We will also use $x_{i,j}(t)$ and $x_{i,j}(*)$ to denote the allocation at time t and at the competitive equilibrium, respectively. When the time is not relevant or could be easily derived from the context, we will use $u_i(k)$ and $u_i(K)$ to denote the utility received by buyer i from seller k and all sellers, respectively.

3 Convergence of Proportional Dynamics

In this section, we justify that the proportional dynamics still converges to the competitive equilibrium with a reasonable rate. The proof centers around a potential function $\Phi(t)$ that generalizes the one used in establishing convergence in the one-seller one-item setting (for linear Fisher markets [30] and linear exchange markets [4]):

$$\Phi(t) = \sum_{i,k} B_i(k,*) \ln \frac{B_i(k,*)}{B_i(k,t)},$$

i.e., the KL-divergence between the budget allocation at the competitive equilibrium and the one at time t. The main task is to show that the potential keeps decreasing over time. We will see in the proof how the pacing equilibrium helps us overcome the difficulties brought by the one-seller multi-item setting.

Theorem 1. *A competitive equilibrium is a fixed point of proportional dynamics, and proportional dynamics always converges to a competitive equilibrium.*

Proof. A fixed point of the dynamics should satisfy that

$$B_i(k,t+1) = \frac{u_i(k,t)}{u_i(K,t)} \cdot B_i = B_i(k,t), \forall i,k.$$

Since the bang-per-bucks of all items bought by a buyer are equalized at competitive equilibrium, the budget allocation $\{B_i(k,*)\}_{k\in K}$ at competitive equilibrium satisfy the above equation, and the first half of the theorem is justified.

To establish convergence, we relate $\Phi(t)$ with $\Phi(t+1)$ as follows:

$$\begin{aligned}
&B_i(k,*) \ln \frac{B_i(k,*)}{B_i(k,t+1)} \\
=&B_i(k,*) \ln \left(\frac{B_i(k,*)}{B_i} \cdot \frac{u_i(K,t)}{u_i(k,t)} \right) \\
=&B_i(k,*) \ln \left(\frac{u_i(k,*)}{u_i(K,*)} \cdot \frac{u_i(K,t)}{u_i(k,t)} \right)
\end{aligned}$$

$$
\begin{aligned}
=&B_i(k,*)\ln\left(\frac{B_i(k,*)}{B_i(k,*)}\cdot\frac{B_i(k,t)}{B_i(k,t)}\cdot\frac{u_i(k,*)}{u_i(K,*)}\cdot\frac{u_i(K,t)}{u_i(k,t)}\right)\\
=&B_i(k,*)\ln\frac{B_i(k,*)}{B_i(k,t)}-B_i(k,*)\ln\frac{u_i(K,*)}{u_i(K,t)}-B_i(k,*)\ln\left(\frac{B_i(k,*)}{B_i(k,t)\cdot\frac{u_i(k,*)}{u_i(k,t)}}\right),
\end{aligned}
$$

where the second equality holds since the competitive equilibrium is a fixed point.[7] Summing over buyers and sellers gives

$$
\Phi(t)-\Phi(t+1)=\sum_{i,k}B_i(k,*)\ln\frac{u_i(K,*)}{u_i(K,t)}+\sum_{i,k}B_i(k,*)\ln\left(\frac{B_i(k,*)}{B_i(k,t)\cdot\frac{u_i(k,*)}{u_i(k,t)}}\right).
$$

For the first term, it holds that

$$
\sum_{i,k}B_i(k,*)\ln\frac{u_i(K,*)}{u_i(K,t)}=\sum_i B_i\ln\frac{u_i(K,*)}{u_i(K,t)}\geq 0,
$$

since $u=u(K,*)$ maximizes the EG-objective $\sum_i B_i\ln u_i$. For the second term,[8] let $\alpha_i(k,t)=\frac{B_i(k,t)}{u_i(k,t)}$ be the pacing multiplier for buyer i at the pacing equilibrium of sub-market k. We have

$$
\begin{aligned}
\tilde{B}:=&\sum_{i,k}B_i(k,t)\cdot\frac{u_i(k,*)}{u_i(k,t)}=\sum_k\sum_i\alpha_i(k,t)\sum_{j\in J_k}v_{i,j}x_{i,j}(*)\\
\leq&\sum_k\sum_i\alpha_i(k,t)\sum_{j\in J_k}v_{i,j}x_{i,j}(t)=\sum_k\sum_i B_i(k,t)=B,
\end{aligned}
$$

where the inequality holds since, at the pacing equilibrium, every item is sold to the buyer with the maximum bid $\alpha_i v_{i,j}$ (see Sect. 2.2). Then

$$
\sum_{i,k}B_i(k,*)\ln\left(\frac{B_i(k,*)}{B_i(k,t)\cdot\frac{u_i(k,*)}{u_i(k,t)}}\right)\geq\sum_{i,k}B_i(k,*)\ln\left(\frac{B_i(k,*)}{\frac{B}{\tilde{B}}\cdot B_i(k,t)\cdot\frac{u_i(k,*)}{u_i(k,t)}}\right)\geq 0,
$$

where the second inequality holds due to the non-negativity of KL-divergence.

Now we have shown that $\Phi(t)$ is monotonically decreasing. By the non-negativity of KL divergence, $\Phi(t)$ is lower bounded. As a result, $\Phi(t)$ converges and $\Phi(t)-\Phi(t+1)$ goes to zero as t goes to infinity. Recall that

[7] Regarding the risk of division by zero, note that $B_i(k,t)$ and $u_i(k,t)$ can become zero only if all items owned by seller k have a zero value for buyer i. In this case, $B_i(k,*)$ must also be zero, and the corresponding terms can be safely ignored.

[8] In the one-seller one-item setting [30], based on the per-item proportional rule, the second term can be directly simplified to the KL-divergence between $p(*)$ and $p(t)$, which denotes the item prices at the competitive equilibrium and at time t, respectively. This cannot be done here. We instead use the property of pacing equilibrium to bound the term, and thus justify that it is a non-trivial generalization of the per-item proportional rule.

$\Phi(t) - \Phi(t+1) \geq \sum_{i,k} B_i(k,*) \ln \frac{u_i(K,*)}{u_i(K,t)} \geq 0$ and the equality only happens when $u_i(K,t) = u_i(K,*), \forall i$ by the uniqueness of utilities at the competitive equilibrium. Therefore $u_i(K,t) \to u_i(K,*)$ as $t \to \infty$. KL-divergence equals zero only if two distributions are identical, so we have $B_i(k,t) \cdot \frac{u_i(k,*)}{u_i(k,t)} \to B_i(k,*)$, i.e., the bang-per-bucks $\frac{u_i(k,t)}{B_i(k,t)}$ converge to $\frac{u_i(k,*)}{B_i(k,*)}$, the bang-per-bucks at the competitive equilibrium, and so do the item prices.

To see the convergence of $B_i(k,t)$, first take the limiting budget allocation $\{B_i(k,**)\}_{i\in I,k\in K}$ of some convergent sub-sequence of $\{B_i(k,t)\}_{t=0}^{+\infty}$ (which exists as the sequence is bounded). We have shown that the bang-per-bucks and item prices converge to a competitive equilibrium, therefore $\{B_i(k,**)\}_{i\in I,k\in K}$ should also be the budget allocation at some competitive equilibrium. Recall that in the definition of Φ we can use any budget allocation at the competitive equilibrium for $\{B_i(k,*)\}_{i\in I,k\in K}$, so we can replace it with $\{B_i(k,**)\}_{i\in I,k\in K}$. The analysis in previous paragraphs still holds, and now we have $\Phi(t)$ converges to 0 and thus $B_i(k,t)$ converges to $B_i(k,**)$. (If $\Phi(t)$ is defined with some $\{B_i(k,*)\}_{i\in I,k\in K}$ differing from the limit point, it will rest at some value larger than zero. By replacing it with $\{B_i(k,**)\}_{i\in I,k\in K}$, we rule out the possibility that $B_i(k,t)$ oscillates among several different equilibrium values and thus justify its convergence.) □

For the convergence rate, by summing both sides of the inequality $\Phi(t) - \Phi(t+1) \geq \sum_{i,k} B_i(k,*) \ln \frac{u_i(K,*)}{u_i(K,t)}$ over t, we have

$$\Phi(1) - \Phi(T+1) \geq \sum_{t=1}^{T} \left(\sum_i B_i \ln u_i(K,*) - \sum_i B_i \ln u_i(K,t) \right).$$

Dividing both sides by T and we can see that the average EG-objective converges to the optimum at a rate of $O(1/T)$ (the proof is adapted from the one in [4] to show a similar rate for proportional dynamics in Arrow-Debreu markets).

4 Incentive of Buyers

Considering that sellers implement the proportional rule truthfully, we examine the incentive of buyers to deviate and focus on the buyer-side game defined as follows.

Definition 2. *A* ***budget allocation game*** *of a linear Fisher market is a normal form game among buyers. Each buyer i allocates an amount $B_i(k)$ of money to seller k. The budget allocation should be feasible: $\sum_k B_i(k) \leq B_i, \forall i$. Items in J_k are sold according to the allocation at the pacing equilibrium with budgets $\{B_i(k)\}_{i\in I}$. The utility of each buyer is the total valuation it receives.*

In Appendix A [23],[9] we give an example to show that buyer's optimization problem could be both non-convex and non-smooth. In addition, to learn

[9] Appendices can be found in the full version of this paper [23].

the utility curve, information such as budgets and valuations of other buyers is required. Without such knowledge, the buyer can only access the utility function point-by-point through repeated interaction with sellers, during which other buyers' budget allocation might change as well. As a result, it is impractical for the buyer to manipulate its budget allocation and optimize its utility exactly.[10]

On the other hand, the proportional rule is not only easily implementable, but also provides a satisfying performance guarantee as shown below.

Theorem 2. *Fixing other buyers' budget allocation over sellers, if buyer i's bang-per-bucks* $\{\frac{u_i(k)}{B_i(k)}\}_{k \in K}$ *are equalized among all sellers, then the corresponding budget allocation* $\{B_i(k)\}_{k \in K}$ *achieves a utility of at least* $1/2$ *of the optimal.*

To prove the theorem, we will use the monotonicity of the bang-per-buck and the received utility at the pacing equilibrium with respect to the buyer's submitted budget, which are known results in the literature (see Appendix D [23] for references) and stated as Lemma 1 below.

Lemma 1. *Assume that seller k truthfully implements the pacing equilibrium. Fixing other buyers' submitted budgets to seller k, buyer i's received utility* $u_i(k)$ *is weakly increasing with respect to* $B_i(k)$*, while the bang-per-buck* $\frac{u_i(k)}{B_i(k)}$ *is weakly decreasing with respect to* $B_i(k)$*.*

Proof (Theorem 2). Assume that, except for buyer i, all the other buyers' budget allocations are fixed. Let $\{B_i(k)\}_{k \in K}$ be the budget allocation of buyer i such that its bang-per-bucks are equalized across all sellers, and $u_i(k)$ be the utility received from seller k by submitting $B_i(k)$ to it. Let $u_i = \sum_k u_i(k)$.

Consider any other feasible budget allocation $\{\tilde{B}_i(k)\}_{k \in K}$. Let $\tilde{u}_i(k)$ be the utility received from seller k by submitting $\tilde{B}_i(k)$ to seller k, and $\tilde{u}_i = \sum_k \tilde{u}_i(k)$. Partition K into $K_{\leq}$ and $K_{>}$ such that for each $k \in K_{\leq}$, $\tilde{B}_i(k) \leq B_i(k)$, and for each $k \in K_{>}$, $\tilde{B}_i(k) > B_i(k)$.

By the first result of Lemma 1, for each $k \in K_{\leq}$, buyer i spends no more money, so it receives no more utility: $\tilde{u}_i(k) \leq u_i(k)$. By the second result of Lemma 1, for each $k \in K_{>}$, buyer i spends more money, so its bang-per-buck cannot be higher: $\frac{\tilde{u}_i(k)}{\tilde{B}_i(k)} \leq \frac{u_i(k)}{B_i(k)} = \frac{u_i}{B_i}$. Combining both parts together, we have:

$$\tilde{u}_i = \sum_{k \in K_{\leq}} \tilde{u}_i(k) + \sum_{k \in K_{>}} \tilde{u}_i(k) \leq \sum_{k \in K_{\leq}} u_i(k) + \sum_{k \in K_{>}} \tilde{B}_i(k) \cdot \frac{u_i}{B_i} \leq u_i + u_i = 2u_i.$$

□

5 Incentive of Sellers

In real-world markets, the additive boost mechanism is widely implemented by sellers to subsidize or penalize each buyer in each auction. With additive boosts

[10] In comparison, when each seller owns only one item, a profitable deviation is easily computable for buyers. See Appendix B [23] for a discussion.

c, buyers are ranked by the boosted bids $\alpha_i v_{i,j} + c_{i,j}$, but if it wins, the payment is still the non-boosted bid $\alpha_i v_{i,j}$. Note that adding a constant to all $c_{i,j}$ will not change the auction outcome. Below we assume $c_{i,j} \geq 0, \forall i, j$. A **pacing equilibrium with additive boosts** $c \in \mathbb{R}^{n \times m}$ consists of pacing multipliers α and allocation x such that: (1) $p_j = \max_i \alpha_i v_{i,j} + c_{i,j}, \forall j$; (2) if $x_{i,j} > 0$, $\alpha_i v_{i,j} + c_{i,j} = p_j$; (3) $\sum_j (p_j - c_{i,j}) x_{i,j} = B_i, \forall i$; (4) $\sum_i x_{i,j} = 1, \forall j$. Similar to the original version, pacing equilibrium with additive boosts can be characterized by a modified EG program (the result is formally stated in Appendix C [23]). We then show that, if each seller allocates items via the pacing equilibrium with fixed boosts throughout the dynamics, the convergence result remains to hold.

Theorem 3. *With fixed additive boosts c, proportional dynamics always converges to a (market-wise) pacing equilibrium with additive boosts c.*

The proof is given in Appendix E [23]. Observe that, given the submitted budgets $\{B_i(k)\}_{i \in I}$, seller k can implement *any* feasible allocation as the pacing equilibrium by setting proper additive boosts: with the target allocation in mind, the pacing multiplier is given by the inverse of the bang-per-buck, and then boosts can be calculated such that the winner has the highest boosted bid. Therefore a seller can directly choose a feasible allocation within its submarket to attract buyers' proportionally allocated budgets. The following theorem characterizes how much a seller can gain by deviating from the competitive equilibrium when buyers commit to the proportional rule.

Theorem 4. *Let R_k^{OPT} be the optimal revenue of seller k if all the other sellers choose their competitive equilibrium allocation, and R_k^* be its revenue at any competitive equilibrium. Then $R_k^{OPT} \leq 5R_k^*$.*

Proof. For any $x + c > 0$, it holds that $\frac{x}{x+c} \leq \ln(x + c) - \ln c$. We also have the following inequality: $\frac{x}{x+c} \leq 1 = \ln(x + c) - \ln(x + c) + 1$. For each buyer i, let w_i^* be the total valuation received by it from all sellers other than k at some competitive equilibrium, and u_i^* be the valuation received from seller k. Define $u_i = \sum_{j \in J_k} v_{i,j} x_{i,j}$. If all sellers other than k choose their competitive equilibrium allocation, seller k's revenue can be bounded as follows:

$$
\begin{aligned}
R_k(u, w^*) &= \sum_i B_i \cdot \frac{u_i}{u_i + w_i^*} \\
&\leq \sum_i B_i \ln(u_i + w_i^*) + \sum_i B_i \cdot \min\left(-\ln w_i^*, -\ln(u_i + w_i^*) + 1\right) \\
&\leq \sum_i B_i \ln(u_i^* + w_i^*) + \sum_i B_i \cdot \min\left(-\ln w_i^*, -\ln(u_i + w_i^*) + 1\right) \\
&= \sum_i B_i \cdot \min\left(\ln\left(\frac{u_i^* + w_i^*}{w_i^*}\right), \ln\left(\frac{u_i^* + w_i^*}{u_i + w_i^*}\right) + 1\right) \\
&\leq \sum_i B_i \cdot \min\left(\frac{u_i^*}{w_i^*}, \frac{u_i^* + w_i^*}{u_i + w_i^*}\right),
\end{aligned}
$$

where the second inequality holds since (u^*, w^*) maximizes the EG-objective, and the last one holds since $\ln x \le x - 1$.

Suppose that, for every buyer i, $u_i \ge \frac{u_i^*}{2}$, then $\frac{u_i^*+w_i^*}{u_i+w_i^*} \le 2$. In this case,

$$\begin{aligned} R_k(u, w^*) \le& \sum_i B_i \cdot \min\left(\frac{u_i^*}{w_i^*}, 2\right) \\ =& \sum_i B_i \cdot \frac{u_i^*}{u_i^* + w_i^*} \cdot \min\left(\frac{u_i^* + w_i^*}{w_i^*}, \frac{2(u_i^* + w_i^*)}{u_i^*}\right) \\ \le& 3R_k(u^*, w^*). \end{aligned}$$

Recall that $R_k(u, w)$ is a concave function with respect to u. Let u^{OPT} be the best response of seller k against w^*. Then

$$R_k(u^{\text{OPT}}, w^*) \le 2R_k\left(\frac{u^{\text{OPT}} + u^*}{2}, w^*\right) - R_k(u^*, w^*) \le 5R_k(u^*, w^*).$$

□

Nonetheless, a constant incentive ratio might be insufficient to prevent sellers from deviation. Consider a seller-driven dynamics as follows: at time t, seller k allocates its items to buyers with $\{x_{i,j}(t)\}_{j \in J_k}$ and buyer i receives utilities $\{u_i(k,t) = \sum_j v_{i,j} x_{i,j}(t)\}_{k \in K}$. Buyers still update their budget allocations according to the proportional rule. This determines the revenue of each seller at time $t+1$. From the perspective of online optimization, at time t, seller k receives a convex loss: $\sum_i B_i \cdot \frac{\sum_{k' \neq k} u_i(k',t)}{u_i(k,t) + \sum_{k' \neq k} u_i(k',t)}$. The information required for each seller to optimize it (even in an online fashion) is almost minimal (only $B_i, i \in I$ is needed).[11] Hence sellers are more prone to manipulate the pacing equilibrium.

6 Seller Competition and Fairness Guarantee

In this section, we explore the consequence of deviations from the proportional dynamics. Based on the analysis in previous sections, buyers have a strong tendency to follow the proportional rule. Therefore we put the focus on the strategic competition among sellers. In this section, we assume that there are at least two sellers owning items with $v_{i,j} > 0$ for each buyer i.[12]

---

[11] $u_i(k,t)$ is also required, but it is always known to the seller since in auto-bidding markets the valuations are typically produced by machine learning models of the seller. B_i may be slightly harder to acquire. It could be reported in the financial report of the buyer, or learned from direct communication with its advertising team.

[12] If there is only one seller who provides positive utility for a buyer, the seller can fully absorb the buyer's budget with an infinitesimal amount of items.

Definition 3. *A* ***seller competition game*** *is defined over a linear Fisher market among sellers. The strategy space of seller k is the set of feasible allocations $x_{i,j} \in [0,1], i \in I, j \in J_k$. The revenue of seller k with allocation profile x is given by $R_k(x) = \sum_i B_i \cdot \frac{\sum_{j\in J_k} v_{i,j}x_{i,j}}{\sum_{j\in J_k} v_{i,j}x_{i,j} + \sum_{j\in J\setminus J_k} v_{i,j}x_{i,j}}$. Without loss of generality,[13] we only consider allocation profiles x such that $\forall i, k$, $\sum_{j\in J\setminus J_k} v_{i,j}x_{i,j} \geq \epsilon$ with a sufficiently small $\epsilon > 0$. With this assumption, the revenue is always continuous and the set of allocation profiles is still compact and convex. A* ***pure Nash equilibrium*** *(PNE) of a seller competition game is an allocation profile x such that each seller is best responding, i.e., $R_k(x) \geq R_k(x')$ for all x' that agrees with x for all $j \notin J_k$.*

In addition, note that if two allocation profiles produce the same utility profile $\{u_i(k)\}_{i\in I, k\in K}$, their corresponding game outcomes $\{R_k(x)\}_{k\in K}$ are also identical. As a result, the strategy space of seller k can be equivalently defined as the set of feasible utilities $\{u_i(k)\}_{i\in I}$. The PNE could be defined in the same spirit, and we still restrict our attention on utility profiles such that $\forall i, k$, $\sum_{k'\neq k} u_i(k') \geq \epsilon$ with a sufficiently small $\epsilon > 0$.

Theorem 5. *The seller competition game always admits a unique pure Nash equilibrium.*

The proof is given in Appendix G [23]. In general, the PNE of the seller competition game is not a competitive equilibrium. But we have the following approximation guarantee for the NSW.

Theorem 6. *The NSW at the PNE of the seller competition game is at least $(1 - \Delta)$ of the optimal, where $\Delta := \max_{i,k} \frac{B_i(k,*)}{B_i}$ is the maximum fraction of budget earned by a single seller for a single buyer.*

Proof. Let $u_i(k, \dagger)$ be the utility received by buyer i from seller k at the PNE, and $u_i(k, *)$ be the utility at the competitive equilibrium. For the convenience of notation, in this proof we let $u_i(\dagger) = \sum_k u_i(k, \dagger)$ and $u_i(*) = \sum_k u_i(k, *)$.

By first-order optimality of each seller's revenue at PNE, we have

$$\sum_i B_i \cdot \frac{u_i(\dagger) - u_i(k,\dagger)}{(u_i(\dagger)^2)} \cdot (u_i(k,*) - u_i(k,\dagger)) \leq 0, \forall i. \tag{1}$$

By the definition of Δ, we have

$$\sum_k B_i(k,\dagger) B_i(k,*) \leq \left(\sum_k B_i(k,\dagger)\right) \cdot \max_k \{B_i(k,*)\} \leq \Delta B_i^2. \tag{2}$$

We first consider the arithmetic mean of the buyer-wise ratio $\frac{u_i(*)}{u_i(\dagger)}$, which can be decomposed into two parts:

$$\sum_i B_i \cdot \frac{u_i(*)}{u_i(\dagger)} = \sum_{i,k} B_i \cdot \frac{u_i(k,*)}{u_i(\dagger)}$$

[13] See Appendix F [23] for why the assumption is without loss of generality.

$$= \sum_{k,i} B_i \cdot \frac{u_i(k,*)}{u_i(\dagger)} \cdot \left(1 - \frac{u_i(k,\dagger)}{u_i(\dagger)}\right) + \sum_{i,k} B_i \cdot \frac{u_i(k,*)}{u_i(\dagger)} \cdot \frac{u_i(k,\dagger)}{u_i(\dagger)}.$$

Using inequality (1), the first part can be bounded as follows:

$$\begin{aligned}
\sum_k \sum_i B_i \cdot \frac{u_i(k,*)}{u_i(\dagger)} \cdot \left(1 - \frac{u_i(k,\dagger)}{u_i(\dagger)}\right) &= \sum_k \sum_i B_i \cdot \frac{u_i(\dagger) - u_i(k,\dagger)}{(u_i(\dagger)^2)} \cdot u_i(k,*) \\
&\leq \sum_k \sum_i B_i \cdot \frac{u_i(\dagger) - u_i(k,\dagger)}{(u_i(\dagger)^2)} \cdot u_i(k,\dagger) \\
&\leq \sum_i B_i \sum_k \frac{u_i(k,\dagger)}{u_i(\dagger)} = B.
\end{aligned}$$

For the second part, note that at either equilibrium, buyer's bang-per-bucks are equalized among sellers: $\frac{u_i(k,\dagger)}{u_i(\dagger)} = \frac{B_i(k,\dagger)}{B_i}, \frac{u_i(k,*)}{u_i(*)} = \frac{B_i(k,*)}{B_i}$. Combined with inequality (2), we have

$$\begin{aligned}
\sum_i \sum_k B_i \cdot \frac{u_i(k,*)}{u_i(\dagger)} \cdot \frac{u_i(k,\dagger)}{u_i(\dagger)} &= \sum_k \sum_i \frac{u_i(*)}{B_i u_i(\dagger)} \cdot \sum_k B_i(k,\dagger) B_i(k,*) \\
&\leq \Delta \sum_i B_i \cdot \frac{u_i(*)}{u_i(\dagger)}.
\end{aligned}$$

Putting both parts together and rearranging the terms gives

$$(1 - \Delta) \sum_i B_i \cdot \frac{u_i(*)}{u_i(\dagger)} \leq B.$$

By the weighted AM-GM inequality, the ratio of NSW at each equilibrium is bounded by:

$$\prod_i \left(\frac{u_i(*)}{u_i(\dagger)}\right)^{\frac{B_i}{B}} \leq \frac{1}{B} \cdot \sum_i B_i \cdot \frac{u_i(*)}{u_i(\dagger)} \leq \frac{1}{1 - \Delta}.$$

□

The approximation ratio improves as Δ becomes smaller. Intuitively, Δ describes how *monopolized* the market is. For instance, if $\Delta = 0.9$, then at the competitive equilibrium some buyer i will spend 90% of its budget on a single seller j, and other sellers may not have enough power to compete with seller j for buyer i. As a result, seller j may raise the price for buyer i and spare its items for other buyers' budgets, which hurts the utility of buyer i and the overall fairness. Conversely, with a small Δ, no seller earns a significant portion of buyers' budgets and the market is more balanced and competitive. In this case, even if sellers deviate, the end result will remain close to the competitive equilibrium.

7 Conclusion

This paper extends proportional dynamics to auto-bidding auction markets, a first generalization to the one-seller multi-item scenario. While maintaining a consistent convergence to the competitive equilibrium, we find that buyers adhere more to the proportional rule, while sellers can easily deviate for more revenue. This leads to a novel seller-side game with a unique pure Nash equilibrium, which still ensures good fairness in competitive markets. We hope that our work could help people understand the strategic behaviors in online advertising markets, and provide valuable insights into modern price-forming dynamics.

References

1. Aggarwal, G., Badanidiyuru, A., Mehta, A.: Autobidding with constraints. In: International Conference on Web and Internet Economics, pp. 17–30. Springer (2019)
2. Balseiro, S., Deng, Y., Mao, J., Mirrokni, V., Zuo, S.: Robust auction design in the auto-bidding world. In: Advances in Neural Information Processing Systems, vol. 34, pp. 17777–17788 (2021)
3. Birnbaum, B., Devanur, N.R., Xiao, L.: Distributed algorithms via gradient descent for Fisher markets. In: Proceedings of the 12th ACM Conference on Electronic Commerce, pp. 127–136 (2011)
4. Brânzei, S., Devanur, N., Rabani, Y.: Proportional dynamics in exchange economies. In: Proceedings of the 22nd ACM Conference on Economics and Computation, pp. 180–201 (2021)
5. Brânzei, S., Gkatzelis, V., Mehta, R.: Nash social welfare approximation for strategic agents. Oper. Res. **70**(1), 402–415 (2022)
6. Chen, X., Teng, S.H.: Spending is not easier than trading: on the computational equivalence of fisher and arrow-Debreu equilibria. In: International Symposium on Algorithms and Computation, pp. 647–656. Springer (2009)
7. Cheung, Y.K., Cole, R., Tao, Y.: Dynamics of distributed updating in Fisher markets. In: Proceedings of the 2018 ACM Conference on Economics and Computation, pp. 351–368 (2018)
8. Cheung, Y.K., Hoefer, M., Nakhe, P.: Tracing equilibrium in dynamic markets via distributed adaptation. arXiv preprint arXiv:1804.08017 (2018)
9. Cole, R., Fleischer, L.: Fast-converging tatonnement algorithms for one-time and ongoing market problems. In: Proceedings of the Fortieth Annual ACM Symposium on Theory of Computing, pp. 315–324 (2008)
10. Cole, R., Gkatzelis, V., Goel, G.: Mechanism design for fair division: allocating divisible items without payments. In: Proceedings of the Fourteenth ACM Conference on Electronic Commerce, pp. 251–268 (2013)
11. Conitzer, V., et al.: Pacing equilibrium in first price auction markets. Manage. Sci. **68**(12), 8515–8535 (2022)
12. Conitzer, V., Kroer, C., Sodomka, E., Stier-Moses, N.E.: Multiplicative pacing equilibria in auction markets. Oper. Res. **70**(2), 963–989 (2022)
13. Deng, Y., Mao, J., Mirrokni, V., Zuo, S.: Towards efficient auctions in an auto-bidding world. In: Proceedings of the Web Conference 2021, pp. 3965–3973 (2021)
14. Despotakis, S., Ravi, R., Sayedi, A.: First-price auctions in online display advertising. J. Mark. Res. **58**(5), 888–907 (2021)

15. Dvijotham, K., Rabani, Y., Schulman, L.J.: Convergence of incentive-driven dynamics in Fisher markets. Games Econom. Behav. **134**, 361–375 (2022)
16. Eisenberg, E., Gale, D.: Consensus of subjective probabilities: the pari-mutuel method. Ann. Math. Stat. **30**(1), 165–168 (1959)
17. Feldman, M., Lai, K., Zhang, L.: A price-anticipating resource allocation mechanism for distributed shared clusters. In: Proceedings of the 6th ACM Conference on Electronic Commerce, pp. 127–136 (2005)
18. Gao, Y., Peysakhovich, A., Kroer, C.: Online market equilibrium with application to fair division. In: Advances in Neural Information Processing Systems, vol. 34, pp. 27305–27318 (2021)
19. Goktas, D., Zhao, J., Greenwald, A.: Tâtonnement in homothetic fisher markets. In: Proceedings of the 24th ACM Conference on Economics and Computation, pp. 760–781 (2023)
20. Golrezaei, N., Lobel, I., Paes Leme, R.: Auction design for ROI-constrained buyers. In: Proceedings of the Web Conference 2021, pp. 3941–3952 (2021)
21. Kelly, F.: Charging and rate control for elastic traffic. Eur. Trans. Telecommun. **8**(1), 33–37 (1997)
22. Li, J., Tang, P.: Vulnerabilities of single-round incentive compatibility in auto-bidding: Theory and evidence from ROI-constrained online advertising markets (2023)
23. Li, J., Tang, P.: Proportional dynamics in linear fisher markets with auto-bidding: Convergence, incentives and fairness. arXiv preprint arXiv:2407.11872 (2024)
24. Orlin, J.B.: Improved algorithms for computing Fisher's market clearing prices: Computing Fisher's market clearing prices. In: Proceedings of the Forty-Second ACM Symposium on Theory of Computing, pp. 291–300 (2010)
25. Paes Leme, R., Sivan, B., Teng, Y.: Why do competitive markets converge to first-price auctions? In: Proceedings of The Web Conference 2020, pp. 596–605 (2020)
26. Shapley, L., Shubik, M.: Trade using one commodity as a means of payment. J. Polit. Econ. **85**(5), 937–968 (1977)
27. Shmyrev, V.I.: An algorithm for finding equilibrium in the linear exchange model with fixed budgets. J. Appl. Ind. Math. **3**, 505–518 (2009)
28. Walras, L.: Éléments d'économie politique pure: ou. Théorie de la richesse sociale. F, Rouge (1900)
29. Wu, F., Zhang, L.: Proportional response dynamics leads to market equilibrium. In: Proceedings of the Thirty-ninth Annual ACM Symposium on Theory of Computing, pp. 354–363 (2007)
30. Zhang, L.: Proportional response dynamics in the Fisher market. Theoret. Comput. Sci. **412**(24), 2691–2698 (2011)

Almost Envy-Free Allocation of Indivisible Goods: A Tale of Two Valuations

Alireza Kaviani[1(✉)], Masoud Seddighin[2], and AmirMohammad Shahrezaei[1]

[1] Sharif University of Technology, Tehran, Iran
akaviani05@gmail.com

[2] Tehran Institute for Advanced Studies (TeIAS), Khatam University, Tehran, Iran

Abstract. The existence of EFX allocations stands as one of the main challenges in discrete fair division. In this paper, we present symmetrical results on the existence of EFX and its approximate variations for two distinct valuations: restricted additive valuations and (p, q)-bounded valuations introduced by Christodoulou *et al.* [23]. In a (p, q)-bounded instance, each good has relevance for at most p agents, and any pair of agents shares at most q common relevant goods.

We show that instances with $(\infty, 1)$-bounded valuations admit EF2X allocations and EFX allocations with at most $\lfloor n/2 \rfloor - 1$ discarded goods, mirroring results for the restricted additive setting [3]. We also present $(\sqrt{2}/2)-$EFX algorithms for both restricted additive and $(\infty, 1)$-bounded subadditive settings. The symmetry of these results suggests these valuations share symmetric structures. Building on this, we propose an EFX allocation for restricted additive valuations when $p = 2$ and $q = \infty$.

1 Introduction

Fair division is a longstanding problem in mathematics and economics [13,14,24,30]. It seeks methods to allocate resources so that each party receives a fair share. Modern research on this problem began with Steinhaus in 1945, by introducing the cake-cutting problem [30]. This problem involves dividing a heterogeneous, divisible resource among agents with different preferences, such that each individual believes her share is fair. Since the introduction of cake-cutting, numerous fairness notions have emerged, such as proportionality, envy-freeness, and equity. Fortunately, scientists have significantly advanced in providing guarantees for these criteria; see [14] for an overview.

Over the past two decades, a discrete version of the cake-cutting problem, known as fair allocation, has garnered interest among computer scientists [2,10,15,18,26,27]. In the fair allocation problem, the resource is a set of indivisible goods rather than a divisible cake. Fair allocation of indivisible goods presents many intriguing unresolved questions. In this paper, we investigate one of the most challenging: envy-freeness up to any good, or EFX—referred to by Ariel Procaccia as "*Fair Division's Most Enigmatic Question*" [29]. Consider a set

M. Mavronicolas et al. (Eds.): WINE 2024, LNCS 15534, pp. 261–276, 2026.
https://doi.org/10.1007/978-3-032-08560-3_15

M of indivisible goods and n agents. Each agent i has a monotone valuation function $v_i : 2^M \rightarrow \mathbb{R}^{\geq 0}$ that assigns a non-negative value to each subset of goods. If we allocate bundle X_i to agent i for $1 \leq i \leq n$, the allocation is EFX if, for every $1 \leq i, j \leq n$ and every $X'_j \subset X_j$ we have $v_i(X_i) \geq v_i(X'_j)$.

EFX is indeed, a relaxation of the more stringent notion called *"envy-freeness"*. In an envy-free allocation, each agent prefers their bundle over any other agent's bundle, meaning $v_i(X_i) \geq v_i(X'_j)$ must hold even when $X'_j = X_j$. While examples easily disprove the existence of envy-free allocations, it remains an open question whether EFX allocations are always possible. As of now, EFX allocations are known to exist only in a few limited scenarios. For instance, with two agents having monotone valuations, an EFX allocation can always be found [28]. A similar result holds for three agents where two have monotone valuations, and one has an *MMS-feasible* valuation [2,20]. There are also some guarantees proved for an arbitrary number of agents. For additive valuations, we can always guarantee 0.618-EFX [5,25].[1] Moreover, for general monotone valuations, an EFX allocation exists that discards at most $n-1$ goods [22]. For a detailed overview, refer to Sect. 4.

In recent advances related to EFX allocations, our focus has been drawn to two separate studies that provide EFX guarantees in distinct scenarios. The first study, conducted by Christodoulou *et al.* [23], introduces a family of valuation functions where each good holds relevance (i.e., has a non-zero marginal value) for at most p agents, and any pair of agents share at most q common relevant goods. They present an EFX allocation for the case where $p = 2$ and $q = 1$, leaving the more general cases as open directions:

> "*Unless one can solve the basic question if* EFX *allocations always exist – partial progress can possibly be made for larger values of p and q*" [23].

In the second study, Akrami *et al.* [3] consider EFX allocations for the restricted additive valuations, where the valuations are additive, but the value of every good g for every agent is restricted to be either 0 or v_g. This setting has been extensively studied for different allocation problems, particularly for the maximin fairness notion, which is commonly known as the Santa Claus problem. For a detailed discussion on the significance and applications of the restricted additive valuations, refer to [3]. Akrami *et al.* [3] prove the existence of EF2X allocations for the restricted additive setting. An allocation $\langle X_1, X_2, \ldots, X_n \rangle$ is EF2X, if for every agent i and j and every $X'_j \subset X_j$ such that $|X'_j| \leq |X_j| - 2$, we have $v_i(X_i) \geq v_i(X'_j)$.

In this paper, we uncover intriguing parallels between these two seemingly different settings—the restricted additive setting and (p, q)-bounded valuations (Table 1). The results in this paper, along with those from [23] and [3], form a symmetric set of results:

> when p is either 2 or ∞, every allocation algorithm with EFX, EF2X, or approximate EFX guarantees for $(p, 1)$-bounded valuations has a corre-

[1] See Sect. 2 for a formal definition of β-EFX.

sponding algorithm with the same guarantee for (p, ∞)-bounded valuations with the restricted additive property.

The algorithms we use in our work introduce several techniques and ideas that may be of independent interest. In particular, we revisit the concept of rank introduced by Farhadi *et al.* [25] and, based on this concept, introduce the virtual welfare of the agents for an allocation, which may be of independent interest. You can find a detailed explanation of these techniques in Sect. 3. These results can serve as a building block for providing guarantees for larger values of p and q.

2 Basic Concepts

Every allocation problem instance is represented as a triple $\mathcal{I} = (N, M, \mathcal{V})$, where N is a set of n agents, M is a set of m goods, and $\mathcal{V}$ is the set of valuation functions. We refer to agents by their index from 1 to n and use the variable g to refer to a good. Our results in this paper relate to two types of valuation functions. Let us first define them formally.

Definition 1. *Let $\mathcal{V} = \{v_1, v_2, \dots, v_n\}$ be a set of additive valuation functions where for every i, $v_i : 2^M \to R^{\geq 0}$. We say $\mathcal{V}$ is restricted additive, if for every good g and every $1 \leq i \leq n$, we have $v_i(\{g\}) \in \{0, v_g\}$. For convenience, for every subset S of goods, we define $v(S) = \sum_{g \in S} v_g$.*

The second type of valuations we study is (p, q)-bounded valuations.

Definition 2. *[Christodoulou et al. [23]] Let $\mathcal{V} = \{v_1, v_2, \dots, v_n\}$ be a set of monotone valuation functions, where each $v_i : 2^M \to \mathbb{R}^{\geq 0}$. A good g is not relevant to agent i if $\forall X_i \subseteq M \setminus g, v_i(X_i \cup g) = v_i(X_i)$, and relevant otherwise. Set $\mathcal{V}$ is (p, q)-bounded if each good is relevant to at most p agents, and there are at most q goods relevant to any pair of agents.*

Note that the valuations in $\mathcal{V}$ are not necessarily additive; we only require them to be monotone. Here, we consider a slightly adjusted variant with an additional assumption that the valuations are strictly monotone for relevant goods. That is, for every v_i and a relevant good g, we have $\forall X_i \subseteq M\setminus\{g\}, v_i(X_i \cup \{g\}) - v_i(X_i) > 0$. For a set S of goods, we define $S[i]$ as a subset of S that are relevant to agent i, i.e., $S[i] = \{g \mid g \in S \text{ and } v_i(\{g\}) > 0\}$. Accordingly, we define $S[i, j]$ as the set of goods in S that are relevant both to agent i and agent j. These definitions are applicable to both (p, q)-bounded and restricted additive valuations.

An allocation is a partition of goods into $n + 1$ bundles $\langle X_0, X_1, X_2, \dots, X_n\rangle$, where for every $1 \leq i \leq n$ bundle X_i is the share of agent i, and bundle X_0 is the set of unallocated goods. We sometimes refer to X_0 as the pool. Allocation X is complete, if $X_0 = \emptyset$, and partial otherwise.

Let $\beta \in [1, +\infty]$ be a constant. For a given allocation X, we say that agent i β-envies bundle X_j if $\beta v_i(X_i) < v_i(X_j)$. Additionally, agent i β-strongly envies bundle X_j if, there exists a good $g \in X_j$, such that $\beta v_i(X_i) < v_i(X_j \setminus g)$. An allocation is $(1/\beta)$-EFX if no agent β-strongly envies any other agent. For simplicity, when $\beta = 1$, we drop the prefix and use terms envy, strongly envy, and EFX instead of 1-envy, 1-strongly envy, and 1-EFX. We also consider another relaxation of the EFX known as EF2X. In an EF2X allocation, any envy that an agent i may have towards X_j must disappear after removing any two goods from X_j, for every i, j, we have $\forall_{g,g' \in X_j}, v_i(X_i) \geq v_i(X_j \setminus \{g, g'\})$.

Weighted Envy Graph, Rank, and Virtual Value. The weighted envy graph is a flexible structure that includes several meaningful graphs relating to EFX allocations as a subgraph.

Definition 3. *Given an allocation X, the weighted envy graph G_X is a complete weighted bidirectional graph with n vertices. For each pair of vertices i and j, there is a directed edge from i to j with weight $w_X(i,j) = v_i(X_j)/v_i(X_i)$. If $v_i(X_j) = v_i(X_i) = 0$, we ignore that edge.*

By definition, when $w_X(i,j) > 1$, agent i envies agent j. Note that the edges in G_X might have weight $+\infty$, which occurs when an agent values her bundle as 0. To prevent such situations, we initiate our algorithms with a special allocation, which we refer to as the basic feasible allocation.

Definition 4. *X is a basic feasible allocation, if for every i, $|X_i| = 1$, and $\Pi_i v_i(X_i)$ is maximized.*

Indeed, in the basic feasible allocation, we allocate one item to each agent, such that the Nash welfare is maximized. Lemma 1 states that we can assume without loss of generality that in the basic feasible allocation, every agent has a non-zero value for her allocated item.

Lemma 1. *Assume $\mathcal{I} = (N, M, \mathcal{V})$ is an allocation instance, and let X be a basic feasible allocation. If for some agent i, $v_i(X_i) = 0$, then there exists another instance $\mathcal{I}' = (N', M', \mathcal{V}')$ with $|N'| < |N|$, such that any α-EFX or α-EF2X guarantee for $\mathcal{I}'$ implies the same guarantee for $\mathcal{I}$.*

We also define the **rank** of an agent for an allocation X as follows.

Definition 5. *Given an allocation X, assume $v_i(X_i) > 0$ for every agent i, and that G_X has no cycle with a total product of edge weights greater than 1. The rank of agent i in G_X, denoted by $\mathfrak{R}_i(X)$, is defined as the maximum product of edge weights along any path ending at vertex i, i.e., $\mathfrak{R}_i(X) = \max_{p \in P_i} \prod_{e \in p} w_X(e)$, where P_i is the set of all paths in G_X ending at vertex i. Also, we define the **rankpath** of agent i as $\arg\max_{p \in P_i} \prod_{e \in p} w_X(e)$, and the first vertex in the rankpath of vertex i is referred to as the root of vertex i, denoted by $\mathsf{root}(i, X)$. If there are multiple such paths, we select the path with the fewest number of edges, and if there are still multiple options, we select the one whose root has the smallest index. By definition, for every i, $\mathsf{root}(i, X)$ has rank 1.*

The idea of rank is very helpful in our allocation algorithms. This concept was initially introduced by Farhadi *et al.* [25]. Here, we explore this concept and uncover several intriguing implications of rank. In particular, we employ this concept to define the virtual value of each agent.

Definition 6. *Let X be an allocation such that for every agent i, $v_i(X_i) > 0$ and also G_X has no cycle with a total product of edges greater than* 1*. We define the virtual value of i for allocation X as $\mathfrak{v}_i(X) = v_i(X_i)/\mathfrak{R}_i(X)$.*

Note that the virtual value of each agent is also dependent on the bundle of other agents. Our algorithms are designed in a way that each agent compares their virtual value with the actual values of the other agents. Accordingly, we say agent i **virtually envies** agent j in allocation X, if $\mathfrak{v}_i(X) < v_i(X_j)$. In the same way, we define terms **virtually strong envy** and **virtually** EFX.

Definitions 5 and 6 are well-defined only for allocations where the weighted envy graph has no cycles with a total product of weights greater than 1. Thus, it is essential to ensure that this property holds throughout our algorithms. Here, we provide a necessary and sufficient condition for G_X to satisfy this property. Consider the following linear program:

$$\begin{aligned} \text{Minimize} \quad & \sum_{i=1}^{n} \mathfrak{r}_i \\ \forall_{1\le i,j\le n} \quad & \mathfrak{r}_i w_X(i,j) \le \mathfrak{r}_j \\ \forall_{1\le i\le n} \quad & \mathfrak{r}_i \ge 1 \end{aligned} \tag{1}$$

where $w_X(i,j)$ is the weight of the edge from i to j in G_X. We prove that if LP(1) is feasible, then G_X has no cycle with a weight product strictly greater than 1.

Lemma 2. *Given an allocation X such that $LP(1)$ is feasible for G_X. Then, G_X admits no cycle with a total weight product greater than 1.*

Lemma 3. *Given an allocation X and let $\mathfrak{r}_1, \mathfrak{r}_2, \ldots, \mathfrak{r}_n$ be a feasible solution to $LP(1)$. Then, for every agent i we have $\mathfrak{R}_i(X) \le \mathfrak{r}_i$.*

Observation 1. *Let P be the rankpath of agent i. Then for every edge $j \to k \in P$ we have $\mathfrak{R}_j(X)w_X(j,k) = \mathfrak{R}_k(X)$.*

Indeed, Lemmas 3 and 2 together establish that if LP(1) is feasible for allocation X, then G_X contains no cycle with a weight product exceeding one. Moreover, the solution to this LP corresponds precisely to the ranks of the agents. Specifically, for the restricted additive valuations, a straightforward condition ensures the feasibility of LP (1).

Lemma 4. *Suppose that the set of valuations is restricted additive, and let X be an allocation. If for each agent i, $X_i[i] = X_i$, the product of weights along any cycle in G_X is at most 1.*

Algorithm 1: General Framework for Allocation Algorithms

```
Function Allocate(X_0):
    X ← Basic Feasible Allocation;
    while there exists an applicable update rule do
        // The number of rules is k
        for i ← 1 to k do
            if Rule i is applicable on X then
                Apply Rule i to X to obtain X';
                X ← X';
                break;
            end
        end
    end
    Execute final step on X to obtain X*;
    return X*;
```

In this paper, we use $G_{\beta,X}$ to refer to a specific subgraph of G_X that comprises all the vertices but includes only the edges with weight more than β. In our algorithms we use $G_{1,X}$, $G_{\sqrt{2},X}$, and $G_{0,X}$ as our base graphs. we prove a structural property of $G_{\beta,X}$.

Lemma 5. *Consider restricted additive valuations, and let X be a $1/\beta$-EFX allocation for some $\beta \in [1, +\infty)$. For every edge $i \to j$ in $G_{\beta,X}$, it follows that $v_i(X_j) = v(X_j)$.*

2.1 A General Framework of Our Allocation Algorithms

Although we present algorithms with different guarantees for various valuation functions, they all follow a unified framework, which we describe here to avoid repetition.

Each algorithm consists of a set of updating rules, referred to by their indices (e.g., Rule 1, Rule 2, etc.). Every rule has preconditions that determine when it applies and updates that modify the allocation. The process starts with a basic feasible allocation, and at each step, the lowest applicable rule is selected and applied.

We define several properties and ensure they hold after each rule is applied. Each algorithm also has a potential function ϕ, which increases after each update. With an upper bound on ϕ, the algorithm eventually reaches a state where no further updates can be made. A final step then returns the resulting allocation, satisfying the desired fairness notion.

Algorithm 1 presents the pseudocode for this general framework, using $\mathcal{X}$, $\mathcal{X}'$, and $\mathcal{X}^*$ for allocations before, during, and after updates.

3 Results Overview

We provide fairness guarantees for restricted additive and (p, q)-bounded valuations. These results, alongside existing results for the restricted additive and (p, q)-bounded valuations form a symmetrical pattern:

- Akrami *et al.* [3] show that for (∞, ∞)-bounded valuations with the restricted additive property a complete EF2X allocation, and a partial EFX allocation with less than $n/2$ discarded goods exists. We prove the same guarantees for $(\infty, 1)$-bounded valuations (in the full version).
- Christodoulou [23] show that for $(2, 1)$-bounded valuations we can find an EFX allocation. We prove the same guarantee for $(2, \infty)$-bounded restricted additive valuations (Theorem 3).
- We show that both $(\infty, 1)$-bounded subadditive valuations, and (∞, ∞)-bounded valuations with the restricted additive property admit a $(\sqrt{2}/2)$-EFX allocation (Theorems 1 and 2).

The first result is preliminary and deferred to the full version, while the second and third results form our main contributions.

Table 1. Overview of the results. To show the symmetry of the results for two valuations, we use symbols ♣, ♦, ♠, ♥. Results with the same label are symmetric. Note that for Theorem 2, we also have a subadditivity assumption.

<table>
<tr><th></th><th>Monotone</th><th>Restricted additive</th></tr>
<tr><td>(2,1)-bounded</td><td colspan="2">EFX [23] ♦</td></tr>
<tr><td>(2,∞)-bounded</td><td>Open</td><td>EFX (Theorem 3)♦</td></tr>
<tr><td>(∞,1)-bounded</td><td colspan="2">EF2X (Full version)♣
$\langle$EFX, $\lfloor \frac{n}{2} \rfloor - 1\rangle$ (Full version)♠
$\frac{\sqrt{2}}{2}$-EFX (Theorem 2*)♥</td></tr>
<tr><td>(∞,∞)-bounded</td><td>Open</td><td>EF2X [3]♣
$\langle$EFX, $\lfloor \frac{n}{2} \rfloor - 1\rangle$ [3]♠
$\frac{\sqrt{2}}{2}$-EFX (Theorem 1)♥</td></tr>
</table>

In our algorithms we use various subgraphs of the weighted envy graph, including $G_{0,X}$, $G_{1,X}$, and $G_{\sqrt{2},X}$. For all these algorithms, it is crucial to ensure that G_X does not contain a cycle with a total product of weights strictly greater than 1. For the restricted additive valuations, as shown in Lemma 4, if we allocate only relevant goods to each agent, then G_X does not have such cycles (Properties §(i) and ↑(i)). For (p, q)-bounded valuations, we perform our update rules in a way that keeps LP(1) feasible (Property †(i)). When G_X has no cycle with a total product of weights strictly greater than one, terms such as rank, rankpath, and virtual value are well-defined. Another property that we maintain in our algorithms is that the **allocation is EFX with respect to the virtual values** (Properties ↑(ii) and †(ii)).

As a warmup, in the full version we show that every family of $(\infty, 1)$-bounded valuation functions admit EFX allocations with less than $n/2$ discarded goods. Next, we extend this allocation and obtain a complete EF2X allocation. These two results complement the guarantees of Akrami *et al.* [3] for the restricted additive setting. Hence, most of the rules we use in our allocation algorithms resemble those in Akrami *et al.* [3]. However, the structure of (p, q)-bounded valuations allows us to use a simpler potential function to prove the termination of our algorithms.

An essential part of the algorithm proposed by Akrami *et al.* [3] is an elegant process called envy-elimination. In the envy-elimination process, we eliminate all possible strong envies from the allocation while ensuring that the potential function either remains unchanged or increases. Surprisingly, in the envy-elimination process we only remove goods from the bundles, however, by carefully defining the potential function, we can show that we never fall into an infinite loop trap. The reason that here our potential function is much simpler is that, here we don't use envy-elimination in our algorithm. Instead, we make sure that after each update the allocation is EFX.

The envy-elimination process makes the restricted additive approach a great starting point for deriving more general EFX results: for restricted additive valuations, we needn't worry about the EFX guarantee after each update, as envy-elimination restores this property. This allows for various updates. Here, we refine the envy-elimination process to obtain a $(\sqrt{2}/2)$-EFX allocation for restricted additive valuations. Our algorithm uses two simple update rules, envy-elimination, and a final step.

Interestingly, we achieve the same guarantee for $(\infty, 1)$-bounded subadditive valuations using a simple algorithm with one updating rule and a final step. This rule relies on rank and virtual value. If any agent virtually envies the pool, we select a subset of goods and assign it to an agent based on the ratio of the subset's value to the agent's virtual value, adjusting the allocation through her rankpath. This rule proves very useful.

We later use the same rule to propose an EFX allocation algorithm for $p = 2$ with restricted additive valuations. This algorithm is more technically complex, involving four updating rules and a final step, but the key rule remains similar to the previous one. Since these algorithms don't involve envy-elimination, their potential functions are simple: after each update, Nash welfare increases.

We note that almost all known updating rules for EFX allocations for a general number of agents are designed to increase the utility of agents. However, for special cases, such as when $n = 3$, some update rules might actually decrease social welfare. This discrepancy suggests that for the general case, we may need updating rules that allow for a temporary decrease in agents' utility to perform the algorithm. For the restricted additive setting, Akrami *et al.* [3] employed the envy elimination process to reduce agents' utility. By employing a complex potential function, they show that the algorithm never falls into an infinite loop. In our approach, we leverage virtual value to propose updates that can decrease the value of agents while ensuring that the algorithm does not fall into an infinite loop. This is why we believe that the idea of virtual value has the potential to be applied to more general cases.

Several intriguing directions emerge for exploration following this work. One direction is whether the observed symmetry between restricted additive valuations and $(\infty, 1)$-bounded valuations extends to other problem domains. Notably, there are significant results for the maximin fairness notion and restricted additive valuations, commonly referred to as the Santa Claus problem [6–8]. Investigating whether analogous results can be achieved for the Santa Claus problem with $(\infty, 1)$-bounded valuations would be interesting. Furthermore, it is worthwhile to consider the generalization of our algorithms to more general settings, such as additive and subadditive valuations. One can think of using the concept of rank and virtual value and the idea in the suggested updating rule to potentially achieve a $\frac{\sqrt{2}}{2}$-EFX allocation in the additive setting.

4 Further Related Work

Finding Approximate EFX *Allocations.* Finding approximately fair allocations is common for various fairness notions, including EFX. Beyond the studies mentioned in the introduction, Barman *et al.* [9] explored approximate EFX allocations. They introduced a parameter $\gamma \in (0, 1]$ to bound the multiplicative range of nonzero values for goods across agents and developed an algorithm that yields a $\frac{2\gamma}{\sqrt{5+4\gamma}-1}$-EFX allocation. Amanatidis *et al.* [4] also investigated $(2, \infty)$-bounded valuations and proved the existence of a 2/3-EFX allocation.

Relaxing EFX. Some studies explore relaxed variations of EFX to achieve better guarantees. Farhadi *et al.* [25] examine removing a random good instead of the least valuable one, establishing a 0.74 approximation for this modified notion. Caragiannis *et al.* [16] introduce epistemic EFX, where for each agent, there exists a way to shuffle others' goods so the agent doesn't envy anyone. Another variation is envy-freeness up to one less preferred good, introduced by Barman *et al.* [12], which limits the value of the eliminated good.

EFX *with Charity.* A strand of research has delved into finding partial EFX allocations rather than complete ones. Naturally, we want to keep the number of discarded goods as small as possible. The pioneering work in this area was a partial EFX allocation, with the Nash welfare at least half of the optimal [17]. Subsequently, Chaudhury *et al.* [22] prove the existence of a partial EFX allocation that leaves at most $n-1$ goods unallocated. Berger *et al.* [12] reduced the number of discarded goods to $n - 2$. Chaudhury *et al.* [21] presented a general framework for achieving a partial $(1-\epsilon)$-EFX allocations with a sublinear number of unallocated goods for any ϵ. The number of unallocated goods is then successively reduced to ultimately $O_\epsilon(\sqrt{n \log n})$ [1,11,19].

5 $\frac{\sqrt{2}}{2}$-EFX for Restricted Additive Valuations

In this section, we present a $\frac{\sqrt{2}}{2}$-EFX allocation algorithm for restricted additive valuations. The algorithm uses two updating rules and a final step. After each update, an envy elimination process ensures the allocation remains $\frac{\sqrt{2}}{2}$-EFX. Our algorithm also guarantees that Property §(i) holds at every step before the final step. This property holds trivially for the initial feasible allocation.

§(i) : For every agent i, $\mathcal{X}_i[i] = \mathcal{X}_i$.

Given Lemma 4, Property §(i) implies that $G_{\sqrt{2},\mathcal{X}}$ is always acyclic. [2] We assume that $G_{\sqrt{2},\mathcal{X}}$ contains k source agents. For convenience, we also assume that the agents are re-indexed after each update so that agents $1, 2, \ldots, k$ are the sources of $G_{\sqrt{2},\mathcal{X}}$, and $v(\mathcal{X}_1) \leq v(\mathcal{X}_2) \leq \ldots \leq v(\mathcal{X}_k)$. We define our potential function as $\phi(\mathcal{X}) = [v(\mathcal{X}_1), v(\mathcal{X}_2), \ldots, v(\mathcal{X}_k), \infty]$. We designate our algorithm in this section as C$\sqrt{2}$XRA.

5.1 Envy Elimination

As we mentioned, the envy elimination process is initially introduced by Akrami *et al.* [3]. This process aims to eliminate strong envies among agents while preserving essential allocation properties. Here, we propose a simple alternative for the envy elimination process: Given an allocation $\mathcal{X}$ with potential strong envies, as long as there exist two agents, i and j, where agent i $\sqrt{2}$-strongly envies agent j, we find a good $g \in \mathcal{X}_j$ such that $\sqrt{2}v_i(\mathcal{X}_i) < v_i(\mathcal{X}_j \setminus \{g\})$, and move g from $\mathcal{X}_j$ to $\mathcal{X}_0$.

Given the nature of the process, we know that the resulting allocation is $\frac{\sqrt{2}}{2}$-EFX. As we show in Lemma 6, this process also guarantees that the value of the potential function does not decrease.

Lemma 6. *Let $\mathcal{X}$ be a $\frac{\sqrt{2}}{2}$-EFX allocation satisfying Property §(i), and let $\mathcal{X}'$ be the allocation after applying envy elimination. Then, $\mathcal{X}'$ satisfies Property §(i). Furthermore we have $\phi(\mathcal{X}) \leq \phi(\mathcal{X}')$.*

5.2 Update Rule 1

The first rule takes an allocation $\mathcal{X}$ as input, and if there exists a source agent i and a good $g \in \mathcal{X}_0$ such that good g is relevant to agent i, we allocate g to agent i.

[2] Note that this also holds for the initial allocation.

Lemma 7. *Let $\mathcal{X}$ be an allocation satisfying Property* § *(i), and let $\mathcal{X}'$ be the allocation after applying Rule 1. Then $\mathcal{X}'$ satisfies Property* § *(i). Furthermore, we have $\phi(\mathcal{X}) < \phi(\mathcal{X}')$.*

5.3 Update Rule 2

In the second updating rule, if for some agent i, $v_i(\mathcal{X}_i) < \sqrt{2}v_i(\mathcal{X}_0)$, we update the allocation as follows: since Rule 1 is not applicable, i is not a source. Also, since $G_{\sqrt{2},\mathcal{X}}$ is acyclic, there is a source, let's say j, with a directed path to i in $G_{\sqrt{2},\mathcal{X}}$. We shift the allocation through this path, i.e., for every edge $i_1 \rightarrow i_2$ in this path, we allocate $\mathcal{X}_{i_2}$ to agent i_1. We also give all the goods that are valuable to agent i in $\mathcal{X}_0$ to i, and return the goods in $\mathcal{X}_j$ back to the pool.

Lemma 8. *Let $\mathcal{X}$ be an allocation satisfying Property* § *(i). Assume that Rule 1 is not applicable, and let $\mathcal{X}'$ be the allocation after applying Rule 2. Then, $\mathcal{X}'$ satisfies Property* § *(i). Furthermore, we have $\phi(\mathcal{X}) < \phi(\mathcal{X}')$.*

5.4 Final Step

When none of Rules 1 and 2 is feasible, we give the entire pool to agent 1 and terminate the allocation. In Theorem 1 we show that the resulting allocation is $\frac{\sqrt{2}}{2}$-EFX.

Theorem 1. *Suppose the valuations are restricted additive, and let $\mathcal{X}^*$ be the allocation returned by* $\mathsf{C\sqrt{2}XRA}$*. Then, $\mathcal{X}^*$ is a $\frac{\sqrt{2}}{2}$-EFX allocation.*

6 $\frac{\sqrt{2}}{2}$-EFX for $(\infty, 1)$-Bounded Valuations

In this section, we propose a $\frac{\sqrt{2}}{2}$-EFX allocation algorithm for agents with $(\infty, 1)$-bounded subadditive valuations. Our algorithm consists of a single updating rule, followed by a final step. Throughout the algorithm, we use the Nash social welfare of the allocation as our potential function, that is, $\phi(\mathcal{X}) = \mathsf{NSW}(\mathcal{X})$. We also ensure that the following properties are consistently maintained throughout the algorithm before the final step:

†(i) LP (1) is feasible for allocation $\mathcal{X}$.
†(ii) No agent **virtually strongly envies** another agent. Formally, for any pair of agents i and j, and any good $g \in \mathcal{X}_j$, we have: $\mathfrak{v}_i(\mathcal{X}) \geq v_i(\mathcal{X}_j \setminus g)$.
†(iii) For every agent i, $\mathcal{X}_i[i] = \mathcal{X}_i$.

We designate our algorithm in this section as $\mathsf{C\sqrt{2}XPQ}$.

6.1 Update Rule 1

This update rule is applicable when an agent virtually envies the pool. Let $\hat{\mathcal{X}_0} \subseteq \mathcal{X}_0$ denote a minimal subset of the pool that is virtually envied by at least one agent. Consider agent i with the maximum value of $v_i(\hat{\mathcal{X}_0})/\mathfrak{v}_i(\mathcal{X})$. Note that all goods in $\hat{\mathcal{X}_0}$ are relevant to agent i. Let P denote the rankpath of agent i. We allocate $\hat{\mathcal{X}_0}$ to agent i and shift the bundles along path P. Formally, for every edge $j \to k \in P$, we allocate $\mathcal{X}_k$ to agent j. Then, we return the bundle of the first agent in P to the pool. Note that this update rule does not rely on properties of $(\infty, 1)$-bounded valuations and can be used in the general setting. Thus, we state our lemmas in this section in general form.

To prove that $\mathcal{X}'$ satisfies the required properties, we define $\mathfrak{r}_j$ for each agent j as follows:

$$\mathfrak{r}_j = \begin{cases} \mathfrak{R}_j(\mathcal{X}), & \text{if } j \notin P \\ \mathfrak{R}_k(\mathcal{X}), & \text{if } j \to k \in P \\ \frac{v_j(X_0^*)}{\mathfrak{v}_j(\mathcal{X})}, & \text{if } j = i \end{cases}. \tag{2}$$

We now show that $\mathfrak{r}$ is a feasible solution to LP (1) for allocation $\mathcal{X}'$.

Lemma 9. *Let $\mathcal{X}$ be an allocation satisfying Properties* † *(i) and* † *(ii), and let $\mathcal{X}'$ be the allocation after applying Rule 1. Then, $\mathfrak{r}$ is a feasible solution to LP* (1) *for allocation $\mathcal{X}'$.*

Lemma 10. *Let $\mathcal{X}$ be an allocation satisfying Properties* † *(i) and* † *(ii) and* † *(iii), and let $\mathcal{X}'$ be the allocation after applying Rule 1. Then, $\mathcal{X}'$ satisfies Properties* † *(i),* † *(ii), and* † *(iii). Furthermore, we have $\phi(\mathcal{X}') > \phi(\mathcal{X})$.*

6.2 Final Step

When Rule 1 is not applicable, we perform a final step to achieve a complete $\frac{\sqrt{2}}{2}$-EFX allocation. Let i be an agent with rank equal to one in $\mathcal{X}$, and S be the set of all other agents whose rank is at most $\sqrt{2}$. In the final step, first we iterate through S in an arbitrary order and ask each agent to pick all the goods in the pool that are relevant to her. Then we allocate all the remaining goods in the pool to agent i and return the allocation.

Theorem 2. *Suppose the valuations are restricted additive, and let $\mathcal{X}^*$ be the allocation returned by C$\sqrt{2}$XPQ. Then, $\mathcal{X}^*$ is a $\frac{\sqrt{2}}{2}$-EFX allocation.*

7 EFX for the Restricted Additive Valuations with P = 2

Our final algorithm finds a complete EFX allocation for $(2, \infty)$-bounded and restricted additive valuations. With $p = 2$, each good is relevant to at most two agents, while any pair of agents may share multiple relevant goods. Here, for a set S, $S[i, i]$ refers to goods relevant only to agent i.

Our algorithm consists of four updating rules and a final step. Unlike our previous algorithms, agents may become satisfied and be eliminated from the process. Therefore, here N refers to the set of remaining agents. We also define rank, rankpath, and virtual value based on the allocation of the remaining agents. We use $\phi(\mathcal{X}) = (n - |N|, \mathsf{NSW}(\mathcal{X}))$ as the potential function to prove the termination of our algorithm: Therefore, after each update, either the number of satisfied agents increases or the Nash social welfare of the allocation increases. We designate our algorithm in this section as PQRAX. Furthermore, our updates are designed in a way that the following properties are always satisfied during the algorithm:

↑(i) For every agent $i \in N$, there is an agent j such that $\mathcal{X}_i \subseteq M[i, j]$. We call agent j, the corresponding agent of agent i. Note that the corresponding agent of i can be i itself.

↑(ii) For every agent $i, j \in N$, and every $g \in \mathcal{X}_j$ we have $v_i(\mathcal{X}_j \setminus \{g\}) \leq \mathfrak{v}_i(\mathcal{X})$.

Note that Properties ↑(i) and ↑(ii) hold trivially for the basic feasible allocation. We also prove that after each update, the virtual value of each agent does not decrease. This property also plays a key role in our proofs.

Observation 2. *Let X be an allocation that satisfies Property ↑(i). Then for every $i \in N$, we have $\mathfrak{v}_i(X) = v(X_{root(i,X)})$.*

Observation 3. *Let X be an allocation satisfying Property ↑(i), and $i, j \in N$ be two agents that i has a path to j in $G_{0,X}$. Then we have $\mathfrak{v}_j(X) \leq \mathfrak{v}_i(X)$.*

Lemma 11. *Let X be an allocation satisfying Property ↑(i), and suppose for every $i \in N$ and $g \in X_i$ we have $v(X_i \setminus \{g\}) \leq \mathfrak{v}_i(X)$. Then X satisfies Property ↑(ii).*

Update Rule 1

This rule is applicable when an agent $i \in N$ with $\mathfrak{R}_i(\mathcal{X}) = 1$ and a good $g \in \mathcal{X}_0$ exists such that g has the same type as the goods in $\mathcal{X}_i$, i.e., assuming $\mathcal{X}_i \subseteq M[i, j]$ for some j, then $g \in M[i, j]$. If so, we replace $\mathcal{X}_i$ with a minimal subset of $\mathcal{X}_i \cup \{g\}$ that has a value more than $v(\mathcal{X}_i)$.

Observation 4. *Let $\mathcal{X}'$ be the result of applying Rule 1 to $\mathcal{X}$. Then, all edges in $G_{0,\mathcal{X}}$ remain unchanged in $G_{0,\mathcal{X}'}$.*

Lemma 12. *Let $\mathcal{X}$ be an allocation satisfying Properties ↑(i) and ↑(ii), and let $\mathcal{X}'$ be the result of applying Rule 1 on $\mathcal{X}$. Then, $\mathcal{X}'$ satisfies Properties ↑(i) and ↑(ii). Furthermore, we have $\phi(\mathcal{X}) < \phi(\mathcal{X}')$ and for every remaining agent k we have $\mathfrak{v}_k(\mathcal{X}) \leq \mathfrak{v}_k(\mathcal{X}')$.*

Update Rule 2

This rule is applicable if there exists an agent $i \in N$ such that for some j, $\mathfrak{v}_i(\mathcal{X}) < v(\mathcal{X}_0[i,j])$. If so, we update the allocation as follows: let j be the agent with the minimum virtual value such that $\mathfrak{v}_i(\mathcal{X}) < v(\mathcal{X}_0[i,j])$, and let $P = i_1 \rightarrow i_2 \rightarrow \ldots \rightarrow i_k \rightarrow i$ be the rankpath of agent i. We allocate a minimal subset of $\mathcal{X}_0[i,j]$ with value more than $\mathfrak{v}_i(\mathcal{X})$ to agent i, shift the bundles through P, and return $\mathcal{X}_{i_1}$ back to the pool.

Lemma 13. *Let $\mathcal{X}$ be an allocation satisfying Properties $\uparrow$(i) and$\uparrow$(ii), and let $\mathcal{X}'$ be the result of applying Rule 2 on $\mathcal{X}$. Then, $\mathcal{X}'$ satisfies Properties $\uparrow$(i) and $\uparrow$(ii). Furthermore, we have $\phi(\mathcal{X}) < \phi(\mathcal{X}')$ and for every remaining agent k we have $\mathfrak{v}_k(\mathcal{X}) \leq \mathfrak{v}_k(\mathcal{X}')$.*

Update Rule 3

This rule is applicable when there is an agent $i \in N$ such that for every agent $j \neq i$, $v_i(\mathcal{X}_j) = 0$. If so, we ask agent i to pick either $\mathcal{X}_i$ or $\mathcal{X}_0[i]$ that is more valuable to him and return the other bundle back to the pool. Also, we mark agent i as satisfied and remove her from the process.

Lemma 14. *Let $\mathcal{X}$ be an allocation satisfying Properties $\uparrow$(i) and $\uparrow$(ii), and let $\mathcal{X}'$ be the result of applying Rule 3 on $\mathcal{X}$. Then, $\mathcal{X}'$ satisfies Properties $\uparrow$(i) and $\uparrow$(ii). Furthermore, we have $\phi(\mathcal{X}) < \phi(\mathcal{X}')$ and for every remaining agent k we have $\mathfrak{v}_k(\mathcal{X}) \leq \mathfrak{v}_k(\mathcal{X}')$.*

Lemma 15. *Let i be an agent satisfied by Rule 3. Then, i will never strongly envy any remaining agent, and no remaining agent will ever strongly envy agent i.*

Update Rule 4

First, we prove that when the first three rules are not applicable, then $G_{0,\mathcal{X}}$ has a special structure.

Lemma 16. *Let $\mathcal{X}$ be an allocation satisfying Properties $\uparrow$(i) and $\uparrow$(ii). If the first three rules are not applicable, then $G_{0,\mathcal{X}}$ consists of one or several disjoint cycles.*

In the fourth updating rule, we select and satisfy one agent with rank 1. Let i be an agent with $\mathfrak{R}_i(\mathcal{X}) = 1$ and suppose $i \rightarrow j_1 \rightarrow j_2 \ldots \rightarrow j_k \rightarrow i$ be the cycle in $G_{0,\mathcal{X}}$ containing i. We ask agent i to pick either $\mathcal{X}_i \cup \mathcal{X}_0[i]$ or $\mathcal{X}_{j_1}$, whichever that is more valuable to her. We then mark agent i as satisfied and remove her from the process. Finally, if agent i picked $\mathcal{X}_{j_1}$, we shift the bundles through the cycle, i.e., for every $2 \leq \ell < k$, we give $\mathcal{X}_{\ell+1}$ to agent j_ℓ and give $\mathcal{X}_i$ to agent j_k.

Lemma 17. *Let $\mathcal{X}$ be an allocation satisfying Properties $\uparrow$ (i) and $\uparrow$ (ii) and let $\mathcal{X}'$ be the result of applying Rule 4 on $\mathcal{X}$. Then, $\mathcal{X}'$ satisfies Properties $\uparrow$ (i) and $\uparrow$ (ii). Furthermore, we have $\phi(\mathcal{X}) < \phi(\mathcal{X}')$ and for every remaining agent k we have $\mathfrak{v}_k(\mathcal{X}) \leq \mathfrak{v}_k(\mathcal{X}')$.*

Lemma 18. *Let i be an agent satisfied by Rule 4. Then, i will never strongly envy any remaining agent, and no remaining agent will ever strongly envy agent i.*

Final Step

Finally, if only one agent is left behind, i.e., $|N| = 1$, we allocate the whole set of items in $\mathcal{X}_0$ to the remaining agent and return the allocation.

Theorem 3. *Suppose the valuations are restricted additive and $(2, \infty)$-bounded, and let $\mathcal{X}^*$ be the allocation returned by* PQRAX*. Then, $\mathcal{X}^*$ is an* EFX *allocation.*

References

1. Akrami, H., Alon, N., Chaudhury, B.R., Garg, J., Mehlhorn, K., Mehta, R.: EFX allocations: simplifications and improvements. arXiv preprint arXiv:2205.07638 (2022)
2. Akrami, H., Alon, N., Chaudhury, B.R., Garg, J., Mehlhorn, K., Mehta, R.: EFX: a simpler approach and an (almost) optimal guarantee via rainbow cycle number. In: Proceedings of the 24th ACM Conference on Economics and Computation, EC, pp. 61–61 (2023)
3. Akrami, H., Rezvan, R., Seddighin, M.: An EF2X allocation protocol for restricted additive valuations. In: Proceedings of the 31st International Joint Conference on Artificial Intelligence, IJCAI, pp. 17–23 (2022)
4. Amanatidis, G., Filos-Ratsikas, A., Sgouritsa, A.: Pushing the frontier on approximate EFX allocations. arXiv preprint arXiv:2406.12413 (2024)
5. Amanatidis, G., Markakis, E., Ntokos, A.: Multiple birds with one stone: beating 1/2 for EFX and GMMS via envy cycle elimination. Theor. Comput. Sci. **841**, 94–109 (2020)
6. Annamalai, C., Kalaitzis, C., Svensson, O.: Combinatorial algorithm for restricted max-min fair allocation. ACM Trans. Algor. **13**(3), 1–28 (2017)
7. Asadpour, A., Feige, U., Saberi, A.: Santa claus meets hypergraph matchings. ACM Trans. Algor. **8**(3), 1–9 (2012)
8. Bamas, É., Rohwedder, L.: Better trees for santa claus. In: Proceedings of the 55th Annual ACM Symposium on Theory of Computing, STOC, pp. 1862–1875 (2023)
9. Barman, S., Kar, D., Pathak, S.: Parameterized guarantees for almost envy-free allocations. In: Proceedings of the 23rd International Conference on Autonomous Agents and Multiagent Systems, AAMAS, pp. 151–159 (2024)
10. Barman, S., Krishnamurthy, S.K., Vaish, R.: Finding fair and efficient allocations. In: Proceedings of the 19th ACM Conference on Economics and Computation, EC, pp. 557–574 (2018)

11. Berendsohn, B.A., Boyadzhiyska, S., Kozma, L.: Fixed-point cycles and approximate EFX allocations. In: 47th International Symposium on Mathematical Foundations of Computer Science, MFCS (2022)
12. Berger, B., Cohen, A., Feldman, M., Fiat, A.: Almost full EFX exists for four agents. In: Proceedings of the 36th AAAI Conference on Artificial Intelligence, vol. 36, pp. 4826–4833 (2022)
13. Brams, S.J., Taylor, A.D.: An envy-free cake division protocol. Am. Math. Monthly 9–18 (1995)
14. Brams, S.J., Taylor, A.D.: Fair Division: From Cake-Cutting to Dispute Resolution. Cambridge University Press, Cambridge (1996)
15. Budish, E.: The combinatorial assignment problem: approximate competitive equilibrium from equal incomes. J. Polit. Econ. **119**(6), 1061–1103 (2011)
16. Caragiannis, I., Garg, J., Rathi, N., Sharma, E., Varricchio, G.: New fairness concepts for allocating indivisible items. In: Proceedings of the 32nd International Joint Conference on Artificial Intelligence, IJCAI, pp. 2554–2562 (2023)
17. Caragiannis, I., Gravin, N., Huang, X.: Envy-freeness up to any item with high nash welfare: the virtue of donating items. In: Proceedings of the 20th ACM Conference on Economics and Computation, EC, pp. 527–545 (2019)
18. Caragiannis, I., Kurokawa, D., Moulin, H., Procaccia, A.D., Shah, N., Wang, J.: The unreasonable fairness of maximum nash welfare. In: Proceedings of the 2016 ACM Conference on Economics and Computation, EC, pp. 305–322. ACM (2016)
19. Chashm Jahan, S., Seddighin, M., Seyed-Javadi, S.M., Sharifi, M.: Rainbow cycle number and EFX allocations: (almost) closing the gap. In: Proceedings of the 32th International Joint Conference on Artificial Intelligence, IJCAI, pp. 2572–2580 (2023)
20. Chaudhury, B.R., Garg, J., Mehlhorn, K.: EFX exists for three agents. In: Proceedings of the 21st ACM Conference on Economics and Computation, EC, pp. 1–19 (2020)
21. Chaudhury, B.R., Garg, J., Mehlhorn, K., Mehta, R., Misra, P.: Improving EFX guarantees through rainbow cycle number. In: Proceedings of the 22nd ACM Conference on Economics and Computation, EC, pp. 310–311 (2021)
22. Chaudhury, B.R., Kavitha, T., Mehlhorn, K., Sgouritsa, A.: A little charity guarantees almost envy-freeness. SIAM J. Comput. **50**(4), 1336–1358 (2021)
23. Christodoulou, G., Fiat, A., Koutsoupias, E., Sgouritsa, A.: Fair allocation in graphs. In: Proceedings of the 24th ACM Conference on Economics and Computation, EC, pp. 473–488 (2023)
24. Even, S., Paz, A.: A note on cake cutting. Disc. Appl. Math. **7**(3), 285–296 (1984)
25. Farhadi, A., Hajiaghayi, M., Latifian, M., Seddighin, M., Yami, H.: Almost envy-freeness, envy-rank, and nash social welfare matchings. In: Proceedings of the 35th AAAI Conference on Artificial Intelligence, vol. 35, pp. 5355–5362 (2021)
26. Ghodsi, M., HajiAghayi, M., Seddighin, M., Seddighin, S., Yami, H.: Fair allocation of indivisible goods: improvements and generalizations. In: Proceedings of the 19th ACM Conference on Economics and Computation, EC, pp. 539–556 (2018)
27. Kurokawa, D., Procaccia, A.D., Wang, J.: Fair enough: guaranteeing approximate maximin shares. J. ACM **65**(2), 8 (2018)
28. Plaut, B., Roughgarde, T.: Almost envy-freeness with general valuations. In: Proceedings of the 29th Annual ACM-SIAM Symposium on Discrete Algorithms, EC, pp. 2584–2603 (2018)
29. Procaccia, A.D.: Technical perspective: an answer to fair division's most enigmatic question. Commun. ACM **63**(4), 118–118 (2020)
30. Steinhaus, H.: The problem of fair division. Econometrica **16**(1) (1948)

Best-of-Both-Worlds Fair Allocation of Indivisible and Mixed Goods

Xiaolin Bu[1], Zihao Li[2], Shengxin Liu[3], Xinhang Lu[4](✉), and Biaoshuai Tao[1]

[1] Shanghai Jiao Tong University, Shanghai, China
{lin_bu,bstao}@sjtu.edu.cn

[2] Nanyang Technological University, Singapore, Singapore
zihao004@e.ntu.edu.sg

[3] Harbin Institute of Technology, Shenzhen, Guangdong, China
sxliu@hit.edu.cn

[4] The University of New South Wales, Sydney, Australia
xinhang.lu@unsw.edu.au

Abstract. We study the problem of fairly allocating either a set of indivisible goods or a set of mixed divisible and indivisible goods (i.e., mixed goods) to agents with additive utilities, taking the best-of-both-worlds perspective of guaranteeing fairness properties both ex-ante and ex-post. The ex-post fairness notions considered in this paper are relaxations of envy-freeness, specifically, EFX for indivisible-goods allocation, and EFM for mixed-goods allocation. For two agents, we show that there is a polynomial-time randomized algorithm that achieves ex-ante envy-freeness and ex-post EFX / EFM simultaneously. For n agents with bi-valued utilities, we show there exist randomized allocations that are (i) ex-ante proportional and ex-post EFM, and (ii) ex-ante envy-free, ex-post EFX, and ex-post fractionally Pareto optimal.

Keywords: Fair division · Indivisible goods · Mixed goods · Randomization and approximation

1 Introduction

Fair division studies the problem of fairly allocating scarce resources among agents with heterogeneous preferences over the resources. It is a fundamental problem in society, with real-world applications such as divorce settlement, rent division, course allocation, and more. Two classic fairness notions are *envy-freeness (EF)* and *proportionality (PROP)*. An allocation is said to be envy-free if each agent values her own bundle weakly better than any other bundle in the allocation, and proportional if every agent gets a bundle of value at least $1/n$ times her value for the entire resources, where n is the number of agents.

Our focus in this paper is on the settings of allocating *indivisible* goods as well as a mixture of both divisible and indivisible goods (henceforth referred to as *mixed* goods) when agents have additive utilities, which have received significant attention in recent years [8, 40, 46, 47]. Despite being desirable properties,

M. Mavronicolas et al. (Eds.): WINE 2024, LNCS 15534, pp. 277–294, 2026.
https://doi.org/10.1007/978-3-032-08560-3_16

neither envy-freeness nor proportionality can always be satisfied when (deterministically) allocating indivisible or mixed goods among the agents.

To circumvent the issue, relaxations of the notions have been proposed and studied. For instance, when allocating indivisible goods, *envy-freeness up to any good (EFX)* (resp., *envy-freeness up to one good (EF1)*) requires that an agent's envy towards another agent should be eliminated after the hypothetical removal of any (resp., some) good from the latter agent's bundle [22,25]. While the existence of EFX allocations is only known in special cases (see our Related Work), the weaker notion of EF1 can always be satisfied, even for any number of agents with arbitrary monotonic utilities [39]. With mixed goods, Bei et al. [15] proposed *envy-freeness for mixed goods (EFM)*, which generalizes both envy-freeness and EF1: An agent is envy-free towards any agent whose bundle contains some divisible goods and EF1 towards the rest. They showed that an EFM allocation always exists for any number of agents with additive utilities.

An alternative method to achieve fairness is through randomization [see, e.g., 1,2,19,23,33]. With the power of randomization, envy-freeness and proportionality can both be easily and trivially achieved by giving all goods to a single agent uniformly at random. The realized allocation, however, is patently unfair since all agents but one are left empty-handed.

Aziz et al. [11] timely introduced the *best-of-both-worlds (BoBW)* approach, which combines the two aforementioned methods with the goal of constructing a randomized allocation (i.e., a lottery over deterministic allocations) that is exactly fair ex-ante (before the randomness is realized) and approximately fair ex-post (after the randomness is realized). They showed that ex-ante envy-freeness and ex-post EF1 can be simultaneously achieved when agents have additive utilities. In this paper, our goal is to strengthen the ex-post fairness guarantee to EFX when allocating indivisible goods, and, for the first time, extend the study of best-of-both-worlds fairness to the mixed-goods setting.

Our Results. As is common in the literature of fair division, we assume that agents have additive utilities when allocating either indivisible or mixed goods.

In Sect. 3, we focus on two agents. While the work of Feldman et al. [29] showed that there always exists a randomized allocation that is ex-ante EF and ex-post EFX, we improve upon the result by devising a *polynomial-time* algorithm to compute such a randomized allocation. (An independent work of Garg and Sharma [32] gives a different polynomial-time algorithm to achieve ex-ante EF and ex-post EFX, and in addition, ex-post $\frac{4}{5}$-MMS.) Built upon this result, we then show with mixed goods, a randomized allocation that is simultaneously ex-ante EF and ex-post EFM can be found in polynomial time.

Next, we consider the n-agent cases with mixed-goods and indivisible-goods settings in Sects. 4 and 5 respectively. In both cases, we assume agents have *bi-valued* utilities: each agent's utility for each good is one of two possible values.

In Sect. 4, we devise a polynomial-time algorithm to compute an integral allocation sampled from a randomized allocation that is ex-ante proportional and ex-post EFM. Our result adds to the growing literature revolved around EFM when allocating mixed resources [18,36–38,43].

In Sect. 5, we devise a polynomial-time algorithm to compute an integral allocation sampled from a randomized allocation that is ex-ante EF, ex-post EFX, and ex-post fPO.[1] This generalizes multiple results known in the literature. For instance, the compatibility between ex-ante EF, ex-post EFX, and ex-post PO was only known for binary utilities [13,34].[2] For bi-valued utilities, we only knew the compatibility of ex-post notions between EFX and PO [7] as well as between EFX and fPO [31].

Related Work. The fair allocation of indivisible goods has received extensive attention in the past decades [8,46,47]. Liu et al. [40] overviewed the recent developments of mixed-goods allocations. Below, we first discuss EFX existence results, followed by an overview of best-of-both-worlds fairness in fair division.

EFX Existence. The existence of EFX allocations remains largely open, though progress has been made in special cases where the number of agents or agents' utility functions are restricted. An EFX allocation always exists for two agents with general utilities [44], for three agents [3,28], as well as for any number of agents with restricted utilities such as being identical [44], binary submodular [13] or more general [20], bi-valued [7,31], or when there are two types of agents and agents of the same type have identical additive utilities [41]. There have been lines of work focusing on approximately-EFX allocations [3,6], partial allocations that are EFX [17,24,26], or a combination of both [27].

BoBW Fairness. In addition to the compatibility result between ex-ante EF and ex-post EF1 mentioned above, Aziz et al. [11] showed several impossibility results regarding achieving BoBW fair and economically efficient allocations. For agents with weighted *entitlements*, ex-ante weighted envy-freeness (WEF) is compatible with ex-post weighted transfer envy-freeness up to one good, but not compatible with any stronger ex-post WEF relaxation [9,35]. For agents with subadditive utilities, ex-ante $\frac{1}{2}$-EF, ex-post $\frac{1}{2}$-EFX and ex-post EF1 can be achieved simultaneously [29]. Best-of-both-worlds fairness has also been explored for fair-share-based notions like proportionality and the *maximin share (MMS)* guarantee for agents with additive [5,14] or fractionally subadditive utilities [4]. Babaioff et al. [14] showed ex-ante proportionality and ex-post $\frac{1}{2}$-MMS are compatible. The ex-post MMS approximation ratio was improved in [5], at the cost of weakening ex-ante fairness guarantees.

Slightly further afield, the BoBW paradigm has also been applied to collective choice contexts, e.g., committee voting [10,48] and participatory budgeting [12].

2 Preliminaries

For each positive integer t, let $[t] := \{1, 2, \ldots, t\}$. Denote by $N = [n]$ the set of n agents to whom we allocate resources. In this paper, we consider both indivisible-goods and mixed-goods settings. Let $M \cup D$ be the set of mixed goods, where

[1] Ex-post *fractionally Pareto optimality (fPO)*, which can be found in Defrnition 2.3, is a stronger notion of economic efficiency than ex-post Pareto optimality (PO).

[2] The result in [13] works for *binary submodular* (also known as *matroid-rank*) utilities.

$M = \{g_1, g_2, \ldots, g_m\}$ is the set of m indivisible goods and $D = \{d_1, d_2, \ldots, d_{\overline{m}}\}$ is the set of $\overline{m}$ *homogeneous* divisible goods. When $D = \emptyset$, we are concerned with the indivisible-goods setting.

Allocations. A *fractional allocation* of mixed goods $M \cup D$ to the agents in N is specified by a non-negative $n \times (m+\overline{m})$ matrix $\mathbf{X} = (X_{ig})_{i \in N, g \in M \cup D}$ such that for each $g \in M \cup D$, $\sum_{i \in N} X_{ig} = 1$; here, $X_{ig} \in [0, 1]$ denotes the fraction of good g assigned to agent i. The i-th row X_i of the matrix denotes the goods allocated to agent i in the fractional allocation. When we simply say "an allocation", it will mean a fractional allocation, unless otherwise clear from the context.

A fractional allocation $\mathbf{X}$ is *integral* if $X_{ig} \in \{0, 1\}$ for every $i \in N$ and every indivisible good $g \in M$. Given an integral allocation $\mathbf{X}$, denote by $M_i := \{g \in M \mid X_{ig} = 1\}$ the set of indivisible goods assigned to agent i, and let the $\overline{m}$-dimensional vector $\mathbf{x}_i = (x_{i1}, x_{i2}, \ldots, x_{i\overline{m}}) := (X_{ig})_{g \in D}$ represent the divisible goods received by agent i. We will refer to $A_i = (M_i, \mathbf{x}_i)$ as the *bundle* of mixed goods allocated to agent i. The integral allocation is written as $\mathcal{A} = (A_i)_{i \in N}$. We slightly abuse notation by also using $\mathcal{A}$ as an $n \times (m + \overline{m})$-matrix representation of the integral allocation.

A *randomized allocation* is a probability distribution over integral allocations and specified by a set of $s \in \mathbb{N}$ tuples $\{(p_j, \mathcal{A}_j)\}_{j \in [s]}$, where $p_j \in [0, 1]$, $\sum_{j \in [s]} p_j = 1$, and $\mathcal{A}_j$ is an integral allocation implemented with probability p_j. A randomized allocation $\{(p_j, \mathcal{A}_j)\}_{j \in [s]}$ is called an *implementation* of (or, interchangeably, *implements*) a fractional allocation $\mathbf{X}$ if $\mathbf{X} = \sum_{j \in [s]} p_j \cdot \mathcal{A}_j$.

Utilities. Each agent $i \in N$ is associated with an *additive* utility function u_i, where $u_i(g) \geq 0$ denotes the agent's utility for fully receiving good $g \in M \cup D$. We say that agents have *bi-valued* utilities if for each $i \in N$ and $g \in M \cup D$, $u_i(g) \in \{a, b\}$, where $0 \leq a < b$ are distinct, non-negative real values.[3] When $a = 0$ and $b = 1$, agents are said to have *binary* utilities. Given an allocation $\mathbf{X}$, the utility of agent i is given by $u_i(\mathbf{X}) = u_i(X_i) = \sum_{g \in M \cup D} X_{ig} \cdot u_i(g)$.

Solution Concepts. We now introduce fairness notions. The first two concepts—envy-freeness [30,49] and proportionality [45]—are defined for allocations, fractional or integral. An allocation $\mathbf{X}$ is said to satisfy *envy-freeness (EF)* if for any pair of agents $i, j \in N$, $u_i(X_i) \geq u_i(X_j)$, and *proportionality (PROP)* if for any agent $i \in N$, $u_i(X_i) \geq \frac{u_i(M \cup D)}{n}$. Our next notions are compelling relaxations of envy-freeness when concerning integral allocations of indivisible goods.

Definition 2.1 (EF1 [22,39] and EFX [25,44]). *An integral allocation $(M_i)_{i \in N}$ of indivisible goods M is said to satisfy*

- envy-freeness up to one good (EF1) *if for any pair of agents $i, j \in N$ such that $M_j \neq \emptyset$, we have $u_i(M_i) \geq u_i(M_j \setminus \{g\})$ for some $g \in M_j$;*
- envy-freeness up to any good (EFX) *if for any pair of agents $i, j \in N$ such that $M_j \neq \emptyset$, we have $u_i(M_i) \geq u_i(M_j \setminus \{g\})$ for all $g \in M_j$.*

[3] We will later use an agent's "large good" (resp., "small good") to refer to a good of value b (resp., a) to the agent.

With mixed goods, the following notion generalizes both EF and EF1.

Definition 2.2 (EFM [15]). An integral allocation $(A_i)_{i \in N}$ of mixed goods $M \cup D$ satisfies *envy-freeness for mixed goods (EFM)* if for any pair of $i, j \in N$,

- if agent j's bundle consists of only indivisible goods (i.e., $\boldsymbol{x}_j = \mathbf{0}$) and $M_j \neq \emptyset$, there exists some $g \in M_j$ such that $u_i(A_i) \geq u_i(M_j \setminus \{g\})$;
- otherwise, $u_i(A_i) \geq u_i(A_j)$.

Finally, we introduce an economic efficiency notion.

Definition 2.3 (fPO). *An integral allocation $(M_i)_{i \in N}$ of indivisible goods M is said to satisfy* fractionally Pareto optimal (fPO) *if there is no fractional allocation $\mathbf{Y}$ of indivisible goods M such that $u_i(Y_i) \geq u_i(M_i)$ for all agents $i \in N$ and $u_j(Y_j) > u_j(M_j)$ for some agent $j \in N$.*

We say a randomized allocation satisfies a property ex-ante (resp., ex-post) if the fractional allocation it implements (resp., every integral allocation in its support) satisfies the property. This paper concerns the problem of designing randomized algorithms that simultaneously achieve desirable properties both ex-ante and ex-post. Most of our algorithms do not explicitly output randomized allocations and instead *sample* integral allocations from their supports.

3 Two Agents

In this section, we study the best-of-both-worlds fair allocation of indivisible and mixed goods for two agents. With indivisible goods, there exists a randomized allocation that is ex-ante EF and ex-post EFX [29]. We improve upon their result by presenting a *polynomial-time* algorithm to identify such an allocation. We then leverage this algorithm to provide a randomized allocation of mixed goods that is both ex-ante EF and ex-post EFM in polynomial time.

In Proposition A.2 and Lemma A.1 of Feldman et al. [29], the requirement for ex-post EFX involves initially identifying two allocations, denoted as $\mathcal{A}^1$ and $\mathcal{A}^2$, which minimize the difference between two bundles under utility functions u_1 and u_2 respectively. The cut-and-choose protocol is then employed to determine the owner of each bundle in both allocations. The minimization of differences ensures that, for each $i \in [2]$, the loss in allocation $\mathcal{A}^i$ can be bounded by the gain in allocation $\mathcal{A}^{3-i}$, thereby ensuring ex-ante EF. Their mechanism involves an NP-hard step of finding an allocation that minimizes the difference. In contrast, our focus is solely on identifying two corresponding EFX allocations under utility functions u_1 and u_2, ensuring that the loss in $\mathcal{A}^i$ is bounded by the gain in $\mathcal{A}^{3-i}$ for each agent i. The pseudocode of our algorithm, along with all other omitted proofs and details, can be found in the full version of our paper [21].

In our algorithm, we incorporate a subroutine called `LocalSearch`, which, given an arbitrary allocation (A, B) and a utility function u, returns an integral

EFX allocation with respect to u.[4] The following lemma shows this subroutine produces in polynomial time an EFX allocation with a weakly smaller utility difference between its two bundles compared to the provided allocation (A, B).

Lemma 3.1. `LocalSearch`(A, B, u) *returns in polynomial time an integral EFX allocation* (A', B') *under utility function* u *with* $|u(B') - u(A')| \leq |u(B) - u(A)|$.

By leveraging the subroutine, we now describe our main procedure. For each agent $i \in [2]$, we maintain an allocation $\mathcal{A}^i$ which is an integral EFX allocation under u_i. If one of them is an envy-free allocation under the utility profile (u_1, u_2), we return this allocation directly. Otherwise, either there exists an allocation $\mathcal{A}^i$ which has a smaller utility difference over two bundles than that in $\mathcal{A}^{3-i}$ under u_{3-i}, we update $\mathcal{A}^{3-i}$ from $\mathcal{A}^i$ by using `LocalSearch`, or we find the desired ex-ante EF allocation $\{(0.5, \mathcal{A}^1), (0.5, \mathcal{A}^2)\}$.

Theorem 3.2. *For indivisible goods, an ex-ante EF and ex-post EFX randomized allocation can be found in polynomial time for two agents with additive utilities.*

For mixed goods, the above algorithm (precisely, Algorithm 3 in the full version [21]) can be employed to identify an allocation that is both ex-ante EF and ex-post EFM. Specifically, we first merge all divisible goods in D to a single indivisible good d with an equal total utility and then execute [21, Algorithm 3] on $M \cup \{d\}$. If the output is a single integral allocation, it is an EF (thus EFM) allocation. Otherwise, if there exists an agent $i \in [2]$ such that the good d is in bundle A_2^i (the bundle with the larger utility), we can transfer a fraction of d to A_1^i to reach EF. If no such agent exists, since all divisible goods are consistently in the bundle with the smaller utility, EFM can be reduced to EF1, which is then satisfied by the ex-post EFX property.

Theorem 3.3. *For mixed goods, an ex-ante EF and ex-post EFM randomized allocation can be found in polynomial time for two agents with additive utilities.*

We remark that the above procedure can also ensure a stronger ex-post fairness property called *EFXM* [15,43], which replaces the EF1 criteria by the EFX criteria when comparing to the bundle containing only indivisible goods.

---

[4] While our `LocalSearch` technique and the proof technique used in [44] to show their Theorem 4.2 share similarities, as we both update the current non-EFX allocation by moving some good from a bundle of higher value to a bundle of lower value, there are differences, mainly because we consider different utility classes. For instance, Plaut and Roughgarden showed the *leximin++ solution* is EFX for n agents with general but identical utilities. In addition to bundles' values, leximin++ solution also takes into account of bundles' sizes. But since we focus on two agents with additive utilities, our `LocalSearch` only updates allocations based on bundles' values.

4 Ex-Ante PROP + Ex-Post EFM for Mixed Goods

We focus on the best-of-both-worlds fairness for agents with bi-valued utilities over the set of mixed divisible and indivisible goods. Our main result is below:

Theorem 4.1. *For mixed goods and agents with bi-valued utilities, there exists an algorithm which computes in polynomial time an integral allocation sampled from a randomized allocation that is ex-ante proportional and ex-post EFM.*

4.1 Technical Overview

We devise the following techniques for BoBW fairness with mixed goods.

Baseline Comparison. This technique is designed to show our randomized allocation satisfies ex-ante proportionality. Fix any agent i, we construct a *baseline allocation* $\mathcal{B}^i = (B_1^i, \ldots, B_n^i)$ based on u_i. We partition the outcomes of the randomized allocations to n events $\mathcal{E}_1, \ldots, \mathcal{E}_n$ and show that, for each $k = 1, \ldots, n$, the utility for agent i under each outcome in the event $\mathcal{E}_k$ is at least $u_i(B_k^i)$. The expected utility of agent i is then at least $\sum_{k=1}^n \Pr(\mathcal{E}_k) \cdot u_i(B_k^i)$. If we can show $\mathcal{B}^i$ satisfies

$$\sum_{k=1}^{n} \Pr(\mathcal{E}_k) \cdot u_i(B_k^i) \geq \frac{1}{n} \cdot u_i(M \cup D), \tag{4.1}$$

then the ex-ante proportionality for agent i is proved.

To apply this technique, we need to carefully design the baseline allocation $\mathcal{B}^i$ for each agent i and the partition to the n events such that Eq. 4.1 holds. A natural idea is to design the partition of the event space with $\Pr(\mathcal{E}_1) = \cdots = \Pr(\mathcal{E}_n) = \frac{1}{n}$ such that Eq. 4.1 holds with equality for any baseline allocation $\mathcal{B}^i$. This technique is first illustrated in Sect. 1 where $m \leq n$. In the later part with general m, more sophisticated baseline allocations are constructed.

Minimal Unmatchable Group. Our algorithm starts by allocating indivisible goods, and the divisible goods are allocated by a water-filling process [15] (see also Lemma 4.2). To guarantee BoBW fairness, the indivisible goods must be allocated carefully in order to satisfy some properties that enable the application of the baseline comparison technique described above. We have interpreted the allocation problem as a matching problem in bipartite graphs and identified a key structure, *minimal unmatchable groups*, that is crucial for BoBW fairness. Many other techniques in the bipartite graph matching problem such as *augmenting paths* are also applied in our result. See Sect. 4.4 for more details.

4.2 Preparations

First of all, we justify that we can assume without loss of generality that there is only one divisible good, denoted as d. We also let $u_i(d) := u_i(D) = \sum_{j=1}^{\overline{m}} u_i(d_j)$ for each $i \in N$. Specifically, whenever we say that an ϵ portion of d is allocated to an agent i, we refer to the scenario of allocating an ϵ portion of each divisible good $d_1, \ldots, d_{\overline{m}}$ to the agent, i.e., $x_{i1} = \cdots = x_{i\overline{m}} = \epsilon$. In this way, the set of $\overline{m}$ divisible goods can be treated as the single divisible good d. In the remainder of this section, we thus assume that there is only one divisible good $D = \{d\}$.

Given an EF1 allocation of indivisible goods, Algorithm 1 of Bei et al. [15] specifies a way to allocate the divisible goods and finally obtains an EFM allocation of the mixed goods. We refer to this process of allocating divisible goods on top of an EF1 allocation of indivisible goods as the *water-filling* process. During the water-filling process, agents may swap their bundles; however, no bundle will be split. This is formally stated in the following lemma 4.2.

Lemma 4.2 (Bei et al. [15]). *Given an EF1 allocation $(M_i)_{i \in N}$ of indivisible goods M, an EFM allocation $((M_{\pi(1)}, \mathbf{x}_1), \ldots, (M_{\pi(n)}, \mathbf{x}_n))$ can be computed in polynomial time, where π is a permutation of $[n]$.*

Theorem 4.3. *For agents with additive utilities over mixed goods and $m \leq n$, Algorithm 1 computes in polynomial time an integral allocation sampled from a randomized allocation that is ex-ante proportional and ex-post EFM.*

Proof. Observe that Algorithm 1 runs in polynomial time. Due to Round-Robin, the allocation after the `foreach`-loop is EF1. Then, by Lemma 4.2, the final output allocation is EFM; hence, the randomized allocation is ex-post EFM.

To show the randomized allocation is ex-ante proportional, a key observation is that, for any fixed partition $(X_1, \ldots, X_n)$ of $M \cup D$, if an agent has a probability of $\frac{1}{n}$ to receive each bundle X_i, then her expected utility is exactly $\frac{u_i(M \cup D)}{n}$. As

---

Algorithm 1: Ex-ante PROP+ex-post EFM allocation when $m \leq n$

4.3 A Special Case Where $m \leq n$

We first consider the case $m \leq n$, i.e., the number of indivisible goods is at most the number of agents. This result will be used in later parts. Note, utility functions are not required to be bi-valued.

Our algorithm is shown in Algorithm 1. The Round-Robin algorithm is adopted in the first step to allocate the indivisible goods where the order of the agents π is sampled uniformly at random. The water-filling process is then executed to allocate divisible goods and obtain allocation $\mathcal{A}^\pi$. The required randomized allocation we find is $\{(\frac{1}{n!}, \mathcal{A}^\pi)\}$ and is denoted by $\mathcal{R}$.

Input: Agents N, mixed goods $M \cup D$, and agents' utility functions.

$\forall i \in [n]$, $A_i \leftarrow \emptyset$

Let π be a uniformly random permutation of N.

foreach *agent i according to the order π* **do** // Round-Robin.

    $g^* \leftarrow \arg\max_{g \in M} u_i(g)$

    $A_i \leftarrow A_i \cup \{g^*\}$, $M \leftarrow M \setminus \{g^*\}$

Execute the water-filling process. // See Lemma 4.2.

return *Allocation $\mathcal{A}^\pi$*

---

the permutation of the agents is generated uniformly at random in line 3, the probability that an agent i is ranked k-th in the order is $\frac{1}{n}$. Our goal is to define a *baseline allocation* $\mathcal{B}^i = (B_1^i, \ldots, B_n^i)$ for each agent $i \in N$ where the value of the bundle that agent i receives in the actual allocation when she is ranked k-th is no less than the k-th bundle in the baseline allocation (i.e., B_k^i) regardless of the permutation of other agents.

Given a permutation π, we use $A(\pi, k)$ to denote the bundle allocated to the agent ranked k-th in the allocation output by Algorithm 1. Consider an arbitrary permutation π where agent i is ranked k-th (i.e., $\pi(k) = i$), we will show that $u_i(A(\pi, k)) \geq u_i(B_k^i)$ where $\mathcal{B}^i = (B_1^i, \ldots, B_n^i)$ is the baseline allocation for agent i defined below.

Definition 4.4. *The* baseline allocation $\mathcal{B}^i$ for agent i *is obtained by*

1. *letting agent i partition M into n bundles using the Round-Robin algorithm (in particular, when $m \leq n$, the t-th bundle (i.e., B_t^i) contains the t-th preferred good for $t \leq m$ while the t-th bundle is empty for $t > m$), and*
2. *executing the water-filling process for the divisible goods according to agent i's utility to obtain an EFM allocation for the utility profile $(u_i, u_i, \ldots, u_i)$.*

The following properties of $\mathcal{B}^i$ are straightforward. First, each B_j^i contains at most one indivisible good (i.e., the j-th preferred indivisible good of agent i if $j \leq m$, and no indivisible good if $j > m$). Second, to guarantee EFM, the bundles containing some fraction of the divisible good must have the same value under $u_i(\cdot)$. Let x be that value. Third, due to EFM, $u_i(B_j^i) \geq x$ for each $j \in [n]$.

Next, we show that $u_i(A(\pi, k)) \geq u_i(B_k^i)$ for any permutation π with $\pi(k) = i$. It holds trivially when $B_k^i = \emptyset$, or when B_k^i only contains one indivisible good as agent i will receive at least her k-th preferred good during the Round-Robin phase of Algorithm 1. When B_k^i contains divisible goods, we need to show that $u_i(A(\pi, k)) \geq x$ (recall that x denotes the value of the bundles with divisible goods in $\mathcal{B}^i$). Suppose this is not the case, and $u_i(A_i) < x$ in the actual output allocation $\mathcal{A} = (A_1, \ldots, A_n)$ where $A_i = A(\pi, k)$. Let $\mathcal{I}$ be the set of indices of bundles in $\{A_1, \ldots, A_n\}$, each of which is envied by agent i. It is worth noting that due to that allocation $\mathcal{A}$ satisfies EFM, those bundles do not contain any divisible good. It is also easy to see $x > u_i(A_i) \geq u_i(A_j)$ for all $j \notin \mathcal{I}$. Hence, we reach the following contradiction:

$$\begin{aligned} u_i(M \cup D) &= \sum_{j \in \mathcal{I}} u_i(A_j) + \sum_{j \notin \mathcal{I}} u_i(A_j) \\ &< \sum_{j=1}^{|\mathcal{I}|} u_i(B_j^i) + (n - |\mathcal{I}|) \cdot x \leq \sum_{j=1}^{n} u_i(B_j^i) = u_i(M \cup D). \qquad (4.2) \end{aligned}$$

Finally, we have

$$u_i(\mathcal{R}) = \sum_{k=1}^{n} \sum_{\pi:\pi(k)=i} \frac{u_i(A(\pi, k))}{n!} \geq \sum_{k=1}^{n} \sum_{\pi:\pi(k)=i} \frac{u_i(B_k^i)}{n!} = \sum_{k=1}^{n} \frac{u_i(B_k^i)}{n} = \frac{u_i(M \cup D)}{n},$$

which implies the allocation satisfying ex-ante proportionality. □

Note that the above algorithm can hardly be generalized to the case with more than n indivisible goods. Suppose we define the baseline allocation for each agent in the same way and B_k^i contains divisible goods, then Eq. 4.2 may fail. In more detail, bundle A_j, where $j \in \mathcal{I}$ may contain more than one indivisible good. Hence, it may be that $\sum_{j \in \mathcal{I}} u_i(A_j) > \sum_{j=1}^{|\mathcal{I}|} u_i(B_j^i)$. As a result, Eq. 4.2 does not necessarily hold, leading to the possibility of $u_i(A(\pi, k)) = u_i(A_i) < u_i(B_k^i)$.

4.4 General m with Bi-valued Utilities

We now proceed to the general case with arbitrary numbers of goods and only provide the high-level ideas here. Our algorithm allocates the indivisible goods iteratively and then allocates the divisible goods by the water-filling process. At each iteration, we attempt to allocate each agent a large good of value b. If we construct a bipartite graph $G = (U, V, E)$ where U denotes the set of agents, V denotes the set of indivisible goods, and an edge represents that an agent has value b to a good, then we can find a maximum matching in G at each iteration. If a matching of size n (i.e., a perfect matching) is found, we allocate each agent a "large good" and move on to the next iteration.

At some iteration, a perfect matching may no longer exist. We identify agents $Z \subseteq U$ such that

1. Z cannot be fully matched to large goods; i.e., the Hall's condition fails for Z: $|\Gamma(Z)| < |Z|$, and
2. the remaining agents $U \setminus Z$ can be fully matched; in addition, for each indivisible good that has value b to an agent in $U \setminus Z$, it has value a to any agent in Z.

Intuitively speaking, agents in Z are "mostly finished" with their large goods, while a perfect matching may continue to exist in the future iterations for agents in $U \setminus Z$. We call a minimal set of agents Z satisfying the above requirements a *minimal unmatchable group*. Its precise definition and the algorithm to find such a group can be found in the full version of our paper [21].

At each iteration, agents Z and indivisible goods $\Gamma(Z)$ are temporarily removed. We find a perfect matching for agents in $U \setminus Z$, and allocate a "small good" to each agent in Z. In future iterations, we recursively handle $U \setminus Z$. As a result, more and more agents (and the corresponding "neighboring goods") are removed. For agents that remain, we find a perfect matching and correspondingly allocate each of them a "large good"; for agents that are removed, we allocate each of them a "small good". Thus, in each iteration, exactly n goods are allocated and each agent receives exactly one good. We stop when the number of the remaining indivisible goods is less than n.

Now, the number of unallocated indivisible goods is at most $2n-2$, including at most $n-1$ goods that are temporarily removed by the algorithm (those that are neighbors of removed agents) and at most $n-1$ goods that remain after the last iteration. *It is crucial that the current partial allocation satisfies envy-freeness* (which can be easily verified for each iteration based on the property of minimal unmatchable groups). This allows us to reduce our problem to the case with at most $2n - 2$ indivisible goods, as we can fix the allocation of the indivisible goods allocated by the above procedure. We formally describe this procedure and the reduction to $m \leq 2n - 2$ in the full version of our paper [21].

The most technical part comes to the handling of the case with $m \leq 2n - 2$, whose details can be found in [21]. The case with $m \leq n$ has already been handled in Sect. 1. The case where $n < m \leq 2n - 2$ is handled with a similar idea of comparing to a baseline allocation but requiring a substantial amount of additional effort. We again find a minimal unmatchable group Z of agents, and agents in Z and in $U \setminus Z$ are handled separately in the "last two rounds". Sample a permutation π uniformly at random. In the first round, agents with "higher priority" in Z receive "the last large good", while all agents in $U \setminus Z$ receive large goods. In the second round, we allocate agents with "higher priority" one good in the Round-Robin way. Finally, the water-filling process is applied to allocate the divisible good. The analysis for ex-ante proportionality is the most tricky part. Different baseline allocations are used for agents in Z and agents in $U \setminus Z$.

5 Ex-Ante EF+Ex-Post EFX and fPO for Indivisible Goods

In this section, we focus on the indivisible-goods setting and investigate the best-of-both-world guarantee for both fairness and efficiency when the agents have bi-valued utilities. Here, we will assume $a > 0$ for a clearer demonstration. For the case of $a = 0$, it can be reduced to the binary setting and has been addressed [13]. Our main result is the following:

Theorem 5.1. *In the indivisible-goods setting where agents have bi-valued utilities, Algorithm 2 computes in polynomial time an integral allocation sampled from a randomized allocation that is ex-ante EF, ex-post EFX, and ex-post fPO.*

Theorem 5 of Aziz et al. [11] presented an instance with two goods and two agents with additive utilities to show that no randomized allocation is simultaneously ex-ante EF, ex-post EF1 and ex-post fPO. Their impossibility result continues to hold when replacing ex-ante EF by ex-ante PROP. It is worth noting that the instance consists of three possible values for the goods, which can be considered in the settings allowing tri-valued utilities or *personalized* bi-valued utilities. Their impossibility result indicates that the assumption of bi-valued utilities is *necessary* in order to achieve best-of-both-world fairness *and* economic efficiency. For tri-valued or personalized bi-valued utilities, Aziz et al.'s impossibility result holds for the weaker notions of ex-ante PROP and ex-post EF1, with two agents and two goods. For bi-valued utilities, our Theorem 5.1 indicates that the best-of-both-worlds fairness can hold for the stronger notions of ex-ante EF and ex-post EFX, with an arbitrary number of agents and goods. Our result shows a sharp contrast between bi-valued and tri-valued utilities, and thereby completes the whole picture.

5.1 Technique Discussion and Detailed Descriptions of Algorithm 2

For bi-valued utilities, Garg and Murhekar [31] devised a polynomial-time algorithm to find an EFX and fPO allocation. Their algorithm updates the allocation through consecutive transfers of goods along a path, a process seemingly challenging to implement for achieving an ex-ante fairness guarantee. We instead draw inspiration from the MATCH&FREEZE algorithm of Amanatidis et al. [7], which computes an EFX allocation for bi-valued utilities in polynomial time.

Algorithm 2: Bi-valued indivisible goods: Ex-ante EF + ex-post EFX + ex-post fPO

Input: Agents N, indivisible goods M, and agents' utility functions.

$\forall i \in [n]$, $A_i \leftarrow \emptyset$, $c_i \leftarrow 0$
$t \leftarrow 1$, cnt $\leftarrow 0$
Uniformly generate a random permutation π of the agents.
while $|M| \geq$ [# unfrozen agents] + cnt **do**
    Find the minimal unmatchable group Z_t among N and M via [21, Algorithm 4].
    while $|Z_t| > 0$ *and* $\exists i \in N \setminus Z_t, g \in A_i$ *s.t.* $u_j(g) = b$ *for* $j \in Z_t$ **do**
        Let $g' \in M$ be the good matched to agent i in a perfect matching between $N \setminus Z_t$ and $M \setminus \Gamma(Z_t)$.
        $M \leftarrow M \cup \{g\} \setminus \{g'\}$, $A_i \leftarrow A_i \cup \{g'\} \setminus \{g\}$
        Find the minimal unmatchable group Z_t among N and M via [21, Algorithm 4].
    if $|Z_t| = 0$ **then** Find a perfect matching between N and M; add agent i's matched good to A_i. Remove the matched goods from M.
    else
        foreach $i \in Z_t$ *in the reversed order of* π **do**
            if i *can find an augmenting path to a good in* $\Gamma(Z_t)$ **then**
                Update $\mathcal{A}^\pi$ according to this path.
                Freeze i for the next $\lfloor b/a - 1 \rfloor$ rounds.
            else Set i as quiet.
        Match the corresponding good to each agent in $N \setminus Z_t$ in a perfect matching between $N \setminus Z_t$ and $M \setminus \Gamma(Z_t)$.
        $N \leftarrow N \setminus Z_t$, $M \leftarrow M \setminus \Gamma(Z_t)$, $t \leftarrow t + 1$
    cnt $\leftarrow$ cnt + [# quiet agents]
    $c_i \leftarrow c_i + 1$ for each quiet agent i.
Let $M' \leftarrow M \cup \left(\bigcup_{i \in N} A_i\right)$ and match each good $g \in A_i$ for some $i \in N$ to a copy of agent i initially.
foreach i *in the first* $|M| -$ cnt *of order* π *after removing the frozen agents* **do**
    if i *can find an augmenting path to a good in* M' **then** Update $\mathcal{A}^\pi$ according to this augmenting path.
    else $c_i \leftarrow c_i + 1$
Allocate the remaining goods arbitrarily based on c_i.
return *Allocation* $\mathcal{A}^\pi$

It is worth noting that the MATCH&FREEZE need not return a PO allocation [31, Appendix A.2]. In what follows, we present the first polynomial-time algorithm which can return an ex-ante EF, ex-post EFX, and ex-post fPO allocation when agents have bi-valued utilities. When describing our Algorithm 2, we use the term "round" interchangeably with each execution of the loop in line 4.

We first introduce three possible states for the agents: *active*, *quiet* and *frozen*. Initially, all agents are *active*. When executing line 12 to 18, if an agent $i \in Z_t$ is allocated a large good in line 13, she will become *frozen* for the next $\lfloor b/a - 1 \rfloor$ rounds, indicating that she will not receive any good in these $\lfloor b/a - 1 \rfloor$ rounds. After those rounds, she will become *quiet*. If $i \in Z_t$ does not receive a large good at this round, she will directly become *quiet* in line 16. Intuitively, during the execution of the loop in line 4 to 20, an agent i can only accept small goods after she turns quiet, and the count for these goods is recorded in c_i. We also refer to active and quiet agents as *unfrozen* agents.

We now describe Algorithm 2. Lines 1–3 initialize the algorithm, where c_i records the number of reserved small goods for agent i, crucial for the market clear in line 25. The counter cnt records the total number of reserved goods, precisely the sum of all c_i's during the loop (lines 4–20). A permutation π is generated uniformly at random for subsequent allocations.

The main algorithm proceeds through a multi-round procedure in lines 4–20. In each round, we first check whether the remaining goods in M are sufficient to match one to each unfrozen agent in line 4. If this condition is met, we attempt to find a specific minimal unmatchable group Z_t such that all agents in Z_t have value a over each good matched to agents in $N \setminus Z_t$ in lines 5–9.[5] This is to ensure the efficiency guarantee by allocating more large goods intuitively.

Given Z_t, if $|Z_t| = 0$, indicating that there exists a perfect matching between all agents in N and the current set of goods in M, lines 10–10 are executed to add this matching directly. When Z_t is non-empty, we first find a maximum matching between Z_t and $\Gamma(Z_t)$ using the augmenting path technique in the reversed order of π (lines 12–16). All goods in $\Gamma(Z_t)$ can be allocated to some agent from the construction of Z_t (see [21, Algorithm 4]). The reversed order in line 12 ensures ex-post EFX after the final allocation step in lines 21-24. We then find a perfect matching between $N \setminus Z_t$ and $M \setminus \Gamma(Z_t)$ in lines 17–18 and match these goods accordingly. At the end of each round, we maintain c_i for each quiet agent i and the counter cnt, to allocate a good to those who have not received a good at this round *virtually* (lines 19–20).

If there are not enough goods to allocate one to each unfrozen agent, the `while`-loop in line 4 is terminated and the final allocation step in lines 21–24 is executed. At this step, a large good is allocated to the first $|M| - \text{cnt}$ unfrozen agents under π as far as possible. Line 21 can adjust the previous allocation for the remaining agents in N for a better efficiency guarantee. Line 24 updates the number of reserved small goods for an agent if she cannot be matched to any large good. Line 25 clears the market based on the reserved number c_i for each

[5] The relevant concepts can be found in the full version of our paper [21].

agent i, which allocates these goods *actually*. We can then obtain the desired randomized allocation.

Throughout the algorithm, the randomization of our algorithm (the choice of π) only influences the allocation in lines 12–18 and 21–24, which would not affect the assignment of agents to groups Z_t (resp., goods to $\Gamma(Z_t)$). This observation will be helpful for the analysis of the ex-ante property.

5.2 Fairness

Before presenting the results, we first provide some properties of the allocation returned by Algorithm 2. Let N^t and M^t be the corresponding set N and M at the beginning of round t.

Observation 5.2. *For each agent $i \in N$,*

- *she receives exactly one good at each unfrozen round;*
- *if she belongs to some Z_t which is removed from N in line 18 (assuming this corresponds to round r_i), she values each good allocated to her before round r_i at b and each good in A_j for $j \in N^{r_i} \setminus Z_t$ and $M^{r_i} \setminus \Gamma(Z_t)$ at a;*
- *if she does not belong to any Z_t, she values each good allocated to her before the termination of the* `while`*-loop in line 4 at b.*

According to these observations, we first show that each output integral allocation by our algorithm is EFX. To prove this, we adopt the induction to maintain the following property after the matching at each round: for each agent i, this agent will not envy other agents j except for the case that agents i and j are in the same Z_t and agent i has never been frozen while agent j has been frozen. Under this case, agent i does not envy j using the EFX criteria. Together with the analysis for the final allocation step, we conclude the lemma.

Lemma 5.3. *Every integral allocation returned by Algorithm 2 is EFX.*

We then come to the ex-ante fairness guarantee of our algorithm. The idea is to prove any envy from an agent i to another agent j under π can be eliminated by the gain under another permutation π' which just exchanges i and j in π.

Lemma 5.4. *Algorithm 2 returns an ex-ante EF allocation.*

5.3 Efficiency

We leverage the Fisher market (see [21] for relevant concepts) to provide our efficiency guarantee.

Lemma 5.5. *Every integral allocation returned by Algorithm 2 is fPO.*

Proof. To leverage the First Welfare Theorem [42] for proving this, our goal is to find a proper price vector $\boldsymbol{p}$ for each realized integral allocation $\mathcal{A}$ such that $(\mathcal{A}, \boldsymbol{p})$ is a market equilibrium of a Fisher market. Here, we can directly set the budget e_i for each agent $i \in N$ as $\sum_{g \in A_i} p_g$ to ensure the market clear. It suffices to find $\boldsymbol{p}$ and show that each agent $i \in N$ only receives goods in $\mathrm{MBB_i}$.

Denote the set of agents not in any previous Z_t by N^r. We assume $m \geq n$ here and the analysis for the corner case when $m < n$ can be found in [21]. Define the price vector $\mathbf{p}$ as follows:

- for each agent i in some Z_t, we set $p_g = u_i(g)$ for $g \in A_i$;
- for each agent $i \in N^r$ who receives both large and small goods, set $p_g = u_i(g)$ for each $g \in A_i$.

For the prices of the remaining goods, we will set them iteratively: if there exists one agent i without pricing her goods such that some agent who has priced her goods before values some good in A_i at b, we set $p_g = u_i(g)$ for each $g \in A_i$. If there is no further pricing that can be made, we set the price as a for each remaining good and we call each agents who owns these goods *low-price agent*. We then need to show each agent $i \in N$ only receives goods in $\mathrm{MBB_i}$.

For each agent i in some Z_t, the ratio $u_i(g)/p_g$ for each good $g \in A_i$ is exactly 1. For each good g which is allocated to some agent j in a $Z_{t' \leq t}$, if $p_g = u_j(g) = a$, this good is *actually* allocated at line 25 and g is in $M^{r_i} \setminus \Gamma(Z_t)$, which is valued a by i from the second term in Observation 5.2, whose ratio $u_i(g)/p_g$ is at most 1. For each good g which is allocated to some agent j in $N^{r_i} \setminus Z_t$, we also have $u_i(g) = a$ from Observation 5.2. Thus, each agent i in some Z_t only receives goods in $\mathrm{MBB_i}$.

For each low-price agent, from the third term in Observation 5.2, all goods allocated to her in the multi-round procedure must be large. Because this agent is not priced before the end of the iteration, she receives no small goods. Since the goods she owns are priced at a which is the lowest price we set and all goods in her bundle must be valued at b under her valuation, all goods in her bundle must be in $\mathrm{MBB_i}$.

We then come to some agent $i \in N^r$ who sets $p_g = u_i(g)$ for each $g \in A_i$. The ratio $u_i(g)/p_g$ for each good $g \in A_i$ is exactly 1. If she values a good g in some $A_j, j \in N$ at b (thus j cannot be a low-price agent otherwise it will be priced before the end from the iteration) and $u_j(g) = p_g = a$, we can backward the iteration from agent i and finally achieve an agent $i' \in N^r$ who receives both large and small goods from her perspective. This backward path along with the good presented in the condition of the iteration is a feasible augmenting path at line 23 and agent i' would not receive a small good, which leads to a contradiction. This completes the proof for the case when $m \geq n$. □

We are now ready to prove our main result in this section.

*Proof. Proof of Theorem*5.1 From Lemmas 5.3, 5.4 and 5.5, it suffices to show Algorithm 2 can terminate in polynomial time. From the description of our algorithm, since $|M|$ keeps decreasing which bounds the total number of rounds by m and the only thing we need is to show lines 6-9 can be terminated in polynomial time. From the construction of Z_t according to [21, Algorithm 4], if we keep the perfect matching between $N \setminus Z_t$ and $M \setminus \Gamma(Z_t)$ by just replacing the good g' by g, one edge from $j \in Z_t$ to g' is added, this leads to one of the following cases: either (1) one augmenting path can be performed and the size of the maximum matching between N and M is increased by 1, or (2) there is no additional augmenting path but $|\Gamma(Z_t)|$ is increased by at least 1. After at most n times of the case (2) occurs, case (1) will occur. From the total number of the occurrences of (1) is at most n, lines 6-9 can be terminated in $O(n^2)$ loops. □

6 Conclusion

In this paper, we have studied the best-of-both-worlds fairness for indivisible- and mixed-goods allocations. With indivisible goods, we provided polynomial-time algorithms that achieve ex-ante EF and ex-post EFX allocation for two agents and ex-ante EF, ex-post EFX, and ex-post fPO for n agents with bi-valued utilities. With mixed goods, we showed polynomial-time algorithms that achieve ex-ante EF and ex-post EFM allocation for two agents and ex-ante PROP and ex-post EFM for n agents with bi-valued utilities.

In future research, it would be interesting to further strengthen the results in this paper, the most intriguing of which is perhaps the (in)compatibility between ex-ante EF and ex-post EFM for additive utilities. For mixed goods, another interesting direction is to extend the BoBW-fair study to share-based fairness properties, e.g., maximin share (MMS) guarantee [16].

Acknowledgments. This work was partially supported by ARC Laureate Project FL200100204 on "Trustworthy AI", by the National Natural Science Foundation of China (Grant No. 62102117), by the Shenzhen Science and Technology Program (Grant No. GXWD20231129111306002), by the Guangdong Basic and Applied Basic Research Foundation (Grant No. 2023A1515011188), and by the National Natural Science Foundation of China (No. 62102252).

The authors would like to thank Jiaxin Song for his great suggestions on this paper.

References

1. Abdulkadiroğlu, A., Sönmez, T.: Random serial dictatorship and the core from random endowments in house allocation problems. Econometrica **66**(3), 689–701 (1998)
2. Akbarpour, M., Nikzad, A.: Approximate random allocation mechanisms. Rev. Econ. Stud. **87**(6), 2473–2510 (2020)
3. Akrami, H., Alon, N., Chaudhury, B.R., Garg, J., Mehlhorn, K., Mehta, R.: EFX: a simpler approach and an (almost) optimal guarantee via rainbow cycle number. In: Proceedings of the EC, p. 61 (2023)
4. Akrami, H., Mehlhorn, K., Seddighin, M., Shahkarami, G.: Randomized and deterministic maximin-share approximations for fractionally subadditive valuations. In: Proceedings of the NeurIPS, pp. 58821–58832 (2023)

5. Akrami, H., Garg, J., Sharma, E., Taki, S.: Improving approximation guarantees for maximin share. In: Proceedings of the EC (2024). Forthcoming
6. Amanatidis, G., Markakis, E., Ntokos, A.: Multiple birds with one stone: beating 1/2 for EFX and GMMS via envy cycle elimination. Theoret. Comput. Sci. **841**, 94–109 (2020)
7. Amanatidis, G., Birmpas, G., Filos-Ratsikas, A., Hollender, A., Voudouris, A.A.: Maximum Nash welfare and other stories about EFX. Theoret. Comput. Sci. **863**, 69–85 (2021)
8. Amanatidis, G., et al.: Fair division of indivisible goods: recent progress and open questions. Artif. Intell. **322**, 103965 (2023)
9. Aziz, H., Ganguly, A., Micha, E.: Best of both worlds fairness under entitlements. In: Proceedings of the AAMAS, pp. 941–948 (2023)
10. Aziz, H., Lu, X., Suzuki, M., Vollen, J., Walsh, T.: Best-of-both-worlds fairness in committee voting. In: Proceedings of the WINE, p. 676 (2023)
11. Aziz, H., Freeman, R., Shah, N., Vaish, R.: Best of both worlds: Ex-ante and ex-post fairness in resource allocation. Oper. Res. **72**(4), 1674–1688 (2024)
12. Aziz, H., Lu, X., Suzuki, M., Vollen, J., Walsh,T.: Fair lotteries for participatory budgeting. In: Proceedings of the AAAI, pp. 9469–9476 (2024)
13. Babaioff, M., Ezra, T., Feige, U.: Fair and truthful mechanisms for dichotomous valuations. In: Proceedings of the AAAI, pp. 5119–5126 (2021)
14. Babaioff, M., Ezra, T., Feige, U.: On best-of-both-worlds fair-share allocations. In: Proceedings of the WINE, pp. 237–255 (2022)
15. Bei, X., Li, Z., Liu, J., Liu, S., Lu, X.: Fair division of mixed divisible and indivisible goods. Artif. Intell. **293**, 103436 (2021)
16. Bei, X., Liu, S., Lu, X., Wang, H.: Maximin fairness with mixed divisible and indivisible goods. Auton. Agent. Multi-Agent Syst. **35**(2), 1–21 (2021). https://doi.org/10.1007/s10458-021-09517-7
17. Berger, B., Cohen, A., Feldman, M., Fiat, A.: Almost full EFX exists for four agents. In: Proceedings of the AAAI, pp. 4826–4833 (2022)
18. Bhaskar, U., Sricharan, A.R., Vaish, R.: On approximate envy-freeness for indivisible chores and mixed resources. In: Proceedings of the APPROX, pp. 1:1–1:23 (2021)
19. Bogomolnaia, A., Moulin, H.: A new solution to the random assignment problem. J. Econ. Theory **100**(2), 295–328 (2001)
20. Bu, X., Song, J., Yu, Z.: EFX allocations exist for binary valuations. In: Proceedings of the IJTCS-FAW, pp. 252–262 (2023)
21. Bu, X., Li, Z., Liu, S., Lu, X., Tao, B.: Best-of-both-worlds fair allocation of indivisible and mixed goods. CoRR, abs/2410.06877v1 (2024)
22. Budish, E.: The combinatorial assignment problem: approximate competitive equilibrium from equal incomes. J. Polit. Econ. **119**(6), 1061–1103 (2011)
23. Budish, E., Che, Y.-K., Kojima, F., Milgrom, P.: Designing random allocation mechanisms: theory and applications. Am. Econ. Rev. **103**(2), 585–623 (2013)
24. Caragiannis, I., Gravin, N., Huang, X.: Envy-freeness up to any item with high Nash welfare: The virtue of donating items. In: Proceedings of the EC, pp. 527–545 (2019)
25. Caragiannis, I., Kurokawa, D., Moulin, H., Procaccia, A.D., Shah, N., Wang, J.: The unreasonable fairness of maximum Nash welfare. ACM Trans. Econ. Comput. **7** (3), 12:1–12:32 (2019)
26. Chaudhury, B.R., Kavitha, T., Mehlhorn, K., Sgouritsa, A.: A little charity guarantees almost envy-freeness. SIAM J. Comput. **50**(4), 1336–1358 (2021)

27. Chaudhury, B.R., Garg, J., Mehlhorn, K., Mehta, R., Misra, P.: Improving envy freeness up to any good guarantees through rainbow cycle number. Math. Oper. Res. (2023). Forthcoming
28. Chaudhury, B.R., Garg, J., Mehlhorn, K.: EFX exists for three agents. J. ACM, **71** (1), 4:1–4:27 (2024)
29. Feldman, M., Mauras, S., Narayan, V.V., Ponitka, T.: Breaking the envy cycle: best-of-both-worlds guarantees for subadditive valuations. In: Proceedings of the EC (2024). Forthcoming. See also https://arxiv.org/abs/2304.03706
30. Foley, D.K.: Resource allocation and the public sector. Yale Econ. Essays **7**(1), 45–98 (1967)
31. Garg, J., Murhekar, A.: Computing fair and efficient allocations with few utility values. Theoret. Comput. Sci. **962**, 113932 (2023)
32. Garg, J., Sharma, E.: Best-of-both-worlds fairness of the envy-cycle-elimination algorithm. arXiv: 2410.08986 (2024)
33. Gölz, P., Peters, D., Procaccia, A.D.: In this apportionment lottery, the house always wins. Oper. Res. (2024). Forthcoming
34. Halpern, D., Procaccia, A.D., Psomas, A., Shah, N.: Fair division with binary valuations: one rule to rule them all. In: Proceedings of the WINE, pp. 370–383 (2020)
35. Hoefer, M., Schmalhofer, M., Varricchio, G.: Best of both worlds: agents with entitlements. In: Proceedings of the AAMAS, pp. 564–572 (2023)
36. Li, B., Li, Z., Liu, S., Wu, Z.: Allocating mixed goods with customized fairness and indivisibility ratio. In: Proceedings of the IJCAI, pp. 2868–2876 (2024)
37. Li, Z., Liu, S., Lu, X., Tao B.: Truthful fair mechanisms for allocating mixed divisible and indivisible goods. In: Proceedings of the IJCAI, pp. 2808–2816 (2023)
38. Li, Z., Liu, S., Lu, X., Tao, B., Tao, Y.: A complete landscape for the price of envy-freeness. In: Proceedings of the AAMAS, pp. 1183–1191 (2024)
39. Lipton, R.J., Markakis, E., Mossel, E., Saberi, A.: On approximately fair allocations of indivisible goods. In: Proceedings of the EC, pp. 125–131 (2004)
40. Liu, S., Lu, X., Suzuki, M., Walsh, T.: Mixed fair division: a survey. J. Artif. Intell. Res. **80**, 1373–1406 (2024)
41. Mahara, R.: Existence of EFX for two additive valuations. Discret. Appl. Math. **340**, 115–122 (2023)
42. Mas-Colell, A., Whinston, M.D., Green, J.R., et al.: Microeconomic Theory. Oxford University Press, Oxford (1995)
43. Nishimura, K., Sumita, H.: Envy-freeness and maximum Nash welfare for mixed divisible and indivisible goods. CoRR, abs/2302.13342 (2023)
44. Plaut, B., Roughgarden, T.: Almost envy-freeness with general valuations. SIAM J. Discret. Math. **34**(2), 1039–1068 (2020)
45. Steinhaus, H.: The problem of fair division. Econometrica **16**(1), 101–104 (1948)
46. Suksompong, W.: Constraints in fair division. ACM SIGecom Exchanges **19**(2), 46–61 (2021)
47. Suksompong, W.: Weighted fair division of indivisible items: a review. Inf. Process. Lett. **187**, 106519 (2025)
48. Suzuki, M., Vollen, J.: Maximum flow is fair: a network flow approach to committee voting. In: Proceedings of the EC (2024). Forthcoming
49. Varian, H.R.: Equity, envy, and efficiency. J. Econ. Theory **9**(1), 63–91 (1974)

Tree Splitting Based Rounding Scheme for Weighted Proportional Allocations with Subsidy

Xiaowei Wu(✉) and Shengwei Zhou

IOTSC, University of Macau, Macau, China
{xiaoweiwu,yc17423}@um.edu.mo

Abstract. We consider the problem of allocating m indivisible items to a set of n heterogeneous agents, aiming at computing a proportional allocation by introducing subsidy (money). It has been shown by Wu et al. (WINE 2023) that when agents are unweighted a total subsidy of $n/4$ suffices (assuming that each item has value/cost at most 1 to every agent) to ensure proportionality. When agents have general weights, they proposed an algorithm that guarantees a weighted proportional allocation requiring a total subsidy of $(n-1)/2$, by rounding the fractional bid-and-take algorithm. In this work, we revisit the problem and the fractional bid-and-take algorithm. We show that by formulating the fractional allocation returned by the algorithm as a directed tree connecting the agents and splitting the tree into canonical components, there is a rounding scheme that requires a total subsidy of at most $n/3 - 1/6$.

Keywords: Fair allocation · Chores · Weighted · Proportional · Subsidy

1 Introduction

We study the fair allocation problem of allocating a set of m indivisible items M to a group of n heterogeneous agents N. When all items M give positive values to all agents, the problem refers to the fair allocation of *goods*, which represents the situation of distributing resources or public goods. Analogous but contrary, when all items M give negative values (positive costs) to all agents, the problem refers to the fair allocation of *chores*, which covers the task allocation among employees. In this paper, we mainly focus on the allocation of chores, where each agent i has a cost function $c_i : 2^M \rightarrow \mathbb{R}^+ \cup \{0\}$, but our main result also applies to the allocation of goods. We consider the general weighted setting in which each agent i has a $w_i > 0$ that represents her obligation to undertake the chores. We normalize the weights of agents such that $\sum_{i \in N} w_i = 1$. Traditionally, the fair allocation problem considers how to fairly allocate the items into n bundles $(X_1, \ldots, X_n)$, while each agent receives exactly one bundle that guarantees some

The authors are ordered alphabetically. This work is funded by the Science and Technology Development Fund (FDCT), Macau SAR (file no. 0014/2022/AFJ, 0085/2022/A, 001/2024/SKL and SKL-IOTSC-2024-2026).

M. Mavronicolas et al. (Eds.): WINE 2024, LNCS 15534, pp. 295–313, 2026.
https://doi.org/10.1007/978-3-032-08560-3_17

fairness criteria for her. We say that agent i has cost $c_i(S)$ for bundle $S \subseteq M$. One natural and well-studied fairness notion is *proportionality* (PROP) [33]: an allocation is called weighted proportional (WPROP) [31] if for all agent i, $c_i(X_i) \leq w_i \cdot c_i(M)$, where we refer to $w_i \cdot c_i(M)$ as the proportional share of agent i. Unfortunately, when items are indivisible, PROP allocations are not guaranteed to exist, e.g., considering allocating a single item to two agents. One possible way to circumvent this non-existence result is by introducing money to eliminate the inevitable unfairness. Maskin [30] first proposed the setting called fair allocation with money that allows each agent to receive a subsidy $s_i \geq 0$ to eliminate unfairness, under which the objective is to minimize the total subsidized money. Halpern and Shah [24] considered the fair allocation of goods with money, for another well-known fairness notion, called *envy-freeness* [22]. Assuming that each item has a value of at most 1 to each agent, they conjecture that a total subsidy of $n-1$ suffices to guarantee envy-freeness for the allocation of goods. The conjecture was later verified by Brustle et al. [13]. Very recently, Aziz et al. [4] considered the weighted setting and characterized the property of guaranteeing weighted envy-free allocation with subsidy.

Allocation of Chores with Subsidy. The setting of fair allocation with subsidy can be naturally applied to the allocation of chores. Consider a situation where a factory, such as *Foxconn*, wants to distribute a bunch of production tasks to n different working units. Different working units have different capacities, e.g., their proportional shares of the total tasks, and need to be compensated with money when assigned tasks exceeding their capacity, i.e., for overtime payment. To ensure fairness (for a stable partnership), the mass of tasks each working unit receives, after removing the part that the subsidy offsets, should not exceed its proportional share of the total tasks. On the other hand, to maximize benefits, the factory objects to minimize the total (additional) overtime payments, under the situation that all tasks are completed. In other words, we aim to compute an allocation $(X_1, \ldots, X_n)$ and subsidies $(s_1, \ldots, s_n)$ such that $c_i(X_i) - s_i \leq w_i \cdot c_i(M)$ for all agent $i \in N$, with a small amount of total subsidy $\|\mathbf{s}\|_1 = \sum_{i \in N} s_i$. When all agents have equal weights, Wu et al. [35] proposed an algorithm that computes a proportional allocation with a total subsidy of at most $n/4$. Their algorithm is based on rounding a fractional allocation returned by the Moving Knife Algorithm. However, when agents have general weights, their algorithm can only guarantee an upper bound of $(n-1)/2$ on the total subsidy.

1.1 Our Results

In this paper, we revisit the problem of fair allocation of chores with subsidy. A commonly used framework for this problem is to first compute a fractional proportional allocation and then design a rounding scheme to turn the allocation integral with bounded total subsidy. For the unweighted setting, the algorithm of Wu et al. [35] that guarantees a total subsidy of at most $n/4$ is based on rounding a well-structured fractional allocation returned by the Moving Knife Algorithm. Specifically, the algorithm computes a fractional allocation such that each agent receives a contiguous bundle of items in which at most two of them (the first and the last) are fractional. However, the algorithm fails to generalize to the weighted setting where agents have arbitrary weights. As they pointed out, any algorithm that requires each agent to receive a contiguous bundle of items, fails to compute a (fractional) weighted proportional allocation.

On the other hand, if we do not require any structural property on the fractional allocation, it is easy to compute a weighted proportional allocation. For example, the Eating Algorithm that lets all agents eat their most preferred available item continuously at the same time, with speeds proportional to their weights, guarantees weighted envy-freeness, and thus weighted proportionality. However, it is easy to find instances for which almost all items are shared by multiple agents in the fractional allocation returned by the Eating Algorithm. This complex item-sharing structure introduces major difficulties for devising the rounding scheme and bounding the incurred subsidy. As a comparison, in the fractional allocation returned by the Moving Knife Algorithm, there are at most $n-1$ fractional items and each agent receives at most two such items. Therefore, we consider the Fractional Bid-and-Take algorithm (FBAT) that computes fractional weighted proportional allocations with a certain level of canonical structure. The same fractional algorithm is considered in [35], where it is shown that there are at most $n-1$ fractional items in the returned allocation. However, in this allocation, an agent may receive many fractional items. Consequently, their rounding scheme can only guarantee an upper bound of $(n-1)/2$ on the total subsidy. Whether strictly less subsidy is sufficient to guarantee (weighted) proportionality remains unknown and is proposed as an open problem.

In this paper, we revisit the fractional bid-and-take algorithm and answer this question affirmatively. We characterize the structure of the fractional allocations returned by FBAT, by introducing the *item-sharing* graph where agents are nodes and fractional items are edges. We show that the item-sharing graph of the allocation returned by FBAT is a directed tree[1]. We introduce a general rounding framework based on tree splitting, which allows us to split the directed tree graph into canonical components and round each component independently. We show that there exists a rounding scheme that guarantees weighted proportionality with a total subsidy of at most $n/3 - 1/6$.

Main Result. For the allocation of chores to a group of n agents with general weights, we can compute in polynomial time a weighted proportional allocation with subsidy, where the total subsidy is at most $n/3 - 1/6$.

We remark that rounding the item-sharing graph via tree splitting requires striking a balance between the fineness of the splitting and tightness of the subsidy upper bound. The finer we split the tree, the easier we can upper bound the total subsidy: but this upper bound is usually very loose. For instance, if we take every edge as a component, we can easily recover the simple rounding scheme that guarantees a total subsidy of $(n-1)/2$ [35]. On the other hand, regarding the tree as a whole when rounding items could help give tighter upper

[1] As a comparison, the item-sharing graph defined by the allocation returned by the Moving Knife Algorithm is a path; while that for the Eating Algorithm can be any complex dense multi-graph.

bounds but it would be very difficult to characterize the required subsidy. One notable difficulty in designing the rounding scheme is to handle the fractional items that are shared by multiple agents, which we call the *shattered items*. A shatter item corresponds to multiple edges (that form a path, which we call an *atom-path*) in the item-sharing graph. Since every item should be rounded to exactly one agent, we need to ensure that all edges that correspond to the same shattered item belong to one component after splitting, which puts constraints on the feasible tree splittings. To get around this difficulty, we design an atom-path-centric splitting that recursively splits the tree into a (near) atom-path and a collection of subtrees meeting certain constraints, and use mathematical induction to bound the total subsidy. We further extend our framework to the allocation of goods with some minor modifications. All omitted proofs and the extension to the good setting can be found in our full version.

1.2 Other Related Works

Since the seminal work of Steinhaus [33], the fair allocation problem has received significant attention in the past decades. Other than proportionality, another well-studied fairness notion is *envy-freeness* (EF) [22], that is, no agent wants to exchange her bundle of items with another agent to improve her utility. Since both EF and PROP are not guaranteed to exist when items are indivisible, a line of literature focused on some relaxed fairness notions. *Envy-freeness up to one item* (EF1) [28] and *envy-freeness up to any item* (EFX) [15] are two widely studied relaxations of EF, which require that the envy between any two agents can be eliminated by removing some item; any item respectively. Similarly, we have *proportionality up to one item* (PROP1) [9,18] and *proportionality up to any item* (PROPX) [6] to be two relaxations of PROP. In addition, *maximin share* (MMS) [14] is another popular relaxation of PROP. For a more comprehensive overview of the fair allocation literature, we refer to the recent surveys by Amanatidis et al. [1] and Aziz et al. [5].

Weighted Setting. Motivated by real-world applications where agents are usually not equally obliged, Chakraborty et al. [16] proposed the *weighted* setting. They introduced the *weighted envy-freeness up to one item* (WEF1) for the allocation of goods and show that WEF1 allocations always exist. Lately, WPROP1 allocations have been proved to exist for chores [12], and the mixture of goods and chores [6]. Li et al. [26] showed the existence and computation of WPROPX allocations of chores. Wu et al. [36] and Springer et al. [32] proposed algorithms for the computation of WEF1 allocations for chores. The weighted version of MMS (WMMS) has also been studied for both goods [20] and chores [3]. In addition, generalizing MMS to the weighted setting also inspires the fairness notion of *Anyprice Share* (APS), which has attracted much attention [7,21] We refer to the recent survey [34] for a comprehensive review of weighted fair allocation.

Fair Allocation with Money. Beyond additive valuation functions, Brustle et al. [13] showed that a subsidy of $2(n-1)$ dollars per agent suffices to guarantee envy-freeness for monotonic functions. The result was recently improved by

Kawase et al. [25] to $n-1$ per agent. Barman et al. [8] considered the dichotomous valuations and showed that envy-freeness can be guaranteed with a per-agent subsidy of at most 1. Regarding truthfulness, Goko et al. [23] showed that for submodular functions, there exists a truthful mechanism that guarantees envy-freeness with a subsidy of at most one dollar per agent. The subsidy setting was recently extended to the weighted setting [17,19]. A similar setting is called fair allocation with monetary transfers, introduced by Aziz [2], that allows agents to transfer money to each other. Instead of minimizing the total subsidy, their result focused on the characterization of allocations that are equitable and envy-free with monetary transfers. When considering money as a divisible good, the setting of fair allocation with money is similar to the fair allocation of mixed divisible and indivisible items, which also receives much attention [10,11,27]. For a more detailed review of the existing works on mixed fair allocation, please refer to the recent survey [29].

2 Preliminary

In the following, we introduce the notation and the fairness notions. We consider the problem of allocating m indivisible chores M to n agents N where each agent $i \in N$ has an additive cost function $c_i : 2^M \to \mathbb{R}^+ \cup \{0\}$. We assume that the agents are weighted, i.e., each agent $i \in N$ has a weight $w_i > 0$ (that represents her obligation for undertaking the chores) and $\sum_{i \in N} w_i = 1$. We denote by $\mathbf{w} = (w_1, \dots, w_n)$ the weights of agents. A cost function c_i is said to be *additive* if for any bundle $S \subseteq M$ we have $c_i(S) = \sum_{e \in S} c_i(\{e\})$. For convenience, we use $c_i(e)$ to denote $c_i(\{e\})$. We use $\mathbf{c} = (c_1, ..., c_n)$ to denote the cost functions of agents. We assume w.l.o.g. that each item has cost at most one to each agent, i.e. $c_i(e) \le 1$ for any $i \in N$, $e \in M$. Given an instance $\mathcal{I} = (M, N, \mathbf{w}, \mathbf{c})$, an allocation of $\mathcal{I}$ is represented by an n-partition $\mathbf{X} = (X_1, \dots, X_n)$ of the items M, where $X_i \cap X_j = \emptyset$ for all $i \neq j$ and $\cup_{i \in N} X_i = M$. In allocation $\mathbf{X}$, agent $i \in N$ receives bundle X_i. For convenience of notation, given any set $X \subseteq M$ and $e \in M$, we use $X + e$ and $X - e$ to denote $X \cup \{e\}$ and $X \setminus \{e\}$, respectively.

Definition 1 (WPROP). *An allocation* $\mathbf{X}$ *is called weighted proportional (WPROP) if* $c_i(X_i) \le w_i \cdot c_i(M)$ *for all* $i \in N$.

We use WPROP_i to denote agent i's proportional share, i.e., $\mathsf{WPROP}_i = w_i \cdot c_i(M)$. We denote $s_i \ge 0$ the subsidy we give to agent $i \in N$, $\mathbf{s} = (s_1, \dots, s_n)$ to denote the set of subsidies, and $\|\mathbf{s}\|_1 = \sum_{i \in N} s_i$ to denote the total subsidy.

Definition 2 (WPROPS). *An allocation* $\mathbf{X}$ *with subsidies* $\mathbf{s} = (s_1, \dots, s_n)$ *is called weighted proportional with subsidies (WPROPS) if for any* $i \in N$ *we have* $c_i(X_i) - s_i \le \mathsf{WPROP}_i$.

Given any instance $\mathcal{I} = (M, N, \mathbf{w}, \mathbf{c})$, we aim to find WPROPS allocation $\mathbf{X}$ with a small amount of total subsidy. Unlike envy-freeness with subsidy[2], we

[2] An allocation with subsidies is weighted envy-free with subsidy if $c_i(X_i)/w_i - s_i \le c_j(X_j)/w_j - s_j$ for all $i, j \in N$. We remark that even in the unweighted setting, not

can turn any allocation $\mathbf{X}$ into weighted proportional by introducing subsidies to the agents as follows:

$$s_i = \max\{c_i(X_i) - \mathsf{WPROP}_i, 0\}, \qquad \forall i \in N.$$

Therefore, in the rest of this paper, we mainly focus on computing the allocation $\mathbf{X}$. The subsidy to each agent will be automatically decided by the above equation. Before we dive into our algorithm and analysis, we first introduce a reduction from general instances to identical ordering instances.

Definition 3 (Identical Ordering (IDO) Instances). *An instance is called identical ordering (IDO) if all agents have the same ordinal preference on the items, i.e.,* $c_i(e_1) \geq c_i(e_2) \geq \cdots \geq c_i(e_m)$ *for all* $i \in N$.

In the unweighted setting, Wu et al. [35] showed that if there exists an algorithm for computing PROPS allocations for IDO instances, then it can be converted to an algorithm that works for general instances while preserving the subsidy requirement. We show that the reduction can be naturally generalized to the weighted setting. With this reduction, in the rest of this paper, we only consider IDO instances.

Lemma 1. *If there exists a polynomial time algorithm that given any IDO instance computes a WPROPS allocation with total subsidy at most* α*, then there exists a polynomial time algorithm that given any instance computes a WPROPS allocation with total subsidy at most* α.

3 Technical Overview

Similar to previous results on the unweighted setting, our algorithm has two main steps: we first compute a fractional WPROP allocation, in which a small number of items are fractionally allocated; then we find a way to round the fractional allocation to an integral one. Since some agents may have costs exceeding their proportional share after rounding, we offer subsidies to these agents.

For the unweighted setting, it has been shown that there exists a rounding scheme with a total subsidy of no more than $n/4$, based on the fractional allocation returned by the Moving Knife Algorithm. However, as pointed out by Wu et al. [35], the algorithm fails to compute fractional WPROP allocations when agents have different weights.[3] Hence, generalizing the results for the unweighted setting to the weighted setting is non-trivial. For the weighted setting, they consider the Fractional Bid-and-Take Algorithm (FBTA) and show that there exists a rounding scheme with a total subsidy of no more than $(n-1)/2$. Whether the

all allocations are envy-freeable. In other words, there exist allocations that cannot guarantee envy-freeness even if we introduce unbounded subsidies to the agents.

[3] In fact, any algorithm that tries to allocate each agent a continuous interval (as the Moving Knife Algorithm computes) fails in the weighted setting. See [35] for a counter-example.

fractional WPROP allocation returned by FBTA allows total subsidy strictly smaller than $(n-1)/2$ remains unknown. In this paper, we revisit the algorithm and give a better characterization of the fractional WPROP allocation returned by FBTA. We show that the returned fractional allocation forms a tree structure connecting the agents and turning the allocation integral requires a rounding scheme on the tree. This provides a connection between the fair allocation problem with the graph theory. We believe that this special tree structure, naturally decided by the fractional allocation, might be of independent interest.

3.1 Fractional Bid-and-Take Algorithm

We first introduce the notation for representing a fractional allocation. We denote $x_i(e) \in [0,1]$ the fraction of item e allocated to agent i. We use $\mathbf{x} = (\mathbf{x}_1, \ldots, \mathbf{x}_n)$ to denote the fractional allocation, where $\mathbf{x}_i = \{x_i(e)\}_{e \in M} \in [0,1]^m$ is the fractional bundle allocated to agent i. Note that the allocation is complete iff $\sum_{i \in N} x_i(e) = 1$ for every $e \in M$. In the fractional allocation $\mathbf{x}$, we have $c_i(\mathbf{x}_i) = \sum_{e \in M}(x_i(e) \cdot c_i(e))$ for each agent i. We call $\mathbf{x}$ an integral allocation if $x_i(e) \in \{0,1\}$ for any i and e. To avoid ambiguity, we consistently use $\mathbf{X}$ to refer to an integral allocation. We present the Fractional Bid-and-Take Algorithm which has been shown to compute a fractional WPROP allocation in [35].

Algorithm 1: Fractional Bid and Take Algorithm (FBTA)

Input: An IDO instance $(M, N, \mathbf{w}, \mathbf{c})$ with $c_i(e_1) \le c_i(e_2) \le \cdots \le c_i(e_m)$ for all $i \in N$

$\mathbf{x}_i \leftarrow \mathbf{0}^m, \forall i \in N$, $A \leftarrow N$, $\mathbf{z} \leftarrow \mathbf{1}^m$, $j \leftarrow 1$
while $j \le m$ **do**
  Let $i \leftarrow \operatorname{argmin}_{i' \in A} \frac{c_{i'}(e_j)}{c_{i'}(M)}$; `// break tie lexicographically`
  if $c_i(\mathbf{x}_i) + z_j \cdot c_i(e_j) > \mathsf{WPROP}_i$ **then**
    $x_i(e_j) \leftarrow \frac{\mathsf{WPROP}_i - c_i(\mathbf{x}_i)}{c_i(e_j)}$; `// it is guaranteed that` $x_i(e_j) \le 1$
    $z_j \leftarrow z_j - x_i(e_j)$, $A \leftarrow A \setminus \{i\}$;
  else
    $x_i(e_j) \leftarrow z_j$, $z_j \leftarrow 0$, $j \leftarrow j+1$;

Output: A fractional allocation $\mathbf{x} = (\mathbf{x}_1, \ldots, \mathbf{x}_n)$.

Lemma 2. *The output allocation* $\mathbf{x}$ *is a fractional WPROP allocation.*

Note that in the above WPROP allocation, the number of fractional items is at most $n-1$, since the number of fractional items can increase only when some agent becomes inactive, and there is at least one active agent when the algorithm terminates. However, in the allocation, an agent might receive many fractional items, which introduces difficulties in designing the rounding scheme.

Toward a better understanding of the FBTA, we give an example instance of the chores allocations to weighted agents. Throughout the paper, whenever we use $\mathcal{I}^*$, we mean the instance described as follows.

Example 1. Consider the following instance $\mathcal{I}^*$ given by $(M, N, \mathbf{w}, \mathbf{c})$. The agents are $N = \{1, 2, 3, 4, 5, 6\}$, the items are $M = \{e_1, e_2, e_3, e_4, e_5, e_6\}$ and $w_1 = w_2 = w_3 = 1/12, w_4 = 1/6, w_5 = 1/4, w_6 = 1/3$. The cost functions are additive with the costs of the agents for each item shown in Table 1. Note that the instance is IDO and we have $\mathsf{WPROP}_1 = \mathsf{WPROP}_2 = \mathsf{WPROP}_3 = 0.4$, $\mathsf{WPROP}_4 = 0.9$, $\mathsf{WPROP}_5 = 1.5$, and $\mathsf{WPROP}_6 = 1.7$. After running Algorithm 1 on instance $\mathcal{I}^*$, we obtain the following fractional allocation (see Table 2), where the number in each cell corresponds to the fraction of item the agent receives.

Table 1. Example instance $\mathcal{I}^*$.

	e_1	e_2	e_3	e_4	e_5	e_6
agent 1	0.7	0.7	0.7	0.7	1	1
agent 2	0.8	0.8	0.8	0.8	0.8	0.8
agent 3	0.7	0.8	0.8	0.8	0.8	0.9
agent 4	0.8	0.8	0.8	1	1	1
agent 5	1	1	1	1	1	1
agent 6	0.8	0.8	0.8	1	1	1

Table 2. The fractional allocation.

	e_1	e_2	e_3	e_4	e_5	e_6
agent 1	4/7	0	0	0	0	0
agent 2	0	1/2	0	0	0	0
agent 3	3/7	1/8	0	0	0	0
agent 4	0	3/8	3/4	0	0	0
agent 5	0	0	0	1	1/2	0
agent 6	0	0	1/4	0	1/2	1

3.2 Item-Sharing Graph and Tree Structure

In this section, we introduce a graph called *item-sharing graph* that characterizes the structure of the fractional allocations. Given a fractional allocation, we say that an item is "shared" by agents i and j if $x_i(e) > 0$ and $x_j(e) > 0$. For any item e that is shared by agents i, j, we can view e to be an edge between i and j, while agents are the nodes. Naturally, such a sharing relation would induce a graph in which two agents are connected if they share an item (see Fig. 1 for an example.) However, in such a graph, if an item is shared by many agents, a complete graph will be formed among these agents, which introduces complications to our rounding scheme. Furthermore, we notice that even if an item e is shared by multiple agents, the allocation happens in sequence. For each agent $i \in N$, we denote by e^i the last item that agent i receives. If agent i reaches her (weighted) proportionality and turns inactive before she takes item e^i completely, the algorithm will allocate (the remaining part of) item e^i to some active agent $j \neq i$. We call agent j a *successor* of agent i if the algorithm tries to allocate the remaining part of item e^i to j.

Definition 4 (successor). *If when agent i turns inactive the item e^i is not fully allocated and agent j is the next agent who was chosen to take e^i, we call j a* successor *of agent i with respect to e^i.*

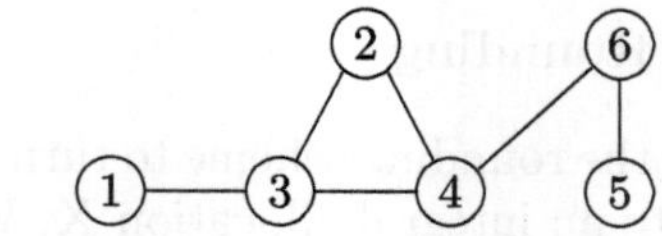

Fig. 1. When simply viewing items as edges between agents, the graph might be complex when an item is shared by more than two agents, e.g., e_2 is (fractional) allocated to agents 2, 3, 4.

Item-sharing Graph. Let nodes be the agents, where each agent has a directed edge to its successor (if any). We abuse the notation slightly and use e^i to refer to the edge from agent i to her successor. We call the resulting graph $G = (N, E)$ the *item-sharing graph* (see Fig. 2 for an example). Note that since each node has at most one outgoing edge, $G = (N, E)$ is a forest and $|E| \leq n - 1$.

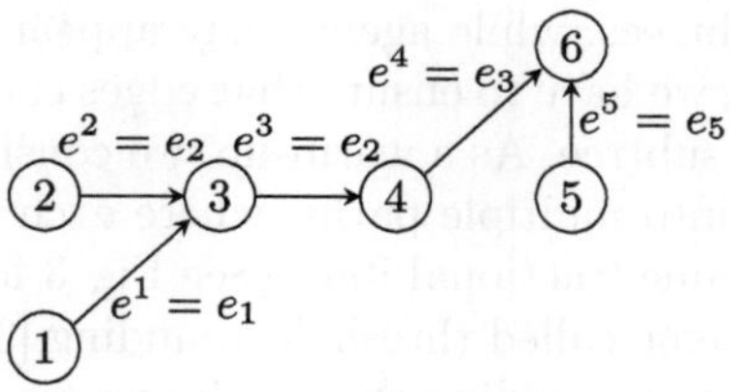

Fig. 2. The item-sharing graph for the returned allocation on instance $\mathcal{I}^*$.

Note that each edge in G corresponds to an item that is shared by the two endpoint agents. Moreover, it is possible that multiple edges correspond to the same item, e.g., $e^2 = e^3 = e_2$ in the instance $\mathcal{I}^*$ (see Fig. 2). Hence the set of edges E forms a (multi)set of fractional items. By regarding the successor of each agent as its parent, we obtain a collection of directed trees, where each agent with no outgoing edge is a root. For example, the graph G^* of $\mathcal{I}^*$ is a tree rooted at agent 6. We remark that an agent might have no outgoing edge for multiple reasons: 1) the agent is active when the algorithm terminates, or 2) the agent reaches her proportionality exactly when she fully consumed the item e^i.

Observation 1 (Tree Structure). *Given an instance $\mathcal{I} = (M, N, \mathbf{w}, \mathbf{c})$, let $\mathbf{x}$ be the (fractional) allocation returned by the FBTA and $G = (N, E)$ be the corresponding item-sharing graph, then G is a forest.*

Remark. We remark that if we construct the item-sharing graph based on an allocation returned by the Moving Knife Algorithm (in the unweighted setting), then the constructed graph G is a path since each agent has at most one outgoing edge and one incoming edge. Hence, for any item-sharing graph that is a path, we can naturally generalize the results of [35] to the weighted setting, which provides an (optimal) rounding scheme that bounds the total subsidy in $n/4$.

3.3 Tree Splitting and Rounding

In this section, we introduce the rounding scheme to turn the fractional allocation $\mathbf{x}$ returned by Algorithm 1 to an integral allocation $\mathbf{X}$. We call a fractional item *rounded* to agent i if, in the integral allocation, the item is fully allocated to agent i. We only round item e to some agent i with $x_i(e) > 0$ in the fractional allocation $\mathbf{x}$. Our goal is to find a rounding scheme with limited total subsidy.

Note that when the item-sharing graph G is a forest, the trees in G are independent to each other. In other words, the set of fractional items of a tree is disjoint from those of any other tree. For any two trees, we can round them independently. Hence, in the remaining part of this paper, we assume w.l.o.g. that the item-sharing graph G is a tree (instead of a forest). We show that by splitting the tree into (edge-disjoint) subtrees and rounding each subtree independently, we can upper bound the total required subsidy to achieve proportionality. In the splitting, we regard each subtree as a collection of edges. We denote the size of each tree by the number of edges it contains. By splitting based on edges, we obtain a collection of subtrees, while agents may appear in multiple trees. Note that during the splitting, we have to ensure that edges corresponding to the same item belong to the same subtree. As a warm-up, we consider the simple splitting that splits the graph G into multiple paths, where each path contains all edges that correspond to the same fractional item (See Fig. 3 for an example). We use the natural rounding scheme called threshold rounding [35] that greedily rounds a fractional item to the agent holding the maximum fraction of the item.

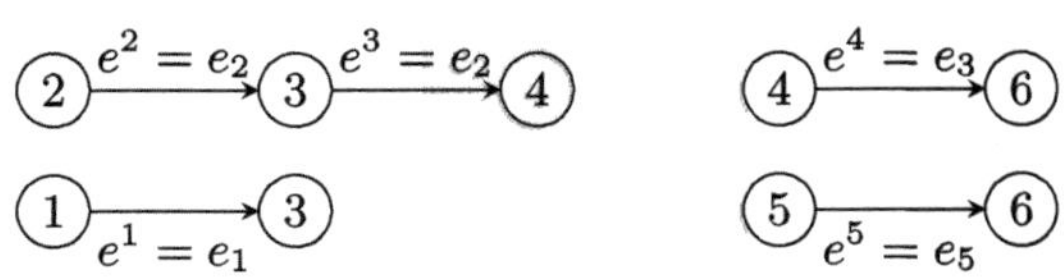

Fig. 3. A simple splitting of the item-sharing graph of instance $\mathcal{I}^*$.

Threshold Rounding. Given $\mathbf{x}$ returned by Algorithm 1, for any $e \in M$, we use $N(e) = \{i \in N : x_i(e) > 0\}$ to denote the set of agents who gets (a fraction of) item e. Note that if $|N(e)| = 1$ then item e is integrally allocated to a single agent. We call an item e *fractional* if $|N(e)| \geq 2$. In the threshold rounding schemes, we round item e to the agent holding the maximum fraction of it.

The following theorem has been shown in [35]. As a warm-up towards our tree-splitting framework, here we give an alternative proof.

Theorem 2. *There exists an algorithm that computes WPROP allocations with subsides at most $(n-1)/2$ for the allocation of chores to a group of weighted agents having general additive cost functions.*

Proof. We first split the tree into multiple paths where all edges of each path correspond to the same fractional item. For each path (each fractional item e), we use the threshold rounding that rounds e to agent $i^* = \text{argmax}_{i \in N} x_i(e)$, which incurs a subsidy $(1 - x_{i^*}(e)) \cdot c_{i^*}(e) \leq 1 - x_{i^*}(e) \leq \frac{|N(e)|-1}{|N(e)|} \leq \frac{|N(e)|-1}{2}$. Since item e corresponds to $|N(e)| - 1$ edges and the total number of edges is $n-1$, by a summation of the above upper bound, we show that the total required subsidy is at most $(n-1)/2$. □

As we will show, if all items are shared by exactly two agents, then it is not very difficult to improve the above splitting (by "pairing" edges) and give a better upper bound of $n/3 - 1/6$ (see Sect. 4). However, the existence of items that are shared by three or more agents complicates the splitting. We call such items *shattered items.* Roughly speaking, every shattered item puts a constraint on how the tree can be split and may cause an increase in the required total subsidy unless we do the splitting in a very careful way. We will discuss the case when shattered items exist in Sect. 5.

4 Rounding Trees Without Shattered Items

Recall that we assume without loss of generality that the item-sharing graph $G = (N, E)$ is a tree, i.e., $|E| = n - 1$. In this section, we consider the case when there is no shattered items. In other words, each of the edges in E corresponds to a unique fractional item that is shared by its two endpoints. We first consider the case of $n = 3$, in which the tree is a path with two edges.

4.1 Three Agents with Two Fractional Items

For the unweighted setting, Wu et al. [35] showed an upper bound of 3/4 for the total subsidy required to round paths with length 2. They also presented an instance with three agents lower-bounding the total subsidy by 2/3. In this section, we close this gap by showing a rounding strategy that guarantees a WPROPS allocation with total subsidy $\leq 2/3$, even for the weighted case.

Theorem 3. *For a tree with size* 2, *among four possible rounding schemes, the one that minimizes the total subsidy always guarantees a WPROPS allocation with a total subsidy of at most* 2/3.

For an instance $\mathcal{I} = (M, N, \mathbf{w}, \mathbf{c})$ where $N = \{1, 2, 3\}$, let e_1 be the item shared by agents 1, 2 and e_2 be the item shared by agents 2, 3 respectively. Ignoring the directions of edges, the item-sharing graph G is as Fig. 4 shows.

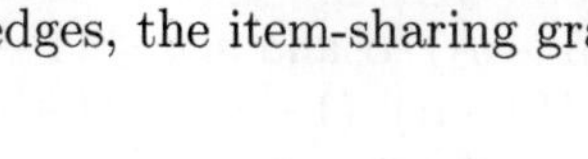

Fig. 4. The item-sharing graph for three agents.

We first introduce four different rounding schemes, that we call LL, RR, LR, and RL. We use LL to denote the scheme that rounds e_1 to agent 1 and e_2 to agent 2; RR to denote the scheme that rounds e_1 to agent 2 and e_2 to agent 3; LR to denote the scheme that rounds e_1 to agent 1 and e_2 to agent 3; RL to denote the scheme that rounds both e_1 and e_2 to agent 2. In the following, we upper bound the required subsidies for the four rounding schemes. We use $s_{\mathsf{LL}}, s_{\mathsf{RR}}, s_{\mathsf{LR}}, s_{\mathsf{RL}}$ to denote the subsidy required in the four rounding strategies respectively. Let $a_1 = c_2(e_1), a_2 = c_2(e_2)$, we assume w.l.o.g. that $a_1 \geq a_2$ and let $\alpha = a_2/a_1 \in (0, 1]$. For any variable x, we use $(x)^+$ to denote $\max\{x, 0\}$.

Lemma 3. *For the four different rounding schemes, the total required subsidy can be upper bounded as shown in the following table, where y_1 and y_2 denote $x_2(e_1)$ and $x_2(e_2)$ respectively.*

Rounding Scheme	Upper Bound for Total Subsidy
LL	$y_1 + ((1 - y_2) \cdot \alpha - y_1)^+$
RR	$y_2 + ((1 - y_1) - y_2 \cdot \alpha)^+$
LR	$y_1 + y_2$
RL	$(1 - y_1) + (1 - y_2) \cdot \alpha$

Proposition 1. *For any* $0 < x, y < 1$, *we have* $\min\{xy, (1 - x)(1 - y)\} \leq 1/4$

Proof of. Theorem 3: To prove Theorem 3, we discuss different cases depending on whether $((1 - y_2) \cdot \alpha - y_1)^+ = 0$ and $((1 - y_1) - y_2 \cdot \alpha)^+ = 0$. Note that $(1 - y_2) \cdot \alpha - y_1 \leq (1 - y_1) - y_2 \cdot \alpha$. Hence there are three different cases to discuss.

When $((1 - y_2) \cdot \alpha - y_1)^+ = 0$ and $((1 - y_1) - y_2 \cdot \alpha)^+ = 0$, we have $s_{\mathsf{LL}} \leq y_1$ and $s_{\mathsf{RR}} \leq y_2$. By summing up $s_{\mathsf{LL}}, s_{\mathsf{RR}}, s_{\mathsf{RL}}$, we have

$$\begin{aligned} s_{\mathsf{LL}} + s_{\mathsf{RR}} + s_{\mathsf{RL}} &\leq y_1 + y_2 + (1 - y_1) + (1 - y_2) \cdot \alpha \\ &\leq y_1 + y_2 + (1 - y_1) + (1 - y_2) = 2, \end{aligned}$$

which leads to $\min\{s_{\mathsf{LL}}, s_{\mathsf{RR}}, s_{\mathsf{RL}}\} \leq 2/3$. When $((1 - y_2) \cdot \alpha - y_1)^+ = 0$ and $((1 - y_1) - y_2 \cdot \alpha)^+ > 0$, we have $s_{\mathsf{LL}} \leq y_1$ and $s_{\mathsf{RR}} \leq (1 - y_1) + y_2 \cdot (1 - \alpha)$. By Proposition 1, we have $\min\{y_2 \cdot (1 - \alpha), (1 - y_2) \cdot \alpha\} \leq 1/4$. Hence we have $\min\{s_{\mathsf{RR}}, s_{\mathsf{RL}}\} \leq (1 - y_1) + \frac{1}{4} = \frac{5}{4} - y_1$. By summing up s_{LL} and $\min\{s_{\mathsf{RR}}, s_{\mathsf{RL}}\}$, we have $s_{\mathsf{LL}} + \min\{s_{\mathsf{RR}}, s_{\mathsf{RL}}\} \leq \frac{5}{4}$, which implies $\min\{s_{\mathsf{LL}}, s_{\mathsf{RR}}, s_{\mathsf{RL}}\} \leq \frac{5}{8} < \frac{2}{3}$. Finally, we consider the case that $((1 - y_2) \cdot \alpha - y_1)^+ > 0$ and $((1 - y_1) - y_2 \cdot \alpha)^+ > 0$. In this case we have $s_{\mathsf{LL}} \leq (1 - y_2) \cdot \alpha$ and $s_{\mathsf{RR}} \leq (1 - y_1) + y_2 \cdot (1 - \alpha)$. Then we have $\frac{2-\alpha}{\alpha} \cdot s_{\mathsf{LL}} + s_{\mathsf{RR}} + s_{\mathsf{LR}} = (2 - \alpha) \cdot (1 - y_2) + (1 + (2 - \alpha) \cdot y_2) = 3 - \alpha$. Hence we have $\min\{s_{\mathsf{LL}}, s_{\mathsf{RR}}, s_{\mathsf{LR}}\} \leq \frac{\alpha \cdot (3-\alpha)}{2+\alpha} \leq \frac{2}{3}$.

Since in all cases there exists a rounding scheme requiring a total subsidy of 2/3, by choosing the one that minimizes the total amount of subsidies among the four possible assignments, the total subsidy required is at most 2/3. ■

Given Theorem 3, we argue that for any instance, there exists a rounding strategy that guarantees a WPROPS allocation with a total subsidy of at most $n/3 - 1/6$. Our algorithm is based on a tree splitting that splits the item-sharing graph G into subtrees with size at most two. For each subtree, we do rounding independently and bound the subsidy by $2/3$.

4.2 Simple Splitting

Recall that during the splitting, we regard each tree as a collection of edges. If we can regard two adjacent edges as a subtree, we can bound the subsidy for the tree by $2/3$, following Theorem 3. Since there are $n-1$ edges, by repeatedly splitting subtrees with size two, we can bound the total subsidy by $\lceil (n-1)/2 \rceil \cdot 2/3 \leq n/3$. In the following, we introduce the Simple Splitting, which we use to split any item-sharing graph to a collection of subtrees with size two.

Algorithm 2: Simple Splitting

Input: An item-sharing graph G.

```
Let i be the deepest (furthest from the root) node in G ;
Let j be i's parent ;
if j has another child l ≠ i then
    Take edges e^i, e^l as a subtree T ;
else if j is the root then
    Take edge e^i as a subtree T ;
else
    Take edges e^i, e^j as a subtree T ;
```

Output: The subtree T and the remaining graph G' (if it is not empty).

Lemma 4. *After repeatedly running Simple Splitting on graph G, we obtain a collection of trees with size two, and at most one of them has size one.*

Given Theorem 3, we can round each subtree independently and bound the subsidy required for each tree by $2/3$. We show that the total subsidy can be upper bounded by $n/3$.

Theorem 4. *Given the fractional allocation $\mathbf{x}$ with $n - 1$ fractional items returned by Algorithm 1, there exists a rounding scheme that returns an integral WPROPS allocation with total subsidy at most $(n-1)/3$ when n is odd; at most $n/3 - 1/6$ when n is even.*

Proof. Considering the item-sharing graph $G = (N, E)$ with $|E| = n - 1$. Recall that in each splitting, we get a subtree with size two. When n is odd, after Simple Splitting, we obtain exactly $(n-1)/2$ trees with size two. By Theorem 3, for each tree, we can round the fractional items with a total subsidy of at most $2/3$. Hence the total subsidy across all subtrees is at most $\frac{2}{3} \cdot \frac{n-1}{2} = \frac{n-1}{3}$.

When n is even, after Simple Splitting, we obtain $(n-2)/2$ trees with size two and one tree with size one. For the tree with size one, we use threshold rounding which incurs a subsidy of at most $1/2$. Hence the total subsidy across all subtrees is at most $\frac{2}{3} \cdot \frac{n-2}{2} + \frac{1}{2} = \frac{n-2}{3} + \frac{1}{2} = \frac{n}{3} - \frac{1}{6}$. □

5 Rounding Trees with Shattered Item

In this section, we consider the case when there is at least one shattered item. Note that the Simple Splitting might not work in the case where multiple edges correspond to the same fractional item. Consider the allocation for instance $\mathcal{I}^*$, returned by Algorithm 1. The item-sharing graph G of the allocation is as Fig. 5 shows, where the two red edges e^2 and e^3 correspond to the same item e_2. Following the Simple Splitting edges e^2 and e^1 would be taken as a subtree, while e^3 and e^4 are put into another subtree. Recall that for each subtree, we apply the rounding scheme as in Theorem 3 independently. However, we notice that items e^2 and e^3 can not be rounded independently. We define the structure like e^2, e^3 as an *atom-path*, which contains edges that cannot be separated.

Definition 5 (Atom-path). *A path $i \to \cdots \to j$ is called an* atom-path *if all edges in the path correspond to the same item, and the path is maximal.*

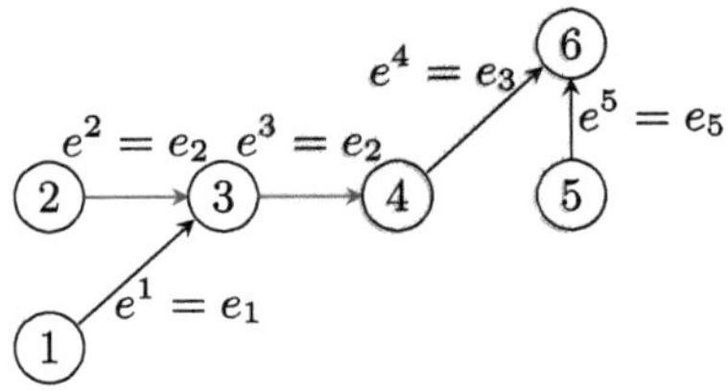

Fig. 5. The item-sharing graph G^* of instance $\mathcal{I}^*$.

We note that any atom-path has a size of at least 2. When there exists an atom-path in G, we need to ensure that the edges in the atom-path are not separated into different subtrees. In the following, we introduce the splitting and bound the total subsidy by $n/3$ as follows.

Lemma 5. *For any tree with size $z \geq 2$ such that there exists an atom-path in the tree, there exists a rounding scheme with total subsidy bounded by $z/3$.*

Theorem 5. *Given the fractional allocation $\mathbf{x}$ with less than $n-1$ fractional items returned by Algorithm 1, there exists a rounding scheme that returns an integral WPROPS allocation with total subsidy at most $(n-1)/3$.*

5.1 Atom-Path Splitting

In this section, we introduce the general tree splitting that we use when there exists at least one atom-path in the item-sharing graph G. A naive idea is to take the atom-path alone (with size at least two) as a subtree and recursively do the splitting on the remaining graph (which might have multiple connected components). For example, in Fig. 5, we can split the tree into the atom-path $\{e^2, e^3\}$, a subtree with edges $\{e^4, e^5\}$, and another subtree with edge e^1. Then we can round the subtrees independently as in Sect. 4. However, notice that rounding a path with size one, e.g., item e^1 in Fig. 5, may cause an increase in subsidy by $1/2$. This may cause a problem when the length of the atom-path is large, as each of these "hanging edges" will incur an increase of $1/2$ in the subsidy. It would be even harder to bound the total subsidy if there are other atom-paths in some connected components of the resulting graph after removing the atom-path. However, we notice that there is an easy fix to the above problem: we can simply attach the "hanging edges" to the atom-path to form a subtree, which we call an *expanded atom-path* (the formal definition will be given later). On a high level, we want to do the splitting such that after taking out an expanded atom-path, each connected component of the resulting graph is *good*.

Definition 6 (Good Subtree). *We say that a subtree is* good *if either it contains at least one atom-path, or it has even size.*

Note that we can bound (using mathematical induction) the total subsidy required to round a good subtree with size z by $z/3$ using Lemma 5 (if it contains an atom-path) or Theorem 4 (if it has even size). To ensure that we can split the tree into an expanded atom-path and a collection of good subtrees, we check each connected component after taking out the atom-path:

- If the component is a good subtree, we keep it directly and bound the required subsidy by $1/3$ of its size, following Lemma 5 or Theorem 4.
- For any connected component with odd size and without an atom-path, we show that there exists an edge in the connected component that can be attached to the atom-path such that after removing the edge the remaining graph is a collection of good subtrees.

By running Atom-path Splitting for a graph with atom-path(s), we obtain a collection of good subtrees and an atom-path with multiple edges attached to it.

Example 2. Consider a tree graph G with an atom-path (e^1, e^2, e^3). After cutting the atom-path we obtain four subtrees T_1, T_2, T_3, T_4 (see Fig. 6a). By definition T_2 and T_4 are good subtrees. Since T_3 has size one, we can simply attach the edge to the atom-path. Observe that T_1 also has an odd size. Hence there must exit an edge (the edge with blue color in Fig. 6a) that is incident to node 1 whose removal results in a collection of even-size subtrees.

Note that each node in the atom-path is attached at most one edge. We call the resulting structure an *expanded atom-path*.

Algorithm 3: Atom-path Splitting

Input: An item-sharing graph G with atom-path.

Pick an arbitrary atom-path and let it be a subtree ;
Remove the edges of the atom-path from the original tree ;
for *each connected component* T **do**
  if T *has an atom-path or has even size* **then**
    Let T be one subtree of the splitting ;
  else T *has odd size and no atom-path*
    Let $(i, j) \in T$ be an edge such that (1) agent i is in the atom-path and (2) the subtree containing j after removing (i, j) has even size;
    Attached (i, j) to the atom-path;
    Do Simple Splitting on each connected component of $T \setminus \{(i, j)\}$;

Output: A collection of subtrees and the remaining graph G'.

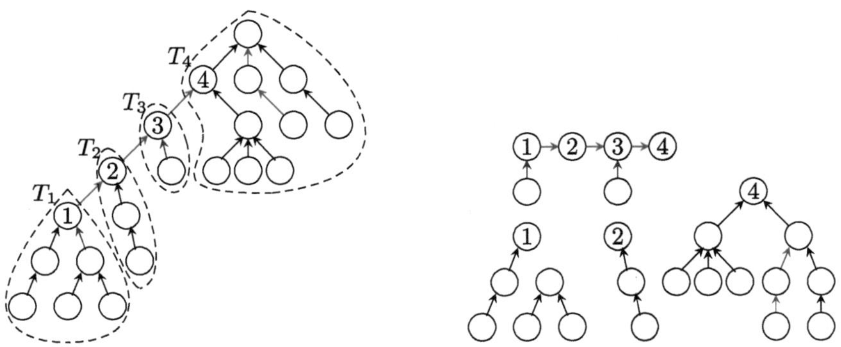

(a) Example with atom-path (e^1, e^2, e^3). (b) The splitting result of Figure 6a.

Fig. 6. An example tree and its splitting result

Definition 7 (Expanded Atom-path). *We call a subtree an* expanded atom-path *if it is composed of an atom-path and a collection of attached edges, where each node in the atom-path is incident to at most one attached edge.*

Lemma 6. *For any graph G containing (at least) one atom-path, Algorithm 3 splits G into an expanded atom-path and a collection of good subtrees.*

Recall that we can upper bound the subsidy required for a good subtree with z edges by $z/3$. To upper bound the total subsidy of a graph with atom-path(s), it remains to bound the subsidy required to round an expanded atom-path.

Lemma 7. *For any expanded atom-path consisting of an atom-path with k edges and $h \leq k + 1$ attached edges, there exists a rounding scheme with total subsidy bounded by $(k + h)/3$.*

Proof of Lemma 5: We prove the lemma by induction on z. We first consider the base case of a tree with $z = 2$ edges, which is an atom-path. In such case, we only have one fractional item that is shared by three agents. Using threshold rounding, we round the item to the agent holding the maximum fraction of it, which incurs a subsidy of at most $2/3$. Hence the lemma holds for the base case.

Now suppose that for any tree with edges $z' < z$ containing at least one atom-path, there exists a rounding scheme with total subsidy at most $z'/3$. We consider a tree with z edges containing at least one atom-path. Following Lemma 6, given any tree with at least one atom-path, the Atom-path splitting returns an expanded atom-path and a collection of subtrees that either contains an atom-path or has even size. For the former case, we can bound its required subsidy by $z'/3$, where z' is the number of edges it contains, by induction hypothesis. For the latter case, we can also bound the required subsidy by $z'/3$ by Simple Splitting and Theorem 4. Finally, by Lemma 7, for the expanded atom-path we can also bound the required subsidy by $1/3$ of its size. Since the summation of the size of all good subtrees and the expanded atom-path is z, the total subsidy required is at most $z/3$, and the lemma follows. ■

6 Conclusion and Open Problems

In this work, we revisit the problem of computing weighted proportional allocations with subsidy, and propose a rounding scheme for the fractional bid-and-take algorithm based on tree splitting. We show that for any instance with n agents with general weights, we can compute a weighted proportional allocation that requires a total subsidy of at most $n/3$. Our work leaves several natural problems open. The most fascinating open problem is to investigate the existence of algorithms that guarantee weighted proportionality requiring a total subsidy of at most $n/4$. We conjecture that the fractional bid-and-take algorithm admits such rounding schemes. It would also be interesting to study other fairness notions like Maximin Share or AnyPrice Share and see if we can further lower the subsidy requirement. Finally, we believe that it would be interesting to apply the tree splitting based framework to other fair allocation settings, e.g., the Best of Both Worlds setting, that requires rounding fractional allocations.

References

1. Amanatidis, G., et al.: Fair division of indivisible goods: recent progress and open questions. Artif. Intell. **322**, 103965 (2023)
2. Aziz, H.: Achieving envy-freeness and equitability with monetary transfers. In: AAAI, pp. 5102–5109. AAAI Press (2021)
3. Aziz, H., Chan, H., Li, B.: Weighted maxmin fair share allocation of indivisible chores. In: IJCAI, pp. 46–52. ijcai.org (2019)
4. Aziz, H., Huang, X., Kimura, K., Saha, I., Sun, Z., Suzuki, M., Yokoo, M.: Weighted envy-free allocation with subsidy. CoRR arxiv:2408.08711 (2024)
5. Aziz, H., Li, B., Moulin, H., Wu, X.: Algorithmic fair allocation of indivisible items: a survey and new questions. SIGecom Exch. **20**(1), 24–40 (2022)

6. Aziz, H., Moulin, H., Sandomirskiy, F.: A polynomial-time algorithm for computing a pareto optimal and almost proportional allocation. Oper. Res. Lett. **48**(5), 573–578 (2020)
7. Babaioff, M., Ezra, T., Feige, U.: Fair-share allocations for agents with arbitrary entitlements. In: EC, p. 127. ACM (2021)
8. Barman, S., Krishna, A., Narahari, Y., Sadhukhan, S.: Achieving envy-freeness with limited subsidies under dichotomous valuations. In: IJCAI, pp. 60–66. ijcai.org (2022)
9. Barman, S., Krishnamurthy, S.K.: On the proximity of markets with integral equilibria. In: AAAI, pp. 1748–1755 (2019)
10. Bei, X., Li, Z., Liu, J., Liu, S., Lu, X.: Fair division of mixed divisible and indivisible goods. Artif. Intell. **293**, 103436 (2021)
11. Bhaskar, U., Sricharan, A.R., Vaish, R.: On approximate envy-freeness for indivisible chores and mixed resources. In: APPROX-RANDOM. LIPIcs, vol. 207, pp. 1:1–1:23. Schloss Dagstuhl - Leibniz-Zentrum für Informatik (2021)
12. Brânzei, S., Sandomirskiy, F.: Algorithms for competitive division of chores. Math. Oper. Res. (2023)
13. Brustle, J., Dippel, J., Narayan, V.V., Suzuki, M., Vetta, A.: One dollar each eliminates envy. In: EC, pp. 23–39. ACM (2020)
14. Budish, E.: The combinatorial assignment problem: approximate competitive equilibrium from equal incomes. In: BQGT, p. 74:1. ACM (2010)
15. Caragiannis, I., Kurokawa, D., Moulin, H., Procaccia, A.D., Shah, N., Wang, J.: The unreasonable fairness of maximum nash welfare. ACM Trans. Econ. Comput. **7**(3), 12:1–12:32 (2019)
16. Chakraborty, M., Igarashi, A., Suksompong, W., Zick, Y.: Weighted envy-freeness in indivisible item allocation. ACM Trans. Econ, Comput. **9**(3), 18:1–18:39 (2021)
17. Choo, D., Ling, Y.H., Suksompong, W., Teh, N., Zhang, J.: Envy-free house allocation with minimum subsidy. Oper. Res. Lett. **54**, 107103 (2024)
18. Conitzer, V., Freeman, R., Shah, N.: Fair public decision making. In: EC (2017)
19. Dai, S., Chen, Y., Wu, X., Xu, Y., Zhang, Y.: Weighted envy-freeness in house allocation. CoRR arxiv:2408.12523 (2024)
20. Farhadi, A., et al.: Fair allocation of indivisible goods to asymmetric agents. J. Artif. Intell. Res. **64**, 1–20 (2019)
21. Feige, U., Huang, X.: On picking sequences for chores. In: EC, pp. 626–655. ACM (2023)
22. Foley, D.: Resource allocation and the public sector. In: Yale Economic Essays, pp. 45–98 (1967)
23. Goko, H., Igarashi, A., Kawase, Y., Makino, K., Sumita, H., Tamura, A., Yokoi, Y., Yokoo, M.: A fair and truthful mechanism with limited subsidy. Games Econ. Behav. **144**, 49–70 (2024)
24. Halpern, D., Shah, N.: Fair division with subsidy. In: Fotakis, D., Markakis, E. (eds.) SAGT 2019. LNCS, vol. 11801, pp. 374–389. Springer, Cham (2019). https://doi.org/10.1007/978-3-030-30473-7_25
25. Kawase, Y., Makino, K., Sumita, H., Tamura, A., Yokoo, M.: Towards optimal subsidy bounds for envy-freeable allocations. In: AAAI, pp. 9824–9831. AAAI Press (2024)
26. Li, B., Li, Y., Wu, X.: Almost (weighted) proportional allocations for indivisible chores. In: WWW, pp. 122–131. ACM (2022)
27. Li, Z., Liu, S., Lu, X., Tao, B.: Truthful fair mechanisms for allocating mixed divisible and indivisible goods. In: IJCAI, pp. 2808–2816. ijcai.org (2023)

28. Lipton, R.J., Markakis, E., Mossel, E., Saberi, A.: On approximately fair allocations of indivisible goods. In: EC, pp. 125–131. ACM (2004)
29. Liu, S., Lu, X., Suzuki, M., Walsh, T.: Mixed fair division: a survey. In: AAAI, pp. 22641–22649. AAAI Press (2024)
30. Maskin, E.S.: On the fair allocation of indivisible goods. In: Arrow and the Foundations of the Theory of Economic Policy, pp. 341–349. Springer, Heidelberg (1987). https://doi.org/10.1007/978-1-349-07357-3_12
31. Robertson, J.M., Webb, W.A.: Cake-cutting algorithms - be fair if you can. A K Peters (1998)
32. Springer, M., Hajiaghayi, M., Yami, H.: Almost envy-free allocations of indivisible goods or chores with entitlements. In: AAAI, pp. 9901–9908. AAAI Press (2024)
33. Steihaus, H.: The problem of fair division. Econometrica **16**, 101–104 (1948)
34. Suksompong, W.: Weighted fair division of indivisible items: a review. Inf. Process. Lett. 106519 (2024)
35. Wu, X., Zhang, C., Zhou, S.: One quarter each (on average) ensures proportionality. In: International Conference on Web and Internet Economics, pp. 582–599. Springer, Heidelberg (2023). https://doi.org/10.1007/978-3-031-48974-7_33
36. Wu, X., Zhang, C., Zhou, S.: Weighted EF1 allocations for indivisible chores. In: EC, p. 1155. ACM (2023)

A Computer-Aided Approach for Approximate Nash Equilibria

Xiaotie Deng[1], Dongchen Li[2], and Hanyu Li[1](✉)

[1] CFCS, School of Computer Science, Peking University, Beijing, China
{xiaotie,lhydave}@pku.edu.cn

[2] The School of Computing and Data Science, The University of Hong Kong, Pokfulam, Hong Kong
dongchen.li@connect.hku.hk

Abstract. Ever since the landmark PPAD-completeness result for Nash equilibria in two-player normal-form games, significant research has focused on developing polynomial-time algorithms for ϵ-approximate Nash equilibria (ϵ-NE). The challenge of establishing the optimal approximation guarantee in polynomial time remains pivotal. While advancements have been made for two-player games, progress in multi-player games is still limited. Difficulties arise due to the increased sophistication of multi-player games and the lack of tools for analyzing approximation bounds. This paper presents a method that allows machines to perform approximation analysis for multi-player games using a domain-specific language called LegoNE. LegoNE enables researchers to design algorithms with only high-level intuitions, while it automatically uncovers the underlying structures and proves the approximation bounds on its own. Using LegoNE, we design a new algorithm for three-player games that achieves a $(0.5+\delta)$-NE, improving the previous best bound $(0.6+\delta)$. This shows that human-machine collaboration allows us to obtain higher-level understandings and better results.

Keywords: approximate Nash equilibria · automation · programming language · formal verification · multi-player games

1 Introduction

1.1 Motivation

Nash equilibrium (NE) is a solution concept in which no player can gain more by unilaterally deviating from their current strategy. Ever since Nash's existence proof of NE in finite normal-form games [27], NE has become a standard solution concept in game theory. The non-constructive nature of Nash's proof leads to the computational consideration of NE, which also becomes a fundamental problem at the intersection of computer science and economics. A series of cornerstone works show that computing exact NE is PPAD-complete [10] for two-player games and FIXP-complete [18] for r-player games with $r \geq 3$. Due

M. Mavronicolas et al. (Eds.): WINE 2024, LNCS 15534, pp. 314–331, 2026.
https://doi.org/10.1007/978-3-032-08560-3_18

to these hardness results, it is well-believed that there is no polynomial-time algorithm for computing exact NE.

This motivates the study of *ϵ-approximate Nash equilibrium* (ϵ-NE), where players can gain at most ϵ additional payoff by deviating from the original strategy. It is shown that computing ϵ-NE in an r-player m-action game has a quasi-polynomial time algorithm with respect to r and m [4, 25]. Such an upper bound result matches the recent breakthrough of the lower bound result by Rubinstein [29], which shows that under a moderate assumption[1], there exists an unknown constant ϵ^*, and computing ϵ-NE even in two-player games requires quasi-polynomial time for any $\epsilon < \epsilon^*$.

Attention thus turns to finding the smallest ϵ that admits a polynomial-time algorithm for ϵ-NE. Ever since the PPAD-hardness results [10, 13], a flourishing line of research has been devoted to improving such an ϵ, most focusing on two-player games. Kontogiannis, Panagopoulou, and Spirakis [24] proposed an algorithm with $\epsilon = 3/4$; Daskalakis, Mehta, and Papadimitriou improved it to $1/2$ [15] and $0.38 + \delta$ [14]; very soon, Bosse, Byrka, and Markakis [5] reached $\epsilon = 0.364$. Then, in the first WINE conference, the renowned work [31] by Tsaknakis and Spirakis introduced a gradient-descent approach to achieve $\epsilon = 0.3393 + \delta$, which remained the state of the art for 15 years. Very recently, the stepping-stone work by Deligkas, Fasoulakis, and Markakis [16] improved the bound to $\epsilon = 1/3 + \delta$ by carefully revising the Tsaknakis-Spirakis algorithm.

While the progress on two-player games is remarkable, it almost remains blank for multi-player games. The only algorithmic result is a polynomial-time algorithm computing $(1/2 + \delta)$-NE for polymatrix games [17], which is a proper generalization of the gradient-descent approach by Tsaknakis and Spirakis [31]. We also know that if we have an algorithm for ϵ-NE in r-player games, then we can easily extend it to compute a $1/(2 - \epsilon)$-NE in $(r + 1)$-player games [6, 21]. Using this extension technique, if we focus on three-player games, the best-known algorithm computes a $(0.6 + \delta)$-NE [16]. For more than 16 years, the design and analysis techniques for multi-player algorithms have stagnated.

1.2 Contributions

The main result of this paper is as follows.

Theorem 1. *There exists a polynomial-time algorithm that computes a* $(0.5+\delta)$-*NE for three-player games.*

We improve the best-known bound for three-player games from $(0.6 + \delta)$ to $(0.5 + \delta)$ by proposing a new algorithm and analyzing its approximation bound.

To make the improvement, we must address two challenges. In the two-player setting, different strategies are constructed and allowed to interplay in an appropriate way to establish an approximation bound. When applying similar ideas to the three-player setting, we face the following challenges.

---

[1] That is, the exponential-time hypothesis (ETH) for PPAD.

- For algorithm design, the interplay between different strategies is much more complicated in three-player games than in two-player games. In the traditional approach for two-player algorithms, results from the approximation analysis are used to guide the selection of parameters in the algorithm. However, it becomes more sophisticated for three-player algorithms because the parameter space grows exponentially with the number of players.
- The algorithm analysis is also more complicated in three-player games. The approximation bound must be represented by tensor products, which are more complicated to analyze than matrix products in two-player games, making matrix-based techniques less feasible. Consequently, for three-player games, we previously had only one simple technique with analyzable approximation bounds, the extension technique, which views the two-player algorithm as a black box and extends it to three-player games.

To overcome these challenges, we develop new techniques for designing and analyzing algorithms for approximate NE, as summarized below.

- For algorithm design, we propose the *optimal mixing* operations. First, it adopts the traditional *mixing* technique in the literature, that is, to "mix" the strategies constructed in different manners to achieve a better approximation bound. Thus, the mixing operation creates interplays between different strategies. Second, it extends the optimization viewpoint in [31] and finds the *optimal* way to mix the strategies.
- For the algorithm analysis, our proof of Theorem 1 introduces a *computer-aided* methodology. It incorporates techniques from *programming languages* and *formal verification.*
  More specifically, we introduce a (domain-specific) programming language called *LegoNE.* LegoNE allows users to define basic operations as Lego bricks and write the pseudo-code of an algorithm with defined basic operations as if playing with Lego bricks. Using Floyd-Hoare semantic [20,22], the LegoNE compiler can automatically turn the pseudo-code into a fixed-size constrained optimization problem, whose optimal value is exactly the approximation bound of the algorithm. An overview of the proof procedure using LegoNE is shown in Fig. 1.

One should note that it is not rare to use an optimization solver in algorithm analysis. Many works also rely on an optimization solver to compute the final approximation/competitive ratio or complexity results [3,7,8,12,19,23,26,28,33]. The contribution of our work is to introduce verification and programming language techniques into the approximation analysis to automate the process of transferring an algorithm into an optimization problem that solves its approximation bound.

Facing the aforementioned challenges, our techniques bring the following benefits.

- When designing an algorithm, we can focus on the design itself and no longer need the instructions from the analysis of the approximation bound. To do

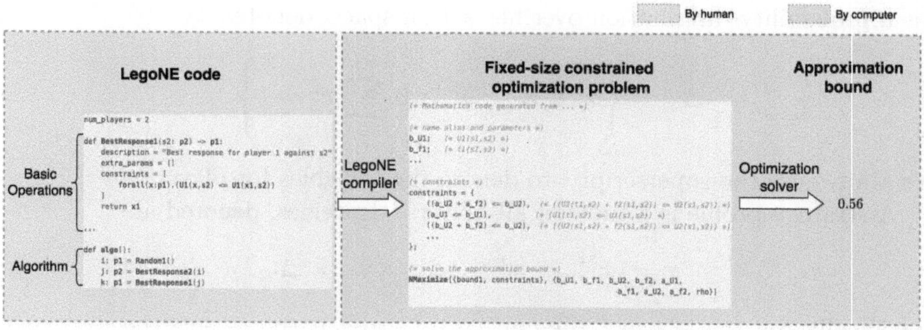

Fig. 1. An overview of the LegoNE framework.

so, we simply use the optimal mixing operation as a subroutine to optimally mix strategies in an algorithm. Then, in the algorithm design phases, the algorithm itself can automatically discover the optimal way to mix strategies.

– A computer program can discover by itself the underlying relationships between various components of an algorithm. Although we have numerous design techniques for approximate NE algorithms, *analyzing* these techniques in multi-player games, particularly their interplays, remains a significant challenge. This is the main obstacle to going beyond the extension technique and opening the black box of two-player algorithms. LegoNE addresses this challenge by automatically discovering these interplays, presenting them as constraints, and producing an approximation analysis with deeper insights than the traditional black-box approach.

With these two advantages, researchers are freed from the complexities of detailed calculations, allowing them to focus on high-level intuition and insights for algorithm design, while the computer handles the discovery of lower-level insights and performs the tedious calculations required for algorithm analysis. Altogether, we are able to design and analyze the new algorithm for three-player games and prove the approximation bound of $0.5 + \delta$.

This paper is organized as follows. In Sect. 2, we introduce the basic definitions and notations related to multi-player (normal-form) games and approximate Nash equilibria. In Sect. 3, we present the LegoNE framework and the optimal mixing operation. In Sect. 4, we show how to use LegoNE to automate the approximation analysis of the new algorithm. Finally, in Sect. 5, we conclude the paper and discuss future work.

2 Preliminaries

In this section, we introduce the basic definitions and notations related to multi-player (normal-form) games and approximate Nash equilibria.

Games and Mixed Strategies. We consider a normal-form game with r players. Each player i has an action space A^i of size n_i. The *(mixed) strategy* $\boldsymbol{x}^i$ of player

i is a probability distribution over her action space, denoted by

$$\Delta_{n_i} = \left\{ \boldsymbol{x}^i \in \mathbb{R}^{n_i} : \boldsymbol{x}^i \geq \mathbf{0}, \sum_{k=1}^{n_i} \boldsymbol{x}_k^i = 1 \right\}.$$

We always use the superscript i to denote the variables for player i.

A strategy profile is a tuple of all players' strategies, denoted as

$$\boldsymbol{x} = (\boldsymbol{x}^1, \ldots, \boldsymbol{x}^r) \in \Delta_{n_1} \times \cdots \times \Delta_{n_r}.$$

Specifically, for each player i, the unit vector $\boldsymbol{e}_k \in \Delta_{n_i}$ represents the strategy of choosing only the kth action, which is called a *pure strategy.* We also write $\boldsymbol{x}^{-i}$ to denote the strategy profile of all players except player i and $\boldsymbol{x}^{-i,j}$ to denote the strategy profile of all players except players i and j.

Each player i has a *payoff* function u_i from $\prod_{j=1}^r A^j$ to $\mathbb{R}$. By adjusting the payoffs through shifting and scaling, we assume that the payoffs are in the range $[0, 1]$. The payoff function can be naturally extended to the mixed strategy space, where $u_i(\boldsymbol{x})$ is the *expected* payoff of player i under strategy profile $\boldsymbol{x}$, i.e., $u_i(\boldsymbol{x}) = \mathsf{E}_{a \sim \boldsymbol{x}}[u_i(a)]$.

Note that the definition of u_i is equivalent to the $[0, 1]$-bounded *multi-linear* function over $\Delta_{n_1} \times \cdots \times \Delta_{n_r}$, that is, $u_i \in [0, 1]$ is a linear function of each player's mixed strategy. Formally, for any $i \in [r]$, any $\boldsymbol{x}^i, \underline{\boldsymbol{x}}^i \in \Delta_{n_i}$, any $\boldsymbol{x}^{-i} \in \Delta_{n_1} \times \cdots \times \Delta_{n_{i-1}} \times \Delta_{n_{i+1}} \times \cdots \times \Delta_{n_r}$, and any $\lambda \in [0, 1]$, we have

$$u_i(\lambda \boldsymbol{x}^i + (1 - \lambda)\underline{\boldsymbol{x}}^i, \boldsymbol{x}^{-i}) = \lambda u_i(\boldsymbol{x}^i, \boldsymbol{x}^{-i}) + (1 - \lambda) u_i(\underline{\boldsymbol{x}}^i, \boldsymbol{x}^{-i}).$$

Approximate Nash Equilibria from the Perspective of Optimization. We generalize [31] to define *ϵ-approximate Nash equilibria* (ϵ-NE) for multi-player games. First, define the *regret* of player i under strategy profile $\boldsymbol{x}$ as

$$f_i(\boldsymbol{x}) := \max_{\underline{\boldsymbol{x}}^i \in \Delta_{n_i}} u_i(\underline{\boldsymbol{x}}^i, \boldsymbol{x}^{-i}) - u_i(\boldsymbol{x}) = \max_{k \in [n_i]} u_i(\boldsymbol{e}_k, \boldsymbol{x}^{-i}) - u_i(\boldsymbol{x}).$$

Intuitively, $f_i(\boldsymbol{x})$ measures the maximum extra payoff that player i can get by deviating from $\boldsymbol{x}^i$ to any other mixed strategy. By the linearity of u_i, the maximum can be achieved by a pure strategy.

Then, define $f(\boldsymbol{x}) := \max_{i \in [r]} f_i(\boldsymbol{x})$. In the literature, a strategy profile $\boldsymbol{x}$ is an ϵ-NE if $f(\boldsymbol{x}) \leq \epsilon$. In other words, $\boldsymbol{x}$ is an ϵ-NE if no player can gain more than ϵ by deviating from their current strategy. $f(\boldsymbol{x})$ is called the *approximation* of $\boldsymbol{x}$. Particularly, a strategy profile is a *Nash equilibrium* (NE) if and only if it is a 0-NE. The global minimum of f over $\Delta_{n_1} \times \cdots \times \Delta_{n_r}$ is always 0 for the existence of NE [27].

3 The New Algorithm Designed in a Lego-Brick Style

In this section, we show how to design our new algorithm for Theorem 1 in a Lego-brick style. As in the Lego world, algorithm design can be viewed as assembling different basic operations (bricks) to form a full algorithm. For example, to sort three real numbers a, b, c, we have to use comparison operations (if-else) and swap operations. When we write a sequence of if-else and swap operations in a specific order, we get the full algorithm for sorting three numbers. We design our new algorithm in the exact same way.

We first present the basic operations and then present the full algorithm for Theorem 1 by assembling these basic operations.

3.1 Basic Operations

In the literature, since all algorithms entangle the algorithm design with the analysis, the basic operations are not explicitly mentioned. However, we can extract the basic operations from these algorithms. We list the polynomial-time basic operations used in our new algorithm as follows.

- **Random strategy**:
$$\boldsymbol{x}^i = \mathsf{Randomi}().$$
Description: Sample a random strategy $\boldsymbol{x}^i$ for player i.
- **Best response** (first occurred in [15]):
$$\boldsymbol{x}^i = \mathsf{BestResponse}(\boldsymbol{x}^{-i}).$$
Description: For a player i, given a strategy profile $\boldsymbol{x}^{-i}$ of the other players, find the best response strategy $\boldsymbol{x}^i = \arg\max_{\boldsymbol{x}^i \in \Delta_{n_i}} u_i(\boldsymbol{x}^i, \boldsymbol{x}^{-i})$.
- **Uniform mixing** (first occurred in [24]):
$$\boldsymbol{x}^i = \mathsf{UniformMixing}(\boldsymbol{x}^i_1, \ldots, \boldsymbol{x}^i_s).$$
Description: For a player i, given strategies $\boldsymbol{x}^i_1, \ldots, \boldsymbol{x}^i_s$, output the strategy $\boldsymbol{x}^i = \frac{1}{s}\sum_{k=1}^{s} \boldsymbol{x}^i_k$.
- **Branching**:
$$\mathsf{IfThenElse}(a, b, \text{branch } 1, \text{branch } 2).$$
Description: Given certain values a, b, go to branch 1 if $a \geq b$, and go to branch 2 otherwise.
- **Stationary point** (first occurred in [31]):
$$\boldsymbol{x}^i, \boldsymbol{x}^j, \boldsymbol{y}^i, \boldsymbol{y}^j = \mathsf{StationaryPoint}(\boldsymbol{x}^{-i,j}).$$
Description: Given a strategy profile $\boldsymbol{x}^{-i,j}$ of the other players, find the stationary point $(\boldsymbol{x}^i, \boldsymbol{x}^j)$ and dual point $(\boldsymbol{y}^i, \boldsymbol{y}^j)$ of function
$$\max\{f_i(\cdot, \cdot, \boldsymbol{x}^{-i,j}), f_j(\cdot, \cdot, \boldsymbol{x}^{-i,j})\}.$$

All but the last basic operation are quite direct. The last basic operation, StationaryPoint, is a key operation in the current state of the art [16] and previous state of the art [31] in two-player games. It views approximate NE as an optimization problem, i.e., finding the minimum of the function f defined in Sect. 2. Then, it computes the local optimum of f using a gradient descent method. To compute the steepest descent direction, it computes a linear program, which also gives a dual solution. After a few iterations, it finds a stationary point $(\boldsymbol{x}^i, \boldsymbol{x}^j)$ (i.e., zero-gradient point), together with a dual solution $(\boldsymbol{y}^i, \boldsymbol{y}^j)$.

Then, we propose a new basic operation, termed *optimal mixing*, which can be used to combine *arbitrary* strategies produced by basic operations. The idea is to simply find the *optimal* convex combination of all occurred strategies that minimizes the function f. Formally, the optimal mixing operation is defined as follows.

Definition 1 (Optimal mixing). *Given a set of strategies* $\boldsymbol{s}_1^1, \dots, \boldsymbol{s}_{t_1}^1 \in \Delta_{n_1}, \dots, \boldsymbol{s}_1^r, \dots, \boldsymbol{s}_{t_r}^r \in \Delta_{n_r}$, *and payoff functions* $u_1, \dots, u_r$, *the optimal mixing operation* OptimalMixing *outputs the strategy profile* $(\boldsymbol{s}_*^1, \dots, \boldsymbol{s}_*^r)$ *that minimizes the function* f *on* $\mathcal{M}$, *the set of all convex combinations of the strategies, i.e.,*

$$\mathcal{M} = \Bigg\{ (\boldsymbol{s}^1, \dots, \boldsymbol{s}^r) \in \Delta_{n_1} \times \dots \times \Delta_{n_r} : \\ \boldsymbol{s}^k = \sum_{i=1}^{t_k} \boldsymbol{\alpha}_i^k \boldsymbol{s}_i^k ; \boldsymbol{\alpha}_j^k \geq 0, j \in [t_k], \sum_{i=1}^{t_k} \boldsymbol{\alpha}_i^k = 1, k \in [r] \Bigg\}.$$

Importantly, since an algorithm can only have a fixed number of basic operations, the optimal mixing operation should only consider a fixed number of strategies.

Then, we show that for fixed r (number of players) and $t_1, \dots, t_r$ (numbers of strategies), the optimal mixing operation has a fully polynomial-time approximation scheme (FPTAS).

Theorem 2. *For fixed* r *and* $t_1, \dots, t_r$, *the optimal mixing operation has an FPTAS.*

More precisely, for any $\varepsilon > 0$, *there is an algorithm that outputs a strategy profile* $(\underline{\boldsymbol{s}}^1, \dots, \underline{\boldsymbol{s}}^r)$ *such that*

$$f(\underline{\boldsymbol{s}}^1, \dots, \underline{\boldsymbol{s}}^r) \leq f(\boldsymbol{s}_*^1, \dots, \boldsymbol{s}_*^r) + \varepsilon$$

in $\mathrm{poly}(n_1, \dots, n_r, 1/\varepsilon)$ *time, where* $(\boldsymbol{s}_*^1, \dots, \boldsymbol{s}_*^r)$ *is the output of the optimal mixing operation and* $n_1, \dots, n_r$ *are the sizes of the action sets* $A_1, \dots, A_r$ *(i.e., the input size).*

Here we only sketch the proof idea. Essentially, we treat the optimal mixing as a continuous constrained optimization problem. The objective function f is C-Lipschitz continuous in the L^∞ norm with respect to coefficients $(\boldsymbol{\alpha}_i^k)_{k,i}$.

Formally, for any $x, y \in \mathcal{M}$ with coefficients $\boldsymbol{\alpha} = (\boldsymbol{\alpha}_i^k)_{k,i}$ and $\boldsymbol{\beta} = (\boldsymbol{\beta}_i^k)_{k,i}$,

$$|f(x) - f(y)| \leq C \left\| \boldsymbol{\alpha} - \boldsymbol{\beta} \right\|_{L^\infty},$$

where $C = r \max_{k \in [r]}\{t_k\}$.

Then, we divide the domain $\mathcal{M}$ into a finite number of grids, and then we take the grid point that minimizes the function f as the output of the optimal mixing. It can be proved that the number of grids is constant to n_i, so the running time is polynomial to n_i.

3.2 Full Algorithm for Theorem 1

Now we can present the full algorithm for Theorem 1 by assembling the basic operations. The algorithm is given in Algorithm 1. The design idea of Algorithm 1 is a proper combination of the previous state of the art [31] and the state of the art [16] for two-player games.

Algorithm 1. Our algorithm for three-player games

$\boldsymbol{x}^3 = \mathsf{Random3}()$
$(\boldsymbol{x}_s^1, \boldsymbol{x}_s^2, \boldsymbol{w}^1, \boldsymbol{w}^2) = \mathsf{StationaryPoint}(\boldsymbol{x}^3)$
if $f_1(\boldsymbol{x}_s^1, \boldsymbol{x}_s^2, \boldsymbol{x}^3) \geq f_2(\boldsymbol{x}_s^1, \boldsymbol{x}_s^2, \boldsymbol{x}^3)$ **then**
    $\hat{\boldsymbol{y}}^2 = \mathsf{UniformMixing}(\boldsymbol{x}_s^2, \boldsymbol{w}^2)$
    $\hat{\boldsymbol{y}}^1 = \mathsf{BestResponse}(\hat{\boldsymbol{y}}^2, \boldsymbol{x}^3)$
    $\boldsymbol{z}_s^2, \boldsymbol{z}_s^3, \boldsymbol{w}_p^2, \boldsymbol{w}_p^3 = \mathsf{StationaryPoint}(\boldsymbol{x}_s^1)$
    $\boldsymbol{o}^1, \boldsymbol{o}^2, \boldsymbol{o}^3 = \mathsf{OptimalMixing}(\boldsymbol{x}_s^1, \boldsymbol{w}^1, \hat{\boldsymbol{y}}^1; \boldsymbol{x}_s^2, \boldsymbol{w}^2, \hat{\boldsymbol{y}}^2, \boldsymbol{z}_s^2, \boldsymbol{w}_p^2; \boldsymbol{x}^3, \boldsymbol{z}_s^3, \boldsymbol{w}_p^3)$
else
    $\hat{\boldsymbol{y}}^1 = \mathsf{UniformMixing}(\boldsymbol{x}_s^1, \boldsymbol{w}^1)$
    $\hat{\boldsymbol{y}}^2 = \mathsf{BestResponse}(\hat{\boldsymbol{y}}^1, \boldsymbol{x}^3)$
    $\boldsymbol{z}_s^1, \boldsymbol{z}_s^3, \boldsymbol{w}_p^1, \boldsymbol{w}_p^3 = \mathsf{StationaryPoint}(\boldsymbol{x}_s^2)$
    $\boldsymbol{o}^1, \boldsymbol{o}^2, \boldsymbol{o}^3 = \mathsf{OptimalMixing}(\boldsymbol{x}_s^1, \boldsymbol{w}^1, \hat{\boldsymbol{y}}^1, \boldsymbol{z}_s^1, \boldsymbol{w}_p^1; \boldsymbol{x}_s^2, \boldsymbol{w}^2, \hat{\boldsymbol{y}}^2; \boldsymbol{x}^3, \boldsymbol{z}_s^3, \boldsymbol{w}_p^3)$
end if
return $(\boldsymbol{o}^1, \boldsymbol{o}^2, \boldsymbol{o}^3)$

As a comparison, we here restate the traditional extension technique and see how the current best algorithm for three-player games is obtained. Suppose an algorithm A computes an α-NE for two-player games. Then, the algorithm A can be extended to a $1/(2-\alpha)$-NE for three-player games [6,21] as in Algorithm 2.

Algorithm 2. Previous Algorithm for Three-Player Games

$\boldsymbol{x}^3 = \mathsf{Random3}()$
$(\boldsymbol{x}_o^1, \boldsymbol{x}_o^2) = A(u_1(\cdot, \cdot, \boldsymbol{x}^3), u_2(\cdot, \cdot, \boldsymbol{x}^3))$ `// A is a two-player algorithm`
$\boldsymbol{x}_o^3 = \mathsf{BestResponse}(\boldsymbol{x}_o^1, \boldsymbol{x}_o^2)$
$\delta = (1-\alpha)/(2-\alpha)$
return $(\boldsymbol{x}_o^1, \boldsymbol{x}_o^2, (1-\delta)\boldsymbol{x}_o^3 + \delta \boldsymbol{x}^3)$

Clearly, Algorithm 1 does not use this extension technique. Instead, it uses the optimal mixing operation to combine the strategies produced by basic operations.

Moreover, for all algorithms in the literature, without optimal mixing, people have to manually choose the combination parameters (e.g., δ in the extension technique) to guarantee a certain approximation bound. In contrast, the optimal mixing operation automatically finds the best combination of strategies to minimize such a bound.

4 Computer-Aided Approximation Analysis Using LegoNE

In Sect. 3, we propose a new algorithm for approximate Nash equilibria in three-player games. However, the main difficulty lies not in the algorithm *design* but the *analysis*. In this section, we introduce a verification approach to analyze the approximation bound of an arbitrary algorithm for approximate Nash equilibria. The main tool we develop is a programming language, *LegoNE*, to write an algorithm and automatically derive its approximation bound. With LegoNE, we can develop Algorithm 1 with only high-level intuitions and then automatically derive its approximation bound.

We first introduce the key idea for the verification approach, the Floyd-Hoare semantics, and its specific form for approximate Nash equilibria. Then, we show how LegoNE can be used to translate an algorithm into a fixed-size constrained optimization problem, whose solution exactly gives the approximation bound of the algorithm.

4.1 Floyd-Hoare Semantics and Logic Encoding

Usually, when we describe an algorithm, we say how it will execute step by step and what the output will be. This is called the *operational semantics* for an algorithm. However, if we want to prove certain properties of an algorithm, we care more about what *properties* a step implies than how it is *executed*. Thus, we can use logic formulas to encode the properties of all steps and then use these formulas as axioms to prove the properties of the algorithm. This is the key idea of the Floyd-Hoare semantics [20,22].

As a concrete example, suppose we want to find the maximum element in a list $\{a_1, a_2, a_3\}$. An algorithm for this purpose is to compare a_1 and a_2, then compare the larger one with a_3. The basic operation is simply

$$s = \mathsf{IfThenElse}(a_1 > a_2, a_1, a_2).$$

Equivalently, this operation finds an s satisfying the logic formula (encoding)

$$[(s = a_1) \wedge (a_1 > a_2)] \vee [(s = a_2) \wedge (a_1 \leq a_2)].$$

Thus, a branch operation can be viewed as the above formula. To prove that the algorithm's output y_{out} is indeed the maximum element in the list, it suffices to show that we can derive the following logic formula from the encodings of the basic operations:

$$(y_{\text{out}} \geq a_1) \wedge (y_{\text{out}} \geq a_2) \wedge (y_{\text{out}} \geq a_3) \wedge$$
$$(y_{\text{out}} = a_1 \vee y_{\text{out}} = a_2 \vee y_{\text{out}} = a_3).$$

For approximate NE, we can use exactly the same idea. We can encode the basic operations in the algorithm as logic formulas, and then use these formulas to prove "This algorithm has an approximation bound of ϵ", which is also another logic formula.

As an illustration, we consider Algorithm 3, a simpler two-player algorithm in the literature [15], whose approximation bound is 0.5. We keep the notations the same as in the original paper.

Algorithm 3. DMP 0.5-NE algorithm

1: $i =$ Random1()
2: $j =$ BestResponse(i)
3: $k =$ BestResponse(j)
4: $r_1, r_2 =$ OptimalMixing($i, k; j$)

This algorithm simply samples a random pure strategy i for player 1, then finds the best response j for player 2 against i, and then finds the best response k for player 1 against j. Finally, it outputs the optimal mixing of i, j, k.[2]

We want to show that this algorithm has an approximation bound of 0.5. To do this, we first need to formally state the conclusion we want to prove in a logic formula. A natural way is to say that for any payoff functions u_1, u_2, whenever i, j, k, r_1, r_2 are produced by Algorithm 3, the approximation of the output strategy profile (r_1, r_2) is at most 0.5, i.e.,

$$\begin{aligned}&(\forall u_1, u_2)(\forall i, j, k, r_1, r_2)\\&\quad(i, j, k, r_1, r_2 \text{are produced by Algorithm 3} \rightarrow f(r_1, r_2) \leq 0.5).\end{aligned} \tag{1}$$

Then, we need to figure out the premises of the implication. Observe line 2 of Algorithm 3, which states that j is the best response to i. Equivalently, for any strategy y of player 2, the payoff of player 2 against i is at most the payoff against j. We can write this as a logic formula:

$$\forall y(u_2(i, y) \leq u_2(i, j)).$$

In this way, we encode step 2 of Algorithm 3 as a logic formula. Similarly, we can encode steps 1 and 3 as logic formulas. We denote such a mapping from one step to a logic formula as

$$\phi[\text{step}] \mapsto [\text{logic formula}].$$

[2] Rigorously speaking, in the original paper, there is no OptimalMixing operation. Instead, the algorithm uses a parameter to mix strategies i and k.

A much more difficult step is to encode the $\mathsf{OptimalMixing}$ operation given in Definition 1. The optimal mixing operation is the solution to a continuous optimization problem, which is not directly expressible in logic formulas. Instead, we provide a formula that is *implied* by the optimal mixing operation, namely, a necessary condition of the optimal mixing operation. Formally, we have the following theorem.

Theorem 3. *For any fixed* r *and* $t_1, \ldots, t_r$*, if* $\boldsymbol{x}_o^1, \ldots, \boldsymbol{x}_o^r$ *are the output of the optimal mixing operation given the input strategies* $\mathbf{s}_i^k$*, then there is a term* L^* *such that*

$$\phi\big[(\boldsymbol{x}_o^1, \ldots, \boldsymbol{x}_o^r) = \mathsf{OptimalMixing}(\boldsymbol{s}_1^1, \ldots, \boldsymbol{s}_{t_r}^r)\big] \mapsto [f(\boldsymbol{x}_o^1, \ldots, \boldsymbol{x}_o^r) \leq L^*]$$

and L^* *is expressed by* f_i *values on the input strategies* $\mathbf{s}_i^k$ *with arithmetic operations* $+, -, \times, \div, <, >,$ $\max, \min$ *operations over finite elements, and branch operations.*

The proof of Theorem 3 is quite intuitive yet technical. We omit it here.

Note that the particular form of the upper bound L^* is not arbitrarily chosen. The form given in the theorem is important for the later computer-aided method to utilize easily.

Finally, we can complete the encoding in (1) by using the above formulas:

$$\begin{aligned}
&\forall u_1, \forall u_2, \forall i, \forall j, \forall k, \forall r_1, \forall r_2\, (&& \\
&\quad (\forall y\, (u_2(i, y) \leq u_2(i, j))\wedge && \text{(encoding of } j = \mathsf{BestResponse}(i)) \\
&\quad \forall x\, (u_1(x, j) \leq u_1(k, j))\wedge && \text{(encoding of } k = \mathsf{BestResponse}(j)) \\
&\quad \phi[(r_1, r_2) = \mathsf{OptimalMixing}(i, k, j)]\wedge && \text{(encoding of optimal mixing)} \\
&\quad \text{inherent formulas}) && \text{(inherent formulas)} \\
&\quad \rightarrow f(r_1, r_2) \leq 0.5). && \text{(approximation bound)}
\end{aligned}$$

Here, "inherent formulas" are the formulas that are not explicitly stated in the algorithm but are implicitly satisfied by the algorithm. This can also be formalized as the domains of the payoff functions and strategies, but we omit them here for simplicity. For example, payoff functions are in $[0, 1]$ and function f_i and u_i have the following relationship:

$$f_i(\boldsymbol{x}^i, \boldsymbol{x}^{-i}) = \max_{\underline{\boldsymbol{x}}^i} u_i(\underline{\boldsymbol{x}}^i, \boldsymbol{x}^{-i}) - u_i(\boldsymbol{x}^i, \boldsymbol{x}^{-i}).$$

The above procedure can be generalized to Algorithm 1. More generally, using LegoNE, we can write the logic encodings of an *arbitrarily defined* basic operation.

Remark 1. In fact, *all* basic operations in the literature can be easily encoded in LegoNE in the following form:

$$\forall \boldsymbol{x}_1^1, \ldots, \boldsymbol{x}_{m_r}^r\, \forall U_1, U_2, \ldots (\alpha\ \mathsf{COMP}\ \beta),$$

where $\boldsymbol{x}_t^s$ is the t-th strategy variable of player s (like i, j, k in (1)), U_i is the payoff variable (which is one of $u_1, u_2, \ldots, u_r$), and α, β are arithmetic expressions over $\boldsymbol{x}_t^s$ and U_i. The COMP is a comparison operator, like $\leq$, $=$, and $\geq$.

4.2 An Overview of Computer-Aided Techniques in LegoNE

Now we describe how computer programs can be used to discover the underlying properties of an algorithm and use them to derive an approximation bound. Again, we consider Algorithm 3 as an illustrative example. The procedure described here can be generalized to Algorithm 1 as well as *any* other algorithms implemented in LegoNE.

The main idea is to reduce the problem of proving an approximation bound to deciding a real-variable arithmetic formula. This is achieved by the following steps.

Step 1: Interplay Analysis and Quantifier Elimination. First, we need to extract the interplay between strategies produced by the algorithm from the logic encodings. This is done by the quantifier elimination technique, more specifically, the *instantiation* technique.

For example, Algorithm 3 produces strategies i, j, k, r_1, r_2. Clearly, these strategies are not arbitrary strategies but have interplays with each other. We need to extract relations from these interplays to derive the approximation bound.

Consider line 3 in Algorithm 3. The logic encoding of this line is

$$\forall x, (u_1(x, j) \leq u_1(k, j)).$$

Let us see two different kinds of relations that can be derived from this logic encoding.

- Importantly, there is a *universal* quantifier $\forall x$. This means that the inequality holds for *any* strategy x of player 1. Specifically, this inequality holds when $x = i$. Thus, we can derive $u_1(i, j) \leq u_1(k, j)$ from the logic encoding of line 3. This reveals a relation between $u_1(i, j)$ and $u_1(k, j)$.
- For another kind of relation, since this inequality holds for any x, it also holds when $u_1(\cdot, j)$ reaches its maximum. Thus, we can derive $\max_x u_1(x, j) \leq u_1(k, j)$. This is a relation between $\max_x u_1(x, j)$ and $u_1(k, j)$.

Both kinds of relations can be automatically derived by computer programs since they are simply instantiations of the universal quantifiers.

Now we show the connections between these two kinds of relations and the approximation bound. Using Theorem 3, we know that $f(r_1, r_2)$ is upper bounded by a term L^*, which is expressed by f_i values on the input strategies i, j, k with arithmetic operations. If we can show that $L^* \leq 0.5$, then we can prove that the algorithm has an approximation bound of 0.5.

To achieve this, recall that function f_i is defined as

$$f_i(\boldsymbol{x}^i, \boldsymbol{x}^{-i}) = \max_{\underline{\boldsymbol{x}}^i} u_i(\underline{\boldsymbol{x}}^i, \boldsymbol{x}^{-i}) - u_i(\boldsymbol{x}^i, \boldsymbol{x}^{-i}).$$

Particularly, in the case of Algorithm 3, L^* contains the following terms: $u_1(i,j)$, $u_1(k,j)$, $u_2(i,j)$, $u_2(k,j)$, $\max_x u_1(x,j)$, $\max_y u_2(i,y)$, and $\max_y u_2(k,y)$. The above two kinds of relations exactly reveal the relations over these terms!

Following the above procedure, we actually deduce from (1) a logic formula in the following form:

$$\begin{aligned}&\forall u_1, \forall u_2, \forall i, \forall j, \forall k, \forall r_1, \forall r_2\, (\dots\\ &\quad \wedge [u_1(i,j) \le u_1(k,j)] \wedge [\max_x u_1(x,j) \le u_1(k,j)] \wedge \quad \text{(derived from line 3)}\\ &\quad \dots \to f(r_1, r_2) \le 0.5).\end{aligned}$$

An important observation is that the above formula only involves variables u_i, i, j, k, r_1, r_2 in a very compact form: they are only used in the following forms: $u_i(x,y)$, $\max_x u_1(x,y)$, $\max_y u_2(x,y)$, $f_i(x,y)$, and $f(x,y)$. Thus, the universal quantifiers can be relaxed over $u_1(i,j)$, $u_1(k,j)$, and so on:

$$\begin{aligned}&\forall u_1(i,j) \forall u_1(k,j) \forall \max_x u_1(x,j) \cdots (\dots\\ &\quad \wedge [u_1(i,j) \le u_1(k,j)] \wedge [\max_x u_1(x,j) \le u_1(k,j)] \wedge \quad \text{(derived from line 3)}\\ &\quad \dots \to f(r_1, r_2) \le 0.5).\end{aligned}$$

In the original goal (1), the universal quantifiers $\forall u_1$, $\forall u_2$, $\forall i$, ... range over all possible payoff functions and strategies. They are infinite-dimensional objects. By the above quantifier elimination, we can reduce the infinite-dimensional objects to finitely many real variables. This is a key step to use computer programs to prove the bound.

Step 2: Decide Real Arithmetic Theories. With Step 1, we have reduced the problem to proving a logic formula involving only finitely many real variables and arithmetic operations. Actually, such formulas form the *theory of first-order real arithmetic* $\mathsf{FOL}_\mathbb{R}$. $\mathsf{FOL}_\mathbb{R}$ contains first-order formulas that involve finitely many real variables, universal or existential quantifiers $(\forall, \exists)$, logic connectives $(\wedge, \vee, \to, \neg)$, and arithmetic operations $(+, -, \times, \div, <, =)$.[3]

$\mathsf{FOL}_\mathbb{R}$ has the following important property.

Theorem 4 (Tarski [30]). *$\mathsf{FOL}_\mathbb{R}$ is decidable. That is, there exists an algorithm, for any $\phi \in \mathsf{FOL}_\mathbb{R}$, it decides whether ϕ is valid (i.e., a true statement).*

In the verification community, various algorithms have been proposed to efficiently decide (subsets of) $\mathsf{FOL}_\mathbb{R}$. The fastest algorithm for general $\mathsf{FOL}_\mathbb{R}$ uses *cylindrical algebraic decomposition* (CAD) [9,11].

Based on this theorem, the formula obtained in Step 1, belonging to $\mathsf{FOL}_\mathbb{R}$, can be directly proved by CAD. This completes the proof of the approximation bound of an algorithm all by computer programs.

[3] Note that the maximum operators can be expressed in $\mathsf{FOL}_\mathbb{R}$: Define $\max\{x,y\}$ by a new variable z satisfying $(x \ge y \to z = x) \wedge (x < y \to z = y)$. Similarly, the minimum operator $\min\{x,y\}$ and absolute value $|x|$ can be expressed in $\mathsf{FOL}_\mathbb{R}$.

One More Step: Find the Approximation Bound by Computer Programs. With the above steps, we can *prove* an approximation bound of an algorithm. However, when we design an algorithm, we cannot know in advance what the approximation bound will be. Thus, we need to *find* the approximation bound by computer programs. This is a much more practical problem.

Actually, this step can be easily achieved by computer programs. For a naive approach, we can simply enumerate all possible approximation bounds from 0 to 1 with a binary search. For each approximation bound ϵ, we can check whether the formula is valid using CAD. If it is valid, then we know that the algorithm has an approximation bound of ϵ. Otherwise, we know that we have to search for a greater ϵ.

However, such a naive approach is not efficient. To determine a bound with an accuracy of 0.01, we need to run CAD at least seven times ($2^{-7} \approx 0.0078$). To make the process more efficient, we need the following observation.

Consider Algorithm 3 again. This time, we do not know the approximation bound of the algorithm. Instead, we set an unknown approximation bound b and rewrite (1) as

$$\begin{aligned}&(\forall u_1, u_2)(\forall i, j, k, r_1, r_2)\\ &\qquad (i, j, k, r_1, r_2 \text{are produced by Algorithm 3} \rightarrow f(r_1, r_2) \leq b).\end{aligned} \tag{2}$$

We need to find a real number b as small as possible such that the above formula is valid.

Similarly, we use quantifier elimination in Step 1 and turn (2) into

$$\begin{aligned}&\forall \max_{y'} u_2(x, y'), \forall \max_{x'} u_1(x', y), \forall u_s(x, y), \forall f_s(x, y)\\ &\qquad (\phi \rightarrow [f(r_1, r_2) \leq b]).\end{aligned} \tag{3}$$

Here, ϕ is a quantifier-free formula.

Using Theorem 3 to construct an upper bound L^* of $f(r_1, r_2)$, we only need to find a real number b such that $L^* \leq b$. Recall that all real variables appearing in L^* are included in universal quantifiers in (3). Since we need $L^* \leq b$ to hold for all values of these real variables, we can compute the maximum value of L^* over all possible values of these real variables and set b to be this maximum value. Clearly, bound b calculated in this way makes formula (2) valid. Thus, we can write down the following constrained optimization problem:

$$\begin{aligned}\text{maximize}\quad & \max\{f_1(r_1, r_2), f_2(r_1, r_2)\}\\ & \qquad \text{over } f_s(x, y), u_s(x, y), \max_{y'} u_2(x, y'), \max_{x'} u_1(x', y),\\ & \qquad b, s \in \{1, 2\}, x \in \{i, k, r_1\}, y \in \{j, r_2\}\\ \text{subject to}\quad & \phi.\end{aligned} \tag{4}$$

Then we show how to solve this optimization problem. First, since constraint ϕ is quantifier-free, we can write the premise into disjunctive normal form (DNF), say $(a_1 \leq a_2 \wedge a_5 \geq a_6) \vee (b_1 = b_2 \wedge b_3 \geq b_4) \vee (c_1 \geq c_2)$. This step can be also performed by computer programs [32].

Second, (4) can be divided into three optimization problems: to maximize $f(r_1, r_2)$ subject to $a_1 \le a_2 \wedge a_5 \ge a_6$, to maximize $f(r_1, r_2)$ subject to $b_1 = b_2 \wedge b_3 \ge b_4$, and to maximize $f(r_1, r_2)$ subject to $c_1 \ge c_2$. All of these optimization problems are fixed-size constrained optimization problem, and thus can be solved by numerical solvers (like Mathematica [2] or Gurobi [1]). Suppose the optimal value of these problems are v_1, v_2, v_3. Then, we actually prove that

- if $b_1 = b_2 \wedge b_3 \ge b_4$, then $f(r_1, r_2) \le v_1$;
- if $a_1 \le a_2 \wedge a_5 \ge a_6$, then $f(r_1, r_2) \le v_2$;
- if $c_1 \ge c_2$, then $f(r_1, r_2) \le v_3$.

Finally, combining these results, we show that under constraint ϕ, $f(r_1, r_2) \le \max\{v_1, v_2, v_3\}$. Thus, we can set $b = \max\{v_1, v_2, v_3\}$ as the final approximation bound.

Note that all above steps can be directly applied to Algorithm 1. Actually, we use mathematica to compute the final $0.5 + \delta$ approximation bound. To show the generality of our method, we also implement most two-player algorithms in the literature in LegoNE and reprove their approximation bounds by the same method.

5 Conclusion

This paper introduces LegoNE, a framework for designing algorithms for approximate NE. We demonstrate that LegoNE allows for a modular, Lego-brick style approach to algorithm design and that the approximation analysis can be fully performed by machine. Using this method, we improve the state-of-the-art algorithm for approximate NE in three-player games from $(0.6 + \delta)$ to $(0.5 + \delta)$.

LegoNE establishes a new paradigm of human-machine collaboration for algorithm design and analysis. In this model, humans concentrate on high-level creative design, contributing diverse ideas, while machines take care of the detailed, accurate analysis by following predefined patterns. This division allows both to excel in their strengths, fostering more efficient and innovative algorithm development.

Acknowledgments. This work is supported by National Science and Technology Major Project (No. 2022ZD0114904) and Natural Science Foundation of China (Grant No. 6212290003). The authors would like to thank Ruyi Ji and Yuhao Li for helpful discussions. The authors would also like to thank the anonymous reviewers for their valuable comments and suggestions.

Disclosure of Interests. The authors have no competing interests to declare that are relevant to the content of this article.

References

1. Gurobi Optimization. https://www.gurobi.com/

2. Wolfram Mathematica: Modern Technical Computing. https://www.wolfram.com/mathematica/
3. Abolhassani, M., Ehsani, S., Esfandiari, H., HajiAghayi, M., Kleinberg, R., Lucier, B.: Beating 1-1/e for ordered prophets. In: Proceedings of the 49th Annual ACM SIGACT Symposium on Theory of Computing, Montreal, Canada, pp. 61–71. ACM (2017). https://doi.org/10.1145/3055399.3055479
4. Babichenko, Y., Barman, S., Peretz, R.: Empirical distribution of equilibrium play and its testing application. Math. Oper. Res. **42**(1), 15–29 (2017). https://doi.org/10.1287/moor.2016.0794
5. Bosse, H., Byrka, J., Markakis, E.: New algorithms for approximate nash equilibria in bimatrix games. In: Deng, X., Graham, F.C. (eds.) WINE 2007. LNCS, vol. 4858, pp. 17–29. Springer, Heidelberg (2007). https://doi.org/10.1007/978-3-540-77105-0_6
6. Bosse, H., Byrka, J., Markakis, E.: New algorithms for approximate nash equilibria in bimatrix games. Theor. Comput. Sci. **411**(1), 164–173 (2010). https://doi.org/10.1016/j.tcs.2009.09.023
7. Bubna, A., Chiplunkar, A.: Prophet inequality: order selection beats random order. In: Leyton-Brown, K., Hartline, J.D., Samuelson, L. (eds.) Proceedings of the 24th ACM Conference on Economics and Computation, EC 2023, London, United Kingdom, 9–12 July 2023, pp. 302–336. ACM (2023). https://doi.org/10.1145/3580507.3597687
8. Cai, Y., Wu, J.: On the optimal fixed-price mechanism in bilateral trade. In: Proceedings of the 55th Annual ACM Symposium on Theory of Computing, STOC 2023, pp. 737–750. Association for Computing Machinery, New York (2023). https://doi.org/10.1145/3564246.3585171
9. Caviness, B.F., Johnson, J.R.: Quantifier Elimination and Cylindrical Algebraic Decomposition. Springer (2012)
10. Chen, X., Deng, X., Teng, S.H.: Settling the complexity of computing two-player Nash equilibria. J. ACM **56**(3), 14:1–14:57 (2009). https://doi.org/10.1145/1516512.1516516
11. Collins, G.E.: Quantifier elimination for real closed fields by cylindrical algebraic decompostion. In: Brakhage, H. (ed.) GI-Fachtagung 1975. LNCS, vol. 33, pp. 134–183. Springer, Heidelberg (1975). https://doi.org/10.1007/3-540-07407-4_17
12. Correa, J.R., Saona, R., Ziliotto, B.: Prophet secretary through blind strategies. Math. Program. **190**(1), 483–521 (2021). https://doi.org/10.1007/S10107-020-01544-8
13. Daskalakis, C., Goldberg, P.W., Papadimitriou, C.H.: The complexity of computing a nash equilibrium. SIAM J. Comput. **39**(1), 195–259 (2009). https://doi.org/10.1137/070699652
14. Daskalakis, C., Mehta, A., Papadimitriou, C.: Progress in approximate nash equilibria. In: Proceedings of the 8th ACM Conference on Electronic Commerce, San Diego, California, USA, pp. 355–358. ACM (2007). https://doi.org/10.1145/1250910.1250962
15. Daskalakis, C., Mehta, A., Papadimitriou, C.: A note on approximate nash equilibria. In: Spirakis, P., Mavronicolas, M., Kontogiannis, S. (eds.) WINE 2006. LNCS, vol. 4286, pp. 297–306. Springer, Heidelberg (2006). https://doi.org/10.1007/11944874_27
16. Deligkas, A., Fasoulakis, M., Markakis, E.: A polynomial-time algorithm for 1/3-approximate nash equilibria in bimatrix games. In: Chechik, S., Navarro, G., Rotenberg, E., Herman, G. (eds.) 30th Annual European Symposium on Algorithms,

ESA 2022, 5–9 September. LIPIcs, vol. 244, pp. 41:1–41:14. Schloss Dagstuhl - Leibniz-Zentrum für Informatik, Berlin/Potsdam, Germany (2022). https://doi.org/10.4230/LIPIcs.ESA.2022.41
17. Deligkas, A., Fearnley, J., Savani, R., Spirakis, P.: Computing approximate nash equilibria in polymatrix games. Algorithmica **77**(2), 487–514 (2017). https://doi.org/10.1007/s00453-015-0078-7
18. Etessami, K., Yannakakis, M.: On the complexity of nash equilibria and other fixed points. SIAM J. Comput. **39**(6), 2531–2597 (2010). https://doi.org/10.1137/080720826
19. Fahrbach, M., Huang, Z., Tao, R., Zadimoghaddam, M.: Edge-weighted online bipartite matching. J. ACM **69**(6), 45:1–45:35 (2022). https://doi.org/10.1145/3556971
20. Floyd, R.W.: Assigning meanings to programs. In: Colburn, T.R., Fetzer, J.H., Rankin, T.L. (eds.) Program Verification: Fundamental Issues in Computer Science, pp. 65–81. Springer, Dordrecht (1993). https://doi.org/10.1007/978-94-011-1793-7_4
21. Hémon, S., de Rougemont, M., Santha, M.: Approximate nash equilibria for multi-player games. In: Monien, B., Schroeder, U.-P. (eds.) SAGT 2008. LNCS, vol. 4997, pp. 267–278. Springer, Heidelberg (2008). https://doi.org/10.1007/978-3-540-79309-0_24
22. Hoare, C.A.R.: An axiomatic basis for computer programming. Commun. ACM **12**(10), 576–580 (1969). https://doi.org/10.1145/363235.363259
23. Jaillet, P., Lu, X.: Online stochastic matching: new algorithms with better bounds. Math. Oper. Res. **39**(3), 624–646 (2014). https://doi.org/10.1287/moor.2013.0621
24. Kontogiannis, S.C., Panagopoulou, P.N., Spirakis, P.G.: Polynomial algorithms for approximating nash equilibria of bimatrix games. In: Spirakis, P., Mavronicolas, M., Kontogiannis, S. (eds.) WINE 2006. LNCS, vol. 4286, pp. 286–296. Springer, Heidelberg (2006). https://doi.org/10.1007/11944874_26
25. Lipton, R.J., Markakis, E., Mehta, A.: Playing large games using simple strategies. In: Proceedings of the 4th ACM Conference on Electronic Commerce, pp. 36–41 (2003)
26. Liu, Z., Ren, Z., Wang, Z.: Improved approximation ratios of fixed-price mechanisms in bilateral trades. In: Proceedings of the 55th Annual ACM Symposium on Theory of Computing, STOC 2023, pp. 751–760. Association for Computing Machinery, New York (2023). https://doi.org/10.1145/3564246.3585160
27. Nash, J.: Non-cooperative games. Ann. Math. **54**(2), 286–295 (1951)
28. Peng, B., Tang, Z.G.: Order selection prophet inequality: from threshold optimization to arrival time design. In: 63rd IEEE Annual Symposium on Foundations of Computer Science, FOCS 2022, Denver, CO, USA, 31 October–3 November 2022, pp. 171–178. IEEE (2022). https://doi.org/10.1109/FOCS54457.2022.00023
29. Rubinstein, A.: Settling the complexity of computing approximate two-player nash equilibria. In: 2016 IEEE 57th Annual Symposium on Foundations of Computer Science (FOCS), pp. 258–265 (2016). https://doi.org/10.1109/FOCS.2016.35
30. Tarski, A., McKinsey, J.C.C.: A Decision Method for Elementary Algebra and Geometry. University of California Press, dgo - digital original, 1 edn (1951). https://doi.org/10.2307/jj.8501420
31. Tsaknakis, H., Spirakis, P.G.: An optimization approach for approximate nash equilibria. In: Deng, X., Graham, F.C. (eds.) WINE 2007. LNCS, vol. 4858, pp. 42–56. Springer, Heidelberg (2007). https://doi.org/10.1007/978-3-540-77105-0_8
32. Van Dalen, D.: Logic and Structure. Universitext, Springer, London (2013). https://doi.org/10.1007/978-1-4471-4558-5

33. Williams, V.V., Xu, Y., Xu, Z., Zhou, R.: New bounds for matrix multiplication: from alpha to omega. In: Proceedings of the 2024 Annual ACM-SIAM Symposium on Discrete Algorithms (SODA), pp. 3792–3835. Proceedings, Society for Industrial and Applied Mathematics (2024). https://doi.org/10.1137/1.9781611977912.134

The Complexity of Symmetric Bimatrix Games with Common Payoffs

Abheek Ghosh(✉) and Alexandros Hollender

Department of Computer Science, University of Oxford, Oxford OX1 3QG, U.K.
{abheek.ghosh,alexandros.hollender}@cs.ox.ac.uk

Abstract. We study symmetric bimatrix games that also have the common-payoff property, i.e., the two players receive the same payoff at any outcome of the game. Due to the symmetry property, these games are guaranteed to have symmetric Nash equilibria, where the two players play the same (mixed) strategy. While the problem of computing such symmetric equilibria in general symmetric bimatrix games is known to be intractable, namely PPAD-complete, this result does not extend to our setting. Indeed, due to the common-payoff property, the problem lies in the lower class CLS, ruling out PPAD-hardness. In this paper, we show that the problem remains intractable, namely it is CLS-complete. On the way to proving this result, as our main technical contribution, we show that computing a Karush-Kuhn-Tucker (KKT) point of a quadratic program remains CLS-hard, even when the feasible domain is a simplex.

Keywords: Computational Complexity · Quadratic Programming · Total Search Problems · Bimatrix Games

1 Introduction

The Nash equilibrium notion is an extremely natural solution concept for games. First of all, it has a very simple definition: every player should be happy with what they are playing at equilibrium, given what the other players are playing. In particular, this is something that each player can easily check for themselves. A second, arguably equally important, property is that every game is guaranteed to have a Nash equilibrium [26,27]. As a result, Nash equilibria have been used as the de facto standard solution concept in a vast array of economic settings.

One crucial question, which was somewhat overlooked for a long time, is whether Nash equilibria are easy to find. The advent of computer science made this question precise and formal: *is there a polynomial-time algorithm for computing a Nash equilibrium of a game*? Given that, despite significant efforts, no such efficient algorithm had been found, it was natural to suspect that the problem is in fact intractable. The theory of NP-completeness, although very successful, can unfortunately not be used for this problem. Interestingly, it is precisely the two core properties mentioned above – efficient verifiability and guaranteed existence – which make it essentially impossible for the problem to

M. Mavronicolas et al. (Eds.): WINE 2024, LNCS 15534, pp. 332–347, 2026.
https://doi.org/10.1007/978-3-032-08560-3_19

be NP-hard. Indeed, the problem lies in the class TFNP of total NP search problems, and no TFNP problem can be NP-hard, unless NP = co-NP [22]. The complexity class PPAD, the correct class – as it turned out – to capture the complexity of computing Nash equilibria, was only defined in the early 90 s [29], and the celebrated results of Daskalakis, Goldberg, and Papadimitriou [8], and Chen, Deng, and Teng [4] finally proved that the problem is PPAD-complete.

The PPAD-completeness of computing Nash equilibria provides strong evidence that the problem cannot be solved in polynomial time. Indeed, many different problems with no known efficient algorithms are known lie in this class. More recently, the hardness of PPAD has also been shown assuming various cryptographic assumptions [3,5,16]. This intractability result brings into question whether the Nash equilibrium notion is in fact a reasonable solution concept for these games. Indeed, the PPAD-completeness result [4] even applies to bimatrix games, namely simple normal-form games between just two players.

In the past two decades, an important research direction has consisted in investigating whether Nash equilibria can at least be computed efficiently for special restricted classes of games. A very natural such restriction of bimatrix games are *symmetric* bimatrix games. These are games in which nothing changes if we swap the labels of the two players. Formally, the payoff matrix of the second player is just the transpose of the payoff matrix of the first player. This is a very natural condition, which is in particular satisfied by many standard games such as the prisoner's dilemma, the stag hunt, or the game of chicken. Nash proved that symmetric games always admit *symmetric* equilibria [26], where both players play the same mixed strategy. As a result, it is natural to only consider equilibria that have this desirable property in this setting. Unfortunately, this problem remains PPAD-complete [30, Theorem 2.4]. Other notable negative results for bimatrix games include the PPAD-hardness of sparse games [4], win-lose games [4], and rank-3 games [23]. Some positive results have been obtained for restrictions such as, of course, zero-sum games [28], but also rank-1 bimatrix games [1]. Interestingly, all these results have shown either PPAD-completeness or polynomial-time solvability. This raises the question of whether there are natural classes of bimatrix games for which the problem is neither PPAD-complete, nor in P.

Our contribution. In this work, we identify for the first time a natural class of bimatrix games for which the equilibrium problem has *intermediate* complexity. Namely, we show that the problem of computing a Nash equilibrium in such games is CLS-complete, meaning that it is unlikely to be PPAD-complete or polynomial-time solvable.

The class of games we consider are *symmetric bimatrix games with common payoffs*, meaning that the two players have the same payoff for any given outcome of the game. More formally, these are symmetric games $(A, A^\intercal)$ that also satisfy $A = A^\intercal$, and we consider symmetric equilibria in such games.[1] These games,

[1] Since these games have common payoffs, a pure Nash equilibrium is guaranteed to exist and can be easily located by simply picking the maximum entry in the payoff matrix A. However, note that such pure equilibria will usually *not* be symmetric.

studied by Emmons et al. [10], can be used to model settings in multiagent reinforcement learning and cooperative AI, although mostly with a larger number of players. Since our hardness result already holds for only two players, it also extends to these settings.

By using a well-known connection [21,25], our result immediately also implies the CLS-completeness of computing (not necessarily symmetric) Nash equilibria in imitation games (A, I) with symmetric matrix A.[2]

Our result is obtained by leveraging a known connection between equilibria of these games and Karush-Kuhn-Tucker (KKT) points of quadratic programs of a particular form. This connection was already observed and used by McLennan and Tourky [21] in their work on NP-hardness of various decision problems related to Nash equilibria. In our context, it immediately yields CLS-membership for our equilibrium computation problem. Our main technical contribution is to also establish CLS-hardness of the problem. This is achieved by showing that the problem of computing a KKT point of a quadratic program *with simplex constraints* is CLS-complete. Recent work of Fearnley et al. [14] had established CLS-completeness of the problem *with box constraints*, and had left open the case of simplex constraints. We present a direct reduction from box constraints to simplex constraints, thus resolving this case as well.

Continuous Local Search (CLS). The class CLS, introduced by Daskalakis and Papadimitriou [9], lies both in PPAD and in PLS. The class PLS captures the complexity of various problems in discrete local optimization, such as the local max-cut problem [12,17,19,31]. Fearnley et al. [13] proved that CLS = PPAD $\cap$ PLS, and that CLS exactly captures the complexity of finding gradient descent fixed points, i.e., KKT points of general smooth functions given by circuits. Babichenko and Rubinstein [2] further showed CLS-completeness of computing mixed Nash equilibria of congestion games with many players, as well as of finding KKT points of polynomials of degree five. Recently, Fearnley et al. [14] showed that this remains the case even for polynomials of degree two, i.e., quadratic programming. Their result only holds for box constraints.

A CLS-completeness result indicates that the problem is unlikely to lie in P, since the cryptographic hardness results mentioned above for PPAD also apply to CLS. It is also believed that CLS $\neq$ PPAD. Indeed, it is known that no black-box reduction from PPAD to CLS exists [24], and, so far, all collapses of classes in the context of TFNP have used black-box reductions.

Further related work. Kontogiannis and Spirakis [18] use KKT points of an appropriately defined quadratic formulation of symmetric bimatrix games to obtain almost 1/3-approximate Nash equilibria. It is known that computing Nash equilibria in imitation games is PPAD-complete [4,6]. Furthermore, Murhekar and Mehta [25] show that although the problem does not admit a FPTAS (Fully Polynomial Approximation Scheme), there is a PTAS. Various decision problems about the existence of Nash equilibria with certain properties in bimatrix games have been proved NP-complete, even for symmetric or imitation games [7,15,21].

[2] Here I denotes the identity matrix.

Outline. In Sect. 2, we present some basic definitions and standard results about bimatrix games, and we introduce the KKT problem for quadratic programs. In Sect. 3, we explain the connection between the two, and state our results for games. Finally, in Sect. 4, we present our main technical result, namely the CLS-hardness of KKT points in quadratic programs with simplex constraints.

2 Preliminaries

Let $[n] = \{1, 2, \ldots, n\}$. Let $\Delta^{n-1} \subset \mathbb{R}^n$ denote the $(n-1)$-dimensional simplex in $\mathbb{R}^n$, i.e., $\boldsymbol{x} \in \Delta^{n-1}$ if $x_i \geq 0$ for all $i \in [n]$ and $\sum_i x_i = 1$.

2.1 Game Theory: Bimatrix Games

Definition 1. *Let $n, m \in \mathbb{N}$. A bimatrix game (A, B) is given by two matrices $A = (a_{ij}) \in [0,1]^{n\times m}$ and $B = (b_{ij}) \in [0,1]^{n\times m}$, with the interpretation that player 1 (the row player) and player 2 (the column player) have n and m pure strategies respectively. If the row player plays its ith strategy, and the column player plays its jth strategy, then the payoff to the row player is a_{ij}, and the payoff to the column player is b_{ij}.*

A bimatrix game is

- symmetric, *if $B = A^\intercal$ (in particular, $n = m$).*
- common-payoff, *if $B = A$.*
- *an* imitation *game, if $n = m$ and $B = I_n$ (the $n \times n$ identity matrix).*

Definition 2. *Let $\varepsilon \geq 0$. A* mixed strategy profile *$(\boldsymbol{x}, \boldsymbol{y}) \in \Delta^{n-1} \times \Delta^{m-1}$ is an ε-Nash equilibrium if*

$$\boldsymbol{x}^\intercal A\boldsymbol{y} \geq \hat{\boldsymbol{x}}^\intercal A\boldsymbol{y} - \varepsilon, \quad \text{for all } \hat{\boldsymbol{x}} \in \Delta^{n-1},$$
$$\boldsymbol{x}^\intercal B\boldsymbol{y} \geq \boldsymbol{x}^\intercal B\hat{\boldsymbol{y}} - \varepsilon, \quad \text{for all } \hat{\boldsymbol{y}} \in \Delta^{m-1}.$$

The equilibrium is symmetric, *if $n = m$ and $\boldsymbol{x} = \boldsymbol{y}$.*

Fact 1. (Nash [26]). *Every bimatrix game has a Nash equilibrium. Every symmetric bimatrix game has a symmetric Nash equilibrium.*

The following alternative, stronger way of defining approximation of Nash equilibria will be useful in the paper.

Definition 3. *Let $\varepsilon \geq 0$. A mixed strategy profile $(\boldsymbol{x}, \boldsymbol{y}) \in \Delta^{n-1} \times \Delta^{m-1}$ is an* ε-well-supported *Nash equilibrium if for all $i \in [n]$*

$$x_i > 0 \implies (A\boldsymbol{y})_i \geq \max_j (A\boldsymbol{y})_j - \varepsilon$$

and for all $i \in [m]$

$$y_i > 0 \implies (\boldsymbol{x}^\intercal B)_i \geq \max_j (\boldsymbol{x}^\intercal B)_j - \varepsilon.$$

It is easy to see that any *ε-well-supported* Nash equilibrium is also an ε-Nash equilibrium, but the converse is not necessarily true. Nevertheless, the notions are known to be computationally equivalent in the following sense.

Lemma 1. (Chen et al. [4, Lemma 3.2]**)** *In a bimatrix game* (A, B) *with* $A, B \in [0,1]^{n \times m}$, *given any* $\varepsilon^2/8$*-Nash equilibrium, we can find an* ε*-well-supported Nash equilibrium in polynomial time.*

This continues to hold in the case of *symmetric* equilibria in symmetric bimatrix games.[3]

2.2 Optimization: The KKT Problem for Quadratic Programs

Definition 4. (Karush–Kuhn–Tucker (KKT) Conditions) *Consider the following constrained optimization problem in* $\mathbb{R}^n$

$$\min_{\boldsymbol{x} \in \mathbb{R}^n} \quad f(\boldsymbol{x}) \tag{1}$$

$$\textit{subject to} \quad g_j(\boldsymbol{x}) \le 0, \quad \textit{for } j \in [m]. \tag{2}$$

We assume that the function f *is continuously differentiable and the functions* g_j *for* $j \in [m]$ *are continuously differentiable and convex. The Karush–Kuhn–Tucker (KKT) conditions are:*

$$\nabla f(\boldsymbol{x}) + \sum_j u_j \nabla g_j(\boldsymbol{x}) = 0, \tag{3}$$

$$u_j g_j(\boldsymbol{x}) = 0, \quad \textit{for } j \in [m], \tag{4}$$

$$u_j \ge 0, \quad \textit{for } j \in [m], \tag{5}$$

and $\boldsymbol{x}$ *must also satisfy the original constraints in* (2)*. The* u_j *variables are called the dual variables. A point* $\boldsymbol{x}$ *that satisfies the KKT conditions above is called a KKT point.*

A KKT point is stable with respect to gradient descent and vice-versa. In gradient descent, we move towards the direction of decreasing gradient of $f(\boldsymbol{x})$, i.e., towards a direction that has positive inner-product with $-\nabla f(\boldsymbol{x})$, unless by doing so we violate one of the constraints. The $u_j g_j(\boldsymbol{x}) = 0$ condition in (4) ensures that $u_j > 0$ only if $g_j(\boldsymbol{x}) = 0$, i.e., u_j can take positive values only if the corresponding constraint is a tight one. Then, $\sum_j u_j \nabla g_j(\boldsymbol{x})$ corresponds to the convex cone of the gradients of the tight constraints, and the condition (3) ensures that $-\nabla f(\boldsymbol{x})$ is in this cone; therefore, by moving towards a direction that decreases $f(\boldsymbol{x})$, we violate a constraint.

[3] It suffices to inspect the proof of [4, Lemma 3.2] to see that for symmetric bimatrix games, the symmetry of a strategy profile is retained in the polynomial-time transformation.

Definition 5. (ε-approximate KKT Conditions) *The ε-approximate KKT (or simply ε-KKT) conditions relax the KKT conditions in* (3)–(5) *(for the optimization problem in* (1)–(2)*):*

$$\nabla f(\boldsymbol{x}) + \sum_j u_j \nabla g_j(\boldsymbol{x}) \in [-\varepsilon, \varepsilon], \tag{6}$$

$$u_j g_j(\boldsymbol{x}) = 0, \qquad \textit{for} j \in [m], \tag{7}$$

$$u_j \geq 0, \qquad \textit{for} j \in [m]. \tag{8}$$

In this paper, we primarily focus on quadratic programs (QPs) with linear constraints. In such a QP, the objective function $f(\boldsymbol{x})$ in (1) is a quadratic function and the functions $g_j(\boldsymbol{x})$ in the constraints in (2) are linear functions. In particular, we look at box constraints and simplex constraints.

Definition 6. (QP with Box Constraints) *A quadratic program with box constraints is given as*

$$\min_{\boldsymbol{x}} \quad \frac{1}{2}\boldsymbol{x}^\intercal A \boldsymbol{x} + \boldsymbol{b}^\intercal \boldsymbol{x} \tag{9}$$

$$\textit{subject to} \quad 0 \leq x_i \leq 1, \qquad \textit{for} i \in [n], \tag{10}$$

where $A = (a_{ij}) \in \mathbb{R}^{n\times n}$ is symmetric without loss of generality, $a_{ij} = a_{ji}$, and $\boldsymbol{b} = (b_i) \in \mathbb{R}^n$.

The ε-KKT conditions for a QP with box constraints can be simplified to the following form:

Definition 7. (ε-KKT Conditions for a QP with Box Constraints) *A feasible point $\boldsymbol{x}$ satisfies the ε-KKT Conditions if*

$$\textit{for} i \in [n] : \begin{cases} \sum_j a_{ij}x_j + b_i \geq -\varepsilon, & \textit{if} x_i = 0, \\ \sum_j a_{ij}x_j + b_i \in [-\varepsilon, \varepsilon], & \textit{if} 0 < x_i < 1, \\ \sum_j a_{ij}x_j + b_i \leq \varepsilon, & \textit{if} x_i = 1. \end{cases} \tag{11}$$

It is known that computing an ε-KKT point of a QP with box constraints is CLS-complete [14].

Definition 8. (QP with a Simplex Constraint) *A quadratic program with simplex constraints is given as*

$$\min_{\boldsymbol{x}} \quad \frac{1}{2}\boldsymbol{x}^\intercal A \boldsymbol{x} + \boldsymbol{b}^\intercal \boldsymbol{x} \tag{12}$$

$$\textit{subject to} \quad \sum_i x_i = 1, \tag{13}$$

$$x_i \geq 0, \qquad \textit{for} i \in [n]. \tag{14}$$

where $A = (a_{ij}) \in \mathbb{R}^{n\times n}$ is symmetric without loss of generality, $a_{ij} = a_{ji}$, and $\boldsymbol{b} = (b_i) \in \mathbb{R}^n$.

Note that a QP with a scaled version of the simplex constraint of the form

$$\sum_i x_i = s \tag{15}$$

for any fixed $s > 0$ is equivalent to the canonical form: we can change variables from x_i to $x'_i = x_i/s$ and the coefficients from a_{ij} to $a'_{ij} = a_{ij}s^2$ and from b_i to $b'_i = b_i s$ to recover the canonical form. We shall use this scaled version in some of our proofs for convenience.

Definition 9. (ε-KKT Conditions for a QP with a Simplex Constraint) *A feasible point $\boldsymbol{x}$ satisfies the ε-KKT Conditions if there exists $u \in \mathbb{R}$ such that*

$$\text{for} i \in [n] : \begin{cases} \sum_j a_{ij}x_j + b_i \geq u - \varepsilon & \text{if} x_i = 0, \\ \sum_j a_{ij}x_j + b_i \in [u - \varepsilon, u + \varepsilon] & \text{if} x_i > 0. \end{cases} \tag{16}$$

$u \in \mathbb{R}$ is the dual variable corresponding to the $\sum_i x_i = 1$ constraint.

One of the main contributions of our paper is to show that computing an ε-KKT point for simplex-constrained QPs is CLS-complete through a reduction from the same computational problem for box-constrained QPs.

3 The Complexity of Symmetric Common-Payoff Games

The main result of this section is the following.

Theorem 1. *The problem of computing a symmetric ε-Nash equilibrium in a symmetric bimatrix game with common payoffs is CLS-complete.*

Remark 1. Here, and everywhere else in this paper, when we talk about ε-Nash equilibria (or ε-well-supported Nash equilibria, or ε-KKT points), we always consider the setting where ε is provided as part of the input, in standard binary representation. In particular, this means that all our hardness results apply to the setting where ε is *inverse-exponential* in the size of the input. Naturally, our results also apply to exact solutions,[4] namely to the case where $\varepsilon = 0$. Finally, when ε is inverse-polynomial, an ε-equilibrium can be computed in polynomial time by using gradient descent on the optimization problem given in Lemma 2. In other words, the problem admits a FPTAS and thus our hardness result for inverse-exponential ε is optimal.

[4] For the games and KKT problems we consider, it can be shown using standard techniques (see, e.g., [11]) that they admit exact rational solutions with polynomial-length representations, and that the exact versions are equivalent to the inverse-exponential ε versions.

Proof. By Lemma 1, it suffices to establish this result for approximate well-supported equilibria of the game. By Lemma 2, stated and proved below, the problem is equivalent to computing an approximate KKT point of a quadratic program with a simplex constraint. Finally, by Theorem 3, our main technical result which is stated and proved in the next section, this problem is CLS-complete. Note that we can ensure that all the payoffs lie in [0, 1] by a standard rescaling and offset argument.

The following lemma was proved by McLennan and Tourky [21] for exact equilibria. Here we show that the result continues to hold if we introduce approximation.

Lemma 2. *Let* (A, B) *be a symmetric bimatrix game with common payoffs, i.e.,* $B = A^\mathsf{T} = A$*. Then, a strategy profile* $(\boldsymbol{x}, \boldsymbol{x})$ *is an* ε*-well-supported Nash equilibrium, if and only if* $\boldsymbol{x}$ *is an* ε*-KKT point of*

$$\begin{aligned} \min_{\boldsymbol{x}} \quad & -\boldsymbol{x}^\mathsf{T} A \boldsymbol{x} \\ \textit{subject to} \quad & \sum_i x_i = 1, \\ & x_i \geq 0, \qquad \textit{for} i \in [n]. \end{aligned}$$

Proof. Using the fact that the matrix $A = (a_{ij})$ is symmetric, we can rewrite

$$-\boldsymbol{x}^\mathsf{T} A \boldsymbol{x} = -2 \sum_{i<j} a_{ij} x_i x_j - \sum_i a_{ii} x_i^2.$$

and thus

$$\frac{\partial}{\partial x_i}(-\boldsymbol{x}^\mathsf{T} A \boldsymbol{x}) = -2 \sum_j a_{ij} x_j = -2(A\boldsymbol{x})_i.$$

As a result, $\boldsymbol{x}$ satisfies the ε-KKT conditions if any only if there exists $u \in \mathbb{R}$ such that

$$\text{for} i \in [n] : \begin{cases} -2(A\boldsymbol{x})_i \geq u - \varepsilon & \text{if} x_i = 0, \\ -2(A\boldsymbol{x})_i \in [u - \varepsilon, u + \varepsilon] & \text{if } x_i > 0, \end{cases}$$

or equivalently, if and only if there exists $v \in \mathbb{R}$ such that

$$\text{for } i \in [n] : \begin{cases} (A\boldsymbol{x})_i \leq v + \varepsilon/2 & \text{if } x_i = 0, \\ (A\boldsymbol{x})_i \in [v - \varepsilon/2, v + \varepsilon/2] & \text{if } x_i > 0. \end{cases}$$

Now, it is not hard to see that such $v \in \mathbb{R}$ exists if and only if

$$(A\boldsymbol{x})_i \geq \max_j (A\boldsymbol{x})_j - \varepsilon \qquad \text{for all} i \text{with} x_i > 0.$$

But this is exactly the definition of $(\boldsymbol{x}, \boldsymbol{x})$ being an ε-well-supported Nash equilibrium of the game (A, A).

3.1 Consequences for Imitation Games

The CLS-completeness for symmetric common-payoff games immediately yields the following result.

Theorem 2. *The problem of computing an ε-Nash equilibrium in an imitation game (A, I) with $A^\top = A$ is CLS-complete.*

Proof. As before, by Lemma 1, it suffices to establish this result for approximate well-supported equilibria of the game. By Lemma 3, stated and proved below, this problem is equivalent to that of finding a symmetric approximate well-supported Nash equilibrium in the symmetric game $(A, A^\top)$ with $A^\top = A$, i.e., with common payoffs. The result then follows by Theorem 1.

The following lemma is well-known for exact equilibria, see e.g. [21]. We present a version that also holds for approximate equilibria.

Lemma 3. *Let A be a square $n \times n$ matrix, and let $\varepsilon < 1/n$. Then the following are equivalent for all $\boldsymbol{y}$:*

(1) $(\boldsymbol{y}, \boldsymbol{y})$ is a symmetric ε-well-supported Nash equilibrium of the symmetric game $(A, A^\top)$,
(2) there exists $\boldsymbol{x}$ such that $(\boldsymbol{x}, \boldsymbol{y})$ is an ε-well-supported Nash equilibrium of the imitation game (A, I).

Proof. First, assume that $(\boldsymbol{y}, \boldsymbol{y})$ is a symmetric ε-well-supported Nash equilibrium of the symmetric game $(A, A^\top)$. Then, we have that for all $i \in [n]$

$$y_i > 0 \implies (A\boldsymbol{y})_i \geq \max_j (A\boldsymbol{y})_j - \varepsilon.$$

Now, define strategy $\boldsymbol{x}$ to be the uniform distribution on $S := \{i \in [n] : y_i > 0\}$. Then, we have that for all $i \in [n]$

$$x_i > 0 \implies y_i > 0 \implies (A\boldsymbol{y})_i \geq \max_j (A\boldsymbol{y})_j - \varepsilon$$

and

$$y_i > 0 \implies x_i = 1/|S| = \max_j x_j \implies (I\boldsymbol{x})_i = \max_j (I\boldsymbol{x})_j$$

i.e., $(\boldsymbol{x}, \boldsymbol{y})$ is an ε-well-supported Nash equilibrium of the imitation game (A, I).

Conversely, assume that there exists $\boldsymbol{x}$ such that $(\boldsymbol{x}, \boldsymbol{y})$ is an ε-well-supported Nash equilibrium of the imitation game (A, I). Then, we have that for all $i \in [n]$

$$y_i > 0 \implies (I\boldsymbol{x})_i \geq \max_j (I\boldsymbol{x})_j - \varepsilon \implies x_i \geq \max_j x_j - \varepsilon \geq 1/n - \varepsilon > 0$$

where we used the fact that $\max_j x_j \geq 1/n$, since $\boldsymbol{x}$ is a probability distribution. Thus, we obtain

$$y_i > 0 \implies x_i > 0 \implies (A\boldsymbol{y})_i \geq \max_j (A\boldsymbol{y})_j - \varepsilon$$

which means that $(\boldsymbol{y}, \boldsymbol{y})$ is a symmetric ε-well-supported Nash equilibrium of the symmetric game $(A, A^\top)$.

4 CLS-Hardness of the KKT Problem for QPs on the Simplex

We show that computing approximate KKT points for simplex-constrained QPs is CLS-complete through a reduction from the same computational problem for box-constrained QPs.

Theorem 3. *The problem of computing an ε-KKT point of a quadratic program with a simplex constraint and no first-order terms in the objective function*

$$\min_{\boldsymbol{x}} \quad \frac{1}{2}\boldsymbol{x}^\intercal A\boldsymbol{x} \tag{17}$$

$$\textit{subject to} \quad \sum_i x_i = 1, \tag{18}$$

$$x_i \geq 0, \qquad \textit{for} \, i \in [n], \tag{19}$$

where $A = (a_{ij}) \in \mathbb{R}^{n\times n}$ is symmetric without loss of generality, $a_{ij} = a_{ji}$, is CLS-complete.

Proof. Notice that the simplex-constrained QP in the theorem statement does not have any first-order terms of the form $b_i x_i$. This is without loss of generality because $b_i x_i$ can be replaced by $b_i x_i(\sum_j x_j)$ to make all terms second-order (as $\sum_j x_j = 1$). So, the two formulations are equivalent. Similarly, the version where the matrix A is not symmetric is equivalent to the version where it is symmetric, since we can just replace A by $(A + A^\intercal)/2$, without changing the objective function.

Further, as discussed in Sect. 2.2, a scaled simplex constraint of the form $\sum_j x_j = s$ for some $s > 0$ instead of $\sum_j x_j = 1$ also keeps the problem effectively unchanged (by suitable change of variables). So, for convenience, we will use QPs that include first-order terms and have scaled simplex constraints in our reduction.

We start with an arbitrary instance of a QP with box constraints; we assume that $A = (a_{ij}) \in \mathbb{R}^{n\times n}$ and $\boldsymbol{b} = (b_i) \in \mathbb{R}^n$ are the coefficients of this box-constrained QP, the same notation as (9). We want to compute an ε-approximate KKT point of this QP. We reduce this to the problem of computing a δ-approximate KKT point of a polynomially larger simplex-constrained QP and where $\delta = \Theta(\varepsilon)$.

Let $M = \max(1, \max_i(|b_i| + \sum_j |a_{ij}|))$ and $\delta = \varepsilon/(4 + 4nM)$. We construct the following QP with $2n + 1$ variables $\boldsymbol{x} = (x_1, \ldots, x_n)$, $\boldsymbol{y} = (y_1, \ldots, y_n)$, and z:

$$\min_{\boldsymbol{x},\boldsymbol{y},z} \quad \sum_{i<j} a_{ij}x_i x_j + \sum_i \frac{a_{ii}}{2}x_i^2 + \sum_i b_i x_i + \frac{M}{2\delta}\sum_i (x_i + y_i - 1)^2 \tag{20}$$

$$\text{subject to} \quad \sum_i x_i + \sum_i y_i + z = 2n, \tag{21}$$

$$x_i, y_i, z \geq 0, \qquad \text{for} \, i \in [n]. \tag{22}$$

The coefficients a_{ij} and b_i are the same ones as the original box-constrained QP. The intuition for this constructed simplex-constrained QP is as follows: The x_i variables will approximately correspond to the original n variables of the box-constrained QP. The $(x_i + y_i - 1)^2$ term in the objective tries to push the variables towards satisfying $x_i + y_i = 1$. The $y_i \geq 0$ variable occurs only in this term and allows x_i to take values ≤ 1. So, overall the $(x_i + y_i - 1)^2$ term acts as a soft $x_i \leq 1$ constraint. The variable z is to satisfy the simplex condition if $\sum_i x_i + \sum_i y_i < 2n$.

A δ-approximate KKT point $(\boldsymbol{x}, \boldsymbol{y}, z)$ of the above QP (20)–(22) satisfies the following conditions:

$$0 \geq u - \delta, \qquad \text{if} z = 0, \tag{23}$$

$$0 \in [u - \delta, u + \delta], \qquad \text{if} z > 0, \tag{24}$$

$$\frac{M}{\delta}(x_i + y_i - 1) \geq u - \delta, \qquad \text{if } y_i = 0, \tag{25}$$

$$\frac{M}{\delta}(x_i + y_i - 1) \in [u - \delta, u + \delta], \qquad \text{if } y_i > 0, \tag{26}$$

$$\sum_j a_{ij}x_j + b_i + \frac{M}{\delta}(x_i + y_i - 1) \geq u - \delta, \qquad \text{if } x_i = 0, \tag{27}$$

$$\sum_j a_{ij}x_j + b_i + \frac{M}{\delta}(x_i + y_i - 1) \in [u - \delta, u + \delta], \qquad \text{if } x_i > 0, \tag{28}$$

for all $i \in [n]$ and where $u \in \mathbb{R}$.

Lemma 4. *$x_i + y_i < 2$ for all $i \in [n]$, which also implies $z > 0$ and $u \in [-\delta, \delta]$.*

Proof. Irrespective of whether $z = 0$ or $z > 0$, from (23) and (24), we have $u \leq \delta$. For contradiction, let $x_i + y_i \geq 2$ for some $i \in [n]$. Then either $x_i > 0$ or $y_i > 0$. If $y_i > 0$, then using (26), we have

$$\frac{M}{\delta}(x_i + y_i - 1) \leq u + \delta \implies \frac{M}{\delta} \leq 2\delta \implies \delta^2 \geq \frac{M}{2},$$

which is a contradiction because $\delta \leq 1/4$ and $M \geq 1$ by definition. Similarly, if $x_i > 0$, then using (28), we have

$$\begin{aligned}
&\sum_j a_{ij}x_j + b_i + \frac{M}{\delta}(x_i + y_i - 1) \leq u + \delta \\
&\implies \sum_j (-|a_{ij}|)x_j - |b_i| + \frac{M}{\delta} \leq 2\delta \quad \text{because} u \leq \delta, x_i + y_i \geq 2, \text{and} x_j \geq 0, \\
&\implies -2nM + \frac{M}{\delta} \leq 2\delta \qquad \text{because} x_j \leq 2n forall j, \\
&\implies M \leq 2\delta^2 + \delta 2nM \leq \frac{1}{8} + \frac{1}{2} \qquad \text{because} \delta \leq \frac{1}{4 + 4nM},
\end{aligned}$$

which is a contradiction because $M \geq 1$. So, $x_i + y_i < 2$ for all $i \in [n]$, which implies that $\sum_i (x_i + y_i) < 2n \implies z > 0$ using (21), which implies $u \in [-\delta, \delta]$ using (24).

From Lemma 4, we know that $u \in [-\delta, \delta]$, using this to get rid of u in (25)–(28), we get the following conditions

$$\frac{M}{\delta}(x_i + y_i - 1) \geq -2\delta, \qquad \text{if } y_i = 0, \tag{29}$$

$$\frac{M}{\delta}(x_i + y_i - 1) \in [-2\delta, 2\delta], \qquad \text{if } y_i > 0, \tag{30}$$

$$\sum_j a_{ij}x_j + b_i + \frac{M}{\delta}(x_i + y_i - 1) \geq -2\delta, \qquad \text{if } x_i = 0, \tag{31}$$

$$\sum_j a_{ij}x_j + b_i + \frac{M}{\delta}(x_i + y_i - 1) \in [-2\delta, 2\delta], \qquad \text{if } x_i > 0. \tag{32}$$

We now try to get rid of the dependency on y_i in the KKT conditions (29)–(32), albeit making these inequalities slightly weaker.

Lemma 5. *For all $i \in [n]$, we have*

$$\sum_j a_{ij}x_j + b_i \geq -4\delta, \qquad \textit{if} x_i = 0, \tag{33}$$

$$\sum_j a_{ij}x_j + b_i \in [-4\delta, 4\delta], \qquad \textit{if } x_i \in (0,1), \tag{34}$$

$$\sum_j a_{ij}x_j + b_i + \frac{M}{\delta}(x_i - 1) \leq 2\delta, \qquad \textit{if } x_i \geq 1. \tag{35}$$

Proof. We consider the following two cases:

1. $x_i + y_i < 1$. As $x_i + y_i - 1 < 0$, we have $\frac{M}{\delta}(x_i + y_i - 1) < 0$. Plugging this in (31) and (32), we have for all $x_i \geq 0$ (and $x_i < 1$ by the case assumption)

$$\sum_j a_{ij}x_j + b_i + \frac{M}{\delta}(x_i + y_i - 1) \geq -2\delta \implies \sum_j a_{ij}x_j + b_i \geq -2\delta.$$

   Further, from (29) and (30), we know that $\frac{M}{\delta}(x_i+y_i-1) \geq -2\delta$ for all $y_i \geq 0$. Plugging this into (32), if $x_i > 0$, we have

$$\sum_j a_{ij}x_j + b_i + \frac{M}{\delta}(x_i + y_i - 1) \leq 2\delta$$

$$\implies \sum_j a_{ij}x_j + b_i \leq 2\delta - \frac{M}{\delta}(x_i + y_i - 1) \leq 4\delta.$$

2. $x_i + y_i \geq 1$. If $y_i = 0$, we know that $x_i \geq 1$, and (32) reduces to

$$\sum_j a_{ij}x_j + b_i + \frac{M}{\delta}(x_i - 1) \in [-2\delta, 2\delta].$$

If $y_i > 0$, from (30) we have $\frac{M}{\delta}(x_i + y_i - 1) \in [-2\delta, 2\delta]$. Plugging this into (31) and (32), we get

$$\sum_j a_{ij}x_j + b_i \geq -4\delta, \qquad \text{if } x_i = 0, \tag{36}$$

$$\sum_j a_{ij}x_j + b_i \in [-4\delta, 4\delta], \qquad \text{if } x_i > 0. \tag{37}$$

Further, as $y_i > 0$, from (32) we also trivially have for $x_i \geq 1$

$$\begin{aligned}\sum_j a_{ij}x_j + b_i + \frac{M}{\delta}(x_i - 1) &= \sum_j a_{ij}x_j + b_i + \frac{M}{\delta}(x_i + y_i - 1) - \frac{M}{\delta}y_i \\ &\leq 2\delta - \frac{M}{\delta}y_i \leq 2\delta.\end{aligned}$$

We propose $\boldsymbol{x}' = (x_1', \ldots, x_n')$, where $x_i' = \min(1, x_i)$, as an ε-approximate KKT point of the original box-constrained QP.

If $x_i \leq 1$ for all i, then from Lemma 5, we are already at a $4\delta \leq \varepsilon$ approximate KKT point.

Let us now consider the case where $x_i > 1$ for some i. As $x_i > 1$, from Lemma 5, we have

$$\begin{aligned}2\delta &\geq \sum_j a_{ij}x_j + b_i + \frac{M}{\delta}(x_i - 1) \geq -\sum_j |a_{ij}|x_j - |b_i| + \frac{M}{\delta}(x_i - 1) \\ &\geq -\sum_j 2|a_{ij}| - |b_i| + \frac{M}{\delta}(x_i - 1) \qquad \text{as } x_j < 2 \text{ from Lemma 4} \\ &\geq -2M + \frac{M}{\delta}(x_i - 1) \\ \implies x_i &\leq 1 + 2\delta\left(1 + \frac{\delta}{M}\right) \leq 1 + 4\delta.\end{aligned}$$

This implies that for all i, $x_i' \in [x_i - 4\delta, x_i]$. So, for all i, $\sum_j a_{ij}(x_j - x_j') \leq \sum_j |a_{ij}||x_j - x_j'| \leq 4\delta M$. Similarly, for all i, $\sum_j a_{ij}(x_j - x_j') \geq -4\delta M$. Putting this together with (33)–(35) of Lemma 5, we get

$$\sum_j a_{ij}x_j' + b_i \geq \sum_j a_{ij}x_j - 4\delta M + b_i \geq -4\delta(1 + M), \qquad \text{if } x_i' = x_i = 0,$$

$$\sum_j a_{ij}x_j' + b_i \in \sum_j a_{ij}x_j + [-4\delta M, 4\delta M] + b_i \in [-4\delta(1 + M), 4\delta(1 + M)],$$
$$\text{if } x_i' = x_i \in (0, 1),$$

$$\sum_j a_{ij}x_j' + b_i \leq \sum_j a_{ij}x_j + 4\delta M + b_i + \frac{M}{\delta}(x_i - 1) \leq 4\delta(1 + M),$$
$$\text{if } x_i' = 1 \iff x_i \geq 1,$$

where on the last line we also used the fact that $\frac{M}{\delta}(x'_i - 1) = 0$ when $x'_i = 1$ and $\frac{M}{\delta}(x_i - 1) \geq 0$ when $x_i \geq 1$. As $4\delta(1 + M) \leq \varepsilon$, $\boldsymbol{x}'$ satisfies the required box-constraint KKT conditions, and we are done.

5 Open Problems

Although the computation of equilibria in general bimatrix games is known to be PPAD-complete, there are some interesting special cases for which the complexity remains open. For example, rank-1 bimatrix games are easy [1], and rank-3 games are hard [23], but the complexity of rank-2 games remains open. In another direction, it is known that *win-lose* games, where all payoffs are 0 or 1, are hard [4]. Separately, it is also known that *sparse* bimatrix games, where the number of non-zero entries in each row and column is constant, are also hard [4]. However, the complexity of bimatrix games that are both sparse and win-lose remains open [20]. In a certain sense, these are the simplest bimatrix games, and understanding their complexity is an intriguing question.

Acknowledgments. The authors wish to thank the reviewers for suggestions that helped improve the presentation of the paper. A.G. was supported by EPSRC Grant EP/X040461/1 "Optimisation for Game Theory and Machine Learning". A.H. thanks Koosha Samieefar for pointing out the connection with KKT points of quadratic programming shown in [21].

References

1. Adsul, B., Garg, J., Mehta, R., Sohoni, M.: Rank-1 bimatrix games: a homeomorphism and a polynomial time algorithm. In: Proceedings of the 43rd ACM Symposium on Theory of Computing (STOC), pp. 195–204 (2011). https://doi.org/10.1145/1993636.1993664
2. Babichenko, Y., Rubinstein, A.: Settling the complexity of Nash equilibrium in congestion games. In: Proceedings of the 53rd ACM Symposium on Theory of Computing (STOC), pp. 1426–1437 (2021). https://doi.org/10.1145/3406325.3451039
3. Bitansky, N., Paneth, O., Rosen, A.: On the cryptographic hardness of finding a nash equilibrium. In: Proceedings of the 56th Symposium on Foundations of Computer Science (FOCS), pp. 1480–1498 (2015). https://doi.org/10.1109/focs.2015.94
4. Chen, X., Deng, X., Teng, S.H.: Settling the complexity of computing two-player Nash equilibria. J. ACM **56**(3), 14:1–14:57 (2009). https://doi.org/10.1145/1516512.1516516
5. Choudhuri, A.R., Hubáček, P., Kamath, C., Pietrzak, K., Rosen, A., Rothblum, G.N.: Finding a nash equilibrium is no easier than breaking Fiat-Shamir. In: Proceedings of the 51st ACM Symposium on Theory of Computing (STOC), pp. 1103–1114 (2019). https://doi.org/10.1145/3313276.3316400
6. Codenotti, B., Štefankovič, D.: On the computational complexity of Nash equilibria for bimatrix games. Inf. Process. Lett. **94**(3), 145–150 (2005). https://doi.org/10.1016/j.ipl.2005.01.010

7. Conitzer, V., Sandholm, T.: New complexity results about Nash equilibria. Games Econom. Behav. **63**(2), 621–641 (2008). https://doi.org/10.1016/j.geb.2008.02.015
8. Daskalakis, C., Goldberg, P.W., Papadimitriou, C.H.: The complexity of computing a nash equilibrium. SIAM J. Comput. **39**(1), 195–259 (2009). https://doi.org/10.1137/070699652
9. Daskalakis, C., Papadimitriou, C.: Continuous local search. In: Proceedings of the 22nd ACM-SIAM Symposium on Discrete Algorithms (SODA), pp. 790–804 (2011). https://doi.org/10.1137/1.9781611973082.62
10. Emmons, S., Oesterheld, C., Critch, A., Conitzer, V., Russell, S.: For learning in symmetric teams, local optima are global Nash equilibria. In: Proceedings of the 39th International Conference on Machine Learning (ICML). pp. 5924–5943 (2022). https://proceedings.mlr.press/v162/emmons22a.html
11. Etessami, K., Yannakakis, M.: On the complexity of nash equilibria and other fixed points. SIAM J. Comput. **39**(6), 2531–2597 (2010). https://doi.org/10.1137/080720826
12. Fabrikant, A., Papadimitriou, C., Talwar, K.: The complexity of pure Nash equilibria. In: Proceedings of the 36th Annual ACM Symposium on Theory of Computing (STOC), pp. 604–612. ACM (2004). https://doi.org/10.1145/1007352.1007445
13. Fearnley, J., Goldberg, P., Hollender, A., Savani, R.: The complexity of gradient descent: CLS = PPAD ∩ PLS. J. ACM **70**(1), 7:1–7:74 (2022). https://doi.org/10.1145/3568163
14. Fearnley, J., Goldberg, P.W., Hollender, A., Savani, R.: The complexity of computing KKT solutions of quadratic programs. In: Proceedings of the 56th ACM Symposium on Theory of Computing (STOC), pp. 892–903 (2024). https://doi.org/10.1145/3618260.3649647
15. Gilboa, I., Zemel, E.: Nash and correlated equilibria: Some complexity considerations. Games Econom. Behav. **1**(1), 80–93 (1989). https://doi.org/10.1016/0899-8256(89)90006-7
16. Jawale, R., Kalai, Y.T., Khurana, D., Zhang, R.: SNARGs for bounded depth computations and PPAD hardness from sub-exponential LWE. In: Proceedings of the 53rd ACM Symposium on Theory of Computing (STOC), pp. 708–721 (2021). https://doi.org/10.1145/3406325.3451055
17. Johnson, D.S., Papadimitriou, C.H., Yannakakis, M.: How easy is local search? J. Comput. Syst. Sci. **37**(1), 79–100 (1988). https://doi.org/10.1016/0022-0000(88)90046-3
18. Kontogiannis, S., Spirakis, P.: Approximability of symmetric bimatrix games and related experiments. In: Proceedings of the 10th International Symposium on Experimental Algorithms, pp. 1–20 (2011). https://doi.org/10.1007/978-3-642-20662-7_1
19. Krentel, M.W.: Structure in locally optimal solutions. In: Proceedings of the 30th Annual Symposium on Foundations of Computer Science (FOCS), pp. 216–221 (1989). https://doi.org/10.1109/SFCS.1989.63481
20. Liu, Z., Sheng, Y.: On the approximation of Nash equilibria in sparse win-lose games. Proceedings of the 32nd AAAI Conference on Artificial Intelligence (AAAI), pp. 1154–1160 (2018). https://doi.org/10.1609/aaai.v32i1.11439
21. McLennan, A., Tourky, R.: Simple complexity from imitation games. Games Econom. Behav. **68**(2), 683–688 (2010). https://doi.org/10.1016/j.geb.2009.10.003
22. Megiddo, N., Papadimitriou, C.H.: On total functions, existence theorems and computational complexity. Theoret. Comput. Sci. **81**(2), 317–324 (1991). https://doi.org/10.1016/0304-3975(91)90200-L

23. Mehta, R.: Constant rank two-player games are PPAD-hard. SIAM J. Comput. **47**(5), 1858–1887 (2018). https://doi.org/10.1137/15m1032338
24. Morioka, T.: Classification of search problems and their definability in bounded arithmetic. Master's thesis, University of Toronto (2001). https://www.collectionscanada.ca/obj/s4/f2/dsk3/ftp04/MQ58775.pdf
25. Murhekar, A., Mehta, R.: Approximate Nash equilibria of imitation games: algorithms and complexity. In: Proceedings of the 19th International Conference on Autonomous Agents and Multiagent Systems (AAMAS), pp. 887–894 (2020). https://doi.org/10.5555/3398761.3398865
26. Nash, J.: Non-cooperative games. The Annals of Math. **54**(2), 286–295 (1951). https://doi.org/10.2307/1969529
27. Nash, J.F.: Equilibrium points in n-person games. Proc. Natl. Acad. Sci. **36**(1), 48–49 (1950). https://doi.org/10.1073/pnas.36.1.48
28. von Neumann, J.: Zur Theorie der Gesellschaftsspiele. Mathematische Annalen **100**(1), 295–320 (1928). https://doi.org/10.1007/bf01448847
29. Papadimitriou, C.H.: On the complexity of the parity argument and other inefficient proofs of existence. J. Comput. Syst. Sci. **48**(3), 498–532 (1994). https://doi.org/10.1016/S0022-0000(05)80063-7
30. Papadimitriou, C.H.: The complexity of finding Nash equilibria. In: Nisan, N., Roughgarden, T., Tardos, É., Vazirani, V.V. (eds.) Algorithmic Game Theory, pp. 29–52. Cambridge University Press (2007). https://doi.org/10.1017/cbo9780511800481.004
31. Schäffer, A.A., Yannakakis, M.: Simple local search problems that are hard to solve. SIAM J. Comput. **20**(1), 56–87 (1991). https://doi.org/10.1137/0220004

Logarithmic Comparison-Based Query Complexity for Fair Division of Indivisible Goods

Xiaolin Bu[1], Zihao Li[2], Shengxin Liu[3](✉), Jiaxin Song[1], and Biaoshuai Tao[1]

[1] Shanghai Jiao Tong University, Shanghai, China
{lin_bu,sjtu_xiaosong,bstao}@sjtu.edu.cn
[2] Nanyang Technological University, Singapore, Singapore
zihao004@e.ntu.edu.sg
[3] Harbin Institute of Technology Shenzhen, Shenzhen, China
sxliu@hit.edu.cn

Abstract. We study the problem of fairly allocating m indivisible goods to n agents, where agents may have different preferences over the goods. In the traditional setting, agents' valuations are provided as inputs to the algorithm. In this paper, we adopt the query model, which has been widely considered for other similar problems (such as matching [28], graph isomorphism [30], and equilibrium in game [7]), to our fair division problem. In particular, we consider a new *comparison-based query model* where the algorithm presents two bundles of goods to an agent and the agent responds by telling the algorithm which bundle she prefers. We investigate the query complexity for computing allocations with several fairness notions including *proportionality up to one good* (PROP1), *envy-freeness up to one good* (EF1), and *maximin share* (MMS). Our main result is an algorithm that computes an allocation that satisfies both PROP1 and $\frac{1}{2}$-MMS within $O(\log m)$ queries with a constant number of n agents. For identical and additive valuation, we present an algorithm for computing an EF1 allocation within $O(\log m)$ queries with a constant number of n agents. To complement the positive results, we show that the lower bound of the query complexity for any of the three fairness notions is $\Omega(\log m)$ even with two agents.

Keywords: Fair Division · Comparison-based Query · Indivisible Goods

1 Introduction

Fair division of indivisible resources concerns how to fairly allocate a set of heterogeneous indivisible goods/items to a set of agents with varying preferences for these resources [2]. The concept of the fair division was initially introduced

Shengxin Liu and Biaoshuai Tao are corresponding authors. The full version of the paper can be found in [14].

M. Mavronicolas et al. (Eds.): WINE 2024, LNCS 15534, pp. 348–365, 2026.
https://doi.org/10.1007/978-3-032-08560-3_20

by Steinhaus [37,38] and has been extensively studied in the fields of economics, computer science, and social science. The primary objective of fair division is to achieve *fairness* given that agents have different preferences over the resources.

Query-Based Fair Allocation. In most existing literature, agents' preferences are typically represented by a set of *valuation functions*, each of which assigns a non-negative value to every item subset, reflecting how much an agent likes it.In the classical setting [1,13,15,24,34], a fair division algorithm has explicit knowledge of each agent's valuation function: the valuation functions are provided as inputs to the algorithm. This is called *the direct revelation.*

The direct revelation simplifies the burden of theoretical analysis. However, many real-world scenarios may not fit this neat model since the valuation functions may be unavailable upfront. For example, to enhance user experience, many platforms adopt indirect and gentle approaches to understand user's preferences. Ridesharing platforms (Lyft and Uber) collect users' ratings after each ride to gain insights into users' preferences for different types of cars. Similarly, online homestay platforms (Ctrip and Airbnb) employ a recommendation system that suggests rooms on their main page. They learn about users' preferences by analyzing the history of clicks and interactions.

To model these scenarios, a line of works [27,29,31] assumes the allocation algorithm does not know the preferences explicitly. Instead, a *value-based query* is adopted to access an agent's value for a specific bundle. This line of research focuses on optimizing the query complexity to compute a fair allocation.

Comparison-based Queries. In this paper, we propose a new *comparison-based query model.* In this model, the agents retain their *intrinsic* valuation functions. However, instead of requiring them to disclose their valuation functions directly or querying their explicit value for specific bundles, we can only learn about each agent's preferences by querying for her value relationship among different bundles. In each comparison-based query, we present two bundles to an agent, and then let her specify which bundle she prefers. Under this model, we study the problem of optimizing query complexity to obtain a fair allocation.

Compared to the value-based query model, the comparison-based query model has several advantages in many practical scenarios. It is much more convenient for agents to report their ordinal preferences than values. When providing feedback or expressing preferences, most users tend to prefer a simplified process. The comparison-based query enables users to only answer "yes" or "no" without the need to provide concrete values. It also protects agents' privacy to a large extent. From the perspective of information theory, each response conveys only one bit of information, which is minimal. Our algorithmic results based on this new query model achieve fair allocations while making agents reveal as little information as possible.

Fairness Criteria. Various notions of fairness have been proposed over the years. Among them, *envy-freeness* (EF) [23,41] and *proportionality* (PROP) [38] are the most natural ones, stating respectively that no agent envies another and each agent receives at least an average share of the total value.

Unfortunately, such allocations may not always exist for indivisible items (e.g., allocating one unit item to two agents). The arguably most prominent relaxation of envy-freeness is *envy-freeness up to one item* (EF1) [16,27], which states that, for any pairs of agents i and j, i will not envy j if one item is (hypothetically) removed from j's bundle. As for proportionality, two different relaxations have been proposed in the past literature: *proportionality up to one item* (PROP1) [3,18] and *maximin share* (MMS) [15,34]. Similar to EF1, PROP1 requires each agent i's proportionality if one item is (hypothetically) added to i's bundle. For MMS, each agent i maintains a threshold MMS_i that is defined by the maximum value of the least preferred bundles among all n-partition of the goods (n is the number of the agents). An allocation is MMS if each agent's received value is at least this threshold.

For additive valuations, EF1 and PROP1 allocations are known to always exist [15,27], and an MMS allocation may not exist [34]. Therefore, researchers look for the approximation version of MMS: α-MMS, indicating each agent's received value is at least $\alpha\cdot$ MMS_i. The current best-known result is that an α-MMS allocation always exists for constant $\alpha = 3/4 + 3/3836$ [1]. Although an MMS allocation always satisfies PROP1 for additive valuations, there exists a constant $\alpha < 1$ such that PROP1 and α-MMS are incomparable. When the goods have similar values, PROP1 is close to PROP, whereas α-MMS, being close to α-proportionality, is a weaker criterion. On the other hand, when there is a good with more than $1/n$ of the value of the entire bundle set, an allocation where an agent receives an empty bundle can be PROP1; in this case, α-MMS provides a much better fairness guarantee.

In this paper, we design an algorithm that finds allocations that simultaneously satisfy PROP1 and $\frac{1}{2}$-MMS.

Sublinear Complexity in the Number of Goods. A challenge in many typical fair division models is that the number of goods m may be extremely large (e.g., computational resources, housing resources, transportation resources, etc.). Our comparison-based query model offers a solution for handling such cases. We seek algorithms with fewer queries sublinear in the number of items m. Note that direct adaptations of existing algorithms [27,29] require the query numbers linear in m to compute an EF1 allocation.

1.1 Our Results and Technical Overviews

In this paper, we focus on additive valuations and deterministic algorithms. As the main result of the paper, we propose an algorithm that uses $O(n^4 \log m)$ comparison-based queries to compute an allocation that is simultaneously PROP1 and $\frac{1}{2}$-MMS. For EF1, we present algorithms that use $O(\log m)$ queries for up to three agents, and $O(n^2 \log m)$ queries for n agents with identical valuations. The results are complemented by the fact that $\Omega(\log m)$ queries are necessary to achieve any of the three mentioned fairness notions: PROP1, EF1, and α-MMS (for any $\alpha > 0$), which holds even for any fixed constant number of agents with identical valuations. This matches the query complexity of our algorithms for constant numbers of agents.

Our main result is built upon three techniques: the subroutine `MinBundleKept`, the subroutine `PROP1-PROP`, and the *Hall matching*. The `MinBundleKept` subroutine aims to find a partial allocation that is approximately balanced with most of the goods allocated. It iteratively finds a tentative n-partition, finds the bundle with the minimum value by comparison-based queries, and keeps this bundle. This subroutine is the key building block in all our algorithms. By this subroutine, we can design algorithms for computing PROP1 and EF1 allocations when agents' valuations are identical.

Roughly speaking, the `PROP1-PROP` subroutine decides if a bundle approximately meets the proportional requirement of an agent. Specifically, as discussed at the beginning of Sect. 3, it is impossible to check with comparison-based queries whether an allocation satisfies the PROP1 condition for an agent or whether a bundle is worth at least $1/n$ of this agent's value on the entire item set (it is perhaps interesting to see that computing a PROP1 allocation is possible while checking if an allocation is PROP1 is impossible). The `PROP1-PROP` subroutine takes a bundle and an agent as inputs, and outputs "yes" if the bundle satisfies the PROP1 requirement for the agent, and "no" if the bundle fails to achieve PROP. Notice that a bundle may satisfy both "yes" and "no" conditions; in this case, both outputs "yes" and "no" are acceptable.

With the subroutine `PROP1-PROP`, given n bundles, we can construct a bipartite graph $G = (A, B, E)$ where A contains n vertices representing the agents, B contains n vertices representing the bundles, and an edge indicates a bundle surely satisfies the PROP1 requirement for an agent. To find a PROP1 and $\frac{1}{2}$-MMS allocation, we iteratively find a matching from $X \subseteq A$ to $Y \subseteq B$ such that there is no edge from an agent in $A \setminus X$ to a bundle in B, and allocate the bundles in Y to the agents in X. This property guarantees that the agents in $A \setminus X$ believe the remaining unallocated items (the union of the bundles in $B \setminus Y$) are still worthy enough so that they will receive a bundle satisfying the PROP1 and $\frac{1}{2}$-MMS conditions in the future. Such matching can be found by removing all the agents in A that fail the Hall condition $|\Gamma(S)| \geq |S|$, and we call the matching from X to Y a *Hall matching*.

The relationship between our techniques and our results is shown in our full version [14].In Sect. 3, we present an algorithm with $O(n^4 \log m)$ queries to find a PROP1 allocation.In Sect. 4, we discuss our results on EF1 allocations.Finally, in Sect. 5, we present our main result, an algorithm with $O(n^4 \log m)$ queries that finds an allocation simultaneously satisfying PROP1 and $\frac{1}{2}$-MMS.Note that although our result of PROP1 in Sect. 3 is weaker than our main result in Sect. 5 and the latter is not built upon the former, we discuss the PROP1 algorithm first because it illustrates the application of our three techniques in a simpler way.

1.2 Related Work

Value-based Query Complexity for Indivisible Items. There have been several studies on value-based query complexity for computing fair allocations of indivisible items [29,31], which allows the algorithm to acquire agent i's specific value for a bundle X of items, i.e., $u_i(X)$. Observe that each comparison-based query can be simulated by two value-based queries, which implies that the lower bound of comparison-based query complexity is no less than that of value-based query complexity asymptotically. For value-based queries, Plaut and Roughgarden [31] showed that, even for two agents with identical and monotone valuations, the query complexity of computing an EFX (envy-free up to any item) allocation is exponential in terms of the number m of indivisible items. Oh et al. [29] explored the value-based query complexity for more fairness notions. For EFX with two agents, Oh et al. [29] provided a lower bound of $\Omega(m)$ and a matching upper bound of $O(m)$ queries for additive valuations. For EF1 with no more than three agents, the query complexity is $\Theta(\log m)$ for additive valuations and $O(\log^2 m)$ for monotone valuations. With more than three agents, a careful analysis of the classic *envy-graph* procedure is provided, which requires $O(nm)$ queries in the worst case.

Communication Complexity. When there is no restriction on the queries, the communication complexity (i.e., the number of bits exchanged during interaction) of finding a fair allocation has also attracted great interest in existing papers [11,22,32]. Plaut and Roughgarden [32] considered the communication complexity for α-EF/α-PROP beyond additive valuations. Our primary positive result directly implies that the communication complexity of finding a PROP1 and $\frac{1}{2}$-MMS allocation requires at most $O(n^4 \log m)$ bits. Feige's recent work [22] further improved the upper bound of finding a PROP1 allocation to $O(n \log m \log n)$ and showed that this bound also applies for finding an APROP (PROP1 and $\frac{n}{2n-1}$-TPS) allocation. However, the above results cannot yield a better upper bound of the query complexity for finding a PROP1+$\frac{1}{2}$-MMS allocation within the comparison-based query model. Specifically, the former result requires each agent to compute a PROP1 allocation offline according to her own valuation function (thus no information exchange is needed), whose complexity is not included in the communication model yet is considered in our query model, and the latter result requires value-based queries. Feige [22] also provided a lower bound $\Omega(n \log \frac{m}{n})$ of the communication complexity (even for randomized protocols) for finding a PROP1 allocation, which implies the lower bound under our query model for a constant number of agents.

Query Complexity for Divisible Items. In the context of fair division of *divisible* resources (i.e., cake-cutting), a valuation function can be a general real function on an interval, which makes it impossible to be succinctly encoded. Thus, the query complexity has been extensively studied in the cake-cutting literature [5,6,10,12,17,26,35,36,40]. The most well-known query model is the Robertson-Webb (RW) query model [36], where two types of queries, namely **Eval** and **Cut**, are allowed. In particular, $\mathbf{Eval}_i(x, y)$ asks agent i the value of interval $[x, y]$ and $\mathbf{Cut}_i(x, r)$ asks agent i for a point y such that agent i's value to interval $[x, y]$ is r. The moving-knife [19] and Even-Paz [21] algorithms compute a proportional allocation with connected pieces within the complexity of $O(n^2)$ and $O(n \log n)$ RW queries respectively. On the other hand, a lower bound of

$\Omega(n \log n)$ RW queries is given by Edmonds and Pruhs [20]. Computing an envy-free allocation is more challenging. An envy-free allocation can be found by the *cut-and-choose* protocol for two agents. For a general number of agents, Aziz and Mackenzie [5] proposed an algorithm within a finite number of RW queries while the best known lower bound is $\Omega(n^2)$ by Procaccia [33].

Other Aspects. Our problem is also implicitly related to the fair division of indivisible goods using only ordinal or partial information about the preferences (e.g., [4,8,25]). We refer to the excellent survey for an overview of recent progress on fair allocations of indivisible goods [2].

2 Preliminaries

Let $[k]$ denote $\{1, \ldots, k\}$. There are $m \in \mathbb{N}^+$ indivisible goods, denoted by $M = \{g_1, \ldots, g_m\}$, required to be allocated to $n \in \mathbb{N}^+$ agents, denoted by $N = \{1, \ldots, n\}$. A bundle is a subset of M. An allocation $\mathcal{A} = \{A_1, A_2, \ldots, A_n\}$ is a n-partition of the set of goods M, where A_i is the bundle assigned to agent i. We say an allocation is a *partial allocation* if the union of these bundles is a proper set of M (i.e., $\cup_{i=1}^n A_i \subsetneq M$).

Assume each agent $i \in [n]$ has an *intrinsic* and *non-negative* valuation function $u_i : \{0,1\}^{[m]} \to \mathbb{R}_{\geq 0}$. The set of all valuation functions is called value profile $\mathcal{U}$, which is not given as input. To learn about an agent's preference, an algorithm is only allowed to select an agent, present her with two bundles, and ask which one she prefers. Note that even if we could get the comparison result, we still do not know the explicit values of the two bundles. Our query model is referred to as the *comparison-based query model.*

Definition 1. *We adopt the following query in the comparison-based query model:*

- ***Comp***$_i(X, Y)$*: given two bundles* $X, Y \subseteq M$*, let agent* i *compare* $u_i(X)$ *and* $u_i(Y)$*, and return the bundle with a higher value (if* $u_i(X) = u_i(Y)$*, she could return an arbitrary one).*

Throughout this paper, we assume all the valuation functions are *additive*, i.e., $u_i(S) = \sum_{g \in S} u_i(\{g\})$ for any $i \in [n]$ and $S \subseteq M$. For simplicity, we refer to $u_i(\{g\})$ as $u_i(g)$. We say the valuation functions are *identical* if $u_1 = u_2 = \cdots = u_n$.

To measure fairness, the most natural notions are *envy-freeness* (EF) and *proportionality* (PROP). An allocation $\mathcal{A}$ is said to satisfy *envy-freeness (EF)* if no agent prefers the bundle of another agent to her own, i.e., $u_i(A_i) > u_i(A_j)$ for all $i, j \in [n]$ An allocation $\mathcal{A}$ is said to satisfy *proportionality (PROP)* if $u_i(A_i) \geq \frac{1}{n} u_i(M)$ for any agent i.

Since EF and PROP may not always be satisfied, we consider the following relaxations, EF1 and PROP1. For a verbal description, EF1 requires that no agent envies another agent following the removal of some single good from the otherâĂŹs bundle. PROP1 ensures that, for every agent i, there exists an item g such that the total value of agent i's allocation is at least $\frac{1}{n}$ of the overall value (i.e., $u_i(M)$) if g is included in agent i's bundle.

Definition 2. (EF1). *An allocation $\mathcal{A}$ is said to satisfy* envy-freeness up to one item (EF1) *if, for any two agents i and j, either $u_i(A_i) \geq u_i(A_j)$ or there exists an item $g \in A_j$ such that $u_i(A_i) \geq u_i(A_j \setminus \{g\})$.*

Definition 3. (PROP1). *An allocation $\mathcal{A}$ is said to satisfy* proportionality up to one item (PROP1)*, if for any agent i, there exists an item $g \in M \setminus A_i$ such that $u_i(A_i \cup \{g\}) \geq u_i(M)/n$.*

To find an EF1 allocation for monotonic valuation functions, Lipton et al. [27] proposed *envy-graph*, where each vertex represents an agent and there is a directed edge from vertex i to vertex j if agent i envies agent j. The *envy-graph procedure* starts from an empty allocation. When there is an unallocated item, it selects a source agent (an agent whose corresponding vertex in the envy-graph has no incoming edges) and assigns that item to her, and then runs the *cycle-elimination* algorithm until there are no cycles in the envy-graph.

Definition 4. (Cycle-elimination). *If there is a cycle $i_1 \rightarrow i_2 \rightarrow \dots i_k \rightarrow i_1$ (also written as i_{k+1}) on the envy-graph, shift the bundles along this path (i.e., agent i_j receives agent i_{j+1}'s bundle for any $j \in [k]$).*

Besides, *maximin share* (MMS) is also extensively studied for indivisible items. The maximin share of agent i, denoted by MMS_i, is defined as the maximum value of the least preferred bundle among all allocations under valuation function u_i. By the definition of MMS, we can observe that $\text{MMS}_i \leq \frac{1}{n}u_i(M)$. An allocation is said to be MMS-fair if the value of the bundle received by each agent i is at least MMS_i, which may not always exist [34]. An allocation is said to be α-maximin share fair (α-MMS) if each agent i receives a bundle with at least α-fractional of their maximin share.

Definition 5. (α-MMS). *An allocation $\mathcal{A}$ satisfies α-maximin share fair (α-MMS), if for any agent i, $u_i(A_i) \geq \alpha \cdot \text{MMS}_i = \alpha \cdot \max_{\mathcal{A} \in \Pi(M)} \min_{j \in [n]} u_i(A_j)$.*

In this paper, we study the problem of optimal query complexity (defined as the *minimum number of comparison-based queries* required) to find a fair allocation. Note that the algorithms in this paper are restricted to be deterministic. We first show a lower bound of the query complexity for finding allocations satisfying the above fairness notions. As illustrated in Sect. 1.2, the effect of a single comparison-based query can be achieved by two valued-based queries. This directly implies that the lower bound for comparison-based query complexity is at least half of the one for the value-based query. We next present a counter-example (using similar ideas of the construction in Theorem 5.5 in Oh et al. [29]) showing the value-based query complexity for PROP1 will be $\Omega(\log m)$, which directly implies Theorem 6. Its proof can be found in our full version [14].

Theorem 6. *The comparison-based query complexity of computing a PROP1/EF1/α-MMS allocation is $\Omega(\log m)$ even for a constant number of agents with identical valuations. Here, α is an arbitrary real number in $(0, 1]$.*

3 Query Complexity of PROP1

In this section, we focus on the query complexity of finding a PROP1 allocation. A straightforward idea is to adapt the moving-knife method [9,15,39], which costs about $O(\text{poly}(n)\cdot \log m)$ queries. However, in our model, it is impossible to determine whether a bundle is PROP1 (or PROP) for an agent. Below we provide two counter-examples.

- $m = n+1$ and for each $i \in [n]$, $u_i(g_1) = \ldots = u_i(g_n) = 1$ and $u_i(g_{n+1}) = x$, where $x \in (1, 2)$. Determining whether an empty bundle $\emptyset$ is PROP1 for an agent i requires judging whether $x \geq \frac{n+x}{n}$, which is equivalent to $x \geq \frac{n}{n-1}$. However, this cannot be achieved by the comparison-based query.
- $n = m = 3$ and $u_i(g_1) = 0, u_i(g_2) = x, u_i(g_3) = 1$, where $0 < x < 1$. To check whether $\{g_2\}$ is PROP, one must figure out whether $x \geq \frac{1}{2}$, which is impossible to achieve by the comparison-based query.

To address this issue, we define a subroutine `PROP1-PROP-Subroutine` in Theorem 8, which allows the overlapping of 'yes' bundles (satisfying PROP1) and 'no' bundles (not satisfying PROP). We show that the subroutine is sufficient for finding a PROP1 for non-identical additive valuations in Theorem 9.

Algorithm 1 considers a special case where the valuations are identically u. The first phase of Algorithm 1 introduces a subroutine called `MinBundleKept`. It begins with an empty allocation $\mathcal{A} = (\emptyset, \ldots, \emptyset)$. Then in each round, it evenly partitions the unallocated items into n bundles $X_1, \ldots, X_n$ and tries to append the X_i to agent i's bundle A_i. For the one holding the minimal utility, it updates its bundle to $A_i \cup X_i$. For others, it sends back X_i to the pool of unallocated items and keeps A_i unchanged. When there are less than n items remaining, `MinBundleKept` may not change the allocation anymore: there may be an agent i such that $A_i \cup X_i$ has the smallest value with $X_i = \emptyset$.

Next, when there are less than n items unallocated, the second phase (Lines 8-14) is conducted. By sorting all the bundles, we assume $u(A_1) \leq \ldots \leq u(A_n)$ without loss of generality. Next, in each round, assume there are r remaining unallocated items $g_1, \ldots, g_r$. We respectively add them to the r bundles $A_1, \ldots, A_r$ with the smallest values. If the smallest bundle is still located in the r bundles, we assign the new item to this bundle, return other new items to the pool, and swap bundles to maintain $u(A_1) \leq \ldots \leq u(A_n)$. Otherwise, if the smallest bundle is no longer among the r bundles, we directly assign each item g_k to bundle A_k for each agent $k \in [r]$ and terminate our algorithm.

The correctness of Algorithm 1 is stated in Theorem 7.

Theorem 7. *For identical and additive valuation, a PROP1 allocation can be found by Algorithm 1 through $O(n^2 \log m)$ queries with running time $O(n^2 \log m)$.*

Here we provide a sketch of proof. More details can be found in our full version [14].For simplification, assume $u(M) = 1$. Our proof is mainly built upon two observations.First, before triggering the "if"-condition at Line 11, $u(A_i) \leq 1/n$

Algorithm 1: Algorithm for computing PROP1 allocations for identical and additive valuation

// Phase I: Subroutine MinBundleKept $([n], M)$:
1–6 [redacted]
7 // Phase II: Allocate the remaining items in P
8–15 [redacted]
16 **return** *the allocation* $\mathcal{A} = (A_1, A_2, \ldots, A_n)$

for any $i \in [n]$.In the first phase, as we append X_i to A_i only if $B_i = A_i \cup X_i$ has the minimum utilities among all newly-created bundles $\{B_k\}_{k\in[n]}$, hence, $u(A_i) \le 1/n$.Afterward, when the "if"-condition is not triggered, the operations after Line 11 also ensure this invariant. Second, at the last round of the first phase, one bundle B_k must exist such that $u(B_k) \ge 1/n$. Based on the two observations, we demonstrate that the "if"-condition at Line 11 must be triggered during the second phase. When the "if"-condition at Line 11 is triggered, according to the first observation, there must exist one bundle A_j satisfying $u(A_j \cup \{g_j\}) \ge 1/n$ with $j \in [r]$. Since $u(A_j \cup \{g_j\}) \ge u(A_i \cup \{g_i\}) \ge u(A_{r+1}) \ge u(A_j)$ ($A_i \cup \{g_i\}$ is defined at Line 10), the property of PROP1 is satisfied.

We then introduce a subroutine PROP1-PROP(n, M, B, u): given a bundle $B \subseteq M$, it verifies whether $u(B) < 1/n$ or $u(B \cup \{g\}) \ge 1/n$ for some item g outside B. To do that, an intuitive approach is to adapt Algorithm 1 with $O(n^2 \log m)$ queries, where the detailed algorithm and proof can be found in our full version [14]. In Algorithm 2, we implement the subroutine with a lower complexity of $O(n \log m)$ (proved in Theorem 8).

Specifically, in Algorithm 2, given a valuation function u and a bundle B, we partition the items into a sequence $B, A_1, I_1, \ldots, A_{n-1}, I_{n-1}, R$ such that each bundle $A_\ell \cup \{I_\ell\}$ is weakly more valuable than B. Some of the latter items and bundles, however, may be empty because of the lack of items. It will not affect the correctness of our algorithm, and we can directly ignore the corresponding

Algorithm 2: PROP1-PROP(n, M, B, u)

Line up the items in the order $(B, M \setminus B)$;
Partition the item sequence into $B, A_1, I_1, A_2, I_2, \dots, A_{n-1}, I_{n-1}, R$ (some of the latter bundles may be empty), such that for each $i \in [n-1]$, $A_i \neq \emptyset$ and $I_i \neq \emptyset$, we have $u(A_i) \leq u(B)$ and $u(A_i \cup \{I_i\}) > u(B)$, where each I_i is a single item and R is the set of remaining items;
if $R = \emptyset$ **return** *yes*;
return *no*;

inequalities for these terms. If there are no remaining items in the sequence, i.e., $R = \emptyset$, this implies PROP1 and the algorithm will output 'yes'. Otherwise, it will output 'no', indicating B is not PROP.

Note that the range of 'yes' and 'no' may overlap. If a bundle satisfies PROP1 and its value is no more than the proportional value, the algorithm may output either 'yes' or 'no'. Its correctness is stated in Theorem 8 and its proof can be found in our full version [14].

Theorem 8. *PROP1-PROP(n, M, B, u) in Algorithm 2 determines if a bundle is PROP1 or not PROP: Given a constant number of agents n, a bundle $B \subseteq M$, for a valuation function u, B is PROP1 if 'yes' is returned and is not PROP if 'no' is returned via at most $O(n \log m)$ queries and the same running time.*

Based on Theorem 7 and PROP1-PROP-Subroutine, for additive valuation functions, we provide Algorithm 3 that is able to find a PROP1 allocation in $O(n^4 \log m)$ queries as stated in Theorem 9, with proof in our full version [14].

Theorem 9. *For non-identical additive valuations, a PROP1 allocation can be found by Algorithm 3 via $O(n^4 \log m)$ queries with running time $O(n^4 \log m)$.*

Algorithm 3 works as follows. We first fix an arbitrary agent k and then apply Algorithm 1 with respect to agent k's valuation function u_k. Let the output partition be $\mathcal{B} = (B_1, \dots, B_n)$, where every bundle is PROP1 to agent k. At Line 5, a bipartite graph is constructed where the two sets of vertices are, respectively, the n agents and the n bundles in $\mathcal{B}$. The edges in the graph are constructed through PROP1-PROP-Subroutine, where if agent i says 'yes' to bundle B_j, we add an undirected edge (i, B_j), and we further add edges between agent k and each bundle. We can observe that an agent will not answer 'no' to all bundles since the value of each bundle cannot be less than $1/n$ at the same time. Hence, the number of created edges will be non-zero.

Then, we find a maximal subset of the agents such that its size is larger than its neighbors' size greedily. Denote the subset by Z and its neighbors by $\Gamma(Z)$, and the complementary of them by $\bar{Z}$ and $\bar{\Gamma}(Z)$ respectively. As agent k is adjacent to all bundles from Line 5, $k \notin Z$. Thus $|\bar{Z}| > 0$. Lemma 10 shows a perfect matching can be found between $\bar{Z} = Q \setminus Z$ and a subset of $\bar{\Gamma}(Z) = \mathcal{B} \setminus \Gamma(Z)$, where $Q = N$ at this point. The proof of Lemma 10 can be found in our full version [14].

Algorithm 3: Algorithm for computing PROP1 allocations for non-identical additive valuations

---

1 Let $A_i \leftarrow \emptyset$ for $i \in [n]$;
2 Let $P \leftarrow M$ be the set of unallocated items and $Q \leftarrow [n]$ be the set of agents not receiving items;
3 **while** $|Q| > 0$ **do**
4   Let an agent $k \in Q$ decide a PROP1 partition $\mathcal{B} = (B_1, \ldots, B_{|Q|})$ of P according to u_k by Algorithm 1;
      `// Subroutine BipartiteConstruction`$(k, Q, P, \mathcal{B})$:
5   [illegible]
6   Let $Z \subseteq Q$ be the maximal subset such that $|Z| > |\Gamma(Z)|$, where $\Gamma(Z)$ denotes all the neighbors of Z;
7   Find a perfect matching between $Q \setminus Z$ and a subset of $\mathcal{B} \setminus \Gamma(Z)$;
8   **foreach** *agent* $i \in Q \setminus Z$ *that is matched to* $B_j \in \mathcal{B} \setminus \Gamma(Z)$ **do**
9   Update $A_i \leftarrow B_j$, $Q \leftarrow Q \setminus \{i\}$ and $P \leftarrow P \setminus B_j$;
10 **return** *the allocation* $\mathcal{A} = (A_1, A_2, \ldots, A_n)$

Lemma 10. *The perfect matching in Line 7 of Algorithm 3 always exists between* $\bar{Z} = Q \setminus Z$ *and a subset of* $\bar{\Gamma}(Z) = \mathcal{B} \setminus \Gamma(Z)$.

In perfect matching, each agent in $\bar{Z}$ is matched to an adjacent bundle, while some bundles in $\bar{\Gamma}(Z)$ may remain unmatched. Each matched bundle is then allocated to the respective matched agent. For the remaining agents in Z and the unallocated items, we repeat the above process with fewer agents and items until each agent receives a bundle.

Lemma 11 shows the allocation obtained by Algorithm 3 satisfies PROP1. The high-level idea is that if a bundle B_j is allocated to one agent $i \in Q$, then B_j is PROP1 to agent i. On the other hand, for any other agent $k \in Q$ not receiving any bundle, $v_k(B_j)$ is less than the proportional share of $v_i(P)$ since there exists no edge between k and B_j. Proof of Lemma 11 can be found in our full version [14].

Lemma 11. *The allocation returned by Algorithm 3 is PROP1.*

4 Query Complexity of EF1

This section studies the query complexity of computing an EF1 allocation. For a constant number of agents, the well-known Round-Robin algorithm that guarantees EF1 requires $O(m \log m)$ queries as each agent needs to sort the items in descending order, and the envy-graph procedure requires $O(m)$ queries as each round of the procedure allocates one item and costs a constant number of queries. There is still a gap between the existing algorithms and the lower bound

$\Omega(\log m)$ to compute an EF1 allocation under the comparison-based query model for a constant number of agents. In this section, we endeavor to bridge this gap and improve the query complexity to compute an EF1 allocation.

When there are two agents, the algorithm *cut-and-choose* outputs an EF1 allocation. The query complexity mainly depends on the first agent's division step, which can be implemented by a line binary search with $O(\log m)$ queries, and the second agent's choosing step only requires one query. Thus, we have the following theorem and the detailed proof can be found in our full version [14].

Theorem 12. *For two agents with additive valuations, the query complexity and running time of computing an EF1 allocation are $O(\log m)$.*

We now come to our main result in this section that for identical valuation, the query complexity for EF1 is $O\left(n^2 \log m\right)$, which provides a tight $\Theta(\log m)$ bound for a constant number of agents. The case of $m < n$ is trivial, and we focus on the case of $m \geq n$.

Theorem 13. *An EF1 allocation can be found through $O(n^2 \log m)$ queries with running time $O(n^2 \log m)$ for identical and additive valuation functions.*

Our algorithm follows the steps below, which is illustrated in our full version [14] for a better understanding.

Step 1: We run the subroutine `MinBundleKept` $([n], M)$ in Algorithm 1. The algorithm returns a partial allocation $(A_1, A_2, \ldots, A_n)$ and a set of unallocated items $P = \{g_1, \ldots, g_{n-1}\}$.[1] Without loss of generality, assume $u(g_1) \leq \cdots \leq u(g_{n-1})$ and $u(A_1) \leq \cdots \leq u(A_n)$.

Step 2: We go back to the round in which A_n was last updated in `MinBundleKept`. In that round, we have a temporary allocation $\mathcal{B} = (B_1, \ldots, B_{n-1}, A_n)$ such that A_n is the one with the smallest utility among all. Let B_{n-1} be the bundle containing g_{n-1} (if not, we switch it to B_{n-1}). Line up the bundles from the left to the right in the order of $B_1, \ldots, B_{n-1}, A_n$ and, in particular, arrange g_{n-1} in the right of B_{n-1} such that g_{n-1} is adjacent to A_n.

Step 3: Update the first $n-1$ bundles in a moving-knife manner. We find another $n-1$ bundles $C_1, C_2, \ldots, C_{n-1}$ from left to right such that each C_i for $i \in [n-1]$ is weakly larger than A_n and smaller than A_n after removing C_i's rightmost item, which ensures the EF1 conditions from C_i to A_n and from A_n to C_i. As A_n is not larger than B_i with $i \in [n-1]$, the right endpoint of bundle C_i is to the left of (or at most the same as) B_i, which means such an allocation (maybe partial) $\mathcal{C} = (C_1, \ldots, C_{n-1}, A_n)$ exists. If there is no unallocated item between C_{n-1} and A_n, an EF1 allocation is already found.

---

[1] We can verify that there will always be exactly $n-1$ unallocated items when `MinBundleKept` terminates. When $m > n$, there will be a round where $n < |P| \leq 2n$. In this situation, at most two items are allocated each round until $|P| \leq n$. Therefore, there must exist a point where $|P| = n$ or $n-1$. If the former case occurs, it will then turn to $|P| = n-1$ after one more round.

Step 4: If there are still unallocated items, denote them by R. We have $g_{n-1} \in R$ as it is placed directly to the left of A_n. If g_{n-1} is the only item in R, we proceed to Step 5.

Otherwise, we proceed with the following operations in the order of g_{n-2}, g_{n-3}, ..., g_1 and exchange each of them with items of R. As shown in Appendix of our full version [14], if the bundle C_2 contains g_k, g_k is exchanged to R, while some items from R are exchanged to C_2 so that the new bundle D is still weakly larger than or equal to A_n (to guarantee EF1). We refer to the new bundle after exchange as D_k for each C_k. Note that the unallocated items are also arranged in a line. During each exchanging operation, the items in R are taken from the leftmost end, and the returned item is arranged such that the rightmost items are $g_j, g_{j+1}, \ldots, g_{n-1}$ for some $j \in [n-1]$. Repeat this process until no more exchange can be made or $R \subseteq P$. Then, we obtain a partial allocation $\mathcal{D} = (D_1, \ldots, D_{n-1}, A_n)$ and a set of unallocated items R.
Step 5: Allocate $S = R \setminus P$ to an arbitrary agent. For the remaining items in R, run the envy-graph procedure to allocate them based on $\mathcal{D}$.

Proof of Theorem 13 can be found in our full version [14].

We further consider the query complexity for three agents with additive valuations. This algorithm is almost a straightforward adaption of Algorithm 3 of [29], where the algorithm for identical valuation in Theorem 13 is adapted. The detailed algorithm and its proof can be found in our full version [14].

Theorem 14. *For three agents with additive valuations, the query complexity and running time of computing an EF1 allocation are $O(\log m)$.*

5 Query Complexity of MMS+PROP1

We have already shown that a PROP1 allocation can be found via $O(\log m)$ queries for a constant number of agents with additive valuations. Besides PROP1, MMS is also a well-studied notion of fairness in the fair division of indivisible goods. In this section, we will provide an algorithm that can achieve PROP1 and $\frac{1}{2}$-MMS simultaneously via $O\left(n^4 \log m\right)$ queries, which is shown in Algorithm 4 in our full version [14].

In the following detailed descriptions of our algorithm, we utilize two concepts: the bundles of the *first type* and the bundles of the *second type*. Roughly speaking, the first type of bundle is a singleton containing a valuable item while the second type of bundle may not consist of only one item. The specific bundles of the *first type* will be presented in the following descriptions, while the remaining bundles are of the *second type*.

Similar to Algorithm 3, we perform a multi-round algorithm. For each round, according to the set Q of agents not receiving any items and the set P of unallocated items, we first choose an arbitrary agent $k \in Q$ and find a PROP1 and $\frac{1}{2}$-MMS allocation subject to agent k's valuation u_k. In this part, we first use the subroutine `MinBundleKept` (Q, P) defined in Algorithm 1 to find the partial allocation $\mathcal{B}$ and a set P' of unallocated items whose size is less than $|Q|$. For this, there exist two possible cases. For the first case, the two most valuable bundles among all bundles in $\mathcal{B}$ and all items in P' are both items in P'. In this case, we can directly allocate the most valuable item in P' to agent k and restart the main procedure for the remaining items and agents by ignoring all previous bundles of the second type. Here, the bundle allocating this item to agent k is a bundle of the *first type*. Otherwise, we can run the procedure described in Theorem 13 to obtain an EF1 allocation $\mathcal{C}$ under u_k, which is the wanted PROP1 and $\frac{1}{2}$-MMS allocation under valuation u_k.

We then construct a bipartite graph G between all remaining agents in Q and all bundles in the above EF1 allocation $\mathcal{C}$, similar to the `BipartiteConstruction` function in Algorithm 3. Here, an edge between an agent i and a bundle C_j is constructed only if allocating C_j to agent i can meet agent i's requirement for both PROP1 and $\frac{1}{2}$-MMS. For the above agent k, we directly add an edge (k, C_j) for each bundle C_j in $\mathcal{C}$. For each remaining agent i, we first execute the `PROP1-PROP-Subroutine` in Algorithm 2 to identify the bundles that satisfy PROP1 to agent i. However, different from that used in Algorithm 3, we also need to ensure $\frac{1}{2}$-MMS here. Thus, we consider the allocation in two subcases. For the first case, agent i can receive only one item in $\{I_1, \ldots, I_{|Q|-1}\}$ to reach both PROP1 and $\frac{1}{2}$-MMS. Under this case, we can again allocate the most valuable item to agent i and restart the main procedure with only the bundles of the first type allocated. Here, the bundle allocating this item to agent i is a bundle of the *first type*. Otherwise, we can claim that agent i can meet the requirement for PROP1 and $\frac{1}{2}$-MMS after receiving C_j.

We now perform a similar reduction procedure as in Algorithm 3, where we find a maximal subset violating Hall's marriage theorem. We then find a perfect matching of the remaining agents to some bundles and reduce the original instance to a smaller one after removing these agents and their matched bundles.

We now present our main theorem in this section. The analysis of query complexity and running time can be found in our full version [14].

Theorem 15. *For additive valuations, a PROP1 and $\frac{1}{2}$-MMS allocation can be found by Algorithm 4 via $O\left(n^4 \log m\right)$ queries with running time $O\left(n^4 \log m\right)$.*

Proof Structure. Similar to the proof of Theorem 9, our high-level idea is to iteratively reduce the input instance's size and demonstrate the recursion's correctness. As we specified, there are two types of bundles: one only consists of a single large item for some agent, and the other may consist of multiple items and is found by the perfect-matching procedure. In Lemma 21, we show that every allocated first type of bundle satisfies the required fairness constraints for the agent who receives the bundle. Further, Lemma 16 demonstrates that removing a first type bundle will not damage the fairness constraints for the remaining agents.

Therefore, it suffices to show every allocated second type of bundle also satisfies PROP1 and $\frac{1}{2}$-MMS. In Lemma 19, we establish that any bundle in the computed allocation $\mathcal{C}$ satisfies PROP1 and $\frac{1}{2}$-MMS for one fixed agent k. Lemma 20 then verifies each allocated bundle also satisfies PROP1 and $\frac{1}{2}$-MMS for the agent receiving the bundle. For the agents who do not receive any bundle and will be handled in the next iteration, Lemma 17 and Lemma 18 demonstrate that the PROP1 and $\frac{1}{2}$-MMS constraints can be satisfied on the “smaller” fair division instance where the allocated second type of bundles are removed.

We are ready to begin our proof. The following lemma shows that removing the bundles of the first type will not damage the PROP1 and MMS conditions, whose proof can be found in our full version [14].

Lemma 16. *Given a problem instance I of allocating a set M items to N agents, fixing an agent i' and a good o, we can define* MMS_i *and* MMS'_i *for each agent $i \neq i'$ as the maximum share of agent i under the original instance I and the instance I' removing agent i' and the good o from I. We also assume a PROP1 allocation $\mathcal{A} = \{A_i : i \in N \setminus \{i'\}\}$ for I'. For each $i \in N \setminus \{i'\}$, we have:*

- $\text{MMS}'_i \geq \text{MMS}_i$;
- *there exists an item $g \in M \setminus A_i$ such that $u_i(A_i \cup \{g\}) \geq \frac{1}{|N|}u_i(M)$.*

Since the bundles of the first type only contain exactly one item, if the allocation containing only the bundles of the second type is PROP1 and $\frac{1}{2}$-MMS under the reduced instance, by Lemma 16, these bundles can also meet the conditions of PROP1 and $\frac{1}{2}$-MMS under the original instance. Thus, we will first assume there is no bundle of the first type and check the properties for the bundles of the second type in the following.

Then we present two following lemmas that help us to keep the PROP1 and MMS properties during the reduction of the problem. Proof of Lemma 17 can be found in our full version [14] and Lemma 18 will be proved later.

Lemma 17. *For the set P of unallocated items and the set Q of agents not receiving items during the procedure, for any $o \in P$ and any $i \in Q$, we have $u_i(P \setminus \{o\})/(|Q| - 1) \geq u_i(M \setminus \{o\})/(|N| - 1)$.*

Lemma 18. *For the set P of unallocated items and the set Q of agents not receiving items during the procedure, for any $i \in Q$, we have $\frac{1}{|Q|}u_i(P) \geq \frac{1}{|N|}u_i(M)$.*

Lemma 19. *The EF1 allocation $\mathcal{C}$ under u_k is PROP1 and the valuation of each bundle is at least $\frac{1}{2}\text{MMS}_k$ under u_k for the instance which allocates the set P of items to the set Q of agents.*

From Lemma 19 (whose proof can be found in our full version [14]), we get the correctness of the added edge (k, C_j). We next show all added edges (i, C_j) also satisfy the corresponding conditions for PROP1 and $\frac{1}{2}$-MMS, with proof deferred to Appendix of our full version [14].

Lemma 20. *For each added edge (i, C_j), allocating C_j to agent i can meet the conditions for PROP1 and $u_i(C_j) \geq \frac{1}{2}\text{MMS}_i$ under the instance that allocates the set P of items to the set Q of agents.*

From the previous two lemmas, we can conclude that the bundles of the second type can satisfy the PROP1 and $\frac{1}{2}$-MMS conditions under the corresponding instance of items P and agents Q. From the existence of the set Z and the perfect matching procedure which can be shown in the same way as in Lemma 10, it suffices to show: the feasibility of the PROP1 conditions under the reduced instance can induce the feasibility of the PROP1 conditions in the original instance and Lemma 17 holds.

For the induction of the PROP1 conditions, this can be directly derived from the correctness of Lemma 18. According to Theorem 8, in any round that allocates a set P of unallocated items to a set Q of agents, each matched C_j in this round is valued less than $\frac{1}{|Q|}u_i(P)$ under u_i for each $i \in Z$. From this, we observe that the term $\frac{1}{|Q|}u_i(P)$ for each remaining agent i becomes larger after this round. Summarizing over all rounds, we can conclude the statement in Lemma 18.

We have already shown the bundles of the second type satisfy the conditions for PROP1 and $\frac{1}{2}$-MMS above, it suffices to show they also hold for the bundles of the first type to prove the correctness of our algorithm. Proof of Lemma 21 can be found in our full version [14].

Lemma 21. *For a bundle A_k of the first type allocated to an agent k, $u_k(A_k) \geq \frac{1}{2n}u_k(M) \geq \frac{1}{2}\text{MMS}_k$ and there exists a good $g \in M \setminus A_k$ such that $u_k(A_k \cup \{g\}) \geq \frac{1}{n}u_k(M)$.*

Acknowledgments. The research of Biaoshuai Tao was supported by the National Natural Science Foundation of China (No. 62102252). The research of Shengxin Liu was partially supported by the National Natural Science Foundation of China (No. 62102117), by the Shenzhen Science and Technology Program (No. GXWD20231129111306002), and by the Guangdong Basic and Applied Basic Research Foundation (No. 2023A1515011188).

References

1. Akrami, H., Garg, J.: Breaking the 3/4 barrier for approximate maximin share. In: Proceedings of the ACM-SIAM Symposium on Discrete Algorithms (SODA), pp. 74–91 (2024)
2. Amanatidis, G., et al.: Fair division of indivisible goods: recent progress and open questions. Artif. Intell. **322**, 103965 (2023)
3. Aziz, H., Caragiannis, I., Igarashi, A., Walsh, T.: Fair allocation of indivisible goods and chores. Auton. Agent. Multi-Agent Syst. **36**(1), 3:1–3:21 (2022)
4. Aziz, H., Li, B., Xing, S., Zhou, Y.: Possible fairness for allocating indivisible resources. In: Proceedings of the International Conference on Autonomous Agents and Multiagent Systems (AAMAS), pp. 197–205 (2023)
5. Aziz, H., Mackenzie, S.: A discrete and bounded envy-free cake cutting protocol for any number of agents. In: Proceedings of the Annual IEEE Symposium on Foundations of Computer Science (FOCS), pp. 416–427 (2016)

6. Aziz, H., Mackenzie, S.: A discrete and bounded envy-free cake cutting protocol for four agents. In: Proceedings of the Annual ACM Symposium on Theory of Computing (STOC), pp. 454–464 (2016)
7. Babichenko, Y.: Query complexity of approximate Nash equilibria. J. ACM **63**(4) (2016)
8. Benadè, G., Halpern, D., Psomas, A.: Dynamic fair division with partial information. In: Proceedings of the Annual Conference on Neural Information Processing Systems (NeurIPS), pp. 3703–3715 (2022)
9. Bilò, V., et al.: Almost envy-free allocations with connected bundles. Games Econom. Behav. **131**, 197–221 (2022)
10. Brams, S.J., Taylor, A.D.: An envy-free cake division protocol. Am. Math. Mon. **102**(1), 9–18 (1995)
11. Brânzei, S., Nisan, N.: Communication complexity of cake cutting. In: Proceedings of the ACM Conference on Economics and Computation (EC), p. 525 (2019)
12. Brânzei, S., Nisan, N.: The query complexity of cake cutting. In: Proceedings of the Annual Conference on Neural Information Processing Systems (NeurIPS), pp. 37905–37919 (2022)
13. Bu, X., Li, Z., Liu, S., Song, J., Tao, B.: On the complexity of maximizing social welfare within fair allocations of indivisible goods. arXiv preprint arXiv:2205.14296 (2022)
14. Bu, X., Li, Z., Liu, S., Song, J., Tao, B.: Logarithmic comparison-based query complexity for fair division of indivisible goods. arXiv preprint arXiv:2404.18133 (2024)
15. Budish, E.: The combinatorial assignment problem: approximate competitive equilibrium from equal incomes. J. Polit. Econ. **119**(6), 1061–1103 (2011)
16. Caragiannis, I., Kurokawa, D., Moulin, H., Procaccia, A.D., Shah, N., Wang, J.: The unreasonable fairness of maximum Nash welfare. ACM Trans. Econ. Comput. **7**(3), 1–32 (2019)
17. Cechlárová, K., Pillárová, E.: On the computability of equitable divisions. Discret. Optim. **9**(4), 249–257 (2012)
18. Conitzer, V., Freeman, R., Shah, N.: Fair public decision making. In: Proceedings of the ACM Conference on Economics and Computation (EC), pp. 629–646 (2017)
19. Dubins, L.E., Spanier, E.H.: How to cut a cake fairly. Am. Math. Monthly **68**(1P1), 1–17 (1961)
20. Edmonds, J., Pruhs, K.: Cake cutting really is not a piece of cake. In: Proceedings of the ACM-SIAM Symposium on Discrete Algorithms (SODA), pp. 271–278 (2006)
21. Even, S., Paz, A.: A note on cake cutting. Discret. Appl. Math. **7**(3), 285–296 (1984)
22. Feige, U.: Low communication protocols for fair allocation of indivisible goods. arXiv preprint arXiv:2407.07641 (2024)
23. Foley, D.K.: Resource allocation and the public sector. Yale Econ. Essays **7**(1), 45–98 (1967)
24. Ghodsi, M., Hajiaghayi, M.T., Seddighin, M., Seddighin, S., Yami, H.: Fair allocation of indivisible goods: improvement. Math. Oper. Res. **46**(3), 1038–1053 (2021)
25. Halpern, D., Shah, N.: Fair and efficient resource allocation with partial information. In: Proceedings of the International Joint Conference on Artificial Intelligence (IJCAI), pp. 224–230 (2021)
26. Kurokawa, D., Lai, J., Procaccia, A.: How to cut a cake before the party ends. In: Proceedings of the AAAI Conference on Artificial Intelligence (AAAI), pp. 555–561 (2013)

27. Lipton, R.J., Markakis, E., Mossel, E., Saberi, A.: On approximately fair allocations of indivisible goods. In: Proceedings of the ACM Conference on Electronic Commerce (EC), pp. 125–131 (2004)
28. Nisan, N.: The demand query model for bipartite matching. In: Proceedings of the Annual ACM-SIAM Symposium on Discrete Algorithms (SODA), pp. 592–599 (2021)
29. Oh, H., Procaccia, A.D., Suksompong, W.: Fairly allocating many goods with few queries. SIAM J. Discret. Math. **35**(2), 788–813 (2021)
30. Onak, K., Sun, X.: The query complexity of graph isomorphism: bypassing distribution testing lower bounds. In: Proceedings of the Annual ACM SIGACT Symposium on Theory of Computing (STOC), pp. 165–171 (2018)
31. Plaut, B., Roughgarden, T.: Almost envy-freeness with general valuations. SIAM J. Discret. Math. **34**(2), 1039–1068 (2020)
32. Plaut, B., Roughgarden, T.: Communication complexity of discrete fair division. SIAM J. Comput. **49**(1), 206–243 (2020)
33. Procaccia, A.D.: Thou shalt covet thy neighbor's cake. In: Proceedings of the International Joint Conference on Artificial Intelligence (IJCAI), pp. 239–244 (2009)
34. Procaccia, A.D., Wang, J.: Fair enough: Guaranteeing approximate maximin shares. In: Proceedings of the ACM Conference on Economics and Computation (EC), pp. 675–692 (2014)
35. Procaccia, A.D., Wang, J.: A lower bound for equitable cake cutting. In: Proceedings of the ACM Conference on Economics and Computation (EC), pp. 479–495 (2017)
36. Robertson, J., Webb, W.: Cake-Cutting algorithms: be fair if you can. Peters/CRC Press (1998)
37. Steinhaus, H.: The problem of fair division. Econometrica **16**(1), 101–104 (1948)
38. Steinhaus, H.: Sur la division pragmatique. Econometrica **17**, 315–319 (1949)
39. Stromquist, W.: How to cut a cake fairly. Am. Math. Mon. **87**(8), 640–644 (1980)
40. Stromquist, W.: Envy-free cake divisions cannot be found by finite protocols. Electr. J. Comb. **15**(1), #R11 (2008)
41. Varian, H.R.: Equity, envy, and efficiency (1973)

Ex-Post Individually Rational Bayesian Persuasion

Jiahao Zhang[1(✉)], Shuran Zheng[2], Renato Paes Leme[3], and Zhiwei Steven Wu[1]

[1] Carnegie Mellon University, Pittsburgh, PA 15213, USA
{jiahaozhang,zstevenwu}@cmu.edu
[2] Tsinghua University, Beijing, China
shuranzheng@mail.tsinghua.edu.cn
[3] Google Research, New York, NY 10011, USA
renatoppl@google.com

Abstract. In the Bayesian persuasion model, a sender can convince a receiver to choose an alternative action to the one originally preferred by the receiver. A crucial assumption in this model is the sender's commitment to a predetermined information disclosure policy (signaling scheme) and the receiver's trust in this commitment. However, in practice, it is difficult to monitor whether the sender adheres to the disclosure policy, and the receiver may refuse to follow the recommendation due to a lack of trust. Trust becomes particularly strained when the receiver knows that the sender will incur obvious losses when truthfully following the protocol. In this work, we propose the notion of *ex-post individually rational (ex-post IR)* Bayesian persuasion: after observing the state, the sender is never asked to send a signal that is less preferred than no information disclosure. An ex-post IR Bayesian persuasion policy is more likely to be truthfully followed by the sender, thereby providing stronger incentives for the receiver to trust the sender. Our contributions are threefold. First, we demonstrate that the optimal ex-post IR persuasion policy can be efficiently computed through a linear program, while also offering its geometric characterization. Second, we show that surprisingly, for non-trivial classes of games, the requirement of ex-post IR constraints does not incur any cost to the sender's utility. Finally, we compare ex-post IR Bayesian persuasion to other information disclosure models that ensure different notions of credibility.

Keywords: Bayesian persuasion · Ex-post · Credibility

1 Introduction

In the wake of the seminal work by [21] and [11], there has been a significant surge in research interests over the past decade concerning the design of optimal information disclosure policies, especially Bayesian persuasion. In the standard formulation of Bayesian persuasion, a sender strategically discloses information

M. Mavronicolas et al. (Eds.): WINE 2024, LNCS 15534, pp. 366–382, 2026.
https://doi.org/10.1007/978-3-032-08560-3_21

to influence a receiver's belief and chosen action. A crucial assumption in this model is that the sender will commit to a signaling policy and the receiver will trust this commitment. However, in most practical scenarios, this commitment is difficult to monitor, which may lead to the receiver refusing to follow the persuasion due to a lack of trust. The receiver's trust is especially fragile when there are clear deviations that could benefit the sender in certain situations. Consider the following scenario:

Example 1 (Lending). There is a credit reporting agency (the sender) recommending borrowers to a lender (the receiver). The lender makes loaning decisions based on the borrowers' trustworthiness, i.e., whether the loan will be repaid. Suppose that the lender will reject the loan application if the probability of repayment is below 0.3. The lender will issue a small loan if the probability is in $[0.3, 1)$ and grant a huge loan if the probability is 1. Suppose the lender's prior about a borrower's repayment probability is 0.5, which means that without any recommendation from the sender, the lender will issue a small loan. Suppose the sender knows whether a borrower will repay the loan based on their credit history, and the sender can send a signal to the lender. The sender's goal is to persuade the lender to approve loans. The sender's utility is 0 if the loan is rejected, 1 if a small loan is approved, and 10 if a huge loan is approved. Then the optimal Bayesian persuasion policy that maximizes the sender's expected utility is to fully reveal whether the borrower will repay the loan, which gives an expected utility of 5.

The optimal information disclosure policy in Example 1 has its Achilles' heel. The sender may hesitate to truthfully reveal low-credit borrowers who will not repay the loans, as it will yield a lower utility than no information disclosure, which guarantees a small loan for the borrower. Consequently, the receiver might cast doubt on the reliability of the entire protocol, as the sender incurs a clear loss when truthfully sending the signal. In the absence of trust, this policy may fail to effectively persuade the receiver. Motivated by this tension between incentives and trust, our work considers the notion of *ex-post individually rational* (ex-post IR) Bayesian persuasion: after observing the state (the borrower's trustworthiness), the sender never sends a signal that yields a lower payoff for the sender than no information disclosure. We provide three types of results.

1. First, we study how to compute the optimal signaling scheme for ex-post IR Bayesian persuasion. We show that the optimal signaling scheme can still be solved by linear programming even if we add the ex-post IR constraint. In addition, we provide a geometric characterization of the optimal ex-post IR signaling scheme by extending the concave closure representation by [11] to incorporate the ex-post IR constraint. In the binary state setting, our geometric result enables a clean visualization of the optimal ex-post IR signaling policy.
2. Second, we establish conditions where the ex-post IR constraint imposes no cost to the sender. While there are instances in which the ex-post IR constraint incurs a cost to the optimal sender utility, a natural question is when

ex-post IR incurs no cost. Building on our aforementioned geometric result, we give a complete geometric characterization in the binary state case (Theorem 2) of when the sender achieves the same optimal utilities whether or not ex-post IR constraint is required. Our proof technique links the optimal ex-post IR signaling scheme with the *quasiconcave closure* of the sender's expected utility function. As a corollary, the geometric characterization enables a more computationally efficient algorithm than simply running linear programming to determine whether the ex-post IR constraint incurs a sender utility gap. However, our geometric approach for the binary state setting does not extend to the general state space setting. Going beyond binary states, we focus on characterizing the utility gap for two families of Bayesian persuasion problems that are motivated by real-world applications.

- We define a broad class of problems called *trading games* (Definition 8), which include *bilateral trade* and first-price auction with reserve price as special cases. We show that the optimal signaling schemes of all trading games are ex-post IR.
- We identify a general class of games that meet two conditions called *cyclical monotonicity* and *weak logarithmic supermodularity*, with the *credence goods* market as a notable special case. We introduce a new notion of the *greedy signaling scheme* and show that it is ex-post IR for this class. Then in the special case of the credence goods market, we show that there exists a greedy signaling scheme is optimal, which implies that ex-post IR constraint incurs no cost to the sender's utility.

3. Third, we compare the sender's optimal expected utility under varying degrees of commitment power. At one extreme is the standard Bayesian persuasion model, where the sender has full commitment power, i.e., they can commit any Bayes-plausible signaling scheme as they like. At the other extreme is the cheap talk model, where the sender has no commitment power at all. As a result, the sender's optimal expected utility is always higher in Bayesian persuasion than in cheap talk [13]. We show that in the ex-post IR Bayesian persuasion model, the sender's optimal utility is always between the optimal utilities of Bayesian persuasion and cheap talk *if and only if* the sender has an ordered preference over actions (see Assumption 1). Finally, we compare with a model called *credible persuasion* studied by [12], which also studied the credibility of the sender. We show that the sender's optimal utility in credible persuasion is also between the optimal utilities of Bayesian persuasion and cheap talk. However, we provide two examples showing that the utility of the optimal ex-post IR signaling scheme could be higher or lower than that of optimal credible persuasion.

1.1 Related Literature

A substantial body of previous research has studied the credibility of the sender in Bayesian persuasion. They consider the notion of credibility in different ways, which are all different from ours. In the work by [20], the sender sends a signal to the receiver at some cost that depends on the signal and the realized world state, which makes the sender more credible. [15] considers the sender to communicate with the receiver through an intermediary which can be influenced by the sender with some probability. [12] characterizes the notion of credibility by restricting the sender to only commit to a marginal distribution of signals rather than the whole signaling scheme. There are more works about the limitation of the commitment power via repeated interactions [4,18], verifiable information [2,9], information control [17], and mediation [22].

Our paper considers two applications of Bayesian persuasion: bilateral trade and credence goods market. The information transmission models about these two applications has also been extensively studied in previous works. For the bilateral trade, [1] considers the welfare of the sender (consumer) in both monopolistic and competitive markets, and [8] considers the voluntary disclosure in the bilateral trade rather than the Bayesian persuasion. For the credence goods market, [7] compares the market outcome with and without commitment and their model is extended by [10].

Technically, at a very high level, our work relies on two classic tools, which have also been successfully used in previous works. The first is quasiconcave closure. For example, [14] shows that in cheap talk, when the sender's utility is state-independent, the optimal sender utility can be characterized by the quasiconcave closure of the sender's expected utility function. The second is the greedy algorithm. For example, [16] studies mechanisms that use greedy allocation rules and show that all such mechanisms obtain a constant fraction of the optimal welfare.

1.2 Organization

The rest of the paper is organized as follows: Sect. 2 introduces the model of Bayesian persuasion and the definition of ex-post IR Bayesian persuasion. Section 3 provides results about the optimality of ex-post IR Bayesian persuasion in both algebraic and geometric views, which is similar to [11]. Section 4 provides results of when there is no sender utility gap between the Bayesian persuasion and the ex-post IR Bayesian persuasion. Section 5 compares the optimal sender's utilities under different information transmission models that ensure different notions of credibility. All omitted proofs are in the appendix, which can be found in the full version of our paper[1].

2 Model

We introduce the standard Bayesian persuasion problem with finite actions and states, where a sender chooses a signal to reveal information to a receiver who then takes an action. The sender's utility $v(a, \theta)$ and the receiver's utility $u(a, \theta)$

[1] https://arxiv.org/abs/2312.04973.

depend on the receiver's action $a \in A$ and the state of world $\theta \in \Theta$. Throughout this paper, we impose two standard assumptions. First, the sender has a preference over the actions regardless of the state. See Assumption 1 for the formal description. Second, there is no dominated action for the receiver, i.e. for any action a, there exists a posterior belief μ that a is the best action for the receiver under μ. This is without loss of generality because we can drop dominated actions.

Assumption 1. *Let $|A| = n$. The sender's preference over action is $a_1 \geq a_2 \geq \ldots \geq a_n$. In other words, for any $\theta \in \Theta$ and any $i, j \in [n]$ that $i \leq j$, we have $v(a_i, \theta) \geq v(a_j, \theta)$.*

The sender and receiver share a common prior $\mu_0 \in \Delta(\Theta)$ about the state. The sender commits to a *signaling scheme* ϕ, which is a randomized mapping from the state to signals. Formally, the signaling scheme is defined as $\phi : \Theta \mapsto \Delta(S)$, where S is the set of all possible signals. For convenience, let $\phi_\theta \in \Delta(S)$ be the conditional signal distribution when the state is $\theta \in \Theta$ and we denote by $\phi_\theta(s)$ the probability of sending signal $s \in S$ when the state is θ. For ease of notation, we use ϕ to also denote the distribution over signals induced by the signaling scheme ϕ and the prior μ_0 and denote $\phi(s)$ the probability of sending signal $s \in S$.

The interaction between the sender and the receiver goes on as follows:

1. the sender commits to a signaling scheme ϕ and the receiver observes it;
2. the sender observes a realized state of nature $\theta \sim \mu_0$;
3. the sender draws a signal $s \in S$ according to the distribution ϕ_θ and sends it to the receiver;
4. the receiver observes the signal s and updates her belief about the state according to the *Bayes rule*;
5. the receiver chooses an action that maximizes her expected utility.

The receiver's posterior belief after seeing $s \in S$ is denoted by μ_s. We have $\mu_s(\theta) = \frac{\mu_0(\theta)\phi_\theta(s)}{\phi(s)}$.

For any posterior belief μ_s, denote the receiver's best response $a^*(\mu_s) \in \arg\max_{a \in A} \sum_{\theta \in \Theta} u(a, \theta)\mu_s(\theta)$. For convenience, let the index of the best response under prior $a^*(\mu_0)$ be k^*, i.e. $a^*(\mu_0) = a_{k^*}$. As is customary in the literature, we assume that the receiver breaks ties in favor of the sender. The sender's expected utility when sending s will be equal to $\hat{v}(\mu_s) := \sum_{\theta \in \Theta} v\big(a^*(\mu_s), \theta\big)\mu_s(\theta)$. An outcome is a pair of the realized world state and the receiver's action. Denote $\pi \in \Delta(A \times \Theta)$ the probability distribution of outcomes given the signaling scheme is ϕ and the prior μ_0. We have

$$\pi(a, \theta) = \sum_{s \in S : a^*(\mu_s) = a} \sum_{\theta \in \Theta} \phi_\theta(s)\mu_0(\theta).$$

Sometimes we use π to represent a signaling scheme. Then, we introduce the definition of the ex-post individually rational (IR) signaling scheme.

Definition 1 (Ex-post individual rationality). *A signaling scheme* ϕ *is ex-post IR for the sender if for any state* $\theta \in \Theta$ *and any signal* $s \in S$, $v(a^*(\mu_s), \theta) \geq v(a^*(\mu_0), \theta)$.

The definition of ex-post IR requires that compared to no communication, the sender will not regret sending the signal s in any case. It can also mean that after the sender sees the state and draws a signal in Step 3, he will not hesitate to send the signal. Next, we give the definition of the Bayesian persuasion instance or *game* for short.

Definition 2 (game). *A game (Bayesian persuasion instance) is a quadruple* (A, Θ, v, u) *where* A *is the action set,* Θ *is the state set,* v *is the sender's utility function and* u *is the receiver's utility function.*

3 Warmup: Optimal Ex-Post IR Signaling Schemes

It is well known that the optimal signaling scheme of a standard Bayesian persuasion problem with a finite number of actions and states can be solved by linear programming. In addition, the optimal expected Sender utility can be characterized by the concave closure of the expected Sender utility function $\hat{v} : \Delta(\theta) \mapsto R$, which maps a posterior belief to the expected Sender utility. In this section, we extend these two standard results to the setting of ex-post IR Bayesian persuasion. We first show that the ex-post IR constraints are linear, which implies that the optimal ex-post IR signaling scheme can also be solved by linear programming. Then we provide a concavefication characterization of the optimal ex-post IR signaling scheme.

3.1 Linear Programming

The linear programming for Bayesian persuasion is

$$\max_{\pi} \sum_{a \in A} \sum_{\theta \in \Theta} v(a, \theta) \pi(a, \theta)$$

subject to

$$\sum_{\theta \in \Theta} (u(a, \theta) - u(a', \theta)) \pi(a, \theta) \leq 0 \quad \forall a, a' \in A \tag{1}$$

$$\sum_{a \in A} \pi(a, \theta) = \mu_0(\theta) \quad \forall \theta \in \Theta, \tag{2}$$

where Eq. (1) is the incentive compatible constraints for the Receiver and Eq. (2) is to make sure the signaling scheme is *Bayes-plausible*. To solve the optimal ex-post IR signaling scheme, we only need to add the following ex-post IR constraints

$$\pi(a, \theta) = 0 \quad \forall a \in A, \theta \in \Theta, u(a, \theta) < u(a^*(\mu_0), \theta). \tag{3}$$

3.2 Concave Closure

Let V be the concave closure of the expected utility function $\hat{v}$.

Fact 1. *The value of the optimal signaling scheme is $V(\mu_0)$. The sender benefits from persuasion if and only if $V(\mu_0) > \hat{v}(\mu_0)$*

Now we consider Bayesian persuasion with ex-post IR constraints. Instead of computing the concave closure on the whole domain of posterior belief, we define a closure only on the posterior belief μ that $a^*(\mu) \geq a^*(\mu_0)$. Let A^+ be the set of actions preferred by Sender over $a^*(\mu_0)$. Formally we have

$$V_{\text{EX-POST}}(\mu_0) = \max_{\tau \in \Delta(\Delta(\Theta))} \left\{ \sum_{\mu \in \text{supp}(\tau)} \tau(\mu)\hat{v}(\mu) : \sum_{\mu \in \text{supp}(\tau)} \tau(\mu)\mu = \mu_0, \quad a^*(\mu) \in A^+, \forall \mu \in \text{supp}(\tau) \right\}. \tag{4}$$

Theorem 1. *The value of the optimal ex-post IR signaling scheme is $V_{\text{EX-POST}}(\mu_0)$. The sender benefits from ex-post IR persuasion if and only if $V_{\text{EX-POST}}(\mu_0) > \hat{v}(\mu_0)$.*

The proof directly follows from the Eq. (4) and [11], which is straightforward and omitted. Recall Example 1 in the introduction, Fig. 1 visualize the concave closure V and the concave closure with ex-post IR constraints $V_{\text{EX-POST}}$ in this example.

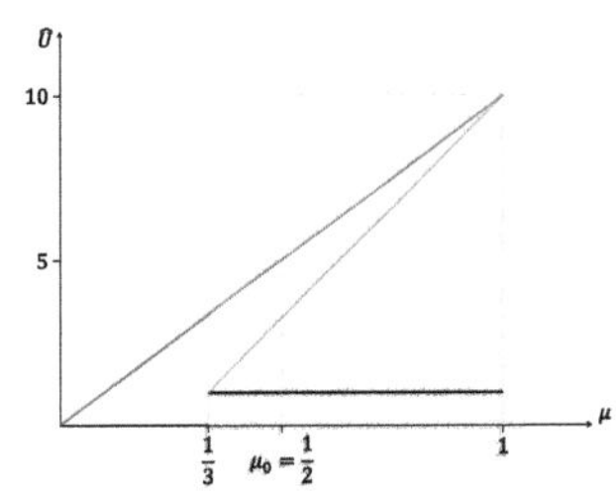

Fig. 1. The black line is the utility curve of the sender's expected utility function $\hat{v}$ in Example 1. The red line is the concave closure V. The blue line is the concave closure with ex-post IR constraints $V_{\text{EX-POST}}$. (Color figure online)

4 The Effect of the Ex-Post IR Constraint

Now that we know how to compute the optimal ex-post IR signaling scheme and its concave closure representation. In this section, we compare the sender's optimal signaling schemes with and without the ex-post IR constraint, and examine how the ex-post IR requirement affects the sender's expected utility and provide more explicit results. We focus on the following question: under what conditions does the ex-post IR constraint incurs no cost to the sender's utility?

Firstly, in Sect. 4.1, we characterize the impact of the ex-post IR constraint by a simple function: *smoothed quasiconcave closure*. We show that the ex-post IR constraint imposes no costs if and only if the smoothed quasiconcave closure is concave. This function is simple: it is basically a refined *quasiconcave closure* of the sender's expected utility function, smoothed by connecting the adjacent endpoints. See Theorem 2 for details. For the general cases, our results are driven by applications. We identify two non-trivial classes of games with real-world applications, demonstrating that ex-post IR constraint incurs no costs in these games. The first class of games is *trading game* in Sect. 4.2. The second class of games is in Appendix A with credence goods market as its application.

4.1 Geometric Characterization in the Binary State Case

In this section, we consider a binary state set $\Theta = \{\theta_1, \theta_2\}$ and a finite action set $A = \{a_1, a_2, ...a_n\}$. In this case, it's known that the optimal signaling scheme can be visualized from the concave closure of the sender's expected utility function. In our paper, instead of considering the concave closure, we show how to determine whether there is a gap between the optimal signaling scheme and optimal ex-post IR signaling scheme through the quasiconcave closure of the sender's expected utility function. For convenience, we first define the *partition* of $[0, 1]$ and the thresholds of a partition.

Definition 3 (partition). *A family of intervals P is a partition of $[0, 1]$ if and only if*

1. *the family P does not contain $\emptyset$,*
2. *the union of intervals in P is equal to $[0, 1]$,*
3. *the intersection of any two distinct intervals is $\emptyset$.*

For a partition $\{I_i\}_{i=1}^n$ of $[0, 1]$, let $\alpha_0 = 0$ and α_i be the right endpoint of interval I_i. We can always sort $\{I_i\}_{i=1}^n$ such that $0 = \alpha_0 < \alpha_1 < \cdots < \alpha_n = 1$. The threshold of I_i and I_{i+1} is α_i for any $1 \le i < n$. Putting it all together, we denote a partition P with its thresholds as $P = (\{I_i\}_{i=1}^n, \{\alpha_i\}_{i=0}^n)$. Sometimes we will omit thresholds $\{\alpha_i\}_{i=0}^n$ if they are not needed.

In the binary state case, say $\mu(\theta_1)$ is the independent variable of the function $\hat{v}$, the sender's expected utility function is a piecewise linear function that divides the domain $[0, 1]$ into n disjoint *intervals* $I_1, I_2, \cdots, I_n$. Each interval I_i corresponds to a action $a^*(I_i)$ that the receiver's best response under μ is $a^*(I_i)$ for any posteriors μ that $\mu(\theta_1) \in I_i$. In particular, let the interval corresponding to the sender's favorite action a_1 be I_{i^*}. Therefore, the sender's expected utility function $\hat{v}$ corresponds to a partition $\hat{P} = (\{I_i\}_{i=1}^n, \{\alpha\}_{i=0}^n)$ of $[0, 1]$. When $\mu(\theta_1) \in (\alpha_{i-1}, \alpha_i)$ for all $1 \le i \le n$, the best response for the receiver is $a^*(I_i)$. When $\mu(\theta_1) = \alpha_i$ for some i, the receiver breaks the tie in favor of the sender. We then present the definition of *quasiconcave function* and *quasiconcave closure*.

Definition 4 (quasiconcave function). *A function* $f : [0,1] \to \mathbb{R}$ *defined on [0,1] is quasiconcave if for all* $x, y \in [0,1]$ *and* $\lambda \in [0,1]$*, we have* $f(\lambda x + (1-\lambda)y) \geq \min\{f(x), f(y)\}$.

Definition 5 (quasiconcave closure). *The quasiconcave closure of a function* $f : [0,1] \to \mathbb{R}$ *defined on* $[0,1]$ *is the pointwise lowest quasiconcave and upper semicontinuous function that is no less than* f *at every point on* $[0,1]$.

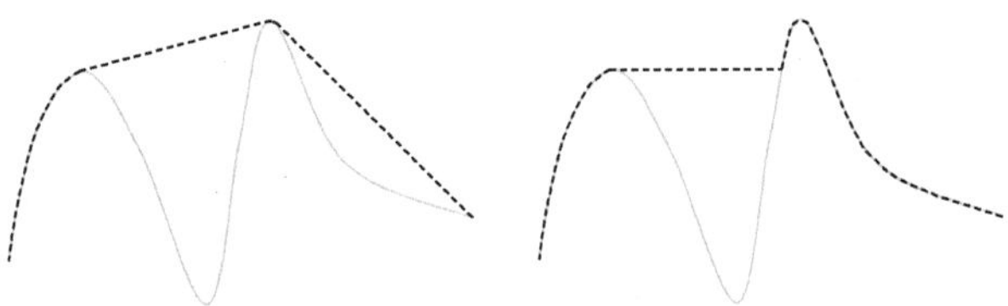

Fig. 2. A function with its concave (left) and quasiconcave (right) closures.

Next, we present a useful lemma of the quasiconcave closure of the sender's expected utility function. It says the quasiconcave closure of $\hat{v}$ is also a piecewise function and shows the monotonicity of the quasiconcave closure. Above we say the function $\hat{v}$ corresponds to the partition $\hat{P}$. It is actually true that any piecewise function f on $[0,1]$ corresponds to a partition P of [0,1] consisting of all subdomains of f. The proof is simple and omitted. Now we use this fact to formally define the monotonicity of the piecewise function with respect to intervals (Fig. 2).

Definition 6 (monotone with respect to intervals). *Consider a one-dimensional piecewise function* $f : [0,1] \to \mathbb{R}$ *corresponding to a partition* $\{I_i\}_{i=1}^n$ *of* $[0,1]$. f *is increasing (decreasing) w.r.t intervals if for all* $i, j \in [n]$ *that* $i < j$ *and for all* x, y *that* $x \in I_i, y \in I_j$, $f(x) \leq (\geq) f(y)$.

Then we formally present our lemma of the quasiconcave closure as follows.

Lemma 1. *Let* $\overline{V}$ *be the quasiconcave closure of the sender's utility function* $\hat{v}$. *Then we have*

1. *The quasiconcave closure* $\overline{V}$ *is a piecewise function corresponding to a partition* $\overline{P} = (\{J_j\}_{j=1}^m, \{\beta_j\}_{j=0}^m)$ *of* $[0,1]$ *such that* $\overline{V}$ *is continuous on each interval* J_j *for all* $j \in [m]$ *and* β_j *is a discontinuity for* $1 \leq j < m$.
2. *There exists* $i^* \in [n]$ *and* $j^* \in [m]$ *such that* $I_{i^*} \subset J_{j^*}$, *i.e. there exists an interval in* $\overline{P}$ *such that it contains the interval* I_{i^*} *corresponding the sender's favorite action* a_1.
3. *For all* $j < j^*$, $\overline{V}$ *is increasing w.r.t intervals and for all* $j > j^*$, $\overline{V}$ *is decreasing w.r.t intervals.*

Now we are ready to describe our main result in this section. It shows how to construct a smoothed quasiconcave closure directly from the quasiconcave closure and there is no gap between the optimal signaling scheme and optimal ex-post IR signaling scheme if and only if such a smoothed quasiconcave closure is concave.

Definition 7 (smoothed quasiconcave closure). *Let the quasiconcave closure $\overline{V}$ corresponds to the partition $\overline{P} = (\{J_j\}_{j=1}^m, \{\beta_j\}_{j=0}^m)$ of $[0,1]$ mentioned in Lemma 1. The smoothed quasiconcave closure Γ is a function defined on [0,1] that for all $1 \le j \le m$, if $x \in [\beta_{j-1}, \beta_j]$,*

$$\Gamma(x) = \frac{\overline{V}(\beta_j) - \overline{V}(\beta_{j-1})}{\beta_j - \beta_{j-1}}(x - \beta_{j-1}) + \overline{V}(\beta_{j-1}).$$

That is, we smoothen the quasiconcave closure by connecting adjacent endpoints $(\beta_{j-1}, \overline{V}(\beta_{j-1}))$ and $(\beta_j, \overline{V}(\beta_j))$ for all $1 \le j \le m$. We provide a simple example in Fig. 3 to illustrate the definition of smoothed quasiconcave closure.

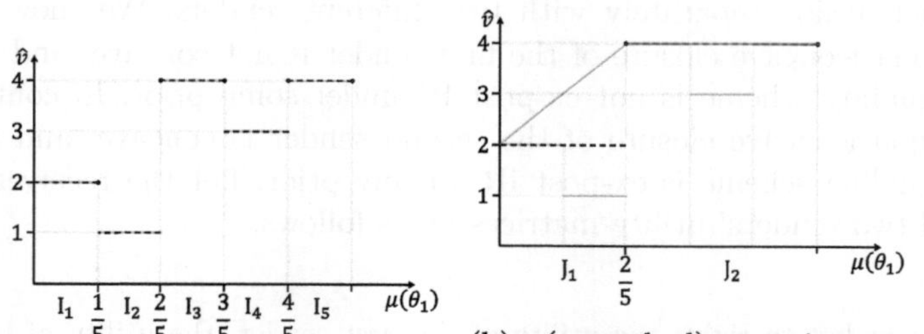

(a) sender's utility function

(b) (smoothed) quasiconcave closure

Fig. 3. A simple example of smoothed quasiconcave closure. Left: the curve of the sender's expected utility function $\hat{v}$ corresponding to a partition $\hat{P} = (\{I_i\}_{i=1}^5, \{\alpha_i\}_{i=0}^5)$ of $[0,1]$ that $\alpha_i = \frac{i}{5}$ for all $0 \le i \le 5$. Right: the dashed line is the quasiconcave closure $\overline{V}$ corresponding to a partition $\overline{P} = (\{J_1, J_2\}, \{\beta_0 = 0, \beta_1 = \frac{2}{5}, \beta_2 = 1\})$ and the red line is the smoothed quasiconcave closure. (Color figure online)

Theorem 2. *The optimal signaling scheme is ex-post IR if and only if Γ is concave.*

Remark 1. Given $|A| = n$, Theorem 2 provides an algorithm to determine if the optimal signaling scheme is ex-post IR with the $O(n \log n)$ complexity, which is better than simply running a linear program to solve the optimal signaling scheme and check whether it is ex-post IR. In detail, we can sort each row of the receiver's utility matrix according to the first entry of each row $u(a_i, \theta_1)$ for all $i \in [n]$ with complexity $O(n \log(n))$. Then we can solve the sender's expected utility function $\hat{v}$ with complexity $O(n)$. Because when the row of the receiver's utility matrix is sorted, we can compute the threshold of each piece of $\hat{v}$ by inequality $\mu(\theta_1)u(a_i, \theta_1) + \mu(\theta_2)u(a_i, \theta_2) \leq \mu(\theta_1)u(a_{i+1}, \theta_1) + \mu(\theta_2)u(a_{i+1}, \theta_2)$ for all $1 \leq i < n$. At last, we can construct Γ and check whether it is concave in $O(n)$ complexity according to Theorem 2.

The intuition behind Theorem 2 lies in the two fundamental properties of the quasiconcave closure $\overline{V}$: (1) $\overline{V}(\mu_2) \geq \min\{\overline{V}(\mu_1), \overline{V}(\mu_3)\}$ for all $\mu_1 \leq \mu_2 \leq \mu_3 \in [0,1]$ (quasiconcavity); (2) $\overline{V}$ is increasing (decreasing) w.r.t intervals on the left (right) of Interval J_{j^*} (monotonicity). On one hand, if Γ is concave, it is the concave closure of the expected utility function. Then for any prior, the optimal signaling scheme will induce better actions for the sender than the best response under prior because of the monotonicity, which means the optimal signaling scheme is ex-post IR. On the other hand, if the optimal signaling scheme is ex-post IR, we can prove that the concave closure is actually the smoothed quasiconcave closure by the definition of quasiconcavity. Hence the smoothed quasiconcave closure is concave.

Finally, we provide a toy example for that can help us better interpret the result of Theorem 2 through visualization. In this example, a fixed receiver plays Bayesian persuasion separately with two different senders. We show that the smoothed quasiconcave closure of the first sender is not concave, and thus, the optimal signaling scheme is not ex-post IR under some prior. In contrast, the smoothed quasiconcave closure of the second sender is concave, and thus, the optimal signaling scheme is ex-post IR for any prior. Let the receiver's utility matrix and two senders' utility matrices be as follows.

Table 1. From left to right: the utility of the first sender, the utility of the second sender, the utility of the receiver.

v	$\theta_1 = 0$	$\theta_2 = 1$
a_1	4	4
a_2	3	3
a_3	2	2
a_4	1	1

v	$\theta_1 = 0$	$\theta_2 = 1$
a_1	4	4
a_2	7/2	7/2
a_3	2	2
a_4	1	1

v	$\theta_1 = 0$	$\theta_2 = 1$
a_1	8	0
a_2	7	3
a_3	0	8
a_4	3	7

The utility functions of these senders are both state independent. The first sender's utility is $4, 3, 2, 1$ when the receiver's best response is a_1, a_2, a_3, a_4 respectively. The second sender's utility is $4, 7/2, 2, 1$ when the receiver's best response is a_1, a_2, a_3, a_4 respectively. The receiver takes action a_3 when the posterior probability of state being zero $\mu_s(0) \leq \frac{1}{4}$, takes action a_4 when $\frac{1}{4} \leq \mu_s(0) \leq \frac{1}{2}$, takes action a_2 when $\frac{1}{2} \leq \mu_s(0) \leq \frac{3}{4}$, and takes action a_1 otherwise. The figures of the senders' expected utility functions are Figs. 4a and 5a. The optimal signaling schemes can be visualized from Figs. 4b and 5b. In the first case, when the prior is $\frac{1}{2} \leq \mu_s(0) \leq \frac{3}{4}$, the optimal disclosure policy is divided the prior into two posterior s_1, s_2, $\mu_{s_1}(0) = 0$ and $\mu_{s_2}(0) = \frac{3}{4}$ with the best response being a_3 and a_1 respectively. But the best response under the prior is $a_2 > a_3$. So the sender is worse off under s_1, which means the optimal signaling scheme is not ex-post IR. In contrast, the sender is never worse off under any signal of the optimal disclosure policy. Meanwhile, the figures of (smoothed) quasiconcave closure Γ are in Figs. 4c and 5c. The difference is that the smoothed quasiconcave closure is not concave in the first case while is concave in the second case.

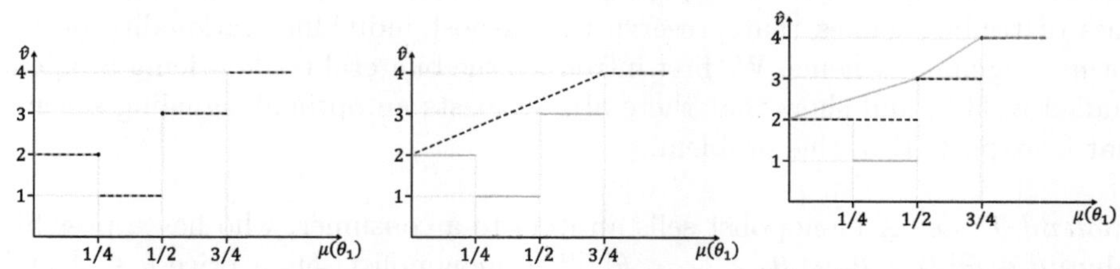

(a) sender's utility function (b) concave closure (c) (smoothed) quasiconcave

Fig. 4. The example corresponding to the first sender (Table 1) that the optimal signaling scheme is not ex-post IR. From left to right: the curve of the first sender's utility function (dashed line in Fig. 4a), the concave closure of the first sender's utility function (dashed line in Fig. 4b), the quasiconcave closure (dashed line in Fig. 4c) and the smoothed quasiconcave closure (red line in Fig. 4c) of the first sender's expected utility function. The smoothed quasiconcave closure is not concave. (Color figure online)

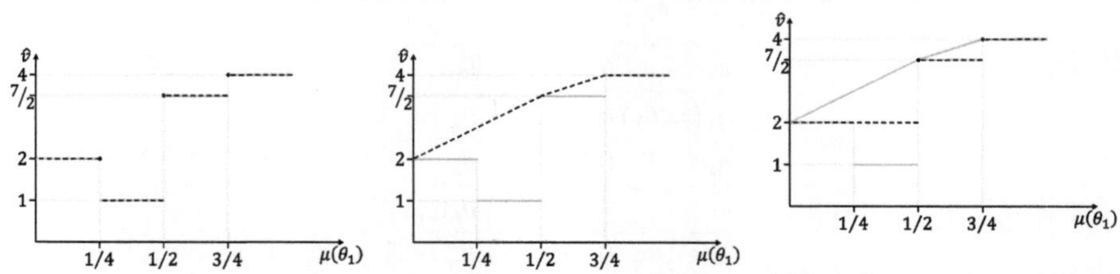

(a) sender's utility function (b) concave closure (c) (smoothed) quasiconcave

Fig. 5. The example corresponding to the second sender (Table 1) that the optimal signaling scheme is ex-post IR. From left to right: the curve of the second sender's expected utility function (dashed line in Fig. 5a), the concave closure of the second sender's utility function (dashed line in Fig. 5b), the quasiconcave closure (dashed line in Fig. 4c) and the smoothed quasiconcave closure (red line in Fig. 4c) of the second sender's utility function. The smoothed quasiconcave closure is concave. (Color figure online)

4.2 No Gap in the Bilateral Trade Market

In this section, we introduce a class of Bayesian persuasion problems in which the sender will not be affected by the ex-post IR constraint, which we term as *trading games*. The discovery of trading games is inspired by the bilateral trade persuasion problem studied in [3,8]. We first observe that the bilateral trade persuasion problem always has an optimal signaling scheme that is ex-post IR. Drawing from this insight, we identify three key properties that guarantee the ex-post IR of the optimal signaling scheme. These three properties yield a broader class of trading games that preserve the ex-post individual rationality of the optimal signaling scheme. We first introduce the bilateral trade pricing problem studied in [3,8] and show that there always exists an optimal signaling scheme that is ex-post IR in this problem.

Bilateral Trade. A monopolist sells an item to a consumer, who has n possible values $\theta \in \Theta$: $0 < \theta_1 < \theta_2 < ... < \theta_n$. The monopolist sets a price $a \in A$ and the trade only happens when the consumer's value is higher than the price. The monopolist will not set a price between any two closest values θ_i and θ_{i+1} for some i because the monopolist can increase the price to θ_{i+1} without decreasing the probability of making a deal. Therefore, we can let $A = \Theta$ without loss of generality. The monopolist doesn't know the consumer's value but shares a prior μ_0 of the value with the consumer. Before the trade, the consumer commits to a signaling scheme ϕ. Then the monopolist sets the price according to the posteriors μ_s for any realized signal $s \in S$. The utility of the consumer is $v(a, \theta) = (\theta - a) \cdot \mathbb{1}_{\theta \geq a}$ and the utility of the monopolist is $u(a, \theta) = a \cdot \mathbb{1}_{\theta \geq a}$.

Table 2. The utility of receiver in the bilateral trade

v	θ_1	θ_2	$\cdots$	θ_n
$a_1(=\theta_1)$	θ_1	θ_1	$\cdots$	θ_1
a_2	0	θ_2	$\cdots$	θ_2
$\cdots$	$\cdots$	$\cdots$	$\cdots$	θ_{n-1}
a_n	0	0	$\cdots$	θ_n

Proposition 1. *In the bilateral trade persuasion problem defined in [3], there exists an optimal signaling scheme that is ex-post IR.*

We then extend the bilateral trade game to a general class of games called *trading games*, in which the ex-post IR constraint does not affect the sender's expected utility. We first make a few observations about the bilateral trade problem: (1) the utility of the receiver is upper-triangular (2) the non-zero part (see Table 2) corresponding columns of the receiver utility is increasing from top to bottom (3) the sum of the non-zero part corresponding columns of the sender utility and the receiver utility is a constant (see formal definition in Definition 8). In the context of bilateral trade games, the first feature means that the buyer won't buy the item if the price is higher than its value; the second feature means that when the price is higher, the buyer's utility is lower; the third feature means that once the trade happens, the social welfare is a constant equals to the buyer's value. We extend the bilateral trade game to a general class of games called *trading games*. As long as a game satisfies the aforementioned properties, we call it a trading game (see Definition 8).

Definition 8. *(Trading game) A game (A, Θ, v, u) is a trading game if $|A| = |\Theta| = n$ for any integer n and the following conditions hold*

1. *u is upper-triangular and non-negative: $u(a_i, \theta_k) \geq 0$ for all $i, k \in [n]$ and $u(a_i, \theta_k) = 0$ if $i > k$.*
2. *the non-zero part corresponding columns of the receiver utility is increasing: $u(a_i, \theta_k) \leq u(a_j, \theta_k)$ for all $i, j, k \in [n]$ that $i < j \leq k$.*
3. *$v(a_i, \theta_k) + u(a_i, \theta_k) = c_k$ where c_k is a positive constant for all $i, k \in [n]$ that $i \leq k$.*

These three conditions generally hold in trading problems with a buyer and a seller that do not have to be a bilateral trade problem. The following example is about the Bayesian persuasion problem in the first price auction (FPA) with a reserve price, which we show is also a trading game.

Example 2 (First-price auction). Consider the first-price auction with multiple bidders, when we set reserve prices, we only care about the maximum bid. It is useful to abstract bidders in the auction as a single "mega-bidder" whose value is the maximum of values and who always bids the maximum of all bids (see [6]). Now we can consider the pricing problem between the seller(receiver) and the representative winning bidder(sender). The state is the value of the bidder $\theta_1 \leq \theta_2 \leq ... \leq \theta_n$. The action is the reserve price set by the seller. Notice that the seller will never set a reserve price between some θ_{i-1} and θ_i because he has the incentive to increase the reserve price to exact θ_i. Then the receiver has exactly n actions $a_1, ..., a_n$ which a_i is setting the reserve price at θ_i. The bidder's bidding function $b(r, v)$, which is determined by the reserve price r and the bidder's value v, is increasing w.r.t to the reserve price r. We have $v(a_i, \theta_j) = \theta_j - b(\theta_i, \theta_j)$ if $i \leq j$ otherwise $v(a_i, \theta_j) = 0$ and $u(a_i, \theta_j) = b(\theta_i, \theta_j)$ if $i \leq j$ otherwise $u(a_i, \theta_j) = 0$.

We then show that as long as the game is a trading game, there exists an optimal signaling scheme that is ex-post IR.

Theorem 3. *For all trading games, there exists an optimal signaling scheme that is ex-post IR.*

To illustrate the idea of the proof, we first provide a lemma about the property of the receiver's best response, which relies on the first two properties of Definition 8.

Lemma 2. *Assume the utility matrix u is upper-triangular, non-negative, and the non-zero part corresponding columns of μ is increasing, for any subset $S \subset [n]$, there exists a posterior belief μ_S supported on S that the receiver is indifferent with actions a_k for all $k \in S$ under the posterior μ_S.*

Based on how the receiver best responds, we show that there is a specific structure of ex-post IR signaling schemes. Recall that a signaling scheme induces a set of possible posteriors. We now describe how to construct all posteriors for such signaling schemes step by step. In the first step $t = 1$, let μ_1 be the prior μ_0 and $S_1 = \{1, 2, \cdots, n\}$. We compute a posterior μ_{S_1} such that the receiver is in indifferent among θ_i for all $i \in S_1$. Then we subtract certain proportions of μ_{S_1} from μ_1 until some supports of μ_0 become zero. In detail, let $\mu_2 = \mu_1 - p_1\mu_{S_1}$ such that $p_1 = \min_{k \in S_1} \frac{\mu_1(\theta_k)}{\mu_{S_1}(\theta_k)}$. μ_{S_1} is the first posterior we want and its probability in the signaling scheme is p_1. Let S_2 be the indexes of the support of μ_2, i.e., $S_2 = \{i | \mu_2(\theta_i) > 0\}$. Given μ_2 and S_2, we repeat this process at $t = 2$ and so forth. This procedure will stop in finite steps because at least one component μ_t will become zero for $t \geq 1$. Suppose it stops in m steps. The signaling scheme consist of m posteriors $\mu_{S_1}, \cdots, \mu_{S_m}$ and $\mu_0 = \sum_{t=1}^{m} p_t \mu_{S_t}$. We know that these posteriors exist because of Lemma 2. We will show that $S_m = \{k^*\}$ which means that θ_{k^*} is the last component that becomes zero. Therefore, we end up dividing the prior into several posterior beliefs that a_{k^*} is one of the best responses under each of them. The sender hence always enjoys a utility better than the utility under the prior, which means the signaling scheme is ex-post IR. We then complete the proof by using the third property of Definition 8 to show that such a signaling scheme is optimal. The detailed proof is in Appendix B.

5 Comparison of the Sender's Optimal Utility

In this section, we compare the sender's expected utility under different credibility requirements. We consider four information transmission models with different credibility levels: Bayesian persuasion, ex-post IR Bayesian persuasion, credible persuasion [12], and cheap talk [5]. In many previous works about Bayesian persuasion with partial commitment, such as [12,19], the optimal sender utility achieved by their model is between the Bayesian persuasion and cheap talk. Therefore, it is natural to compare the sender's optimal utility of our ex-post IR Bayesian persuasion with the commitment power of Bayesian persuasion and cheap talk. Besides, since our model considers the credibility of the sender, we compare the ex-post IR Bayesian persuasion to another model called *credible persuasion* in the paper by [12], which also considers the credibility of the sender. Given the sender's utility v, receiver's utility u, the optimal sender utilities gained in these four models is the functions of prior μ_0 and denoted as $V_{\text{BP}}(\mu_0)$, $V_{\text{EXPOST}}(\mu_0)$, $V_{\text{CREDIBLE}}(\mu_0)$, $V_{\text{CHEAP}}(\mu_0)$. We formally give the definition that the optimal sender utility achieved in one information transmission model is better than the optimal sender utility achieved in another information transmission model.

Definition 9. *Consider two information transmission model X, Y. If for any v, u, μ_0, we have $V_X(\mu_0) \geq V_Y(\mu_0)$, we say the optimal sender utility achieved in X is better than the optimal sender utility achieved in Y, which is denoted by $V_X \succeq V_Y$.*

Next, we compare the optimal sender utility of the aforementioned four information transmission models. We give a brief introduction of credible persuasion in Appendix C.

Theorem 4. *The ranking of the optimal sender's utility is as follows:*

1. $V^{\text{BP}} \succeq V^{\text{credible}}, V^{\text{expostBP}} \succeq V^{\text{cheaptalk}}$.
2. *There exists v, u such that for any μ_0, $V^{\text{credible}}(\mu_0) \geq V^{\text{expostBP}}(\mu_0)$.*
3. *There exists v, u such that for any μ_0, $V^{\text{expostBP}}(\mu_0) \geq V^{\text{credible}}(\mu_0)$.*

The first part of Theorem 4 that ex-post IR Bayesian persuasion and credible persuasion are partial commitment models between Bayesian persuasion and cheap talk. The second part of Theorem 4 that the commitment power of ex-post IR Bayesian persuasion and credible persuasion differs in different cases based on a series of lemma that show when the optimal sender utility of ex-post IR Bayesian persuasion or credible persuasion is the same as the optimal sender utility of Bayesian persuasion. Then we can construct examples when one model's optimal sender utility is equal to the optimal sender utility of Bayesian persuasion and the other model's optimal sender utility is strictly smaller than the optimal sender utility of Bayesian persuasion.

6 Conclusion

In this paper, we studied ex-post IR Bayesian persuasion, a novel Bayesian persuasion model with restricted commitment power. We show that the ex-post IR constraint is a linear constraint; hence, linear programming can still solve the optimal ex-post IR signaling scheme. Our main results give broad conditions when there is no utility gap for the sender between the optimal signaling scheme and the optimal ex-post IR signaling scheme. At last, we show that, in the sense of the sender's optimal utility, our model and credible persuasion is an information transmission model between Bayesian persuasion and cheap talk. We further show when the optimal sender utility of our model is higher than the optimal sender utility of credible persuasion and vice versa.

References

1. Ali, S.N., Lewis, G., Vasserman, S.: Voluntary disclosure and personalized pricing. In: Proceedings of the 21st ACM Conference on Economics and Computation, pp. 537–538 (2020)
2. Ben-Porath, E., Dekel, E., Lipman, B.L.: Mechanisms with evidence: commitment and robustness. Econometrica **87**(2), 529–566 (2019)

3. Bergemann, D., Brooks, B., Morris, S.: The limits of price discrimination. Am. Econ. Rev. **105**(3), 921–957 (2015)
4. Best, J., Quigley, D.: Honestly dishonest: a solution to the commitment problem in Bayesian persuasion. Technical report, Mimeo (2016)
5. Crawford, V.P., Sobel, J.: Strategic information transmission. Econometrica J. Econom. Soc. 1431–1451 (1982)
6. Feng, Z., Lahaie, S., Schneider, J., Ye, J.: Reserve price optimization for first price auctions. arXiv preprint arXiv:2006.06519 (2020)
7. Fong, Y.F., Liu, T., Wright, D.J.: On the role of verifiability and commitment in credence goods markets. Int. J. Ind. Organ. **37**, 118–129 (2014)
8. Glode, V., Opp, C.C., Zhang, X.: Voluntary disclosure in bilateral transactions. J. Econ. Theory **175**, 652–688 (2018)
9. Hart, S., Kremer, I., Perry, M.: Evidence games: truth and commitment. Am. Econ. Rev. **107**(3), 690–713 (2017)
10. Hu, X., Lei, H.: Information transmission in monopolistic credence goods markets. arXiv preprint arXiv:2303.13295 (2023)
11. Kamenica, E., Gentzkow, M.: Bayesian persuasion. Am. Econ. Rev. **101**(6), 2590–2615 (2011)
12. Lin, X., Liu, C.: Credible persuasion. arXiv preprint arXiv:2205.03495 (2022)
13. Lipnowski, E.: Equivalence of cheap talk and Bayesian persuasion in a finite continuous model (2020)
14. Lipnowski, E., Ravid, D.: Cheap talk with transparent motives. Econometrica **88**(4), 1631–1660 (2020)
15. Lipnowski, E., Ravid, D., Shishkin, D.: Persuasion via weak institutions. J. Polit. Econ. **130**(10), 2705–2730 (2022)
16. Lucier, B., Syrgkanis, V.: Greedy algorithms make efficient mechanisms. In: Proceedings of the Sixteenth ACM Conference on Economics and Computation, pp. 221–238 (2015)
17. Luo, Z., Rozenas, A.: Strategies of election rigging: trade-offs, determinants, and consequences. Q. J. Polit. Sci. **13**(1), 1–28 (2018)
18. Mathevet, L., Pearce, D., Stacchetti, E.: Reputation for a degree of honesty. Technical report, Working paper (2022)
19. Min, D.: Bayesian persuasion under partial commitment. Econ. Theor. **72**(3), 743–764 (2021)
20. Nguyen, A., Tan, T.Y.: Bayesian persuasion with costly messages. J. Econ. Theory **193**, 105212 (2021)
21. Rayo, L., Segal, I.: Optimal information disclosure. J. Polit. Econ. **118**(5), 949–987 (2010)
22. Salamanca, A.: The value of mediated communication. J. Econ. Theory **192**, 105191 (2021)

Optimal Scoring Rule Design under Partial Knowledge

Yiling Chen[1] and Fang-Yi Yu[2(✉)]

[1] Harvard University, Cambridge, MA 02138, USA
yiling@seas.harvard.edu
[2] George Mason University, Fairfax, VA 22030, USA
fangyiyu@gmu.edu

Abstract. This paper studies the design of optimal proper scoring rules when a principal has partial knowledge of an agent's signal distribution. Recent work [23] characterizes the proper scoring rules that maximize the increase of an agent's payoff when the agent chooses to access a costly signal to refine a posterior belief from his prior prediction, under the assumption that the agent's signal distribution is fully known to the principal. In our setting, the principal only knows about a set of distributions where the agent's signal distribution belongs. We formulate the scoring rule design problem as a max-min optimization that maximizes the worst-case increase in payoff across the set of distributions.

We propose an efficient algorithm to compute an optimal scoring rule when the set of distributions is finite, and devise a fully polynomial-time approximation scheme that accommodates various infinite sets of distributions. We further remark that widely used scoring rules, such as the quadratic and log rules, as well as previously identified optimal scoring rules under full knowledge [23], can be far from optimal in our partial knowledge settings.

Keywords: Proper scoring rule · Crowdsourcing · Optimization

1 Introduction

Proper scoring rules are scoring functions that incentivize truthful information elicitation: an agent with a subjective belief about an uncertain event maximizes his expected score by making a prediction according to his belief. If the agent can acquire a costly signal to refine his belief, which proper scoring rules maximally incentivize the agent's information acquisition? Recent work [23] explores this question under the assumption that the principal has full knowledge about the agent's information structure. In contrast, we investigate this optimal scoring rule design question when the principal has only partial knowledge about the agent's information structure, only knowing the set of information structures that the agent's belongs to.

Incentivizing information acquisition is crucial in many real-world applications. For example, in conference reviewing, a reviewer could spend little to

M. Mavronicolas et al. (Eds.): WINE 2024, LNCS 15534, pp. 383–400, 2026.
https://doi.org/10.1007/978-3-032-08560-3_22

no effort and form his prior assessment about a paper primarily based on the length of the paper, the amount of grammatical errors in the introduction, or the references cited. But a conference chair would aspire for a high-effort review where the reviewer carefully reads the paper and then forms his, more informed, posterior assessment. As another example, in crowdsourced prediction for the reproducibility of scientific studies [24,27], an expert could form a prediction on the replication outcome of a study based on his general knowledge about the study's topic, the reputation of the publication venue, and the authors' affiliations. But a more valuable and accurate prediction requires the expert to examine the study's methodology and evaluation procedures carefully. In these applications, the principal ideally hopes to devise a proper scoring rule that maximizes the expected increase in score if the agent acquires the information, to maximally incentivize information acquisition. The principal however only has limited knowledge about the agent's information structure.

More formally, an agent's information structure consists of two parts: the agent's prior belief about the random variable of interest (the prior) and the distribution of the agent's signal conditioned on every realization of the random variable (an experiment). We formulate the optimal scoring rule design problem as a max-min optimization that maximizes the agent's worst-case increase in score across the set of possible information structures. We explore the problem for four settings of principal's knowledge, ranging from the special full-knowledge case to varying degree of partial knowledge:

1. The principal knows both the prior and the experiment and hence the set of information structures is a singleton. We reprove [23]'s results and show that the optimal scoring rules are v-shaped in Proposition 1. In addition, Proposition 1 also provides a close-form expression for an optimal scoring rule.
2. The principal knows the prior but is uncertain about the experiment. The set of information structures shares the same prior, as introduced in Example 2. Interestingly, we show in Theorem 2 that the same v-shaped scoring rule in Proposition 1 is also optimal.
3. The principal knows the experiment but is uncertain about the agent's prior. The set of information structures shares the same experiment, as introduced in Example 1. We present a fully polynomial time approximation scheme (FPTAS) that outputs approximately optimal piece-wise linear scoring rules in Theorem 4.
4. The principal is uncertain about both the prior and the experiment. In Theorem 3, we propose an efficient algorithm to compute an optimal scoring rule when the set of distributions is finite. Additionally, for ρ-correlated information structures defined in Example 3, we develop an FPTAS to find approximately optimal scoring rules in Theorem 4. ρ-correlated information structures have interesting connections to Beta-Bernoulli model [28] and noise operator in Boolean function analysis [31].

We then run simulations to evaluate the performance of two frequently used proper scoring rules (quadratic scoring rule and log scoring rule), the v-shaped scoring rule (which is optimal in the first two settings) and the piece-wise linear scoring rules obtained by our FPTAS for the ρ-correlated information structures. The simulations show that our algorithm's piecewise linear scoring rules perform well, and provide the most uniform incentive when the signal and the state have a large correlation. However, when the correlation is small, log scoring rule outperforms our piecewise linear scoring rules and other scoring rules. In particular, as the correlation decreases, we observe that the piecewise scoring rule approaches the log scoring rule which may suggest an interesting connection between ρ-correlated information structure and the log scoring rule. The v-shaped scoring rule empirically performs worst in our partial knowledge setting.

Organization and Contribution of Technical Results. Using Savage's representation of proper scoring rules [18,25,36], we convert variable space of our max-min optimization problem from scoring rules to convex functions where the increase in payoff becomes the Jensen's gap of the associated convex function as Problem 1.

In Sect. 3.1, we derive a geometric interpretation of the optimization problem. Lemma 1 shows that the information gain amounts to how *curved* the associated convex function is at the prior of the information structure. This interpretation is critical as it leads to Theorem 2 that shows v-shaped scoring rule is optimal for known prior setting (Example 2).

In Sects. 3.2 and 3.3, we delve into the scenario of unknown prior settings. We first use linear program to find an optimal piecewise linear scoring rule when the collection of information structures is finite in Sect. 3.2. Then we venture into the domain of infinite information, and provide an FPTAS (Theorem 4) for various infinite sets (Examples 1 and 3). Informally, our FPTAS runs the linear program on a finite subset of the infinite set of information structures, and provides approximation guarantees as long as the finite subset is an ϵ-covering of the original set under earth mover's distance. To this end, we relate the earth mover's distance to our optimization problem in Lemma 3. Additionally, we design a novel coupling that can bound the earth mover's distance of two posterior predictions by the total variation distance of their joint distributions on signal and state in Lemma 4. This coupling argument may be of independent interest for designing approximation algorithms for information aggregation and elicitation.

Related Work. Our problem can be seen as purchasing prediction from strategic people. The work on this topic can be roughly divided into two categories according to whether agents can misreport their signal or prediction. Below we focus on the relationship of our work to the most relevant technical scholarship.

In the first category, to ensure agents reporting their signals or predictions truthfully, there are two settings according to whether money is used for incentive alignment. In the first setting, the analyst uses monetary payments to incentivize agents to reveal their data truthfully. The challenge is to ensure truth-telling gets the highest payments. Existing works verify agents' reports by using either an observable ground truth (proper scoring rules) or peers' reports (peer prediction). For the second setting, individuals' utilities directly depend on an inference or learning outcome (e.g. they want a regression line to be as close to their own data point as possible) and hence they have incentives to manipulate their reported data to influence the outcome [8,11,22,26,34].

Our setting generalizes [23]'s. Our max-min optimization formulation captures the principal's partial knowledge about agents' information structure, while theirs focuses more on known information structure. We consider both ex-ante and ex-post setting which allow us to compare our result to the log scoring rule which is arguably one of the most important proper scoring rules but cannot be ex-post bounded. They show the optimal scoring rule is v-shaped when the information structure is known. We generalize the result to known prior setting in Theorem 2.

[29] also study optimal scoring rule design problem, but is more related to sequential method [17]. Instead of imposing bounded payment conditions, their objective comprises both payment and accuracy. They consider a special case of Beta-Bernoulli information structures where the prior is uninformative and design scoring rule for an agent to sequentially acquire samples from a Bernoulli distribution.

Finally, [33] study an information acquisition problem: When the information structure and the cost of information are known, the principal chooses a proper scoring rule (menu of contracts) to incentivize information acquisition as cheaply as possible subject to limited liability. Informally, their formulation can be seen as the dual of our problem. Instead of maximizing information gain subjected to bounded payment conditions, they want to minimize payment with a lower bound on the information gain. Their optimal scoring rules are also v-shaped and similar to our known information structure setting.

In the second category, agents cannot misreport their signal or prediction. The problem of purchasing data from people has been investigated with different focuses, e.g. privacy concerns [10,12,15,16,30,37], effort and cost of data providers [1,5,7,9,35,38], and reward allocation [2,14].

More generally, our problem is also related to contract theory [19]. Recent work also studies optimal contracts where the principal does not fully know the agent's cost. [3,21] Moreover, [4] study delegation problem that tries to optimize the efforts of others. However, the critical difference that sets ours apart from these works is the principal's preference. Most of the work in contract theory aims to maximize the principal's utility, but ours treats the principal's preference as a budget constraint and optimize the efforts of the agents. Our formulation may be more suitable for complicated problems, e.g., peer grading or conference review, that consists of multiple sub-problems, and the principal's utility can not be easily decomposed according to sub-problems. For the modelling choice, [13] studies auction design where the unknown value distribution is from a known collection, without assuming a prior over the collection.

2 Model and Preliminaries

In Sect. 2.1, we first introduce proper scoring rules and two boundedness notions. Then we define information structures that formalize the relationship between the costly signal and the event of interest, and we specify the principal's partial knowledge and introduce three motivating examples. Finally, we define the value of information and our scoring rule design problem. We further simplify our problem by connecting scoring to convex function in Sect. 2.2.

2.1 Optimal Scoring Rule for Costly Information

This paper studies the design of scoring rules for binary state[1] which maps an agent's reported prediction x and the realized ground state $\omega \in \{0,1\}$ to a score for the agents, $PS(\omega, x)$. As a principal designs a scoring rule for the agent, one desirable scoring rule should elicit truthful predictions.

Definition 1. *A scoring rule is* proper *if for all predictions $x, x' \in [0,1]$,*

$$\mathbb{E}_{\omega \sim x}[PS(\omega, x)] \geq \mathbb{E}_{\omega \sim x}[PS(\omega, x')].$$

In other words, if an agent's prediction for $\omega = 1$ is x, he cannot gain a higher score by misreporting x'. Here $\omega \sim x$ denote random variable $\omega = 1$ with probability x, and 0 otherwise.

Another desirable property of a scoring rule is boundedness as the principal has a finite budget for the reward. Here we consider the following two notions.

Definition 2. *A scoring rule PS is* ex-post bounded *by $B > 0$ if for all $\omega \in \{0,1\}$ and $x \in [0,1]$, $PS(\omega, x) \in [0, B]$. Alternatively, PS is* ex-ante bounded *by B if for all $x \in [0,1]$ $\mathbb{E}_{\omega \sim x}[PS(\omega, x)] \in [0, B]$.*

In addition to properness and boundedness, the principal often wants to design proper scoring rules that incentivize effort. Specifically, when the agent can refine his prediction by exerting a binary effort for a costly signal, how can the principal design a scoring rule that maximizes the agent's perceived gain from exerting effort without complete knowledge of the prior and posterior distributions?

Prior, Posterior, and Information Structures. The agent can access a costly signal $S \in \mathcal{S}$ with a finite support that improves his prediction of the unknown binary state of the world $W \in \{0,1\}$. Specifically, the agent has an ***information structure*** P which is a joint distribution on the state and signal. An information structure P consists of a *prior* $\pi \in [0,1]$ of the state and an *experiment* $\sigma : \{0,1\} \to \Delta(\mathcal{S})$ which is a conditional distribution of the signal given the state, so that for all state $\omega \in \Omega$ and signal $s \in \mathcal{S}$,

$$\pi = P(W=1) \text{ and } \sigma(s|\omega) = P(S = s \mid W = \omega).$$

We will refer P by the pair of prior and experiment (π, σ).

Given P with (π, σ), if the agent ignores the signal, his truthful prediction of the state is the *prior*, $P(W=1) = \pi$. If the agent accesses the signal and

[1] We discuss the d-dimensional setting in the appendix.

sees $S = s$, his truthful prediction becomes the *posterior* $P(W = 1 \mid S = s) = \frac{\pi\sigma(s|1)}{\pi\sigma(s|1)+(1-\pi)\sigma(s|0)}$. The posterior prediction $X(s) := P(W = 1 \mid S = s)$ is a random variable over the randomness of signal. We will omit s and write the random variable of posterior prediction as X. For instance, the expectation of posterior equals prior $\mathbb{E}X = \pi$ by Bayes' rule.

Partial Knowledge of Information Structures. In our partial knowledge setting, the principal only knows a collection of information structures $\mathcal{P}$ that the agent's information structure falls into. Below are three examples of collections of information structures that model the principal's partial knowledge. We will use these as running examples throughout the paper.

First the principal may know the signal distribution given ground state (experiment) but does not know the agent's background (prior). Intuitively, this captures online crowdsourcing setting, e.g., image annotation, where the background of the agent is unknown, but quality of signal can be controlled. Note that the priors are bounded away from zero and one by $\delta > 0$, because if the prior is zero or one, the posterior is identical to the prior.

Example 1. Given an experiment σ and $0 < \delta \leq 1/2$, a collection of information structures with homogeneous experiment is $\mathcal{P}_\sigma = \{(\pi, \sigma) : \pi \in [\delta, 1 - \delta]\}$.

To another extreme, the principal may know agent's prior but does not know the agent's experiment. This can model peer grading in classroom where the student's prior is known but the quality of his work is uncertain.

Example 2. Given a prior $\pi \in (0, 1)$ and a set of experiments σ_i where $i \in \mathcal{I}$, a collection of information structures with homogeneous prior is $\{(\pi, \sigma_i) : i \in \mathcal{I}\}$.

The first example has the same experiment for all information structures, and the second has the same prior. We provide an example where the prior and experiment both vary.

Example 3. Given $\rho \in [0, 1]$, $0 < \delta \leq 1/2$ and a binary signal space $\mathcal{S} = \{0, 1\}$, a ρ-correlated experiment produces signal that equals the ground state with probability ρ and samples from prior independently otherwise. A collection of ρ-correlated information structures $\mathcal{P}_\rho$ is

$$\{(\pi, \sigma) : \pi \in [\delta, 1 - \delta], \sigma(1|1) = \rho + (1 - \rho)\pi, \sigma(1|0) = (1 - \rho)\pi\}.$$

Note that the value ρ controls the correlation between the signal and the ground state. In particular, if $\rho = 0$, the signal is independent of the ground state, and if $\rho = 1$, the signal perfectly agrees with the ground state. Similar to Example 1, the priors are bounded away from zero and one.

There are several interesting interpretations of ρ-correlated experiments. First, the information structure is the posterior predictive distribution of Beta-Bernoulli model [28] given an additional sample, and ρ captures the strength of prior. Second, the distribution of the signal and the state is also known as ρ-correlated pair in Boolean function analysis [31]. We formalize this connection in the appendix.

Information Gain and Max-Min Optimization Problem. Now we formalize the agent's perceived gain from exerting effort under a proper scoring rule PS. Let $PS(x) := \mathbb{E}_{\omega \sim x} PS(\omega, x)$ be the expected score of truthful reporting a prediction x. Given PS and P with (π, σ), the expected score of the agent truthfully reporting his initial prediction is $PS(\pi)$. Alternatively, if the agent accesses the costly signal $S = s$, he gets score $PS(X(s))$. Thus, his expected score of accessing the costly signal before seeing it is $\mathbb{E}\, PS(X)$ over the randomness of X. Since $\mathbb{E}\, X = \pi$, the agent's gain from exerting effort is

$$\mathbb{E}\, PS(X) - PS(\pi) = \mathbb{E}\, PS(X) - PS(\mathbb{E}\, X).$$

We call the above ***information gain*** on P under the proper scoring rule PS. Because the information gain is fully determined by the posterior distribution random variable, we can use P and X interchangeably. Additionally, we let $\mathcal{X}$ be the collection of random variables of prediction induced from information structures P in $\mathcal{P}$.

To maximize the agent's gain from exerting effort for any possible information structure in $\mathcal{X}$, the principal finds a bounded proper scoring rule $PS : \{0,1\} \times [0,1] \to \mathbb{R}$ which maximizes the worst-case information gain

$$\max_{PS} \min_{X \in \mathcal{X}} \mathbb{E}\, PS(X) - PS(\mathbb{E}\, X).$$

We will simplify our optimization in Sect. 2.2 as Problem 1.

2.2 Savage's Representation of Proper Scoring Rules

We characterize the space of bounded proper scoring rules. First, Savage's representation of proper scoring rules connects proper scoring rules and convex functions.

Theorem 1 ([25,36]). *When the state space $\Omega = \{0,1\}$ is binary, for every proper scoring rule PS, there exists a convex function $H : [0,1] \to \mathbb{R}$ so that for all $x \in [0,1]$ and $\omega \in \{0,1\}$*

$$PS(\omega, x) = \begin{cases} H(x) + \partial H(x) \cdot (1-x) & \text{when } \omega = 1, \\ H(x) - \partial H(x) \cdot x & \text{when } \omega = 0 \end{cases}$$

where $\partial H(x)$ is a subgradient of H at x.

Conversely, for every convex function $H : [0,1] \to \mathbb{R}$, there exists a proper scoring rule such that the above condition hold.

With above characterization, we can specify proper scoring rules by their associated convex functions. For instance, a *quadratic scoring rule* has a convex function $2(x^2+(1-x)^2)-1$, and a *log scoring rule* has a convex function $x\log_2 x + (1-x)\log_2(1-x)+1$. We will study ***piecewise-linear scoring rules*** whose associated convex function is piecewise linear $\max\{f_1(x), \ldots, f_m(x)\}$ with affine functions f_i, $i = 1, \ldots, m$. Finally, ***v-shaped scoring rules*** in [23] are special cases of piecewise linear scoring rules, whose corresponding convex function is v-shaped with parameter (a, b, c, x_0) so that $H_{(a,b,c,x_0)}(x) = \max\{a(x - x_0) + c, b(x - x_0) + c\}$ with the *vertex* at x_0. Figure 3a shows examples of quadratic and log scoring rules, Fig. 1 is for v-shaped scoring rules, and Fig. 2 is for piecewise scoring rules.

Now we reformulate our optimization problem in terms of convex functions with the following lemmas by simple applications of Theorem 1. The first lemma converts *the information gain as the gap of Jensen's inequality of the associated convex function.* The second lemma shows the bounded conditions can be checked by the value and sub-gradient of the convex function. The proofs are in the appendix.

Lemma 1. *For any proper scoring rule PS with convex function H, the expected score of truthfully reporting x is $PS(x) = H(x)$, and the information gain of information structure X under the proper scoring rule PS is $\mathbb{E}\,PS(X) - PS(\mathbb{E}\,X) = \mathbb{E}[H(X)] - H(\mathbb{E}\,X)$.*

We will define the information gain of information structure X under the proper scoring rule with H as

$$\mathfrak{I}_X(H) := \mathbb{E}[H(X)] - H(\mathbb{E}\,X). \tag{1}$$

Lemma 2. *Given $B > 0$, for any proper scoring rule PS with convex function H, PS is ex-post bounded by B if and only if $H(x) + \partial H(x)(1 - x)$ and $H(x) - \partial H(x)x$ are in $[0, B]$ for all $x \in [0, 1]$.*

PS is ex-ante bounded by B if and only if $H(x) \in [0, B]$ for all $x \in [0, 1]$.

Let $\mathcal{B}_{\text{expost}}$ denote the set of ex-post bounded convex functions and $\mathcal{B}_{\text{exante}}$ for ex-ante bounded convex function as Lemma 2. By Lemma 2, $\mathcal{B}_{\text{expost}} \subseteq \mathcal{B}_{\text{exante}}$. We now derive the simplified program for our max min optimization problem over the space of convex functions.

Problem 1. Given a set of information structures $\mathcal{X}$ and a set of convex functions $\mathcal{B}$, find a convex function $H : [0, 1] \to \mathbb{R}$ which maximizes the worst-case information gain

$$\max_{H \in \mathcal{B}} \min_{X \in \mathcal{X}} \mathfrak{I}_X(H).$$

We will focus on $\mathcal{B}$ being $\mathcal{B}_{\text{expost}}$ in ex-post bounded setting and $\mathcal{B}_{\text{exante}}$ in ex-ante setting.

3 Main Results

In Sect. 3.1, we explore the setting of known information structure and show v-shaped scoring rules are optimal in Proposition 1. We further show that the same v-shaped scoring rule is also optimal in the setting of known prior but uncertain about the experiment in Theorem 2. We shift our focus to the unknown prior setting. We consider finite collections of information structures in Sect. 3.2, and devise an efficient algorithm that solves for the optimal scoring rules in Theorem 3. Finally, we design an FPTAS in Theorem 4 for infinite collections of information structures, Examples 1 and 3.

We provide outlines of proofs and intuitions, while complete proofs are deferred to the appendix.

3.1 Singleton and Homogeneous Prior Information Structures

As a warm-up, let's consider the principal exactly knows the agent's information structure so that $\mathcal{X} = \{X\}$ is a singleton. We show that the optimal H can be v-shaped (defined in Sect. 2.2) with the vertex at the prior in both ex-ante and ex-post bounded settings.

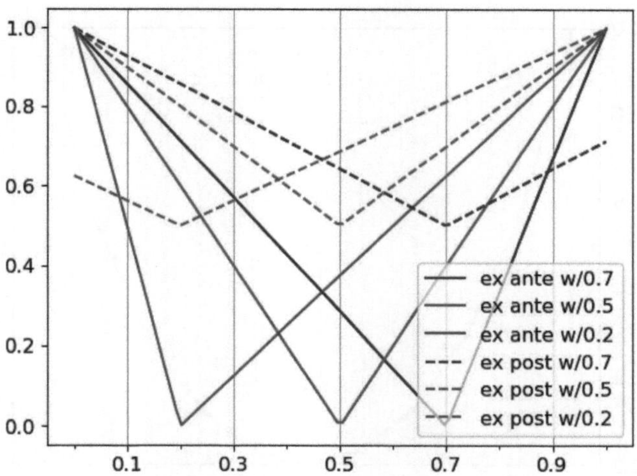

Fig. 1. The optimal scoring rules for the homogeneous prior setting. With Theorem 2, the dashed lines are the optimal v-shaped convex functions for three different priors (0.7, 0.5, and 0.2) in the ex-post setting, and the solid ones are in the ex-ante setting.

Proposition 1 (singleton). *If $\mathcal{X} = \{X\}$ is singleton in ex-post or ex-ante bounded settings, there exists an optimal scoring rule associated with a v-shaped convex function for Problem 1.*

1. *A v-shaped convex function with $x_0 = \pi$, $a = \frac{-B}{\pi}, b = \frac{B}{1-\pi}$ and $c = 0$ is optimal in ex-ante bounded setting with B.*
2. *A v-shaped convex function with $x_0 = \pi$, $a = \frac{-B}{2\max\{\pi,1-\pi\}}$, $b = \frac{B}{2\max\{\pi,1-\pi\}}$, and $c = \frac{B}{2}$ is optimal in ex-post bounded setting with B.*

While the above result is already proved in [23], we provide an explicit closed-form expression of the optimal solution and present an alternative proof in the appendix. We generalize the proof and show that v-shaped scoring rule is optimal for any collection of information structure when they share the same prior (defined in Example 2).

Theorem 2 (homogeneous prior). *Given any collection of information structures with homogeneous prior π (Example 2), there exists an optimal scoring rule associated with a v-shaped convex function with the vertex at π in both ex-ante and ex-post bounded settings respectively.*

The proof follows from the fact that the optimal v-shaped scoring rules only depend on the prior. When multiple information structures have the same prior, the same v-shaped scoring rule is still optimal.

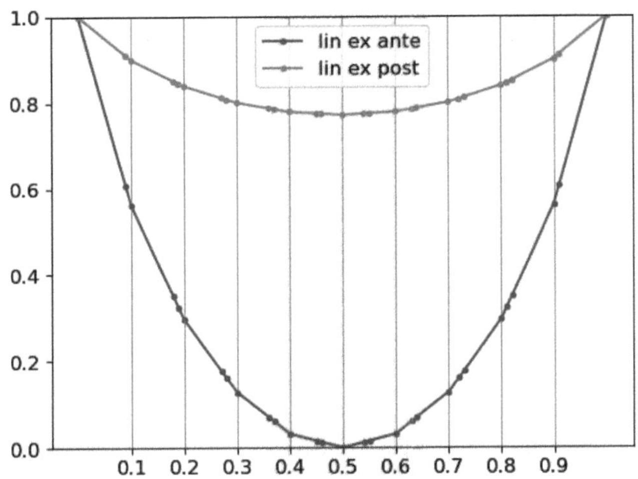

Fig. 2. The optimal piecewise linear functions for a finite collection of information structures $\mathcal{P}_{\rho,N} = \{(\pi,\sigma) \in \mathcal{P}_\rho : N\pi \in \mathbb{N}\}$ with $\rho = 0.1$ and $N = 10$ (Sect. 4). In both ex-post and ex-ante setting, the vertex of the function is at the prior of the collection of information structures which is aligned with our intuition in Sect. 3.3 which suggests maximizing curvature at the prior.

These results suggest that the principal should choose H that is "curved" at the prior in order to incentivize the agent to derive the signal and move away from the prior as Fig. 1.

3.2 Finite Collections of Information Structures

If the collection of information structures is *finite* so that $|\mathcal{X}| < \infty$, we give an efficient algorithm to compute an optimal piecewise linear scoring rule. Finite collections of information structures are natural when there are finite types of agents, and is useful to approximate some infinite collections of information structures as shown in the next section.

Theorem 3. *If $\mathcal{X}$ is finite, for both bounded settings, there exists an optimal proper scoring rule that is piecewise linear and can be derived by solving a linear program in time polynomial in $|\mathcal{X}| \cdot |\mathcal{S}|$.*

The main idea is that when $\mathcal{X}$ is finite $\mathfrak{J}_X(H)$ in Eq. (1) only depends on the evaluations of H on the support of $\mathcal{X}$ $\overline{\text{supp}}(\mathcal{X}) := \cup_{X\in\mathcal{X}}(\text{supp}(X) \cup \{\mathbb{E}\, X\})$. Thus, instead of searching all possible bounded convex functions, we can reduce the dimension of Problem 1 and use a linear program whose variables contain the evaluations of H in $\overline{\text{supp}}(\mathcal{X})$ and linear constraints ensure that the evaluations can be extended to a convex piecewise linear function. Figure 2 presents an example of piece-wise linear convex function outputted by our linear program. This observation allows us to solve the problem in weakly polynomial time in $|\overline{\text{supp}}(\mathcal{X})| \leq |\mathcal{X}| \cdot |\mathcal{S}|$. We present the formal proof in the appendix.

3.3 Infinite Collections of Information Structures

For a non-finite collection of information structures, solving Problem 1 exactly is generally infeasible, because even evaluating the objective value $\min_{X\in\mathcal{X}} \mathfrak{J}_X(H)$ may depend on evaluating all $X \in \mathcal{X}$, which could require unbounded time. In particular, this limitation also applies to Examples 1 and 3. Below, we extend our algorithms for finite collections of information structures to more general collections. We demonstrate the idea by approximately solving the optimal scoring rule for Examples 1 and 3 in Theorem 4. Remark 1 discusses the potential and limitation of our method for abstract collections of information structures.

Theorem 4. *Given any $\epsilon > 0, \delta > 0$ and $B > 0$, in ex-post bounded setting $\mathcal{B} = \mathcal{B}_{expost}$ (or ex-ante setting $\mathcal{B} = \mathcal{B}_{exante}$), there exists an efficient algorithm that outputs an ϵ-optimal scoring rule H on information structures with a homogeneous experiment (in Example 1) so that,*

$$\min_{X\in\mathcal{X}} \mathfrak{J}_X(H) \geq \max_{H'\in\mathcal{B}} \min_{X\in\mathcal{X}} \mathfrak{J}_X(H') - \epsilon,$$

with running time polynomial in B and $1/\epsilon$ (or B, $1/\epsilon$ and $1/\delta$ respectively). The same results hold for ρ-correlated information structures in Example 3.

Our FPTAS computes a finite collection of information structures, and runs our linear program in Theorem 3. Recall that $\mathcal{P}_\sigma$ is the collection of information structures with homogeneous experiment (in Example 1), our algorithm picks a finite set information structures

$$\mathcal{P}_{\sigma,N} = \{(\pi, \sigma) \in \mathcal{P}_\sigma : N\pi \in \mathbb{N}\}$$

with $N = \lfloor B/(2\epsilon) \rfloor$ and outputs an optimal piecewise linear scoring rule $H_{\sigma,N}$ for $\mathcal{P}_{\sigma,N}$ using Theorem 3. Similarly, for ρ-correlated information structures, let

$\mathcal{P}_{\rho,N} = \{(\pi, \sigma) \in \mathcal{P}_\rho : N\pi \in \mathbb{N}\}$. Our algorithm outputs an optimal piecewise linear scoring rule $H_{\rho,N}$ for $\mathcal{P}_\rho$.

The main challenge is to show the approximation guarantees. We observe that given a pair of information structures, the difference between information gains should be small if their posterior distribution is close. Formally, given two posteriors X and X' induced by two information structures P and P' respectively, the ***earth mover's distance*** (EMD) [6] between posteriors is $d_{EM}(X, X') = \sup_{f:\|f\|_{Lip}\leq 1} \mathbb{E}[f(X)] - \mathbb{E}[f(X')]$ where $\|f\|_{Lip}$ is the Lipschitz constant of f. The following lemma relates the earth mover's distance to our optimization problem.

Lemma 3. *Consider two posteriors X and X', and $B > 0$. For any $H \in \mathcal{B}_{expost}$, $|\mathfrak{I}_X(H) - \mathfrak{I}_{X'}(H)| \leq 2B \cdot d_{EM}(X, X')$. Similarly, if* $\text{supp}(\mathcal{X})$ *and* $\text{supp}(\mathcal{X}')$ *are contained in* $[1/L, 1 - 1/L]$ *for some $L > 0$, for any $H \in \mathcal{B}_{exante}$, $|\mathfrak{I}_X(H) - \mathfrak{I}_{X'}(H)| \leq 2LB \cdot d_{EM}(X, X')$.*

To use the above results, we need to show $\mathcal{P}_{\sigma,N}$ is a good *covering* for $\mathcal{P}_\sigma$ so that for all information structures in $\mathcal{P}_\sigma$ there exists one in $\mathcal{P}_{\sigma,N}$ that has a small earth mover's distance on the posteriors. Here, we design a novel coupling that can bound the earth mover's distance of two posterior predictions by the total variation distance on signal and state space. Given two information structures P and P' on $\Omega \times \mathcal{S}$, the ***total variation distance*** (TVD) [6] between these information structures is $d_{TV}(P, P') = \frac{1}{2}\sum_{w,s} |P(w, s) - P'(w, s)|$.

Lemma 4. *Given two information structures P and P' with induced posterior predictions X and X', $d_{EM}(X, X') \leq d_{TV}(P, P')$.*

Lemma 4 upper bound the EMD of posterior predictions by the TVD of information structures. The proof use the maximal coupling of P and $P'$to couple X and X' using the maximal coupling of P and P'. Then we show the expected difference of X and X' in our coupling can be converted to the total variation distance. Note that the EMD between two random variables is always smaller than the TVD between them, but Lemma 4 uses the TVD of the information structures, which can be much smaller than the TVD of posterior.[2]

Lemma 5. *Given $\delta > 0$, N, and σ, for all $P \in \mathcal{P}_\sigma$, there exists $P' \in \mathcal{P}_{\sigma,N}$ so that $d_{TV}(P, P') \leq \frac{1}{N}$. Similarly, given $\rho > 0$, for all $P \in \mathcal{P}_\rho$, there exists $P' \in \mathcal{P}_{\rho,N}$ so that $d_{TV}(P, P') \leq \frac{2-\rho}{N}$. $\overline{\text{supp}}(\mathcal{P}_{\sigma,N}) = O(N|\mathcal{S}|)$, and $\overline{\text{supp}}(\mathcal{P}_{\rho,N}) = O(N)$.*

Lemma 5 shows $\mathcal{P}_{\sigma,N}$ and $\mathcal{P}_{\rho,N}$ are good coverings for $\mathcal{P}_\sigma$ and $\mathcal{P}_\rho$ in TVD respectively, by direct computation.

[2] For instance, the TVD of posterior distribution between two ρ-correlated information structures with prior π and π' respectively is 1, but the TVD of information structures is less than $2|\pi - \pi'|$ as shown in Lemma 5.

Proof (Proof of Theorem 4). Given $\delta, \epsilon > 0$, and $\mathcal{P}_\sigma = \{(\pi, \sigma) : \pi \in [\delta, 1-\delta]\}$ in Example 1, we run our algorithm in Theorem 3 on a finite collection of information structures $\mathcal{P}_{\sigma,N} = \{(\pi, \sigma) : \pi \in [\delta, 1-\delta], N\pi \in \mathbb{N}\}$ with an integer $N = \lceil 2B/\epsilon \rceil$. Because $|\mathcal{P}_{\sigma,N}| = O(B/\epsilon)$, the running time is polynomial in B/ϵ and $|\mathcal{S}|$. For the approximation guarantee, given any ex-post bounded $H' \in \mathcal{H}_{expost}$ and $X \in \mathcal{P}_\sigma$, there exists $X' \in \mathcal{P}_{\sigma,N}$ so that

$$d_{EM}(X, X') \leq \frac{\epsilon}{2B}$$

by Lemmas 4 and 5. Moreover, by Lemma 3 and $N \geq 2B/\epsilon$

$$|\mathfrak{J}_X(H') - \mathfrak{J}_{X'}(H')| \leq 2B \cdot d_{EM}(X, X') \leq \epsilon.$$

Therefore, the optimal H for $\mathcal{P}_{\sigma,N}$ satisfies that

$$\min_{X \in \mathcal{X}} \mathfrak{J}_X(H') \leq \min_{X \in \mathcal{X}'} \mathfrak{J}_X(H') + \epsilon \leq \min_{X \in \mathcal{X}'} \mathfrak{J}_X(H) + \epsilon$$

which completes the proof for the ex-post bounded case. The proof for ex-ante bounded case is similar to the above.

Remark 1. Our FPTAS runs the linear program on a finite subset of an infinite set of information structures and provides approximation guarantees as long as the finite subset is an ϵ-covering of the original set under earth mover's distance by Lemma 3. The approach easily extends to any set with efficiently computable small coverings.[3] Furthermore, the optimization Problem 1 can be seen as zero-sum games between a designer choosing a scoring rule and the nature picking an information structure. [20] Hence, if a best response oracle on Θ or an ϵ-covering is available, various no-regret and best response algorithms can be employed to solve it.

However, the main challenge of solving Problem 1 lies in understanding the set Θ. Unlike standard optimization problems, Θ is often non-convex, including Examples 1 and 3. Consequently, a general efficient algorithm assuming the existence of efficiently computable small coverings or best response oracle may appear vacuous. Instead, we believe our method of finding and proving the small covering property could be a valuable tool for future research in analyzing their own Θ.

4 Simulations

We compare the performance of common scoring rules (Sect. 2.2) within the context of Problem 1. We will focus on ρ-correlated information structures (Example 3) in the ex-ante setting. This collection of information structures offers a platform to evaluate these scoring rules under the uncertainty of prior (and the experiment to less extent). The other combinations (homogeneous experiment, or ex-post setting) yield comparable outcomes and are deferred to the appendix.

[3] By [32, Theorem 2.2.11], there always exists an ϵ-covering with size $O(1/\epsilon)$.

We consider the correlation of signal and the state is either small $\rho = 0.025$ or large $\rho = 0.25$, and the prior π is in $[0.01, 0.99]$. For proper scoring rules, we use 1) a quadratic scoring rule, 2) a log scoring rule, 3) a v-shaped scoring rule with $(a, b, c, x_0) = (-2, 2, 0, 0.5)$ which is optimal when prior is 0.5 as Theorem 2, and 4) the piecewise linear scoring rule $H_{\rho,N}$ for $\mathcal{P}_{\rho,N}$ derived by our algorithm in Theorem 4 with $N = 50$. Figure 3a shows the associated convex functions for these five proper scoring rules. We include these scoring rules to investigate three questions: How good are classic scoring rules in our setting (the quadratic and log scoring rules)? How does misspecified prior harm the optimal scoring rule in the known prior setting (v-shaped scoring rule)? How does our FPTAS generalizes (piecewise linear scoring rules)?

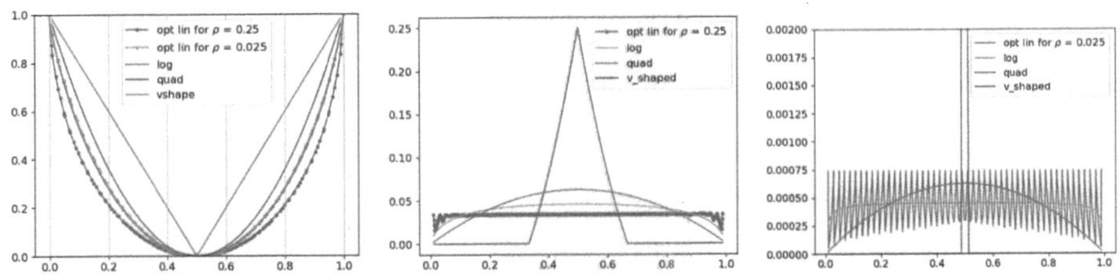

(a) Associated convex functions (b) Information gain with $\rho = 0.25$. (c) Information gain with $\rho = 0.025$.

Fig. 3. Figure 3a shows the five associated convex functions in Sect. 4. By Lemma 2 all scoring rules are ex-ante bounded by $B = 1$. Figures 3b and 3c present the information gain on each ρ-correlated information structures. x axis is the prior and y axis is the information gain.

To begin with, we evaluate all our scoring rules with the information gain on each information structure in $\mathcal{P}_{\rho,N'}$ with $N' = 1000 > N$ that serves as a proxy of the infinite collection $\mathcal{P}_\rho$ and is a superset of $\mathcal{P}_{\rho,N}$. Figures 3b and 3c present the outcomes where the y-axis represents the information gain, while the x-axis denotes the position of the prior. Note that our piecewise linear scoring rules are optimal for $\mathcal{P}_{\rho,N}$, but only ϵ-optimal for $\mathcal{P}_{\rho,N'}$ with $\epsilon \leq \frac{2-\rho}{N} \leq 0.04$ by Theorem 4. The difference of information gains between $\mathcal{P}_{\rho,N}$ and $\mathcal{P}_{\rho,N'}$ measure how well our method generalizes.

Figure 3b shows the information gain on ρ-correlated information structures with large correlation $\rho = 0.25$. The piecewise linear scoring rule for $\mathcal{P}_{\rho,N}$ has the worst-case information gain 0.0341 on the original set $\mathcal{P}_{\rho,N}$, and generalizes well on the superset $\mathcal{P}_{\rho,N'}$ with the worst-case information gain 0.0149. Comparatively, log scoring rule gets 0.0094, the quadratic scoring rule gets 0.0024, and the v-shaped scoring rule gets 0 worst-case information gain on $\mathcal{P}_{\rho,N'}$. The v-shaped scoring rule performs varies significantly: It achieves the highest information gain on information structures with prior centered around 0.5 aligning with the implications of Theorem 2, but also gets zero information gain when prior is above 0.7 or below 0.3. This is because when the prior is far away from the vertex (0.5), information structure can have a support contained in a flat area $(0, 0.5)$ or $(0.5, 1)$ and leading to zero information gain because the Jensen's inequality is tight on affine functions.

Figure 3c shows the information gains when the correlation is small $\rho = 0.025$. The piecewise linear scoring rule has the worst-case information gain $5.77 \cdot 10^{-5}$ compared to $2.76 \cdot 10^{-4}$ under the log scoring rule, $2.48 \cdot 10^{-5}$ on quadratic scoring rule, and 0 under the v-shaped scoring rule. First, the information gains are much small in the small correlation setting, because the posterior is barely move away from prior.[4] Second, information gains under the piecewise linear scoring rule on $\mathcal{P}_{\rho,N'}$ has several periodic peaks each peak. Note that the piecewise linear can be seen as several small v-shaped scoring rules. If an information structure's prior is at a vertex, the information gain is large. However, if the support of the information structure is contained in a flat area, the information gain is near zero. Noteworthy, the number of vertices of our optimal scoring rule equal $N + 1 = 51$ by Theorem 3 which is also the number of peaks in the information gains. On the other hand, the log and quadratic scoring rule are strictly convex which do not have any sharp transition between flat area and vertex, so the information gains change smoothly as the priors change. Finally, observe that the piecewise linear scoring rules are surprisingly close to the log scoring rule in Fig. 3a. In Fig. 4, we conduct an additional simulations and observe that difference between the piecewise scoring rule for $\mathcal{P}_{\rho,N}$ and the log scoring rule

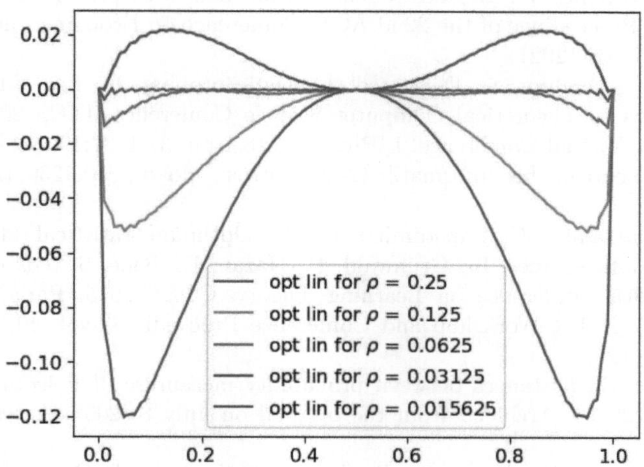

Fig. 4. The difference between log scoring rule and the piecewise optimal scoring rules for ρ-correlated information structures $\mathcal{P}_{\rho,N}$ as ρ ranging from 1/4 to 1/64.

[4] Indeed, we can show that a correlated signal is Blackwell dominated by a correlated signal with larger correlation, and thus has a smaller information gain under any proper scoring rules.

uniformly decreases as ρ decreases. This results suggest an interesting potential connection between ρ-correlated information structure and the log scoring rule.

5 Conclusion

We propose the problem of optimal proper scoring rules design when the principal has partial knowledge of an agent's signal distribution. As shown in our simulations, this approach may serve as a benchmark for various scoring rules. We devise efficient algorithms for four principal knowledge settings, and design a novel coupling to bound the earth mover's distance of posterior by the total variation distance of the signal and state space.

Acknowledgments. We thank Grant Schoenebeck and anonymous reviewers for their valuable feedback on restructuring the paper. This work is supported by NSF under award No. IIS-2147187.

References

1. Abernethy, J., Chen, Y., Ho, C.J., Waggoner, B.: Actively purchasing data for learning. In: ACM EC 2015 (2015)
2. Agarwal, A., Dahleh, M.A., Sarkar, T.: A marketplace for data: an algorithmic solution, pp. 701–726 (2019). https://doi.org/10.1145/3328526.3329589
3. Alon, T., Dütting, P., Talgam-Cohen, I.: Contracts with private cost per unit-of-effort. In: Proceedings of the 22nd ACM Conference on Economics and Computation, pp. 52–69 (2021)
4. Bechtel, C., Dughmi, S.: Delegated stochastic probing. In: Lee, J.R. (ed.) 12th Innovations in Theoretical Computer Science Conference, ITCS 2021, 6–8 January 2021, Virtual Conference. LIPIcs, vol. 185, pp. 37:1–37:19. Schloss Dagstuhl - Leibniz-Zentrum für Informatik (2021). https://doi.org/10.4230/LIPICS.ITCS.2021.37
5. Cai, Y., Daskalakis, C., Papadimitriou, C.H.: Optimum statistical estimation with strategic data sources. In: Grünwald, P., Hazan, E., Kale, S. (eds.) Proceedings of The 28th Conference on Learning Theory, COLT 2015, Paris, France, 3–6 July 2015. JMLR Workshop and Conference Proceedings, vol. 40, pp. 280–296. JMLR.org (2015)
6. Chatterjee, S.: Distances between probability measures. PDF lecture notes, UC Berkeley (2008). Archived from the original on July 8, 2008. Accessed 21 June 2013
7. Chen, Y., Immorlica, N., Lucier, B., Syrgkanis, V., Ziani, J.: Optimal data acquisition for statistical estimation. In: Tardos, É., Elkind, E., Vohra, R. (eds.) Proceedings of the 2018 ACM Conference on Economics and Computation, Ithaca, NY, USA, 18–22 June 2018, pp. 27–44. ACM (2018). https://doi.org/10.1145/3219166.3219195
8. Chen, Y., Podimata, C., Procaccia, A.D., Shah, N.: Strategyproof linear regression in high dimensions. In: Tardos, É., Elkind, E., Vohra, R. (eds.) Proceedings of the 2018 ACM Conference on Economics and Computation, Ithaca, NY, USA, 18–22 June 2018, pp. 9–26. ACM (2018). https://doi.org/10.1145/3219166.3219175

9. Chen, Y., Zheng, S.: Prior-free data acquisition for accurate statistical estimation. In: Karlin, A.R., Immorlica, N., Johari, R. (eds.) Proceedings of the 2019 ACM Conference on Economics and Computation, EC 2019, Phoenix, AZ, USA, 24–28 June 2019, pp. 659–677. ACM (2019). https://doi.org/10.1145/3328526.3329564
10. Cummings, R., Ioannidis, S., Ligett, K.: Truthful linear regression. In: Grünwald, P., Hazan, E., Kale, S. (eds.) Proceedings of The 28th Conference on Learning Theory, COLT 2015, Paris, France, 3–6 July 2015. JMLR Workshop and Conference Proceedings, vol. 40, pp. 448–483. JMLR.org (2015)
11. Dekel, O., Fischer, F.A., Procaccia, A.D.: Incentive compatible regression learning. J. Comput. Syst. Sci. **76**(8), 759–777 (2010). https://doi.org/10.1016/J.JCSS.2010.03.003
12. Fleischer, L., Lyu, Y.: Approximately optimal auctions for selling privacy when costs are correlated with data. In: Faltings, B., Leyton-Brown, K., Ipeirotis, P. (eds.) Proceedings of the 13th ACM Conference on Electronic Commerce, EC 2012, Valencia, Spain, 4–8 June 2012, pp. 568–585. ACM (2012). https://doi.org/10.1145/2229012.2229054
13. Fu, H., Haghpanah, N., Hartline, J., Kleinberg, R.: Optimal auctions for correlated buyers with sampling. In: Proceedings of the Fifteenth ACM Conference on Economics and Computation, pp. 23–36 (2014)
14. Ghorbani, A., Zou, J.Y.: Data shapley: equitable valuation of data for machine learning. In: Chaudhuri, K., Salakhutdinov, R. (eds.) Proceedings of the 36th International Conference on Machine Learning, ICML 2019, 9–15 June 2019, Long Beach, California, USA. Proceedings of Machine Learning Research, vol. 97, pp. 2242–2251. PMLR (2019)
15. Ghosh, A., Ligett, K., Roth, A., Schoenebeck, G.: Buying private data without verification. In: Babaioff, M., Conitzer, V., Easley, D.A. (eds.) ACM Conference on Economics and Computation, EC 2014, Stanford , CA, USA, 8–12 June 2014, pp. 931–948. ACM (2014). https://doi.org/10.1145/2600057.2602902
16. Ghosh, A., Roth, A.: Selling privacy at auction. In: Shoham, Y., Chen, Y., Roughgarden, T. (eds.) Proceedings 12th ACM Conference on Electronic Commerce (EC-2011), San Jose, CA, USA, 5–9 June 2011, pp. 199–208. ACM (2011). https://doi.org/10.1145/1993574.1993605
17. Ghosh, M., Mukhopadhyay, N., Sen, P.K.: Sequential Estimation, vol. 904. Wiley (2011)
18. Gneiting, T., Raftery, A.E.: Strictly proper scoring rules, prediction, and estimation. J. Am. Stat. Assoc. **102**(477), 359–378 (2007)
19. Grossman, S.J., Hart, O.D.: An analysis of the principal-agent problem. In: Foundations of Insurance Economics, pp. 302–340. Springer (1992)
20. Guo, Y., Hartline, J.D., Huang, Z., Kong, Y., Shah, A., Yu, F.: Algorithmic robust forecast aggregation. CoRR abs/2401.17743 (2024). https://doi.org/10.48550/ARXIV.2401.17743
21. Guruganesh, G., Schneider, J., Wang, J.R.: Contracts under moral hazard and adverse selection. In: Biró, P., Chawla, S., Echenique, F. (eds.) EC 2021: The 22nd ACM Conference on Economics and Computation, Budapest, Hungary, 18–23 July 2021, pp. 563–582. ACM (2021). https://doi.org/10.1145/3465456.3467637
22. Hardt, M., Megiddo, N., Papadimitriou, C.H., Wootters, M.: Strategic classification. In: Sudan, M. (ed.) Proceedings of the 2016 ACM Conference on Innovations in Theoretical Computer Science, Cambridge, MA, USA, 14–16 January 2016, pp. 111–122. ACM (2016). https://doi.org/10.1145/2840728.2840730

23. Li, Y., Hartline, J.D., Shan, L., Wu, Y.: Optimization of scoring rules. In: Pennock, D.M., Segal, I., Seuken, S. (eds.) EC 2022: The 23rd ACM Conference on Economics and Computation, Boulder, CO, USA, 11–15 July 2022, pp. 988–989. ACM (2022). https://doi.org/10.1145/3490486.3538338
24. Liu, Y., et al.: Replication markets: Results, lessons, challenges and opportunities in AI replication. CoRR abs/2005.04543 (2020)
25. McCarthy, J.: Measures of the value of information. Proc. Natl. Acad. Sci. U.S.A. **42**(9), 654 (1956)
26. Meir, R., Procaccia, A.D., Rosenschein, J.S.: Algorithms for strategyproof classification. Artif. Intell. **186**, 123–156 (2012). https://doi.org/10.1016/J.ARTINT.2012.03.008
27. Michael, G., et al.: Are replication rates the same across academic fields? Community forecasts from the darpa score programme. Roy. Soc. Open Sci. **7**(200566) (2020)
28. Murphy, K.P.: Machine Learning - A Probabilistic Perspective. Adaptive Computation and Machine Learning Series. MIT Press (2012)
29. Neyman, E., Noarov, G., Weinberg, S.M.: Binary scoring rules that incentivize precision. In: Biró, P., Chawla, S., Echenique, F. (eds.) EC 2021: The 22nd ACM Conference on Economics and Computation, Budapest, Hungary, 18–23 July 2021, pp. 718–733. ACM (2021). https://doi.org/10.1145/3465456.3467639
30. Nissim, K., Vadhan, S.P., Xiao, D.: Redrawing the boundaries on purchasing data from privacy-sensitive individuals. CoRR abs/1401.4092 (2014)
31. O'Donnell, R.: Analysis of Boolean Functions. Cambridge University Press (2014)
32. Panaretos, V.M., Zemel, Y.: An Invitation to Statistics in Wasserstein Space. Springer (2020)
33. Papireddygari, M., Waggoner, B.: Contracts with information acquisition, via scoring rules. In: Pennock, D.M., Segal, I., Seuken, S. (eds.) EC 2022: The 23rd ACM Conference on Economics and Computation, Boulder, CO, USA, 11–15 July 2022, pp. 703–704. ACM (2022). https://doi.org/10.1145/3490486.3538261
34. Perote, J., Perote-Peña, J.: Strategy-proof estimators for simple regression. Math. Soc. Sci. **47**(2), 153–176 (2004). https://doi.org/10.1016/S0165-4896(03)00085-4
35. Roth, A., Schoenebeck, G.: Conducting truthful surveys, cheaply. CoRR abs/1203.0353 (2012)
36. Savage, L.J.: Elicitation of personal probabilities and expectations. J. Am. Stat. Assoc. **66**(336), 783–801 (1971)
37. Waggoner, B., Frongillo, R.M., Abernethy, J.D.: A market framework for eliciting private data. In: Cortes, C., Lawrence, N.D., Lee, D.D., Sugiyama, M., Garnett, R. (eds.) Advances in Neural Information Processing Systems 28: Annual Conference on Neural Information Processing Systems 2015, 7–12 December 2015, Montreal, Quebec, Canada, pp. 3510–3518 (2015)
38. Zheng, S., Waggoner, B., Liu, Y., Chen, Y.: Active information acquisition for linear optimization, pp. 167–176 (2018)

Strategic Facility Location via Predictions

Qingyun Chen[1](✉), Nick Gravin[2], and Sungjin Im[1]

[1] University of California, Merced, CA, USA
{qchen41,sim3}@ucmerced.edu

[2] Key Laboratory of Interdisciplinary Research of Computation and Economics, Shanghai University of Finance and Economics, Shanghai, China
nikolai@mail.shufe.edu.cn

Abstract. The facility location with strategic agents is a canonical problem in the literature on mechanism design without money. Recently, Agrawal et al. [1] considered this problem in the context of machine learning augmented algorithms, where the mechanism designer is also given a prediction of the optimal facility location. An ideal mechanism in this framework produces an outcome that is close to the social optimum when the prediction is accurate (consistency) and gracefully degrades as the prediction deviates from the truth, while retaining some of the worst-case approximation guarantees (robustness). The previous work only addressed this problem in the two-dimensional Euclidean space providing optimal trade-offs between robustness and consistency guarantees for deterministic mechanisms.

We consider the problem for *general* metric spaces. Our only assumption is that the metric is continuous, meaning that any pair of points must be connected by a continuous shortest path. We introduce a novel mechanism that in addition to agents' reported locations takes a predicted optimal facility location $\hat{o}$. We call this mechanism Harmonic, as it selects one of the reported locations $\tilde{\ell}_i$ with probability inversely proportional to $d(\hat{o}, \tilde{\ell}_i) + \Delta$ for a constant parameter Δ. While Harmonic mechanism is not truthful, we can *characterize the set of undominated strategies* for each agent i as solely consisting of the points on a shortest path from their true location ℓ_i to the predicted location $\hat{o}$. We further derive *consistency and robustness guarantees on the Price of Anarchy (PoA)* for the game induced by the mechanism. Specifically, assuming that $\Delta = \frac{\varepsilon}{2n} \cdot \sum_{i\in[n]} d(\hat{o}, \ell_i)$ is closely related to the average distance to the predicted location $\hat{o}$, our consistency guarantee is arbitrarily close to optimum (PoA is $1+\varepsilon$), while having a constant robustness guarantee (PoA is $O(1+1/\varepsilon^3)$ in general and is $O(1+1/\varepsilon^2)$ for strictly convex metric spaces). We also show that for a constant number of agents $n = O(1)$, Harmonic(Δ) mechanism with $\Delta = 0$ attains 1-consistency and $O(1)$-robustness.

1 Introduction

The facility location is a canonical problem attracting a lot of interest in many different fields such as operation research, artificial intelligence, social choice, and

M. Mavronicolas et al. (Eds.): WINE 2024, LNCS 15534, pp. 401–419, 2026.
https://doi.org/10.1007/978-3-032-08560-3_23

economics [9]. Facility location with strategic customers serves as an exemplary setting of the approximate mechanism design without payments that has been extensively studied in the past fifteen years [3,4,12–16,20,24,25,28,30].

In the case of a single facility, the problem involves n agents residing in a metric space $\mathcal{M}$, each agent i with their own most preferred location $\ell_i \in \mathcal{M}$ of the single facility $f \in \mathcal{M}$. The agents are strategic and may misreport their location hoping to influence the final choice of the facility $f \in \mathcal{M}$ and minimize their personal cost given by the distance $d(\ell_i, f)$. The central planner collects reported locations from the agents and then decides where to open the facility. For the most common objective, a.k.a., social cost minimization, the goal is two-fold: first, the outcome should be a good approximation for minimizing the average distance from the facility to all agents if not the optimum; and second, no agents should have an incentive to misreport their true location.

The traditional approach to this problem aims to find mechanisms with worst-case performance guarantees on all possible inputs. This approach has been criticized in a broader context of algorithm design for its often pessimistic bounds and its focus on adversarial instances, which are not frequently encountered in practice [26]. To overcome the practical shortcomings of traditional algorithms, Lykouris and Vassilvitskii [21] have recently introduced a powerful framework of *machine learning augmented algorithms*. ML augmented algorithms strive to surpass the limitations of traditional algorithms by incorporating predictions as supplementary input. These algorithms are supposed to do very well when furnished with accurate predictions (consistency guarantee), while gracefully degrading in performance as prediction quality declines. Ideally, they also uphold robust guarantees for all predictions, ensuring reliability even in the face of imperfect forecasts (robustness guarantee). ML augmented algorithms have found applications in strengthening online and streaming algorithms with predictions about future unknown input, as well as in accelerating algorithms by utilizing knowledge gleaned from past inputs. See [23] for a survey of ML augmented algorithms.

Very recently, there has been a surge of interest in applying ML augmented framework to mechanism design [5,6,10,17,31] pioneered by Agrawal et al. [1]. Their work specifically addresses the strategic facility location problem in two-dimensional Euclidean space $\mathcal{M} = \mathbb{R}^2$. To illustrate the setting, consider the following examples:

Example 1. A group of $n = 100$ people (say alumni from the same school) want to decide on a meeting place (e.g., for the school reunion). Before asking people about their preferred meeting location, the meeting planner may already have a good idea about how many people still live near the school and how many will be coming from a greater city area; thus the planner could predict which spot might work well for everyone (either somewhere close to the school, or a well accessible place in the city center).

Example 2. A city wants to open a new government office to serve the public. The city likely possesses extensive data on residents, thus can find a good candidate location. Alternatively, it might opt for a central location as a likely near-optimal solution, even without in-depth analysis.

These kinds of scenarios are neatly captured by the model of [1], which assumes access to a prediction of the optimal location obtained via machine-learning algorithm, or other means. This assumption about the output (and not the input)[1] is particularly well suited for the mechanism design scenario in which it is usually difficult to accurately predict the input from all n agents.

Agrawal et al. [1] demonstrate that it is possible to design a strategy-proof mechanism that is arbitrarily close to the optimal solution in the two-dimensional setting, provided access to a quality prediction of the facility location. This is a significant advancement, as previously only a $\sqrt{2}$-approximate strategy-proof mechanism was known for this setting without predictions. Their mechanism builds upon a well known coordinate-wise median-point mechanism for low dimensional Euclidean spaces $\mathbb{R}^d$: the predicted location $\hat{o} \in \mathbb{R}^2$ gets a certain weight $c \in [0, 1]$ and is added to the agents' reports, the output is given by the coordinate-wise median. The parameter c is based on the expected quality of predictions: if the mechanism designer anticipates high-quality predictions, they can set the parameter c close to 1, resulting in a nearly 1-consistent (nearly optimal) algorithm. Specifically, the algorithm is $(1 + r)$- consistent and $g(r)$-robust[2], where $r(c) \in (0, \infty)$ is a control-parameter that governs a trade-off between consistency and robustness and $g(r) \approx \frac{1}{\sqrt{2r}}$ is a function tending to infinity as $r \to 0$. I.e., when prediction is extremely inaccurate, the algorithm maintains a constant approximation guarantee for any fixed value of r, ensuring a robust solution quality. These consistency and robustness guarantees are best possible among all anonymous deterministic mechanisms [1].

In this paper, we seek to investigate if the facility location problem continues to benefit from predictions even in the *general* metric setting. We ask

How to use ML augmented advice in general metric spaces $\mathcal{M}$?

1.1 Challenges

We discuss two primary approaches considered in the literature and the difficulties in extending them to the general metric space with predictions.

Median-Point Like Mechanisms. The median-point mechanism is perhaps the most common approach for the facility location game. It assumes that the locations are in the (low-dimensional) Euclidean space and computes the median of the reported locations along each coordinate. The algorithm is strategyproof, and it is optimum for the single dimensional Euclidean space [25] and $\sqrt{2}$-approximate for the two dimensional Euclidean space [18]. However, the only

[1] It is common in the ML augmented algorithms literature to have predictions about algorithm's input rather than output.

[2] In general, an algorithm is α-consistent and β-robust if it is an α-approximation when the prediction is perfect and a β-approximation for arbitrary predictions.

known approximation guarantee of the coordinate-wise median is $O(\sqrt{d})$ for the Euclidean norm in $\mathbb{R}^d$ spaces, and the mechanism is inapplicable in general metric spaces. In fact, it is known that any deterministic mechanism is $\Omega(n)$ factor off the optimum even in a circle metric [27]. Hence, the approach of Agrawal et al. [1] is only limited to low dimensional Euclidean spaces and does not extend to general metrics.

Random Dictatorship. Understanding mechanism design in general metric spaces presents significant challenges, as evidenced by the limited research conducted in this area (see a survey [9]). Indeed, any mechanism with a sublinear in n approximation guarantee must be randomized [27]. The only known $O(1)$-approximation strategyproof mechanism that works for general $\mathcal{M}$ is *Random Dictatorship* (RD), which chooses one of the reported locations uniformly at random. RD is a $2(1 - 1/n)$-approximation [2]. There are a few characterization results of strategyproof mechanisms for special metrics $\mathcal{M}$ (such as trees, or cycles) [11,22]. Also there is a characterization of group-strategyproof mechanisms [29] in strictly convex spaces, which can only give $\Omega(n)$ approximation to the optimal social cost. Unfortunately, these results do not offer any new mechanisms that can work in general metric spaces. On the positive side, the ML augmented framework offers a richer space of possible mechanisms by adding extra information about the predicted location to the input. On the negative side, there are still fundamental roadblocks to having good strategyproof ML augmented mechanisms, which are discussed below.

Possible Mechanisms. First, it is reasonable to restrict attention to randomized mechanisms that select facility f from a distribution over reported locations $\boldsymbol{\ell}$ and the predicted location $\hat{o}$, as we may not even know which other points general $\mathcal{M}$ may or may not have. This is still a rich family of mechanisms as each selection probability $\{g_i(\cdot)\}_{i\in[n+1]}$ may depend on $O(n^2)$ pairwise distances. Unfortunately, we do not know any strategyproof mechanisms of such form beyond RD and for a good reason: each agent controls n variables with many degrees of freedom in general $\mathcal{M}$; manipulations by a single agent may affect all g_j for $j \in [n+1]$ in a rather complex way even for simple and well-behaved functions $\{g_i(\cdot)\}_{i\in[n+1]}$. One can still have a strategyproof mechanism by choosing f from a fixed distribution over reported locations $\boldsymbol{\ell}$ and predicted optimal location $\hat{o}$. However, if $\mathbf{Pr}[f \leftarrow \hat{o}]$ is not a constant independent of n, then we do not get any interesting consistency guarantees. On the other hand, if $\hat{o}$ is very far from each reported location ℓ_i, then we do not get any robustness guarantee.

A more reasonable attempt to balance consistency and robustness is to combine RD mechanism with a selection of $\hat{o}$ by letting agents vote on two alternative outcomes: RD($\boldsymbol{\ell}$) or $\hat{o}$. This approach, however, has its own problems that can be illustrated with our previous example of finding a meeting place for $n = 100$ alumni (Example 1). Imagine that a single person $i^* \in [n]$ is on vacation at a remote tropical island in another country at the time of the meeting, while the remaining 99 people are all in town. Although the RD mechanism selects i^* only with a small 1% probability, it is still incredibly inconvenient for anyone else. We can think of i^* as an adversarial agent, who does not care about the outcome, but still has a unilateral power to sabotage the outcome of $\texttt{RD}(\boldsymbol{\ell})$ for everyone else and make it arbitrarily worse than the predicted location $\hat{o}$. Thus any approach involving voting on two alternatives $\texttt{RD}(\boldsymbol{\ell})$ or $\hat{o}$ must give each agent a unilateral power to impose $\hat{o}$ outcome. This leads to *Consensus Mechanism*: the outcome is a predicted location $\hat{o}$, unless all agents agree on $\texttt{RD}(\boldsymbol{\ell})$ outcome. The consensus mechanism has perfect 1-consistency, since at least one agent must prefer $\hat{o}$ over $\texttt{RD}(\boldsymbol{\ell})$ when $\hat{o}$ is the optimal location. It also has a constant robustness guarantee for a small number of agents $n = O(1)$ (e.g., $n = 3$), since (i) in the case when at least one agent prefers $\hat{o}$ over $\texttt{RD}(\boldsymbol{\ell})$, the prediction is a constant approximation to the optimum, and otherwise (ii) $\texttt{RD}(\boldsymbol{\ell})$ is a $2(1 - 1/n)$-approximation to the optimum [2].

Untruthful Mechanisms. It turns out that the consensus mechanism is not strategyproof. E.g., if one agent i slightly prefers $\hat{o}$ over $\texttt{RD}(\boldsymbol{\ell})$ and everyone else strongly prefers $\texttt{RD}(\boldsymbol{\ell})$, then one of the agents $j \neq i$ may greatly benefit by reporting a closer location $\widetilde{\ell}_j$ to ℓ_i and thus convince i to choose $\texttt{RD}(\boldsymbol{\ell}_{\text{-}j}, \widetilde{\ell}_j)$ rather than $\hat{o}$. On the other hand, any selfish deviation $\widetilde{\ell}_i \neq \ell_i$ by agent i may only improve the outcome for all other agents[3]. I.e., non truthful mechanisms do not necessarily lead to bad outcomes in facility location games and thus should not be excluded from consideration. Furthermore, while the literature on strategic facility location is primarily concerned with truthful mechanisms, other areas of mechanism design have encountered a fair number of non truthful mechanisms, e.g., first-price or position auctions in revenue maximization settings. For such mechanisms, a standard approach is to analyse the equilibria of the ensuing game and establish good Price of Anarchy (PoA) bounds (e.g., [8]).

1.2 Our Contributions

While the consensus mechanism suggests that positive results are possible in ML augmented framework, it does not provide a satisfactory solution for the most interesting case, when the number of agents n is large. Indeed, its robustness guarantee of $\Omega(n)$ scales linearly with n (e.g., when one agent i is located exactly at the prediction $\ell_i = \hat{o}$ and all other agents stay at the same far away location $\ell_j = P \in \mathcal{M}$ for all $j \neq i$) and it does not allow for any trade-offs between consistency and robustness guarantees.

In this paper, we propose a novel mechanism, which we term `Harmonic` and which to the best of our knowledge has not been considered before. The mechanism chooses a reported location $\widetilde{\ell}_i \in \{\widetilde{\ell}_1, \ldots, \widetilde{\ell}_n\}$ with a probability *inversely* proportional to its distance $d(\widetilde{\ell}_i, \hat{o})$ to the predicted location $\hat{o}$ plus a constant Δ, i.e., $\mathbf{Pr}[f \leftarrow \widetilde{\ell}_i]$ is proportional to $1/(d(\hat{o}, \widetilde{\ell}_i) + \Delta)$. The intuition for the inversely proportional selection rule is to penalize agents for reporting remote locations to be more robust against manipulations by adversarial agents[4]. The

[3] Any such deviation would make everyone prefer $\texttt{RD}(\boldsymbol{\ell}_{\text{-}i}, \widetilde{\ell}_i)$ over $\hat{o}$.

[4] Recall a major issue of $\texttt{RD}(\boldsymbol{\ell})$: a single adversarial agent can make everyone's expected cost arbitrarily large. With an inversely proportional rule, reporting a remote location has a bounded impact on the expected cost of any agent.

parameter Δ allows us to balance consistency and robustness guarantees: smaller Δ means better consistency and makes `Harmonic` resemble the consensus mechanism, while larger Δ brings `Harmonic` closer to the `RD` mechanism. The hope is that if the predicted optimal location $\hat{o}$ is close to accurate, then many agents should benefit by shifting their reports closer to $\hat{o}$ (agent i gets a higher chance of being selected at the reported location $\widetilde{\ell}_i$ at the cost of higher distance to their true location $d(\ell_i, \widetilde{\ell}_i)$). On the other hand, when prediction $\hat{o}$ is highly inaccurate, the hope is that only a few agents decide to move closer to $\hat{o}$ and we get a good robustness guarantee. This is also the situation where the constant parameter Δ helps to reduce the impact of a single agent on the mechanism's outcome.

Results. First, while `Harmonic` is not strategy-proof, we show that it significantly limits the agents' undominated action space—in a good direction. Specifically, the best response of each agent i is to report her preferred location, ℓ_i or the predicted location, $\hat{o}$, or any point on the shortest path between ℓ_i and $\hat{o}$, when she is indifferent between ℓ_i and $\hat{o}$. In other words, each agent can only choose from the shortest paths between her preferred location and the predicted location. We then prove that Pure Nash Equilibrium (PNE) always exists via Kakutani's fixed-point theorem (Sect. 3.2). The only assumption we use is that every pair of points in the metric space is connected by a continuous shortest path. This is a rather mild assumption, as one can easily embed a given discrete metric into a continuous one by taking a convex combination over the points in the space. In particular, for any discrete metric $\mathcal{M}$, one can simply extend the metric to $\overline{\mathcal{M}}$ by adding a continuous shortest path between each pair of discrete points.

Second, we analyse the Price of Anarchy (PoA) of the ensuing game induced by the `Harmonic(`Δ`)` mechanism. We assume that the trade-off parameter $\Delta = c \cdot a$ is related to the average distance $a \stackrel{\text{def}}{=} \frac{1}{n}\sum_{i\in[n]} d(\hat{o}, \ell_i)$ to the prediction $\hat{o}$ for a constant c. When predicted location $\hat{o}$ is γ-approximation to the optimum, `Harmonic` is $\gamma(1+2c)$-consistent (Theorem 2). On the robustness side, we show that `Harmonic` is $O(1+1/c^3)$-robust (Theorem 4) in general, and give a better robustness guarantee of $O(1+1/c^2)$, if the metric is strictly convex (Theorem 3). I.e., for a small constant $c = \varepsilon/2$ we achieve $(1+\varepsilon)$-consistency that smoothly degrades with the prediction's accuracy, and respectively $O(1+1/\varepsilon^3)$ and $O(1+1/\varepsilon^2)$ robustness[5]. Furthermore, if $\Delta = 0$, we obtain $O(n)$-robustness (Theorem 5) and 1-consistency, which translates into a constant robustness and 1-consistency for a constant $n = O(1)$ number of agents (see the discussion in the full version).

[5] Note that our consistency/robustness guarantees rely on the trade-off parameter Δ. I.e., we rely on the approximately correct estimation of the average distance a, which is arguably not a very strong assumption on the setting (a is a single number with simple dependencies on each agent's individual location ℓ_i). Also note that, if Δ is underestimated ($\frac{\Delta}{a} = o(\varepsilon)$), then we get better consistency and worse robustness; and if Δ is overestimated ($\frac{\Delta}{a} = \omega(\varepsilon)$), then we get better robustness.

1.3 Other Related Work

Mechanism design without payments often falls short of producing the optimal solution, leading to potentially suboptimal results. Agrawal et al. [1] initiated the application of machine learning-augmented algorithms in this field to overcome these limitations. For the utilitarian objective of minimizing the average distance of agents to the facility, their algorithm achieves $\sqrt{2c^2+2}/(1+c)$-consistency and $\sqrt{2c^2+2}/(1-c)$-robustness in the two-dimensional space, where $c \in [0,1)$ is a parameter to be chosen by the mechanism. As mentioned before, a $\sqrt{2}$-approximate strategy-proof mechanism is known and it is the best possible without predictions [18]. For the egalitarian social cost of minimizing the maximum distance to the facility, Agrawal et al. [1] presented an algorithm with 1-consistency and $1+\sqrt{2}$-robustness. For this setting, a 2-approximate strategy-proof mechanism is known without predictions [3,18]. A very recent work [5] studies randomized algorithms in one-dimensional and two-dimensional settings. For the obnoxious version of maximizing the total distance to the obnoxious facility with predictions, Istrate and Bonchis recently considered various metrics such as circles, trees, and one-dimensional and two-dimensional hypercubes [19].

Very recently, an ML augmented mechanism was developed for the k-facility location game in general metric spaces [7]. However, it assumes predictions for each point, requires assigning points to facilities in a balanced manner, and does not provide an absolute robustness guarantee. We defer more related work to the full version of this paper.

1.4 Roadmap

In Sect. 2, we define the problem and our `Harmonic` mechanism. In Sect. 3, we discuss the undominated strategies and equilibria concepts. In Sect. 4 we give analysis of our mechanism, providing consistency and robustness bounds. In Sect. 5 we conclude our paper. Due to the space constraints, most proofs are deferred to the full version of this paper.

2 Preliminaries

In a (single) facility location problem the goal is to place a facility f in a given metric space $\mathcal{M}$ with a distance function $d: \mathcal{M} \times \mathcal{M} \to \mathbb{R}$ to serve a set of n agents. Each agent $i \in [n]$ has a most preferred location $\ell_i \in \mathcal{M}$ and incurs a cost of $d(f, \ell_i)$, when the facility is placed at $f \in \mathcal{M}$. The objective is to minimize the social cost $\mathsf{SC}(f, \boldsymbol{\ell}) \stackrel{\text{def}}{=} \sum_{i\in[n]} d(f, \ell_i)$. The optimal cost is achieved by facility o, i.e., $\mathsf{opt}(\boldsymbol{\ell}) = \min_{o' \in \mathcal{M}} \sum_{i\in[n]} d(o', \ell_i) = \sum_{i\in[n]} d(o, \ell_i)$. The facility location f is an α-approximation if $\mathsf{SC}(f, \boldsymbol{\ell}) \le \alpha \cdot \mathsf{opt}(\boldsymbol{\ell})$.

In the strategic setting, the location ℓ_i is private information of each agent $i \in [n]$, i.e., i may report a different location $\widetilde{\ell}_i$ to the mechanism. The mechanism $f: \mathcal{M}^n \to \mathcal{M}$ takes the reported location vector $\widetilde{\boldsymbol{\ell}} = (\widetilde{\ell}_1, \widetilde{\ell}_2, \dots, \widetilde{\ell}_n)$ and outputs the facility $f(\widetilde{\boldsymbol{\ell}})$. For notational convenience, we may let f denote the mechanism's outcome. If the mechanism is randomized, then the outcome is a distribution $\pi(\mathcal{M})$ of the facilities in $\mathcal{M}$. We assume that agents are selfish and rational, i.e., each agent i tries to minimize their own expected cost, $\mathsf{cost}_i \stackrel{\text{def}}{=\!=} \mathbb{E}[d(f,\ell_i)]$ for the chosen facility $f(\widetilde{\boldsymbol{\ell}}_{-i},\widetilde{\ell}_i)$, when i reports $\widetilde{\ell}_i$ and the remaining $n-1$ agents report $\widetilde{\boldsymbol{\ell}}_{-i}$. The mechanism is called *dominant strategy incentive compatible* (DSIC) if $\mathbb{E}[d(f(\widetilde{\ell}_i,\boldsymbol{\ell}_{-i}),\ell_i)] \geq \mathbb{E}[d(f(\boldsymbol{\ell}),\ell_i)]$ for all $i \in [n]$ and all $\boldsymbol{\ell} \in \mathcal{M}^n$. E.g., a random dictatorship mechanism that selects an agent i uniformly at random and lets i to place the facility at their preferred location ℓ_i, i.e., $f : \mathcal{M}^n \to \pi(\mathcal{M})$ is given by $f \sim \text{UNI}\{\ell_1,\ldots,\ell_n\}$, is a DSIC mechanism [2]. On the other hand, a non-truthful mechanism induces a game among n agents. Similar to DSIC mechanisms, we assume that agents only employ undominated strategies[6] (see Claim 1). We further assume that agents reach Nash Equilibrium $\widetilde{\boldsymbol{\ell}} \in \mathtt{NE}(\boldsymbol{\ell})$, i.e., at a given reported profile $\widetilde{\boldsymbol{\ell}} = (\widetilde{\ell}_1,\ldots,\widetilde{\ell}_n)$ no agent i attains smaller expected cost by a unilateral deviation $(\widetilde{\ell}'_i,\widetilde{\boldsymbol{\ell}}_{-i})$. The Price of Anarchy (PoA) of the mechanism $f : \mathcal{M}^n \to \pi(\mathcal{M})$ is the worst-case ratio of the expected social cost attained at equilibrium and the optimal social cost.

$$\mathsf{PoA}(f) \stackrel{\text{def}}{=\!=} \max_{\boldsymbol{\ell}\in\mathcal{M}^n} \max_{\widetilde{\boldsymbol{\ell}}\in\mathtt{NE}(\boldsymbol{\ell})} \frac{\mathbb{E}[\mathsf{SC}(f(\widetilde{\boldsymbol{\ell}}),\boldsymbol{\ell})]}{\mathsf{opt}(\boldsymbol{\ell})} \tag{1}$$

ML Augmented Mechanism Design for Facility Location Problem. The mechanism designer in addition to the reports $\widetilde{\boldsymbol{\ell}}$ is given a prediction $\hat{o}$ of the optimum location o. Formally, the mechanism is $f(\widetilde{\boldsymbol{\ell}},\hat{o}) : \mathcal{M}^{n+1} \to \pi(\mathcal{M})$ (to emphasize the value of Δ used, we may put it as a subscript; we have $f_\Delta(\cdot,\cdot)$ in our case, where $\Delta = \frac{c}{n}\cdot\mathsf{SC}(\hat{o},\boldsymbol{\ell})$). Consistency and robustness are two standard measures that describe the performance of an algorithm with predictions [21].

The *consistency* captures how well the algorithm performs, when prediction is correct ($\hat{o} = o(\boldsymbol{\ell})$): an algorithm (or a truthful mechanism) is α-consistent if it achieves an α-approximation to the optimum. In case of non truthful mechanisms, we say that a mechanism $f(\widetilde{\boldsymbol{\ell}},\hat{o})$ is α-consistent if its price of anarchy (1) is at most α on any instance $\boldsymbol{\ell} \in \mathcal{M}^n$ when $\hat{o} = o(\boldsymbol{\ell})$. It is often unreasonable to expect that the predicted solution $\hat{o}$ is 100% accurate. Thus it is important to understand how approximation guarantee degrades with the quality of the prediction. To this end, we measure the quality of the prediction $\hat{o}$ as the ratio of $\mathsf{SC}(\hat{o},\boldsymbol{\ell})$ and $\mathsf{opt}(\boldsymbol{\ell})$.[7] Namely, we say that prediction is γ-*accurate* if $\frac{\mathsf{SC}(\hat{o},\boldsymbol{\ell})}{\mathsf{opt}(\boldsymbol{\ell})} \leq \gamma$.

In contrast, *robustness* captures how much of the worst-case guarantees the mechanism retains, when the prediction is arbitrarily wrong. Namely, mechanism $f(\widetilde{\boldsymbol{\ell}},\hat{o})$ is β-robust, if the price of anarchy of the game induced by f is at most

---

[6] No agent i plays a strategy $\widetilde{\ell}_i$ with a higher expected cost than another strategy $\widetilde{\ell}'_i$ on every profile $\boldsymbol{\ell}_{-i} \in \mathcal{M}^{n-1}$.

[7] One can also use $d(\hat{o},o)$ as a measure of the accuracy of the prediction. A drawback of this approach is that o might not be unique, or that a pair of far away points may yield similar social cost.

β, i.e.,

$$\beta \geq \max_{\boldsymbol{\ell},\hat{o}} \max_{\widetilde{\boldsymbol{\ell}} \in \mathtt{NE}(\boldsymbol{\ell}, f_\Delta)} \frac{\mathbb{E}[\mathsf{SC}(f_\Delta(\widetilde{\boldsymbol{\ell}}, \hat{o}), \boldsymbol{\ell})]}{\mathsf{opt}(\boldsymbol{\ell})}.$$

The ML augmented framework studies which α-consistency and β-robustness guarantees are attainable. Typically, there is a trade off between feasible α and β. We provide guarantees on the robustness and consistency as functions of the prediction accuracy γ and the ratio $c = \frac{\Delta}{a}$ between given parameter Δ and average cost $a \stackrel{\text{def}}{=\!=} \frac{1}{n}\mathsf{SC}(\hat{o}, \boldsymbol{\ell})$ for the predicted location $\hat{o}$.

Harmonic Mechanism. We analyze the following `Harmonic` mechanism, which installs facility f at $\widetilde{\ell}_i$ with probability inversely proportional to $d(\widetilde{\ell}_i, \hat{o}) + \Delta$ for a constant Δ. This mechanism is not dominant strategy incentive compatible

ALGORITHM 1: Harmonic mechanism $\mathtt{Harmonic}(\Delta)$

Data: Reported locations $\widetilde{\boldsymbol{\ell}} = (\widetilde{\ell}_1, \ldots, \widetilde{\ell}_n)$, prediction $\hat{o} \in \mathcal{M}$
Result: Facility $f \sim \pi\{\widetilde{\ell}_1, \ldots, \widetilde{\ell}_n\}$
for $i \in [n]$ **do** let $d_i = d(\widetilde{\ell}_i, \hat{o})$ Choose $\mathbf{Pr}[f \leftarrow \widetilde{\ell}_i] \stackrel{\text{def}}{=\!=} \frac{1/(d_i+\Delta)}{\sum_{j=1}^{n} 1/(d_j+\Delta)}$ for $i \in [n]$;
// $\mathbf{Pr}[f \leftarrow \widetilde{\ell}_i] \propto \frac{1}{d_i+\Delta}$

(DSIC) in general. However, as we will show shortly, the set of undominated strategies for each agent is rather limited. In other words, `Harmonic` mechanism is not too far from strategy-proof mechanisms. We further analyse how far the social cost at a Nash equilibrium is from the optimum social cost, namely the price of anarchy under `Harmonic` selection rule f.

3 Strategies and Equilibria in Harmonic Mechanism

We first describe the set of strictly dominant strategies of each agent $i \in [n]$ under `Harmonic` mechanism. For notational brevity, we use ℓ_j for $j \neq i$ in place of $\widetilde{\ell}_j$, as we consider a fixed agent i's strategy for all possible reports of the other agents and one can simply pretend that $\boldsymbol{\ell}_{\text{-}i}$ is a variable vector to agent i.

3.1 Undominated Strategies

We consider the expected $\mathsf{cost}_i(\widetilde{\ell}_i)$ of a fixed agent i as a function of her report $\widetilde{\ell}_i$, while assuming that the reports $\boldsymbol{\ell}_{\text{-}i}$ of other agents follow a distribution[8] $\mathcal{F}_{\text{-}i}$. Then

[8] The main focus of our paper is on pure Nash equilibria, but here we allow randomization in other agents' strategies.

$$\mathsf{cost}_i(\widetilde{\ell}_i) = \mathop{\mathbb{E}}_{\boldsymbol{\ell}_{\text{-}i}\sim\mathcal{F}_{\text{-}i}}\left[\frac{\sum_{j\neq i}\frac{d(\ell_i,\ell_j)}{d(\ell_j,\hat{o})+\Delta}+\frac{d(\ell_i,\widetilde{\ell}_i)}{d(\widetilde{\ell}_i,\hat{o})+\Delta}}{\sum_{j\neq i}\frac{1}{d(\ell_j,\hat{o})+\Delta}+\frac{1}{d(\widetilde{\ell}_i,\hat{o})+\Delta}}\right] = \mathop{\mathbb{E}}_{\boldsymbol{\ell}_{\text{-}i}}\left[\frac{C_1(\boldsymbol{\ell}_{\text{-}i})+\frac{x}{y+\Delta}}{C_2(\boldsymbol{\ell}_{\text{-}i})+\frac{1}{y+\Delta}}\right], \quad (2)$$

where agent i can control only two parameters $x \stackrel{\text{def}}{=} d(\ell_i,\widetilde{\ell}_i) \geq 0$ and $y \stackrel{\text{def}}{=} d(\widetilde{\ell}_i,\hat{o}) \geq 0$ by possibly misreporting her true location, while the remaining terms $C_1 \geq 0$ and $C_2 > 0$ in the numerator and denominator respectively only depend on the reports of other agents $\boldsymbol{\ell}_{\text{-}i}$. Thus, we can study agent i's cost as a function of only $x, y \in \mathbb{R}_+$, and each fixed realization of $\boldsymbol{\ell}_{\text{-}i} \sim \mathcal{F}_{\text{-}i}$. For convenience we continue to use the same notations C_1, C_2, and cost_i for each fixed realization of $\boldsymbol{\ell}_{\text{-}i}$: $\mathsf{cost}_i(x,y,\boldsymbol{\ell}_{\text{-}i}) = \frac{C_1+x/(y+\Delta)}{C_2+1/(y+\Delta)}$. A few simple observations are in order.

Observation 1. *The cost $\mathsf{cost}_i(x,y,\boldsymbol{\ell}_{\text{-}i})$ is a strictly increasing in x for any $C_1, C_2, y, \Delta \geq 0$.*

By triangle inequality $x+y = d(\ell_i,\widetilde{\ell}_i)+d(\widetilde{\ell}_i,\hat{o}) \geq d(\ell_i,\hat{o})$ and $x+d(\ell_i,\hat{o}) \geq y$. Hence,

Observation 2. $x \geq |d(\ell_i,\hat{o}) - y|$.

We further note that player i would never choose y larger than $d(\ell_i,\hat{o})$.

Claim 1. *The strategy of agent i with $y > d(\ell_i,\hat{o})$ is strictly dominated by the strategy $x = 0$, $y = d(\ell_i,\hat{o})$ for any $C_1, C_2 \geq 0$.*

Therefore, $d(\widetilde{\ell}_i,\hat{o}) = y \in [0, d(\ell_i,\hat{o})]$. By Observation 1, agent i chooses minimal possible x for any given y and any $\boldsymbol{\ell}_{\text{-}i}$. Now, $x \geq d(\ell_i,\hat{o}) - y$ by Observation 2. If $\mathcal{M}$ is a *continuous space*[9], then there always exists a $\widetilde{\ell}_i \in \mathcal{M}$ with $x \stackrel{\text{def}}{=} d(\widetilde{\ell}_i,\ell_i) = d(\ell_i,\hat{o}) - y$. I.e., player i's undominated strategies in the continuous metric space are $\widetilde{\ell}_i$ on a shortest path $\mathsf{S\text{-}Path}(\hat{o},\ell_i)$ between $\hat{o}$ and ℓ_i.

Claim 2. *If $\mathcal{M}$ is a continuous metric space, then reports on the shortest paths $\widetilde{\ell}_i \in \mathsf{S\text{-}Path}(\hat{o},\ell_i)$ strictly dominate other strategies of agent i.*

3.2 Equilibria Concepts

Many solution concepts may be used to describe the outcome of a game, such as Pure Nash (PNE), Mixed Nash (MNE), Correlated (CE), and Coarse Correlated (CCE) equilibria. In case of the game induced by the `Harmonic` mechanism f, the solution concepts that involve randomization are not ideal. Indeed, an agent i might find it difficult to compute their expected cost (2) for a fixed reported location $\widetilde{\ell}_i$ let alone finding the best response over all possible $\widetilde{\ell}_i \in \mathsf{S\text{-}Path}(\hat{o},\ell_i)$.

[9] Formally, for any two points $P_1, P_2 \in \mathcal{M}$ and $x \in [0, d(P_1,P_2)]$ there exists a point $P \in \mathcal{M}$ such that $d(P,P_1) = x$ and $d(P,P_2) = d(P_1,P_2) - x$.

E.g., if $n-1 = 10$ agents randomize in $\boldsymbol{\ell}_{\text{-}i}$ between two pure strategies in a mixed Nash Equilibrium, then (2) would already have more than 1000 fraction terms.

Thus we adopt the Pure Nash Equilibrium (PNE) solution concept to describe the outcome of `Harmonic` mechanism f. It is important to keep in mind that, unlike MNE, PNE may not exist. We show below that PNE always exists in a continuous metric space $\mathcal{M}$, i.e., in a metric where every pair of points has a continuous shortest path between them. The proof refers to the full version.

Theorem 1. *Let $\mathcal{M}$ be continuous (any pair of points has a shortest path), then for any initial positions $\boldsymbol{\ell} = (\ell_i)_{i\in[n]}$ of $[n]$ agents there is a Pure Nash Equilibrium under* `Harmonic` *mechanism.*

For discrete (non continuous) metric spaces $\mathcal{M}$ to allow randomness, we can let agents to explicitly report their locations as a distribution over a finite number of points in $\mathcal{M}$. I.e., we consider an extension of the metric space $\mathcal{M}$ to a larger metric space $\overline{\mathcal{M}}$ over all finite convex combinations $\overline{\mathcal{M}} = \{\boldsymbol{\alpha} = \sum_{i\in[k]} \alpha_i \cdot \ell_i \mid \sum_{i\in[k]} \alpha_i = 1,\ \alpha_i \geq 0,\ \ell_i \in \mathcal{M}\}$, where a naturally induced metric in $\overline{\mathcal{M}}$ is given by the earth mover's distance between two probability distributions $\boldsymbol{\alpha}, \boldsymbol{\beta}$ over points in $\mathcal{M}$. Then $\overline{\mathcal{M}}$ is a continuous metric space and thus PNE does exist in $\overline{\mathcal{M}}$. Furthermore, when many agents participate in `Harmonic` mechanism, a reasonable approximation and/or simplification of their expected costs (due to good concentration) can be done by calculating expectations of the numerator and denominator in (2) separately and then using the ratio of expectations instead of the expected ratio. This approach corresponds exactly to agents playing pure strategies in $\overline{\mathcal{M}}$.

Best Response in Pure Nash Equilibria We next show in Claim 3 which locations on the shortest paths could be the best response of agent i for a fixed profile of other agents' reports $\boldsymbol{\ell}_{\text{-}i}$: i only needs to decide between two choices $\widetilde{\ell}_i \in \{\ell_i, \hat{o}\}$ and only when she is indifferent, then i may also play any $\widetilde{\ell}_i \in \mathsf{S\text{-}Path}(\hat{o}, \ell_i)$.

Claim 3. *Given a profile $\boldsymbol{\ell}_{\text{-}i}$ of agents $[n] \setminus \{i\}$ locations, agent i's best response is to only report*

- *her true location $\widetilde{\ell}_i = \ell_i$ when $\mathsf{cost}_i(\ell_i, \boldsymbol{\ell}_{\text{-}i}) < d(\ell_i, \hat{o}) + \Delta$;*
- *predicted location $\widetilde{\ell}_i = \hat{o}$ when $\mathsf{cost}_i(\ell_i, \boldsymbol{\ell}_{\text{-}i}) > d(\ell_i, \hat{o}) + \Delta$; then $\mathsf{cost}_i(\widetilde{\ell}_i, \boldsymbol{\ell}_{\text{-}i}) \geq d(\ell_i, \hat{o}) + \Delta$;*
- *any point $\widetilde{\ell}_i \in \mathsf{S\text{-}Path}(\hat{o}, \ell_i)$ when $\mathsf{cost}_i(\ell_i, \boldsymbol{\ell}_{\text{-}i}) = d(\ell_i, \hat{o}) + \Delta$; then $\mathsf{cost}_i(\ell_i, \boldsymbol{\ell}_{\text{-}i}) = \mathsf{cost}_i(\widetilde{\ell}_i, \boldsymbol{\ell}_{\text{-}i})$.*

4 Price of Anarchy

Here we study the Price of Anarchy (PoA) of the PNE of `Harmonic` mechanism f in metric $\mathcal{M}$ or $\overline{\mathcal{M}}$. We will make extensive use of Claims 2, 3 to prove PoA guarantees for the consistency and robustness of `Harmonic` mechanism. We derive two different PoA bounds for robustness: (a) when $\mathcal{M}$ is a strictly convex space (b) when $\mathcal{M}$ is a general metric space.

We first introduce some simplified notations that would help us to use Claim 3. Let $\widetilde{\boldsymbol{\ell}} = (\widetilde{\ell}_i)_{i\in[n]}$ be a PNE. Let $t_i \stackrel{\text{def}}{=} d(\hat{o}, \ell_i)$ denote the distance between predicted location $\hat{o}$ and the true location ℓ_i of agent $i \in [n]$. In equilibrium, some of the agents may report different locations $\widetilde{\ell}_i$, we denote by $d_i \stackrel{\text{def}}{=} d(\hat{o}, \widetilde{\ell}_i)$ the distance from $\hat{o}$ to the reported location of $i \in [n]$. Note that player i will always have $t_i \geq d_i$ in an equilibrium. To write down the $\mathsf{cost}_i(\widetilde{\ell}_i)$ of agent i we will use $c_{ij} \stackrel{\text{def}}{=} d(\ell_i, \widetilde{\ell}_j)$ (note that $c_{ij} \neq c_{ji}$) that denotes the distance between true location of $i \in [n]$ and the reported location of $j \in [n]$. In particular, $c_{ii} = d(\ell_i, \widetilde{\ell}_i) = t_i - d_i$. Then

$$\mathsf{cost}_i(\widetilde{\ell}_i) = \frac{\sum_{j\in[n]} c_{ij}/(d_j + \Delta)}{\sum_{j\in[n]} 1/(d_j + \Delta)} \tag{3}$$

By Claim 3 each agent either reports $\widetilde{\ell}_i = \hat{o}$, or $\widetilde{\ell}_i = \ell_i$, or $\widetilde{\ell}_i \in \mathsf{S\text{-}Path}(\hat{o}, \ell_i)$ in a PNE. Let

$$S \stackrel{\text{def}}{=} \{i \in [n] \mid \widetilde{\ell}_i = \hat{o}\}, \quad T \stackrel{\text{def}}{=} \{i \in [n] \mid \widetilde{\ell}_i = \ell_i\}, \quad U \stackrel{\text{def}}{=} [n] \setminus (S \cup T)$$

and also use $\overline{S} = [n] \setminus S = U \cup T$. We also use $D \stackrel{\text{def}}{=} \sum_{j\in[n]} 1/(d_j + \Delta)$ to denote the denominator of each $\mathsf{cost}_i(\widetilde{\boldsymbol{\ell}})$ for $i \in [n]$.

4.1 Consistency

In this section, we show that the $\texttt{Harmonic}(\Delta)$ mechanism can achieve a solution arbitrarily close to the optimum depending on the quality of the prediction. Specifically, when the predicted location $\hat{o}$ gives an almost optimum estimation of the true optimum ($\sum_{i\in[n]} d(\ell_i, \hat{o}) \approx \sum_{i\in[n]} d(\ell_i, o)$) and $\frac{\Delta}{\mathsf{opt}/n} = O(\varepsilon)$, we are able to show that PoA of $\texttt{Harmonic}(\Delta)$ mechanism is close to 1. The following theorem refers to the full version.

Theorem 2. *If $\hat{o}$ is γ-accurate, i.e., $\frac{\mathsf{SC}(\hat{o},\boldsymbol{\ell})}{\mathsf{opt}(\boldsymbol{\ell})} \leq \gamma$, and $c \stackrel{\text{def}}{=} \Delta/a$ for an average cost $a = \frac{1}{n}\mathsf{SC}(\hat{o}, \boldsymbol{\ell})$ of the prediction $\hat{o}$, then $\texttt{Harmonic}(\Delta)$ mechanism is $\gamma \cdot (1 + 2c)$-consistent.*

4.2 Robustness

Here, we show robustness guarantees for arbitrary bad predicted location $\hat{o}$ for the $\texttt{Harmonic}$ mechanism. Our guarantees depend on $c = \Delta/a$ (the larger c, the better) that equals to the ratio between the parameter Δ of the $\texttt{Harmonic}$ mechanism and $a = \frac{1}{n}\mathsf{SC}(\hat{o}, \boldsymbol{\ell})$. We will first derive a few useful lemmas and then prove two PoA guarantees: first, for the case when $\mathcal{M}$ is strictly convex space, and, second, for the general metric.

Recall that Harmonic mechanism chooses each location $\widetilde{\ell}_i$ for $i \in [n]$ with probability $\frac{1/(d_i+\Delta)}{D}$ where $D \stackrel{\text{def}}{=} \sum_{i\in[n]} 1/(d_i+\Delta)$. The expected social cost $\mathsf{SC} = \sum_{j\in[n]} \frac{1/(d_j+\Delta)}{D} \sum_{i\in[n]} c_{ij}$, which we partition into three terms by dividing $j \in [n]$ into three sets S, U, and T according to the equilibrium conditions from Claim 3 as follows

$$\mathsf{SC} = \frac{|S|/\Delta}{D} \sum_{i\in[n]} t_i + \sum_{j\in T} \frac{1/(t_j+\Delta)}{D} \sum_{i\in[n]} d_{ij} + \sum_{j\in U} \frac{1/(d_j+\Delta)}{D} \sum_{i\in[n]} c_{ij}. \quad (4)$$

We first consider the term corresponding to the set of agents T, who reported their true location $\widetilde{\ell}_j = \ell_j$. It is useful to keep in mind that the random dictatorship mechanism is a 2-approximation to the optimal social cost [2].

Lemma 1 (Theorem 3.1 in [2]). *The random dictatorship mechanism (i.e., choosing location ℓ_j of agent $j \in [n]$ with probability $1/n$) is 2-approximation.*

This Lemma 1 is a useful comparison point for estimating the contribution to the social cost by agents in T. Indeed, if we can show that the probabilities of selecting $\widetilde{\ell}_i = \ell_i$ are at most a constant factor away from the uniform sampling, then agents in T contribute no more than a constant factor compared to the optimum social cost. The next Lemma 2 proves that and gives useful probability estimates for selecting agents $i \in U$ under Harmonic mechanism.

Lemma 2. *For all $i \in T \cup U$, the probability that location $\widetilde{\ell}_i$ is selected is at most*

$$\frac{1/(d_i+\Delta)}{\sum_{j\in[n]} 1/(d_j+\Delta)} \le \frac{2}{n} \cdot \frac{t_i+\Delta}{d_i+\Delta}. \quad (5)$$

The Lemma 2 implies that the Harmonic mechanism chooses each location $\widetilde{\ell}_j = \ell_j$ for agents $j \in T$ with probability at most $2/n$. I.e., the contribution of agents in T to the expected social cost is at most $O(1)$ of the uniform sampling and, hence, is $O(\mathsf{opt})$.

Corollary 1. $\sum_{j\in T} \frac{1/(t_j+\Delta)}{D} \sum_{i\in[n]} d_{ij} \le 4 \cdot \mathsf{opt}$.

The other two terms corresponding to agents $j \in S$ and $j \in U$ are the main source of Harmonic mechanism's inefficiency. Namely, it is impossible to achieve $O(1)$ robustness guarantee, when Δ is significantly smaller than $a = \frac{1}{n}\sum_{i\in[n]} t_i$. Indeed, consider for example an instance, in which $\Delta = 0$, a single agent j stays exactly at the predicted location $\ell_j = \hat{o}$, and all other agents $i \in [n] \setminus \{j\}$ are all situated at the same spot in $\mathcal{M}$ far away from $\hat{o}$. In the unique equilibrium agent j reports $\widetilde{\ell}_j = \hat{o}$, while all other agents $i \neq j$ report ℓ_i; the mechanism picks $\widetilde{\ell}_j = \hat{o}$ with large probability resulting in an $\Omega(n)$ inefficient placement of the facility. Still, when the parameter Δ is not too small compared to a, we are able to show a constant approximation guarantee regardless of the predicted location $\hat{o}$.

We begin our analysis with the term in (4) corresponding to agents $j \in S$. There we have a factor $\frac{1}{D \cdot \Delta}$ and our next lemma relates this term to the parameter $c = \Delta/a$.

Lemma 3. *Let* $\Delta = \frac{c}{n} \cdot \sum_{i \in [n]} d(\ell_i, \hat{o})$ *in* Harmonic(Δ). *Then* $\frac{1}{\Delta \cdot D} \le \max\left\{\frac{1/c+1}{n}, \frac{1/c+1}{n/c}\right\}$.

The Lemmas 1, 2, 3 and Corollary 1 are mostly enough to get a robustness guarantee of $O(1 + 1/c^2)$, if the set of agents U in PNE $\widetilde{\boldsymbol{\ell}}$ was empty. Unfortunately, the presence of set U significantly complicates the analysis. Next, we derive two different robustness guarantees: first for an easier case, when metric $\mathcal{M}$ is a strictly convex space, and later for the general metric.

Robustness in Strictly Convex Spaces. Here, we show robustness guarantee for arbitrary bad predicted location $\hat{o}$ for the `Harmonic` mechanism, assuming that $\mathcal{M}$ is a strictly convex space, i.e., the distance between two vectors $\mathbf{v_1}, \mathbf{v_2} \in \mathcal{M}$ is given by $d(\mathbf{v_1}, \mathbf{v_2}) = \|\mathbf{v_1} - \mathbf{v_2}\|$, and $\|\mathbf{v_1} + \mathbf{v_2}\| = \|\mathbf{v_1}\| + \|\mathbf{v_2}\|$ for $\mathbf{v_1}, \mathbf{v_2} \neq 0$ implies that $\mathbf{v_1} = c \cdot \mathbf{v_2}$ for a $c > 0$.

Theorem 3. Harmonic(Δ) *is a* $O(1 + 1/c^2)$*-robust in strictly convex space* $\mathcal{M}$, *if* $\Delta = \frac{c}{n} \cdot \mathsf{SC}(\hat{o}, \boldsymbol{\ell})$.

To prove the theorem, we first bound bound c_{ij} in `Harmonic`, which follows from strict convexity:

Claim 4. *For strictly convex space* $\mathcal{M}$, *we have* $c_{ij} \le \frac{t_j - d_j}{t_j} t_i + \frac{d_j}{t_j} d_{ij}$ *for* $j \in U$.

Applying this upper bound on c_{ij} for all $j \in U$ and $i \in [n]$ in the expected social cost (4) gives

$$\begin{aligned}\mathsf{SC} &\overset{\text{Claim 4}}{\le} \frac{|S|}{\Delta \cdot D} \sum_{i \in [n]} t_i + \sum_{j \in U} \frac{1/(d_j + \Delta)}{D} \sum_{i \in [n]} \left(t_i + \frac{d_j}{t_j} d_{ij}\right) + \sum_{j \in T} \frac{1/(t_j + \Delta)}{D} \sum_{i \in [n]} d_{ij} \\ &= \underbrace{\left(\frac{|S|}{\Delta \cdot D} + \sum_{j \in U} \frac{1}{D(d_j + \Delta)}\right) \cdot \sum_{i \in [n]} t_i}_{(*)} + \underbrace{\sum_{j \in U} \frac{d_j/t_j}{(d_j + \Delta) D} \sum_{i \in [n]} d_{ij} + \sum_{j \in T} \frac{1/(t_j + \Delta)}{D} \sum_{i \in [n]} d_{ij}}_{(**)}.\end{aligned}$$

We first shall get an upper bound on the second term $(**)$ above in a very similar way to Lemma 2 and Corollary 1.

$$(**) \le \sum_{j \in U} \frac{2}{n} \cdot \frac{t_j + \Delta}{d_j + \Delta} \cdot \frac{d_j}{t_j} \sum_{i \in [n]} d_{ij} + \frac{2}{n} \sum_{j \in T} \sum_{i \in [n]} d_{ij} \le \frac{2}{n} \sum_{\substack{j \in U \cup T, \\ i \in [n]}} d_{ij} \le \frac{2}{n} \sum_{i,j \in [n]} d_{ij} \le 4 \cdot \mathsf{opt}, \quad (6)$$

where the first inequality holds by Lemma 2; the second inequality holds as $\frac{t_j + \Delta}{d_j + \Delta} \frac{d_j}{t_j} = \frac{1 + \Delta/t_j}{1 + \Delta/d_j} \le 1$ for all $j \in U$; the last inequality holds by Lemma 1 similar to Corollary 1. Next, we get an upper bound on the $(*)$ term. We first note that

$$(*) \le \left(\frac{|S|}{\Delta \cdot D} + \sum_{j\in U} \frac{1}{D\cdot\Delta}\right) \cdot \sum_{i\in[n]} t_i = (|S|+|U|)\Delta \cdot \frac{1}{\Delta D} \cdot \frac{\sum_{i\in[n]} t_i}{\Delta} \le$$

$$(|S|+|U|)\Delta \cdot \max\left\{\frac{1/c+1}{n}, \frac{1/c+1}{n/c}\right\} \cdot \frac{n}{c}, \quad (7)$$

where we got the first inequality by substituting $\frac{1}{d_j+\Delta}$ with $\frac{1}{\Delta}$; used definition of $c = \frac{\Delta}{a} = \frac{\Delta \cdot n}{\sum_{i\in[n]} t_i}$ to simplify $\frac{\sum_{i\in[n]} t_i}{\Delta}$ and Lemma 3 to get an upper bound on $\frac{1}{\Delta D}$ in the second inequality. Finally, we give an upper bound on $(|S|+|U|)\Delta$ in the following Claim 5. The proof relies on equilibrium conditions from Claim 3 and Claim 4.

Claim 5. *For strictly convex space $\mathcal{M}$, $(|S|+|U|)\Delta \le \frac{2}{n}\sum_{i,j\in[n]} d_{ij}$.*

We get the following bound on $(*)$ by applying Claim 5 to (7).

$$(*) \le \frac{2}{n}\sum_{i,j\in[n]} d_{ij} \cdot \max\left\{\frac{1/c+1}{n}, \frac{1/c+1}{n/c}\right\} \cdot \frac{n}{c} = \max\left\{\frac{2+2c}{c^2}, \frac{2+2c}{c}\right\} \cdot \frac{1}{n}\sum_{i,j\in[n]} d_{ij}.$$

I.e., $(*) = O(1+\frac{1}{c^2}) \cdot \mathsf{opt}$. We conclude the proof of Theorem 3 by combining this bound on $(*)$ with the constant upper bound (6) on $(**)$ in (4.2).

Robustness in General Metric Spaces. The goal of this section is to extend our robustness result for strictly convex spaces to general metric (Theorem 4). However, since general metric $\mathcal{M}$ may have rather different geometry than strictly convex spaces[10], we lose an additional factor $1/c$ in our robustness guarantee.

Theorem 4. *Harmonic(Δ) is a $O(1+1/c^3)$-robust in metric space, if $\Delta = \frac{c}{n}\cdot \mathsf{SC}(\hat{o}, \boldsymbol{\ell})$.*

4.3 Constant Number of Agents

We conclude the PoA analysis of `Harmonic` mechanism with a special case of a small number of agents $n = O(1)$. In this regime, we are able to simplify the mechanism by setting the parameter $\Delta = 0$, i.e., we ignore the estimate $\hat{a}$ for the average distance $a = \frac{1}{n}\sum_{i\in[n]} d(\ell_i, \hat{o})$, and get the following `Harmonic`(0) mechanism.

[10] Consider a two-dimensional space $L_1(\mathbb{R}^2)$, and consider a point $(0,1) \in \mathsf{S\text{-}Path}((0,0),(1,1))$. The distances from another point $(1,0)$ to the endpoints are $d((1,0),(0,0)) = d((1,0),(1,1)) = 1$, while $d((1,0),(0,1)) = 2$.

ALGORITHM 2: Harmonic mechanism `Harmonic`(0)

Data: Reported locations $\widetilde{\boldsymbol{\ell}} = (\widetilde{\ell}_1, \ldots, \widetilde{\ell}_n)$, prediction $\hat{o} \in \mathcal{M}$
Result: Facility $f \sim \pi\{\widetilde{\ell}_1, \ldots, \widetilde{\ell}_n\}$
for $i \in [n]$ **do** let $d_i = d(\widetilde{\ell}_i, \hat{o})$ Choose $\mathbf{Pr}[f \leftarrow \widetilde{\ell}_i] \stackrel{\text{def}}{=} \frac{1/d_i}{\sum_{j=1}^{n} 1/d_j}$ for $i \in [n]$;
`//` $\mathbf{Pr}[f \leftarrow \widetilde{\ell}_i] \propto \frac{1}{d_i}$ `(proportional to)`

Theorem 5. *`Harmonic`(0) is 1-consistent (γ-consistent for a γ-accurate prediction $\hat{o}$) in general metric $\mathcal{M}$. It is $O(n)$-robust, when $\mathcal{M}$ is a strictly convex space.*

Remark 1. The assumption that $\mathcal{M}$ is strictly convex space is crucial for deriving $O(1)$-robustness bound. This is because each agent $i \in U$ may increase the distance to everyone else compared to $\hat{o}$ and ℓ_i. The analysis for sets S and T can be carried in the same way as for strictly convex spaces.

To see this, consider the following example of Nash equilibrium in a circle metric space $\mathcal{M}$ with just $n = 2$ agents with arbitrary large PoA. Specifically, let $d(\ell_1, \ell_2) = 1$, $d(\hat{o}, \ell_1) = d(\hat{o}, \ell_2) = M$ while the length of the whole circle is $2M+1$ for a large constant $M \in \mathbb{R}$. Then the following $\widetilde{\boldsymbol{\ell}}$ is a Nash equilibrium: $\widetilde{\ell}_1$ and $\widetilde{\ell}_2$ are on their respective shortest paths from ℓ_1 and ℓ_2 to $\hat{o}$ with $d(\widetilde{\ell}_1, \hat{o}) = d(\widetilde{\ell}_2, \hat{o}) = 0.5$. It is easy to verify that each of the agents $i = 1$ and $i = 2$ is indifferent between reporting $\widetilde{\ell}_i = \hat{o}$ or $\widetilde{\ell}_i = \ell_i$. I.e., by Claim 3 $\widetilde{\boldsymbol{\ell}}$ is a Nash equilibrium with the social cost $\mathsf{SC}(\widetilde{\boldsymbol{\ell}}) = 2M$. The optimal location is, e.g., at $f = \ell_1$ with the social cost of 1.

5 Conclusions

In this paper we study a canonical problem of strategic single-facility location in general metric spaces under new lenses of ML augmented mechanism design framework. This framework not only allows to circumvent worst-case analysis limitations, but also enriches the design space of mechanisms in interesting new ways. It naturally led us to consider new type of non-truthful mechanisms (such as `Harmonic` mechanism) that have not appeared in the prior literature. We got useful insights about undominated strategies and equilibria structure for this mechanism. We proved that `Harmonic` mechanism has a $1 + \varepsilon$ price of anarchy bound when predictions are (nearly) accurate, while retaining a constant PoA of $O(1 + 1/\texttt{poly}(\varepsilon))$ in the worst-case, when $\hat{o}$ is arbitrary bad (given that our mechanism's parameter $\Delta = \varepsilon \cdot \mathsf{SC}(\hat{o}, \boldsymbol{\ell})$).

Our PoA analysis of consistency and especially robustness significantly deviates from a typical PoA analysis, as (i) the general smoothness argument does not help in breaking 2-approximation barrier and (ii) we have to use metric conditions in a non-trivial way (e.g., our PoA bounds are different for strictly convex and general metric spaces, when $n = O(1)$). The tightness of our PoA bounds for `Harmonic` remains an open problem. Another interesting open question is to find a mechanism that does not depend on a parameter Δ, but admits similar $(1+\varepsilon)$-consistency and $O(1+1/\texttt{poly}(\varepsilon))$-robustness PoA guarantees.

Acknowledgments. Gravin's research is supported by National Key R & D Program of China (2023YFA1009500), by NSFC grant 61932002, and by "the Fundamental Research Funds for the Central Universities" in China. Chen and Im are supported in part by NSF grants CCF-1844939, CCF-2121745, and CCF-2423106.

References

1. Agrawal, P., Balkanski, E., Gkatzelis, V., Ou, T., Tan, X.: Learning-augmented mechanism design: leveraging predictions for facility location. In: Proceedings of the 23rd ACM Conference on Economics and Computation, pp. 497–528 (2022)
2. Alon, N., Feldman, M., Procaccia, A.D., Tennenholtz, M.: Strategyproof approximation mechanisms for location on networks. arXiv preprint arXiv:0907.2049 (2009)
3. Alon, N., Feldman, M., Procaccia, A.D., Tennenholtz, M.: Strategyproof approximation of the minimax on networks. Math. Oper. Res. **35**(3), 513–526 (2010)
4. Aziz, H., Lam, A., Suzuki, M., Walsh, T.: Random rank: the one and only strategyproof and proportionally fair randomized facility location mechanism. In: Koyejo, S., Mohamed, S., Agarwal, A., Belgrave, D., Cho, K., Oh, A. (eds.) Advances in Neural Information Processing Systems 35: Annual Conference on Neural Information Processing Systems 2022, NeurIPS 2022, New Orleans, LA, USA, 28 November–9 December 2022 (2022)
5. Balkanski, E., Gkatzelis, V., Shahkarami, G.: Randomized strategic facility location with predictions (2024). https://arxiv.org/abs/2409.07142
6. Balkanski, E., Gkatzelis, V., Tan, X.: Strategyproof scheduling with predictions. In: Kalai, Y.T. (ed.) 14th Innovations in Theoretical Computer Science Conference, ITCS 2023, 10–13 January 2023, MIT, Cambridge, Massachusetts, USA. LIPIcs, vol. 251, pp. 11:1–11:22 (2023)
7. Barak, Z., Gupta, A., Talgam-Cohen, I.: Mac advice for facility location mechanism design (2024). https://arxiv.org/abs/2403.12181
8. Caragiannis, I., et al.: Bounding the inefficiency of outcomes in generalized second price auctions. J. Econ. Theory **156**, 343–388 (2015)
9. Chan, H., Filos-Ratsikas, A., Li, B., Li, M., Wang, C.: Mechanism design for facility location problems: a survey. In: Zhou, Z. (ed.) Proceedings of the Thirtieth International Joint Conference on Artificial Intelligence, IJCAI 2021, Virtual Event/Montreal, Canada, 19–27 August 2021, pp. 4356–4365. ijcai.org (2021)
10. Christodoulou, G., Sgouritsa, A., Vlachos, I.: Mechanism design augmented with output advice (2024). https://arxiv.org/abs/2406.14165
11. Dokow, E., Feldman, M., Meir, R., Nehama, I.: Mechanism design on discrete lines and cycles. In: Faltings, B., Leyton-Brown, K., Ipeirotis, P. (eds.) Proceedings of the 13th ACM Conference on Electronic Commerce, EC 2012, Valencia, Spain, 4–8 June 2012, pp. 423–440. ACM (2012)

12. Escoffier, B., Gourvès, L., Kim Thang, N., Pascual, F., Spanjaard, O.: Strategy-proof mechanisms for facility location games with many facilities. In: Brafman, R.I., Roberts, F.S., Tsoukiàs, A. (eds.) ADT 2011. LNCS (LNAI), vol. 6992, pp. 67–81. Springer, Heidelberg (2011). https://doi.org/10.1007/978-3-642-24873-3_6
13. Feldman, M., Wilf, Y.: Strategyproof facility location and the least squares objective. In: Kearns, M.J., McAfee, R.P., Tardos, É. (eds.) Proceedings of the fourteenth ACM Conference on Electronic Commerce, EC 2013, Philadelphia, PA, USA, 16–20 June 2013, pp. 873–890. ACM (2013)
14. Fotakis, D., Tzamos, C.: Winner-imposing strategyproof mechanisms for multiple facility location games. In: Saberi, A. (ed.) WINE 2010. LNCS, vol. 6484, pp. 234–245. Springer, Heidelberg (2010). https://doi.org/10.1007/978-3-642-17572-5_19
15. Fotakis, D., Tzamos, C.: On the power of deterministic mechanisms for facility location games. ACM Trans. Economics and Comput. **2**(4), 15:1–15:37 (2014)
16. Fotakis, D., Tzamos, C.: Strategyproof facility location for concave cost functions. Algorithmica **76**(1), 143–167 (2016)
17. Gkatzelis, V., Kollias, K., Sgouritsa, A., Tan, X.: Improved price of anarchy via predictions. In: Pennock, D.M., Segal, I., Seuken, S. (eds.) EC 2022: The 23rd ACM Conference on Economics and Computation, Boulder, CO, USA, 11–15 July 2022, pp. 529–557. ACM (2022)
18. Goel, S., Hann-Caruthers, W.: Optimality of the coordinate-wise median mechanism for strategyproof facility location in two dimensions. Soc. Choice Welfare **61**(1), 11–34 (2023)
19. Istrate, G., Bonchis, C.: Mechanism design with predictions for obnoxious facility location. arXiv preprint arXiv:2212.09521 (2022)
20. Lu, P., Sun, X., Wang, Y., Zhu, Z.A.: Asymptotically optimal strategy-proof mechanisms for two-facility games. In: Parkes, D.C., Dellarocas, C., Tennenholtz, M. (eds.) Proceedings 11th ACM Conference on Electronic Commerce (EC-2010), Cambridge, Massachusetts, USA, 7–11 June 2010, pp. 315–324. ACM (2010)
21. Lykouris, T., Vassilvitskii, S.: Competitive caching with machine learned advice. J. ACM **68**(4), 24:1–24:25 (2021)
22. Meir, R., Lang, J., Lesca, J., Mattei, N., Kaminsky, N.: A market-inspired bidding scheme for peer review paper assignment. In: Thirty-Fifth AAAI Conference on Artificial Intelligence, AAAI 2021, Thirty-Third Conference on Innovative Applications of Artificial Intelligence, IAAI 2021, The Eleventh Symposium on Educational Advances in Artificial Intelligence, EAAI 2021, Virtual Event, 2–9 February 2021, pp. 4776–4784. AAAI Press (2021)
23. Mitzenmacher, M., Vassilvitskii, S.: Algorithms with predictions. Commun. ACM **65**(7), 33–35 (2022)
24. Procaccia, A., Wajc, D., Zhang, H.: Approximation-variance tradeoffs in facility location games. In: Proceedings of the AAAI Conference on Artificial Intelligence, vol. 32 (2018)
25. Procaccia, A.D., Tennenholtz, M.: Approximate mechanism design without money. ACM Trans. Econ. Comput. (TEAC) **1**(4), 1–26 (2013)
26. Roughgarden, T.: Beyond the Worst-Case Analysis of Algorithms. Cambridge University Press (2021)
27. Schummer, J., Vohra, R.V.: Strategy-proof location on a network. J. Econ. Theory **104**(2), 405–428 (2002)
28. Serafino, P., Ventre, C.: Heterogeneous facility location without money. Theor. Comput. Sci. **636**, 27–46 (2016)

29. Tang, P., Yu, D., Zhao, S.: Characterization of group-strategyproof mechanisms for facility location in strictly convex space. In: Biró, P., Hartline, J.D., Ostrovsky, M., Procaccia, A.D. (eds.) EC 2020: The 21st ACM Conference on Economics and Computation, Virtual Event, Hungary, 13–17 July 2020, pp. 133–157. ACM (2020)
30. Walsh, T.: Strategy proof mechanisms for facility location in euclidean and manhattan space. arXiv preprint arXiv:2009.07983 (2020)
31. Xu, C., Lu, P.: Mechanism design with predictions. In: Raedt, L.D. (ed.) Proceedings of the Thirty-First International Joint Conference on Artificial Intelligence, IJCAI 2022, Vienna, Austria, 23–29 July 2022, pp. 571–577. ijcai.org (2022)

Beyond the Worst Case: Distortion in Impartial Culture Electorates

Ioannis Caragiannis and Karl Fehrs(✉)

Aarhus University, Åbogade 34, 8200 Aarhus N, Denmark
{iannis,karl}@cs.au.dk

Abstract. *Distortion* is a well-established notion for quantifying the loss of social welfare that may occur in voting. As voting rules take as input only ordinal information, they are essentially forced to neglect the exact values the agents have for the alternatives. Thus, in worst-case electorates, voting rules may return low social welfare alternatives and have high distortion. Accompanying voting rules with a small number of cardinal queries per agent may reduce distortion considerably.

To explore distortion beyond worst-case conditions, we use a simple stochastic model according to which the values the agents have for the alternatives are drawn independently from a common probability distribution. This gives rise to so-called *impartial culture electorates*. We refine the definition of distortion so that it is suitable for this stochastic setting and show that, rather surprisingly, all voting rules have high distortion *on average*. On the positive side, for the fundamental case where the agents have random *binary* values for the alternatives, we present a mechanism that achieves approximately optimal average distortion by making a *single* cardinal query per agent. This enables us to obtain slightly suboptimal average distortion bounds for general distributions using a simple randomized mechanism that makes one query per agent. We complement these results by presenting new tradeoffs between the distortion and the number of queries per agent in the traditional worst-case setting.

Keywords: Computational social choice · Voting · Distortion · Impartial culture

1 Introduction

Voting has been the subject of social choice theory for centuries. Traditionally, having elections as the main application area in mind, social choice theorists have made a lot of progress in understanding the axiomatic properties of *voting rules* (e.g., see [39] for an introduction to basic voting axioms). However, the use of voting goes far beyond elections, and today, it provides a compelling way of collective decision-making. So, in addition to the traditional axiomatic treatment of voting rules, other approaches that have an optimization flavour have attracted a lot of attention, starting with the work of Young (see [38] and references therein).

M. Mavronicolas et al. (Eds.): WINE 2024, LNCS 15534, pp. 420–437, 2026.
https://doi.org/10.1007/978-3-032-08560-3_24

Among others, the utilitarian approach [9] assumes that voters have values for the alternatives. The preferences expressed in the ballot of a voter are a very short summary of these values, e.g., a ranking of the candidates in terms of their values, a set of candidates with values exceeding a threshold, or just the alternative with the highest value for the voter. A voting rule takes as input the voters' ballots and computes an outcome (e.g., a winning alternative). As it has no access to the underlying values of the voters for the alternatives, the outcome depends only on the short summaries in the voters' ballots. Rather unsurprisingly, the outcome of the voting rule will be sub-optimal if evaluated in terms of the voters' values for the alternatives.

How far can the outcome be from the optimal alternative? To answer this question, we need to specify a measure to assess the quality of alternatives. Following a rather standard approach in the literature, we use the *social welfare*—the total value the agents have for an alternative—as our quality measure. Then, the following question arises naturally. How far from the optimum can the outcome of a voting rule be in terms of social welfare? The notion of *distortion*, introduced by Procaccia and Rosenschein [34], comes to answer this question. The distortion of a voting rule is defined as the worst-case ratio, among all voting profiles on a set of alternatives, between the maximum social welfare among all alternatives and the social welfare of the winning alternative returned by the voting rule.

The distortion of voting rules has been the subject of a long list of papers in computational social choice for more than ten years now; see the nice survey of the key results in the area by Anshelevich et al. [5]. For example, under mild assumptions about the valuations, the ubiquitous plurality rule has a distortion of $O(m^2)$, where m is the number of alternatives [13]. Nevertheless, the distortion is inherently high. Even when the values for the alternatives are restricted (e.g., the total value of each agent for all alternatives is normalized to 1), the distortion of any (possibly randomized) voting rule can be as bad as $\Theta(\sqrt{m})$ [9,20].

A recent approach by Amanatidis et al. [1] aims to bypass such lower bounds by making very limited use of the underlying cardinal information. Here, besides the ranking submitted as a ballot by an agent, the voting rule (or, better, the *mechanism*) can pose queries to the agent regarding her value for particular alternatives. Even though optimal distortion results are now possible (by naively querying the values of all alternatives in each agent's ranking), the important problem to be solved is to design mechanisms that achieve low distortion by making a limited number of queries per agent. Among other results, Amanatidis et al. [1] present an algorithm that achieves constant distortion by making at most $O(\log^2 m)$ queries per agent. On the negative side, they show that any mechanism that achieves constant distortion must make at least $\Omega\left(\frac{\log m}{\log\log m}\right)$ per agent. Both results refer to deterministic mechanisms.

By the definition of distortion, the results discussed above are of a worst-case flavour. Voting rules and mechanisms are evaluated on a profile for which they have the worst possible performance, no matter how frequently such a profile may appear in practice. As such, they may not be suitable to explain the success of certain voting rules in practice.

1.1 Overview of Our Contribution

We follow the utilitarian framework but—motivated by the recent trend of analyzing algorithms from non-worst-case perspectives [35]—attempt an evaluation of voting rules on an *average-case* basis. To this end, we consider a simple stochastic setting in which each agent draws a random value for each alternative according to a common probability distribution. The draws of each agent for each alternative are independent. Naturally, *impartial culture electorates* [33, 37]—i.e., profiles with uniformly random rankings of alternatives—emerge in this way. At the conceptual level, we introduce the notion of *average distortion* for such stochastic preferences. The average distortion of a mechanism is defined as the expected maximum social welfare among the alternatives over the expected social welfare of the alternative returned by the mechanism. To distinguish between the two notions, we use the term *worst-case distortion* to refer to the traditional definition. It is not hard to see that the average distortion is always upper-bounded by the worst-case distortion.

We warm up by showing (in Sect. 3) that, perhaps surprisingly, any (potentially randomized) voting rule has average distortion $\Omega(m)$, implying that the trivial voting rule that selects one of the alternatives uniformly at random (ignoring the profile) has an almost best possible average distortion. Our lower bound uses a very simple *binary* probability distribution (i.e., one that returns values of 1 and 0). In light of this seemingly discouraging result, our primary objective is to understand the additional power a limited number of value queries can give to a voting rule by seeking answers to the following questions:

> Can we design mechanisms that use a few (ideally, a constant number of) queries per agent and have low (ideally, constant) average distortion? What conditions do we need to impose on the distribution from which the agents' valuations are drawn in order to achieve this?

Notice that we do not intend to introduce other restrictions on the electorates, such as requiring the number of agents n to be large compared to the number of alternatives m. Instead, we would like our mechanisms to work for *any* choice of n and m.

Our first positive result is for the family of binary distributions. In Sect. 4, we present a deterministic mechanism, called MEAN, which makes a *single query* per agent and achieves *constant* average distortion. This is our most technically involved result and indicates that using queries (even in a minimal way) can lead to a significant improvement in average distortion.

Binary distributions are, of course, an exceedingly crude model for the agents' valuations. However, we demonstrate the usefulness of our result by employing the mechanism MEAN as a building block for a randomized mechanism, which works for a general distribution F. This mechanism, which we call RTMEAN, still requires only one query per agent and achieves an average distortion of $O\left(\log m + \log \frac{\sigma^2}{\mu^2}\right)$, where μ and σ^2 are the mean and variance of the distribution F. For many distributions of interest, the average distortion bound obtained is only logarithmic in m. A similar idea is used to define our randomized mechanism RtSearch for the traditional model of worst-case distortion. RtSearch uses a logarithmic number of queries per agent and achieves logarithmic worst-case distortion. Such an upper bound is not known for deterministic mechanisms. These results are presented in Sect. 5. To the best of our knowledge, this is the first analysis of randomized mechanisms that make value queries in the distortion literature.

Our upper bound on the average distortion of mechanism Mean is not attainable by deterministic mechanisms in the traditional worst-case model. We prove that no deterministic mechanism that makes a single query per agent can achieve distortion better than $\Omega(\sqrt{m})$, even when the valuations are binary. More importantly, for general valuations, we present a new lower bound of $\Omega(\log m)$ on the number of queries per agent that allows for constant worst-case distortion. This improves the previously best-known lower bound of Amanatidis et al. [1] by a sublogarithmic factor. These results appear in Sect. 6.

Due to lack of space, most proofs have been omitted. They can be found in the full version of our paper [11].

1.2 Further Related Work

Using different methodological approaches, utilities in voting have been considered in social choice theory since the work of Bentham [8] in the 18th century; for recent work on the topic see [6,32]. The stochastic model we follow was introduced by Boutilier et al. [9]. Among other investigations, Boutilier et al. aimed at identifying the optimal voting rule in terms of the expected social welfare of the winning alternative. Their main conclusion is that this voting rule is a positional scoring rule with parameters depending on the probability distribution the agents' values are drawn from. They do not consider notions related to average distortion and do not present related bounds. Gonczarowski et al. [24] define a version of distortion (slightly different from ours) in the model of Boutilier et al. as follows. For a particular profile, they examine the ratio between the expected maximum social welfare and the expected social welfare of the winning alternative, both terms conditioned on the random values of the alternatives being consistent with the profile. They define the expected distortion as the worst-case ratio over all profiles. It can be verified that, for every voting rule, the value of expected distortion according to [24] is bounded above and below by the worst-case distortion and our average distortion, respectively. The main result of [24] is that binomial voting (which, in the terminology of [9], is essentially the average-case optimal voting rule for a fair Bernoulli trial) approximates the rule with minimum expected distortion for values drawn from distributions supported on $[0, 1]$. Both Boutilier et al. [9] and Gonczarowski et al. [24] restrict their attention to voting rules and do not consider value queries to the agents.

In the current paper, we assume that agents have non-negative values for the alternatives. In a series of recent papers on *metric voting*, originating from the work of Anshelevich et al. [4], agents and alternatives are assumed to be located in a metric space, and the preferences of each agent reflect her relative distance from the alternatives. In that different setting, the distortion quantifies the suboptimality of the outcome of voting rules in terms of the *social cost*. This line of research has led to challenging algorithmic problems with beautiful solutions [15,16,23,25]. Cheng et al. [18,19] and Ghodsi et al. [22] consider variants of average-case distortion that, due to the metric setting and additional modelling assumptions, are only distantly related to ours.

The study of the utilitarian framework in the recent CS literature goes beyond the study of single-winner voting. Caragiannis et al. [12,14] and Benadè et al. [7] consider multiwinner voting and participatory budgeting settings, respectively. Filos-Ratsikas et al. [21] investigate voting in distributed settings. The utilitarian approach has also been used in settings that are more general than voting. The main idea is to explore how well algorithms that use only ordinal information about the underlying input (as opposed to cardinal values) can approximate the optimal solutions of combinatorial optimization problems. The survey of Anshelevich et al. [5] covers early work on this hot topic, as well as on the other directions discussed above. The benefits of allowing for limited access to the cardinal information were more recently explored for matching [2,3,26,27] and clustering [10] problems.

We remark that stochastic models are otherwise ubiquitous in the EconCS literature, including a long line of research in mechanism design and auctions originating from the seminal paper of Myerson [30], and a rich menu of other settings like matchings [17], fair division [28], kidney-exchange [36], and many more.

2 Preliminaries

Throughout the paper, we denote by N and A the sets of agents and alternatives and reserve n and m for their cardinalities, respectively. Each agent $i \in N$ has a *ranking* $\succ_i$ of the alternatives, i.e., a strict ordering of the elements in A. A *profile* $P = \{\succ_i\}_{i\in N}$ is just a collection of the agents' rankings. We denote by $\mathcal{P}$ the set of all possible profiles with n agents and m alternatives. A voting rule $\mathcal{M} : \mathcal{P} \rightarrow A$ takes as input a profile and returns a single winning alternative.

We assume that the ranking of each agent results from underlying hidden non-negative *values* that the agent has for each alternative. For $i \in N$, the *valuation function* $\text{val}_i : A \rightarrow \mathbb{R}_{\geq 0}$ returns the values of agent i for the alternatives in A. Then, agent i's ranking $\succ_i$ is *consistent* with val_i. This means that $a \succ_i a'$ only if $\text{val}_i(a) \geq \text{val}_i(a')$ for every pair of alternatives $a, a' \in A$. Given a ranking $\succ$, the function $\text{pos}_\succ : A \rightarrow [m]$ returns the position of a given alternative in the ranking. We say that alternative a is the *top-ranked* alternative of agent i if $\text{pos}_{\succ_i}(a) = 1$.

Consider a set of valuation functions $v = \{\text{val}_i\}_{i \in N}$. We typically refer to the set v as the agents' *valuations*. The *social welfare* of an alternative $a \in A$ is defined as

$$\text{SW}(a, v) = \sum_{i \in N} \text{val}_i(a).$$

Let $\mathcal{V}$ be the set of all possible valuations the agents in N can have for the alternatives in A. We denote by $\mathcal{P}(\text{val}_i)$ the set of rankings that are consistent with the valuation function val_i of agent i. We say that a profile $P = \{\succ_i\}_{i \in N}$ is consistent with the valuations v if the ranking $\succ_i$ of every agent $i \in N$ is consistent with her valuation function val_i, i.e., $\succ_i \in \mathcal{P}(\text{val}_i)$. Then, $\mathcal{P}(v)$ represents the set of profiles in $\mathcal{P}$ which are consistent with the valuations v.

Besides voting rules, we consider *mechanisms* that have, in addition to the profile, access to the valuations. Such a mechanism $\mathcal{M} : \mathcal{P} \times \mathcal{V} \rightarrow A$ takes as input the profile and the valuations and returns a winning alternative. We are particularly interested in mechanisms that use the whole profile on input but only a small part of the valuations by making a limited number of *queries* per agent. A query for the value of agent $i \in N$ for alternative $a \in A$ simply returns the value of $\text{val}_i(a)$.

The *worst-case distortion* of a mechanism $\mathcal{M}$ applied on profiles consistent with valuations from $\mathcal{V}$ is defined as

$$\text{dist}(\mathcal{M}) = \sup_{\substack{v \in \mathcal{V} \\ P \in \mathcal{P}(v)}} \frac{\max_{a \in A} \text{SW}(a, v)}{\text{SW}(\mathcal{M}(P, v), v)},$$

i.e., it is the worst-case ratio—among all valuations and consistent profiles—between the maximum social welfare and the social welfare of the alternative returned by the mechanism. For randomized mechanisms, in which the alternative returned is a random variable, we use the expectation $\mathbb{E}[\text{SW}(\mathcal{M}(P, v), v)]$ (taken over the random choices of $\mathcal{M}$) instead of $\text{SW}(\mathcal{M}(P, v), v)$ in the denominator.

We extend the notion of distortion to *stochastic environments* with random valuations and consistent profiles. We assume that the values of the agents for the alternatives are drawn from a *known* common distribution F. In particular, for each agent $i \in N$ and alternative $a \in A$, the value $\text{val}_i(a)$ is drawn independently from distribution F. Given valuations v selected in this way, a consistent profile P is selected uniformly at random among all profiles in $\mathcal{P}(v)$ (essentially, in her ranking, each agent breaks ties among alternatives in terms of value uniformly at random). This gives rise to uniformly random profiles, which are known as *impartial culture electorates* in the social choice theory literature.

For valuations v, we use $v \sim F$ to denote that, for every alternative $a \in A$ and every agent $i \in N$, the value $\text{val}_i(a)$ is drawn independently from F. We use the notation $\succ_i \sim \mathcal{P}(\text{val}_i)$ to refer to a ranking that is selected uniformly at random among all rankings that are consistent with the valuation function val_i of agent i. Similarly, we use $P \sim \mathcal{P}(v)$ for a profile that is selected uniformly at random among all profiles that are consistent with valuations v. Then, the *expected social welfare* of the winning alternative picked by a deterministic mechanism $\mathcal{M}$ is

$$\mathbb{E}_{\substack{v \sim F \\ P \sim \mathcal{P}(v)}} \left[\mathrm{SW}(\mathcal{M}(P, v), v)\right].$$

When mechanism $\mathcal{M}$ is randomized, the expectation is taken over the randomness of $\mathcal{M}$ as well.

We now adapt the notion of distortion to this stochastic setting. The *average distortion* of a mechanism $\mathcal{M}$ on a family of distributions $\mathcal{F}$ is

$$\mathrm{avdist}(\mathcal{M}, \mathcal{F}) = \sup_{F \in \mathcal{F}} \frac{\mathbb{E}_{v \sim F}\left[\max_{a \in A} \mathrm{SW}(a, v)\right]}{\mathbb{E}_{\substack{v \sim F \\ P \sim \mathcal{P}(v)}}\left[\mathrm{SW}(\mathcal{M}(P, v), v)\right]}.$$

We remark that the trivial voting rules that return a fixed alternative or an alternative selected uniformly at random have average distortion at most m for every distribution F. Denoting them by $\mathcal{M}$, this is due to the simple fact that

$$\begin{aligned}\mathbb{E}_{\substack{v \sim F \\ P \sim \mathcal{P}(v)}}\left[\mathrm{SW}(\mathcal{M}(P), v)\right] &= \frac{1}{m} \cdot \mathbb{E}_{v \sim F}\left[\sum_{a \in A} \mathrm{SW}(a, v)\right] \\ &\geq \frac{1}{m} \cdot \mathbb{E}_{v \sim F}\left[\max_{a \in A} \mathrm{SW}(a, v)\right].\end{aligned}$$

For particular distributions, the average distortion can be much lower. As an example, consider the uniform distribution over the interval $[a, b]$ with $a, b \geq 0$. Then, the expected social welfare of the alternative returned by both trivial rules is $n \cdot (a+b)/2$ while the expected maximum social welfare (over all alternatives) cannot exceed $n \cdot b$. Hence, the average distortion is at most 2 in this case.

A fundamental family that is of central importance to our work is the family $\mathcal{F}^{0/1}$ of *binary distributions*, consisting of the set of probability distributions $\{F_p\}_{p \in [0,1]}$, where F_p is such that the random variable z that is drawn according to F is equal to 1 with probability p and to 0 with probability $1-p$.

3 Average Distortion Can Be High

We begin our technical exposition with a lower bound of $\Omega(m)$ on the average distortion. Essentially, Theorem 1 implies that the trivial voting rules that return a fixed alternative or an alternative selected uniformly at random on every profile have approximately optimal average distortion.

In our proof, we make use of a particular distribution F from the family $\mathcal{F}^{0/1}$. Furthermore, we exploit a class of voting rules called *positional scoring rules.* A positional scoring rule g is defined by a scoring vector $\langle \alpha_1, \alpha_2, ..., \alpha_m \rangle$ with $\alpha_1 \geq \alpha_2 \geq ... \geq a_m \geq 0$. On input a profile P, the voting rule g assigns a score $\mathrm{sc}_g(a, P)$ to every alternative $a \in A$ and the winning alternative is the one with the highest score (breaking ties according to some tie-breaking rule). The scores of the alternatives is computed as follows. For $j = 1, ..., m$, alternative a takes α_j points each time it appears in the j-th position in an agent's ranking, i.e., $\text{sc}_g(a, P) = \sum_{i \in N} \alpha_{\text{pos}_{\succ_i}(a)}$. For example, *plurality* is the positional scoring rule that uses the m-entry scoring vector $\langle 1, 0, ..., 0 \rangle$. The plurality winner on profile P, denoted by $\text{plu}(P)$, is the alternative that appears most often in the top position of the agents' rankings.

Theorem 1. *For every (possibly randomized) mechanism* $\mathcal{M}$, $\text{avdist}(\mathcal{M}, \mathcal{F}^{0/1}) \in \Omega(m)$.

Proof. Consider impartial culture electorates with m alternatives, $n \geq 6m \ln m$ agents, and binary values drawn from the probability distribution $F \in \mathcal{F}^{0/1}$ with $p = \frac{1}{nm}$. Then, the probability that some alternative has positive social welfare is $1 - \left(1 - \frac{1}{nm}\right)^{nm} \geq 1 - e^{-1}$. Thus, $\mathbb{E}_{v \sim F}[\max_{a \in A} \text{SW}(a, v)] \geq 1 - e^{-1}$. Now consider any voting rule $\mathcal{M}$; we will complete the proof by showing that

$$\mathop{\mathbb{E}}_{\substack{v \sim F \\ P \sim \mathcal{P}(v)}} [\text{SW}(\mathcal{M}(P), v)] \leq \frac{4}{m}. \tag{1}$$

Below, we prove inequality (1), assuming that $\mathcal{M}$ is deterministic. For randomized $\mathcal{M}$, the expectation in the LHS of (1) should also be taken over the randomness of $\mathcal{M}$ as well. In that case, inequality (1) follows trivially by our arguments, interpreting $\mathcal{M}$ as a probability distribution over deterministic voting rules.

Let g be the positional scoring rule that uses the scoring vector $\langle \alpha_1, \alpha_2, ..., \alpha_m \rangle$ with

$$\alpha_j = \mathop{\mathbb{E}}_{\substack{\text{val}_i \sim F \\ \succ_i \sim \mathcal{P}(\text{val}_i)}} [\text{val}_i(a) | \text{pos}_{\succ_i}(a) = j],$$

i.e., score α_j is equal to the expected value according to F given by each agent to the alternative ranked at position j.[1] Now, observe that for a given profile $P = \{\succ_i\}_{i \in N}$ and alternative $a \in A$, we have

$$\begin{aligned}\mathbb{E}_{v \sim F}[\text{SW}(a, v) | P \in \mathcal{P}(v)] = \sum_{i \in N} \mathbb{E}_{\text{val}_i \sim F}[\text{val}_i(a) | \succ_i \in \mathcal{P}(\text{val}_i)] &= \sum_{i \in N} \alpha_{\text{pos}_{\succ_i}(a)} \\ &= \text{sc}_g(a, P),\end{aligned}$$

where $\text{sc}_g(a, P)$ denotes the score of alternative a in profile P according to positional scoring rule g. Thus,

$$\begin{aligned}\mathop{\mathbb{E}}_{\substack{v \sim F \\ P \sim \mathcal{P}(v)}} [\text{SW}(\mathcal{M}(P), v)] &= \sum_{P \in \mathcal{P}} \mathbb{E}_{v \sim F}[\text{SW}(\mathcal{M}(P), v) | P \in \mathcal{P}(v)] \cdot \Pr_{v \sim F}[P \in \mathcal{P}(v)] \\ &= \sum_{P \in \mathcal{P}} \text{sc}_g(\mathcal{M}(P), P) \cdot \Pr_{v \sim F}[P \in \mathcal{P}(v)] \\ &= \mathop{\mathbb{E}}_{\substack{v \sim F \\ P \sim \mathcal{P}(v)}} [\text{sc}_g(\mathcal{M}(P), P)].\end{aligned} \tag{2}$$

[1] Boutilier et al. [9] refer to this positional scoring rule g as an average-case optimal social choice function.

Notice that, by the definitions of voting rules plu and g, for any profile P and alternative $a \in A$, it holds

$$\begin{aligned} \mathrm{sc}_g(a,P) &\leq \alpha_1 \cdot \mathrm{sc}_{\mathrm{plu}}(a,P) + n \cdot \alpha_2 \\ &\leq \alpha_1 \cdot \mathrm{sc}_{\mathrm{plu}}(\mathrm{plu}(P),P) + n \cdot (mp - \alpha_1). \end{aligned} \tag{3}$$

The first inequality follows since g gives α_1 points to alternative a for each of its $\mathrm{sc}_{\mathrm{plu}}(a,P)$ appearances in the top position in the agents' rankings and at most α_2 points for its remaining appearances. The second inequality follows since the plurality winner $\mathrm{plu}(P)$ has the highest plurality score (and, hence, $\mathrm{sc}_{\mathrm{plu}}(a,P) \leq \mathrm{sc}_{\mathrm{plu}}(\mathrm{plu}(P),P)$ for every alternative $a \in A$) and since $\alpha_2 \leq \sum_{j=1}^{m} \alpha_j - \alpha_1$ and $\sum_{j=1}^{m} \alpha_j = mp$ (as the expected total value given to all alternatives by an agent).

We will also need the following lemma.

Lemma 1. $\mathbb{E}_{\substack{v \sim F \\ P \sim \mathcal{P}(v)}}[\mathrm{sc}_{\mathrm{plu}}(\mathrm{plu}(P),P)] \leq \frac{3n}{m}$.

Proof. The proof of the lemma will follow by a simple application of the Chernoff bound, e.g., see [29].

Lemma 2 (Chernoff bound, upper tail). *For every binomial random variable Q and any $\delta > 0$, we have*

$$\Pr[Q \geq (1+\delta)\,\mathbb{E}[Q]] \leq \exp\left(-\frac{\delta^2\,\mathbb{E}[Q]}{2+\delta}\right).$$

Consider an alternative $a \in A$ and denote by X_i the random variable indicating whether agent i ranks a first ($X_i = 1$; this happens with probability $1/m$) or not ($X_i = 0$). Observe that $\mathrm{sc}_{\mathrm{plu}}(a,P) = \sum_{i \in N} X_i$, i.e., $\mathrm{sc}_{\mathrm{plu}}(a,P)$ is a binomial random variable with expectation $\frac{n}{m}$. By applying Lemma 2 with $\delta = 1$, we obtain that $\Pr[\mathrm{sc}_{\mathrm{plu}}(a,P) \geq \frac{2n}{m}] \leq \exp\left(-\frac{n}{3m}\right) \leq \frac{1}{m^2}$. The last inequality follows since $n \geq 6m \ln m$. Taking the union bound over the m alternatives then gives $\Pr[\mathrm{sc}_{\mathrm{plu}}(\mathrm{plu}(P),P) \geq \frac{2n}{m}] \leq \frac{1}{m}$. Finally, since the plurality score never exceeds n, we have $\mathbb{E}_{\substack{v \sim F \\ P \sim \mathcal{P}(v)}}[\mathrm{sc}_{\mathrm{plu}}(\mathrm{plu}(P),P)] \leq \frac{2n}{m} \cdot \left(1 - \frac{1}{m}\right) + n \cdot \frac{1}{m} \leq \frac{3n}{m}$. □

Notice that the top alternative of an agent has value 0 with probability $(1-p)^m$ and value 1 otherwise. Thus, $\alpha_1 = 1 - (1-p)^m$. Now, observe that $1-(1-p)^m = 1 - \left(1 - \frac{1}{nm}\right)^m \geq 1 - \frac{1}{e^{1/n}} \geq \frac{1}{n+1}$, using the properties $(1-r/t)^t \leq e^{-r}$ for $t > 0$ and $r \geq 0$ and $e^r \geq 1 + r$. Thus, $mp - \alpha_1 \leq \frac{1}{n} - \frac{1}{n+1} < \frac{1}{n^2}$. Also, using the property $(1-r)^t \geq 1 - rt$ for $t \geq 1$, we get $\alpha_1 = 1-(1-p)^m \leq pm = \frac{1}{n}$. Combining these two last inequalities with Eqs. (2) and (3) and Lemma 1, we obtain

$$\begin{aligned} \mathbb{E}_{\substack{v \sim F \\ P \sim \mathcal{P}(v)}}[\mathrm{SW}(\mathcal{M}(P),v)] &= \mathbb{E}_{\substack{v \sim F \\ P \sim \mathcal{P}(v)}}[\mathrm{sc}_g(\mathcal{M}(P),P)] \\ &\leq \alpha_1 \cdot \mathbb{E}_{\substack{v \sim F \\ P \sim \mathcal{P}(v)}}[\mathrm{sc}_{\mathrm{plu}}(\mathrm{plu}(P),P)] + n(mp - \alpha_1) \\ &\leq \frac{3}{m} + \frac{1}{n} \leq \frac{4}{m}, \end{aligned}$$

as desired by (1). The last inequality follows from the relation between n and m. □

4 Constant Average Distortion with a Single Query per Agent

In the previous section, we saw that the average distortion can be high even for binary distributions if the mechanism has no access to the agents' underlying valuations. Still, binary valuations may reveal a lot of information with a single query. Consider a query for the value that an agent has for the alternative at position j of her ranking. If the query returns a value of 1, this means that the agent has a value of 1 for all alternatives in positions higher than j as well. This observation motivates the following definition of *implied social welfare.*

Definition 1 (implied social welfare). *For a distribution $F \in \mathcal{F}^{0/1}$, assume that a mechanism queried each agent for the value of the alternative in position k of her ranking. For each agent $i \in N$, let $\text{val}_{i,k}$ denote the respective value. The* implied social welfare *then is*

$$\text{SW}_{\text{im}}(a, P, v, k) = \sum_{i \in N} \mathbb{I}\{\text{val}_{i,k} = 1 \wedge \text{pos}_{\succ_i}(a) \leq k\}.$$

We use the notion of implied social welfare in the definition of the following mechanism.

Definition 2 (mechanism Mean). *Given a profile P with underlying valuations drawn from a distribution $F_p \in \mathcal{F}^{0/1}$, the mechanism* Mean *queries each agent for the value of the alternative at position $\tau = \max\{1, \lfloor pm \rfloor\}$. The mechanism then returns the alternative that maximizes the implied social welfare, that is,*

$$\textsc{Mean}(P, v) \in \arg\max_{a \in A} \text{SW}_{\text{im}}(a, P, v, \tau),$$

breaking ties arbitrarily.

We show that mechanism Mean has constant average distortion with a single query for the family of binary distributions. Notably, it is impossible for deterministic 1-query mechanisms to achieve similar guarantees in the traditional setting of worst-case distortion. Indeed, even for binary valuations, the worst case distortion of these mechanisms is $\Omega(\sqrt{m})$; see Theorem 6.

Theorem 2. *Mechanism* Mean *has average distortion at most 27 in impartial culture electorates with n agents and m alternatives, and underlying values drawn from any probability distribution in $\mathcal{F}^{0/1}$.*

For our proof of the theorem, we partition the family $\mathcal{F}^{0/1}$ into the subfamilies of probability distributions $\mathcal{F}_1$, $\mathcal{F}_2$, and $\mathcal{F}_3$ defined as follows:

$$\mathcal{F}_1 = \{F_p \in \mathcal{F}^{0/1} : p \geq 1/m\}$$
$$\mathcal{F}_2 = \{F_p \in \mathcal{F}^{0/1} : 1 - (1 - 1/n)^{1/m} \leq p < 1/m\}$$
$$\mathcal{F}_3 = \{F_p \in \mathcal{F}^{0/1} : p < 1 - (1 - 1/n)^{1/m}\}$$

Let $F_p \in \mathcal{F}^{0/1}$; we now distinguish between the three cases $F_p \in \mathcal{F}_1$, $F_p \in \mathcal{F}_2$, and $F_p \in \mathcal{F}_3$. For the latter two cases, we merely state our bounds on the average distortion of the mechanism as Lemma 3.

Lemma 3. *It holds that* $\text{avdist}(\text{MEAN}, \mathcal{F}_2) \leq 2.6$, *and* $\text{avdist}(\text{MEAN}, \mathcal{F}_3) < 2$.

Showing constant average distortion of the mechanism in the first case ($F_p \in \mathcal{F}_1$) is the most challenging and technically involved part in our proof of Theorem 2. Indeed, the proof of the second case ($F_p \in \mathcal{F}_2$) relies on the same machinery that we introduce for the proof of our next lemma and refinements thereof. Together with Lemma 3, the following statement yields the theorem.

Lemma 4. *It holds that* $\text{avdist}(\text{MEAN}, \mathcal{F}_1) \leq 27$.

Proof. Let τ denote the position that mechanism MEAN queried in each agent's ranking, i.e., $\tau = \max\{1, \lfloor mp \rfloor\}$. For valuations v, let $X_i(v)$ be the number of alternatives for which agent i draws a value of 1 and let $X(v) = \sum_{i \in N} X_i(v)$. We will consider valuations v drawn from the probability distribution F_p. Thus, $X_i(v)$ is a random variable following the binomial distribution with m trials and success probability p. Since $X(v)$ is the sum over i.i.d. binomially distributed random variables, $X(v)$ itself follows a binomial distribution with nm trials and success probability p.

Furthermore, we denote by $Z_i(v, \tau)$ the random variable indicating whether the query at position τ of agent i's ranking returned a value of 1 (then, $Z_i(v, \tau) = 1$) or not (then, $Z_i(v, \tau) = 0$). Clearly, $Z_i(v, \tau) = \mathbb{I}\{X_i(v) \geq \tau\}$ such that the random variable $Z(v, \tau) = \sum_{i \in N} Z_i(v, \tau)$ follows a binomial distribution with n trials and success probability $\Pr[X_i(v) \geq \tau]$. We note that the variables X_i are identically distributed for every $i \in N$.

We will need a technical lemma.

Lemma 5. *For integer* $s \in [nm]$, *define the condition* $\text{Balanced}(v, s)$ *to be* $X(v) = s$ *with* $\lfloor s/n \rfloor \leq X_i(v) \leq \lceil s/n \rceil$ *for* $i \in N$. *Then, for any distribution* $F \in \mathcal{F}^{0/1}$, *it holds that*

$$\mathbb{E}_{v \sim F}\left[\max_{a \in A} \text{SW}(a, v) | X(v) = s\right] \leq \mathbb{E}_{v \sim F}\left[\max_{a \in A} \text{SW}(a, v) | \text{Balanced}(v, s)\right].$$

Consider impartial culture electorates with n agents and m alternatives with underlying values drawn from the distribution $F_p \in \mathcal{F}_1$. Denote by $\text{Box}(v)$ the condition $X_i(v) = \tau$ for $i \in S$ and $X_i(v) = 0$ for $i \in N \setminus S$, where S is a subset of the agents of size exactly $\lfloor n/2 \rfloor$. Now, define the quantity

$$B = \mathbb{E}_{v \sim F}\left[\max_{a \in A} \text{SW}(a, v) | \text{Box}(v)\right], \tag{4}$$

i.e., the expected maximum social welfare among all alternatives given that $\lfloor n/2 \rfloor$ agents draw a value of 1 for exactly τ alternatives and the remaining agents draw values of 0 for all alternatives. The quantity B will be a benchmark that will help us compare the expected social welfare of $\textsc{Mean}(P, v)$ to the maximum social welfare.

As stated above, each $X_i(v)$ is a binomial random variable with m trials and success probability p. Hence, its median is either at the value $\lfloor pm \rfloor$ or $\lceil pm \rceil$, which are both at least τ since $p \geq 1/m$. Thus, $\Pr_{v \sim F}[X_i(v) \geq \tau] \geq 1/2$. The random variable $Z(v, \tau)$ then follows a binomial distribution with n trials and success probability at least $1/2$. By the same property of the median of the binomial distribution, it now holds that $\Pr_{v \in F}[Z(v, \tau) \geq \lfloor n/2 \rfloor] \geq 1/2$. With this observation and by applying the law of total expectation, we have

$$\begin{aligned}
&\mathbb{E}_{\substack{v \sim F_p \\ P \sim \mathcal{P}(v)}} [\mathrm{SW}(\textsc{Mean}(P, v), v)] \\
&= \mathbb{E}_{\substack{v \sim F_p \\ P \sim \mathcal{P}(v)}} [\mathrm{SW}(\textsc{Mean}(P, v), v) | Z(v, \tau) \geq \lfloor n/2 \rfloor] \cdot \Pr_{v \in F_p} [Z(v, \tau) \geq \lfloor n/2 \rfloor] \\
&\quad + \mathbb{E}_{\substack{v \sim F_p \\ P \sim \mathcal{P}(v)}} [\mathrm{SW}(\textsc{Mean}(P, v), v) | Z(v, \tau) < \lfloor n/2 \rfloor] \cdot \Pr_{v \in F_p} [Z(v, \tau) < \lfloor n/2 \rfloor] \\
&\geq \frac{1}{2} \cdot \mathbb{E}_{\substack{v \sim F_p \\ P \sim \mathcal{P}(v)}} [\mathrm{SW}(\textsc{Mean}(P, v), v) | Z(v, \tau) \geq \lfloor n/2 \rfloor] \\
&\geq \frac{1}{2} \cdot \mathbb{E}_{\substack{v \sim F_p \\ P \sim \mathcal{P}(v)}} \left[\max_{a \in A} \mathrm{SW}_{\mathrm{im}}(a, P, v, \tau) | Z(v, \tau) \geq \lfloor n/2 \rfloor \right] \\
&\geq \frac{1}{2} \cdot \mathbb{E}_{v \sim F_p} \left[\max_{a \in A} \mathrm{SW}(a, v) | \operatorname{Box}(v) \right] = \frac{1}{2} \cdot B. \qquad (5)
\end{aligned}$$

The first inequality above is obvious. By definition of $\textsc{Mean}$, $\textsc{Mean}(P, v)$ is the alternative that maximizes the implied social welfare when querying each agent in position τ. Hence, the implied social welfare is a lower bound for the social welfare of $\textsc{Mean}(P, v)$, which yields the second inequality. Finally, under the condition that $Z(v, \tau) \geq \lfloor n/2 \rfloor$, there are at least $\lfloor n/2 \rfloor$ agents i for each of which $X_i(v) \geq \tau$. This includes the condition $\operatorname{Box}(v)$ and the third inequality follows. The last equality follows from the definition of B in (4).

We proceed to bound the expected maximum social welfare from above in terms of B. For this purpose, we require another technical lemma.

Lemma 6. *For every distribution $F \in \mathcal{F}^{0/1}$ and any positive integer j, it holds that*

$$\mathbb{E}_{v \sim F} \left[\max_{a \in A} \mathrm{SW}(a, v) | \operatorname{Balanced}(v, jn\tau) \right] \leq 3j \cdot B.$$

For a positive integer j, define the condition $\text{Range}(v, j)$ to be $(j-1)n\tau < X(v) \le jn\tau$. By applying the law of total expectation, we obtain

$$\begin{aligned}
&\mathbb{E}_{v\sim F_p}[\max_{a\in A} \text{SW}(a, v)] \\
&= \sum_{j=1}^{\infty} \mathbb{E}_{v\sim F_p}[\max_{a\in A} \text{SW}(a, v)|\, \text{Range}(v, j)] \cdot \Pr_{v\sim F_p}[\text{Range}(v, j)] \\
&\le \sum_{j=1}^{\infty} \mathbb{E}_{v\sim F_p}[\max_{a\in A} \text{SW}(a, v)|X(v) = jn\tau] \cdot \Pr_{v\sim F_p}[\text{Range}(v, j)] \\
&\le \sum_{j=1}^{\infty} \mathbb{E}_{v\sim F_p}[\max_{a\in A} \text{SW}(a, v)|\, \text{Balanced}(v, jn\tau)] \cdot \Pr_{v\sim F_p}[\text{Range}(v, j)] \\
&\le 3B \cdot \sum_{j=1}^{\infty} j \cdot \Pr_{v\sim F_p}[\text{Range}(v, j)]. \qquad (6)
\end{aligned}$$

The first inequality follows since the condition $\text{Range}(v, j)$ includes the condition $X(v) = jn\tau$ and since the quantity $\mathbb{E}[\max_{a\in A} \text{SW}(a, v)|X(v) = t]$ is non-decreasing in terms of t. The second inequality follows from Lemma 5 while the third one follows from Lemma 6. We now bound the term $\sum_{j=1}^{\infty} j \cdot \Pr_{v\sim F_p}[\text{Range}(v, j)]$.

$$\begin{aligned}
&\sum_{j=1}^{\infty} j \cdot \Pr_{v\sim F_p}[\text{Range}(v, j)] \\
&= \sum_{j=1}^{\infty} j \sum_{k=(j-1)n\tau}^{jn\tau} \Pr_{v\sim F_p}[X(v) = k] \\
&= \frac{1}{n\tau} \cdot \sum_{j=1}^{\infty} \sum_{k=(j-1)n\tau}^{jn\tau} jn\tau \cdot \Pr_{v\sim F_p}[X(v) = k] \\
&\le \Pr_{v\sim F_p}[X(v) \le n\tau] + \frac{2}{n\tau} \cdot \sum_{j=2}^{\infty} \sum_{k=(j-1)n\tau}^{jn\tau} k \cdot \Pr_{v\sim F_p}[X(v) = k] \\
&\le \frac{1}{2} + \frac{2}{n\tau} \cdot \mathbb{E}_{v\sim F_p}[X(v)] \le 4.5.
\end{aligned}$$

The first inequality follows from the fact that $jn\tau/2 \le (j-1)n\tau \le k$ for $j \ge 2$. The second inequality follows from the definition of the expectation of random variable $X(v)$. Since $X(v)$ follows the binomial distribution with nm trials and success probability p, we have $\mathbb{E}[X(v)] = nmp$ which is at most $2n\lfloor mp \rfloor = 2n\tau$ since $p \ge 1/m$.

Now, inequality (6) implies that $E_{v\sim F}[\max_{a\in A} \text{SW}(a, v)] \le 13.5B$. Combining this bound with inequality (5) lets us conclude that $\text{avdist}(\textsc{Mean}, \mathcal{F}_1) \le 27$, as desired. □

5 Randomized Mechanisms

We now present two randomized mechanisms for impartial culture electorates with underlying valuations drawn from a *general* probability distribution and worst-case electorates, respectively. Both mechanisms are nevertheless similar in spirit. They randomly pick a single threshold from a suitably defined set of thresholds and query each agent to determine a set of alternatives that have value above the threshold. This information is then used to compute an approximation of the alternatives' respective social welfare, which is used to decide the winning alternative.

Impartial Culture Electorates. Our first "random threshold" mechanism RtMean uses mechanism Mean as a building block.

Definition 3 (mechanism RtMean). *The mechanism RtMean uses k thresholds $\ell_1, \ell_2, ..., \ell_k$ with $0 < \ell_1 < ... < \ell_k$ as parameters. Given a profile P with underlying valuations drawn from a probability distribution F, RtMean selects an integer t uniformly at random from $[k]$, and sets $p = \Pr_{z \sim F}[z \geq \ell_t]$. It then simulates an execution of Mean on the distribution $F_p \in \mathcal{F}^{0/1}$ by making the same value queries as Mean for F_p, but interpreting the answer $\mathrm{val}_i(a)$ to a query as 1 if $\mathrm{val}_i(a) \geq \ell_t$ and 0 otherwise. RtMean returns as output the alternative that Mean selects.*

Notice that mechanism RtMean uses exactly one query per agent.

Theorem 3. *Let F be a probability distribution over non-negative, real-valued outcomes with mean μ and variance σ^2. There is a set of thresholds $\ell_1, \ell_2, ..., \ell_k$ such that the average distortion of mechanism RtMean is at most $O(\log m + \log \frac{\sigma^2}{\mu^2})$ when applied to impartial culture electorates with n agents and m alternatives, and underlying values drawn according to F.*

As an immediate consequence of Theorem 3, we obtain $O(\log m)$ bounds on the average distortion of RtMean for fundamental families of distributions. The relevant properties of these distributions can be found in, e.g., Papoulis and Unnikrishna Pillai [31, table 5-2, p. 162].

Corollary 1. *The average distortion of mechanism RtMean is at most $O(\log m)$ when applied to impartial culture electorates with n agents and m alternatives, and underlying values drawn according to*

- *the exponential distribution $E(\lambda)$ where $\mu = \frac{1}{\lambda}, \sigma^2 = \frac{1}{\lambda^2}$,*
- *the chi-squared distribution $\chi^2(k)$ where $\mu = k, \sigma^2 = 2k$, or*
- *the Erlang-k distribution $E_k(\lambda)$ where $\mu = \frac{1}{\lambda}, \sigma^2 = \frac{1}{k\lambda^2}$.*

In the latter two cases, k denotes a positive integer.

We remark that distributions for which our analysis does not give logarithmic average distortion are the γ-distribution $G(\alpha, \beta)$ where $\mu = \alpha\beta$, $\sigma^2 = \alpha\beta^2$ and the β-distribution $\beta(\alpha, \beta)$ where $\mu = \frac{\alpha}{\alpha+\beta}$, $\sigma^2 = \frac{\alpha\beta}{(\alpha+\beta)^2(\alpha+\beta+1)}$ for respective α-parameters very close to zero.

Worst-Case Electorates. Our next random threshold mechanism RtSearch achieves low worst-case distortion.[2]

Definition 4 (mechanism RtSearch). *For a given profile P with underlying valuations v over m alternatives, mechanism RtSearch picks an integer r uniformly at random from the set* $\{1, 2, \ldots, \lceil \log 2m \rceil\}$. *For every agent* $i \in N$, *mechanism RtSearch queries the value* ν_i *of the agent's top-ranked alternative, and finds the set of alternatives* $S_{i,r}$, *for each of which the agent has value more than* $\nu_i/2^r$ *using binary search. The mechanism then returns an alternative* $\text{RtSearch}(P, v) \in \arg\max_{a \in A} \sum_{i \in N} \nu_i \mathbb{I}\{a \in S_{i,r}\}$.

Theorem 4. *Mechanism RtSearch achieves worst-case distortion at most* $O(\log m)$ *with* $O(\log m)$ *queries per agent.*

6 Worst-Case Distortion Lower Bounds

We conclude our technical exposition by presenting two lower bounds on the worst-case distortion.

Lower Bounding the Number of Queries for Constant Distortion. Our first lower bound on the number of queries per agent that are necessary to get constant worst-case distortion improves the previously best bound of Amanatidis et al. [1] by a sublogarithmic factor. More specifically, Amanatidis et al. present a lower bound construction and show that any mechanism that uses up to λ queries per agent must have worst-case distortion at least $\Omega\left(\frac{1}{\lambda} \cdot m^{\frac{1}{2(\lambda+1)}}\right)$. Thus, in order to get constant worst-case distortion, at least $\Omega\left(\frac{\log m}{\log \log m}\right)$ queries per agent are necessary. In their follow-up work, Amanatidis et al. [2] present an improved construction which yields a higher distortion of $\Omega(m^{1/\lambda})$. Unfortunately, λ is now required to be a constant. Therefore, their new construction does not provide any lower bound on the number of queries per agent necessary to get constant worst-case distortion. We prove the next theorem using a considerably different construction which shows that mechanisms making at most λ queries per agent have worst-case distortion at least $\frac{1}{8} \cdot m^{\frac{1}{3\lambda}}$ for values of λ that are allowed to be logarithmic in m.

Theorem 5. *Any deterministic mechanism that achieves a constant worst-case distortion must make* $\Omega(\log m)$ *queries per agent.*

Lower Bounding the Distortion of 1-Query Mechanisms. In Sect. 4, we saw that a single query per agent is sufficient to guarantee constant average distortion when the agents draw their valuations according to a binary distribution. Such a guarantee is not attainable in the traditional setting of worst-case distortion.

[2] We remark that RtSearch is similar in spirit to a mechanism proposed by Benadè et al. [7] in a slightly different participatory budgeting context.

Indeed, Amanatidis et al. [1] proved that any deterministic 1-query mechanism must have distortion $\Omega(m)$. However, their lower bound construction uses valuations that are more complex than binary. Our next result states that, even in the case of binary valuations, the worst-case distortion of deterministic 1-query mechanisms must still be high.

Theorem 6. *Every deterministic* 1*-query mechanism has a worst-case distortion of at least* $\Omega(\sqrt{m})$. *This is true even if the agents have binary values for each of the alternatives.*

7 Discussion and Open Problems

We have initiated the study of average distortion in a simple stochastic setting that creates impartial culture electorates. The main open problem is whether constant average distortion is possible with a small number of queries per agent for general probability distributions of valuations. Throughout the paper, we assume that the distribution is given as part of the input, and this information is crucial to make our mechanisms work. It would be interesting to explore whether this—admittedly strong—requirement can be removed. Other natural extensions of our model include different distributions per alternative or distributions that produce random valuations satisfying the unit-sum or unit-range assumption. These latter assumptions are clearly beyond the reach of our current analysis techniques, as they necessarily imply correlations between the random values an agent has for the alternatives.

For the worst-case setting, we have improved the previously best-known lower bound on the number of queries per agent that are necessary for constant worst-case distortion by deterministic mechanisms. Still, the conjecture of Amanatidis et al. [1] that constant worst-case distortion is possible with $\Theta(\log m)$ deterministic queries per agent is wide open. Furthermore, we have also demonstrated that the use of randomization can yield worst-case bounds that the known deterministic mechanisms cannot achieve. Exploring whether there is a separation between deterministic and randomized mechanisms in terms of their worst-case distortion for a given number of queries per agent is another challenging problem that deserves investigation.

References

1. Amanatidis, G., Birmpas, G., Filos-Ratsikas, A., Voudouris, A.A.: Peeking behind the ordinal curtain: improving distortion via cardinal queries. Artif. Intell. **296**, 103488 (2021)
2. Amanatidis, G., Birmpas, G., Filos-Ratsikas, A., Voudouris, A.A.: Don't roll the dice, ask twice: the two-query distortion of matching problems and beyond. In: Proceedings of the 36th Annual Conference on Neural Information Processing Systems (NeurIPS) (2022)

3. Amanatidis, G., Birmpas, G., Filos-Ratsikas, A., Voudouris, A.A.: A few queries go a long way: information-distortion tradeoffs in matching. J. Artif. Intell. Res. **74**, 227–261 (2022)
4. Anshelevich, E., Bhardwaj, O., Postl, J.: Approximating optimal social choice under metric preferences. In: Proceedings of the 39th AAAI Conference on Artificial Intelligence (AAAI), pp. 777–783 (2015)
5. Anshelevich, E., Filos-Ratsikas, A., Shah, N., Voudouris, A.A.: Distortion in social choice problems: the first 15 years and beyond. In: Proceedings of the 30th International Joint Conference on Artificial Intelligence (IJCAI), pp. 4294–4301 (2021)
6. Apesteguia, J., Ballester, M.A., Ferrer, R.: On the justice of decision rules. Rev. Econ. Stud. **78**(1), 1–16 (2011)
7. Benadè, G., Nath, S., Procaccia, A.D., Shah, N.: Preference elicitation for participatory budgeting. Manage. Sci. **67**(5), 2813–2827 (2021)
8. Bentham, J.: An Introduction to the Principles of Morals and Legislation. Clarendon Press (1907)
9. Boutilier, C., Caragiannis, I., Haber, S., Lu, T., Procaccia, A.D., Sheffet, O.: Optimal social choice functions: a utilitarian view. Artif. Intell. **227**, 190–213 (2015)
10. Burkhardt, J., Caragiannis, I., Fehrs, K., Russo, M., Schwiegelshohn, C., Shyam, S.: Low-distortion clustering with ordinal and limited cardinal information. In: Proceedings of the 38th AAAI Conference on Artificial Intelligence (AAAI) (2024)
11. Caragiannis, I., Fehrs, K.: Beyond the worst case: distortion in impartial culture electorates. CoRR abs/2307.07350 (2023)
12. Caragiannis, I., Nath, S., Procaccia, A.D., Shah, N.: Subset selection via implicit utilitarian voting. J. Artif. Intell. Res. **58**, 123–152 (2017)
13. Caragiannis, I., Procaccia, A.D.: Voting almost maximizes social welfare despite limited communication. Artif. Intell. **175**(9–10), 1655–1671 (2011)
14. Caragiannis, I., Shah, N., Voudouris, A.A.: The metric distortion of multiwinner voting. Artif. Intell. **313**, 103802 (2022)
15. Charikar, M., Ramakrishnan, P.: Metric distortion bounds for randomized social choice. In: Proceedings of the 33rd ACM-SIAM Symposium on Discrete Algorithms (SODA), pp. 2986–3004 (2022)
16. Charikar, M., Ramakrishnan, P., Wang, K., Wu, H.: Breaking the metric voting distortion barrier. J. ACM (forthcoming)
17. Chen, N., Immorlica, N., Karlin, A.R., Mahdian, M., Rudra, A.: Approximating matches made in heaven. In: Proceedings of the 36th International Colloquium on Automata, Languages and Programming (ICALP), Part I. Lecture Notes in Computer Science, vol. 5555, pp. 266–278. Springer (2009)
18. Cheng, Y., Dughmi, S., Kempe, D.: Of the people: voting is more effective with representative candidates. In: Proceedings of the 2017 ACM Conference on Economics and Computation (EC), pp. 305–322 (2017)
19. Cheng, Y., Dughmi, S., Kempe, D.: On the distortion of voting with multiple representative candidates. In: Proceedings of the AAAI Conference on Artificial Intelligence (AAAI) (2018)
20. Ebadian, S., Kahng, A., Peters, D., Shah, N.: Optimized distortion and proportional fairness in voting. In: Proceedings of the 23rd ACM Conference on Economics and Computation (EC), pp. 563–600 (2022)
21. Filos-Ratsikas, A., Micha, E., Voudouris, A.A.: The distortion of distributed voting. Artif. Intell. **286**, 103343 (2020)
22. Ghodsi, M., Latifian, M., Seddighin, M.: On the distortion value of the elections with abstention. In: Proceedings of the 33rd AAAI Conference on Artificial Intelligence (AAAI), pp. 1981–1988 (2019)

23. Gkatzelis, V., Halpern, D., Shah, N.: Resolving the optimal metric distortion conjecture. In: Proceedings of the 61st IEEE Annual Symposium on Foundations of Computer Science (FOCS), pp. 1427–1438 (2020)
24. Gonczarowski, Y.A., Kehne, G., Procaccia, A.D., Schiffer, B., Zhang, S.: The distortion of binomial voting defies expectation. In: Proceedings of the 37th International Conference on Neural Information Processing Systems (NeurIPS) (2024)
25. Kizilkaya, F.E., Kempe, D.: Plurality veto: a simple voting rule achieving optimal metric distortion. In: Proceedings of the 31st International Joint Conference on Artificial Intelligence (IJCAI), pp. 349–355 (2022)
26. Latifian, M., Voudouris, A.A.: The distortion of threshold approval matching (2024)
27. Ma, T., Menon, V., Larson, K.: Improving welfare in one-sided matchings using simple threshold queries. In: Proceedings of the Thirtieth International Joint Conference on Artificial Intelligence (IJCAI), pp. 321–327 (2021)
28. Manurangsi, P., Suksompong, W.: When do envy-free allocations exist? SIAM J. Discret. Math. **34**(3), 1505–1521 (2020)
29. Motwani, R., Raghavan, P.: Randomized Algorithms. Cambridge University Press (1995)
30. Myerson, R.B.: Optimal auction design. Math. Oper. Res. **6**(1), 58–73 (1981)
31. Papoulis, A., Unnikrishna Pillai, S.: Probability, Random Variables and Stochastic Processes. McGraw-Hill Higher Education (2002)
32. Pivato, M.: Asymptotic utilitarianism in scoring rules. Social Choice Welfare **47**(2), 431–458 (2016)
33. Pritchard, G., Wilson, M.C.: Asymptotics of the minimum manipulating coalition size for positional voting rules under impartial culture behaviour. Math. Soc. Sci. **58**(1), 35–57 (2009)
34. Procaccia, A.D., Rosenschein, J.S.: The distortion of cardinal preferences in voting. In: Proceedings of the 10th International Workshop on Cooperative Information Agents (CIA). Lecture Notes in Computer Science, vol. 4149, pp. 317–331. Springer (2006)
35. Roughgarden, T. (ed.): Beyond the Worst-Case Analysis of Algorithms. Cambridge University Press (2020)
36. Toulis, P., Parkes, D.C.: Design and analysis of multi-hospital kidney exchange mechanisms using random graphs. Games Econ. Behav. **91**, 360–382 (2015)
37. Tsetlin, I., Regenwetter, M., Grofman, B.: The impartial culture maximizes the probability of majority cycles. Social Choice Welfare **21**(3), 387–398 (2003)
38. Young, P.: Optimal voting rules. J. Econ. Perspect. **9**(1), 51–64 (1995)
39. Zwicker, W.S.: Introduction to the theory of voting. In: Brandt, F., Conitzer, V., Endriss, U., Lang, J., Procaccia, A.D. (eds.) Handbook of Computational Social Choice, pp. 23–56. Cambridge University Press (2016)

Fair Ordering in Replicated Systems via Streaming Social Choice

Geoffrey Ramseyer(✉) and Ashish Goel

Stanford University, Stanford, CA 94305, USA
geoff.ramseyer@cs.stanford.edu, ashishg@stanford.edu

Abstract. How can we order transactions "fairly" in a replicated state machine (of which today's blockchains are a prototypical example)? In the model of prior work [4,18–20,31], each of n replicas observes transactions in a different order, and the system aggregates these observed orderings into a single order. We argue that this problem is best viewed directly through the lens of the classic preference aggregation problem of social choice theory (instead of as a distributed computing problem), in which rankings on candidates are aggregated into an election result. Two features make this problem novel and distinct. First, the number of transactions is unbounded, and an ordering must be defined over a countably infinite set. And second, decisions must be made quickly and with only partial information. Additionally, some faulty replicas might misreport their observations; the influence of faulty replicas on the output should be well understood.

Prior work studies a "γ-batch-order-fairness" property, which divides an ordering into contiguous batches. If a γ fraction of replicas receive a transaction τ before another transaction τ', then τ' cannot be in an earlier batch than τ. This definition holds vacuously, so we strengthen it to require batches of minimal size, while accounting for faulty replicas. This lens gives a protocol with both strictly stronger fairness and better liveness properties than prior work. We specifically adapt the Ranked Pairs [30] method to this streaming setting. This algorithm can be applied on top of any of the communication protocols (in various network models) of prior work for immediate liveness and fairness improvements. Prior work relies on a fixed choice of γ and a bound on the number of faulty replicas f, but we show that Ranked Pairs satisfies our definition for every $\frac{1}{2} < \gamma \leq 1$ simultaneously and for any f, where fairness guarantees degrade as f increases.

1 Introduction

We study the problem of ordering transactions in the widely-used replicated state machine architecture (e.g., [22,28,32]). In the standard setting, each of a set of n distinct *replicas* maintains a copy of a *state machine*. Replicas communicate to agree on a totally-ordered log of transactions $\tau_1 \prec \tau_2 \prec \tau_3 \prec \ldots$, and then each replica applies the transactions in this order to its local copy of the

M. Mavronicolas et al. (Eds.): WINE 2024, LNCS 15534, pp. 438–456, 2026.
https://doi.org/10.1007/978-3-032-08560-3_25

state machine. An (unbounded) set of clients create and broadcast new transactions. This architecture famously underlies many of today's blockchains, but also underlies many traditional (centralized) services.

This architecture requires that all replicas apply transactions in the same order. For example, suppose a user A with a \$1 balance creates two transactions, one which sends \$1 to user B and one which sends \$1 to user C. Only one of these payments can execute successfully, so if replicas apply these transactions in different orders, then they will disagree on the balances of B and C.

There are many communication protocols through which replicas can agree on *some* total order. Yet in many systems, most notably within today's public blockchains, significant financial value can be derived from ordering transactions in specific ways [10]. These systems must therefore agree on not just any ordering but an "optimal" one, for some notion of optimal.

The key observation for this work is that this problem is a novel streaming variation of the classic preference aggregation problem of social choice theory [2,7]. Prior work on this problem [4,17–19] use different network communication protocols, but ultimately each produces agreement on a "vote" from each replica on the ordering of a set of transactions (in social choice parlance, a "ranking" over "candidates"). Agnostic to the network protocol, how should a system aggregate a set of proposed orders ("rankings") into a single, total order?

Two key differences separate this problem from the classic social choice setting. First, the number of transactions ("candidates") is countably infinite (as clients can continually send new transactions for an unbounded time). And second, the system must produce the output ranking in a streaming fashion. It cannot wait to see the reported orders over the entire (infinite) set of transactions before making a decision on the relative ordering of two transactions. Instead, the system must produce as output an append-only, totally-ordered log of transactions (to be given to the state machine). Furthermore, the system should minimize the delay between when a client sends a transaction and when that transaction is appended to the output log (i.e. the system needs "liveness").

While we believe this problem to be interesting in its own right, studying it directly through the lens of classic social choice theory, enables us to develop an ordering algorithm with both much stronger order "fairness" guarantees (the key property studied in prior work, discussed below) and stronger liveness guarantees than all of the prior work. This algorithm could be deployed on top of any of the network protocols of prior work.

One additional consideration is that replicas might strategically adjust their reported orderings. Prior work [4,18,19] writes that "honest" replicas must report transactions in the order in which they arrive over a network, and any other behavior is "faulty." One desirable property of an order aggregation rule is that the influence of a (colluding) subset of faulty replicas is precisely bounded.

There is a wide body of social choice literature on this underlying aggregation problem. Our goal with this work is to demonstrate an application for our streaming version of the problem, and to demonstrate the value of using social choice results in this application by targeting the precise desiderata raised in prior work. There are many other natural desiderata, notions of fairness, and aggregation rules that may be practically useful. The application of social-choice style aggregation rules in this streaming setting poses a number of interesting open questions. For example, if replicas have distinct financial motivations, or accept bribes from clients to order transactions in specific ways, is there an aggregation rule that maximizes (perhaps approximately) social welfare? For a reader coming from social choice theory, "transaction" could be replaced by "candidate", and "an ordering vote" by "a ranking."

1.1 Our Results

We first observe that for an ordering method to be well-defined on a countably infinite set, it suffices for there to exist a "monotonic" and "asymptotically live" algorithm that implements the ordering method on finite, initial segments of the input. This algorithm must not undo a decision when its (finite) input is extended ("monotonicity") and must eventually output every transaction ("asymptotic liveness"). The core challenge in building such an algorithm is in determining when an algorithm has enough information to make a decision that is consistent with its hypothetical output on any extension of its input.

Prior work of Kelkar et al. [19] defines a notion of "γ-batch-order-fairness" (Definition 4). The output ordering is divided into batches, and if γn replicas receive a transaction τ before another transaction τ', then τ cannot be in a later batch than τ'. However, transactions within a batch can be ordered arbitrarily, which means this definition is vacuously satisfiable by an arbitrarily large batch. We first strengthen this definition in Sect. 3 to precisely capture the intuition that batches should be *minimal*. Minimality must be defined carefully in the presence of faulty replicas; we show that γ-batch-order-fairness cannot be satisfied with faulty replicas and exact notions of batch minimality.

We call our definition (γ, δ)-minimal-batch-order-fairness (Definition 7). Simply put, if a $\frac{1}{2} < \gamma \leq 1$ fraction of replicas vote for τ before τ' (henceforth; $\tau \prec \tau'$) then the output ordering should include $\tau \prec \tau'$ unless there is a sufficiently strong reason to put $\tau' \prec \tau$—that is, a sequence of transactions $(\tau_1, \tau_2, \ldots, \tau_{k-1}, \tau_k)$, with $\tau_1 = \tau$ and $\tau_k = \tau'$, where at least a $\gamma - 2\delta$ fraction of replicas vote for $\tau_i \prec \tau_{i+1}$ for $1 \leq i \leq k-1$. Motivating this definition is the fact that a δ fraction of faulty replicas can, at most, increase or decrease the fraction of replicas that report $\tau \prec \tau'$ by δ.

As an example, given f faulty replicas out of n total replicas and a fixed γ, the protocol of Kelkar et al. [19] satisfies $(\gamma, \frac{f}{n})$-minimal-batch-batch-order-fairness (Lemma 3.3). However, that protocol relies on an explicit choice of parameter γ and an explicit bound on f, and it is not clear whether a higher or lower γ gives a stronger guarantee (see the online appendix for an example). By contrast, the algorithm we give here achieves $(\gamma, \frac{f}{n})$-minimal-batch-order-fairness for every $\frac{1}{2} < \gamma \leq 1$ *simultaneously*, and for any number of faulty replicas. Fairness guarantees smoothly degrade as the number of faulty replicas increases.[1]

This work adapts the ordering known as Ranked Pairs [30] to our streaming setting. Informally, Ranked Pairs operates on a weighted, directed graph, where each vertex is a transaction and an edge from τ to τ' has weight equal to the fraction of replicas that report $\tau \prec \tau'$. Prior work [4, 19], by contrast, operates on the unweighted, directed graph that is derived by taking the weighted graph which our work uses, removing all edges below weight γ, and dropping the weights on the remaining edges. This technical difference is the key to our stronger results.

Analyzing this graph gives an important lemma about what pieces of information Ranked Pairs uses to determine its output. This lemma allows a streaming version of Ranked Pairs to track how uncertainty about as-yet-unseen transactions propagates through the algorithm. Ranked Pairs iterates over edges in order of weight. An interesting observation is that the tiebreaking rule in this ordering critically determines protocol liveness. A worst-case (fixed) rule could cause our streaming algorithm to never output any transactions. However, we show in Sect. 6 how to construct a tiebreaking rule in a streaming fashion that guarantees not only asymptotic liveness but also an explicitly bounded liveness delay. If, for example, at most Δ time elapses between when a transaction τ is sent and when every honest replica votes on it, our algorithm outputs τ after a delay of at most $O(n\Delta)$. This can be reduced to $O(k\Delta)$, by rounding edge weights to the nearest $\frac{1}{k}$— although this rounding reduces the fairness guarantee to $(\gamma, \frac{f}{n} + \frac{1}{2k})$-minimal-batch-order-fairness (again for all $\frac{1}{2} < \gamma \leq 1$ simultaneously, and any f).

Proofs and additional details are deferred to the online appendix, available at https://arxiv.org/abs/2304.02730.

2 Preliminaries and System Model

We consider a model in which there are n *replicas* cooperating to develop a total ordering of transactions. Transactions are received over the network from *clients*. We say that the order that replica receives transactions is its observed ordering.

Definition 1 (Ordering Preference). *An Ordering Preference on a (finite or countably-infinite) set of transactions is a total ordering*[2] $\hat{\sigma}_i = (\tau_{i_1} \prec \tau_{i_2} \prec \tau_{i_3} \prec \ldots)$

However, at any finite time, each replica can have received only a finite number of transactions. Replicas periodically submit these finite observations to a ranking algorithm.

[1] Our algorithms do not require knowledge of a bound on the number of faulty replicas. This notion of "faulty" is distinct from notions of "faulty" in Byzantine-fault tolerant communication protocols.

[2] Isomorphic to a subset of ω.

Definition 2 (Ordering Vote). *A replica i submits to a ranking algorithm an ordering vote on a set of k transactions, $\sigma = (\tau_{i_1}, ... \tau_{i_k})$. (where $\tau_{i_{j_1}} \prec \tau_{i_{j_2}}$ for $j_1 < j_2$). σ' extends σ if σ is an initial segment of σ' (σ extends itself).*

Definition 3 (Ranking Algorithm). *A ranking algorithm $\mathcal{A}(\sigma_1, \ldots, \sigma_n)$ takes as input an ordering vote from each replica and outputs an ordering σ on a subset of the transactions in its input.*

The output need not include every transaction in the input.

A replica is *honest* if its true, observed ordering always extends its ordering vote, and if its vote always contains all transactions that the replica has observed prior to voting. Otherwise, the replica is *faulty*. We denote the number of faulty replicas as f (out of n), and do not assume a bound or knowledge of f.[3]

In the rest of this work, we assume that every replica eventually votes on every transaction. This may not necessarily hold for faulty replicas. A precise choice of response to this type of faulty behavior will depend on the network model and communication protocol, but, informally, if a replica fails to vote on a transaction within a "sufficient" time, it suffices for the honest replicas to fabricate a vote on behalf of the faulty replica. We give an example construction that satisfies this property in the online appendix, but note that prior work on this problem (e.g., [4,18,19]) implicitly use the same ideas to handle unresponsive replicas.

3 Defining Fair Ordering

3.1 Batch Order Fairness

Our discussion here follows the notion of "fairness" discussed in prior work (e.g., [4,18–20]). This notion of "fairness" is that if "most" replicas report a transaction τ before another transaction τ', then the protocol outputs τ before τ'. Making this notion precise requires accounting for what are known as Condorcet cycles. A (super)majority might report τ before τ', another majority reports τ' before τ'', and a third majority reports τ'' before τ. Aequitas [19] addresses this problem by outputting these Condorcet cycles in "batches," which leads to the following definition (paraphrased from [19]).

Definition 4 (γ-batch-order-fairness). *Suppose that τ and τ' are received by all nodes. If γn nodes received τ before τ', then a ranking algorithm never outputs τ' in a later batch than τ.*

In other words, τ and τ' could be in the same batch. Transactions within a batch are ordered *arbitrarily*. Subsequent work [4,18,20] follows this same definition (or a restatement of it).

There are two key problems with this definition. First, it does not preclude a ranking algorithm from putting all transactions into one batch (therefore ordering transactions arbitrarily). For this definition to be meaningful, it needs a

[3] $(\gamma, \frac{f}{n})$-minimal-batch-order-fairness (Definition 7) is only meaningful if $f < \frac{n}{2}$.

notion of "minimality" to batch sizes. We address this problem by strengthening the definition to include a notion of "minimality" (Definition 7). Some of the prior work [4,19,20], but not all [18], implicitly satisfies this definition.

Second, it is not clear whether a higher or lower γ gives stronger fairness properties, yet all of the prior work requires a fixed parameter choice. Indeed, there are simple examples (with no faulty replicas) where Definition 7 only implies ordering restrictions at high values of γ, others at only low values, and still others at only intermediate values (see the online appendix for details). It is not the case that satisfying Definition 7 for a low value of γ implies satisfying the definition for a high value of γ. We address this problem by constructing a protocol that satisfies Definition 7 for all $\frac{1}{2} \leq \gamma \leq 1$ simultaneously.

3.2 Batch Minimality

To construct a precise definition, we start with a weighted dependency graph over a set of transactions.

Definition 5 (Ordering Graph). *Suppose that replicas report ordering votes* $(\sigma_1, ..., \sigma_n)$ *on a set of transactions* V*. The Ordering Graph* $G(\sigma_1, ...\sigma_n)$ *is a complete, weighted directed graph with vertex set* V *and, for each transaction pair* (τ, τ')*, an edge of weight* $w(\tau, \tau') = \alpha$ *if* αn *replicas report* $\tau \prec \tau'$*.*

Any notion of "minimal" should at least require that when there are no cycles in ordering dependencies, the output is consistent with every ordering dependency (as in Theorem IV.1 part 1, Themis [18]). Exact γ-minimality generalizes this notion to the case where cycles in ordering dependencies exist.

Definition 6 (Exact γ-minimality). *For any pair of transactions* τ, τ'*, if* γn *replicas receive* $\tau \prec \tau'$ *but* τ' *is output before* τ*, then there exists a sequence of transactions* $\tau' = \tau_1, ..., \tau_k = \tau$ *where* γn *replicas receive* $\tau_i \prec \tau_{i+1}$ *and* τ_i *is output before* τ_{i+1} *for all* i*.*

However, this notion of minimality is *impossible* to achieve (for any γ) in the face of *any* faulty replicas. The condition on f in Lemma 3.1 is a technical artifact of a construction in the proof. For some parameter settings, such as $n = 7$ with $\gamma = \frac{2}{3}$, the condition admits $f = 1$.

Lemma 3.1. *No protocol can achieve* γ*-batch-order-fairness with exactly* γ*-minimal output batches for any* n *and* f *greater than* $n \mod (n - \lceil \gamma n \rceil + 1)$ *(for* $\gamma > 1/2$*).*

In fact, the construction in the proof contains no cycles in the ordering dependency graph of weight γ, thereby applying even to a weaker version of Definition 6 that applies only to the case where the ordering graph, restricted to edges of weight at least γ, has no cycles.

The core difficulty is that faulty replicas can cause the weight on an edge observed by any algorithm to be different from the true, ground truth weight.

Intuitively, if a protocol has a reliable communication layer (i.e. every honest replica is able to submit a vote on its ordering preference), then the worst that a faulty replica can do is to misreport its ordering preferences. If there are f faulty replicas, then the faulty nodes can artificially reduce or increase the fraction of replicas that vote for $\tau \prec \tau'$ by at most $\frac{f}{n}$. As a protocol cannot distinguish a faulty replica from an honest one by its votes, we therefore must take some error in edge weights into account in our definition of minimality.

Definition 7 ((γ, δ)-minimal-batch-order-fairness). *An ordering is (γ, δ)-minimally-batch-order-fair if, for any transaction pair (τ, τ') that is received in that order by at least γn replicas but output by the protocol in the reverse ordering, then there is a sequence of transactions $\tau' = \tau_1, ..., \tau_k = \tau$ where at least $(\gamma - 2\delta)n$ replicas receive $\tau_i \prec \tau_{i+1}$ and τ_i is output before τ_{i+1}.*

Definition 7 captures the notion that a protocol cannot distinguish between network delays and a δ-fraction of replicas misreporting information. The protocol outputs minimally-sized batches, up to this indistinguishable fraction. Definition 6 corresponds to the case of $\delta = 0$. This definition does not mention "batches" or "minimality," but approximately minimal batches (the strongly connected components of Lemma 3.2) can be recovered from any ordering satisfying it. The second condition of Lemma 3.2 limits the size of output batches.

Lemma 3.2. *Suppose that an output ordering satisfies (γ, δ)-minimal-batch-order-fairness. Compute the ordering graph $\hat{G}$, drop all edges with weight below $\gamma - \delta$, and compute the strongly connected components of the remainder.*

- *If γn replicas received $\tau \prec \tau'$, then either τ and τ' are in the same strongly connected component, or all transactions in the component containing τ are output before any transactions in the component containing τ'.*
- *If γn replicas receive $\tau \prec \tau'$ and there is no sequence of transactions $\tau' = \tau_1, ..., \tau_k = \tau$ where at least $(\gamma - 2\delta)n$ replicas receive $\tau_i \prec \tau_{i+1}$, then all transactions in the component containing τ are output before any transactions in the component containing τ'.*

There is one subtle difference between this definition and the requirement for an explicit sequence of output batches in prior work (e.g., in [18,19]). When there are two disjoint strongly connected components with no dependencies (of strength at least γ) from one to the other, Definition 7 allows the output to interleave these components. This may be strictly required (see the online appendix).

3.3 Comparison to Prior Work

We first apply Definition 7 to related work. Aequitas [19] is instantiated with an explicit choice of parameter γ and bound f on the number of faulty replicas.

Lemma 3.3. *Aequitas [19] with parameters (γ, f) achieves $(\gamma, \frac{f}{n})$-minimal-batch-order-fairness.*

Aequitas requires $\gamma - \frac{1}{2} > \frac{2f}{n}$. Lemma 3.3 strengthens Theorem 6.6 of [19].

Broadly speaking, Aequitas, along with the protocols of Cachin et al. [4] and Kiayias et al. [20], follow the same pattern. Through (very different) communication protocols, they come to agreement on, for each transaction pair τ and τ', the fraction of replicas that report τ before τ'. These protocols choose (in advance) a fixed parameter γ, and then drop from consideration ordering dependences of strength less than γ. This is equivalent to constructing the ordering dependency graph of Definition 5, but then dropping all edges of weight less than γ and then forgetting all of the weights on the remaining edges. The strongly connected components of this graph form the batches output in Aequitas.

Aequitas (as well as the protocols of Cachin et al. [4] and Kiayias et al. [20]) also is not asymptotically live (Definition 9). The strongly connected components of this unweighted dependency graph can be arbitrarily large, and these protocols wait until they observe the entirety of a component before outputting any transaction in it. An exception is one version of the protocol of Kiayias et al. [20], which adds timestamps to transactions and (given precise network assumptions) can output a component in a streaming fashion.

By contrast, Ranked Pairs (Theorem 4.1) achieves $(\gamma, \frac{f}{n})$-minimal-batch-fairness for every γ simultaneously, and for any f (our work does not require knowledge of f). Additionally, Ranked Pairs guarantees a bounded liveness delay (although end-to-end results depend also on network conditions).

Themis [18] claims (Theorem IV.1) that its output can be partitioned into "minimal" batches, which appears to contradict Lemma 3.1. However, Themis uses a much weaker notion of "minimal."

Where γ-batch-order-fairness would create a dependency for τ not after τ' if at least γn replicas receive τ before τ', the relaxation implied by Definition III.1 of [18] creates such a dependency if at least $n(1-\gamma)+1$ receive τ before τ' (that is, it cannot be the case that more than γn receive τ' before τ). Theorem IV.1 of [18] considers its output batches "minimal" if they are singletons when there are no cycles of these dependencies. This notion of minimality is much weaker than what we discuss here, and does not provide any guarantee when there is significant disagreement between replicas on transaction orderings. Lemma 3.4 gives an example highlighting this distinction.

Lemma 3.4. *In Themis [18], even if all nodes report τ before τ' and there are no faulty replicas nor cycles in ordering dependencies (as defined in Definition 4) for any $\gamma > 1/2$, the output may put τ' before τ.*

This kind of counterintuitive behavior is a possible output of Themis's protocol. When two transactions τ, τ' receive $n - f$ votes in one round of Themis, it determines whether to consider as an ordering dependency $\tau \prec \tau'$ or $\tau' \prec \tau$ by *majority vote*. Equivalently, Themis operates in the regime of $\gamma = \frac{1}{2}$, regardless of a chosen parameter value (γ affecs how Themis considers transactions reported by some but not all replicas, which does not affect this discussion).

4 Ranked Pairs Voting

Our protocol is based on the Ranked Pairs method [30]. Ranked Pairs (reproduced in the online appendix) first builds the ordering graph (Definition 5), then iterates through edges in order of edge weight (breaking ties arbitrarily). It builds an acyclic set of edges greedily, adding each new edge so long as it does not create a cycle with edges already in the set. The output ordering is the result of topologically sorting the resulting acyclic graph.

Theorem 4.1. *Given a ordering vote (on every transaction in a finite set) from every replica, Ranked Pairs Voting simultaneously satisfies* $(\gamma, \frac{f}{n})$*-minimal-batch-order-fairness for every* γ*, and does not depend on any fixed bound on* f*.*

Lemma 3.2 and Theorem 4.1 together imply an important property. Ranked Pair's output can be divided into batches consistent with γ-batch-order-fairness for *any* γ (subject to the interleaving discussed in Sect. 3.2). There is no need to choose a γ or arbitrarily order transactions within a batch.

5 Streaming Ranked Pairs

We now turn to the streaming setting that is the focus of our work. There may be an infinite set of transactions to consider, but at any finite time, an algorithm must compute an ordering on a subset of the transactions that have been seen thus far. In a real-world system, such an algorithm would be run on some periodic frequency, and replicas would append to their ordering votes.

An algorithm must be *monotonic* (Definition 8)—if replicas extend their ordering votes, the algorithm, when run again on the extensions, can only extend its prior output.

Definition 8 (Monotonicity). *A ranking algorithm* $\mathcal{A}(\cdot, \ldots, \cdot)$ *is monotonic if, given two sets of ordering votes* $(\sigma_1, \ldots, \sigma_n)$ *and* $(\sigma'_1, \ldots, \sigma'_n)$*, such that* σ'_i *extends* σ_i *for all* $i \in [n]$*,* $\mathcal{A}(\sigma'_1, \ldots, \sigma'_n)$ *extends* $\mathcal{A}(\sigma_1, \ldots, \sigma_n)$*.*

If a ranking algorithm is monotonic, it implies a well-formed definition for aggregating a set of orderings on countably infinite sets of transactions, not just on finite sets. τ comes before τ' in the infinite case if there exists a finite subset of the input orderings such that the algorithm puts $\tau \prec \tau'$ in its output.

However, the vacuous algorithm that never outputs anything is monotonic.

Definition 9 (Asymptotic Liveness). *A ranking algorithm* $\mathcal{A}(\cdot, \ldots, \cdot)$ *is asymptotically live if, given any set of countably infinite ordering votes* $(\hat{\sigma}_1, \ldots, \hat{\sigma}_n)$ *and any transaction* τ *in those votes, there exists* N *such that when each* $\hat{\sigma}_i$ *is trunctated to the first* N *elements to produce* σ_i*,* τ *is included in* $\mathcal{A}(\sigma_1, \ldots, \sigma_n)$*.*

We present our algorithm (Algorithm 1) in two stages. Ranked Pairs iterates through edges of equal weight in an arbitrary order. We start by analyzing a version of our algorithm (which we call here Algorithm 1$*$, specified in full in the online appendix) that, like Ranked Pairs, uses a fixed, arbitrary iteration order. This algorithm is monotonic but not asymptotically live. We then show that the modifications to this order present in Algorithm 1 ensure asymptotic liveness.

5.1 Decisions Are Local

Implementing Ranked Pairs voting in this streaming setting requires first understanding when the algorithm accepts an edge in the ordering graph.

Lemma 5.1. *If Ranked Pairs rejects an edge* (τ, τ') *of weight* α*, then it must accept a directed path from* τ' *to* τ*, where each edge has weight at least* α*.*

Observation 5.2. *Ranked Pairs chooses every edge of weight* 1.

Furthermore, the set of vertices that such a path can visit are bounded to a local neighborhood of τ_A and τ_B.

Definition 10. *Let* τ *be any transaction.*

1. *Let* P_τ *be the set of preceeding transactions—all* τ' *with* $w(\tau', \tau) = 1$.
2. *Let* Q_τ *be the set of concurrent transactions—all* τ' *with* $0 < w(\tau', \tau) < 1$.
3. *Let* R_τ *be the set of subsequent transactions—all* τ' *with* $w(\tau', \tau) = 0$.

Lemma 5.3. *If ranked pairs chooses all edges in a path from* τ_B *to* τ_A*, no transactions on the path are in* R_{τ_A} *or* P_{τ_B}*.*

Lemma 5.3 lets the streaming algorithm decide whether to choose an edge given only a finite amount of information. Once the algorithm sees a vote for a transaction τ from every replica, it can compute the weight in the ordering graph for every edge (τ, τ') and (τ', τ). Unseen transactions must be in R_τ.

Even so, a streaming algorithm cannot always know whether or not Ranked Pairs would choose an edge. As such, we allow our algorithms to leave an edge in an "*indeterminate*" state, and then account for indeterminate edges when considering subsequent edges.

We first define a graph that records the information available at a finite time.

Definition 11 (Streamed Ordering Graph). *Suppose that each replica submits an ordering vote* $(\sigma_1, ..., \sigma_n)$*. Let* V *be the set of all transactions that appear in each vote, and let* $\hat{v}$ *be a new vertex (the "future") The Streamed Ordering Graph is a weighted, directed, complete graph* $\hat{G} = (V', E)$ *on vertex set* $V \cup \{\hat{v}\}$*.*

For each τ *and* τ'*, set weights* $w(\tau, \tau')$ *and* $w(\tau', \tau)$ *as in Definition 5. Set* $w(\tau, \hat{v}) = 1$ *for all* τ*. If there exists* τ' *that appears in the votes of some but not all replicas and which preceeds* τ *in at least one replica's vote, set* $w(\hat{v}, \tau) = 1$*, and otherwise* $w(\hat{v}, \tau) = 0$*.*

Conceptually, the "future" vertex $\hat{v}$ represents all transactions that have not received votes from all replicas. Implicitly, all edge weights not computable from the available information are upper-bounded by 1. An implementation could use a tighter bound, but subsequent arguments require only that the assigned weight on these edges is an upper bound of the true weight.

5.2 A Streaming Algorithm

We can now construct a streaming algorithm to compute Ranked Pairs. Instead of either including or rejecting every edge, Algorithm 1∗ has the option to declare an edge as *indeterminate*; that is, the algorithm cannot determine whether to include or exclude the edge with the information in the input.

The algorithm proceeds identically to the classic Ranked Pairs algorithm, except regarding how it treats *indeterminate* edges. For each new edge, it first checks whether it can accept the edge (without creating a cycle of accepted edges), supposing that all currently *indeterminate* edges are later accepted. Conversely, it also checks whether it can reject the edge, which means checking whether accepting the edge would always create a cycle of accepted edges (assuming the currently *indeterminate* edges are later rejected). The algorithm operates on the modified input graph (Definition 11), where the "future" vertex bounds the information the algorithm has to consider about as-yet-unseen transactions. Informally, all edges between as-yet-unseen transactions are *indeterminate*, so from the perspective of directed connectivity analysis, the unseen transactions form a single, strongly-connected component.

To prove correctness of this algorithm, it suffices to show that its output is monotonic and that the output is consistent with the non-streaming version of Ranked Pairs. By consistency, we mean that at any point, if clients were to stop sending new transactions and all replicas eventually voted on every replica, Ranked Pairs and Algorithm 1∗ would produce the same output.

Lemma 5.4. *Consider a counterfactual scenario where clients stop sending transactions, all replicas eventually receive every transaction, and every replica includes every transaction in its output vote.*

Whenever Algorithm 1∗ includes a determinate edge in H (resp. excludes an edge), that edge is included (resp. excluded) in the output of the non-streaming Ranked Pairs on the counterfactual input.

The same argument shows that Algorithm 1∗ is monotonic.

Corollary 5.5. *Algorithm 1∗ is monotonic.*

Lemma 5.4 implies that Algorithm 1∗'s output matches that of Ranked Pairs.

Theorem 5.6. *The output of Algorithm 1∗ matches an initial segment of the ordering output by the non-streaming Ranked Pairs (on the counterfactual of Lemma 5.4).*

Ranked Pairs has the property that, for any set of inputs $(\sigma_1, \ldots, \sigma_n)$, if it is run on a restriction of the inputs to the transactions that appear as an initial segment of its output on $(\sigma_1, \ldots, \sigma_n)$, its output is unmodified on the restricted inputs (for completeness, we give a proof in the online appendix). As Algorithm 1$*$ is monotonic, its output exactly matches the Ranked Pairs ordering when the input votes are restricted in this way.

6 Liveness and Efficiency

6.1 Efficient Marginal Runtime

Instantiating this protocol means running a ranking algorithm repeatedly (as replicas report additional information). Rather than recompute the streamed ordering graph and the status of every edge in each invocation, Algorithm 1$*$ could use the output of a past invocation, restricted to the decisions that were *determinate*, as an oracle. Corollary 5.5 implies that *determinate* choices are consistent with any extension of the input, so this oracle does not change the output. However, the runtime now depends only on the number of new edges in the streamed ordering graph (see the online appendix).

6.2 Protocol Liveness

Algorithm 1$*$ is not asymptotically live. We give a worst-case example in the online appendix. The problem is the tiebreaking rule for how Ranked Pairs iterates over edges of equal weight. That construction uses a worst-case tiebreaking rule, but Ranked Pairs requires only some tiebreaking rule. Our algorithm can generate a tiebreaking rule dynamically. Specifically, when the algorithm visits edges of weight $\gamma = \frac{k}{n}$, it defers edges that would become indeterminate to the end of the ordering among edges of that weight.

Algorithm 1 thus constructs a (partial) ordering between edges that breaks ties between edges of the same weight. The output of this algorithm is, by the same arguments as in Sect. 5.2, consistent with the output of non-streaming Ranked Pairs, when using an tiebreaking rule consistent with this edge ordering. The oracle mechanism of Sect. 6.1 passes this tiebreaking information from one run of Algorithm 1 to the next.

Lemma 6.1. *Whenever Algorithm 1 includes a determinate edge in H, or excludes an edge from H, the non-streaming Ranked Pairs, when using the implied ordering, would make the same decision.*

This small change enables the following structural lemma.

Definition 12. *An edge $(\hat{\tau}, \hat{\tau'})$ is contemporary to an edge (τ, τ') if and only if $\hat{\tau}$ and $\hat{\tau'}$ are contained in $U_{\tau,\tau'}$.*

Lemma 6.2. *If Algorithm 1 marks an edge (τ_i, τ_j) with weight $\frac{k}{n}$ as indeterminate, there must be a path from τ_j to τ_i contained in U_{τ_i,τ_j} that contains an indeterminate edge of weight at least $\frac{k+1}{n}$.*

ALGORITHM 1: Streaming Ranked Pairs

Input: An ordering vote from each replica $\{\sigma_i\}_{i\in[n]}$
Input: $\hat{H}$, the graph computed in the preceeding invocation of Algorithm 1
$\hat{G} = (V, E, w) \leftarrow \hat{G}(\sigma_1, ..., \sigma_n)$; `/* Compute streamed ordering graph */`
$H \leftarrow (V, \emptyset)$; `/* Initialize empty graph on same vertex set */`
foreach *edge* $(\tau, \hat{v})$ *and* $(\hat{v}, \tau)$ *with* $w(e) = 1$ **do**
  Add e to H, and mark it as *indeterminate*
Add all *determinate* edges of $\hat{H}$ to H, marked *determinate*
foreach γ *in* $(1, \frac{n-1}{n}, \frac{n-2}{n}, \ldots, 0)$ **do**
  $E_\gamma \leftarrow \{e \in E \mid w(e) = \gamma\}$, ordered arbitrarily
  repeat
    $e = (\tau_i, \tau_j) \leftarrow$ first element of E_γ
    $U_{\tau_i,\tau_j} \leftarrow (V \setminus (R_{\tau_i} \cup P_{\tau_j})) \cup \{\hat{v}\}$
    If there is a directed path in H restricted to U_{τ_i,τ_j} from τ_j to τ_i where every edge is *determinate*, then remove the edge from E_γ.
    Else if there is no directed path in H restricted to U_{τ_i,τ_j} from τ_j to τ_i where every edge is either *determinate* or *indeterminate*, add (τ_i, τ_j) to H and mark it as *determinate*. Remove the edge from E_γ.
    Otherwise, defer (τ_i, τ_j) to the end of the ordering on E_γ.
  until E_γ *is empty or no progress is made in one full loop over* E_γ;
  Mark all remaining edges in E_γ as *indeterminate*
Output: The topological sort of H, up to but not including the first transaction bordering an *indeterminate* edge

By contrast, the best bound in Algorithm 1$*$ is that there exists a contemporary indeterminate edge of weight $\frac{k}{n}$. Lemma 6.2 implies that for any indeterminate edge, there is sequence of contemporary indeterminate edges of strictly increasing weights that causes the edge to be indeterminate.

Lemma 6.3. *Suppose Algorithm 1 marks an edge e_1 of weight γ as indeterminate. Then there exists a sequence of indeterminate edges $e_1, \ldots e_k$ of strictly increasing weight such that for each pair of edges $e_i = (\tau_i, \tau_j)$ and $e_{i+1} = (\tau_i', \tau_j')$, e_{i+1} is contemporary to e_i, and the source of e_k is $\hat{v}$.*

A chain cannot be longer than n edges, as each step increases the weight by at least $\frac{1}{n}$. This bound implies asymptotic liveness.

Theorem 6.4. *Algorithm 1 is asymptotically live.*

Additionally, given a network assumption like Assumption 6.5, Lemma 6.3 bounds the temporal delay between when a transaction is sent by a client and when Algorithm 1 adds it to its output.

Assumption 6.5 (Network Delay Bound). *If a client sends a transaction τ at time t, all replicas include τ in their ordering votes before time $t + \Delta$.*

As such, replicas disagree on the ordering between τ and τ' only if they were sent at roughly the same time. Note that Algorithm 1 does not require knowledge of Δ, and does not depend on Assumption 6.5 for correctness.

Observation 6.6. *Consider any two transactions τ_i and τ_j sent at times t_i and t_j, respectively. If $t_i + \Delta < t_j$, then the weight of the edge (τ_i, τ_j) is 1.*

Lemma 6.7. *Consider any three transactions τ, τ_i, and τ_j received by all replicas, that were sent at times t, t_i, and t_j, respectively. Suppose $\tau \in U_{\tau_i,\tau_j}$, and that the weight of (τ_i, τ_j) is not 0. Then $t_j - \Delta \leq t \leq t_j + 2\Delta$.*

This lemma and Lemma 6.3 give an overall time bound.

Theorem 6.8. *A transaction τ is contained in the output of the algorithm of Algorithm 1 after at most $(n+1)\Delta$ time.*

6.3 Trading Accuracy for Liveness

The $O(n\Delta)$ bound in Theorem 6.8 comes from the fact that a chain of indeterminate edges can have $O(n)$ length, which, in turn, follows from the fact that weights have a granularity of $\frac{1}{n}$. Reducing this granularity by rounding weights reduces the maximum length of a chain of indeterminate edges.

Lemma 6.9. *If edge weights are rounded to the nearest $\frac{1}{k}$ in the streamed ordering graph (before Algorithm 1 is applied) then a transaction τ is contained in the output after at most $(k+2)\Delta$ time.*

However, rounding weakens fairness guarantees.

Lemma 6.10. *Rounding edge weights to the nearest $\frac{1}{k}$ before applying Algorithm 1 achieves $(\gamma, \frac{f}{n}+\frac{1}{2k})$-minimal-batch-order-fairness for all γ and f faulty replicas.*

7 Related Work

7.1 Order-Fairness

Closest to this work are Aequitas (Kelkar et al., [19]) and Themis (Kelkar et al., [18]), which propose the definition of γ-batch-order-fairness and instantiate ordering protocols based on aggregating ordering votes into a total order. Aequitas achieves $(\gamma, \frac{f}{n})$-minimal-batch-order-fairness for a fixed choice of γ, but is not asymptotically live. Kelkar et al. [17] instantiate Aequitas in a setting where the number of replicas is not known, and similarly points out the connection between this streaming problem and the classical ranking aggregation problem. Section 3.3 gives a more detailed comparison to these works.

Our definition of (γ, δ)-minimal-batch-order-fairness (Definition 7) strengthens the (vacuously satisfiable) γ-batch-order-fairness definition used in prior work (see Sect. 3 for details). Cachin et al. [4] give a notion of "κ-differential-order-fairness;" Theorem D.1 of [18] shows equivalence of this definition and γ-batch-order-fairness, for a choice of γ (as a function of κ).

Concurrently with this work, Vafadar and Khabbazian [31] observe that an adversary can influence the output of protocols like those above by sending extra transactions, which can force batches to be combined together. Combining batches eliminates the constraints on the output that Definition 4 gives. Lemma 3.4 identifies a similar problem. This observation is a corollary of the fact that Aequitas (as well as the Ranked Pairs algorithm studied here) does not satisfy the Irrelevance of Independent Alternatives axiom [2].

Vafadar and Khabbazian [31] also propose using Ranked Pairs to sort transactions within batches (as, e.g., those output by a protocol like Themis [18]). However, no modification to the ordering of transactions within a batch output in Aequitas [19] can give the Ranked Pairs ordering, because Ranked Pairs might need to interleave batches that Aequitas considers incomparable (as demonstrated by counterexample in the online appendix). This strategy also cannot mitigate Aequitas's lack of asymptotic liveness. This mitigation, if applied to Aequitas, is akin to dropping all edges of weight less than γ before running Ranked Pairs. This strategy, therefore, would achieve $(\gamma', \frac{f}{n})$-minimal-batch-order-fairness for all $\gamma' \geq \gamma$ (a weaker guarantee than what our protocol provides).

Also concurrently with this work, Kiayias et al. [20] propose a different method for ordering transactions within batches (but rely on a specific choice of a parameter γ to determine those batches). A version of that protocol adds timestamps on reported orderings to improve liveness.

As discussed in Sect. 3.3, the batches in Themis [18] are defined slightly differently than in Definition 4. Themis, when computing a graph of ordering dependencies, adds a directed edge between every pair of vertices, choosing the direction of an edge by majority vote (ignoring the parameter γ, which is considered only on edges adjacent to transactions that some replicas have not yet reported observing). In so doing, it merges together any two batches unless, for any transaction τ in one and τ' in the other, a majority of replicas vote $\tau \prec \tau'$ (which bypasses the counterexample mentioned above).

Additionally, Themis's [18] round-based sequencing protocol presents a different problem insurmountable by any process like the one suggested by Vafadar and Khabbazian [31] that sorts batches separately. Namely, Themis may finalize the composition of a batch before has enough information to guarantee that, according to the Ranked Pairs ordering, all transactions in the batch must come before all those not in the batch (and not already in the output). Thus, it may output two batches in sequence, such that running Ranked Pairs on each batch separately produces a different output than running Ranked Pairs on the union of the two batches together. We give a example of this phenomenon in the online appendix.

Zhang et al. [35] add timestamps to ordering votes and sorts transactions by median timestamp to provide a linearizability guarantee, which is incomparable to (γ, δ)-minimal-batch-order-fairness. Mamageishvili et al. [27] propose letting users pay to reduce their reported timestamps.

7.2 Social Choice

Classical work of social choice studies the problem of choosing an order between a fixed, finite set of candidates given a complete set of preferences from each voter [2]. Prior work studies related models in a limited-information setting. Lu and Boutilier [25], Ackerman et al. [1], and Cullinan et al. [9] study the setting where a social choice rule has incomplete access (a partial ordering) to a voter's preferences over a finite set of candidates. Conitzer and Sandholm [8] study the communication complexity of voting rules. Fain et al. [12] study a randomized ordering rule that needs only a constant number of queries to voter preferences.

Fishburn [14] showed that Arrow's impossibility theorem does not hold when the set of voters is infinite, although Kirman and Sondermann [21] find "dictatorial sets" of voters. Grafe and Grafe [15] extend these results to the case of infinitely many alternatives, subject to a continuity condition on the space of ranking preferences. Chichilnisky and Heal [5] and Efimov and Koshevoy [11] study the types of rules admissible in an infinite voters setting.

7.3 Front-Running

Order-fairness is often motivated by the goal of limiting an adversary's ability to order transactions for profit ("front-running, e.g., Daian et al. [10]). Another approach is to commit to an ordering before transaction contents are made public, using threshold encryption or commit-reveal schemes [6,26,33]. Li et al. [23] study the problem of verifiably computing an ordering within a hardware enclave. These could be used on top of a streaming ordering algorithm.

Kavousi et al. [16] randomly shuffle blocks of encrypted transactions. Some financial exchanges [3,29] process transactions in unordered batches, eliminating the need for an ordering within a block (but not eliminating all order manipulations [34]). Ferreira and Parkes [13] and Li et al. [24] construct sequencing rules specific to decentralized exchanges.

8 Conclusion

We study here the problem of ordering transactions in a replicated state machine as a streaming instance of the preference aggregation problem from classical social choice theory. This viewpoint enables us to strengthen the notions of "order-fairness" used in prior work, and then construct algorithms solving this problem with both strictly stronger "fairness" guarantees and strictly stronger liveness properties than all of the prior work. To be specific, our streaming variant of Ranked Pairs satisfies $(\gamma, \frac{f}{n})$-minimal-batch-order-fairness for every $\frac{1}{2} < \gamma \leq 1$ simultaneously, and for any number of faulty replicas f. Fairness guarantees smoothly weaken as the number of faulty replicas increases. For comparison, prior work must fix a choice of γ and a bound on the number of faulty replicas in advance, and can only satisfy γ-batch-order-fairness for that γ.

These notions of "order-fairness" are not the only desiderata with practical significance. Different contexts in which a system is deployed pose different constraints and financial incentives. We believe that social-choice style methods for preference aggregation raises many interesting open questions of practical import. For example, what aggregation rules lead to high social welfare? If clients bribe replicas to prefer certain orderings, what rules maximize (or minimize) this fee revenue? How do incentives distort the output ordering?

References

1. Ackerman, M., Choi, S.Y., Coughlin, P., Gottlieb, E., Wood, J.: Elections with partially ordered preferences. Public Choice **157**, 145–168 (2013)
2. Arrow, K.J.: A difficulty in the concept of social welfare. J. Polit. Econ. **58**(4), 328–346 (1950)
3. Budish, E., Cramton, P., Shim, J.: The high-frequency trading arms race: Frequent batch auctions as a market design response. Q. J. Econ. **130**(4), 1547–1621 (2015)
4. Cachin, C., Mićić, J., Steinhauer, N., Zanolini, L.: Quick order fairness. In: Financial Cryptography and Data Security: 26th International Conference, FC 2022, Grenada, 2–6 May 2022, Revised Selected Papers, pp. 316–333. Springer (2022)
5. Chichilnisky, G., Heal, G.: Social choice with infinite populations: construction of a rule and impossibility results. Soc. Choice Welfare **14**, 303–318 (1997)
6. Cline, D., Dryja, T., Narula, N.: Clockwork: an exchange protocol for proofs of non front-running (2020)
7. Condorcet, M.D., Caritat, M.D.: An essay on the application of analysis to the probability of decisions rendered by a plurality of votes. Classics of Social Choice, pp. 91–112 (1785)
8. Conitzer, V., Sandholm, T.: Communication complexity of common voting rules. In: Proceedings of the 6th ACM Conference on Electronic Commerce, pp. 78–87 (2005)
9. Cullinan, J., Hsiao, S.K., Polett, D.: A borda count for partially ordered ballots. Soc. Choice Welfare **42**, 913–926 (2014)
10. Daian, P., et al.: Flash boys 2.0: frontrunning in decentralized exchanges, miner extractable value, and consensus instability. In: 2020 IEEE Symposium on Security and Privacy (SP), pp. 910–927. IEEE (2020)
11. Efimov, B.A., Koshevoy, G.A.: A topological approach to social choice with infinite populations. Math. Soc. Sci. **27**(2), 145–157 (1994)
12. Fain, B., Goel, A., Munagala, K., Prabhu, N.: Random dictators with a random referee: constant sample complexity mechanisms for social choice. In: Proceedings of the AAAI Conference on Artificial Intelligence, vol. 33, pp. 1893–1900 (2019)
13. Ferreira, M.V., Parkes, D.C.: Credible decentralized exchange design via verifiable sequencing rules. arXiv preprint arXiv:2209.15569 (2022)
14. Fishburn, P.C.: Arrow's impossibility theorem: concise proof and infinite voters. J. Econ. Theory **2**(1), 103–106 (1970)
15. Grafe, F., Grafe, J.: On arrow-type impossibility theorems with infinite individuals and infinite alternatives. Econ. Lett. **11**(1–2), 75–79 (1983)
16. Kavousi, A., Le, D.V., Jovanovic, P., Danezis, G.: Blindperm: efficient mev mitigation with an encrypted mempool and permutation. Cryptology ePrint Archive (2023)
17. Kelkar, M., Deb, S., Kannan, S.: Order-fair consensus in the permissionless setting. In: Proceedings of the 9th ACM on ASIA Public-Key Cryptography Workshop, pp. 3–14 (2022)
18. Kelkar, M., Deb, S., Long, S., Juels, A., Kannan, S.: Themis: fast, strong order-fairness in byzantine consensus. In: Proceedings of the 2023 ACM SIGSAC Conference on Computer and Communications Security, pp. 475–489 (2023)

19. Kelkar, M., Zhang, F., Goldfeder, S., Juels, A.: Order-fairness for byzantine consensus. In: Micciancio, D., Ristenpart, T. (eds.) CRYPTO 2020. LNCS, vol. 12172, pp. 451–480. Springer, Cham (2020). https://doi.org/10.1007/978-3-030-56877-1_16
20. Kiayias, A., Leonardos, N., Shen, Y.: Ordering transactions with bounded unfairness: definitions, complexity and constructions. In: Annual International Conference on the Theory and Applications of Cryptographic Techniques, pp. 34–63. Springer (2024)
21. Kirman, A.P., Sondermann, D.: Arrow's theorem, many agents, and invisible dictators. J. Econ. Theory **5**(2), 267–277 (1972)
22. Lamport, L.: Paxos made simple. ACM SIGACT News (Distributed Computing Column) 32, 4 (Whole Number 121, December 2001), pp. 51–58 (2001)
23. Li, R., Hu, X., Wang, Q., Duan, S., Wang, Q.: Transaction fairness in blockchains, revisited. Cryptology ePrint Archive (2023)
24. Li, Y., Zhang, M., Li, J., Chen, E., Chen, X., Deng, X.: MEV makes everyone happy under greedy sequencing rule. In: Proceedings of the 2023 Workshop on Decentralized Finance and Security, pp. 9–15 (2023)
25. Lu, T., Boutilier, C.: Multi-winner social choice with incomplete preferences. In: Twenty-Third International Joint Conference on Artificial Intelligence (2013)
26. Malkhi, D., Szalachowski, P.: Maximal extractable value (MEV) protection on a DAG. arXiv preprint arXiv:2208.00940 (2022)
27. Mamageishvili, A., Kelkar, M., Schlegel, J.C., Felten, E.W.: Buying time: latency racing vs. bidding for transaction ordering. In: 5th Conference on Advances in Financial Technologies (AFT 2023). Schloss-Dagstuhl-Leibniz Zentrum für Informatik (2023)
28. Oki, B.M., Liskov, B.H.: Viewstamped replication: a new primary copy method to support highly-available distributed systems. In: Proceedings of the Seventh Annual ACM Symposium on Principles of Distributed Computing, pp. 8–17 (1988)
29. Ramseyer, G., Goel, A., Mazières, D.: SPEEDEX: a scalable, parallelizable, and economically efficient decentralized exchange. In: 20th USENIX Symposium on Networked Systems Design and Implementation (NSDI 2023), pp. 849–875 (2023)
30. Tideman, T.N.: Independence of clones as a criterion for voting rules. Soc. Choice Welfare **4**(3), 185–206 (1987)
31. Vafadar, M.A., Khabbazian, M.: Condorcet attack against fair transaction ordering. In: 5th Conference on Advances in Financial Technologies (2023)
32. Yin, M., Malkhi, D., Reiter, M.K., Gueta, G.G., Abraham, I.: Hotstuff: BFT consensus with linearity and responsiveness. In: Proceedings of the 2019 ACM Symposium on Principles of Distributed Computing, PODC 2019, pp. 347–356. Association for Computing Machinery, New York (2019)
33. Zhang, H., Merino, L.H., Estrada-Galinanes, V., Ford, B.: Flash freezing flash boys: countering blockchain front-running. In: 2022 IEEE 42nd International Conference on Distributed Computing Systems Workshops (ICDCSW), pp. 90–95. IEEE (2022)

34. Zhang, M., Li, Y., Sun, X., Chen, E., Chen, X.: Computation of optimal MEV in decentralized exchanges
35. Zhang, Y., Setty, S., Chen, Q., Zhou, L., Alvisi, L.: Byzantine ordered consensus without byzantine oligarchy. In: 14th USENIX Symposium on Operating Systems Design and Implementation (OSDI 2020), pp. 633–649 (2020)

An Impossibility Result for Strongly Group-Strategyproof Multi-winner Approval-Based Voting

Ioannis Caragiannis[1], Rob LeGrand[2], Evangelos Markakis[3,4](✉), and Emmanouil Pountourakis[5]

[1] Aarhus University, Aarhus, Denmark
[2] Angelo State University, San Angelo, TX, USA
[3] Athens University of Economics and Business, Athens, Greece
markakis@gmail.com
[4] Input Output Global (IOG), Athens, Greece
[5] Drexel University, Philadelphia, PA, USA

Abstract. Multi-winner approval-based voting has received considerable attention recently, as an election format. A voting rule in this setting takes as input ballots in which each agent approves a subset of the available alternatives and outputs a committee of alternatives of a given size k. We consider the scenario when a coalition of agents can act strategically and alter their ballots so that the new outcome is strictly better for some coalition member and at least as good for anyone else in the coalition. Voting rules that are robust against this strategic behaviour are called strongly group-strategyproof. We prove that, for $k \in \{1, 2, \ldots, m-2\}$, strongly group-strategyproof multi-winner approval-based voting rules which furthermore satisfy the minimum efficiency requirement of unanimity do not exist, where m is the number of available alternatives. Our proof builds a connection to single-winner voting with ranking-based ballots and exploits the infamous Gibbard-Satterthwaite theorem to reach the desired impossibility result. Our result has implications for paradigmatic problems from the area of approximate mechanism design without money and indicates that strongly group-strategyproof mechanisms for minimax approval voting, variants of facility location, and classification can only have an unbounded approximation ratio.

Keywords: approval voting · computational social choice · mechanism design

1 Introduction

Approval voting offers a simple and easy-to-use format for running elections on multiple issues with binary domains. This may involve either committee elections among a set of candidates (which is the viewpoint adopted in this work), or elections for a set of topics that need to be decided upon simultaneously, often

M. Mavronicolas et al. (Eds.): WINE 2024, LNCS 15534, pp. 457–475, 2026.
https://doi.org/10.1007/978-3-032-08560-3_26

referred to as multiple referenda. Under such a voting format, the voters are allowed to express approval preferences, i.e., each voter can specify an approval set, with as many candidates as she likes, with the interpretation that these are the candidates she is happy with, whereas all other candidates are disapproved by her. In the last decades, many scientific societies and organizations have adopted approval voting for their council elections, including, among many others, the American Mathematical Society and the Game Theory Society.

We are interested in the strategic behavior of voters in approval-based elections. Namely, our work is centered around the design of mechanisms for selecting a committee of a given (a priori fixed) size, which falls under the broader umbrella of *mechanism design without money*. To begin with, the first goal one would pursue is to obtain *strategyproof mechanisms*, where no voter has an incentive to unilaterally misreport her approval preferences to her own benefit. The famous Gibbard-Satterthwaite theorem [12,26] is not applicable in the approval voting domain, as it requires a different space of preferences that express a strict ranking over the set of candidates. Some known extensions of this theorem to weak rankings—e.g., a result by Beja [6]—are not applicable to approval voting either. Thus, it is feasible to have strategyproof mechanisms (that are not necessarily dictatorial), as already pointed out by Caragiannis et al. [9]. As an example, for committees of size k, the classic *minisum rule* (commonly also referred to as the approval voting rule), which selects the candidates with the k highest approval scores, is strategyproof subject to appropriate tie-breaking.

Going beyond strategyproofness, several works have also examined coalitional manipulations, where a group of voters may benefit from a joint deviation. This gives rise to two distinct notions of resilience, depending on how we interpret what it means for a coalition to benefit from a joint deviation. In particular, a standard class is that of *weakly group-strategyproof* (in short, weakly GSP, or sometimes simply referred to as GSP) mechanisms, where no deviation can make all members of a coalition strictly better off. A stronger definition is that of strongly GSP mechanisms where no deviation makes some member of the coalition better off without worsening the other members. Our interest here is in the latter notion of *strong group-strategyproofness.* Even though the definitions of weak and strong group-strategyproofness may not seem to differ significantly at first sight, the current literature conveys a different picture. There are positive results on the existence of weakly GSP mechanisms with desirable properties (e.g., satisfying some form of economic efficiency or approximating some social cost objective), both in voting [9], but also in other fields of mechanism design without money, such as in facility location problems [24]. On the contrary, many of these results do not generalize for strongly GSP mechanisms, and in some cases it has remained an open problem whether comparable results are possible with strongly GSP mechanisms. This is precisely the focus of our work.

Our Results. We consider the problem of selecting a committee of a given fixed size, in elections where voters submit approval ballots. Our main result is in Sect. 3, where we show that unanimous, strongly GSP mechanisms do not exist. Unanimity is a very simple and minimal form of efficiency, requiring that if all voters approve the same committee, this should also be the outcome of the election. We view this as a severe impossibility result for strongly GSP mechanisms. Our proof is based on a construction that exploits the Gibbard-Satterthwaite theorem; even though we are interested in approval-based elections, we essentially provide a reduction from appropriately defined elections with ranking preferences, where the non-existence of strategyproof mechanisms implies the impossibility of strongly GSP mechanisms in our setting. We feel that our proof technique is of independent interest and could have further applications. As a corollary, our result also yields an incompatibility between Pareto-efficiency and strong group-strategyproofness under the approval voting format. Finally, we stress that our results are not based on any computational complexity assumption and, hence, hold for exponential-time mechanisms as well.

We proceed in Sect. 4, by demonstrating some further implications of our main result. Most importantly, we resolve one of the open questions posed by Caragiannis et al. [9], concerning the *minimax approval voting* rule, proposed by Brams et al. [7] as an alternative to the minisum rule. Namely, we prove that there is no strongly GSP mechanism that achieves a finite approximation for the minimax solution. Within the voting domain, we also obtain an impossibility result for *participatory budgeting* under approval ballots. Beyond voting, we also have implications for two other related problems. The first one is a special case of *binary classification*, where we seek a classifier that labels a fixed number of points as positive and the remaining ones as negative. Finally, the last one is a version of the *constrained facility location*, where the facilities can be placed only in particular locations.

Further Related Work. For a more complete exposition of approval voting elections and their applications, we refer to the book of Lackner and Skowron [15]. Regarding resistance to manipulation, there are several positive results for strategyproof mechanisms that generalize to weakly GSP mechanisms. In fact, this connection has been studied more thoroughly by Le Breton and Zaporozhets [16] and Barberà et al. [3], where sufficient conditions have been identified for domains in which strategyproof mechanisms coincide with weakly GSP mechanisms. On a different line of works, there have been some impossibility results when combining strategyproofness with proportional representation axioms and efficiency guarantees, as in [23]. Follow up works on this include also [10,13].

Regarding strongly GSP mechanisms in settings without monetary payments, the literature is rather scarce. We are aware of characterization results for the case of two alternatives by Manjunath [19] and by Barberà et al. [4]. In particular, for approval elections with two alternatives, Manjunath identifies essentially two voting rules that are strongly GSP and Pareto-efficient. On a slightly different direction, the work of Barberà et al. [4] establishes characterizations based on certain monotonicity properties. Finally, in terms of impossibility results, Filos-Ratsikas et al. [11] show that there are no anonymous, strongly GSP mechanisms for a facility location problem, where the goal is to place a facility on a line, under double-peaked preferences. This is a quite different domain however, and does not have any implications for our voting setting.

2 Definitions

We begin by introducing our notation. We let N denote a set of n voters (or *agents*) and A a set of m candidates (or *alternatives*). We consider elections where both the preferences and the ballots are approval-based, i.e., an agent's ballot expresses his approval for each alternative in the subset under consideration. Hence, a voting profile P is a tuple $P = (P_1, \ldots, P_n)$, where $P_i \subseteq A$ denotes the declared preference of agent i. Under this notation, an *approval election* is specified by a tuple (N, A, P). Our focus is on multi-winner elections for selecting a committee of some predetermined size. We let k denote the committee size. So, for any voting rule in our setting, the outcome must be some set $S \subseteq A$, with $|S| = k$. To distinguish with the rules discussed later in Sect. 2.2, we refer to them as *multi-winner approval-based voting rules.*

We extend the notion of *Hamming distance* to subsets of A as follows. We say that the distance between two sets Q and T is the total number of alternatives in which they differ, i.e.,

$$d(Q,T) = |Q \setminus T| + |T \setminus Q| = |Q| + |T| - 2|Q \cap T|.$$

Note that this is precisely the Hamming distance of the sets when represented as binary vectors, where the ith coordinate of each vector equals 1 if the ith alternative belongs to the set and equals 0 otherwise.

2.1 Resistance to Manipulation and Other Properties

We consider strategic agents who may misreport their preferences to the voting rule if this can be in their interest. We assume that each agent prefers committees that include as many alternatives from their approval preference as possible. Equivalently, agents evaluate a committee in terms of its Hamming distance to their approval preference (with the smaller Hamming distance, the better).

Hence, it is desirable for voting rules to be resistant to possible manipulations by the agents. Given a profile P and a voting rule R, we denote by $R(P)$ the outcome of the voting rule on profile P. We also denote by P_{-i} the preferences of all agents besides i. Hence, we can also write P as (P_i, P_{-i}). Our first property indicating resistance to manipulation is *strategyproofness*, which requires that no agent i has an incentive to unilaterally change her preference so as to reduce the distance of P_i from the outcome of the voting rule.

Definition 1. *A voting rule R is* strategyproof *(SP) if for any election (N, A, P), any agent $i \in N$, and any set of alternatives $P'_i \subseteq A$, it holds that*

$$d(P_i, R(P_i, P_{-i})) \leq d(P_i, R(P'_i, P_{-i})).$$

Moving on, we can also consider deviations by coalitions of agents and define two analogous notions of resistance to manipulation. In particular, the first such notion says that there is no deviation that can make all deviating agents strictly better off. Given a profile P, in analogy to P_i and P_{-i}, we let P_S and P_{-S} denote the preferences of a set of agents S and of all agents except S, respectively.

Definition 2. *A voting rule R is* weakly group-strategyproof *(weakly GSP) if for any election (N, A, P) and any coalition $S \subseteq N$ of agents, there is no profile P'_S of the agents in S such that*

$$d(P_i, R(P_S, P_{-S})) > d(P_i, R(P'_S, P_{-S})) \quad \forall i \in S$$

The notion that is the main focus of our work is a stronger requirement, defined as follows.

Definition 3. *A voting rule R is* strongly group-strategyproof *(strongly GSP) if for any election (N, A, P) and any coalition $S \subseteq N$ of agents, there is no profile P'_S of the agents in S such that*

$$d(P_i, R(P_S, P_{-S})) \geq d(P_i, R(P'_S, P_{-S}))$$

for every agent $i \in S$, with strict inequality for at least one agent of S.

The rationale behind this last concept is that we demand the voting rule to be resistant to coalitions in which some of the agents may change their preference profile in order to help other members of the coalition (without necessarily gaining something for themselves). Clearly, strong group-strategyproofness is a stronger notion than weak group-strategyproofness.

Apart from resistance to manipulation, we will also examine two properties related to efficiency. The first one is a very natural axiom for our setting.

Definition 4. *We say that a voting rule R for selecting a committee of size k is* unanimous *if whenever all agents approve the same set S of k alternatives (and nothing else), R outputs S as the selected committee.*

Definition 5. *A voting rule R is* Pareto-efficient *if, for every input profile P, it outputs a set S of k alternatives, such that there is no other committee S' of size k, with at least one agent being closer to S' than S, and all other agents not being further off.*

2.2 Voting by Ranking Ballots

Although our main focus is on approval voting, the proof of our main result is based on a construction that involves elections where the ballot of each agent is a strict ranking of the alternatives in decreasing order of preference. Hence, a preference profile in ranking-based voting is given by a tuple $(\succ_1, \cdots, \succ_n)$, where $\succ_i$ is the ranking of agent i on the set of alternatives A. For brevity, we use $\succ$ to denote the entire preference profile $(\succ_1, \cdots, \succ_n)$. Furthermore, for two alternatives $p, q \in A$, we will use $p \succ_i q$ to denote that agent i prefers p to q (and, hence, p appears higher than q in the ranking of i).

Given a set of agents N, a set of alternatives A, and a preference profile $\succ$, a *ranking-based election* is specified by a tuple $(N, A, \succ)$. We will consider *single-winner ranking-based voting rules*, which return a single alternative as an outcome when applied to a ranking profile. Hence, for a voting rule T, we let

$T(\succ)$ denote the winning alternative of T when given the preference profile $\succ$ as input.

In analogy to Definition 1, we can also define the notion of strategyproofness in this setting as well. In the definition below, for a profile $\succ$, we denote by $\succ_{-i}$ the preference profile of all agents except i.

Definition 6. *A ranking-based voting rule T is strategyproof (SP) if for any ranking-based election $(N, A, \succ)$, any agent $i \in N$, and any ranking $\succ'_i$ of the alternatives in A, it holds that*

$$T(\succ) \succ_i T(\succ'_i, \succ_{-i}).$$

The classic impossibility result due to Gibbard and Satterthwaite, states that any SP and *onto* single-winner ranking-based voting rule for elections with at least three alternatives is a dictatorship. Here, a voting rule T is onto if, for every alternative $p \in A$, there exists a profile $\succ$ such that $T(\succ) = p$. A voting rule T is a dictatorship if there exists an agent i such that for every profile $\succ$, the outcome $T(\succ)$ is the top choice of agent i.

Theorem 1 (Gibbard [12], Satterthwaite [26]). *Any single-winner ranking-based voting rule for ranking-based elections with at least three alternatives, which is SP and onto, must be dictatorial.*

3 Our Impossibility Result

Before we embark on our impossibility result, it is instructive to start with a discussion on the existence of SP and weakly GSP voting rules. If we care only about strategyproofness, the minisum voting rule (with appropriate deterministic tie-breaking) is SP. When it comes to coalitional manipulations, it would be expected that some form of dictatorial mechanisms would be weakly, or even strongly GSP. One crucial issue here with defining a dictatorial rule is that the outcome cannot always coincide exactly with the dictator's preferences, due to the constraint on the size of the committee being exactly k. Therefore, a voting rule is dictatorial in our setting if it falls within the following class of voting rules (referred to as k-completion by LeGrand et al. [17]).

> **Dictatorial k-completion:** Pick an agent i as the dictator. If $|P_i| < k$, then output the union of P_i together with $k - |P_i|$ alternatives from $A \setminus P_i$ (selected according to some tie-breaking order); else if $|P_i| > k$, then output a subset of P_i of size k (again, according to some tie-breaking order); else return P_i.

The k-completion rules do indeed satisfy some form of resistance to coalitional manipulation, but they are still not strongly GSP.

Theorem 2 (Caragiannis et al. [9]). *Any k-completion voting rule is weakly GSP but not strongly GSP.*

The reason a k-completion rule is not always strongly GSP is that, dependent on the tie-breaking rule used, the dictator could help some other agent by changing his approval ballot and steer the rule to select a committee that makes some agents better off, while the dictator is not worse off.

A natural question, posed already a while ago by Caragiannis et al. [9], is whether non-trivial strongly GSP voting rules exist. Our main result is the following theorem, where we obtain a negative answer, as long as we demand the additional and seemingly harmless property of unanimity.

Theorem 3. *Consider approval elections with m alternatives, where the outcome must always be a committee of size k, with $k \in [m-2]$, and $n \geq 3(m-k)$ agents. Then, there is no multi-winner approval-based voting rule that is both strongly GSP and unanimous.*

The remaining section is devoted to the proof of this result. We start with the following lemma, which connects unanimity, strong group-strategyproofness, and Pareto-efficiency.

Lemma 1. *Any strongly GSP and unanimous multi-winner approval-based voting rule is Pareto-efficient.*

Proof. Assume that the voting rule is not Pareto-efficient. Then, there exists an approval election (N, A, P), for which the voting rule returns a k-sized committee of alternatives K, while there exists another k-sized committee K' such that $d(K', P_{i^*}) < d(K, P_{i^*})$ for some agent $i^* \in N$ and $d(K', P_i) \leq d(K, P_i)$ for any other agent $i \in N$. But then, all agents have an incentive to form a coalition and misreport their preferences by simply reporting the set of alternatives K'. In that case, by the unanimity property, K' should be selected as the outcome. The agent i^* would then become strictly better off, and the rest of the agents would not be worse off. This would contradict the fact that our voting rule is strongly GSP. □

Using Lemma 1, we will prove Theorem 3 by showing that no Pareto-efficient multi-winner approval-based voting rule is strongly GSP. We first provide a high-level outline of our proof. We believe that the technique can be of independent interest, potentially useful for establishing other impossibility results as well. The backbone of the proof is a reduction that transforms a ranking-based election (as defined in Sect. 2.2) into an approval election, so that any Pareto-efficient multi-winner approval-based voting rule for the latter is mapped naturally to a single-winner ranking-based voting rule for the former. We show that if the multi-winner approval-based voting rule is furthermore strongly GSP, then the single-winner ranking-based voting rule is SP and onto. Then, the Gibbard-Satterthwaite theorem (Theorem 1) implies certain "dictatorship-like" properties for the multi-winner voting rule, eventually leading to a contradiction.

Definition 7. *[Construction of an approval election from a ranking-based election] Given a parameter k and a ranking-based election $I' = (N', A', \succ)$ with $|N'| = n'$ and $|A'| = m'$, we construct a corresponding approval election $I = (N, A, P(\succ))$, as follows:*

- *For every agent $i \in N'$, we introduce $m' - 1$ different copies of agents, indexed as $(i, 1), \ldots, (i, m'-1)$. Hence, the approval-based election has a total of $n = (m' - 1) \cdot n'$ agents.*
- *The set of alternatives is $A = A' \cup D$, where D consists of $k - 1$ dummy alternatives. Hence, the approval election has a total of $m = |A| = m' + k - 1$ alternatives.*
- *For $i \in [n']$ and $j \in [m'-1]$, agent (i, j) has the approval ballot $P_{(i,j)}$ defined as*

$$P_{(i,j)} = \{\textit{top}\, j \textit{ ranked alternatives in } \succ_i\} \cup D$$

Hence, the approval profile is given by $P(\succ) = \{P_{(i,j)}\}_{i \in [n'], j \in [m'-1]}$.

Example 1. Before we proceed, we demonstrate the construction with a simple example, when n' and m' are small. Consider the ranking-based election $(N', A', \succ)$ with $N' = \{1, 2\}$, $A' = \{x, y, z\}$, and with $\succ$ defined as $x \succ_1 y \succ_1 z$ and $y \succ_2 z \succ_2 x$. The reduction of Definition 7 for $k = 2$ gives an approval election with four agents $N = \{(1, 1), (1, 2), (2, 1), (2, 2)\}$, and four candidates $A = \{x, y, z, d_1\}$. Here, we have only one dummy candidate. The approval ballots are: $P_{(1,1)} = \{x, d_1\}$, $P_{(1,2)} = \{x, y, d_1\}$, $P_{(2,1)} = \{y, d_1\}$, and $P_{(2,2)} = \{y, z, d_1\}$.

Now, let $(N', A', \succ)$ be a ranking-based election and let $(N, A, P(\succ))$ be the corresponding approval election defined by our reduction in Definition 7. Also, let R be a Pareto-efficient multi-winner approval-based voting rule which outputs the committee $R(P(\succ))$ of size k when applied on the profile $P(\succ)$. We will show that $R(P(\succ)) \setminus D$ is a singleton and will use it to define a corresponding single-winner ranking-based voting rule for the ranking-based election.

Lemma 2. *If the multi-winner approval-based voting rule R is Pareto-efficient, then $R(P(\succ)) \setminus D$ is a singleton.*

Proof. For the sake of contradiction, assume that $|R(P(\succ)) \setminus D| \geq 2$. Note that for every agent $i \in N'$, agent $(i, 1)$ of N approves exactly one alternative outside D. Hence, there is at least one alternative in the outcome $R(P(\succ))$, say a^*, that this agent does not approve. Moreover, each alternative in $D \setminus R(P(\succ))$ is approved by all agents. Now, consider replacing a^* in the outcome of R by some alternative in $D \setminus R(P(\succ))$. Agent $(i, 1)$ is strictly better off now, and all other agents are not worse off. Hence, R is not Pareto-efficient, a contradiction. □

Given Lemma 2, we now define the single-winner ranking-based voting rule T, which returns the alternative contained in $R(P(\succ)) \setminus D$ on input the profile of rankings $\succ$. We refer to T as the rule induced by the approval-based rule R. We will show that if the multi-winner approval-based voting rule R is both Pareto-efficient and strongly GSP, the rule T satisfies the premises of the Gibbard-Satterthwaite theorem (Theorem 1).

Lemma 3. *If R is a strongly GSP and Pareto-efficient multi-winner approval-based voting rule, the induced single-winner ranking-based voting rule T is onto and SP.*

Proof. We first prove that T is onto, i.e., for every alternative $a \in A'$, there exists $\succ$ such that $T(\succ) = a$. We construct $\succ$ such that a is the top alternative in every agent's ranking. Note that, in the corresponding approval voting profile $P(\succ)$ defined by our reduction in Definition 7, the only Pareto-efficient committee of size k is $D \cup \{a\}$. Hence, this must be the outcome of R, and this means that T is equal to the single alternative in $R(\succ) \setminus D$, i.e., alternative a.

Second, we show that T is SP. Assume towards a contradiction that T is not SP, i.e., there exists a preference profile $\succ$, some agent i, and some ranking of alternatives $\succ_i'$ such that

$$T(\succ_i', \succ_{-i}) \succ_i T(\succ_i, \succ_{-i}). \tag{1}$$

By our reduction, the approval ballots of the profiles $P(\succ_i, \succ_{-i})$ and $P(\succ_i', \succ_{-i})$ differ only in the agents $(i, 1), \ldots, (i, m' - 1)$. Hence, the main idea is to show that the unilateral deviation to $\succ_i'$ by agent i in the ranking-based election corresponds to successful deviations by the coalition of agents $(i, 1), \ldots, (i, m' - 1)$ in the approval-based election, from profile $P(\succ)$ to profile $P(\succ_i', \succ_{-i})$.

Let $a = T(\succ_i, \succ_{-i})$ and $b = T(\succ_i', \succ_{-i})$. Suppose that according to $\succ_i$, alternative a is in position r and b is in position r'. Observe that Equation (1) implies $r' < r$. Since we assumed that, under rule T, b is the winner in profile $(\succ_i', \succ_{-i})$ and a is the winner in profile $\succ$, this means that the outcome of rule R in the approval voting election is $D \cup \{a\}$ for profile $P(\succ)$ and $D \cup \{b\}$ for profile $P(\succ_i', \succ_{-i})$. Note that the agents (i, j) with $r' \leq j < r$ approve alternative b but do not approve alternative a. Hence, their distance from the outcome of R decreases in profile $P(\succ_i', \succ_{-i})$ compared to profile $P(\succ)$. The remaining agents in the coalition are indifferent, since either they approve both a and b (this holds for agents (i, j) with $j \geq r$, if any), or they approve neither of them (this holds for agents (i, j) with $j < r'$, if any). Hence, we have constructed a deviation, where at least one member of the coalition is better off, and some are indifferent. This contradicts the assumption that the multi-winner approval-based voting rule R is strongly GSP. Therefore, the single-winner ranking-based voting rule T is SP. □

Notice that, according to our reduction in Definition 7, if the number of alternatives in the approval election is at least $k + 2$, then the number of alternatives in the ranking-based election is at least 3. Thus, by Lemmas 2, 3, and Theorem 1, we get the following corollary.

Corollary 1. *Let R be a Pareto-efficient and strongly GSP multi-winner approval-based voting rule for approval elections with $m \geq 3$ alternatives, $k \in [m - 2]$, and $n \geq 3(m - k)$ agents. Then, the induced single-winner ranking-based voting rule T is a dictatorship, i.e., there exists an agent i^* such that for every profile $\succ$, $T(\succ)$ is the top preference of agent i^*.*

The fact that T is a dictatorship implies that R also has some dictatorship-like attributes. We will show that this contradicts the fact that R is strongly GSP, which implies that the intersection of strongly GSP and Pareto-efficient multi-winner approval-based voting rules is empty.

Consider a unanimous (and, by Lemma 1, Pareto-efficient) and strongly GSP multi-winner approval-based voting rule R which outputs committees of size $k \in [m-2]$ for elections with $m \geq 3$ alternatives. We use the reduction of Definition 7, with parameters[1] k, $m' = m - k + 1 \geq 3$, and $n' = 3$, and let T be the single-winner ranking-based voting rule induced by R. By Corollary 1, T is a dictatorship, and without loss of generality, let us assume that the dictator is agent 1.

Now, let x and y be two specific alternatives of A' and consider the rankings $\succ^x$ and $\succ^y$ of the alternatives of A', which differ only in the two top positions, defined as follows. Ranking $\succ^x$ has alternative x first, alternative y second, and then the alternatives of $A' \setminus \{x, y\}$ in some arbitrary order. Ranking $\succ^y$ has alternative y first, alternative x second, and then the alternatives of of $A' \setminus \{x, y\}$ in the same order with $\succ^x$. Based on these, we will construct three distinct approval profiles and argue about their outcome under R, that will eventually lead to a contradiction.

Consider first the ranking profiles $(\succ^x, \succ^x, \succ^y)$ and $(\succ^y, \succ^x, \succ^y)$. The corresponding approval profiles $P_x = P(\succ^x, \succ^x, \succ^y)$ and $P_y = P(\succ^y, \succ^x, \succ^y)$ differ only in the ballot of agent $(1, 1)$, which is $\{x\} \cup D$ in P_x and $\{y\} \cup D$ in P_y. Among the remaining agents in profiles P_x and P_y, the ones that are of interest for the arguments below are agent $(2, 1)$ with approval ballot $\{x\} \cup D$ and agent $(3, 1)$ with approval ballot $\{y\} \cup D$. By the definition of our reduction, the fact that T is a dictatorship of agent 1, and the relation between the voting rules R and T, we have that $R(P_x) = \{x\} \cup D$ and $R(P_y) = \{y\} \cup D$.

Now, consider a third approval profile P_{xy}, differing from P_x and P_y only in the ballot of agent $(1, 1)$ which is now $\{x, y\} \cup D$. For convenience, the three profiles are depicted in Table 1, when $m' = 3$. We remark that this profile is not produced by our reduction. Still, the Pareto-efficiency of voting rule R, implies that $D \subset R(P_{xy})$, since the alternatives in D appear in the approval ballot of all agents in P_{xy}. We distinguish between two cases regarding the single alternative in $R(P_{xy}) \setminus D$.

First, if $R(P_{xy}) \setminus D \neq \{x\}$, then the agents $(1, 1)$ and $(2, 1)$, with ballots $\{x, y\} \cup D$ and $\{x\} \cup D$ under P_{xy}, have Hamming distance at least 1 and exactly 2 respectively, from the k-sized committee $R(P_{xy})$. The deviation of these two agents to approval ballots $\{x\} \cup D$ for both yields the approval profile P_x, for which rule R outputs the committee $\{x\} \cup D$. Its Hamming distance from the approval ballots of agents $(1, 1)$ and $(2, 1)$ in profile P_{xy} is only 1 and 0, respectively, implying the existence of a successful deviating coalition that contradicts the assumption that voting rule R is strongly GSP.

It remains to consider the case $R(P_{xy}) = \{x\} \cup D$. Then, the agents $(1, 1)$ and $(3, 1)$, with ballots $\{x, y\} \cup D$ and $\{y\} \cup D$ under P_{xy}, have Hamming distance 1 and 2 from the committee $R(P_{xy})$ respectively. Their deviation to ballot $\{y\} \cup D$ for both yields the approval profile P_y, for which the rule R outputs the k-sized

[1] These parameters introduce the restriction of $n \geq 3(m-k)$ in the statement of Theorem 3. Actually, our construction yields exactly $3(m-k)$ agents in the approval-based profile. Additional indifferent agents may be added to get higher values of n.

committee $\{y\} \cup D$. Hence, this will yield a Hamming distance of 1 and 0 from the approval ballots of agents $(1,1)$ and $(3,1)$ in the approval profile P_{xy}. This again implies the existence of a successful deviating coalition, contradicting the assumption that R is strongly GSP.

We conclude that R cannot be Pareto-efficient (and, hence, unanimous) and strongly GSP, completing the proof of Theorem 3.

Table 1. An example of the profiles used in the final step of the proof of Theorem 3, when the ranking-based election we start from has $m' = 3$ alternatives. The profiles differ only in the approval ballot of agent $(1,1)$ and any Pareto-efficient outcome of the rule R (i.e., either $\{x\} \cup D$ or $\{y\} \cup D$) on profile P_{xy} violates strong group-strategyproofness.

agent	P_x	P_y	P_{xy}
$(1,1)$	$\{x\} \cup D$	$\{y\} \cup D$	$\{x,y\} \cup D$
$(1,2)$	$\{x,y\} \cup D$	$\{x,y\} \cup D$	$\{x,y\} \cup D$
$(2,1)$	$\{x\} \cup D$	$\{x\} \cup D$	$\{x\} \cup D$
$(2,2)$	$\{x,y\} \cup D$	$\{x,y\} \cup D$	$\{x,y\} \cup D$
$(3,1)$	$\{y\} \cup D$	$\{y\} \cup D$	$\{y\} \cup D$
$(3,2)$	$\{x,y\} \cup D$	$\{x,y\} \cup D$	$\{x,y\} \cup D$
outcome	$\{x\} \cup D$	$\{y\} \cup D$	?

4 Implications

We now discuss some implications of our impossibility result in other settings. We begin by answering a question by Caragiannis et al. [9] about the "approximability" of the minimax approval voting rule. Then, we discuss three problems, which are generalizations of multi-winner approval-based voting (even though some may seem different at first glance).

4.1 Approximating Minimax Approval Voting

Minimax approval voting has been proposed by Brams et al. [7], who argued about its use in comparison to the minisum rule that is mostly adopted in practice. In particular, the k-minimax approval voting rule takes as input a profile of approval ballots and returns a k-sized committee of alternatives that minimizes the maximum Hamming distance from the ballots. Unfortunately, as proved by LeGrand et al. [17], the rule has two important drawbacks. The first is that computing the winning committee is an NP-hard problem, and the second is that the rule is not SP.

As a relaxation, LeGrand et al. [17], Caragiannis et al. [9], and Byrka and Sornat [8] studied *approximate mechanisms* for k-minimax approval voting. Given an approval election (N, A, P) and a k-sized committee C, let

$D(C, P) = \max_{i \in N} d(C, P_i)$ denote the maximum Hamming distance of C from the approval ballots of P. Also, let $C^* \in \arg\min_K D(K, P)$ be a k-sized committee that would be returned by the k-minimax approval voting rule on input the election (N, A, P). A voting rule R approximates the k-minimax solution within a factor of $\rho \geq 1$, if $D(R(P), P) \leq \rho \cdot D(C^*, P)$ for every profile P. The quantity ρ is called the approximation ratio of mechanism R.

The papers by LeGrand et al. [17] and Caragianis et al. [9] present SP and weakly GSP voting rules, which approximate k-minimax approval voting within a factor of $3 - \frac{2}{k+1}$. Among them, the minisum rule with appropriate tie-breaking is SP, whereas the class of k-completion voting rules, presented in Sect. 3, is weakly GSP. The question of whether such an approximation factor can be achieved by a strongly GSP mechanism was left open.

To resolve this open question using our impossibility result, we make first the following claim.

Lemma 4. *A non-unanimous voting rule has an infinite approximation ratio for k-minimax approval voting.*

Proof. Consider such a non-unanimous rule. Non-unanimity implies that there is a profile P in which all agents approve the same k-sized committee of alternatives, say C^*, but the rule returns some different committee, say C. Clearly, $D(C^*, P) = 0$, i.e., the minimax solution has a Hamming distance of 0 to all voters, whereas $D(C, P) > 0$ in this case. This means that the rule has an infinite approximation ratio. □

The following statement now follows due to Theorem 3.

Theorem 4. *Any strongly GSP approval-based multi-winner voting rule in elections with $m \geq 3$ alternatives, $k \in [m-2]$, and $n \geq 3(m-k)$ agents has an infinite approximation ratio for k-minimax approval voting.*

4.2 Participatory Budgeting with an Approval Voting Format

An important application of approval-based voting is in *participatory budgeting* [2,25]. In the most typical scenario, a municipality has a fixed budget and considers implementing several projects. Each project has a known cost, but the total cost of all projects exceeds the available budget. Thus, only a subset of projects can be selected for implementation. Participatory budgeting is used to delegate the selection of projects to the citizens. Each participating citizen casts a ballot with their preferences, and the municipality has to select, using a predefined mechanism, a subset of projects that are within the budget. Among other formats that have been used for eliciting the citizens' preferences, approval ballots seem to be very popular, not only within municipalities, but also in programs offered by blockchain communities or decentralized autonomous organizations (DAOs); e.g., see the recent related survey by Talmon [29].

Let us focus now on the case of projects with equal costs, so that for some integer k, all subsets of k projects are the maximal sets of projects that are within the budget[2]. This setting can be easily seen to be equivalent to multi-winner approval voting. In particular, we say that a mechanism is maximal if it always returns a maximally feasible set of projects for funding (the addition of any other project violates the budget constraint). For the case we are considering, a maximal mechanism would have to select a set of exactly k projects (which can be viewed as a committee of size k in the approval voting election). Assuming that each citizen aims to have as many of her favourite projects selected, our impossibility result implies that unanimity and strong group-strategyproofness are not compatible for participatory budgeting mechanisms that are maximal.

Theorem 5. *Unanimous and strongly GSP maximal mechanisms for participatory budgeting with an approval ballot format do not exist.*

We remark that Theorem 5 applies to mechanisms which allow citizens to cast ballots with arbitrarily many projects, even if their total cost exceeds the available budget. This is not an unusual practice today.[3] Whether Theorem 5 carries over for inputs, where each agent's ballot respects the budget restriction is an interesting open problem.

4.3 Classification with Shared Inputs

An important problem in *supervised learning* is, given a set of training examples that maps points from an input space to labels, to select a classifier from a concept class that is as close to the data in the training example as possible. The work of Meir et al. [21] studies strategic issues in *binary classification* by considering the following multi-agent setting with an input space $\mathcal{X}$, a concept class $\mathcal{H}$ and n agents. The concept class contains classifiers; each classifier $h \in \mathcal{H}$ is a function mapping each point of $\mathcal{X}$ to a label from $\{+, -\}$. The training set consists of a set of points $X \subseteq \mathcal{X}$ and data provided by n agents, where agent i provides a labeling $Y_i : X \rightarrow \{+, -\}$ of the points in X.

A classifier is evaluated for each agent $i \in [n]$ by her loss function ℓ_i, returning the number of data points for which the labeling of the classifier and the labeling provided by agent i differ. I.e., for a classifier $h \in \mathcal{H}$ and agent i,

$$\ell_i(h, Y_i) = \sum_{x \in X} \mathbb{I}\{h(x) \neq Y_i(x)\}.$$

Furthermore, each agent $i \in [n]$ has a positive weight w_i, so that $\sum_{i \in [n]} w_i = 1$, indicating the importance of the agent for the classification task. A classification mechanism takes as input the dataset $Y = (Y_1, Y_2, ..., Y_n)$, which consists of the labelings provided by the agents, along with their weight vector

[2] In fact, for our purposes, the project costs do not need to be the same, as long as any set of k projects is budget feasible, and any superset is infeasible.

[3] The site https://en.wikipedia.org/wiki/List_of_participatory_budgeting_votes has information about the format of such elections, held in the last ten years.

$\mathbf{w} = (w_1, w_2, ..., w_n)$, and returns a classifier from the concept class $\mathcal{H}$. The outcome of the classification mechanism, say $h \in \mathcal{H}$, is evaluated using the global risk $L(h, Y, \mathbf{w})$, defined as

$$L(h, Y, \mathbf{w}) = \sum_{i \in [n]} w_i \cdot \ell_i(h, Y_i).$$

The well-known *empirical risk minimization* (ERM) algorithm selects a classifier minimizing the global risk, i.e.,

$$\text{erm}(Y, \mathbf{w}) \in \arg\min_{h \in \mathcal{H}} L(h, Y, \mathbf{w}).$$

Meir et al. [21] assume that agents are strategic and may decide to misreport their private data to the classification mechanism. Adapting the notions of strategyproofness and strong group-strategyproofness to this setting, we say that a classification mechanism M taking as input the weight vector $\mathbf{w}$ and the private labelings of the agents, $Y = (Y_1, ..., Y_n)$, for the set of points $X \subseteq \mathcal{X}$, is SP if there is no agent $i \in [n]$ and labeling $Y_i' : X \to \{+, -\}$ so that $\ell_i(M(Y, \mathbf{w}), Y_i) > \ell_i(M((Y_{-i}, Y_i'), \mathbf{w}), Y_i)$. The mechanism is strongly GSP if there is no subset of agents $S \subseteq [n]$, agent $i^* \in S$, and labelings $Y_i' : X \to \{+, -\}$ for $i \in S$, so that $\ell_i(M(Y, \mathbf{w}), Y_i) \geq \ell_i(M((Y_{-S}, Y_S'), \mathbf{w}), Y_i)$, for each $i \in S$ and $\ell_{i^*}(M(Y, \mathbf{w}), Y_{i^*}) > \ell_{i^*}(M((Y_{-S}, Y_S'), \mathbf{w}), Y_{i^*})$. A universally SP (resp., strongly GSP) randomized mechanism is a probability distribution over deterministic SP (resp., deterministic strongly GSP) mechanisms.

Meir et al. [21] observe that ERM is not SP and aim to design SP mechanisms which approximate the global risk of ERM. A (possibly randomized) ρ-approximate mechanism M in this context satisfies $\mathbb{E}[L(M(Y, \mathbf{w}), Y, \mathbf{w})] \leq \rho \cdot L(\text{erm}(Y, \mathbf{w}), Y, \mathbf{w})$ for every dataset Y. They present an $O(n)$-approximate SP mechanism, which is also proved to be almost best possible among all deterministic SP mechanisms. A randomized version of this mechanism, which first selects an agent proportionally to the agents' weights and then selects the classifier of minimum loss to the selected agent, is universally SP and has an approximation ratio of $3 - \frac{2}{n}$. This bound was proved to be best possible for randomized SP mechanisms by Meir et al. [20].

We now demonstrate the relation between binary classification with shared inputs and approval voting. Consider a set of m points $X \subseteq \mathcal{X}$ and the concept class $\mathcal{H}$ consisting of the classifiers that label k points of X with $+$ and the remaining with $-$, where $k \in [m-2]$. Such classifiers can make sense when there is a fixed number of points that can receive one of the two labels, due to capacity constraints. E.g., classifiers for school admission or bank loan applications fit under this framework (where only a predetermined number of the applicants will be admitted). Another natural example can be seen with taking $k = n/2$, where the goal is to separate the points into an upper and lower half. We can then think of each point as an alternative, classifiers as k-sized committees (containing the alternatives that the classifier labels with $+$), and input data provided by the agents as approval ballots, where the approval set corresponds to the alternatives labeled as + by the agent. Under this definition, the loss $\ell_i(h, Y_i)$ of agent i for classifier h is the Hamming distance between the k-sized committee corresponding to the classifier h and the approval ballot corresponding to agent i. Thus, classification for this particular concept class with agents aiming at minimizing their loss is equivalent to multi-winner approval voting with agents aiming at minimizing their Hamming distance to the winning committee.

We will use now an argument similar to Lemma 4 in Sect. 4.1, so as to obtain a negative result that extends also for randomized mechanisms. Consider a universally strongly GSP randomized mechanism M which has a strongly GSP deterministic mechanism M_R (corresponding to the multi-winner voting rule R) in its support, i.e., it calls mechanism M_R with positive probability. Then, Theorem 3 implies that R is not unanimous, and hence, there is an approval ballot profile in which all agents agree on k alternatives, but the rule R returns another k-sized committee. Equivalently, this means that there exists a dataset $Y = (Y_1, ..., Y_1)$, consisting of n copies of the same labeling Y_1, so that when the mechanism M_R takes as input Y, it returns a classifier h, producing a labeling that is different from Y_1, i.e., $\ell_i(M_R(h, Y, \mathbf{w}), Y_i) > 0$ and $L(M_R(Y, \mathbf{w}), Y, \mathbf{w}) = \sum_{i \in [n]} w_i \cdot \ell_i(M_R(Y, \mathbf{w}), Y_1) > 0$, thus, $\mathbb{E}[L(M(Y, \mathbf{w}), Y, \mathbf{w})] > 0$. In contrast, the concept class $\mathcal{H}$ contains the classifier h^* which agrees with the labeling Y_1 on the points of X and thus, $L(\text{erm}(Y, \mathbf{w}), Y, \mathbf{w}) = 0$. Therefore, the mechanism M has infinite approximation ratio.

Theorem 6. *Any universally strongly GSP randomized mechanism for binary classification with shared inputs has an infinite approximation ratio.*

We remark that we have silently assumed that the classification mechanism may take as input datasets that are not necessarily *realizable*. This means that the labeling of the points in X provided by some agent may not coincide with the labeling of any classifier from the concept class. This is crucial for proving Theorem 6. If the labeling provided by each agent is realizable by a classifier in the concept class, then the two mechanisms presented by Meir et al. [21] are strongly GSP and universally strongly GSP, respectively. A realizable dataset corresponds to an approval voting setting in which the outcome can be *any* committee of alternatives. Clearly, any dictatorship is strongly GSP in this setting. For realizable datasets, the mechanisms in [21] correspond to such dictatorships.

4.4 Constrained Facility Location in Networks

Facility location has played a key role in the field of approximate mechanism design with money [24]. The simplest version of the problem aims to locate a single facility on the line when strategic agents report their private location and have *single-peaked* preferences. These results usually exploit a famous result by Moulin [22], which characterizes the class of SP mechanisms. The results on the line have been extended to multiple dimensions [5], multiple facilities [18], facility location in networks [1,27], and have addressed constrained versions of the problem; e.g., see [14,28,30].

We focus on the constrained version as well, regarding the allowed locations for placing a facility. The problem we consider is how to locate a facility at a node of a graph, when agents report their private locations in the graph. Our objective is to locate a facility at a node selected from a predefined subset of allowable nodes so that the maximum or the total shortest-path distance of the facility from all agents is minimized. A strongly GSP mechanism here means that if a misreport by any coalition of agents results in an agent from the coalition coming closer to the facility, some other agent in the coalition will be strictly further off. We evaluate the outcome of such a mechanism in terms of the maximum or the total shortest-path distance compared against the optimal solution. We have the following negative result.

Theorem 7. *Strongly GSP mechanisms for constrained facility location in graphs have infinite approximation ratio with respect to both the maximum and the total cost objectives.*

Proof. Again, we resort to an argument analogous to Lemma 4. Consider instances, where the graph is the m-dimensional hypercube and all nodes with k 1s ($k \in [m-2]$) in their binary representation is the allowable set of nodes for placing the facility. Then, the m dimensions correspond to alternatives, each node in the allowable set corresponds to a k-sized committee, and each agent's location in the graph corresponds to an approval ballot. Clearly, the shortest-path distance between two nodes in the hypercube is equal to the Hamming distance of the corresponding approval ballots to committees. Theorem 3 then implies that every strongly GSP mechanism for locating the facility is non-unanimous. Hence, there exists an instance in which all agents have the same preference, i.e., they are all located in the same node of the allowable set, but the mechanism returns another node of the allowable set as an outcome. Hence, the optimal solution has zero cost, both for the maximum and for the total cost objective, while the mechanism returns a solution with positive cost. □

Clearly, it is again the restriction of the allowable set of nodes that leads to the impossibility. When the allowable set contains all graph nodes, dictatorships are strongly GSP and have a finite approximation ratio.

5 Discussion

Let us conclude by mentioning very briefly the few cases that escape from our impossibility. First, notice that our construction has at least six agents. On the other hand, we can easily see that the serial dictatorship mechanism—which returns a k-sized committee that has minimum distance from agent 2 among the k-sized committees of minimum distance from agent 1—is strongly GSP for approval elections with two agents. The case of three, four, and five agents is left open. Furthermore, our construction creates profiles with $m \geq 3$ alternatives and uses committee size $k \in [m-2]$. Also, recall that, for two alternatives, Manjunath [19] provides a positive result with his consensus voting rules. So, the case $m \geq 3$ and $k = m - 1$ is left mysteriously open. Exploring whether our impossibility carries over even for the simplest among these cases with three alternatives and committees of size 2 would require a different construction.

Regarding some further directions for future research, note that the implications of our impossibility results in Sect. 4 is for the existence of finite multiplicative approximations for the problems that we study there. This does not exclude the existence of strongly GSP mechanisms with good additive approximation guarantees. Yet another, very interesting problem is the characterization of weakly GSP mechanisms. This however appears to be more challenging. We note that in contrast to our impossibility result, weakly GSP and unanimous voting rules do exist, such as the k-completion rules described in Sect. 3.

Acknowledgments. Research supported by the framework of the H.F.R.I call "Basic research Financing (Horizontal support of all Sciences)" under the National Recovery and Resilience Plan "Greece 2.0" funded by the European Union âĂŞ NextGenerationEU (H.F.R.I. Project Number: 15877).

References

1. Alon, N., Feldman, M., Procaccia, A.D., Tennenholtz, M.: Strategyproof approximation of the minimax on networks. Math. Oper. Res. **35**(3), 513–526 (2010)
2. Aziz, H., Shah, N.: Participatory budgeting: models and approaches. In: Rudas, T., Péli, G. (eds.) Pathways Between Social Science and Computational Social Science: Theories, Methods, and Interpretations, pp. 215–236. Springer (2021)
3. Barberà, S., Berga, D., Moreno, B.: Individual versus group strategy-proofness: when do they coincide? J. Econ. Theory **145**(5), 1648–1674 (2010)
4. Barberà, S., Berga, D., Moreno, B.: Group strategy-proof social choice functions with binary ranges and arbitrary domains: characterization results. Internat. J. Game Theory **41**, 791–808 (2011)
5. Barberà, S., Gul, F., Stacchetti, E.: Generalized median voter schemes and committees. J. Econ. Theory **61**(2), 262–289 (1993)
6. Beja, A.: Arrow and gibbard-satterthwaite revisited: extended domains and shorter proofs. Math. Soc. Sci. **25**(3), 281–286 (1993)
7. Brams, S.J., Kilgour, D.M., Sanver, M.R.: A minimax procedure for electing committees. Public Choice **132**(3/4), 401–420 (2007)

8. Byrka, J., Sornat, K.: PTAS for minimax approval voting. In: Proceedings of the 10th International Conference on Web and Internet Economics, WINE 2014, pp. 203–217 (2014)
9. Caragiannis, I., Kalaitzis, D., Markakis, E.: Approximation algorithms and mechanism design for minimax approval voting. In: Proceedings of the 24th AAAI Conference on Artificial Intelligence, AAAI 2010, pp. 737–742 (2010)
10. Delemazure, T., Demeulemeester, T., Eberl, M., Israel, J., Lederer, P.: Strategyproofness and proportionality in party-approval multiwinner elections. In: 37th AAAI Conference on Artificial Intelligence, AAAI 2023, pp. 5591–5599 (2023)
11. Filos-Ratsikas, A., Li, M., Zhang, J., Zhang, Q.: Facility location with double-peaked preferences. Auton. Agents Multi Agent Syst. **31**(6), 1209–1235 (2017)
12. Gibbard, A.: Manipulation of voting schemes. Econometrica **41**(4), 587–602 (1973)
13. Kluiving, B., de Vries, A., Vrijbergen, P., Boixel, A., Endriss, U.: Analysing irresolute multiwinner voting rules with approval ballots via SAT solving. In: 24th European Conference on Artificial Intelligence, ECAI 2020, vol. 325, pp. 131–138. IOS Press (2020)
14. Kyropoulou, M., Ventre, C., Zhang, X.: Mechanism design for constrained heterogeneous facility location. In: Proceedings of the 12th International Symposium on Algorithmic Game Theory, SAGT 2019, pp. 63–76 (2019)
15. Lackner, M., Skowron, P.: Multi-Winner Voting with Approval Preferences. Springer Briefs in Intelligent Systems. Springer (2023)
16. Le Breton, M., Zaporozhets, V.: On the equivalence of coalitional and individual strategy-proofness properties. Soc. Choice Welfare **33**(2), 287–309 (2009)
17. LeGrand, R., Markakis, E., Mehta, A.: Some results on approximating the minimax solution in approval voting. In: Proceedings of the 6th International Joint Conference on Autonomous Agents and Multiagent Systems, AAMAS 2007, pp. 1193–1195 (2007)
18. Lu, P., Sun, X., Wang, Y., Zhu, Z.A.: Asymptotically optimal strategy-proof mechanisms for two-facility games. In: Proceedings of the 11th ACM Conference on Electronic Commerce, EC 2010, pp. 315–324 (2010)
19. Manjunath, V.: Group strategy-proofness and voting between two alternatives. Math. Soc. Sci. **63**(3), 239–242 (2012)
20. Meir, R., Almagor, S., Michaely, A., Rosenschein, J.S.: Tight bounds for strategyproof classification. In: Proceedings of the 10th International Conference on Autonomous Agents and Multiagent Systems, AAMAS 2011, pp. 319–326 (2011)
21. Meir, R., Procaccia, A.D., Rosenschein, J.S.: Algorithms for strategyproof classification. Artif. Intell. **186**, 123–156 (2012)
22. Moulin, H.: On strategy-proofness and single peakedness. Public Choice **35**(4), 437–455 (1980)
23. Peters, D.: Proportionality and strategyproofness in multiwinner elections. In: Proceedings of the 17th International Conference on Autonomous Agents and Multi-Agent Systems, AAMAS 2018, pp. 1549–1557 (2018)
24. Procaccia, A.D., Tennenholtz, M.: Approximate mechanism design without money. ACM Trans. Econ. Comput. **1**(4), 18:1–18:26 (2013)
25. Rey, S., Maly, J.: The (computational) social choice take on indivisible participatory budgeting. CoRR abs/2303.00621 (2023)
26. Satterthwaite, M.A.: Strategy-proofness and arrow's conditions: existence and correspondence theorems for voting procedures and social welfare functions. J. Econ. Theory **10**(2), 187–217 (1975)
27. Schummer, J., Vohra, R.V.: Strategy-proof location on a network. J. Econ. Theory **104**(2), 405–428 (2002)

28. Sui, X., Boutilier, C.: Approximately strategy-proof mechanisms for (constrained) facility location. In: Proceedings of the 14th International Conference on Autonomous Agents and Multiagent Systems, AAMAS 2015, pp. 605–613 (2015)
29. Talmon, N.: Social choice around decentralized autonomous organizations: on the computational social choice of digital communities. In: Proceedings of the 22nd International Conference on Autonomous Agents and Multiagent Systems, AAMAS 2023, pp. 1768–1773 (2023)
30. Walsh, T.: Strategy proof mechanisms for facility location at limited locations. In: Proceedings of the 18th Pacific Rim International Conference on Artificial Intelligence, PRICAI 2021, pp. 113–124 (2021)

LP-Based Control for Network Revenue Management Under Markovian Demands

Haixiang Lan[1], Guillermo Gallego[2], Zizhuo Wang[2(✉)], and Yinyu Ye[2,3]

[1] Columbia University, New York, NY, USA
hl3725@columbia.edu
[2] The Chinese University of Hong Kong, Shenzhen, Guangdong, China
{gallegoguillermo,wangzizhuo,yinyuye}@cuhk.edu.cn
[3] Shanghai Jiao Tong University, Shanghai, China

Abstract. We study a network revenue management (NRM) problem, where the seller irrevocably accepts or rejects customers' requests to maximize the total revenue over a finite time horizon subject to limited resources. We consider the setting where customer arrivals are dependent and follow a Markov chain choice model. We propose a deterministic linear program (DLP) which serves as an upper bound of this problem, based on which randomized allocation policies are designed. Specifically, when the customer's arrival distribution is known, we design a static probabilistic allocation (SPA) policy and show that the regret, measured by the difference between its expected revenue and the DLP's optimal value, scales as the square root of the time horizon. When the customer's arrival distribution is unknown, we design an online probabilistic allocation (OPA) policy by sequentially learning the parameters, achieving a similar regret under mild conditions. We show that the asymptotic regrets of our algorithms are optimal in the respective settings. Furthermore, we conduct numerical experiments to validate the superior performance of the proposed algorithms. Finally, we extend our model when customers have finite patience and design an adaptive probabilistic allocation (APA) policy with similar theoretical guarantees.

Keywords: Network Revenue Management · Discrete Choice Models · Online Learning

1 Introduction

The *network revenue management* (NRM) problem is a classical challenge in dynamic decision-making, with a wide range of applications in airlines, hotels, and online platforms. The problem involves maximizing the revenue with limited resources and time. Specifically, multiple types of resources have finite capacity over a finite time horizon. Customers arrive sequentially over time and make requests for some products. Each product consumes a certain combination of resources. For each period, a seller must irrevocably decide whether to accept or reject each customer's request subject to the availability of current resources.

M. Mavronicolas et al. (Eds.): WINE 2024, LNCS 15534, pp. 476–493, 2026.
https://doi.org/10.1007/978-3-032-08560-3_27

If the seller accepts the request for a product, then the corresponding resources are consumed and the customer pays a price for the product. Otherwise, no resource is consumed, and no revenue is gained. The remaining resources have no salvage value at the end of the time horizon. The seller's goal is to maximize the expected revenue during the time horizon.

In classical NRM problems, it is usually assumed that each customer is only interested in one product. If the seller rejects the customer's request, then the customer will leave without purchase. Such a model is usually referred to as the independent demand model. However, in practice, a customer may switch to request another product if an earlier request is rejected. For example, in an airline's context, if a customer wants to buy a certain multi-leg flight for her vacation and finds that the flight is not available (or a certain fare class is not available), then she[1] may change to search for an alternative date or route. In an online advertisement context, if an advertiser is informed that a certain combination of keywords at a certain price is unavailable, then the advertiser may change her set of keywords to bid or change her price (or change the time to display). In either of such contexts, customers' demands are no longer independent but follow a certain stochastic process. We next present a real-world instance that reflects the above-mentioned scenario.

Railway Ticket. In this example, each "product" is a consecutive combination of multiple legs. The platform will display all products' information online. Customers have preferences for products based on departure and arrival times. After clicking the product, which is regarded as sending a request, the customers can see further availability information. The platform can decide to accept or reject customers' requests for products. If the platform accepts the request, then the product will be available (see "Book" status in Fig. 1). Otherwise, the request is rejected and the product will be made unavailable (see "Sold out" status in Fig. 1). The unavailable products are not always sold out since the platform may strategically save resources for potential customers who pay more and travel for longer distances. By rejecting certain requests, the platform can implicitly lead customers to transition to other products such that the resources can be utilized more efficiently. In this example, different fare classes of the two trains, Z223 and Z8981, can be considered as products, and customers can transition among these products. The platform needs to carefully decide whether to accept or reject customers' requests for different products based on the time horizon, remaining resources and customer behaviors.

In this work, we aim to study such a network revenue management problem under a dependent demand model, where the seller's decisions are intertwined with customers' choice behavior. Particularly, we would like to shed light on the following questions:

1. How do we model customers' dependent demands in the NRM problem?
2. How can we design practical policies to maximize the expected revenue with strong theoretical guarantees when customers choice model is known?

[1] We refer to the seller as "he/his" and the customer as "she/her".

(a) All information (b) Product 1 information (c) Product 2 information

Fig. 1. Railway Ticket Example.

3. When the customer's demand model is unknown *a priori*, how to design a policy that can learn the parameters of the problem while optimizing the revenue?

To answer the first question, we adopt the Markov chain choice model proposed by [7]. In the Markov chain choice model, each customer arrives with an initial request for a certain product. If the customer's request for the product is rejected by the seller, then the customer will transition to request another product according to a Markov chain or she may leave the marketplace without making a purchase. In this way, customers transition among products until their requests are accepted or they choose the no-purchase option. Note that this model is a strict generalization of the independent demand model.

Under the aforementioned model, we study the optimal decision problem for the seller. To that end, we consider a deterministic linear program (DLP), which can be viewed as a fluid approximation of the original problem and provide an upper bound for the expected revenue of the optimal policy. Based on the DLP, we design a simple randomized policy which accepts the request for each product with a certain probability and show that the randomized policy is asymptotically optimal. Specifically, we define the notion of *regret* as the performance measure of our policy, which is the difference between our policy's expected revenue and the DLP's optimal value. We show that the regret of the randomized policy scales in the square root of the time horizon.

In reality, the parameters of our model, i.e., the initial arrival probabilities of each product and the transition probabilities between products, may be unknown *a priori*. In such a setting, the seller needs to simultaneously (i) learn customers' preferences based on the collected purchase and transition data, and (ii) maximize the cumulative revenue by making online acceptance or rejection decisions while satisfying the resource constraints. To achieve such an exploration-exploitation tradeoff, we design an online estimate-then-calibrate randomized policy which achieves an optimal order of regret up to a logarithmic factor under some mild conditions. Furthermore, we consider a case when customers have limited patience levels, i.e., they can only make a certain number of transitions before they leave the marketplace. Using similar ideas, we propose an allocation policy that achieves near-optimal performance. We also conduct numerical experiments to validate the performance of the proposed policies.

Before proceeding, we introduce notations to be used throughout. We use the symbol $[n]$ as a shorthand for $\{1, 2, ..., n\}$ and $|\cdot|$ to denote the absolute value of a number or the cardinality of a set. For non-negative sequences $\{a_n\}_{n=1}^{\infty}$ and $\{b_n\}_{n=1}^{\infty}$, we write $a_n \lesssim b_n$ or $a_n = \mathcal{O}(b_n)$ or $b_n = \Omega(a_n)$ if there exists a universal positive constant C such that $a_n \leq C b_n$ for all n. We write $a_n = \tilde{\mathcal{O}}(b_n)$ if $a_n = \mathcal{O}(b_n \text{poly}(\log n))$ where $\text{poly}(x)$ is a polynomial function of x. For two vectors or matrices $\boldsymbol{a}$ and $\boldsymbol{b}$, we use $\boldsymbol{a} \leq \boldsymbol{b}$ to denote component-wise inequalities. We use $\|\cdot\|$ to denote 2-norm of a vector.

The rest of the paper is organized as follows. Section 2 reviews the literature that is relevant to our work. Based on the model and formulation in Sect. 3, we propose a policy under known parameters in Sect. 4 and a policy under unknown parameters in Sect. 5 with near-optimal performance guarantees. We conduct numerical experiments in Sect. 6. In Sect. 7, we extend our model and analysis to the case where customers have finite patience levels. We conclude our work in Sect. 8.

2 Literature Review

Our research is related to discrete choice models, assortment optimization, network revenue management, and online learning. In the following, we review related works in each of these areas.

Markov Chain Choice Model. We adopt the Markov chain choice model in this work. The Markov chain choice model has gained interests in the literature in recent years as an emerging way to capture customer choice behavior. [38] initially mentions such customer choice process in an airline revenue management context. [7] formally introduces the Markov chain choice model and shows that the assortment optimization under the model can be efficiently solved in polynomial time. [13] further solves the assortment problem, as well as the single resource/network revenue management problems by reducing the problem to efficient linear programs.

In this paper, we also adopt the Markov chain choice model as our base model. However, unlike in the previous literature where the transitions of the customers and the decisions of the seller are separated, in our model, the transition of customer is moderated by the decision of the seller, and we study an online learning setting in which the parameters of the model need to be learned while making decisions.

Parameter Estimation in Markov Chain Choice Model. In reality, the initial arrival probability and transition probability of customers may not be known in advance and may need to be estimated by empirical data. There has been much literature addressing this estimation problem. For example, [7] proposes an estimation algorithm by offering all-product assortment and all-but-one assortments, and estimates parameters by observing purchased products each time. [20] extends the algorithm to offering more flexible and realistic assortments. [35] designs an expectation-maximization (EM) algorithm to estimate the parameter by maximizing the log-likelihood function, which is not guaranteed to be statistically consistent nor even converge to a limit. [28] designs an assortment-based estimation algorithm with high probability concentration bounds under mild parameter space assumptions. [17] uses a forward and backward greedy heuristic to directly solve the assortment optimization based on direct comparisons of revenues without parameter estimation. However, the parameters and corresponding optimal policies are estimated by offering assortments and observing purchasing probabilities. Besides offering assortments, [21] estimates the transition matrix from a single sample path with statistical guarantees. In our paper, we use similar ideas to estimate parameters of our model based on customers' transition paths. However, we focus on the online decision making problem with possible moderations of customer's choices.

Online Assortment Optimization. As a topic in online decision making, online assortment optimization has been extensively studied in the recent literature. The information of customer's behavior may not be known in advance and need to be learned over the whole time horizon, incurring an exploration-exploitation trade-off. Furthermore, decisions are made sequentially and irrevocably, thus requiring adaptive strategies. The problem under the multinomial logit (MNL) model is first studied and the corresponding problem is called the MNL-bandit problem. For example, [32] and [34] adopt explore-then-commit policies to show asymptotic $\mathcal{O}(N^2 \log^2 T)$ and $\mathcal{O}(N \log T)$ regret bounds respectively. [1] uses Thompson sampling algorithm to get an $\mathcal{O}(\sqrt{NT} \log T)$ regret bound. [2] further uses the upper confidence bound (UCB) algorithm to get an $\mathcal{O}(\sqrt{NT \log(NT)})$ regret bound. [11] proves a tight $\Omega(\sqrt{NT})$ lower bound for capacitated MNL-bandit problems. [12] finally proposes an optimal policy with a $\Theta(\sqrt{T})$ regret (independent of N) for uncapacitated MNL-bandit problems.

Two closely related works about online assortment optimization under Markov chain choice model are [17] and [28]. Both adopt the explore-then-commit strategy. Specifically, [17] designs a learning algorithm with a problem-dependent regret $\mathcal{O}(\log T)$ for unconstrained assortment optimization. [28] designs a learning algorithm with a problem-independent $\mathcal{O}(T^{\frac{2}{3}} \log T)$ regret when the exact assortment is computable and an $\mathcal{O}(\log T)$ regret otherwise. Compared to the above literature, which do not consider resource capacity, we further integrate online learning under Markov chain choice model with inventory constraints and derive new regret bounds.

Network Revenue Management (NRM). NRM has been introduced by [18,19] in the context of airline industries. An extensively studied performance measure is the revenue loss, i.e., the difference between the expected revenue of the proposed policy and the optimal policy. A common approach is to consider the fluid approximation LP and propose a policy based on its optimal solution. [36] designs a static bid-price control policy to guarantee an $\mathcal{O}(\sqrt{T})$ revenue loss. [31] gets an $o(\sqrt{T})$ revenue loss by re-solving the fluid approximation. [23] further improves the revenue loss to $\mathcal{O}(1)$ under the non-degeneracy assumption. [24] also points out the limitation of re-solving techniques in booking limit and bid price control in terms of re-solving time and frequency. Recently, [9] and [37] use different approaches to achieve $\mathcal{O}(1)$ upper bound on the revenue loss. The latter's approach is further generalized to other online decision making problems, e.g. dynamic pricing, knapsack, assignment and bin-packing [6,15, 16]. When information of the demand and resource is revealed online and the decisions are made sequentially and irrevocably, the problem is also referred as online linear programming, for example [3,29]. Several works focus on unknown underlying distributions for NRM. For example, [22] derives an $\mathcal{O}(\log T)$ regret under unknown demand parameters with a non-degeneracy assumption. [10] improves the regret to $\mathcal{O}(1)$ under the same assumption.

Another stream of works explore correlated customer arrivals. For example, [4] and [5] investigate the co-existence of high expectation and high variance demand, which is absent in independent arrivals. [26] presents a new linear programming formulation for NRM under Markovian correlated customer arrivals and derives a new bid-price policy with constant approximation theoretical guarantee. Beyond NRM, Markovian demands are also studied in other revenue management literature, e.g., pricing [27,33], inventory control [8,14], and online resource allocation [25]. We also study NRM under dependent customer demands but present a new model by capturing customers' Markovian choice behavior.

3 Model and Formulation

Consider a network revenue management problem over a time horizon of T periods. Let $\boldsymbol{r}$ denote the revenue vector $[r_1, r_2, ..., r_n]^\top$, where r_j is the revenue of product j. There are m resources indexed by $i \in [m]$, where resource i has initial capacity B_i. Let $\boldsymbol{B}$ denote the capacity vector $[B_1, B_2, ..., B_m]^\top$. Each product j consumes a_{ij} units of resource i for each i. Let $\boldsymbol{A}_j := [a_{1j}, a_{2j}, ..., a_{mj}]^\top \in \{0,1\}^m$ denote the resource that product j uses. Let $\boldsymbol{A} \in \{0,1\}^{m\times n}$ denote the product incidence matrix $[\boldsymbol{A}_1, \boldsymbol{A}_2, ..., \boldsymbol{A}_n]$.

The purchase process of each customer follows a Markov chain choice model: For each time period $t \in T$, a customer, indexed by t, arrives and initially requests for product j with probability λ_j. Let $\boldsymbol{\lambda}$ denote the arrival probability vector $[\lambda_1, \lambda_2, ..., \lambda_n]^\top$ with $\lambda_j \geq 0$ for all j and $\sum_{j=1}^n \lambda_j \leq 1$. If the request for product j is accepted, then the seller gains revenue r_j and the resources are consumed by $\boldsymbol{A}_j$. If the request for product j is rejected, then the customer transitions to another product k with probability ρ_{jk}. The customer may also transition to a no-purchase option with probability $\rho_{j0} := 1 - \sum_{k\in[n]} \rho_{jk} > 0$. In that case, the customer leaves the system and the seller gains no revenue. We assume ρ_{i0} to be lower bounded by $\underline{\rho} > 0$. We also assume the customer does not request the same product again after an immediate rejection, i.e., $\rho_{ii} = 0$ for $i \in [n]$. Let $\boldsymbol{\rho} \in \mathbb{R}^{n\times n}$ denote the Markov transition matrix. Let $\boldsymbol{\Sigma}$ denote the set of transition matrices $\boldsymbol{\Sigma} := \{\rho \in \mathbb{R}_+^{n\times n} : \rho_{ii} = 0, i \in [n], 1 - \sum_{j\in[n]} \rho_{ij} \geq \underline{\rho} > 0\}$. In conclusion, each customer's requests transition among different products according to a Markov chain until her request is accepted or she transitions to the no-purchase option and leaves the system. We assume the process among different customers are independent. An illustration of the above process is shown in Fig. 2.

Fig. 2. Illustration of Customer Choice Process.

In our problem, the seller needs to decide whether to accept or to reject each request of each customer. The decisions are made in an online fashion and are irrevocable. The seller's decision must satisfy the capacity constraints. Unused resources at the end of the horizon have no value. The goal of the seller is to determine a feasible policy to maximize the expected revenue over the whole time horizon.

In the following, we formalize the above description. We start by formally defining the stochastic process of customers' transitions. To avoid confusion, we use s to denote the time indices *within* the transition process of each single customer and use t to denote the indices of customers, which scales with the horizon T. For each customer t, let $\{Z_s^t\}_{s=1}^\infty$ denote the transition process of the customer t. It has discrete states $\{1, 2, ..., n\}$ and two *additional* special states: W and F, where W means the customer successfully purchases a product and F means the customer leaves the system with no purchase. These two states are absorbing states, i.e., for any $s \in \mathbb{N}$, $\mathbb{P}(Z_{s+1}^t = F | Z_s^t = F) = 1$, $\mathbb{P}(Z_{s+1}^t = W | Z_s^t = W) = 1$. Let $\tau_t := \inf\{s \in \mathbb{N} : Z_{s+1}^t = W \text{ or } Z_{s+1}^t = F\}$, i.e., $\tau_t + 1$ is the time when the customer t leaves the system. We denote an allocation control policy by π. For each customer t, for $s \in [\tau_t]$, $\pi_s^t = 0$ if the seller rejects the customer t's request at time s and $\pi_s^t = 1$ if the seller accepts the customer's request at time s. For each $t \in [T]$, let $\boldsymbol{X}^t(\pi) \in \{0,1\}^n$ denote the purchase vector where $X_i^t(\pi) = 1$ if customer t purchases product i and $X_i^t(\pi) = 0$ otherwise. Similarly, let $\boldsymbol{Y}^t(\pi) \in \mathbb{Z}_{\geq 0}^n$ denote the rejection vector: $Y_i^t(\pi)$ is the number of rejections of customer t's requests for product i, which can be larger than 1 since the seller can reject a customer's request for the same product multiple times. The expected revenue under policy π is $v_\pi := \mathbb{E}(\boldsymbol{r}^\top \sum_{t=1}^T \boldsymbol{X}^t(\pi))$.

In the following, we define the consistency, feasibility and admissibility of a policy π.

Definition 1 (Consistency). *We call an allocation control policy π consistent if for all $t \in [T]$, for all $s_1, s_2 \in [\tau_t]$, the implication $Z_{s_1}^t = Z_{s_2}^t \Rightarrow \pi_{s_1}^t = \pi_{s_2}^t$ holds.*

The consistency of the policy requires that the seller cannot change the decision of a product to a single customer. In other words, if the seller rejects a customer's request for product i, he should always reject the customer's request for product i in the subsequent process. Note that in the presence of consistency, our model is *equivalent* to that the seller offers an assortment to each customer and that the customer purchases a product according to a Markov chain choice model, which is studied in [7] and [13].

Definition 2 (Feasibility and Admissibility). *We call an allocation control policy π feasible if it is consistent, non-anticipating and satisfies the capacity constraints $\sum_{j=1}^{n} \boldsymbol{A}_j \sum_{t=1}^{T} X_j^t(\pi) \leq \boldsymbol{B}$ almost surely. We call an allocation control policy π admissible if it is consistent, non-anticipating and satisfies the capacity constraints in expectation, i.e. $\sum_{j=1}^{n} \boldsymbol{A}_j \sum_{t=1}^{T} \mathbb{E}(X_j^t(\pi)) \leq \boldsymbol{B}$. In the following, we use Π and $\mathcal{A}$ to denote the set of all feasible policies and admissible policies, respectively. By definition, we have $\Pi \subset \mathcal{A}$.*

The seller's problem is to find an optimal feasible policy, which can be written as

$$
\begin{aligned}
\max_{\pi} \quad & \mathbb{E}(\boldsymbol{r}^\top \sum_{t=1}^{T} \boldsymbol{X}^t(\pi)) \\
\text{s.t.} \quad & \boldsymbol{A} \sum_{t=1}^{T} \boldsymbol{X}^t(\pi) \leq \boldsymbol{B} \quad \text{a.s.}
\end{aligned} \tag{1}
$$

In the following, we denote v^* as the optimal value of (1) and π^* as an optimal solution.

3.1 Deterministic Linear Programming (DLP) Formulation

To study the seller's optimal allocation control policy π^*, we consider a DLP formulation of this problem:

$$
\begin{aligned}
v_{\text{DLP}} := \max_{\boldsymbol{x}_t \geq \boldsymbol{0}, \boldsymbol{y}_t \geq \boldsymbol{0}, t \in [T]} \quad & \boldsymbol{r}^\top \sum_{t=1}^{T} \boldsymbol{x}_t \\
\text{s.t.} \quad & \boldsymbol{A} \sum_{t=1}^{T} \boldsymbol{x}_t \leq \boldsymbol{B} \\
& \boldsymbol{x}_t + \boldsymbol{y}_t = \boldsymbol{\lambda} + \boldsymbol{\rho}^\top \boldsymbol{y}_t, \quad \forall t \in [T].
\end{aligned} \tag{DLP-1}
$$

Now we provide some explanation for (DLP-1). In (DLP-1), $\boldsymbol{x}_t$ is a vector representing the purchase probability of each product by customer t, which can be viewed as the expectation of $\boldsymbol{X}^t(\pi)$. Similarly, $\boldsymbol{y}_t$ is a vector representing the expected times of rejecting customer t's requests for each product, which can be viewed as the expectation of $\boldsymbol{Y}^t(\pi)$. Thus, $\boldsymbol{r}^\top \boldsymbol{x}_t$ is the expected revenue of customer t. The objective function is the expected revenue over the entire time horizon. The first set of constraints ensures the capacity constraints must be satisfied in expectation. The second set of constraints follows from a balance equation of any allocation policy. They can be explained as follows: Without loss of generality, we focus on $t = 1$ and $i = 1$, which corresponds to product 1 of customer 1. The left hand side is the summation of accepted and rejected requests of product 1. As for the right hand side, λ_1 is the expected number of requests for product 1 at the first stage, while $\sum_{j=1}^{n} \rho_{j1} y_{1j}$ is the expected number of requests for product 1 starting from the second stage. The two quantities both stand for the total expected number of requests for product 1, hence they should be equal. In the following, we show that (DLP-1) is well defined and provides an upper bound for the expected revenue under the optimal policy π^*.

Proposition 1. (DLP-1) *has a non-empty and bounded feasible region. Moreover,* $v^* \leq v_{\text{DLP}}$.

Considering v_{DLP} as a benchmark, we now define the *regret* of a policy π.

Definition 3. *(Regret). For any policy* π*, we define the revenue loss by* $v^* - v_\pi$ *and the regret of a policy* π *by* $v_{\text{DLP}} - v_\pi$.

By noting that $v^* \leq v_{\text{DLP}}$, we know $v^* - v_\pi \leq v_{\text{DLP}} - v_\pi$, i.e., the revenue loss is upper bounded by the regret. In the following, we focus on analyzing the regret of a policy.

To evaluate the performance of a policy, we adopt an asymptotic regime. We assume there is a sequence of problems indexed by T, and $\boldsymbol{B}$ is linear in T, i.e., $\boldsymbol{B} = T\boldsymbol{b}$ for some $\boldsymbol{b} \in \mathbb{R}_+^m$.

Under the asymptotic regime, we can equivalently reformulate the DLP as follows:

$$\begin{aligned} \max_{\boldsymbol{x} \geq \mathbf{0}, \boldsymbol{y} \geq \mathbf{0}} \quad & \boldsymbol{r}^\top \boldsymbol{x} \\ \text{s.t.} \quad & \boldsymbol{A}\boldsymbol{x} \leq \boldsymbol{b} \\ & \boldsymbol{x} + \boldsymbol{y} = \boldsymbol{\lambda} + \boldsymbol{\rho}^\top \boldsymbol{y}. \end{aligned} \tag{DLP-2}$$

The equivalence is as follows: let $(\boldsymbol{x}_t^*, \boldsymbol{y}_t^*)_{t=1}^T$ be the optimal solution to (DLP-1), then $\boldsymbol{x}^* := \frac{1}{T}\sum_{t=1}^T \boldsymbol{x}_t^*$ and $\boldsymbol{y}^* := \frac{1}{T}\sum_{t=1}^T \boldsymbol{y}_t^*$ are the optimal solution to (DLP-2). Conversely, if $\boldsymbol{x}^*, \boldsymbol{y}^*$ are the optimal solution to (DLP-2), then $\boldsymbol{x}_t^* = \boldsymbol{x}^*/T, \boldsymbol{y}_t^* = \boldsymbol{y}^*/T$ for all T are the optimal solution to (DLP-1). Moreover, $T\boldsymbol{r}^\top \boldsymbol{x}^* = \boldsymbol{r}^\top \sum_{t=1}^T \boldsymbol{x}_t^*$. We remark that the formulation technically coincides with Sect. 5 in [13] but they are different in the problem setting. Our control policies are completely different from [13] and apply to online learning settings when parameters are unknown *a priori*.

4 Static Probabilistic Allocation

In this section, we assume $\boldsymbol{\lambda}$ and $\boldsymbol{\rho}$ are known. In the following, we design an algorithm based on the DLP formulation and analyze its performance.

Let $\boldsymbol{x}^*$ and $\boldsymbol{y}^*$ be an optimal solution to (DLP-2). The idea of the policy is to interpret $x_i^*/(x_i^*+y_i^*)$ as the allocation probability of the request for product i. Note that x_i^* is the expected number of accepted requests at state i. Similarly, y_i^* is the expected number of rejected requests at state i. We can then interpret $x_i^*/(x_i^*+y_i^*)$ as the proportion of accepted requests in total requests for product i. We can therefore choose the accepting probability of requests for product i as $x_i^*/(x_i^*+y_i^*)$, thus achieving the proportion in expectation. Due to capacity constraints, we also need to reject the requests for product i when the remaining resource $\boldsymbol{B}'$ does not satisfy $\boldsymbol{A}_i \leq \boldsymbol{B}'$. We call this policy the Static Probabilistic Allocation (SPA) policy. The detail of the policy is given in Algorithm 1.

Algorithm 1 Static Probabilistic Allocation (SPA)

Input: Model parameters $\boldsymbol{A}, \boldsymbol{r}, \boldsymbol{B}_1 := \boldsymbol{B}, T, \boldsymbol{\lambda}, \boldsymbol{\rho}$, optimal solution $\boldsymbol{x}^*, \boldsymbol{y}^*$ of (DLP-2).
for $t \in [T]$ **do**
  Include product i in the assortment S_t with probability $x_i^*/(x_i^*+y_i^*)$.
  Customer t arrives with first choice i.
  while Customer t does not leave **do**
    Customer t requests for product i.
    if $i \in S_t$ and $\boldsymbol{A}_i \leq \boldsymbol{B}_t$ **then**
      Seller accepts the request and customer t buys product i and leaves.
    else if $i \in S_t$ but $\boldsymbol{A}_i \nleq \boldsymbol{B}_t$ **then**
      Seller rejects the request and all subsequent requests until customer t leaves with no purchase.
    else if $i \notin S_t$ **then**
      Seller rejects the request and customer t transitions to product j
      $i := j$ ($j = 0$ means the customer transitions to the no purchase option and leaves.)
  $\boldsymbol{B}_{t+1} = \boldsymbol{B}_t - \boldsymbol{A}_i$ (or $\boldsymbol{B}_{t+1} = \boldsymbol{B}_t$ if no purchase)

Let v_{SPA} denote the expected revenue of the SPA policy. We have the following results about the regret of the SPA policy in Theorem 1.

Theorem 1. *The regret of the SPA policy* $v_{\text{DLP}} - v_{\text{SPA}}$ *is* $\mathcal{O}(\sqrt{T})$.

5 Online Probabilistic Allocation

In reality, the product and resource information $\boldsymbol{r}, \boldsymbol{A}$ is typically known in advance but the customer arrival probability $\boldsymbol{\lambda}$ and transition probability $\boldsymbol{\rho}$ may be unknown. We call $\boldsymbol{\theta} := (\boldsymbol{\lambda}; \boldsymbol{\rho})$ the parameter in our problem, written as a vector. In this section, we develop a learning-while-doing policy when $\boldsymbol{\theta}$ is unknown in the beginning. We adopt a similar notion of regret to evaluate the performance of our policy. Specifically, we still consider the regret to be $v_{\text{DLP}} - v_\pi$ where v_{DLP} is the optimal DLP for the clairvoyant model under the true parameter, and v_π is the expected revenue of a policy π. For our algorithm design, there are three main challenges: (i) how to estimate parameter $\boldsymbol{\theta}$ using available data; (ii) how to balance the exploration-exploitation trade-off; and (iii) how to control the resource over the entire time horizon without the exact parameter $\boldsymbol{\theta}$.

First, we consider the estimation of customers' arrival probability $\boldsymbol{\lambda}$ and transition probability $\boldsymbol{\rho}$. Over the selling horizon, we can observe the transition paths the customers follow by accepting and rejecting their requests. We propose a parameter estimation method for $\boldsymbol{\theta}$ based on customers' transition paths with consistency guarantee and sub-Gaussian concentration bounds. We can construct an estimator by (i) collecting their first requests to estimate $\boldsymbol{\lambda}$ and (ii) collecting their second requests to estimate $\boldsymbol{\rho}$.

Then, we combine our estimation techniques with the probabilistic allocation policy proposed in Sect. 4. In particular, in the initial τ periods (called the exploration phase), the seller rejects all customers' requests and observes customers' first and second requests (the second request can be the no-purchase option). At the end of this phase, we can get an estimator $\tilde{\boldsymbol{\theta}}$ based on the collected data. We can determine the accepting probability $\tilde{\boldsymbol{p}}$ for each product based on $\tilde{\boldsymbol{\theta}}$. At the following epochs, we use the previously calibrated parameters to make the acceptance and rejection decision for the customer. In the meantime, we update the estimation of $\tilde{\boldsymbol{\theta}}$ using the observations from the current epoch, which we will use in the following epoch. The detailed steps of the algorithm is given in Algorithm 2. In the following, we analyze the performance of the algorithm. We first make two assumptions:

Assumption 1 (Non-degeneracy). *The optimal solution to* (DLP-2) *is non-degenerate and unique.*

Assumption 1 is without loss of generality. In fact one can make an infinitesimal perturbation to the revenue vector $\boldsymbol{r}$ so that this assumption is satisfied, while the change to the objective value is arbitrarily small [30]. Under the assumption, the optimal solution to (DLP-2) is a basic feasible solution.

Assumption 2 (Parameter Space). *The true parameter $\boldsymbol{\theta}$ lies in $\Theta := \Theta_\lambda \times \Theta_\rho$, where*

$$\Theta_\lambda := \{\boldsymbol{\lambda} : 0 < \underline{\lambda} \leq \lambda_i \leq \overline{\lambda} < 1, i \in [n], \sum_{i\in[n]} \lambda_i \leq 1\} \text{ and } \Theta_\rho := \{\boldsymbol{\rho} \in \boldsymbol{\Sigma} : 0 < \underline{\rho} \leq \rho_{ij} \leq \overline{\rho} < 1\}$$

for some known $\underline{\lambda}, \bar{\lambda}, \underline{\rho}$ and $\bar{\rho}$.

Assumption 2 only requires that the true parameters are bounded away from 0 or 1. Specifically, we want the arrival probability to be lower bounded for the convenience of estimation. Otherwise, if λ_i approaches to 0, then we may not obtain enough samples to estimate ρ_{ij} for $j \in [n] \cup \{0\}$.

In the next theorem, we show the regret of Algorithm 2:

Theorem 2 *Suppose Assumptions 1 and 2 hold. There exists μ_1, independent of T, such that for any $C_6 > \mu_1$, if we set $\tau := \lceil C_6\sqrt{T}\log T\rceil$, then the OPA policy has a regret $\tilde{O}(\sqrt{T})$.*

Finally, we show that the regret in Theorem 2 is tight for this problem. In particular, we show that even if we know the parameters $\boldsymbol{\lambda}$ and $\boldsymbol{\rho}$, the gap between v_{DLP} and v^* is $\Omega(\sqrt{T})$. We have the following proposition.

Algorithm 2 Online Probabilistic Allocation (OPA)

Input: Model parameters $T, \boldsymbol{A}, \boldsymbol{r}, \tau, \boldsymbol{B}_{\tau+1} := \boldsymbol{B}$.
(Let K be such that $T = \sum_{k=0}^{K} 4^k \tau$ and epoch k to be the time interval $[\sum_{j=0}^{k} 4^j \tau, \sum_{j=0}^{k+1} 4^j \tau])$

for $t \in [\tau]$ **do** ▷ Exploration Phase (epoch 0)

Reject all customers' all requests, collect arrival and transition data $\boldsymbol{g}^0$ and $\boldsymbol{\rho}^0$, defined as

g_i^0 : the number of customers whose first choice is product i in $[\tau]$.

h_{ij}^0: the number of customers whose first choice is product i and second choice is product j in $[\tau]$ (the second choice includes the no-purchase option). Let

$$\lambda_i^0 := \frac{g_i^0}{\tau}, \qquad \rho_{ij}^0 := \frac{h_{ij}^0}{g_i^0} \; (\frac{0}{0} := 0), \qquad \hat{\boldsymbol{\theta}}^0 := \text{vec}(\{\lambda_i^0\}_{i\in[n]} \cup \{\rho_{ij}^0\}_{i,j\in[n]}). \tag{E-1}$$

$$\tilde{\boldsymbol{\theta}}^0 := \arg\min_{\boldsymbol{\theta}'\in\Theta} \|\boldsymbol{\theta}' - \hat{\boldsymbol{\theta}}^0\|, \quad \text{where } \Theta \text{ is the parameter space in Assumption 2.} \tag{E-2}$$

for epoch $k \in [K]$ **do** ▷ Learning-while-doing phase

Solve the following linear programming and get the optimal solution $\boldsymbol{x}_k$ and $\boldsymbol{y}_k$.

$$\begin{aligned} \max_{\boldsymbol{x}\geq \boldsymbol{0}, \boldsymbol{y}\geq \boldsymbol{0}} \quad & \boldsymbol{r}^\top \boldsymbol{x} \\ \text{s.t.} \quad & \boldsymbol{A}\boldsymbol{x} \leq \boldsymbol{b} \\ & \boldsymbol{x} + \boldsymbol{y} = (\tilde{\boldsymbol{\lambda}}^{k-1}) + (\tilde{\boldsymbol{\rho}}^{k-1})^\top \boldsymbol{y}. \end{aligned} \tag{2}$$

for $t \in$ epoch k **do**

Include product i in the offered assortment S_t with probability $\tilde{p}_{ki} = \frac{x_{ki}}{x_{ki}+y_{ki}}$.

Accept customers' request for product i if $i \in S_k$ and resources permit.

$\boldsymbol{B}_{t+1} = \boldsymbol{B}_t - \boldsymbol{A}_i$ if customer t finally purchases product i (or $\boldsymbol{B}_{t+1} = \boldsymbol{B}_t$ if no purchase)

Collect arrival and transition data $\boldsymbol{g}^k$ and $\boldsymbol{h}^k$. Define:

g_i^k : the number of customers whose first choice is product i in epoch k.

h_{ij}^k: the number of customers whose first choice is product i and second choice is product j in epoch k (the second choice includes the no-purchase option). Let

$$\lambda_i^k := \frac{g_i^k}{\tau}, \qquad \rho_{ij}^k := \frac{h_{ij}^k}{g_i^k} \; (\frac{0}{0} := 0), \qquad \hat{\boldsymbol{\theta}}^k := \text{vec}(\{\lambda_i^k\}_{i\in[n]} \cup \{\rho_{ij}^k\}_{i,j\in[n]}). \tag{E-3}$$

$$\tilde{\boldsymbol{\theta}}^k := \arg\min_{\boldsymbol{\theta}'\in\Theta} \|\boldsymbol{\theta}' - \hat{\boldsymbol{\theta}}^k\|. \tag{E-4}$$

Proposition 2 *The gap between the optimal value of DLP,* v_{DLP} *and the revenue under optimal policy* v^* *is bounded below by* $v_{\text{DLP}} - v^* = \Omega(\sqrt{T})$.

By combining Theorems 1, 2 and Proposition 2, we know that the SPA and the OPA policy have achieved near optimal performances in their respective settings.

6 Numerical Experiments

In this section, we conduct numerical experiments to demonstrate the performance of the SPA policy when $\boldsymbol{\theta}$ is known and the performance of the OPA policy when $\boldsymbol{\theta}$ is unknown.

In our numerical experiments, we consider the following setting of the base problem. We set $m = 10$ and $n = 20$. For the arrival probabilities, we generate n independent random numbers Λ_j, each uniformly distributed on $[0, 1]$ and set $\lambda_j := \Lambda_j / \sum_{j=1}^{n} \Lambda_j$. For the transition probabilities, for each i, we generate n independent random numbers ϕ_{ij} uniformly distributed on $[0, 1]$. We further set $\phi_{ii} = 0$ for all i. We set $\rho_{ij} := \phi_{ij} / \sum_{j=0}^{n} \phi_{ij}$. As for Θ, we set $\underline{\lambda} := 0.01$, $\overline{\lambda} := 0.99$, $\underline{\rho} := 0.001$ and $\overline{\rho} := 0.999$. Let $\hat{\boldsymbol{\theta}} := \text{vec}(\{\lambda_i\}_{i \in [n]} \cup \{\rho_{ij}\}_{i,j \in [n]})$ and let $\boldsymbol{\theta} := \arg\min_{\tilde{\boldsymbol{\theta}} \in \Theta} \|\tilde{\boldsymbol{\theta}} - \hat{\boldsymbol{\theta}}\|$. The revenue vector $\boldsymbol{r}$ is generated uniformly on $[0, 1]$. Each term in the consumption matrix $\boldsymbol{A}$ is generated independently by Bernoulli distribution with probability 0.3. Under the asymptotic regime, we set the selling time horizon T ranging from 10^2 to 10^4. For each T, we set $\boldsymbol{B} = T\boldsymbol{b}$ where each term of $\boldsymbol{b}$ is identically and independently generated according to a uniform distribution on $[0, \frac{1}{2}]$. We replicate the experiments under each T for 100 times and record the average revenue, and finally get the regret by comparing the average revenue and the benchmark DLP. The relationship between the time horizon and the regret under known and unknown parameters, with the SPA and the OPA policies, are presented in Fig. 3a and Fig. 3b respectively. The results after taking the logarithm of the regret and the time horizon are presented in Fig. 3c.

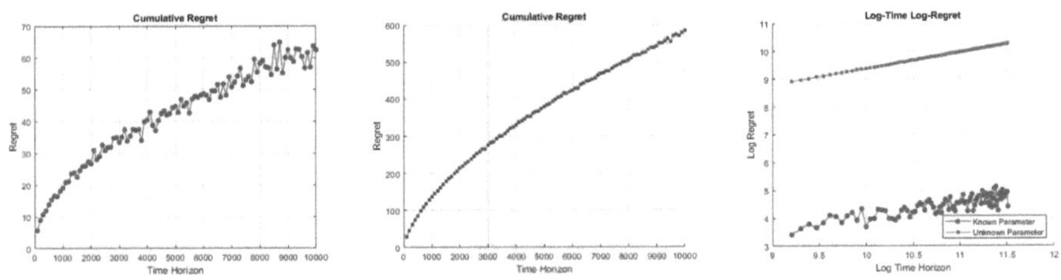

(a) Regret of SPA (b) Regret of OPA (c) Log time v.s. log regret

Fig. 3. Results of Numerical Experiments.

From the figures, we can see that both the SPA policy and the OPA policy show sublinear regret with respect to the time horizon. The regret under the OPA policy is significantly larger than the SPA policy, due to the cost of learning unknown $\boldsymbol{\theta}$. We further perform a linear regression between the logarithm of the regret and the logarithm of T. For the SPA policy, the slope of the regression parameter is 0.532 while for the OPA policy, the slope of the regression parameter is 0.504. Therefore, the results validate the $\mathcal{O}(\sqrt{T})$ and $\tilde{\mathcal{O}}(\sqrt{T})$ regret bounds respectively in Theorems 1 and 2.

7 Model Extension: Finite Patience Level of Customers

In the previous discussion, we assume that customers' purchase process follows a Markov process. In reality, customers may have limited patience and may leave the system after a certain number of transitions. In this section, we assume the number of transitions of a customer is upper bounded by a constant K, which characterizes the patience level of a customer. In the following, we derive a new formulation of our problem and propose an adaptive probabilistic allocation policy (APA) with a provable $\mathcal{O}(\sqrt{T})$ regret bound. An illustration of the above process is shown in Fig. 4.

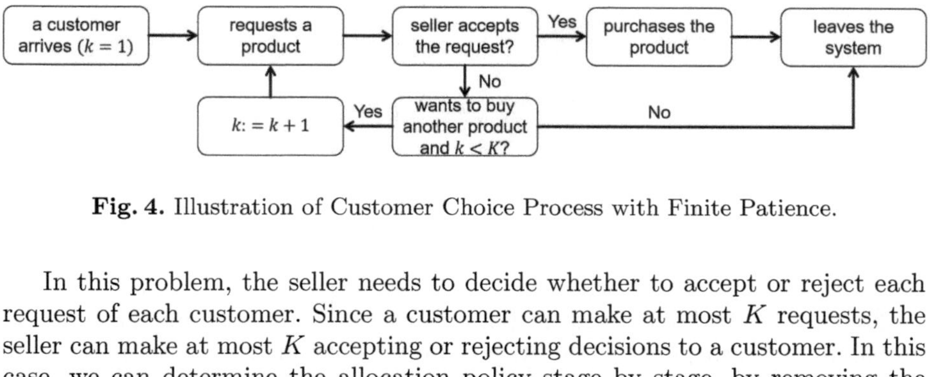

Fig. 4. Illustration of Customer Choice Process with Finite Patience.

In this problem, the seller needs to decide whether to accept or reject each request of each customer. Since a customer can make at most K requests, the seller can make at most K accepting or rejecting decisions to a customer. In this case, we can determine the allocation policy stage by stage, by removing the consistency assumption of the policy.

In the following, we formalize the above description and formulate the corresponding DLP. We denote an allocation control policy by π. For each $t \in [T]$, let $\boldsymbol{X}^t(\pi) \in \{0,1\}^{n\times k}$ denote the purchase matrix where $\boldsymbol{X}^t_{ij}(\pi) = 1$ if the request for product i in stage j of customer t is accepted and $\boldsymbol{X}^t_{ij}(\pi) = 0$ otherwise. Similarly, let $\boldsymbol{Y}^t(\pi) \in \mathbb{Z}^{n\times k}_{\geq 0}$ denote the rejection matrix where $\boldsymbol{Y}^t_{ij}(\pi) = 1$ if the request for product i in stage j of customer t is rejected and $\boldsymbol{Y}^t_{ij}(\pi) = 0$ otherwise. We can write $\boldsymbol{X}^t(\pi) = [\boldsymbol{X}^t_1(\pi), \boldsymbol{X}^t_2(\pi), ..., \boldsymbol{X}^t_K(\pi)]$ and $\boldsymbol{Y}^t(\pi) = [\boldsymbol{Y}^t_1(\pi), \boldsymbol{Y}^t_2(\pi), ..., \boldsymbol{Y}^t_K(\pi)]$. The expected revenue under policy π is $v_\pi := \mathbb{E}(\boldsymbol{r}^\top(\sum_{t=1}^T \sum_{k=1}^K \boldsymbol{X}^t_k(\pi)))$. The seller's problem can be written as

$$
\begin{aligned}
v^*_{\mathrm{F}} := \max_{\pi} \quad & \mathbb{E}(\boldsymbol{r}^\top(\sum_{t=1}^{T}\sum_{k=1}^{K} \boldsymbol{X}^t_k(\pi))) \\
\text{s.t.} \quad & \boldsymbol{A}\sum_{t=1}^{T}\sum_{k=1}^{K} \boldsymbol{X}^t_k(\pi) \leq \boldsymbol{B} \;\; \text{a.s.}
\end{aligned}
\tag{3}
$$

We now present a corresponding fluid approximation of (3):

$$
\begin{aligned}
v_{\mathrm{FDLP}} := \max_{\boldsymbol{x}_{tk},\boldsymbol{y}_{tk}\in\mathbb{R}^n_+,k\in[K],t\in[T]} \quad & \boldsymbol{r}^\top \sum_{t=1}^{T}\sum_{k=1}^{K}\boldsymbol{x}_{tk} \\
\text{s.t.} \quad & \boldsymbol{A}\sum_{t=1}^{T}\sum_{k=1}^{K}\boldsymbol{x}_{tk} \leq \boldsymbol{B} \\
& \boldsymbol{x}_{t1} + \boldsymbol{y}_{t1} = \boldsymbol{\lambda} \quad \forall t \in [T] \\
& \boldsymbol{x}_{tk} + \boldsymbol{y}_{tk} - \boldsymbol{\rho}^\top \boldsymbol{y}_{tk-1} = 0 \quad \forall 2 \leq k \leq K, t \in [T].
\end{aligned} \tag{FDLP}
$$

We still consider an asymptotic regime so that we can consider an equivalent LP in a single period, i.e., with respect to one customer (without loss of generality, $t = 1$). We can formulate a DLP similar to the previous problem,

$$
\begin{aligned}
v_{\text{FDLP-2}} := \max_{\boldsymbol{x}_{k},\boldsymbol{y}_{k}\in\mathbb{R}^n_+,k\in[K]} \quad & \boldsymbol{r}^\top \sum_{k=1}^{K}\boldsymbol{x}_{k} \\
\text{s.t.} \quad & \boldsymbol{A}\sum_{k=1}^{K}\boldsymbol{x}_{k} \leq \boldsymbol{b} \\
& \boldsymbol{x}_{1} + \boldsymbol{y}_{1} = \boldsymbol{\lambda} \\
& \boldsymbol{x}_{k} + \boldsymbol{y}_{k} - \boldsymbol{\rho}^\top \boldsymbol{y}_{k-1} = 0 \quad \forall k \in \{2, ..., K\}.
\end{aligned} \tag{FDLP-2}
$$

For each $k \in [K]$ and $i \in [n]$, $\boldsymbol{x}_k$ is a vector with x_{ki} representing the purchase probability of each product i at stage k by the customer, while $\boldsymbol{y}_k$ is a vector with y_{ki} representing the probability of the event that the customer requests for product i at stage k but is rejected. In other words, $\boldsymbol{x}_k$ and $\boldsymbol{y}_k$ are the expectation of $\boldsymbol{X}_k^t(\pi)$ and $\boldsymbol{Y}_k^t(\pi)$ respectively. We have a proposition similar to Proposition 1.

Proposition 3 (FDLP) *provides an upper bound for* (3)*, i.e.,* $v_{\mathrm{F}}^* \leq v_{\mathrm{FDLP}}$.

In the following, we develop an adaptive allocation policy (APA), by accepting the request for product i at stage k with probability $p_{ki} := \frac{x_{ki}}{x_{ki}+y_{ki}}$. Different from the SPA policy, which accepts the requests for the same product with the same probability at each stage, we can use an adaptive strategy to accept the requests depending on the current stage of the customer. The details of this algorithm are given in Algorithm 3.

We have the following result about the regret of Algorithm 3.

Theorem 3. *The regret of the APA policy* $v_{\mathrm{FDLP}} - v_{\mathrm{APA}}$ *is* $\mathcal{O}(\sqrt{T})$.

Therefore, even when customers have finite patience levels, we can still have a policy that achieves asymptotically optimal performance.

Algorithm 3 Adaptive Probabilistic Allocation (APA)

Input: Model parameters $\boldsymbol{A}, \boldsymbol{r}, \boldsymbol{B}_1 := \boldsymbol{B}, T, \boldsymbol{\lambda}, \boldsymbol{\rho}$, optimal solution x^*, y^* of (FDLP-2)
for $t \in [T]$ **do**
  Customer t arrives with first choice i. Current stage $k = 1$.
  while Customer t does not leave and current stage $k < K$ **do**
    Customer t requests for product i
    Simulate a random variable $X \sim \text{Bernoulli}(x^*_{ki}/(x^*_{ki} + y^*_{ki}))$.
    if $X = 1$ and $\boldsymbol{A}_i \leq \boldsymbol{B}_t$ **then**
      Seller accepts the request and customer t buys product i and leaves.
    else if $X = 1$ and $\boldsymbol{A}_i \not\leq \boldsymbol{B}_t$ **then**
      Seller rejects the request and all subsequent requests until customer t leaves with no purchase.
    else if $X = 0$ **then**
      Seller rejects the request and customer t transitions to product j
      $i := j$ ($j = 0$ means the customer transitions to the no purchase option and leaves.)
      $k := k + 1$.
  $\boldsymbol{B}_{t+1} = \boldsymbol{B}_t - \boldsymbol{A}_i$ (or $\boldsymbol{B}_{t+1} = \boldsymbol{B}_t$ if no purchase)

8 Conclusion

In this paper, we study the NRM problem under Markovian demands. We use the Markov chain choice model to capture the dependence of customer demands. To analyze the decision problem, we build a simple stochastic model and develop a DLP as its fluid approximation. We then propose an SPA policy based on the DLP and show its asymptotic optimality. When the parameters of the Markov chain choice model are unknown in the beginning, we design an OPA policy and show its performance compared to the optimal policy by tight regret bounds. We also extend our model to the impatient customer setting and propose an APA policy with similar theoretical guarantees. Due to space limit, the proofs are omitted. The full version of this paper is on SSRN[2].

Acknowledgments. The research of Guillermo Gallego is supported by grant GRF project 16215280 from HK. The research of Zizhuo Wang is partially supported by the National Natural Science Foundation of China (NSFC) [Grant NSFC-72394361, 72425013], and the Guangdong Key Lab of Mathematical Foundations for Artificial Intelligence.

References

1. Agrawal, S., Avadhanula, V., Goyal, V., Zeevi, A.: Thompson sampling for the MNL-bandit. In: Conference on Learning Theory, pp. 76–78. PMLR (2017)
2. Agrawal, S., Avadhanula, V., Goyal, V., Zeevi, A.: MNL-bandit: a dynamic learning approach to assortment selection. Oper. Res. **67**(5), 1453–1485 (2019)

[2] https://papers.ssrn.com/sol3/papers.cfm?abstract_id=4824587.

3. Agrawal, S., Wang, Z., Ye, Y.: A dynamic near-optimal algorithm for online linear programming. Oper. Res. **62**(4), 876–890 (2014)
4. Aouad, A., Ma, W.: A nonparametric framework for online stochastic matching with correlated arrivals. In: Proceedings of the 24th ACM Conference on Economics and Computation, EC 2023, p. 114. Association for Computing Machinery, New York (2023)
5. Bai, Y., El Housni, O., Jin, B., Rusmevichientong, P., Topaloglu, H., Williamson, D.P.: Fluid approximations for revenue management under high-variance demand. Manage. Sci. **69**(7), 4016–4026 (2023)
6. Banerjee, S., Freund, D.: Uniform loss algorithms for online stochastic decision-making with applications to bin packing. In: Abstracts of the 2020 SIGMETRICS/Performance Joint International Conference on Measurement and Modeling of Computer Systems, pp. 1–2 (2020)
7. Blanchet, J., Gallego, G., Goyal, V.: A Markov chain approximation to choice modeling. Oper. Res. **64**(4), 886–905 (2016)
8. den Boer, A., Perry, O., Zwart, B.: Dynamic pricing policies for an inventory model with random windows of opportunities. Nav. Res. Logist. **65**(8), 660–675 (2018)
9. Bumpensanti, P., Wang, H.: A re-solving heuristic with uniformly bounded loss for network revenue management. Manage. Sci. **66**(7), 2993–3009 (2020)
10. Chen, G., Li, X., Ye, Y.: An improved analysis of LP-based control for revenue management. Oper. Res. **72**(3), 1124–1138 (2024)
11. Chen, X., Wang, Y.: A note on a tight lower bound for capacitated MNL-bandit assortment selection models. Oper. Res. Lett. **46**(5), 534–537 (2018)
12. Chen, X., Wang, Y., Zhou, Y.: Optimal policy for dynamic assortment planning under multinomial logit models. Math. Oper. Res. **46**(4), 1639–1657 (2021)
13. Feldman, J.B., Topaloglu, H.: Revenue management under the Markov chain choice model. Oper. Res. **65**(5), 1322–1342 (2017)
14. Feng, Y., Gallego, G.: Perishable asset revenue management with Markovian time dependent demand intensities. Manage. Sci. **46**(7), 941–956 (2000)
15. Freund, D., Banerjee, S.: Good prophets know when the end is near. Manage. Sci. (2024)
16. Freund, D., Zhao, J.: Overbooking with bounded loss. In: Proceedings of the 22nd ACM Conference on Economics and Computation, pp. 477–478 (2021)
17. Gallego, G., Lu, W.: An optimal greedy heuristic with minimal learning regret for the Markov chain choice model. Available at SSRN 3810470 (2021)
18. Gallego, G., van Ryzin, G.: Optimal dynamic pricing of inventories with stochastic demand over finite horizons. Manage. Sci. **40**(8), 999–1020 (1994)
19. Gallego, G., van Ryzin, G.: A multiproduct dynamic pricing problem and its applications to network yield management. Oper. Res. **45**(1), 24–41 (1997)
20. Gupta, A., Hsu, D.: Parameter identification in Markov chain choice models. Theoret. Comput. Sci. **808**, 99–107 (2020)
21. Hsu, D.J., Kontorovich, A., Szepesvári, C.: Mixing time estimation in reversible Markov chains from a single sample path. In: Advances in Neural Information Processing Systems, vol. 28 (2015)
22. Jasin, S.: Performance of an LP-based control for revenue management with unknown demand parameters. Oper. Res. **63**(4), 909–915 (2015)
23. Jasin, S., Kumar, S.: A re-solving heuristic with bounded revenue loss for network revenue management with customer choice. Math. Oper. Res. **37**(2), 313–345 (2012)
24. Jasin, S., Kumar, S.: Analysis of deterministic LP-based booking limit and bid price controls for revenue management. Oper. Res. **61**(6), 1312–1320 (2013)

25. Jia, J., et al.: Online resource allocation in Markov chains. In: Proceedings of the ACM Web Conference 2023, pp. 3498–3507 (2023)
26. Jiang, J.: Constant approximation for network revenue management with Markovian-correlated customer arrivals. arXiv preprint arXiv:2305.05829 (2023)
27. Keskin, N.B., Li, M.: Selling quality-differentiated products in a Markovian market with unknown transition probabilities. Oper. Res. **72**(3), 885–902 (2024)
28. Li, S., Luo, Q., Huang, Z., Shi, C.: Online learning for constrained assortment optimization under Markov chain choice model. Oper. Res. (2024)
29. Li, X., Ye, Y.: Online linear programming: dual convergence, new algorithms, and regret bounds. Oper. Res. **70**(5), 2948–2966 (2022)
30. Megiddo, N., Chandrasekaran, R.: On the ε-perturbation method for avoiding degeneracy. Oper. Res. Lett. **8**(6), 305–308 (1989)
31. Reiman, M.I., Wang, Q.: An asymptotically optimal policy for a quantity-based network revenue management problem. Math. Oper. Res. **33**(2), 257–282 (2008)
32. Rusmevichientong, P., Shen, Z.J.M., Shmoys, D.B.: Dynamic assortment optimization with a multinomial logit choice model and capacity constraint. Oper. Res. **58**(6), 1666–1680 (2010)
33. Rustichini, A., Wolinsky, A.: Learning about variable demand in the long run. J. Econ. Dyn. Control **19**(5–7), 1283–1292 (1995)
34. Sauré, D., Zeevi, A.: Optimal dynamic assortment planning with demand learning. Manuf. Serv. Oper. Manage. **15**(3), 387–404 (2013)
35. Şimşek, A.S., Topaloglu, H.: An expectation-maximization algorithm to estimate the parameters of the Markov chain choice model. Oper. Res. **66**(3), 748–760 (2018)
36. Talluri, K., van Ryzin, G.: An analysis of bid-price controls for network revenue management. Manage. Sci. **44**(11-part-1), 1577–1593 (1998)
37. Vera, A., Banerjee, S.: The Bayesian prophet: a low-regret framework for online decision making. Manage. Sci. **67**(3), 1368–1391 (2021)
38. Zhang, D., Cooper, W.L.: Revenue management for parallel flights with customer-choice behavior. Oper. Res. **53**(3), 415–431 (2005)

Pandora's Box Problem Over Time

Georgios Amanatidis[1,2], Federico Fusco[3], Rebecca Reiffenhäuser[4], and Artem Tsikiridis[5(✉)]

[1] University of Essex, Colchester, UK
[2] Archimedes/Athena RC, Athens, Greece
[3] Sapienza University of Rome, Rome, Italy
[4] University of Amsterdam, Amsterdam, The Netherlands
[5] Centrum Wiskunde and Informatica (CWI), Amsterdam, The Netherlands
artem.tsikiridis@cwi.nl

Abstract. The Pandora's Box problem models the search for the best alternative when evaluation is costly. In its simplest variant, a decision maker is presented with n boxes, each associated with a cost of inspection and a distribution over the reward hidden within. The decision maker inspects a subset of these boxes one after the other, in a possibly adaptive ordering, and obtains as utility the difference between the largest reward uncovered and the sum of the inspection costs. While this version of the problem is well understood (Weitzman 1979), there is a flourishing recent literature on variants of the problem. In this paper, we introduce a general framework—the Pandora's Box Over Time problem—that captures a wide range of variants where time plays a role, e.g., as it might constrain the schedules of exploration and influence both costs and rewards. In our framework, each box is characterized by time-dependent rewards and costs, and inspecting it might require a box-specific processing time. Once a box is inspected, its reward may deteriorate over time, possibly differently for each box. Our main result is an efficient 21.3-approximation to the optimal strategy, which generally is NP-hard to compute. We further obtain improved results for the natural special cases where boxes have no processing time or when costs and reward distributions are time-independent (but rewards may deteriorate after inspecting).

1 Introduction

In the classic version of Pandora's Box problem, introduced in the seminal paper by Weitzman [29], a decision maker is presented with n boxes, each containing a hidden reward drawn from a known (independent) distribution, but also incurs a fixed cost for inspecting the box and learning the value of the reward. A (possibly adaptive) strategy consists of sequentially inspecting a set of boxes and then picking the largest observed reward. In the end, the decision maker gains the largest reward observed minus the sum of the incurred inspection costs. Surprisingly, Weitzman provides a simple procedure to compute the optimal strategy. Pandora's Box problem naturally models situations like hiring, buying a house, or choosing a school: the decision maker has some prior on the value

M. Mavronicolas et al. (Eds.): WINE 2024, LNCS 15534, pp. 494–512, 2026.
https://doi.org/10.1007/978-3-032-08560-3_28

of each alternative but only a costly inspection (e.g., interviewing a candidate, visiting a house or a school) reveals their actual values. Motivated by these applications, in recent years, many variants of the original problem have been considered (we refer the interested reader to Beyhaghi and Cai [6], and the Related Work). In this work, we study the impact of time on the problem as we propose a unified framework that captures and generalizes existing models.

Time can impact in many different ways, especially given the sequential nature of the Pandora's Box problem. For instance, it is possible that a company has a strict deadline for hiring a new worker, so only a limited number of candidates can be interviewed. Moreover, good candidates may only remain on the market for a short period, or they expect a definite answer within a certain amount of time after the interview. Conversely, in the housing market, it is common that houses that remain too long on the market decrease their price (i.e., their cost) so that the corresponding reward increases or that some days are more convenient to arrange a visit (thus decreasing the cost of inspection). As a further example, when picking a yearly gym membership, or sampling perishable goods like food, picking an option only some time after having tested it often reduces the leftover value one can extract.

We introduce a general framework, the Pandora's Box Over Time problem, which captures these phenomena. Here, each box is associated with a processing time, time-dependent inspection cost, and random reward, and with a non-increasing discounting function that specifies how the value of the realized reward deteriorates in the interval of time between inspection and selection. Our model captures all of these aspects, and, somewhat surprisingly, it still allows for good approximation ratios, despite its generality. However, it is clear from the above examples that many applications do not exhibit all types of time dependency at once. Therefore, we also focus on natural restrictions, for which we show that approximation guarantees can be significantly improved.

Our Contribution. Beyond providing a general framework for Pandora's Box Over Time problem, we provide the following results:

- We show how to efficiently compute a 21.3-approximation to the optimal strategy for Pandora's Box Over Time (Theorem 1). As our model captures the Free Order Prophet problem [1], we also observe that finding the optimal solution is NP-hard in general.
- We give improved approximation results for two special cases, namely when processing times are all zero (Pandora Over Time With Instant Inspection) and when boxes still have processing times and the values of inspected boxes may degrade over time, but both the costs and the distributions of the rewards are not time-dependent (Pandora With Value Discounting). Leveraging the reduced difficulty compared to our very rich, general model, we obtain an $(8 + \varepsilon)$-approximation for the instant inspection variant (Theorem 3) and a 1.37-approximation for the variant with value discounting only (Theorem 4).

Our framework is general enough to capture, and provide results for, a number of models already studied in the literature.

- The original paper introducing the optimal solution to the Pandora's Box problem by Weitzman [29] also encompasses a specific discounting factor for costs and rewards, as well a possibly non uniform processing time. These are captured as special cases of time-dependent costs and rewards, as well as value discounting functions (which regulate the deterioration of inspected rewards) in our general model.
- Berger et al. [4] study a model where boxes can be inspected and chosen only in predetermined time intervals. Their model can be captured by setting the inspection cost for "non-available" time slots to infinity, while the value discounting function captures the fact that inspected boxes cannot be chosen after the deadline.
- In the Pandora's Box with Commitment problem (introduced as an open problem for the first time in the extended version of Olszewsk and Weber [25] and then studied by Fu et al. [15], and Segev and Singla [27]), boxes can be inspected in any order, but the reward can only be collected immediately upon inspection. We capture this model with a value discounting function that drops to zero after the processing time has elapsed, so that a box retains value only at the round it is inspected.
- Our model also captures "offline" constraints that do not seem time-related at first, such as combinatorial constraints on the set of boxes that can be inspected [28]. Using processing times and time-varying cost functions, we capture both cardinality and knapsack constraints. Note that the idea of using value discounting with respect to time, thus allowing the flexibility to address these types of constraints, was already suggested by Weitzman [29], albeit in a more restricted way.
- Finally, our model strictly generalizes Online Pandora [14], where the order in which the boxes have to be processed is fixed in advance. This can be achieved in our model by setting the cost function to infinity for all the time steps but the one proposed by the online ordering.

Technical Challenges and Techniques. The crucial difficulty at the heart of the classic Pandora's Box problem lies in finding a balance between exploration and exploitation: inspecting new boxes improves the reward, but is more costly. This trade-off is solved by Weitzman [29] via the notion of *reservation value*, an index of the intrinsic value of each box. However, such a technical tool is not robust to modifications in the model, as the introduction of time-varying parameters implies that the importance of a box crucially depends on the time step in which it is inspected and, possibly, chosen. The first step of our approach consists in creating a copy for each box b_i for each time step t, so that b_i^t is a proxy for the strategy inspecting box i at time t. This procedure allows us to associate with each box a fixed cost and random reward but creates three problems: (i) we want to avoid that two proxies of the same boxes are inspected together, (ii) we want to force the strategy to inspect a box b_i^t exactly at time t, and (iii) we want to make sure that each inspected box has enough time to be processed. The second step of our approach consists then of imposing compatibility constraints on the boxes that our strategy inspects, so to enforce (i) to (iii). We do so by combining a submodular maximization routine with an adaptivity gap result [9], as in Singla [28]. In particular, the constraints induced by our model are captured by matching constraints in a block bipartite hypergraph, for which we derive an approximation algorithm. This leaves our strategy with a list of boxes that can be inspected one after the other while respecting (i) to (iii). Finally, our last step exploits a reduction to prophet inequality (see also Esfandiari et al. [14]) to find a good stopping rule. Notably, since we are reducing the problem to prophet inequality (where options are immediately discarded or accepted), we can ignore the deterioration of the reward in inspected boxes. Overall, our approach has similarities with Berger et al. [4], but with the crucial differences that boxes have time-varying parameters and non-uniform processing times (so that a simple transversal matroid is not enough to capture the incompatibility between boxes).

For the Pandora Over Time With Instant Inspection problem, we exploit the simplified structure of the hypergraph constraint (which collapses to a graph) to achieve a better approximation result. On the other hand, for the Pandora With Value Discounting, we exploit a reduction to free-order prophet, which directly provides a better approximation guarantee [10,20].

Further Related Work. Many other versions of Pandora's Box problem have been investigated in recent years, including problems with non-obligatory inspection [5,7,13,16], with interdependent valuations [11,12], combinatorial costs [3], precedence constraints [8], and contexts [2]. The Pandora's Box problem has also been investigated from a contract design perspective [21], and in the learning setting [17–19].

2 Model and Preliminaries

We study a generalization of the Pandora's Box problem in which time plays a crucial role for the decision maker. We call this variant the *Pandora's Box Over Time* problem, or simply *Pandora Over Time*. There is a set of boxes $[n] = \{1, \dots, n\}$. A *strategy* π inspects boxes sequentially and (possibly) adaptively. To be more precise, at each round t (starting from $t = 1$), the strategy π will be allowed to do (some of) the following: (a) inspect a box $i \in [n]$ that has not been considered before (if such boxes remain) and proceed to round $t + 1$, (b) stay idle[1] at round t and proceed to round $t + 1$, or (c) *halt*. Given a strategy π, we denote by T_π the random round it halts and by $S(\pi)$ the random ordered tuple of inspected boxes at rounds $1, 2, \dots, T_\pi$.

Unlike the classic Pandora's Box problem, where each box $i \in [n]$ is associated with a scalar cost, here i is associated with a *cost function* $\bar{c}_i : \mathbb{Z}_{>0} \to \mathbb{R}_{\geq 0}$. This extension allows us to model the changing cost of a box based on the time it is inspected. In general, we do not impose any assumptions on the form of the cost

[1] In fact, a strategy π will simulate option (b) by inspecting a "dummy" box that can be processed instantly and has a deterministic reward and cost of 0. It is without loss of generality to assume that such boxes are readily available.

functions. Additionally, each box has a *processing time* $p_i \in \mathbb{N}$, which represents the number of rounds a strategy must wait before it can inspect another box. Specifically, if a strategy π chooses to inspect box $i \in [n]$ at round t, the soonest π may inspect the next box is at round $t' = t + 1 + p_i$.

Our model also captures scenarios in which both the value sampled from each inspected box and the final claimed value depend on time. Specifically, the reward of each box $i \in [n]$ inspected at time t is drawn from a probability distribution D_{it}. We denote by V_{it} the random variable representing the reward drawn from D_{it}. We assume that all distributions D_{it} for $i \in [n]$ and $t \in \mathbb{Z}_{>0}$ are independent. Additionally, the reward in each box $i \in [n]$ is not constant and may degrade over time from the moment the box is inspected to the moment it is (potentially) chosen. Formally, each box is associated with a *value discounting* function $\bar{v}_i : \mathbb{R}_{\geq 0} \times \mathbb{N} \to \mathbb{R}_{\geq 0}$, known to the decision maker. Consider a strategy π that halts at time T_π. Suppose π inspects box $i \in [n]$ at time $t \leq T_\pi$ and samples a reward of V_{it} from D_{it}. If the strategy chooses to collect the reward from box i after $\tau = T_\pi - t$ rounds, it may collect a reward of $\bar{v}_i(V_{it}, \tau) \in \mathbb{R}_{\geq 0}$ rather than V_{it}. Note that we assume $\bar{v}_i(V_{it}, \tau) \leq \bar{v}_i(V_{it}, 0) = V_{it}$ for all $\tau \in \mathbb{N}$, i.e., $\bar{v}_i$ is non-increasing in the variable measuring the time passed since inspection.

We represent an instance of Pandora Over Time by $\mathcal{I} = \left(\bar{c}_i, p_i, (V_{it})_{t\in[H]}, \bar{v}_i\right)_{i\in[n]}$, where H (which is at least $n + \sum_{i=1}^{n} p_i$) denotes the *time horizon* of the instance. That is, there are nH reward probability distributions to which access is given as part of the input. Given a strategy π for $\mathcal{I}$, we use $t_i(\pi) \in \{1, \ldots, T_\pi\}$ to denote the random round at which strategy π inspects box $i \in S(\pi)$. Furthermore, we define the random utility of the decision maker for strategy π as

$$u_{\mathcal{I}}(\pi) := \max_{i\in S(\pi)} \bar{v}_i\left(V_{it_i(\pi)}, T_\pi - t_i(\pi)\right) - \sum_{i\in S(\pi)} \bar{c}_i\left(t_i(\pi)\right) \tag{1}$$

i.e., we assume that when a strategy halts, *the best available reward is always collected* at time T_π. We use π^* to define an optimal strategy, i.e., $\pi^* \in \arg\max_\pi \mathbb{E}_{\boldsymbol{v}\sim\boldsymbol{D}}[u_{\mathcal{I}}(\pi)]$, and we say that a strategy π is an α-approximation of an optimal strategy if $\alpha \cdot \mathbb{E}_{\boldsymbol{v}\sim\boldsymbol{D}}[u_{\mathcal{I}}(\pi)] \geq \mathbb{E}_{\boldsymbol{v}\sim\boldsymbol{D}}[u_{\mathcal{I}}(\pi^*)]$, for $\alpha \geq 1$.

It is not hard to observe that both the Pandora's Box problem and Pandora's Box with Commitment (where, recall, only the last reward can be collected) are special cases of our problem. To see this, consider an instance $(c_i, V_i)_{i\in[n]}$ of Pandora's Box. We can construct an instance $\mathcal{I}$ of Pandora Over Time such that for each $i \in [n]$, $p_i = 0$, $\bar{c}_i(t) = c_i$, and $V_{it} = V_i$ for all $t \in \mathbb{Z}_{>0}$. Then, by setting $\bar{v}_i(V_i, t) = V_i$ for all $t \in \mathbb{N}$ (respectively, $\bar{v}_i(V_i, 0) = V_i$ and $\bar{v}_i(\cdot, t) = 0$ for all $t \neq 0$), the utility obtained by a strategy π for $(c_i, V_i)_{i\in[n]}$ in Pandora's Box (respectively, Pandora's Box with Commitment) coincides with (1) for $(\mathcal{I}, \pi)$.

2.1 A Class of Related Stochastic Optimization Problems

In this section, we present a related class of Pandora's Box problems with added constraints on the sequences of inspected boxes. We call this variant *Constrained*

Pandora. This class of stochastic optimization problems, ever since being proposed by Singla [28], has sparked a rich line of work (see Related Work). Our purpose is to relate Pandora Over Time to instances of Constrained Pandora, to leverage certain known results. In this scenario, as in the classic Pandora's Box, there is a set $[n] = \{1, \dots, n\}$ of boxes, with each box $i \in [n]$ containing a random variable V_i drawn from a publicly known, non-negative distribution D_i. The distributions $D_1, \dots, D_n$ are independent. Moreover, each box $i \in [n]$ is associated with a known cost $c_i \geq 0$. As in our setting, a strategy π at each round t (starting from $t = 1$) may choose to inspect a box $i \in [n]$ (or stay idle/inspect a dummy box) and proceed to round $t+1$, or halt. Crucially, not all uninspected boxes can be inspected at a given time t. In Constrained Pandora, there is a predetermined collection of *feasible sequences* of inspected boxes, denoted by $\mathcal{F}$, and we say that a strategy is *feasible* if it adheres to $\mathcal{F}$. In other words, for any feasible strategy π, and for any random ordered tuple $(i_1, \dots, i_k)$ that is a subsequence of $S(\pi)$ with $k \in [T_\pi]$, it holds that $(i_1, \dots, i_k) \in \mathcal{F}$. We say that $\mathcal{F}$ is *prefix-closed* if for any tuple $(i_1, \dots, i_k) \in \mathcal{F}$ and any $j \in [k]$, it holds that $(i_1, \dots, i_j) \in \mathcal{F}$. We denote an instance of this problem as $\hat{\mathcal{I}} = \big((c_i, V_i)_{i \in [n]}, \mathcal{F}\big)$; when $\mathcal{F}$ contains any possible sequence, i.e., when the problem is the unconstrained Pandora's Box problem, we just write $\hat{\mathcal{I}} = \big((c_i, V_i)_{i \in [n]}\big)$ instead. Let π be a feasible strategy for $\hat{\mathcal{I}}$. For each $i \in [n]$, let $I_i(\pi)$ and $A_i(\pi)$ be the indicator random variables that signify whether box i is inspected by π and whether the reward of box i is collected, respectively. Note that $A_i(\pi) \leq I_i(\pi)$ always holds because a box must be inspected by π before its reward is collected. Additionally, since π is a feasible strategy, the indicator random variables respect the exploration constraints imposed by $\mathcal{F}$. Finally, $A_i(\pi) = 1$ holds for the box with the maximum reward observed among those in $S(\pi)$. Therefore, we can write the random utility of this strategy π as:

$$u_{\hat{\mathcal{I}}}(\pi) := \sum_{i=1}^{n} A_i(\pi) V_i - \sum_{i=1}^{n} I_i(\pi) c_i = \max_{i \in S(\pi)} V_i - \sum_{i \in S(\pi)} c_i. \tag{2}$$

Characterization of Optimal Strategies via Reservation Values. The work of Kleinberg et al. [23] revived interest in the Pandora's Box problem and drew the attention of the economics and computation community towards it. Among other contributions, they provided a new proof of the optimality of Weitzman's rule for the Pandora's Box problem. Later, Singla [28] applied this to more general problems with constraints and different optimization objectives.

Let $\hat{\mathcal{I}}$ be an instance of Constrained Pandora. We define the *reservation value* r_i of each box $i \in [n]$ to be the (unique) solution to the equation $\mathbb{E}_{V_i \sim D_i}[(V_i - r_i)^+] = c_i$ (where $(x)^+$ is a shortcut for $\max(x, 0)$). Moreover, for each box $i \in [n]$, let $Y_i := \min(V_i, r_i)$. We now state a result of Kleinberg et al. [23].

Lemma 1 (rf. Lemma 1 of Kleinberg et al. [23]**).** *Let* $\hat{\mathcal{I}} = \big((c_i, V_i)_{i \in [n]}, \mathcal{F}\big)$. *Then:*

$$\mathbb{E}\left[u_{\hat{\mathcal{I}}}(\pi)\right] \leq \mathbb{E}\left[\sum_{i=1}^{n} A_i(\pi) Y_i\right], \tag{3}$$

for every strategy π. This becomes an equality if π has the following property: whenever it inspects a box i and samples a value $V_i \geq r_i$, it accepts the box. In other words, the probability that $I_i(\pi) = 1$, $A_i(\pi) = 0$, and $V_i > r_i$ is 0.

Bounding the Adaptivity Gap. Consider an instance $\hat{\mathcal{I}} = \big((0, V_i)_{i\in[n]}, \mathcal{F}\big)$ with $\mathcal{F}$ being prefix-closed. Even though inspecting boxes incurs no cost in this stochastic optimization problem, finding the optimal adaptive strategy may be challenging due to the constraints imposed by $\mathcal{F}$. However, Singla [28] showed that the expected utility of the optimal *non-adaptive* strategy is a constant approximation of the optimal adaptive strategy, with the ratio being referred to as the *adaptivity gap*. The constant was later improved by Bradac et al. [9].

Lemma 2 (due to[9]). *Consider an instance $\hat{\mathcal{I}} = \big((0, V_i)_{i\in[n]}, \mathcal{F}\big)$ with $\mathcal{F}$ being prefix-closed. Then, $\max_{S\in\mathcal{F}} \mathbb{E}[\max_{i\in S} V_i] \geq \frac{1}{2} \cdot \max_{\pi\in\Pi} \mathbb{E}[\max_{i\in S(\pi)} V_i]$, where Π is the set of all adaptive strategies for $\hat{\mathcal{I}}$.*

2.2 Prophet Inequality Problems

A final class of stochastic optimization problems useful to our work is Prophet Inequality problems. In such problems, there is a set of random variables $Y_1, \ldots, Y_n$, each of which is drawn from a publicly known, non-negative distribution D_i. The distributions $D_1, \ldots, D_n$ are independent, and in the standard Prophet Inequality setting, the order in which the random variables arrive is adversarial but known. That is, let σ be a permutation of $(1, \ldots, n)$ given to the decision maker (also known as "gambler"). At round $i \in [n]$, the decision maker samples $Y_{\sigma(i)} \sim D_{\sigma(i)}$ and may either accept $Y_{\sigma(i)}$ or move on to the next random variable. Her goal is to approximate $\mathbb{E}[\max_{i\in[n]} Y_i]$, i.e., to compete with a prophet who has access to all random realizations beforehand. Samuel-Cahn [26] showed that there is a simple threshold-based rule, i.e., a strategy that accepts the first sampled value that exceeds a predetermined threshold τ, which achieves a 2-approximation. Let $\mathcal{I}_{\text{PR}} = (V_i)_{i\in[n]}$ be an instance of the problem. We denote by $i^* \in [n]$ the random index for which $Y_{\sigma(i^*)} \geq \tau$ and $Y_{\sigma(j)} < \tau$ for $j < i^*$. An important variant of the problem is the *Free Order Prophet* problem in which the decision maker is free to choose the permutation σ; note that this is the same as Pandora with Commitment and zero costs. The results we need are summarized in the next lemma.

Lemma 3 (due to[22,26], and[10]). *Let $\mathcal{I}_{\text{PR}} = (Y_i)_{i\in[n]}$ be an instance of the Prophet Inequality problem and σ be any given permutation of the random variables. For threshold $\tau = \frac{1}{2}\mathbb{E}\big[\max_{j\in[n]} Y_j\big]$, it holds that $\mathbb{E}\left[Y_{\sigma(i^*)}\right] \geq \frac{1}{2}\mathbb{E}\left[\max_{i\in[n]} Y_i\right]$. Moreover, there exists a permutation $\hat{\sigma}$ of the random variables and a threshold τ, so that $\mathbb{E}\left[Y_{\hat{\sigma}(i^*)}\right] \geq 0.7258\,\mathbb{E}\left[\max_{i\in[n]} Y_i\right]$.*

2.3 Monotone Submodular Functions

We now define monotone submodular functions, as part of our problem (similarly as many related problems) will reduce to maximizing a monotone submodular function subject to combinatorial constraints.

Definition 1. *Given a ground set* $[n]$, *a function* $f : 2^{[n]} \to \mathbb{R}$ *is* monotone submodular *if:* (i) $f(S) \leq f(T)$, *for every* $S \subseteq T \subseteq [n]$ *(monotonicity), and* (ii) $f(S) + f(T) \geq f(S \cup T) + f(S \cap T)$, *for every* $S, T \subseteq [n]$ *(submodularity).*

Let $Y_1, \ldots, Y_n$ be a collection of random variables, with each Y_i being drawn from a non-negative distribution D_i. Consider the set function defined as $f(S) := \mathbb{E}[\max_{i \in S} Y_i]$ for each $S \subseteq [n]$ (using the convention that $\max_{i \in \emptyset} Y_i = 0$). It is not hard to show that f is monotone submodular.

Lemma 4. *The function* f *above is normalized (i.e.,* $f(\emptyset) = 0$*), non-negative, monotone submodular.*

3 An $O(1)$ Approximation for Pandora Over Time

Next we show how to derive a constant approximation to the guarantee of the optimal strategy for Pandora Over Time in its full generality. Note that the problem is NP-hard, as it is a generalization of the Free Order Prophet problem which is known to be NP-hard, even for distributions with support of size 3 [1]. The same is true even for the special cases studied in Sect. 4.

Theorem 1. *There exists a strategy,* π_{MAIN}, *for Pandora Over Time which can be computed efficiently and provides a* 21.3*-approximation to an optimal strategy.*

The section is dedicated to the proof of Theorem 1 and is structured as follows. In Sect. 3.1, for each Pandora Over Time instance, $\mathcal{I}$, we identify a constrained instance, $\hat{\mathcal{I}}$, with a carefully constructed prefix-closed constraint. We refer to $\hat{\mathcal{I}}$ as the *proxy* instance for $\mathcal{I}$. This proxy instance will be central to devising a "good" strategy for our problem in Sect. 3.2, where we present our main result: a strategy that approximates the guarantee of the optimal strategy. To this end, we argue about how the performance of our strategy crucially depends on the performance of an algorithm for maximizing a monotone submodular function under a particular feasibility constraint. Finally, in the full version of our paper, we obtain a constant factor approximation algorithm for the submodular maximization problem in question. Then Theorem 1 directly follows by Corollary 1 and Theorem 2.

We often work with two instances (typically of different variants of the problem) and their respective strategies within the same proof. So, when needed for clarity, we write, e.g., $S(\mathcal{I}, \pi)$ and $S(\hat{\mathcal{I}}, \hat{\pi})$ instead of $S(\pi)$ and $S(\hat{\pi})$, respectively.

3.1 Reduction to a Constrained Pandora Problem

Here we show that each Pandora Over Time instance $\mathcal{I}$ has an equivalent representation as a Constrained Pandora instance $\hat{\mathcal{I}}$ with certain exploration constraints. Moreover, we define a class of strategies for each such instance, each of which has a one-to-one correspondence with a strategy for the underlying Pandora Over Time instance $\mathcal{I}$. The properties of $\hat{\mathcal{I}}$ and its associated strategy will be critical in our analysis. We begin with defining a hypergraph that we will later associate with $\hat{\mathcal{I}}$.

Definition 2. *Given a Pandora Over Time instance $\mathcal{I}$, its associated bipartite hypergraph $\mathcal{H}(\mathcal{I}) = (L, R, E)$ is defined as follows:*

- *For each $i \in [n]$, there is a vertex $b_i \in L$.*
- *For each timeslot $j \in [H]$, there is a vertex $t_j \in R$.*
- *For each $(i, j) \in [n] \times [H]$, there is a hyperedge $e(i, j) := \{b_i\} \cup \{t_k \mid k = j, \dots, j + p_i\}$ in E.*

To simplify the notation, when $\mathcal{I}$ is clear, we will sometimes write $\mathcal{H}$ instead of $\mathcal{H}(\mathcal{I})$. We use $\mathcal{M}(\mathcal{I})$ to denote the set of *matchings* of $\mathcal{H}(\mathcal{I})$, i.e., all the collections of disjoint hyperedges of $\mathcal{H}(\mathcal{I})$. Note that while $\mathcal{H}(\mathcal{I})$ is indeed a bipartite hypergraph (i.e., its vertices are partitioned into two sets, L and R, such that $|L \cap e| \leq 1$, for every hyperedge $e \in E$), it has even more structure. In particular, for every hyperedge $e \in E$, the set $R \cap e$ consists of *consecutive* (with respect to their index) vertices of R. We call such hypergraphs *block bipartite*. This structure on the hyperedges of $\mathcal{H}(\mathcal{I})$ will allow us to approximate the optimal solution to a submodular maximization problem on $\mathcal{M}(\mathcal{I})$ in polynomial time; see the end of this section for the definition of the *Submodular Block Matching* problem.

Proxy Instance and Proxy Strategy. For each instance $\mathcal{I}$, we construct a Constrained Pandora instance denoted by $\hat{\mathcal{I}}_{\text{PROXY}}$. The set of boxes of $\hat{\mathcal{I}}_{\text{PROXY}}$ is $[n] \times [H]$, i.e., for each box $i \in [n]$ and each timeslot $t \in [H]$, we add a box labeled (i, t) to $\hat{\mathcal{I}}_{\text{PROXY}}$. Then, for each box (i, t) we set the reward $V_{(i,t)}$ to follow the distribution D_{it} and the cost of the box (i, t) to be $c_{(i,t)} = \bar{c}_i(t)$. Finally, we construct the collection of ordered tuples $\mathcal{F}$ as follows: for each matching $\{e(i_1, j_1), \dots, e(i_k, j_k)\} \subseteq \mathcal{M}(\mathcal{I})$, we add to $\mathcal{F}$ the ordered tuple of the boxes $\left(\left(i_{\sigma(1)}, j_{\sigma(1)}\right), \dots, \left(i_{\sigma(k)}, j_{\sigma(k)}\right)\right)$ where σ is a permutation of the boxes so that $j_{\sigma(1)} < \cdots < j_{\sigma(k)}$. Furthermore, for each strategy π of $\mathcal{I}$, we define below a strategy $\hat{\pi}$ for $\hat{\mathcal{I}}_{\text{PROXY}}$ which we call the *proxy strategy* of π.

Definition 3. *Let $\mathcal{I}$ be an instance of Pandora Over Time and $\hat{\mathcal{I}}_{\text{PROXY}}$ be its proxy Constrained Pandora instance. Given a strategy π for $\mathcal{I}$, we define its* proxy strategy *$\hat{\pi}$ as the strategy that satisfies the following:*

- *$\hat{\pi}$ inspects box $(i, t) \in [n] \times [H]$ if and only if box $i \in S(\pi)$ and $t = t_i(\pi)$.*
- *$\hat{\pi}$ stays idle at time $j \in [T_\pi]$ if and only if strategy π stays idle at time j.*
- *$\hat{\pi}$ halts at time $j \in [T_\pi]$ if and only if strategy π halts at time j.*

The following lemma connects the two problems (the original and the proxy) and plays a key role in our analysis.

Lemma 5. *Let $\mathcal{I}$ be an instance of Pandora Over Time. For every strategy π for $\mathcal{I}$, its proxy strategy $\hat{\pi}$ for $\hat{\mathcal{I}}_{\text{PROXY}}$ is feasible. Furthermore,*

$$\mathbb{E}\left[u_{\hat{\mathcal{I}}_{\text{PROXY}}}(\hat{\pi})\right] = \mathbb{E}\left[\max_{i \in S(\mathcal{I},\pi)} V_{it_i(\pi)} - \sum_{i \in S(\mathcal{I},\pi)} \bar{c}_i\left(t_i(\pi)\right)\right]. \tag{4}$$

Proof. We use $\hat{\mathcal{I}} := \hat{\mathcal{I}}_{\text{PROXY}}$ for ease of notation. Let $(i_1, \ldots, i_k) = S(\mathcal{I}, \pi)$. We first argue that strategy $\hat{\pi}$ is indeed feasible for $\hat{\mathcal{I}}$, i.e., it satisfies the sequence constraint $\mathcal{F}$ of $\hat{\mathcal{I}}$. By the above rules, the sequence of inspected boxes is $S(\hat{\mathcal{I}}, \hat{\pi}) = ((i_1, t_{i_1}(\pi)), \ldots, (i_k, t_{i_k}(\pi)))$. Clearly, the precedence requirement is satisfied, i.e., for any two boxes j, j' with $i_j(\pi) < t_{i_{j'}}(\pi)$, box $(i_j, t_{i_j}(\pi))$ precedes box $(i_{j'}, t_{i_{j'}}(\pi))$ in $S(\hat{\mathcal{I}}, \hat{\pi})$. In Claim 1, we show that the subgraph of $\mathcal{H}(\mathcal{I})$ induced by $(\hat{\mathcal{I}}, \hat{\pi})$ is a matching.

Claim 1. *The set* $M = \{e(i_1, t_{i_1}(\pi)), \ldots, e(i_k, t_{i_k}(\pi))\}$ *is a matching.*

Proof. Suppose, toward a contradiction, that M is not a matching. This means that there is a pair $(i, t_i(\pi))$ and $(j, t_j(\pi))$ with $e(i, t_i(\pi)) \cap e(j, t_j(\pi)) \neq \emptyset$. Assume, without loss of generality, that $t_i(\pi) < t_j(\pi)$. Since each box in $\mathcal{I}$ can only be inspected once by π, by the first bullet of Definition 3, the same is true for $\hat{\mathcal{I}}$ and $\hat{\pi}$. Thus, clearly $i \neq j$, and therefore $b_i \neq b_j$. Furthermore, since strategy π for $\mathcal{I}$ respects the processing time for both $i, j \in S(\pi)$, it holds that $t_j(\pi) \geq t_i(\pi) + p_i + 1$. If $p_i = 0$, we trivially obtain that $e(i, t_i(\pi)) \cap e(j, t_j(\pi)) = \emptyset$, a contradiction. Consider now the case of $p_i \geq 1$. In this case, strategy π stays idle in the interval $[t_i(\pi)+1, t_i(\pi)+p_i]$. However, by the second bullet of Definition 3, strategy $\hat{\pi}$ also stays idle in the interval $[t_i(\pi)+1, t_i(\pi)+p_i]$. Consequently, by the construction of $\mathcal{H}(\mathcal{I})$, there is no hyperedge $e' \in E$ such that $e(i, t_i(\pi)) \cap e' \neq \emptyset$, which implies that $e(i, t_i(\pi)) \cap e(j, t_j(\pi)) = \emptyset$. This is a contradiction. □

Now notice that, by the construction of $(\hat{\mathcal{I}}, \hat{\pi})$, we have

$$\mathbb{E}\left[\max_{i \in S(\mathcal{I},\pi)} V_{it_i(\pi)} - \sum_{i \in S(\mathcal{I},\pi)} \bar{c}_i(t_i(\pi))\right] = \mathbb{E}\left[\max_{(i,j) \in S(\hat{\mathcal{I}},\hat{\pi})} V_{(i,j)} - \sum_{(i,j) \in S(\hat{\mathcal{I}},\hat{\pi})} c_{(i,j)}\right]$$
$$= \mathbb{E}\left[u_{\hat{\mathcal{I}}}(\hat{\pi})\right].$$

which concludes the proof. □

We continue by presenting an upper bound on the optimal expected utility of Pandora Over Time instances, which will be useful in the analysis for our main result in Sect. 3.2. We slightly abuse notation and write $Y(e(i,j)) := Y_{(i,j)} = \min\{V_{(i,j)}, r_{(i,j)}\}$ for each hyperedge $e(i,j) \in E[\mathcal{H}]$ and box $(i,j) \in [n] \times [H]$.

Lemma 6. *Let $\mathcal{I}$ be an instance of Pandora Over Time and π^* be an optimal strategy for $\mathcal{I}$. Then,* $\mathbb{E}[u_{\mathcal{I}}(\pi^*)] \leq 2 \cdot \max_{M \in \mathcal{M}(\mathcal{I})} \mathbb{E}\left[\max_{e(i,j) \in M} Y(e(i,j))\right]$.

Proof. We use $\hat{\mathcal{I}} := \hat{\mathcal{I}}_{\text{PROXY}}$ for brevity. We have:

$$\mathbb{E}[u_{\mathcal{I}}(\pi^*)] = \mathbb{E}\left[\max_{i \in S(\mathcal{I},\pi^*)} \bar{v}_i\left(V_{it_i(\pi^*)}, T_{\pi^*} - t_i(\pi^*)\right) - \sum_{i \in S(\mathcal{I},\pi^*)} \bar{c}_i(t_i(\pi^*))\right]$$
$$\leq \mathbb{E}\left[\max_{i \in S(\mathcal{I},\pi^*)} V_{it_i(\pi^*)} - \sum_{i \in S(\mathcal{I},\pi^*)} \bar{c}_i(t_i(\pi^*))\right] = \mathbb{E}\left[u_{\hat{\mathcal{I}}}(\hat{\pi})\right]$$

$$\leq \mathbb{E}\Bigg[\sum_{(i,j)\in S(\hat{\mathcal{I}},\hat{\pi})} A_{(i,j)}(\hat{\pi}) Y_{(i,j)}\Bigg] \leq \mathbb{E}\Bigg[\sum_{(i,j)\in S(\hat{\mathcal{I}},\hat{\pi})} A_{(i,j)}(\hat{\pi}) \max_{(k,\ell)\in S(\hat{\mathcal{I}},\hat{\pi})} Y_{(k,\ell)}\Bigg]$$

$$\leq \mathbb{E}\Bigg[\max_{(k,\ell)\in S(\hat{\mathcal{I}},\hat{\pi})} Y_{(k,\ell)}\Bigg] \leq \max_{\pi} \mathbb{E}\Bigg[\max_{(k,\ell)\in S(\hat{\mathcal{I}},\pi)} Y_{(k,\ell)}\Bigg]. \tag{5}$$

The first equality follows from the definition of the utility of a Pandora Over Time instance in (1). The first inequality follows from the fact that $\bar{v}_i$ is non-increasing with respect to the time passed since inspection. The second equality follows since, by Lemma 5, strategy $\hat{\pi}$ for the proxy constrained instance $\hat{\mathcal{I}}$ satisfies (4). Then, we apply Lemma 1 for $(\hat{\mathcal{I}}, \hat{\pi})$ and obtain the second inequality due to (3). Finally, the third inequality follows since $\hat{\pi}$ inspects at most one box of $\hat{\mathcal{I}}$ for each random realization.

Observe that the RHS of (5) equals the optimal expected utility of the Constrained Pandora instance with no costs $\big((0, Y_{(i,j)})_{(i,j)\in[n]\times[H]}, \mathcal{F}\big)$. Since $\mathcal{F}$ is a prefix-closed constraint, we can apply Lemma 2 for this instance and obtain:

$$\max_{\pi} \mathbb{E}\Bigg[\max_{(k,\ell)\in S(\hat{\mathcal{I}},\pi)} Y_{(k,\ell)}\Bigg] \leq 2 \cdot \max_{S\in\mathcal{F}} \mathbb{E}\Big[\max_{(i,j)\in S} Y_{(i,j)}\Big]$$

$$= 2 \cdot \max_{M\in\mathcal{M}(\mathcal{I})} \mathbb{E}\Big[\max_{e(i,j)\in M} Y\left(e(i,j)\right)\Big], \tag{6}$$

where the equality directly follows by the construction of the hypergraph $\mathcal{H}$. Combining (5) and (6) completes the proof. □

Given an instance $\mathcal{I}$ and its associated hypergraph $\mathcal{H} = \mathcal{H}(\mathcal{I})$, consider the set function $f : 2^{E[\mathcal{H}]} \to \mathbb{R}_{\geq 0}$ defined as

$$f(M) := \mathbb{E}\Big[\max_{e(i,j)\in M} Y_{(i,j)}\Big], \quad \text{for each} \quad M \in E\left[\mathcal{H}\right]. \tag{7}$$

By Lemma 4, f is a non-negative monotone submodular function with ground set $E\left[\mathcal{H}\right]$. Under this perspective, we can observe that

$$\max_{M\in\mathcal{M}(\mathcal{I})} \mathbb{E}\Big[\max_{e(i,j)\in M} Y\left(e(i,j)\right)\Big] = \max_{M\in\mathcal{M}(\mathcal{I})} f(M)$$

and interpret the inequality of Lemma 6 as follows: the optimal expected utility of a Pandora Over Time instance is upper-bounded by 2 times the optimal solution of an instance of a monotone submodular maximization problem subject to a matching in a block bipartite hypergraph constraint. We conclude the section with the formal statement of this optimization problem.

The Submodular Maximization Problem. Let $\mathcal{H} = (L, R, E)$ be a block bipartite hypergraph, $\mathcal{M} \subseteq 2^E$ be its set of matchings, and $f : 2^E \to \mathbb{R}_{\geq 0}$ be a monotone submodular function. The Submodular Matching on Block Bipartite Hypergraphs problem, or *Submodular Block Matching* for short, asks for a matching $M^* \in \arg\max_{M\in\mathcal{M}} f(M)$. In the full version of our paper we show how to get a polynomial-time 5.32-approximation algorithm for the problem.

Theorem 2. *There is a polynomial-time* 5.32*-approximation algorithm for Submodular Block Matching.*

3.2 Our Order-Non-adaptive Strategy

In this section we present π_{MAIN} (Strategy 1), a strategy for Pandora Over Time which we will show to be a 21.3-approximation of the optimal expected utility.

Strategy 1: π_{MAIN}

▷ **Input:** An instance $\mathcal{I} = \left(\bar{c}_i, p_i, (V_{it})_{t\in[H]}, \bar{v}_i\right)_{i\in[n]}$ and an α-approximation algorithm ALG for Submodular Block Matching.

// Phase 1: determine schedule of inspection and threshold

Construct the bipartite hypergraph $\mathcal{H}(\mathcal{I}) = (L, R, E)$ according to Def. 2.

Let $f(\cdot)$ be the monotone submodular function on 2^E defined in (7)

Find matching $M = \{e(i_1, t_1), \ldots, e(i_k, t_k)\} \subseteq E$ using ALG on instance with objective f and constraints induced by the matchings of $\mathcal{H}(\mathcal{I})$. // $t_1 < \cdots < t_k$

Set $\tau = 0.5 \cdot \mathbb{E}\left[\max_{e(i,t)\in M} Y(e(i,t))\right]$.

// Phase 2: threshold-based strategy based on the schedule Phase 1

for $\ell = 1, \ldots, k$ **do**

  if $r_{(i_\ell, t_\ell)} \geq \tau$ **then**

    Sample $V_{(i_\ell,t_\ell)} \sim D_{(i_\ell,t_\ell)}$ at time t_ℓ.

    if $V_{(i_\ell,t_\ell)} \geq \tau$ **then**

      halt // the reward of the last inspected box is collected

---

Given an instance of Pandora Over Time, the first step of π_{MAIN} (Phase 1) is to determine a preliminary *schedule* of inspection times *before* inspecting a single box. Note that the strategy may eventually not inspect all the boxes in the preliminary set, as it may halt sooner based on a threshold-based stopping rule we specify in Phase 2. Such strategies are called order-non-adaptive in the literature, see, e.g., Beyhaghi and Cai [6]. The stopping rule we specify is inspired by the approach of Esfandiari et al. [14]. The crux of this approach is to relate the expected utility of our Pandora Over Time problem to the expected utility achieved by the "gambler" in an instance $\mathcal{I}_{\text{PR}}$. Once this is achieved, we then relate the performance of our algorithm to the performance of the prophet.

Lemma 7. *Let $\mathcal{I}$ be an Instance of Pandora Over Time and let* ALG *be an α-approximation algorithm for Submodular Block Matching. It holds that*

$$\mathbb{E}[u_{\mathcal{I}}(\pi_{\text{MAIN}})] \geq \frac{1}{2\alpha} \cdot \max_{M\in\mathcal{M}(\mathcal{I})} \mathbb{E}\left[\max_{e(i,j)\in M} Y\left(e(i,j)\right)\right].$$

Proof. Denote $\pi := \pi_{\text{MAIN}}$ and $\hat{\mathcal{I}} := \hat{\mathcal{I}}_{\text{PROXY}}$ for brevity. Moreover, let $\hat{\pi}$ be the proxy strategy of π for $\hat{\mathcal{I}}$. We have:

$$\mathbb{E}[u_{\mathcal{I}}(\pi)] = \mathbb{E}\left[\max_{i\in S(\mathcal{I},\pi)} \bar{v}_i\left(V_{it_i(\pi)}, T_\pi - t_i(\pi)\right) - \sum_{i\in S(\mathcal{I},\pi)} \bar{c}_i\left(t_i(\pi)\right)\right]$$

$$= \mathbb{E}\Big[\max_{i\in S(\mathcal{I},\pi)} \bar{v}_i\left(V_{it_i(\pi)},0\right) - \sum_{i\in S(\mathcal{I},\pi)} \bar{c}_i\left(t_i(\pi)\right)\Big]$$

$$= \mathbb{E}\Big[\max_{i\in S(\mathcal{I},\pi)} V_{it_i(\pi)} - \sum_{i\in S(\mathcal{I},\pi)} \bar{c}_i\left(t_i(\pi)\right)\Big] = \mathbb{E}\left[u_{\hat{\mathcal{I}}}(\hat{\pi})\right]. \quad (8)$$

The first equality follows from the definition of the expected utility of a Pandora Over Time instance in (1). The second equality follows from the fact that π "collects" the reward of the box inspected at time T_π; indeed, π halts at T_π and at that time the last reward is the only one exceeding τ, so it is the one collected. The third equality follows from the definition of the function $\bar{v}_i$ for $i \in [n]$. Finally, the fourth equality holds since, by Lemma 5, strategy $\hat{\pi}$ for $\hat{\mathcal{I}}$ satisfies (4). To continue, we show that $(\hat{\mathcal{I}}, \hat{\pi})$ satisfies the second condition of Lemma 1; in particular, we show that here (3) holds with equality.

Claim 2. *It holds that* $\mathbb{E}\left[u_{\hat{\mathcal{I}}}(\hat{\pi})\right] = \mathbb{E}\big[\sum_{(i,t)\in[n]\times[H]} A_{(i,t)}(\hat{\pi})Y_{(i,t)}\big]$.

Proof. According to Lemma 3, this identity holds if the following condition is satisfied for $(\hat{\mathcal{I}}, \hat{\pi})$: $\left(\exists (i,t) \in [n]\times[H] : I_{(i,t)}(\hat{\pi}) = 1 \wedge V_{(i,t)} \geq r_{(i,t)}\right) \Rightarrow A_{(i,t)}(\hat{\pi}) = 1$. By the first bullet of Definition 3, $I_{(i,t)}(\hat{\pi}) = 1$ if and only if $i \in S(\pi)$ and $t_i(\pi) = t$. This happens only if the condition on Line 6 of π_{MAIN} is evaluated to **true**. Therefore, $r_{(i,t)} \geq \tau$. Since, by assumption, we additionally have that $V_{(i,t)} \geq r_{(i,t)}$, we can conclude that $V_{(i,t)} \geq \tau$. However, this implies that the condition on Line 8 of π_{MAIN} is evaluated to **true** and therefore, strategy π halts (by Line 9). By the third bullet of Definition 3, strategy $\hat{\pi}$ halts as well. Note that, for each box $j \in S(\pi)$ with $t_j(\pi) < t$ it holds that $V_{(j,t_j(\pi))} < \tau \leq V_{(i,t)}$. Therefore, $A_{(i,t)}(\hat{\pi}) = 1$ (the reward of box (i,t) in $\hat{\mathcal{I}}$ is the maximum among inspected boxes), and the claim follows. □

Let M be the matching returned by ALG with input the function f (as in (7)) and the hypergraph $\mathcal{H}(\mathcal{I})$, and let $i^*(\pi)$ be the random variable denoting the box inspected at time T_π. Clearly, by Line 6 and by Line 8, it holds that $Y_{(i^*(\pi),T_\pi)} = \min\left\{r_{(i^*(\pi),T_\pi)}, V_{(i^*(\pi),T_\pi)}\right\} \geq \tau$. Furthermore, $i^*(\pi)$ is the only box in $S(\pi)$ for which this holds. By invoking Claim 2, we can continue (8) as follows:

$$\mathbb{E}[u_{\mathcal{I}}(\pi)] = \mathbb{E}\Big[\sum_{(i,t)\in[n]\times[H]} A_{(i,t)}(\hat{\pi})Y_{(i,t)}\Big] = \mathbb{E}\left[Y_{(i^*,T_\pi)}\right] = \mathbb{E}\left[Y(e(i^*,T_\pi))\right]$$

$$\geq \frac{1}{2}\cdot\mathbb{E}\Big[\max_{e(i,j)\in M} Y_{e(i,j)}\Big] \geq \frac{1}{2\alpha}\cdot \max_{M\in\mathcal{M}(\mathcal{I})} \mathbb{E}\Big[\max_{e(i,j)\in M} Y\left(e(i,j)\right)\Big].$$

The first inequality follows by observing that $\mathbb{E}\left[Y(e(i^*,T_\pi))\right]$ equals the expected value of the gambler for the instance $\mathcal{I}_{\text{PR}} = (Y_{e(i,j)})_{e(i,j)\in M}$; therefore, the inequality holds due to (3). Finally the second inequality follows from that the matching M is the solution of an α-approximation algorithm for this precise objective. The proof follows. □

Combining Lemmata 6 and 7, we directly get the following general result that relates the guarantee of our order-non-adaptive strategy with the quality of approximation we can achieve for Submodular Block Matching.

Corollary 1. *Let $\mathcal{I}$ be an Instance of Pandora Over Time, π^* be an optimal strategy for $\mathcal{I}$, and* ALG *be an α-approximation algorithm for Submodular Block Matching. It holds that* $\mathbb{E}[u_{\mathcal{I}}(\pi_{\text{MAIN}})] \geq \frac{1}{4\alpha} \cdot \mathbb{E}\left[u_{\mathcal{I}}(\pi^*)\right]$.

Of course, Corollary 1 is a conditional version of our Theorem 1. The proof of the latter follows, by using in Corollary 1 the 5.32-approximation algorithm for Submodular Block Matching of Theorem 2.

4 Two Natural Special Cases

Our Pandora Over Time problem is, by design, very general. This is clear from the discussion on all the different variants of related problems it captures as special cases. It is reasonable to expect that there are several other meaningful restricted versions of Pandora Over Time worth studying. Here we turn to two such special cases; one where the processing times are all zero, and another one where the cost and the distribution of each box do not change over time. Both restrictions are in line with the majority of the related literature, and in both cases we are able to significantly improve over Theorem 1.

4.1 Pandora Over Time With Instant Inspection

Although the use of processing times was suggested by Weitzman [29] in the paper that introduced the Pandora's Box problem, their effect has not been studied before. It is true that the varying processing times create technical complications, forcing us to work with matchings in non-uniform hypergraphs. Removing the processing times is a very natural restriction of Pandora Over Time, resulting in what we call *Pandora Over Time With Instant Inspection.* The general instance of this problem is $\mathcal{I} = \left(\bar{c}_i, 0, (V_{it})_{t\in[H]}, \bar{v}_i\right)_{i\in[n]}$.

Theorem 3. *Fix any constant $\varepsilon > 0$. There exists a strategy for Pandora Over Time With Instant Inspection (a variant of π_{MAIN} (Strategy 1)) which can be computed efficiently and provides an $(8+\varepsilon)$-approximation to an optimal strategy.*

Proof. Let $\mathcal{I}$ be an instance of Pandora Over Time With Instant Inspection. The main observation here is that now $\mathcal{H}(\mathcal{I})$ is not a hypergraph any more, but a bipartite graph instead. That is, in this case, Submodular Block Matching becomes the much better understood Submodular Matching on Bipartite Graphs problem. So, instead of Theorem 2, one would rather invoke the $(2+\delta)$-approximation algorithm of Lee et al. [24] (for any fixed $\delta > 0$) for maximizing a submodular function subject to the intersection of two matroids (bipartite matching being a special case of that). That is, π_{MAIN} in Phase 1 would construct the graph $\mathcal{H}(\mathcal{I})$ and run the $(2+\delta)$-approximation algorithm of Lee et al. [24] for $\delta = \varepsilon/4$, to obtain the matching M. From that point on, everything is analogous to before.

Combining Corollary 1 with this $(2 + \varepsilon/4)$ factor, completes the proof. □

4.2 Pandora Over Time With Constant Costs and Distributions

The aspects of Pandora Over Time that give it the most flexibility are probably the time-dependent costs and distributions. Just by varying the cost functions accordingly, it is easy to simulate a large number of scenarios, like deadlines, time windows, and knapsack constraints with respect of time. It is thus reasonable to consider the simpler class of instances where boxes still have processing times and the values of inspected boxes may degrade over time, but both the costs and the distributions of the rewards are not time-dependent. We call this restriction *Pandora With Value Discounting.* Formally, for each $i \in [n]$ and each $t \in [H]$, we have that $D_{it} = D_i$. Furthermore, for each $i \in [n]$ and each $t \in \mathbb{Z}_{>0}$, we have $\bar{c}_i(t) = c_i$. That is, we consider instances of the form $\mathcal{I} = (c_i, p_i, V_i, \bar{v}_i)_{i\in[n]}$.

Theorem 4. *There exists a strategy for Pandora With Value Discounting (see of π_{FIXED} (Strategy 2)) which can be computed efficiently and provides an* 1.37*-approximation to an optimal strategy.*

We begin with upper bounding the optimal expected utility of such instances.

Lemma 8. *Let $\mathcal{I}$ be an instance of Pandora With Value Discounting. Then,* $\mathbb{E}[u_{\mathcal{I}}(\pi^*)] \leq \mathbb{E}[\max_{i\in[n]} Y_i]$.

Proof. Consider the Pandora's Box instance $\hat{\mathcal{I}} = \big((c_i, V_i)_{i\in[n]}\big)$. We have:

$$\begin{aligned}\mathbb{E}[u_{\mathcal{I}}(\pi^*)] &= \mathbb{E}\left[\max_{i\in S(\pi^*)} \bar{v}_i\left(V_i, T_{\pi^*} - t_i(\pi^*)\right) - \sum_{i\in S(\pi^*)} c_i\right] \leq \mathbb{E}\left[\max_{i\in S(\pi^*)} V_i - \sum_{i\in S(\pi^*)} c_i\right]\\ &= \mathbb{E}\left[u_{\hat{\mathcal{I}}}(\pi^*)\right] \leq \mathbb{E}\left[\sum_{i=1}^{n} A_i(\pi^*)Y_i\right] \leq \mathbb{E}\left[\max_{i\in[n]} Y_i\right].\end{aligned}$$

The first equality follows from (1) for $(\mathcal{I}, \pi^*)$ (simplified for this special class of instances). The first inequality follows from the fact that $\bar{v}_i$ is non-increasing with respect to the time passed since inspection. The subsequent equality follows from (2), noting that π^* is a feasible strategy for $\hat{\mathcal{I}}$. Then, we apply Lemma 1 for $(\hat{\mathcal{I}}, \pi^*)$, and therefore the second to last inequality holds due to (3). The lemma follows since π^* may "collect" the reward of at most one box for each observed $\boldsymbol{V} \sim \boldsymbol{D}$, i.e., $\sum_{i=1}^{n} A_i(\pi^*) \leq 1$ holds by the definition of the problem. □

An interpretation of Lemma 8 is that the optimal expected utility of any instance $\mathcal{I}$ is upper bounded by the optimal value achieved by the prophet for the instance $\mathcal{I}_{\text{PR}} = (Y_i)_{i\in[n]}$. Working similarly to Sect. 3, we design an order-non-adaptive threshold-based strategy for the problem. However, unlike the general case, here the "proxy" instance related to our Pandora Over Time instance is simply an instance of (the unconstrained) Pandora's Box. This constraint-free environment allows us to use a free-order prophet inequality to generate an order of the boxes and this drastically improves performance.

We next show that the expected utility of π_{FIXED} for an instance $\mathcal{I} = (c_i, p_i, V_i, \bar{v}_i)_{i\in[n]}$ equals exactly the expected value achieved by the "gambler" for the related instance $\mathcal{I}_{\text{PR}} = (Y_i)_{i\in[n]}$. For brevity, we use $Y_i := Y_i$ for $i \in [n]$.

Strategy 2: π_{FIXED}

▷ **Input:** An instance $\mathcal{I} = (c_i, p_i, V_i, \bar{v}_i)_{i\in[n]}$ of Pandora With Value Discounting

`// Phase 1: determine schedule of inspection and threshold`

1 Construct $\mathcal{I}_{\text{PR}} = (Y_i)_{i\in[n]}$.

2 Obtain a permutation σ of $[n]$ and threshold $\tau > 0$ that achieve the guarantee of Lemma 3 for $\mathcal{I}_{\text{PR}}$.

`// Phase 2: threshold-based strategy based on the schedule Phase 1`

3 Set $t = 1$.

4 **for** $i = 1, \ldots, n$ **do**

5   **if** $r_{\sigma(i)} \geq \tau$ **then**

6   Sample $V_{\sigma(i)} \sim D_{\sigma(i)}$ at time t.

7   **if** $V_{\sigma(i)} \geq \tau$ **then**

8   **halt** `// the reward of the last inspected box is collected`

9   Set $t = t + 1 + p_{\sigma(i)}$. `// wait for` $p_{\sigma(i)}$ `rounds.`

Lemma 9. *Let $\mathcal{I}$ be an instance of Pandora With Value Discounting and $\mathcal{I}_{\text{PR}} = (Y_i)_{i\in[n]}$. Then, it holds that $\mathbb{E}[u_{\mathcal{I}}(\pi_{\text{FIXED}})] = \mathbb{E}[Y_{\sigma(i^*)}]$.*

Proof. Let $\pi := \pi_{\text{FIXED}}$. Consider the Pandora's Box instance $\hat{\mathcal{I}} = \big((c_i, V_i)_{i\in[n]}\big)$:

$$\mathbb{E}[u_{\mathcal{I}}(\pi)] = \mathbb{E}\left[\max_{i\in S(\pi)} \bar{v}_i\left(V_i, T_\pi - t_i(\pi)\right) - \sum_{i\in S(\pi)} c_i\right] = \mathbb{E}\left[\max_{i\in S(\pi)} \bar{v}_i\left(V_i, 0\right) - \sum_{i\in S(\pi)} c_i\right]$$

$$= \mathbb{E}\left[\max_{i\in S(\pi)} V_i - \sum_{i\in S(\pi)} c_i\right] = \mathbb{E}[u_{\hat{\mathcal{I}}}(\pi)] = \mathbb{E}\left[\sum_{i=1}^{n} A_i(\pi)V_i - I_i(\pi)c_i\right] \tag{9}$$

The second equality follows from the fact that π "collects" the box inspected at time T_π. The third equality follows from the definition of $\bar{v}_i$ for $i \in [n]$. Finally, the fourth equality holds by observing that π is a feasible strategy for $\hat{\mathcal{I}}$. To continue, we will show that $(\hat{\mathcal{I}}, \pi)$ satisfies the second condition of Lemma 1. In particular, we show that $(\hat{\mathcal{I}}, \pi)$ are such that (3) holds with equality.

Claim 3. *It holds that $\mathbb{E}\big[\sum_{i=1}^n A_i(\pi)V_i - I_i(\pi)c_i\big] = \mathbb{E}\big[\sum_{i=1}^n A_i(\pi)Y_i\big]$.*

Proof. According to Lemma 3, this identity holds if the following condition is satisfied for $(\hat{\mathcal{I}}, \pi)$: $(\exists i \in [n] : I_i(\pi) = 1 \wedge V_i \geq r_i) \Rightarrow A_i(\pi) = 1$. Indeed, by Line 5 of π_{FIXED}, $I_i(\pi) = 1$ (a box $i \in [n]$ is inspected) if and only if $r_i \geq \tau$. If we additionally have that $V_i \geq r_i$, we can conclude that $V_i \geq r_i$. This implies that the condition on Line 7 is evaluated to **true** and therefore $A_i(\pi) = 1$. The claim follows. □

By invoking Claim 3, we can continue (9) as follows:

$$\mathbb{E}[u_{\mathcal{I}}(\pi)] = \mathbb{E}\left[\sum_{i=1}^{n} A_i(\pi)V_i - I_i(\pi)c_i\right] = \mathbb{E}\left[\sum_{i=1}^{n} A_i(\pi)Y_i\right] = \mathbb{E}[Y_{\sigma(i^*)}].$$

The last equality follows from the fact that $A_i(\pi) = 1$ holds for the *first* (since π halts after accepting) box $\sigma(i)$ for which $r_{\sigma(i)} \geq \tau$ (Line 5) and $V_{\sigma(i)} \geq \tau$ (Line 7). This holds if and only if $Y_{\sigma(i)} = \min\{r_{\sigma(i)}, V_{\sigma(i)}\} \geq \tau$. This is the definition of the accepted box $\sigma(i^*)$ for $\mathcal{I}_{\text{PR}}$. The proof follows. □

Using the two lemmata, it is now easy to show our main result about Pandora With Value Discounting.

Proof (Proof of Theorem 4*).* Fix an arbitrary instance $\mathcal{I}$ of Pandora With Value Discounting. Let $\mathcal{I}_{\text{PR}} = (Y_i)_{i\in[n]}$ be its induced prophet problem with $Y_i = \min\{r_i, V_i\}$ for $i \in [n]$. For $(\mathcal{I}, \pi_{\text{FIXED}})$, the following holds, completing the proof: $\mathbb{E}\left[u_{\mathcal{I}}(\pi_{\text{FIXED}})\right] = \mathbb{E}\left[Y_{\sigma(i^*)}\right] \geq 0.7258 \cdot \mathbb{E}\left[\max_{i\in[n]} Y_i\right] \geq 0.7258 \cdot \mathbb{E}\left[u_{\mathcal{I}}(\pi^*)\right]$. The first equality follows from Lemma 9, the second inequality follows from Lemma 3, and the last inequality follows from the second part of Lemma 8. □

Acknowledgements. GA was supported by project MIS 5154714 of the National Recovery and Resilience Plan Greece 2.0 funded by the European Union under the NextGenerationEU Program. FF was supported by the FAIR (Future Artificial Intelligence Research) project PE0000013, funded by the NextGenerationEU program within the PNRR-PE-AI scheme (M4C2, investment 1.3, line on Artificial Intelligence), the ERC Advanced Grant 788893 AMDROMA "Algorithmic and Mechanism Design Research in Online Markets", the PNRR MUR project IR0000013-SoBigData.it, and the MUR PRIN project "Learning in Markets and Society". AT was supported by the Gravitation Project NETWORKS, grant no. 024.002.003, and the EU Horizon 2020 Research and Innovation Program under the Marie Skłodowska-Curie Grant Agreement, grant no. 101034253.

References

1. Agrawal, S., Sethuraman, J., Zhang, X.: On optimal ordering in the optimal stopping problem. In: EC, pp. 187–188. ACM (2020)
2. Atsidakou, A., Caramanis, C., Gergatsouli, E., Papadigenopoulos, O., Tzamos, C.: Contextual Pandora's box. In: AAAI, pp. 10944–10952. AAAI Press (2024)
3. Berger, B., Ezra, T., Feldman, M., Fusco, F.: Pandora's problem with combinatorial cost. In: EC, pp. 273–292. ACM (2023)
4. Berger, B., Ezra, T., Feldman, M., Fusco, F.: Pandora's problem with deadlines. In: AAAI, pp. 20337–20343. AAAI Press (2024)
5. Beyhaghi, H., Cai, L.: Pandora's problem with nonobligatory inspection: optimal structure and a PTAS. In: STOC, pp. 803–816. ACM (2023)
6. Beyhaghi, H., Cai, L.: Recent developments in Pandora's box problem: variants and applications. SIGecom Exch. **21**(1), 20–34 (2023)
7. Beyhaghi, H., Kleinberg, R.: Pandora's problem with nonobligatory inspection. In: EC, pp. 131–132. ACM (2019)

8. Boodaghians, S., Fusco, F., Lazos, P., Leonardi, S.: Pandora's box problem with order constraints. Math. Oper. Res. **48**(1), 498–519 (2023)
9. Bradac, D., Singla, S., Zuzic, G.: (Near) optimal adaptivity gaps for stochastic multi-value probing. In: APPROX-RANDOM. LIPIcs, vol. 145, pp. 49:1–49:21. Schloss Dagstuhl - Leibniz-Zentrum für Informatik (2019)
10. Bubna, A., Chiplunkar, A.: Prophet inequality: Order selection beats random order. In: EC, pp. 302–336. ACM (2023)
11. Chawla, S., Gergatsouli, E., McMahan, J., Tzamos, C.: Approximating Pandora's box with correlations. In: APPROX/RANDOM. LIPIcs, vol. 275, pp. 26:1–26:24. Schloss Dagstuhl - Leibniz-Zentrum für Informatik (2023)
12. Chawla, S., Gergatsouli, E., Teng, Y., Tzamos, C., Zhang, R.: Pandora's box with correlations: learning and approximation. In: FOCS, pp. 1214–1225. IEEE (2020)
13. Doval, L.: Whether or not to open Pandora's box. J. Econ. Theory **175**, 127–158 (2018)
14. Esfandiari, H., Hajiaghayi, M.T., Lucier, B., Mitzenmacher, M.: Online Pandora's boxes and bandits. In: AAAI, pp. 1885–1892. AAAI Press (2019)
15. Fu, H., Li, J., Xu, P.: A PTAS for a class of stochastic dynamic programs. In: ICALP. LIPIcs, vol. 107, pp. 56:1–56:14. Schloss Dagstuhl - Leibniz-Zentrum für Informatik (2018)
16. Fu, H., Li, J., Liu, D.: Pandora box problem with nonobligatory inspection: hardness and approximation scheme. In: STOC, pp. 789–802. ACM (2023)
17. Gatmiry, K., Kesselheim, T., Singla, S., Wang, Y.: Bandit algorithms for prophet inequality and Pandora's box. In: SODA, pp. 462–500. SIAM (2024)
18. Gergatsouli, E., Tzamos, C.: Online learning for min sum set cover and Pandora's box. In: ICML. Proceedings of Machine Learning Research, vol. 162, pp. 7382–7403. PMLR (2022)
19. Guo, C., Huang, Z., Tang, Z.G., Zhang, X.: Generalizing complex hypotheses on product distributions: auctions, prophet inequalities, and Pandora's problem. In: COLT. Proceedings of Machine Learning Research, vol. 134, pp. 2248–2288. PMLR (2021)
20. Hill, T.: Prophet inequalities and order selection in optimal stopping problems. Proc. Am. Math. Soc. **88**(1), 131–137 (1983)
21. Hoefer, M., Schecker, C., Schewior, K.: Contract design for Pandora's box. CoRR abs/2403.02317 (2024)
22. Kleinberg, R., Weinberg, S.M.: Matroid prophet inequalities. In: STOC, pp. 123–136. ACM (2012)
23. Kleinberg, R.D., Waggoner, B., Weyl, E.G.: Descending price optimally coordinates search. In: EC, pp. 23–24. ACM (2016)
24. Lee, J., Sviridenko, M., Vondrák, J.: Submodular maximization over multiple matroids via generalized exchange properties. Math. Oper. Res. **35**(4), 795–806 (2010)
25. Olszewski, W., Weber, R.: A more general Pandora rule? J. Econ. Theory **160**, 429–437 (2015)
26. Samuel-Cahn, E.: Comparison of threshold stop rules and maximum for independent nonnegative random variables. Ann. Probab. 1213–1216 (1984)
27. Segev, D., Singla, S.: Efficient approximation schemes for stochastic probing and prophet problems. In: EC, pp. 793–794. ACM (2021)
28. Singla, S.: The price of information in combinatorial optimization. In: SODA, pp. 2523–2532. SIAM (2018)

29. Weitzman, M.L.: Optimal search for the best alternative. Econometrica: J. Econometric Soc. 641–654 (1979)

Convergence to Equilibrium of No-Regret Dynamics in Congestion Games

Volkan Cevher[1], Wei Chen[2], Leello Dadi[1], Jing Dong[3], Ioannis Panageas[4], Stratis Skoulakis[1], Luca Viano[1], Baoxiang Wang[3](✉), Siwei Wang[2], and Jingyu Wu[5]

[1] École polytechnique fédérale de Lausanne (EPFL), Lausanne, Switzerland
{volkan.cevher,leello.dadi,stratis.skoulakis,luca.viano}@epfl.ch
[2] Microsoft Research Asia, Beijing, China
{weic,siweiwang}@microsoft.com
[3] The Chinese University of Hong Kong, Shenzhen, China
jingdong@link.cuhk.edu.cn, bxiangwang@cuhk.edu.cn
[4] Computer Science, University of California, Irvine, USA
ipanagea@ics.uci.edu
[5] University of Science and Technology of China, Hefei, China
wujingyu@mail.ustc.edu.cn

Abstract. The congestion game is a powerful model that encompasses a range of engineering systems such as traffic networks and resource allocation. It describes the behavior of a group of agents who share a common set of F facilities and take actions as subsets of k facilities. In this work, we study the online formulation of congestion games, where agents participate in the game repeatedly and observe feedback with randomness. We note that this paper is the result of the merging of [22] arXiv:2306.13673 and [17] arXiv:2401.09628. In [22], we propose CongestEXP, a decentralized algorithm that is based on the classic exponential weights method. By maintaining weights on the facility level, the regret bound of CongestEXP avoids the exponential dependence on the size of possible facility sets, i.e., $\binom{F}{k} \approx F^k$, and scales only linearly with F. Specifically, we show that CongestEXP attains a regret upper bound of $O(kF\sqrt{T})$ for every individual player, where T is the time horizon. If a strict Nash equilibrium exists, we show that CongestEXP can converge to the strict Nash policy almost exponentially fast in $O(F\exp(-t^{1-\alpha}))$, where t is the number of iterations and $\alpha \in (1/2, 1)$. In [17], we present an online learning algorithm in the bandit feedback model that, once adopted by all agents of a congestion game, results in game-dynamics that converge to an ϵ-approximate Nash Equilibrium in a polynomial number of rounds with respect to $1/\epsilon$, the number of players and the number of available resources. The proposed algorithm also guarantees sublinear regret to any agent adopting it. As a result, our work answers an open question from [16] and extends the recent results of [34] to the bandit feedback model.

Keywords: Congestion Game · Convergence to Nash equilibrium · Online Learning

M. Mavronicolas et al. (Eds.): WINE 2024, LNCS 15534, pp. 513–529, 2026.
https://doi.org/10.1007/978-3-032-08560-3_29

1 Introduction

Congestion games are a class of general-sum games that can be used to describe the behavior of agents who share a common set of facilities (resources) [5]. In these games, each player chooses a combination of facilities, and popular facilities will become congested, yielding a lower utility for the players who choose them. Thus, players are incentivized to avoid congestion by choosing combinations that are less popular among the other players. A range of real-world scenarios can be captured by the congestion game model, such as traffic flow, data routing, and wireless communication networks [13,41,42].

In the model of the congestion game, the Nash equilibrium is a popular concept to describe the behavior of selfish players and the dynamics induced by decentralized algorithms. It describes a stable state of the game where no player can improve their utility by unilaterally changing their choice of actions. When a unique Nash equilibrium exists in the congestion game, it can be a reference point for players to coordinate to avoid suboptimal outcomes. Beyond the Nash equilibrium, social welfare is a significant metric, capturing the overall utility or well-being of all players involved. It serves as a crucial benchmark, enabling the evaluation of the efficiency loss incurred when transitioning from centralized to decentralized algorithms.

In the classic one-shot congestion game setting, the Nash equilibrium and the loss of efficiency due to decentralized dynamics have been well studied [37,39]. However, these results do not provide insights into how players arrive at the equilibrium. This motivates the study of congestion games in an online learning framework, where players participate in the game repeatedly at every time step. This framework better models many realistic scenarios, such as the traffic congestion problem in urban areas. In this repeated congestion game setting, players such as drivers in a congested city must choose between different routes to reach their destinations every day. As more drivers use a particular route, the congestion on that route increases, leading to higher travel times and lower utility. In this scenario, players can update their desired route every day to optimize their utility, but the observed utility by each player may be subject to randomness due to uncertainty in the actual congestion situation (e.g., the influence of the weather). All these make it suitable to model the congestion game in an online learning framework.

This paper is the merged version of [17] and [22]. The rest of the paper is summarized as follows. We will first summarize the setting and the contribution of each paper, then we will focus mainly on the results presented in [22]. The full results and proofs of [17] can be found at arXiv:2401.09628.

1.1 Results and Techniques in [17]

Despite the long interest in bandit online learning algorithms for congestion games, the convergence to Nash Equilibrium of such bandit no-regret learning algorithms is not as well explored. The broad question under consideration here is whether or not the uncoordinated selfish behavior of agents can converge to equilibrium. To the best of our knowledge [16] were the first to provide an update rule (performing under bandit feedback) that once adopted by all agents of an atomic congestion game, the resulting strategies converge to an ϵ-approximate Nash Equilibrium with rate polynomial in n, m and $1/\epsilon$. However, their method does not guarantee the no-regret property. As a result, [16] asked the following question:

Is there a no-regret algorithm, in the bandit feedback model, that once adopted by all agents, results in strategies that converge to an ϵ-approximate Nash Equilibrium in $\text{poly}(n, m, 1/\epsilon)$ *rounds?*

The main contribution of our work consists in providing a positive answer to the open question of [16]. More precisely, we provide an online learning algorithm, called *Online Gradient Descent with Caratheodory Exploration* (BGD − CE), that simultaneously provides both regret guarantees and convergence to Nash Equilibrium.

Informal Theorem. *There exists an online learning algorithm (*BGD − CE*) that performs under bandit feedback and guarantees $\mathcal{O}(m^{2.8}T^{4/5})$ regret to any agent that adopts it. Moreover if all agent adopt* BGD − CE, *then the resulting strategies converge to an ϵ-Nash Equilibrium after $\mathcal{O}(n^{13.5}m^{13}/\epsilon^5)$ steps.*

Our proposed online learning algorithm additionally improves the convergence rate of the algorithm of [16].

Informal Theorem. *For Network Congestion games in Acyclic Directed Graphs (DAGs), Online Gradient Descent with Caratheodory Exploration, can be implemented in polynomial time.*

The above result follows from strategy spaces admitting polynomial size descriptions in this setting. We further exploit the specific structure of DAGs to compute exact 1-barycentric-spanners, which as noted in [4,8] are not trivial to obtain for DAGs. We underline that exact spanners are not necessary, and the approximate method of [4] is perfectly suitable. However, our approach is simple, more efficient, and can be of independent interest.

Our Techniques The fundamental difficulty in designing no-regret online learning algorithms under bandit feedback is to guarantee that each resource is sufficiently explored. Unfortunately, standard bandit algorithms such as EXP3 [3] result in regret bounds of the form $\mathcal{O}(2^{m/2}\sqrt{T})$, that scale exponentially with respect to m. However, a long line of research in combinatorial bandits has produced algorithms with a regret polynomially dependent on m [2,4,6,8,18,25,27,31]. These algorithms, in order to overcome the exploration problem, use various techniques that can roughly be categorized two camps, simultaneous exploration versus alternating explore-exploit, as described by [1]. However, to the best of our knowledge, none of these algorithms have been shown to converge to NE in congestion games once adopted by all agents.

Our online learning algorithm, guaranteeing both no-regret and convergence to equilibrium, is based on combining Online Gradient Descent [43] with a novel exploration scheme, much like [23]. Our exploration strategy is based on coupling the notion of barycentric spanners [4] with the notion of Bounded-Away Polytopes proposed by [34] for the semi-bandit case. More precisely, [34] introduced the notion of μ-Bounded Away Polytope which corresponds to the description polytope of the strategy space (convex hull of all pure strategies) with the additional constraint that each resource is selected with probability at least $\mu > 0$. Projecting on this polytope ensures that the variance of the unobserved cost estimators remains bounded. In order to capture bandit estimators, we extend the notion of μ-Bounded Away Polytope to denote the subset of the description polytope for which each point admits a decomposition with least μ weight on a pre-selected barycentric spanner $\mathcal{B}$.

This technique of projecting on μ-Bounded polytopes closely ressembles the *mixing* strategies employed in online learning schemes that have alternating explore-exploit rounds. In those strategies, a fixed measure is added to bias the algorithm's chosen strategy. The projection on μ-Bounded polytopes, however, scales the point before adding a bias, and, in some rounds, does not alter the point. It is therefore a mix of simultaneous and alternating exploration, depending on the round.

Finally, in order to provide a polynomial-time implementation of OGD − CE for Network Congestion Games on Directed Acyclic Graphs we need exploit its well disposed combinatorial structure. In [17], we propose a novel construction of barycentric spanners for DAGs that outputs a 1-barycentric spanner in polynomial time and yields an efficient selfish routing scheme that converges to an equilibrium.

1.2 Results and Techniques in [22]

While there have been various decentralized algorithms that can attain the Nash equilibria efficiently for general online games, they can suffer from a linear dependency on the number of actions when directly applied to the congestion game [12,19,20,24,26,40], which is exponential with k, F. On the other hand, algorithms designed specifically for congestion games either only converge to Nash equilibria asymptotically [28,29,33], on average [16], on best-iterate [35] or require additional assumptions on the structure of the game [9,10]. Moreover, to the best of our knowledge, there is no algorithm that can simultaneously guarantee both low regret and fast convergence to Nash equilibrium for each player. While some online learning algorithms, such as exponential weights, have been shown to converge faster than others due to the specific choice of regularization [24], previous regret results indicate that their guarantees still rely on the exponentially large number of actions, due to their specific form of updates (exponential weighting) [14].

In this paper, we study the online congestion game with semi-bandit and full information feedback. We propose a decentralized algorithm that modifies the celebrated exponential weights algorithm, which can be utilized by each player

Table 1. Summary of previous results. Here F denotes the number of facilities, T denotes the time horizon, and k is the number of facilities in an action. The semi-bandit feedback means that the learner can only observe the reward associated with the facility chosen, and full information refers to the case where the rewards associated with all facilities are revealed. Best iterate convergence means that the algorithm guarantees the existence of an iterate that meets the criteria of a Nash equilibrium. On the other hand, the last iterate convergence indicates that the final output of the algorithm constitutes a Nash equilibrium.

	Nash convergence type	Convergence to Nash	Regret
[9]	Last iterate	Asymptotic Semi-bandit	None
[16]	Best iterate	$O(F^{1.5}T^{-1/6})$ Semi-bandit	None
[35]	Best iterate	$O\left(F^{1.4}T^{-1/5}\right)$ Semi-bandit	$O\left(F^{2}T^{4/5}\right)$ Semi-bandit
Ours	Last iterate	$O\left(F\exp(-T^{1-\alpha})\right)$, $\alpha \in (1/2, 1)$ Full-information	$O\left(kF\sqrt{T}\right)$ Semi-bandit

without additional information about other players' utility. From the individual player's perspective, we show that the algorithm guarantees sublinear individual regret, with respect to the best action in hindsight when holding the other player's strategy fixed. We remark that the regret is also only linear with respect to the number of facilities. As a result of this, we show that the optimal social welfare can be efficiently approximated, up to an error that is only linear with respect to the number of facilities. When a strict Nash equilibrium exists for the congestion game, we also prove that our algorithm is capable of converging to the strict Nash equilibrium fast, with an almost exponentially fast rate that is only linear with respect to the number of facilities.

2 Related Works

2.1 Learning in Games

Online learning has a long history that is closely tied to the development of game theory. The earliest literature can be traced back to Brown's proposal on using fictitious play to solve two-player zero-sum games [5]. It is now understood that fictitious play can converge very slowly to Nash equilibrium [21]. On the other side, it has been shown that if each player of a general-sum, multi-player game experiences regret that is at most $f(T)$, the empirical distribution of the joint policy converges to a coarse correlated equilibrium of the game with a rate of $O(f(T)/T)$ [7]. This implies that a variety of online learning algorithms such as Hedge and Follow-The-Regularized-Leader algorithms can converge to coarse correlated equilibria at a rate of $O(1/\sqrt{T})$.

While the standard no-regret learning dynamic can guarantee convergence to equilibria, it has been shown that more specialized no-regret learning protocols can do better [12,19,20,26,40]. It has also been shown that when strict pure Nash equilibria are present, algorithms that are based on entropic regularization (e.g. exponential weights) can converge fast to the equilibria [14,24]. Moreover, such convergence rate holds for a variety of different feedback models, from full information to bandit feedback.

Though all of the above-mentioned methods are applicable to congestion games, the results usually involve a linear dependency on the number of actions. As each action is a combination of the different facilities (resources) in the congestion games, the results lead to undesirable exponential dependency on the number of facilities.

2.2 Learning in Online Congestion Games

Congestion games were first introduced in the seminal work [36] as a class of games with pure-strategy Nash equilibria. It has then been extensively studied, where its Nash equilibria have been characterized in [39] and a comprehensive introduction has been given in [37].

In the online setting, many works use no-regret learning to develop learning dynamics in this class of games for efficient convergence. [28] are the first to study no-regret learning for congestion games. They showed that the well-known multiplicative weights learning algorithm results in convergence to pure equilibria. Furthermore, they identified a set of mixed Nash equilibria that are weakly stable and showed that the distribution of play converges to this set. Followup works [29] showed that multiplicative weights algorithms converge to the set of Nash equilibria in the sense of Cesàro means, and [33] investigated the effect of learning rate on convergence to Nash equilibria.

With an additional assumption of convex potential functions, [9,10] established a non-asymptotic convergence rate. However, their rate has an exponential dependency on the number of facilities. [16] gave the first non-asymptotic convergence rate under semi-bandit feedback and without an exponential dependency on the number of facilities. However, the convergence is with respect to the averaged-over-time policy and with a $O(T^{-1/6})$ convergence rate. This result is later improved by concurrent work [35], who proposed an online stochastic gradient descent algorithm that converges to an ϵ-approximate Nash equilibrium in $O(\epsilon^{-5})$ time while each individual player enjoys a regret of $O(T^{4/5})$. However, their convergence is only best iterate convergence and only in terms of potential function values (see a detailed comparison in Table 1).

2.3 Combinatorial Bandits and Shortest Path

Combinatorial bandits offer an extension of the classic multi-armed bandit problem where the player must select an action that involves a combination of various resources [8,11,30]. In a special case, the shortest path problem can be viewed as a combinatorial bandit problem where the resources are edges on a graph and the action is a path [25]. Efficient algorithms have been proposed for these problems, and it has been shown that the sublinear regret only linearly depends on the number of resources. However, it is important to note these algorithms are designed for a single player, and as a result, they may not converge to a Nash equilibrium when applied directly to congestion games by allowing each player to execute the algorithm.

3 Problem Formulation

3.1 Congestion Game

A congestion game with n players is defined by $\mathcal{G} = (\mathcal{F}, \{\mathcal{A}_i\}_{i=1}^n, \{r^f\}_{f\in\mathcal{F}})$, where i) $\mathcal{F}$ is the facility set that contains F facilities; ii) $\mathcal{A}_i$ is the action space for player i and contains A actions (we assume that the action space for each player is the same), where each action $a_i \in \mathcal{A}_i$ is a combination of k facilities in $\mathcal{F}$; and iii) $r^f : (\mathcal{A}_1 \times \cdots \times \mathcal{A}_n) \to [0,1]$ is the reward function for facility $f \in \mathcal{F}$, which only depends on the number of players choosing this facility, i.e., $\sum_{i=1}^n \mathbb{I}\{f \in a_i\}$. We denote $a = (a_i, a_{-i})$ as a joint action, where a_{-i} is the actions of all other players except player i. The total reward collected by player i with joint action $a = (a_i, a_{-i})$ is $r_i(a_i, a_{-i}) = \sum_{f\in a_i} r^f(a_i, a_{-i})$. Without loss of generality, we assume that $r^f(a) \in [0,1]$.

Deterministically playing actions $a = (a_i, a_{-i})$ is referred to as a pure strategy. The player can also play a mixture of pure strategy, $\omega_i \in \Delta(\mathcal{A}_i)$, where $\Delta(\mathcal{A}_i)$ denotes the probability simplex of action space $\mathcal{A}_i$. Similarly, we use $\omega = (\omega_i, \omega_{-i})$ to denote a joint randomized policy.

3.2 Nash Equilibrium

One of the commonly used solution concepts in congestion games is Nash equilibrium (NE), and the policies that lead to a Nash equilibrium are referred to as Nash policies. The players are said to be in a Nash equilibrium when no player has an incentive from deviating from its current policy (as described in the definition below).

Definition 1 (Nash equilibrium). *A policy $\omega^* = (\omega_1^*, \ldots, \omega_n^*)$ is called a* ***Nash equilibirum*** *if for all $i \in [n]$, $r_i(\omega_i^*, \omega_{-i}^*) \geq r_i(\omega_i, \omega_{-i}^*), \forall \omega_i \in \Delta(\mathcal{A}_i)$. When ω^* is pure, the equilibrium is called a* ***pure Nash equilibrium****. In addition, when the strategy is pure and the inequality is a strict inequality, the equilibrium is called a* ***strict Nash equilibrium****.*

Fact 1 ([36]). *There exists a pure Nash equilibrium in any congestion game.*

3.3 Social Welfare and Price of Anarchy

Except for Nash equilibrium, another commonly used metric to measure the efficiency of the dynamics between the players is through social welfare. For a given joint action $a = \{a_i\}_{i=1}^n$, the social welfare is defined to be the sum of players' rewards, i.e., $W(a) = \sum_{i=1}^n r_i(a)$, and the optimal social welfare of the game is defined as $\text{OPT} = \max_{a\in\mathcal{A}_1\times\cdots\times\mathcal{A}_n} W(a)$. This optimality is under the case where a central coordinator could dictate each player's strategy, and each player's individual incentives are not considered.

Based on the definition of OPT, We can define smooth games as follows.

Definition 2 (Smooth game [38]). *A game is* (λ, μ)*-smooth if there exists a joint action* a^* *such that for any joint action* a, $\sum_{i \in n} r_i(a_i^*, a_{-i}) \geq \lambda \text{OPT} - \mu W(a)$.

The results in [32] show that congestion games are smooth when the reward function are affine, that is, when $r^f(a)$ is an affine function on the scalar variable $\sum_{i=1}^n \mathbb{I}\{f \in a_i\}$. This property enables certain decentralized no-regret learning dynamics to efficiently approximate the optimal welfare [40].

3.4 Online Congestion Game

In this paper, we study the congestion game in an online setting with a finite time horizon T, where the underlying reward function is unknown. At each time step $t \in [T]$, each player chooses (randomized) policy ω_i^t, from which it forms a joint policy $\omega^t = (\omega_1^t, \ldots, \omega_n^t)$. Then each player i draws a random action $a_i^t \sim \omega_i^t$, plays this action (denote a^t the joint action), and receives overall reward of $\sum_{f \in a_i^t} R^f(a^t)$, where $R^f(a^t)$'s are random variables that satisfy the following assumption.

Assumption 1. *For any facility* $f \in \mathcal{F}$, *any joint action* a^t *and any player* $i \in [n]$, *let* $\mathcal{H}_t$ *be the history up to time step* $t-1$. *Then, 1)* $R^f(a) \in [0, 1]$, *and 2)* $\mathbb{E}\left[R^f(a^t) \mid \mathcal{H}_t\right] = r^f(a^t)$.

The assumption implies that the mean of $R^f(a^t)$ is always $r^f(a^t)$. Hence the Nash equilibrium and expected social welfare of the online congestion game is the same as those of the offline congestion game.

We consider two types of feedback rules in this paper: *semi-bandit feedback*, and *full information feedback*. In the semi-bandit feedback, player i observes all the $R^f(a^t)$'s for any $f \in a_i$ (only the facilities he played); and in full information feedback, player i observes all possible information $R^f(a_i, a_{-i})$, for every $a_i \in \mathcal{A}_i$, $\forall f \in a$.

The efficiency of a sequence of policy $\{\omega_i^t\}_{t=1}^T$ can be measured by the individual regret of all the players (which is defined as follows).

Definition 3 (Individual regret). *The individual regret of player* i *playing policy* $\{\omega_i^t\}_{t=1}^T$ *is defined as the cumulative difference between the received rewards and the rewards incurred by a best-in-hindsight policy, that is* $\text{Regret}_i(T) = \max_{\omega_i \in \mathcal{A}_i, \{\omega_{-i}^t\}_{t=1}^T} \sum_{t=1}^T r_i(\omega_i, \omega_{-i}^t) - r_i(\omega_i^t, \omega_{-i}^t)$.

4 Algorithm

In this section, we introduce CongestEXP, a decentralized algorithm for online congestion games (Algorithm 1).

The algorithm uses the combinatorial nature of the action space. Each player maintains a sampling distribution ω_i^t and a facility-level reward estimator $\tilde{y}_i^t$. At each time step, they first draw a random action $a_i^t \sim \omega_i^t$ and play this action.

Algorithm 1 CongestEXP

1: **Input:** learning rate η.
2: For all $f \in \mathcal{F}$, initialize $\tilde{y}_i^0(f) = 0$, $\omega_i^0(a)$ initialized according to Equation (2).
3: **for** $t = 1, \dots, T$ **do**
4: Players play strategy $a^t \sim \omega^t$, $\omega^t = (\omega_1^t, \dots, \omega_n^t)$.
5: Each player i observe $R^f(a^t) \sim r^f(a^t)$, for each $f \in a_i^t$.
6: Each player $i \in [N]$ computes $\tilde{y}_i^t(f) = 1 - \frac{\mathbb{I}\{f \in a_i^t\}(1 - R^f(a_i^t, a_{-i}^t))}{q_i^t(f)}$, where $q_i^t(f) = \sum_{a_i \in \mathcal{A}_i, f \in a_i} \omega_i^t(a_i)$.
7: Each player $i \in [N]$ updates $\omega_i^{t+1}(a) = \frac{\prod_{f \in a} \tilde{\omega}_i^t(f)}{\sum_{a_i \in \mathcal{A}_i} \prod_{f' \in a_i} \tilde{\omega}_i^t(f')}, \forall a \in \mathcal{A}_i$, where $\tilde{\omega}_i^t(f) = \tilde{\omega}_i^{t-1}(f) \exp\left(\eta \tilde{y}_i^t(f)\right)$.
8: **end for**

Then they use their received information to update $\tilde{y}_i^t(f)$'s (for all $f \in \mathcal{F}$) as follows

$$\tilde{y}_i^t(f) = 1 - \frac{\mathbb{I}\{f \in a_i^t\}(1 - R^f(a^t))}{q_i^t(f)}, \quad q_i^t(f) = \sum_{a_i \in \mathcal{A}_i, f \in a_i} \omega_i^t(a_i), \tag{1}$$

where $q_i^t(f)$ is the probability that player i selects facility f at time t based on its current policy ω_i^t. One can easily check that $\tilde{y}_i^t(f)$ is an unbiased estimator for $r^f(a^t)$, and with these facility-level reward estimators, the players then update ω_i^{t+1} as follows (exponential weighting), and then proceed to the next time step.

$$\omega_i^{t+1}(a) = \& \frac{\prod_{f \in a} \tilde{\omega}_i^t(f)}{\sum_{a_i \in \mathcal{A}_i} \prod_{f' \in a_i} \tilde{\omega}_i^t(f')}, \forall a \in \mathcal{A}_i, \quad \tilde{\omega}_i^t(f) = \exp\left(\eta \sum_{j=1}^{t} \tilde{y}_i^j(f)\right). \tag{2}$$

On the one hand, in the semi-bandit setting, our algorithm leverages this kind of feedback and estimates rewards at the facility level. We note that this algorithm has also been previously utilized to tackle online shortest path problems and combinatorial bandit problems, as documented in the literature [8,11,15,25]. This enables us to achieve a low individual regret (Theorem 1) and guarantee a lower bound for the overall social welfare (Corollary 1). On the other hand, our algorithm constructs exponential weights based on the reward estimation at the action level. This makes sure that the joint policy ω^t can converge to a Nash equilibrium fast when it is nearby (Theorems 2 and 3).

In summary, our results indicate that adopting Algorithm 1 in a congestion game leads to favorable outcomes. Each player enjoys favorable cumulative individual rewards, without compromising the overall social welfare. Moreover, when the joint policy is close to the Nash equilibrium, players can quickly converge to a stable equilibrium state, avoiding inefficient and chaotic dynamics.

5 Theoretical Results

In this section, we present our main theoretical results.

5.1 Sublinear Individual Regret with Linear Dependency on F

Our first theorem shows that each individual player enjoys a sublinear individual regret.

Theorem 1. *Under semi-bandit feedback, Algorithm 1 with* $\eta = \frac{1}{\sqrt{T}}$ *satisfies that for all* $i \in [n]$, $\mathbb{E}\left[Regret_i(T)\right] = O\left(kF\sqrt{T}\right)$.

Remark 1. By Markov's inequality, the above stated result also holds with high probability.

Compared with naively applying exponential weights on the congestion game (with a regret of $\tilde{O}\left(\sqrt{A_i T}\right)$ [3]), we can see that Theorem 1 reduces the factor $\sqrt{A_i}$ to kF. This is a significant improvement since $A \approx F^k$ is exponentially larger than kF.

Though there exist some works to achieve a similar regret upper bound [20], we emphasize that these algorithms only work in the full-information setting, but not the semi-bandit setting. Besides, our algorithm can converge to a strict Nash equilibrium fast, while existing ones can only guarantee to converge to a coarse correlated equilibrium (please see details in Sect. 5.2). We include the proof in the arXiv version of the paper, and we summarize the technical highlights here.

Technical highlight of Theorem 1 In classical proofs of exponential weights algorithms, the regret is closely linked to the quadratic term of the reward estimator, i.e., $\mathbb{E}_t[\sum_{a_i \in \mathcal{A}_i} \omega_i^t(a_i)\left(\tilde{y}_i^t(a_i)\right)^2]$ [3,30], where $\tilde{y}_i^t(a_i)$ is the estimated reward of action a_i at time step t, and $\mathbb{E}_t[\cdot]$ denotes the conditional expectation over all history up to time t. If we can upper bound this term by a constant polynomial with k and F, then we can remove the exponential factor in the individual regret upper bound.

Specifically, one would get a regret decomposition as follows, $\mathbb{E}\left[\text{Regret}_i(T)\right] \leq \mathbb{E}\left[\frac{kF}{\eta} + \eta \sum_{t=1}^{T} \sum_{a_i \in \mathcal{A}_i} \omega_i^t(a_i)\left(\sum_{f \in a_i} \tilde{y}_i^t(f)\right)^2\right]$.

With our facility level estimator (Eq. (1)), $\tilde{y}_i^t(a_i) = \sum_{f \in a_i} \tilde{y}_i^t(f)$, the second term could be upper bounded as $\sum_{a_i \in \mathcal{A}_i} \omega_i^t(a_i)\left(\sum_{f \in a_i} \tilde{y}_i^t(f)\right)^2 \leq k + k\sum_{f \in \mathcal{F}}\left(\frac{\mathbb{I}\{f \in a_i^t\}(1-R^f(a_i^t, a_{-i}^t))}{q_i^t(f)}\right)^2 q_i^t(f)$. Let $\mathbb{E}_t[\cdot]$ denote conditional expectation over all history up to time t. Notice that our estimator $\tilde{y}_i^t(f)$ is unbiased and $(1 - R_i^f(a^t))^2$ is upper bounded by 1. Then, taking expectations on both sides yield

$$\mathbb{E}\left[\sum_{a_i \in \mathcal{A}_i} \omega_i^t(a_i)\left(\sum_{f \in a_i} \tilde{y}_i^t(f)\right)^2\right] \leq \&k + k\mathbb{E}\left[\sum_{f \in \mathcal{F}}\left(\frac{\mathbb{I}\{f \in a_i^t\}\left(1 - R_i^f\left(a_i^t, a_{-i}^t\right)\right)}{q_i^t(f)}\right)^2 q_i^t(f)\right]$$

$$\leq \&k + k\mathbb{E}\left[\mathbb{E}_{t-1}\left[\sum_{f \in \mathcal{F}} \frac{\mathbb{I}\{f \in a_i^t\}}{q_i^t(f)}\right]\right] \leq k + kF.$$

From the above explanation, one can see the necessity of estimating the rewards at the facility level. If the reward estimator is constructed at the action level, that is, an estimator of the form $\tilde{y}_i^t(a_i) = k - \frac{\mathbb{I}\{a_i=a_i^t\}\left(k-\sum_{f\in a_i^t} R_i^f(a_i^t,a_{-i}^t)\right)}{\omega_i^t(a_i)}$. Consider the case that $R_i^f(a_i^t, a_{-i}^t)$ is always 0 and at the beginning $\omega_i^t(a_i) = 1/|\mathcal{A}_i|$, then this quadratic term $\mathbb{E}_t[\sum_{a_i\in\mathcal{A}_i}\omega_i^t(a_i)\left(\tilde{y}_i^t(a)\right)^2]$ is approximately, $\mathbb{E}_t\left[\sum_{a_i\in\mathcal{A}_i}\omega_i^t(a_i)\left(\tilde{y}_i^t(a)\right)^2\right] \approx \sum_{a_i\in\mathcal{A}_i}\omega_i^t(a_i)\cdot\left(\omega_i^t(a_i)\left(\frac{k}{\omega_i^t(a_i)}\right)^2\right) = k^2|\mathcal{A}_i|$, which scales with the number of actions and is thus always exponentially large.

5.2 Tight Approximation to Optimal Welfare

One immediate consequence of Theorem 1 is that our proposed algorithm can achieve a tight approximation to the optimal social welfare.

Corollary 1. *Under semi-bandit feedback, if the congestion game is (λ,μ)-smooth, then Algorithm 1 with $\eta = \frac{1}{\sqrt{T}}$ satisfies $\frac{1}{T}\sum_{t=1}^T W(\omega^t) \geq \frac{\lambda}{1+\mu}\mathrm{OPT} - O\left(\frac{nkF}{\sqrt{T}(1+\mu)}\right)$.*

We remark that $\frac{\lambda}{1+\mu}\mathrm{OPT}$ is shown to be a tight approximation of optimal social welfare possible by offline algorithms that attain Nash equilibrium in congestion games [38]. Therefore, the above result shows that our algorithm is as efficient as any offline Nash policy asymptotically.

5.3 Fast Convergence to Strict Nash Equilibrium

Beyond the low individual regret, we also show that our algorithm can produce a set of policies $\{\omega_i^t\}_{t=1}^T$ that converges fast to a strict Nash equilibrium ω_i^* in the full-information setting.

We first consider a simple case, where each player observes the expected rewards directly (which also take expectation on the randomness of $a_{-i}^t \sim \omega_{-i}^t$, i.e. $\mathbb{E}_{a_{-i}^t\sim\omega_{-i}^t}[r^f(a_i, a_{-i}^t)]$ for any a_i). In addition, we maintain the following assumption on the action.

Assumption 2. *We assume that any k facilities form a valid action.*

We note that the above assumption is only used to simplify our analysis. When the game does not have such a property, we can include dummy actions to meet this requirement, and the analysis in this section almost remains the same.

Theorem 2. *Under Assumption 2, consider the case where each player receives $\mathbb{E}_{a_{-i}^t\sim\omega_{-i}^t}[r^f(a_i^t, a_{-i}^t)], \forall a_i \in \mathcal{A}_i, \forall f \in a_i$ in a game that permits a strict Nash equilibrium $\omega^* = (\omega_1^*, \cdots, \omega_n^*)$, and let $\tilde{y}_i^t(f) = \mathbb{E}_{a_{-i}^t\sim\omega_{-i}^t}[r^f(a_i^t, a_{-i}^t)]$ in Line 6* *of Algorithm 1. Suppose* $\tilde{y}_i^0(f), \forall i \in [n]$ *is initialized such that* $\omega^0 \in U_M$*, where* U_M *is a neighborhood of the strict Nash equilibrium (as defined in Eq. (3)), then for any* $i \in [n]$ *and any* t*, and under Assumption 2, we have* $\|\omega_i^t - \omega_i^*\|_1 \leq 2(kF\exp(-M - \eta\epsilon t))$*, where* $M \geq \left|\log\left(\frac{\epsilon}{2kF}\right)\right|$*, and* ϵ *is a constant that is game-dependent only.*

Remark 2. We note that the convergence rate of the algorithm can be improved by increasing the step size η. This is because when each player receives expected rewards, the player can take greedy steps toward the equilibrium strategy. This agrees with greedy strategies that are previously employed to reach strict Nash equilibrium [14]. However, such greedy policies would not work in the presence of reward uncertainty, as we will discuss in Theorem 3.

It is worth mentioning that the convergence rate of our algorithm does not rely on the number of actions A_i, but rather solely on the number of facilities F. This is an improvement over the previous findings for exponential weights algorithms with non-combinatorial action spaces in the context of a congestion game, where the rate is linearly dependent on the number of actions [14]. The reason for this is the utilization of our facility-level reward estimation technique once again.

Previous studies on the convergence rate of congestion games [9,10] have established a linear convergence rate when the game possesses a convex potential function and the algorithm is given an appropriate initial starting point. The potential function provides a means to capture the incentives of all players to modify their actions and can be used to characterize the dynamics in policy updates. Assuming the convexity of the potential function implies optimization on a simpler policy optimization landscape. In contrast, our algorithm achieves a much faster rate of convergence, and this convergence rate still holds even in the absence of a convex potential function. This is because that we adopt a different approach where we directly argue through the algorithm update rule that the updated policy will always fall within a neighborhood around the Nash equilibrium. This bypasses the need for a smoothness potential function and demonstrates the effectiveness of our approach.

In addition, we remark that though some variants of Mirror Descent (MD) or Follow-the-Regularized-Leader (FTRL) algorithms are also proven to enjoy sublinear regret with logarithmic dependency on action space in the full information setting [20], these results only imply convergence to an approximate coarse correlated equilibrium and do not enjoy convergence to Nash equilibrium. In comparison, Nash equilibrium is much more stable, as the dynamic will remain there unless external factors change, while coarse correlated equilibrium may be more sensitive to small changes in the correlation method, which can lead to deviation from the equilibrium [32].

[16] and [35] have also investigated the convergence to Nash equilibrium for congestion game, under the notion of best-iterate convergence of the rewards value. Their algorithm ensures that with high probability, there exists a $t \in [T]$ such that $r_i(\omega_i^t)$ is close enough to the rewards attained by Nash equilibrium. [35]

also showed their algorithm can attain sublinear individual regret simultaneously. However, these results do not directly guarantee the convergence of the actual action sequence $\{\omega_i^t\}_{t=1}^T$ (as what we do in this paper).

We include the proof in the arXiv version of the paper, and we summarize the technical highlights here. To prove Theorem 2, we first identify that there exists a neighborhood around the strict Nash equilibrium, such that for any player i, his action in the strict Nash equilibrium is the only optimal choice.

Lemma 1. *If there exists a strict Nash equilibrium $a^* = (a_1^*, \dots, a_n^*)$, then there exists $\epsilon > 0$ and a neighborhood U_ϵ of a^*, such that for all $\tilde{\omega} = (\tilde{\omega}_i, \tilde{\omega}_{-i}) \in U_\epsilon$, $r_i(a_i^*, \tilde{\omega}_{-i}) - r_i(a_i, \tilde{\omega}_{-i}) \geq \epsilon, \quad \forall i \in [n], a_i \in \mathcal{A}_i, a_i \neq a_i^*$, where $r_i(a_i, \omega_{-i})$ is defined as $\mathbb{E}_{a_{-i} \sim \omega_{-i}}[r_i(a_i, a_{-i})]$.*

Moreover, if the difference in reward estimator $\tilde{z}_i^t(a_i) = \sum_{j=0}^t \left(\sum_{f \in a_i} \tilde{y}_i^j(f) - \sum_{f' \in a_i^*} \tilde{y}_i^j(f')\right)$ is upper bounded by some small enough constant, then the induced policy of Algorithm 1 falls into the neighborhood set U_ϵ.

Lemma 2. *Let $\tilde{z}_i^t(a_i) = \sum_{j=0}^t \left(\sum_{f \in a_i} \tilde{y}_i^j(f) - \sum_{f' \in a_i^*} \tilde{y}_i^j(f')\right)$, and define*

$$U_M = \& \left\{\omega^t \text{ computed by Algorithm 1} \mid \tilde{z}_i^t(a_i) \leq -M, \forall a_i \neq a_i^*, \forall i \in [n]\right\}. \tag{3}$$

For sufficiently large M, $U_M \subseteq U_\epsilon$. Moreover, following the updates of Algorithm 1, and under Assumption 2, if $\omega^t \in U_M$, then $\omega^{t+1} \in U_M$.

Thus, if ω^0 is in the neighborhood $U_M \subseteq U_\epsilon$, then the reward estimator $\tilde{z}_i^t(a_i)$ can only decrease (by Lemma 1), and hence the algorithm will give an updated policy ω^t that is also within the neighborhood set U_ϵ.

Also, note that ω_i^* is a strict Nash equilibrium, which implies that $|\omega_i^t - \omega_i^*|_1 = 2(1 - \omega_i^t(a_i^*))$. Hence, to establish the convergence rate, we need to lower bound $\omega_i^t(a_i^*) = \frac{\prod_{f \in a_i^*} \tilde{\omega}_i^t(f)}{\sum_{a' \in \mathcal{A}} \prod_{f' \in a'} \tilde{\omega}_i^t(f')} = \frac{\prod_{f \in a_i^*} \exp\left(\eta \sum_{j=0}^t \tilde{y}_i^j(f)\right)}{\sum_{a' \in \mathcal{A}} \prod_{f' \in a'} \exp\left(\eta \sum_{j=0}^t \tilde{y}_i^j(f')\right)}$.

Technical Challenge. We remark that if we directly apply Lemma 2, we can get

$$\omega_i^t(a_i^*) = \& \frac{\prod_{f \in a_i^*} \tilde{\omega}_i^t(f)}{\sum_{a' \in \mathcal{A}} \prod_{f' \in a'} \tilde{\omega}_i^t(f')} \geq \frac{1}{1 + \sum_{a_i \in \mathcal{A}_i, a_i \neq a_i^*} \left(\prod_{f \in a_i} \tilde{\omega}_i^t(f) - \prod_{f' \in a_i^*} \tilde{\omega}_i^t(f')\right)}$$
$$\geq \& 1 - \sum_{a_i \in \mathcal{A}_i, a_i \neq a_i^*} \left(\prod_{f \in a_i} \tilde{\omega}_i^t(f) - \prod_{f' \in a_i^*} \tilde{\omega}_i^t(f')\right).$$

Suppose one can upper bound $\sum_{a_i \in \mathcal{A}_i, a_i \neq a_i^*} \left(\prod_{f \in a_i} \tilde{\omega}_i^t(f) - \prod_{f' \in a_i^*} \tilde{\omega}_i^t(f')\right) \leq \exp(-t)$, then this gives $1 - \sum_{a_i \in \mathcal{A}_i, a_i \neq a_i^*} \exp(-t)$, which yields a convergence rate of $(|\mathcal{A}_i| - 1)\exp(-t)$ as $\|\omega_i^t - \omega_i^*\|_1 = 2(1 - \omega_i^t(a_i^*))$. However, this approach implies that the convergence rate scales linearly with the number of actions (thus exponentially with the number of facilities), and is what we wanted to avoid in the analysis.

To overcome this exponential dependency, we utilize the fact that any k-facility combination is an action, which means that we can order the facility from $f_1, \ldots, f_F$ in decreasing order of $\tilde{\omega}_i^t(f)$ and $f_1, \ldots, f_k$ form the optimal pure Nash action a_i^*.

Then we consider the case that each player observes only $R^f(a_i^t, a_{-i}^t)$, instead of the expected rewards.

Theorem 3. *Under Assumption 2, consider the case where each player receives a stochastic reward under the full information setting and under Assumption 2. Assume the game permits a strict Nash equilibrium* $\omega^* = (\omega_1^*, \cdots, \omega_n^*)$. *Let* $\tilde{y}_i^t(f) = R^f(a_i^t, a_{-i}^t)$ *in Line 6 of Algorithm 1, and set the learning rate to be time-dependent such that* $\sum_{t=0}^{\infty} \eta_t^2 \leq \frac{\delta \cdot M^2}{8kn(F-1)} \leq \sum_{t=0}^{\infty} \eta_t = \infty$. *Suppose* $\tilde{y}_i^0(f), \forall i \in [n]$ *is initalized such that* $\omega^0 \in U_{2M} \subseteq U_\epsilon$, *then for any* $i \in [n]$ *and any* t, *we have* $\|\omega_i^t - \omega_i^*\|_1 \leq 2kF \exp\left(-M - \epsilon \sum_{j=0}^{t} \eta_j\right)$, *with probability at least* $1 - \delta$, *where* $M \geq \left|\log\left(\frac{\epsilon}{2kF}\right)\right|$, *and* ϵ *is a constant that is game-dependent only.*

We remark that in the case of stochastic rewards, the convergence rate of our algorithm cannot be arbitrarily large, as the learning rate η cannot be taken to be arbitrarily large. If we take $\eta_t = \beta t^{-\alpha}$, with β being a small positive constant and $\alpha \in (1/2, 1)$. Then our convergence rate is $\|\omega_i^t - \omega_i^*\|_1 \leq O\left(\exp\left(-\frac{\beta}{1-\alpha} t^{1-\alpha}\right)\right)$, which is close to exponentially fast convergence. When the reward function is smooth, we remark that this can imply each player only experiences constant regret. We include the proof to Theorem 3 in the arXiv version of the paper.

6 Conclusion

We studied the congestion game under semi-bandit feedback and presented a modified version of the well-known exponential weights algorithm. The algorithm ensures sublinear regret for every player, with the regret being linearly dependent on the number of facilities. Additionally, the proposed algorithm can learn a policy that rapidly converges to the pure Nash policy, with the convergence rate also being linearly dependent on the number of facilities. To our best knowledge, these are the first results on congestion games for sublinear individual regret and geometric Nash convergence rate, without an exponential dependency on the number of facilities.

There are several possible directions to further study the online congestion game. First, as our work only considers the semi-bandit feedback model for individual regret, the regret and convergence rate under the full-bandit feedback model remains unclear. For the Nash convergence result, our algorithm only enjoys theoretical guarantees in the full-information setting. It remains future work to extend this result to semi-bandit and full-bandit feedback models. Moreover, it also remains in question whether the results of this work can be extended to the online Markov congestion game proposed by [16].

Acknowledgement. Baoxiang Wang and Jing Dong are partially supported by the National Natural Science Foundation of China (62106213, 72394361) and an extended support project from the Shenzhen Science and Technology Program. Volkan Cevher was supported by Hasler Foundation Program: Hasler Responsible AI (project number 21043), the European Research Council (ERC) under the European Union's Horizon 2020 research and innovation programme (grant agreement n° 725594 - time-data), and the Army Research Office and was accomplished under Grant Number W911NF-24-1-0048. Leello Dadi acknowledges support from Google. Luca Viano is funded (in part) through a PhD fellowship of the Swiss Data Science Center, a joint venture between EPFL and ETH Zurich.

References

1. Abernethy, J.D., Hazan, E., Rakhlin, A.: Competing in the dark: an efficient algorithm for bandit linear optimization (2009)
2. Audibert, J.Y., Bubeck, S., Lugosi, G.: Regret in online combinatorial optimization. Math. Oper. Res. **39**(1), 31–45 (2014)
3. Auer, P., Cesa-Bianchi, N., Freund, Y., Schapire, R.E.: The nonstochastic multiarmed bandit problem. SIAM J. Comput. **32**(1), 48–77 (2002)
4. Awerbuch, B., Kleinberg, R.D.: Adaptive routing with end-to-end feedback: distributed learning and geometric approaches. In: Proceedings of the Thirty-Sixth Annual ACM Symposium on Theory of Computing, pp. 45–53 (2004)
5. Brown, G.W.: Some notes on computation of games solutions. Technical report, Rand Corp Santa Monica CA (1949)
6. Bubeck, S., Cesa-Bianchi, N., Kakade, S.M.: Towards minimax policies for online linear optimization with bandit feedback. In: Mannor, S., Srebro, N., Williamson, R.C. (eds.) COLT 2012 - The 25th Annual Conference on Learning Theory, Edinburgh, Scotland, 25–27 June 2012. JMLR Proceedings, vol. 23, pp. 41.1–41.14. JMLR.org (2012)
7. Cesa-Bianchi, N., Lugosi, G.: Prediction, Learning, and Games. Cambridge University Press (2006)
8. Cesa-Bianchi, N., Lugosi, G.: Combinatorial bandits. J. Comput. Syst. Sci. **78**(5), 1404–1422 (2012)
9. Chen, P.A., Lu, C.J.: Playing congestion games with bandit feedbacks. In: Proceedings of the 2015 International Conference on Autonomous Agents and Multiagent Systems (2015)
10. Chen, P.A., Lu, C.J.: Generalized mirror descents in congestion games. Artif. Intell. **241**, 217–243 (2016)
11. Chen, W., Wang, Y., Yuan, Y.: Combinatorial multi-armed bandit: general framework and applications. In: International Conference on Machine Learning (2013)
12. Chen, X., Peng, B.: Hedging in games: faster convergence of external and swap regrets. In: Advances in Neural Information Processing Systems (2020)
13. Cheung, M.H., Southwell, R., Huang, J.: Congestion-aware network selection and data offloading. In: 2014 48th Annual Conference on Information Sciences and Systems (CISS), pp. 1–6. IEEE (2014)
14. Cohen, J., Héliou, A., Mertikopoulos, P.: Exponentially fast convergence to (strict) equilibrium via hedging. arXiv preprint arXiv:1607.08863 (2016)
15. Combes, R., Talebi Mazraeh Shahi, M.S., Proutiere, A., et al.: Combinatorial bandits revisited. In: Advances in Neural Information Processing Systems (2015)

16. Cui, Q., Xiong, Z., Fazel, M., Du, S.S.: Learning in congestion games with bandit feedback. In: Advances in Neural Information Processing Systems (2022)
17. Dadi, L., Panageas, I., Skoulakis, S., Viano, L., Cevher, V.: Polynomial convergence of bandit no-regret dynamics in congestion games. arXiv preprint arXiv:2401.09628 (2024)
18. Dani, V., Hayes, T.P., Kakade, S.M.: The price of bandit information for online optimization. In: Proceedings of the 20th International Conference on Neural Information Processing Systems, NIPS 2007, pp. 345–352. Curran Associates Inc., Red Hook, NY, USA (2007)
19. Daskalakis, C., Deckelbaum, A., Kim, A.: Near-optimal no-regret algorithms for zero-sum games. In: Proceedings of the Twenty-Second Annual ACM-SIAM Symposium on Discrete Algorithms (2011)
20. Daskalakis, C., Fishelson, M., Golowich, N.: Near-optimal no-regret learning in general games. In: Advances in Neural Information Processing Systems (2021)
21. Daskalakis, C., Pan, Q.: A counter-example to Karlin's strong conjecture for fictitious play. In: IEEE 55th Annual Symposium on Foundations of Computer Science (2014)
22. Dong, J., Wu, J., Wang, S., Wang, B., Chen, W.: Taming the exponential action set: sublinear regret and fast convergence to Nash equilibrium in online congestion games. arXiv preprint arXiv:2306.13673 (2023)
23. Flaxman, A.D., Kalai, A.T., McMahan, H.B.: Online convex optimization in the bandit setting: gradient descent without a gradient. arXiv preprint cs/0408007 (2004)
24. Giannou, A., Vlatakis-Gkaragkounis, E.V., Mertikopoulos, P.: On the rate of convergence of regularized learning in games: from bandits and uncertainty to optimism and beyond. In: Advances in Neural Information Processing Systems (2021)
25. György, A., Linder, T., Lugosi, G., Ottucsák, G.: The on-line shortest path problem under partial monitoring. J. Mach. Learn. Res. **8**, 2369–2403 (2007)
26. Hsieh, Y.G., Antonakopoulos, K., Mertikopoulos, P.: Adaptive learning in continuous games: optimal regret bounds and convergence to Nash equilibrium. In: Conference on Learning Theory (2021)
27. Kalai, A., Vempala, S.: Efficient algorithms for online decision problems. J. Comput. Syst. Sci. **71**(3), 291–307 (2005), learning Theory 2003
28. Kleinberg, R., Piliouras, G., Tardos, É.: Multiplicative updates outperform generic no-regret learning in congestion games. In: Proceedings of the Forty-First Annual ACM Symposium on Theory of Computing (2009)
29. Krichene, W., Drighès, B., Bayen, A.M.: Online learning of Nash equilibria in congestion games. SIAM J. Control. Optim. **53**(2), 1056–1081 (2015)
30. Lattimore, T., Szepesvári, C.: Bandit Algorithms. Cambridge University Press (2020)
31. Neu, G., Bartók, G.: An efficient algorithm for learning with semi-bandit feedback. In: Jain, S., Munos, R., Stephan, F., Zeugmann, T. (eds.) Algorithmic Learning Theory - 24th International Conference, ALT 2013, Singapore, 6–9 October 2013. Proceedings. LNCS, vol. 8139, pp. 234–248. Springer, Cham (2013)
32. Nisan, N., Roughgarden, T., Tardos, E., Vazirani, V.V.: Algorithmic Game Theory. Cambridge University Press (2007)
33. Palaiopanos, G., Panageas, I., Piliouras, G.: Multiplicative weights update with constant step-size in congestion games: convergence, limit cycles and chaos. In: Advances in Neural Information Processing Systems, vol. 30 (2017)

34. Panageas, I., Skoulakis, S., Viano, L., Wang, X., Cevher, V.: Semi bandit dynamics in congestion games: convergence to Nash equilibrium and no-regret guarantees (2023)
35. Panageas, I., Skoulakis, S., Viano, L., Wang, X., Cevher, V.: Semi bandit dynamicsin congestion games: convergence to Nash equilibrium and no-regret guarantees. In: International Conference on Machine Learning (2023)
36. Rosenthal, R.W.: A class of games possessing pure-strategy Nash equilibria. Int. J. Game Theory **2**, 65–67 (1973)
37. Roughgarden, T.: Routing games. Algorithmic Game Theory **18**, 459–484 (2007)
38. Roughgarden, T.: Intrinsic robustness of the price of anarchy. In: Proceedings of the Forty-First Annual ACM Symposium on Theory of Computing, pp. 513–522 (2009)
39. Roughgarden, T., Tardos, É.: How bad is selfish routing? J. ACM (JACM) **49**(2), 236–259 (2002)
40. Syrgkanis, V., Agarwal, A., Luo, H., Schapire, R.E.: Fast convergence of regularized learning in games. In: Advances in Neural Information Processing Systems, vol. 28 (2015)
41. Tekin, C., Liu, M., Southwell, R., Huang, J., Ahmad, S.H.A.: Atomic congestion games on graphs and their applications in networking. IEEE/ACM Trans. Netw. **20**(5), 1541–1552 (2012)
42. Zhang, F., Wang, M.M.: Stochastic congestion game for load balancing in mobile-edge computing. IEEE Internet Things J. **8**(2), 778–790 (2020)
43. Zinkevich, M.: Online convex programming and generalized infinitesimal gradient ascent. In: Fawcett, T., Mishra, N. (eds.) Machine Learning, Proceedings of the Twentieth International Conference (ICML 2003), Washington, DC, USA, 21–24 August 2003, pp. 928–936. AAAI Press (2003)

Enabling Asymptotic Truth Learning in a Social Network

Kevin Lu[1(✉)], Jordan Chong[2], Matt Lu[3], and Jie Gao[1]

[1] Rutgers University, 57 US Highway 1, New Brunswick, NJ 08901-8554, USA
{kl1160,jg1555}@rutgers.edu

[2] New York University, 50 West 4th Street, New York, NY 10012, USA
jhc10007@nyu.edu

[3] Northwestern University, 633 Clark Street, Evanston, IL 60208, USA
mattlu2029@u.northwestern.edu

Abstract. Consider a network of agents that all want to guess the correct value of some ground truth state. In a sequential order, each agent makes its decision using a single private signal which has a constant probability of error, as well as observations of actions from its network neighbors earlier in the order. We are interested in enabling *network-wide asymptotic truth learning* – that in a network of n agents, almost all agents make a correct prediction with probability approaching one as n goes to infinity. In this paper we study both random orderings and carefully crafted decision orders with respect to the graph topology as well as sufficient or necessary conditions for a graph to support such a good ordering. We first show that on a sparse graph of average constant degree with a random ordering asymptotic truth learning does not happen. We then show a rather modest sufficient condition to enable asymptotic truth learning. With the help of this condition we characterize graphs generated from the Erdös Rényi model and preferential attachment model. In an Erdös Rényi graph, unless the graph is super sparse (with $O(n)$ edges) or super dense (nearly a complete graph), there exists a decision ordering that supports asymptotic truth learning. Similarly, any preferential attachment network with a constant number of edges per node can achieve asymptotic truth learning under a carefully designed ordering but not under either a random ordering nor the arrival order. We also evaluated a variant of the decision ordering on different network topologies and demonstrated clear effectiveness in improving truth learning over random orderings.

Keywords: Social learning · information cascade · herding · asymptotic learning

Kevin Lu and Gao are supported by NSF IIS-2229876, DMS-2220271, DMS-2311064, CCF-2208663, CCF-2118953. Part of this work was carried out while JC and ML were participants in the DIMACS REU program at Rutgers University, supported by the NSF award CNS-2150186 and the NSF HDR TRIPODS award CCF-1934924. Due to space limit the full version with all proofs can be found at [12].

M. Mavronicolas et al. (Eds.): WINE 2024, LNCS 15534, pp. 530–547, 2026.
https://doi.org/10.1007/978-3-032-08560-3_30

1 Introduction

Decision making in a social setting is a natural problem that appears in many real world multi-agent settings. In a security application scenario we have a network of agents making predictions on the state of the network—is the network under attack and if so which type of attack? All agents can collect local data and statistics and use it to make inferences. In addition, agents can also observe the behaviors, announcements, predictions from neighboring agents in the network. Depending on the particular application scenario, an agent may be able to share with neighboring agents some raw data or even prediction models. In other settings, the agents may belong to different entities. Due to privacy reasons they may not be able to share any data or model beyond their prediction of the network state âĂŞ a scenario that is considered here.

Suppose there is an unknown ground truth state θ and a network of agents that all want to guess the correct value of θ. Each agent can make a single private measurement which has a constant probability to error. In addition, an agent can observe the actions/predictions of other agents, who have already made their decisions. Intuitively an agent can make use of her own private measurements as well as the actions of other agents to improve her chance of a correct prediction. This is the setup in the classical sequential learning [9,13]. The agents make decisions in a given sequential order σ. The i-th agent has the knowledge of the decisions of all the previous agents. The central question asks if the n-th agent in the sequence chooses the correct action with probability approaching 1 as n goes to infinity, phrased as *asymptotic learning.*

Asymptotic learning does not happen in a sequential learning setting – due to a well known inherent trap called herding or information cascade. An early (small) group of agents make a wrong decision due to observation errors and misfortune, subsequent agents observing the decisions of the early agents ignore their private signal or measurements, in favor of conforming to the actions of the agents they observe [5–7,19,21]. This can happen with perfectly rational agents and rigorous Bayesian analysis. When this happens, all agents make the wrong prediction despite an abundant number of independent observations and signals in the network. In reality, we indeed observe herding when a population makes seemingly irrational decisions following the crowd.

In this paper we consider sequential learning on a social network, i.e., the model used in [4]. The agents still make decisions in a sequential order but an agent v can only see the actions of v's neighbors who have arrived earlier. In this case, the network topology limits the visibility (or information flow). Thus both the network topology and the agent ordering are crucial for successful learning. Further, we are interested in enabling *network-wide asymptotic truth learning* – that in a network of n agents, the average probability of all agents making a correct prediction approaches one as n goes to infinity. Notice that this is a much stronger requirement than asymptotic learning as we need not just the nth agent, rather, all but $o(n)$ agents achieve asymptotic learning.

Enabling network-wide asymptotic learning is much less studied in the literature. One of the main results in [4] is to present one family of graphs that supports network wide asymptotic learning, namely, the *celebrity* graphs: fully connected bipartite graphs of m celebrity nodes and n normal nodes. Under a random ordering of agents on this network or under an adversarial ordering (but the network topology is randomized), there will be roughly n/m normal nodes that arrive before any celebrity nodes, termed the 'guinea pigs'. These early normal nodes are independent of each other and thus make independent decisions. Now, the first celebrity node is able to effectively aggregate information from these independent decisions and arrive at a much more accurate decision. With the Bayesian model, all remaining normal nodes can now effectively learn from this 'knowledgeable' celebrity node and thus essentially copy its decision. This leads to highly accurate predictions for all remaining vertices. In this celebrity graph, in order for effective aggregation to happen with high probability, we need n/m to be at least $\Omega(\log n)$ at the first celebrity node. We also need n/m to be sublinear $o(n)$ – otherwise the network learning rate is dragged down by the guinea pigs. This means that the network is actually very dense, with average degree polynomial in the network size.

In this paper we are interested in the interplay of network topology and the arrival ordering to enable asymptotic truth learning. We would investigate sufficient and necessary conditions on whether there is a good ordering and if so present one, and what happens under a random ordering. Our study will also provide insights or guidance on improving quality of learning in real-world decision making processes.

Our Contribution

Suppose there is a ground truth state $\theta \in \{0, 1\}$ and all agents aim to learn θ. Following a decision order σ, a node v makes a single private observation which matches θ with a constant probability q. v is also able to observe the decisions of neighbors that are earlier than v in order σ. To make a decision, node v can use the Bayesian model, with all information available so far including the decisions from neighbors, the network topology and the ordering σ. Computing the Bayesian state can be intensive, considering the exponentially growing state space. Alternatively, v can use a simple model such as majority rule on the decisions from neighbors and its private signal. Our goal is to achieve network wide asymptotic truth learning – the average success probability of all agents approaches 1 when the number of agents goes to infinity.

To support asymptotic truth learning, both the network topology and the ordering σ matter. If the network is empty, clearly no asymptotic learning is possible – every agent makes decisions using only local private measurements and thus has a constant error probability. If the network is a complete graph, herding happens with a constant probability. Again we do not have asymptotic truth learning. In the two examples, the ordering σ does not matter but in other graphs, the ordering in which the agents make decisions is crucial. The success of truth learning depends on the combination of a carefully crafted ordering tailored to the network topology, under a proper inference model.

(i) Sparse graph with random ordering: no truth learning. We first start with negative results – when truth learning does not happen. First, when the network has an independent set of size $\Omega(n)$, there is an ordering that does not support truth learning, under both the Bayesian model and the majority rule model. Specifically, if these independent nodes are put first in the ordering, they have to make independent decisions, using only private observations, leading to a learning rate of q per node. Since this is a constant fraction of the entire network size, the network learning rate is thus bounded away from 1 by a constant. Using this simple observation we can show that for any network of constant *average* degree, there is an ordering for which truth learning does not happen. In fact, a random ordering does not support truth learning.

(ii) A sufficient condition for truth learning in Bayesian model: Our positive result for enabling truth learning starts from a sufficient condition that concerns both the network structure and the decision ordering. Specifically if there is a vertex v with a set of neighbours S, $|S| = \omega(1)$, that are independent of each other such that removing S will disconnect at most $o(n)$ vertices from v, then there is a decision ordering such that asymptotic truth learning is achieved under the Bayesian model. The decision ordering will put S first before v and then select a decision ordering that propagate the aggregated signal at v to the rest of the vertices. The agents in S is analogous to the notion of *guinea pigs*, agents who make decisions solely based on their private signals [16,18]. We show that a slightly super-constant number of such guinea pigs in the neighborhood of agent v, allows v to learn a high quality signal – a signal which matches θ with probability approaching 1, that can be effectively propagated to the entire network as long as connectivity to v is maintained.

(iii) Characterization of Erdös Rényi model and PA model. Using this sufficient condition we provide characterization of asymptotic truth learning in the Erdös Rényi model – the classical random graph model with each pair connected by the edge independently decided with probability p. When p is too small, $p = O(1/n)$, the graph is too sparse and has a constant fraction of isolated nodes. Thus there is no decision ordering that supports asymptotic truth learning. When p is a constant the graph has quadratically many edges such that any random ordering does not support truth learning since herding happens. Further, when p is almost 1, with a gap of only $O(n^\epsilon)/n$ for any $0 < \epsilon < 1$, one cannot even find an ordering tailored to the specific graph topology to prevent herding. For any other value p we show that the sufficient condition is met and therefore a truth learning ordering exists.

We also analyze networks obtained from preferential attachment models. Notice that in the preferential attachment (PA) network each newcomer selects k edges, typically with k being a constant. Thus the network has constant average degree. A random ordering does not support truth learning despite the existence of high degree vertices with the potential power of effectively aggregating information. Further, if we follow the natural arrival order in the PA model, each node knows k vertices earlier in the order. When $k \geq 2$, herding happens with a constant probability and again there is no truth learning. By using the sufficient condition and analysis of degree distribution of vertices in a PA model, we show there is indeed a decision ordering in the PA graph that supports asymptotic truth learning. In fact, the preferential attachment network has been considered as one of the 'good' networks in the literature that supports truth learning [4]—and now we know it must be accompanied by a carefully selected decision ordering.

(iv) Enabling truth learning on sparse graphs. The sufficient condition requires at least one node with degree $\omega(1)$. On a network of constant degree everywhere, a random ordering does not work. Motivated by the negative results, we ask if there can be networks of constant degree that *does* support truth learning with a particular vertex ordering. We first show that on a 2D grid graph we can carefully choose a subtree of size $O(\log n)$ such that the root has a high quality signal. Then we can propagate it to the rest of the network.

Further, we show that the butterfly network with a specific ordering of the vertices, does provide network-wide truth learning. First, we consider a fully balanced binary tree with an ordering bottom up along the tree. We can calculate the success rate of each node in the tree, which is monotonically improved up the tree and approaches 1 at the root. The butterfly network is essentially packing n fully balanced binary tree with shared vertices at the bottom layers. Again with a bottom up ordering, the learning rate improves along each layer. The network wide learning rate is thus also approaching 1 as the network size goes to infinity. This is the first example of a constant degree network that supports truth learning without using the sufficient condition.

(iv) Simulation results with majority rule on real-world networks. We report experiments using different network topologies including real world graphs and different decision orderings. Since it is computationally expensive to run a full Bayesian model we used the majority rule in all our experiments. We designed a decision ordering inspired by the sufficient condition, which effectively improves truth learning quality over random decision orders.

1.1 Related Work

Learning in a social network is a well studied topic [9,13], so we only review work that is most relevant to this study, namely sequential learning with a social network structure. Almost all prior work examine only asymptotic learning, i.e., the n-th agent has a success rate approaching 1 as $n \to \infty$. Asymptotic learning occurs when the private signals are *unbounded* – when the success probability of a private signal can be arbitrarily close to 1. When the private signals are bounded (which is the parameter regime studied in this paper), asymptotic learning does not happen due to information cascade [19].

To avoid information cascade one suggestion [16,18] is to restrict social interaction among the agents. In particular the agents early in the sequence, referred to as guinea pigs, are not allowed to observe each other. The late decision makers therefore can infer a great deal from the actions of the guinea pigs. [16] also formulated an optimization problem using a payoff model to find the ideal number of guinea pigs. In this paper, the social network structure is fixed and we are looking for a good decision ordering, if it exists. [1] also considers a fixed sequence of

n agents, but assume that edges in the social network follow a stochastic model. They make a distinction between expanding and non-expanding observations. A social network featuring non-expanding observations describes a topology where the decisions of an infinite number of agents are based on a *finite* set of *excessively influential* agents. Such a network precludes asymptotic learning – as there is a positive probability that the excessively influential agents made the wrong decision which was copied by the agents later in the sequence. In the positive direction, they show that unbounded private signals and expanding observations are sufficient for asymptotic learning. A number of other work use a similar setup with [1]. [14] considered position uncertainty and learning robustness. [2] generalized a linear ordering to an m-dimensional lattice.

[3] uses the same model as in [4] and provides interesting insights on bounded degree graphs. They pointed out that with a random decision order on a bounded degree graph, the decision process enjoys locality – the decision of an agent is only influenced by other agents within a certain distance. Further, they also investigated *local learning requirement* that supports the successful learning of a particular agent, without high degree requirements for aggregation. Notice that this result again does not give network-scale asymptotic truth learning.

2 Models and Problem Definition

Consider a network as an undirected graph $G = (V, E)$ modeling the social network of $|V| = n$ agents. Let θ be the ground truth signal which is a Bernoulli variable with parameter $\frac{1}{2}$ and is unknown to the agents. Each agent makes a private noisy measurement of θ and, together with observations of the predictions made by neighbours in G, try to guess the ground truth. Specifically, each agent v_i observes θ privately with a success probability $q \in (\frac{1}{2}, 1)$, which is a constant away from 1. We represent the measurement by a Bernoulli random variable s_i. $\mathbb{P}\{s_i = \theta\} = q$. In the literature this is called the *bounded belief* model [1,19]. For simplicity we use the same probability q for every agent. The results can be extended to the case that every agent has an error probability within (α, β) with $1/2 < \alpha < \beta < 1$.

Next, let σ be a permutation of V, which we call the *decision order*. In the order of σ, we let each $v_i \in V$ decide on an action, $a_i \in \{0, 1\}$, as a prediction of θ. To simplify notation, we will assume that v_i is the ith node in the order σ, i.e., the decision order is $v_1, v_2, ...v_n$. If two vertices v_i, v_j have an edge in G, we write $v_j \sim v_i$. The set of neighbors of v in G is denoted by $N(v)$. With the decision order σ we define the neighbors of a node v_i under σ as the set of neighboring nodes that arrive before i, i.e., $N_\sigma(v_i) = \{v_j | v_j \sim v_i, j < i\}$. When it is v_i's turn to take an action, v_i can use information from its own signal s_i as well as the published actions from nodes in $N_\sigma(v_i)$.

Here are two models that consider different information each agent uses to decide on its action.

- *Global model/Bayesian model:* Every agent v_i has the knowledge of the global network topology, the decision ordering σ of the vertices, its own private

signal s_i, as well as the actions taken by its neighbors $N_\sigma(v_i)$ that arrive before v_i. With knowledge of the network topology, a node uses a Bayesian model to fully utilize the actions of its neighbors in order to find the optimal action. This model is accurate in describing the learning process in the network setting. However, it is also computationally expensive and not practical to implement.
- *Local model/majority rule model:* Here, every agent v_i has only knowledge of its private signal s_i and the actions of its neighbors $N_\sigma(v_i)$. Without any other knowledge, the agent would just match its action to the majority action of its neighbors $N_\sigma(v_i)$ and its own private signal. This model is more suitable for a distributed setting, computationally much more friendly and has been adopted in a number of prior work [4,11,17].

We now define a learning network (G, q, σ) with a graph G, the observation success rate q and a vertex decision order σ. The learning quality for an agent v_i in the learning network (G, q, σ) is the probability that its prediction is correct: $\ell(v_i) = \mathbb{P}(a_i = \theta)$. The *learning quality* for the entire network G is the ratio of the expected number of agents that learn successfully in the network following the decision order σ:

$$L_\sigma(G) = \frac{1}{|V|} E|\{v_i \in V : a_i = \theta\}| = \frac{1}{|V|} \sum_{v_i \in V} \ell(v_i).$$

These definitions follow from the setup in [4] and [3]. With these definitions, we can define two new objectives. Let $\mathcal{F}$ be a family of graphs such that for any $N \in \mathbb{N}$, there exists $G \in \mathcal{F}$ with size at least N. We say that $\mathcal{F}$ achieves *network-wide asymptotic truth learning* if as $n \to \infty$, $\inf\{L_\sigma(G) : |G| \geq n, G \in F\} \to 1$ with each $L_\sigma(G)$ based on the optimal σ for that G. We say that $\mathcal{F}$ achieves *network-wide asymptotic random truth learning* if in addition σ is a random permutation on G. In future discussions, we can shorten to asymptotic (truth) learning and asymptotic random (truth) learning for brevity. Clearly asymptotic random learning implies asymptotic learning. The goal of this work is to examine both of these objectives.

3 Challenges of Asymptotic Truth Learning

Before analyzing specific graph models and decision orders that enable truth learning, we first look at topologies and orderings under which truth learning *does not* happen. We start with two cases that are well-known in the literature.

- **Independent learning**: The simplest example in which truth learning cannot be obtained is an empty graph. Each node makes decisions purely from its own noisy observation s_v, regardless of the decision order. To maximize the chance of guessing right, by Bayesian rule, the best strategy for an agent v is to report the same value as its private measurement s_v, i.e., $a_v = s_v$. The learning quality is $\ell(v) = \mathbb{P}\{s_v = \theta\} = q$. The learning quality of the entire network is q as well. Notice that this is not asymptotic truth learning as $L(G)$ is always a constant away from 1.

- **Information cascade**: When G is a complete graph, each node v knows its own noisy observation s_v as well as all the actions from nodes earlier than v in the order σ. This is exactly the sequential learning scenario in the literature, regardless of the ordering of nodes. It is well known that information cascade can easily happen in this case [5,6]. In particular, if the first two agents take incorrect actions $a = 1-\theta$, then the third agent (and all agents after that) will ignore its own signal and take a as well. This is because the chance that both agents before v are wrong, essentially $(1-q)^2$, is lower than the chance that s_v is wrong $(1-q)$. Therefore with probability at least $(1-q)^2$ all agents take the wrong action. This means that the learning quality of the entire network is at most $1-(1-q)^2$, which is again away from 1 by a constant. Thus asymptotic truth learning does not occur. Notice that the same situation happens if we use the Majority Rule model.

These two examples represent two challenges that forbid asymptotic truth learning. If the graph is too sparse, too many nodes will have to use independent learning, which by itself would already bring down the network learning quality. On the other hand if the graph is too dense such that every node is under heavy influence of the other vertices that arrive before, unfortunate mistakes in observation signals for the first few vertices can propagate throughout the network. This again leads to a large fraction of nodes that end up with the wrong prediction.

3.1 Graphs of $O(1)$ Average Degree

If a graph is sparse, i.e., has constant average degree, in a random ordering σ, a constant fraction of nodes are independent, i.e., with no neighbors that arrive earlier in the ordering. Therefore they will derive their actions purely from individual observation. Since independent nodes have learning quality to be at least a constant away from 1, random asymptotic learning does not happen.

Theorem 1. *Any family of graphs of constant average degree $\Delta = O(1)$ does not achieve asymptotic random truth learning, under both the Bayesian model and the majority vote model.*

The proof is in the Appendix of [12]. This theorem implies that there is no random asymptotic learning for any constant dimensional grid graphs, constant degree expander graphs, and preferential attachment graph with constant average number of edges per node.

In contrast, [4] proposed the *celebrity graph* which is a complete bipartite graph with vertex set $X \cup Y$, $|X| = n$, $|Y| = m$. In order to have network learning rate $1-\varepsilon$ under a random decision order, $N = O(\frac{1}{\varepsilon}\ln\frac{1}{\varepsilon})$ and $M = O(\frac{1}{\varepsilon^2}\ln\frac{1}{\varepsilon})$. Therefore to get $|N| + |M| = n$, ε is in the order of $1/\sqrt{n}$, which means that the network average degree is roughly $\Omega(\sqrt{n})$, in order to support random asymptotic learning.

On the other hand, even for a graph with constant degree everywhere, if we carefully design a decision order it is still possible to achieve asymptotic truth learning. We show such an example called a butterfly network. A *binary butterfly network* consist of $n = (k+1)2^k$ nodes, where $k+1$ is the depth, or *rank*, of the network. Let (i, j) refer to the jth node in the ith rank. For $i > 0$, the node (i, j) forms an edge with 2 nodes in depth $i-1$: $(i-1, j)$ and $(i-1, m)$, where $m \in \mathbb{Z}$ is found by inverting the i most significant bit of j. Note that due to the recursive structures, a butterfly network can be interpreted as the integration of 2^k complete binary trees. See Fig. 1 for an example. We highlight one binary tree with the root being the top leftmost vertex. We define a bottom-up ordering σ such that vertices of depth i arrive after vertices of depths $i-1$. The proof is in the Appendix of [12].

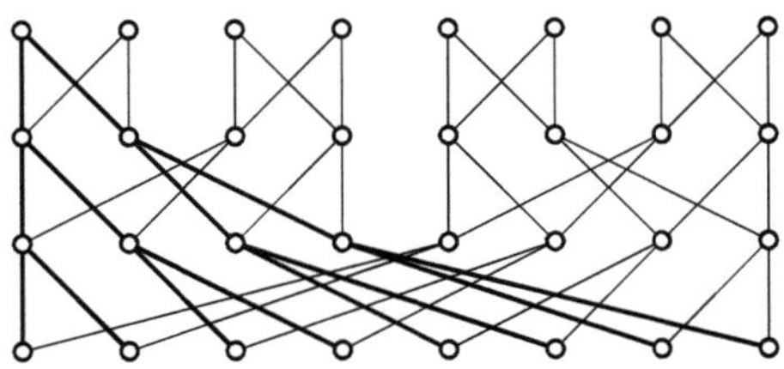

Fig. 1. An example of a butterfly network of $k = 3$.

Theorem 2. *A class of binary butterfly networks under the bottom-up ordering achieves asymptotic truth learning, under both the Bayesian model and the majority vote model.*

4 Enabling Truth Learning Under Bayesian Model

We present the main results in this section. We start with a sufficient condition for asymptotic truth learning under the Bayesian model. Then we use this to analyze Erdös Rényi Graphs, preferential attachment graphs and grid graphs.

Theorem 3. *Let $\mathcal{F}$ be a family of graphs. For each $G = (V, E) \in \mathcal{F}$ with $|V| = n$, if the following holds for some $v \in V$, $S \subset V$:*

1. *$S \subset N(v)$ such that S is independent i.e. $s_1 s_2 \notin E$ for all $s_1, s_2 \in S$;*
2. *$|S| = \omega(1)$, $|S| = o(n)$;*
3. *$G - S$ is a graph where the component containing v has all but $o(n)$ of the remaining vertices,*

then there is an ordering σ under which $\mathcal{F}$ achieves asymptotic truth learning.

Proof. To prove this theorem, there are two steps. First, we will show that v achieves high learning rate by aggregating the decisions of its independent neighbors. Then, we show that under the Bayesian model, we can propagate this learning rate to the rest of the graph.

Obtain a High Quality Signal. We start with the tail bound of a binomial distribution. Suppose X is a random variable with a binomial distribution $\text{Bin}(k, q)$ with an integer k and $0 \leq q \leq 1$,

$$\mathbb{P}\{X = a\} = \binom{k}{a} q^a (1-q)^{k-a}.$$

Lemma 1 (Tail bound of Binomial distribution [20]**).** *Let* $X \sim \text{Bin}(n, p)$*, and let* $\mu = E(X)$*. For any* $0 < \delta < 1$*,*

$$\mathbb{P}\{X \geq (1+\delta)\mu\} \leq \exp(-\frac{\delta^2\mu}{3}), \quad \mathbb{P}\{X \leq (1-\delta)\mu\} \leq \exp(-\frac{\delta^2\mu}{2}).$$

Let v, S be as in the statement of Theorem 3 with $s = |S|$. Then we construct a decision ordering σ. Let all of the vertices in S decide their actions first, in any order within the set. Since all vertices in S are independent, we have $\mathbb{P}\{a_i = \theta\} = q, i \in [s]$. Next, we have $v = v_{s+1}$ make its observation and decision. It is clear that v_{s+1}'s decision would match the majority between its private observation and the decisions of S. Then $\mathbb{P}\{a_{s+1} = \theta\} = \mathbb{P}\{X > (s+1)/2\}$ where $X \sim \text{Bin}(s+1, q)$ is a binomial distribution. By Lemma 1, we have the following probability:

$$\mathbb{P}\{X > \frac{s+1}{2}\} = 1 - \mathbb{P}\{X \leq \frac{s+1}{2}\} \geq 1 - \exp - \frac{(2q-1)^2}{8q}(s+1) = 1 - e^{-O(s)}.$$

Since $s = \omega(1)$, $s \to \infty$ when $n \to \infty$. Thus, $\lim_{n\to\infty} \mathbb{P}\{a_{s+1} = \theta\} = 1$. That is, the learning quality of agent v_{s+1} is asymptotically approaching 1.

Propagate a High Quality Signal. We first establish the property of monotonicity in the Bayesian model. That is, a node's success chance is no worse than the neighbors who have arrived earlier. The proof can be found in the Appendix of [12].

Proposition 1. *Given any learning network* (G, q, σ) *under the Bayesian model,*

$$\ell(v) \geq \max\{\ell(v_i) | v_i \in N_\sigma(v)\}.$$

Now, we will use Proposition 1 to propagate the high quality signal from v_{s+1}. To do this, let the rest of the decision ordering follow the same ordering as a traversal of $V \setminus S$ starting from v_{s+1}, and then the disconnected vertices (if any) follow randomly. Since the rest of the ordering outside of the isolated vertices is a traversal, each vertex that comes after v_{s+1} has a neighbor not in S earlier in the order σ. Thus, by Proposition 1, we have all of those vertices achieve learning rate at least $\ell(v_{s+1})$. Thus the network learning rate $L_\sigma(G)$ is:

$$L_\sigma(G) \geq \frac{1}{n}(1 - e^{-O(s)})(n - o(n)) = 1 - e^{-O(s)} - \frac{o(n)}{n}.$$

Since $s = \omega(1)$, when n goes to infinity, the network learning rate approaches 1.

We remark that the monotonicity property is unique to the Bayesian model. For the majority rule model a node could be misled by the neighbors not having a high quality signal and thus have a dropping learning rate.

4.1 Erdös RényiGraph

An Erdös Rényi graph [8] $G(n,p)$ is defined by $G=(V,E)$, where $|V|=n$, and $p \in (0,1)$, every unordered pair of vertices $v_1, v_2 \in V \times V$ is connected by an edge with probability p. The expected number of edges in an Erdös Rényi graph is $p\binom{n}{2}$. [8] analyzed graph connectivity with different values of p.

- If $p < 1/n$, then a graph in $G(n,p)$ will almost surely have no connected components of size larger than $O(\log n)$.
- If $p = 1/n$, then a graph in $G(n,p)$ will almost surely have a largest component whose size is of order $n^{2/3}$.
- If $np \to c > 1$, where c is a constant, then a graph in $G(n,p)$ will almost surely have a unique giant component containing a constant fraction of the vertices. No other component contains more than $O(\log n)$ vertices.
- If $p < \frac{(1-\varepsilon)\ln n}{n}$ for $0 < \varepsilon < 1$, then a graph in $G(n,p)$ will almost surely contain isolated vertices, and thus is disconnected.
- If $p > \frac{(1+\varepsilon)\ln n}{n}$ for $0 < \varepsilon < 1$, then a graph in $G(n,p)$ will almost surely be connected.

When discussing Erdös Rényi graphs, we consider the families $\mathcal{F} = \{G(n,p(n))\}$ where $p(n)$ is some function of n. In the following discussion, we may use $p = p(n)$ as shorthand. Since Erdös Rényi graphs are random graphs, when considering asymptotic truth learning, we need to define its learning rate differently. Specifically, since an empty graph and a complete each has a non-zero probability of appearing in an Erdös Rényi model, we cannot hope to find a good ordering for each graph in the family. Instead, we take the expected optimal learning rate for each graph G by the probability that G appears in the Erdös Rényi model. Let $F(G)$ be the probability of graph $G \in G_n$, G_n being the set of all graphs of size n, occurring according to the probability distribution of $G(n,p(n))$. We define:

$$L_\sigma(G(n,p(n))) = \sum_{G \in G_n} F(G) L_\sigma(G).$$

For asymptotic truth learning, we always consider the optimal $\sigma(G)$ for each G. For random asymptotic truth learning, σ is a random ordering. We will now show various threshold probabilities where the graph achieves or doesn't achieve truth learning with probability approaching 1 as n approaches infinity. Due to space limit, almost all proofs are in the Appendix of [12].

$p(n) = O(1/n)$: no asymptotic truth learning. When p is in the order of $O(1/n)$, there are a constant fraction of isolated vertices, who have a learning rate of q in any ordering σ. Thus there is no ordering that supports truth learning.

Proposition 2. *The family of Erdös Rényi graphs with parameter $p(n) = \frac{O(1)}{n}$ does not achieve asymptotic truth learning with any ordering σ under both the Bayesian or majority rule model.*

$p(n) = \frac{o(n)}{n}$ and $p(n) = \frac{\omega(1)}{n}$: asymptotic truth learning

Proposition 3. *A family of Erdös Rényi graphs with parameter p satisfying $p = \frac{o(n)}{n}$ and $p = \frac{\omega(1)}{n}$ achieves asymptotic truth learning under the Bayesian model.*

To prove this proposition, we will first prove the following lemma (with proof in the Appendix of [12]) and then use the sufficient condition of Theorem 3.

Lemma 2. *An Erdös Rényi graph $G(n,p)$ with $p = \frac{o(n)}{n}$ and $p = \frac{\omega(1)}{n}$ almost surely contains a giant component of size at least $n - o(n)$.*

Proof (Proof of Proposition 3). Since an edge exists between any two vertices with independent probability p, the degree of each vertex follows the binomial distribution, $\text{Bin}(n-1,p)$, with expectation $d = p(n-1)$, $d = o(n)$ and $d = \omega(1)$. By Lemma 1, for a constant $\delta > 0$, $\mathbb{P}\{\deg(v_i) \notin ((1-\delta)d, (1+\delta)d)\} \leq O(e^{-\delta^2 d})$. Since $d = w(1)$, a randomly selected vertex $v_\alpha \in V$ has degree between $((1-\delta)d, (1+\delta)d)$ with high probability. Next, we select a random subset S of the neighbors of v_α such that $|S| = s = 1/p^{1/3}$ if $p \geq 2/n^{3/4}$ and $d/2$ if not. In the first case, the probability of S being an independent set is $(1-p)^{\binom{s}{2}} \geq 1-p\cdot\binom{s}{2} \geq 1-p^{1/3}$. In the second case, the probability is at least $1 - \frac{pd^2}{4} \geq 1 - 1/n^{1/4}$. In both cases, $s = o(n)$ and $s = \omega(1)$. Thus, with high probability there exists an independent subset of neighbors of v_α of size $\omega(1)$.

Next, we need to show that G remains connected after removing v and S. It is clear that removing a random vertex v from $G(n,p)$ is just $G(n-1,p)$. Removing all of the neighbors of v is equivalent to removing nodes with probability p from $G(n-1,p)$, so the resulting graph is a random graph $G(k,p)$, where k takes values within 0 and $n-1$ with probability following distribution $\text{Bin}(n-1, 1-p)$. We call this resulting graph, G'. From Lemma 1 with high probability, $k = \Theta(n)$.

With our parameter range of p, G' contains with high probability a large component of size $n - o(n)$ by Lemma 2. Now in all cases, we know that $|R| = |N(v_\alpha) \setminus S| \geq (1/2 - \delta)d = \omega(1)$, so we just need R to be connected to all but $o(n)$ vertices of the graph. It is now trivial to see that the probability of at least one edge from R reaching the large component approaches 1, thus satisfying the final necessary condition of Theorem 3.

$p(n) = \Theta(1)$: no random asymptotic truth learning. When $p = \Theta(1)$, we have a dense graph with a quadratic number of edges. If we fix an ordering σ of the vertices, we argue that herding happens on a graph generated randomly from Erdös Rényi model. Thus, if we do not tailor the ordering to the specific graph topology obtained from the Erdös Rényi model we do not have truth learning.

Proposition 4. *A family of Erdös Rényi graphs with $p = \Theta(1)$, does not achieve random asymptotic random truth learning for the Bayesian or majority vote models.*

The above result does not eliminate the possibility of a good ordering chosen with respect to the particular graph topology. Below we present two results depending on how far p is from 1. If p is away from 1 by $\frac{\omega(n^\epsilon)}{n}$ truth learning exists. Otherwise, if p is away from 1 by $\frac{O(n^\epsilon)}{n}$, the graph is nearly a complete graph and there is no ordering for asymptotic truth learning.

$1-p(n)=\frac{\omega(n^\epsilon)}{n}$: asymptotic truth learning. In this scenario, we argue that the conditions of Theorem 3 are met. That is, since p is away from 1 by $\frac{\omega(n^\epsilon)}{n}$, there is an independent set of size $\omega(1)$ in the network. Thus we can choose an ordering that starts from these independent nodes and then propagate the aggregated signal to the rest of the network.

Proposition 5. *A family of Erdös Rényi graphs with parameter p satisfying* $1-p=\frac{\omega(n^\epsilon)}{n}$ *for any* $1>\epsilon>0$ *achieves asymptotic truth learning under the Bayesian model.*

$1-p(n)=\frac{O(n^\epsilon)}{n}$: no asymptotic truth learning. When p is too close to 1, i.e., with gap within $\frac{O(n^\epsilon)}{n}$ for any $1>\epsilon>0$, in any ordering σ, all but a constant number of nodes have at least two neighbors that are placed earlier in σ, therefore herding happens with a constant probability which means no truth learning.

Proposition 6. *A family of Erdös Rényi graphs with parameter p satisfying* $1-p=\frac{O(n^\epsilon)}{n}$ *for any* $\epsilon>0$ *does not achieve asymptotic truth learning under the Bayesian model.*

To prove this, we begin with a lemma:

Lemma 3. *A family of Erdös Rényi graphs with parameter p satisfying* $p=O(n^{-\epsilon})$ *for any constant* $1>\epsilon>0$ *has with high probability* $O(1)$ *maximum induced forest size.*

Proof. First, let N_n^k be the random variable for number of k-cliques in $G(n,p)$. We have:

$$\mathbb{E}(N_n^k)\le\binom{n}{k}(1-p)^{\binom{k}{2}-k+1}p^{k-1}(k+1)^{k-1}$$

from the well-known fact that there can be $(x+1)^{x-1}$ possible rooted forests on x labeled nodes. Plugging in $1-p=O(n^{-\epsilon})$, we get:

$$\mathbb{E}(N_n^k)=O(n^k e^{-k} n^{-k^2} k^k)$$

It's clear that if $k=\omega(1)$, we get $\mathbb{E}(N_n^k)=o(1)$, which means $\mathbb{P}(N_n^k\ge 1)=o(1)$ from Markov's inequality and thus with high probability, the maximum induced forest size is not $\omega(1)$.

Proof (Proof of Proposition 6). Let σ be any decision ordering of $G(n, p)$. Then, let S_σ be the set of vertices such that no more than 1 of their neighbors come before them in σ. We have that S_σ must be an induced forest, so $|S_\sigma| = O(1)$ with high probability by Lemma 3. Now, we argue that if every node in S_σ chooses the wrong decision, then every other node will follow suit regardless of their private signals.

Let's look at the smallest i such that $v_i \notin S_\sigma$. Note that $i \geq 3$. Clearly, every node before it chose incorrectly and v_i is connected to at least 2 of them, so it will just follow the other nodes' decision. Now this is true inductively. For every node not in S_σ, it is connected to at least 2 nodes before it, and both of those nodes chose incorrectly, so the current node will also choose incorrectly. Finally, since $|S_\sigma| = O(1)$ the probability of everything in S_σ choosing incorrectly is also $O(1)$. Thus, we have herding occurs and asymptotic truth learning does not occur.

4.2 Preferential Attachment Graph

A preferential attachment (PA) graph, with positive integer $k = O(1)$ as a parameter, is defined by $G = (V, E)$ where $|V| = n$ and the following process for generating edges:

1. Start with G_{k+1} as a complete graph on $k + 1$ vertices.
2. Given G_t, $t \geq k+1$, generate G_{t+1} by adding a vertex v as well as k undirected edges from v to vertices in G_t. Each edge connects v to a vertex u chosen randomly with probability $\frac{\deg(u)}{D}$ where D is the total current degree in G_t. Continue until G_n is generated and let $G = G_n$.

We use this variation of the PA model to enforce that the graph is connected. Some other models allow self loops. We note that as $n \to \infty$, this variation converges to the same degree distribution as the ones used in proofs in [10], as noted in that book. When k is 1, the graph is a tree. When $k \geq 2$, by using the same argument as in Proposition 4, if the decision ordering is the natural arriving order of the PA graph, there is a constant probability that herding happens. Further, for any constant k, a random ordering does not support truth learning since the average graph degree is a constant, by Theorem 1. In the following we show that for $k \geq 2$ we can achieve asymptotic truth learning by carefully choosing an ordering for graphs following the PA model.

k=$O(1)$: asymptotic truth learning. We first introduce a known theorem about the degree of a vertex as the PA model evolves. Define Γ as the commonly used *Gamma function*:

$$\Gamma(t) = \int_0^\infty x^{t-1} e^{-x} dx \tag{1}$$

Theorem 4 ([10], Exercise 8.14, Equation 8.3.11). *Let v be the vertex inserted to make G_b. Then as $n \to \infty$ and $a = \Theta(n)$, the degree of v,* $\deg(v)$ *in*

G_a converges almost surely to a random variable with expected value:

$$\sum_{s=1}^{k} \frac{\Gamma(ka+\frac{1}{2})}{\Gamma(ka)} \frac{\Gamma(k(b-1)+s)}{\Gamma(k(b-1)+s+\frac{1}{2})} \tag{2}$$

Theorem 5. *A class of preferential attachment graphs for any constant k achieves asymptotic truth learning under the Bayesian model.*

Proof. Our goal is to show that with high probability, the PA graph satisfies the conditions of Theorem 3. First, we know that as $n \to \infty$, the degree distribution of G converges almost surely to the power law distribution with exponent 3 [10]. We may write this distribution as $d(x) = Cx^{-3}$ where $d(x)$ is the proportion of vertices with degree x and $C > 0$ is some constant dependent on k. We have with high probability, the number of vertices of degree exactly k converges to $d(k) = Ck^{-3}n = \Theta(n)$. Let A be the set of vertices in G_t with $t = \log(n)$ and let $0 < \epsilon < Ck^{-3}$. Theorem 4 says for $a \geq \epsilon n$ and for all $v \in A$, $\deg(v)$ in G_a converges almost surely to a random variable, ξ with expectation:

$$\mathbb{E}(\xi) = \sum_{s=1}^{k} \frac{\Gamma(ka+\frac{1}{2})}{\Gamma(ka)} \frac{\Gamma(k(b-1)+s)}{\Gamma(k(b-1)+s+\frac{1}{2})} \tag{3}$$

for some $b \leq \log(n)$. A commonly known fact is $\frac{\Gamma(x+\frac{1}{2})}{\Gamma(x)} = \sqrt{x}(1+O(\frac{1}{x}))$ for $x \to \infty$. Then, taking $n \to \infty$ as well as using $k = O(1)$ gives us the complexity $\mathbb{E}(\xi) = \Omega(\sqrt{\frac{a}{\log(n)}})$. That means $\mathbb{P}(\xi = \Omega(\sqrt{\frac{a}{\log(n)}})) > 0$. Then, as $n \to \infty$, we have with high probability that we can find a vertex $v_\alpha \in A$ such that at $G_{\epsilon n}$, $\deg(v_\alpha) = \Omega(\sqrt{\frac{\epsilon n}{\log(n)}})$. We get that the probability for a new vertex inserted at G_a to be a neighbor of v_α is $\Omega(\sqrt{\frac{\epsilon n}{\log(n)}})/(kn) = \Omega(\frac{1}{\sqrt{n\log(n)}})$. Now by definition of ϵ, there are $\Theta(n)$ vertices of degree k in G that were inserted after $G_{\epsilon n}$. Call the vertices in this set that are also connected to v_α, S. We know $\mathbb{E}(|S|) = \Omega(\frac{1}{\sqrt{n\log(n)}}) \cdot \Theta(n) = \Omega(\sqrt{\frac{n}{\log(n)}})$. We also see that since every vertex in S has degree exactly k, their only neighbors are the vertices they connected to upon insertion, so S is independent. For the same reason, $G - S$ is connected, so we have that v_α and S satisfy the necessary conditions of Theorem 3.

4.3 Grid Graph

The last graph we analyze is the grid graph, i.e., the graph of the lattice points (i, j) with $1 \leq i, j \leq k$ with $n = k^2$ and edges connecting two lattice points (i, j) and (i', j') with $|i - i'| + |j - j'| = 1$. A grid graph has degree at most 4 everywhere. Thus there is no random asymptotic truth learning. Below we show that there is a good ordering with which truth learning happens. The main idea is to find an ordering that builds a high quality signal using $o(n)$ vertices, and then propagate this signal to the rest of the network. Since the grid graph has constant degree we cannot use the first phase in our sufficiency condition (Theorem 3), and have to build that specifically to the grid graph.

Theorem 6. *The class of square grid graphs achieves asymptotic truth learning under the Bayesian model.*

5 Simulations

We will now provide simulated results with different network topologies and orderings. Since the Bayesian model is computationally intensive, in all of our simulations, we utilize the Majority Rule model. Furthermore finding a maximum independent set is NP-hard so we use heuristic ideas to generate an ordering. Note that for each run of the simulation on a single network, a new ordering is created, but the underlying graph structure remains the same. We repeat the simulation on the same network and take the average network learning rate over multiple runs. Finally, in every simulation we utilize a private signal $q = 0.7$.

Ordering. We use random ordering as a baseline and compare with an ordering tailored to the network structure. We first introduce a simple *Two Neighbors* ordering: Choose a node v at random. With an arbitrary order, the nodes in $N(v)$ make their decisions based on the majority rule. After that, v makes a decision. Then nodes with at least two neighbors who have already made decisions will then follow. Repeat the above until all nodes are included.

Furthermore, we introduce an improved *Two Neighbor* ordering to incorporate "high-value" agents. Specifically, we pick the two highest-degree nodes u and v. Using a greedy algorithm, select a size $m < \deg(u)$ independent set of u's neighbors. Have each of the nodes in the subset make a decision and then let u make a decision. Label u as a high-value agent. Do the same for node v. Take all the nodes S' in the network that has at least two neighbors who have made decisions, where the majority of these neighbors are labeled high-value. Have all nodes in S' make a decision. Repeat until termination. If there are no vertices which have at least two high value neighbors, we take a random order of the remaining vertices. Note that the size of the m independent set of neighbors should be large enough to minimize the chance of a negative information cascade, while small enough to allow for further aggregation down the line.

Table 1. email-univ Network

Simulation results $q = 0.7$, $n = 1133$, iterations= 300			
Ordering	Average learning rate		
	email-univ	ER ($p = 0.008$)	PA ($k = 5$)
Random	0.8412	0.8790	0.8778
Two Neighbors	0.8717	0.9443	0.8543
Two Neighbors + High Value	0.9161	0.9633	0.9327

Real-World Network. We run simulations using a dataset of an email network consisting of 1133 nodes and 5451 edges [15]. The network has a density of 0.0085, an average degree of 9, and a maximum degree of 71. We also compare these simulations against an Erdös Rényi model and a PA model with the same number of vertices and density. From Table 1, both *two neighbors* and *two neighbors + high value* ordering outperform the random ordering in the email-univ network. Furthermore, from Fig. 2, the two neighbors + high value ordering performs the best for the email-univ network, with a median learning rate close to 1. While both orderings perform better on average, they are both susceptible to negative information cascades. The shape of the email-univ learning rate distribution resembles the PA model in its polarity of outlier data points. However, the outliers for the email-univ network appear to be more evenly distributed compared to the PA model.

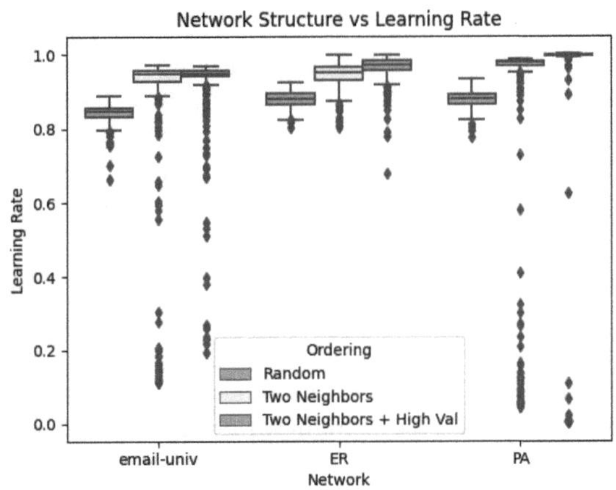

Fig. 2. Simulation results for the email-univ network with $q = 0.7$ over 300 iterations.

Additional simulation results on the Erdös Rényi graph, PA model and low degree graphs with different network parameters can be found in the Appendix of [12].

6 Conclusions

Among all the graphs considered in this paper, except a few cases of very sparse or very dense graphs, the quality of learning in a network can vary drastically with different decision orderings. Given an input graph we ask if there exists a decision ordering that supports asymptotic truth learning. We conjecture that this problem is NP-hard. Furthermore, enabling asymptotic truth learning in graphs of constant degree is generally more challenging. It is also more difficult to analyze the majority vote model although majority vote model is much easier to implement in practice and achieves good performance if the ordering is chosen right. It is future work to further characterize sparse graphs and majority vote model that have asymptotic truth learning.

References

1. Acemoglu, D., Dahleh, M.A., Lobel, I., Ozdaglar, A.: Bayesian learning in social networks. Rev. Econ. Stud. **78**(4), 1201–1236 (2011)
2. Arieli, I., Mueller-Frank, M.: Multidimensional social learning. Rev. Econ. Stud. **86**(3), 913–940 (2018)
3. Arieli, I., Sandomirskiy, F., Smorodinsky, R.: On social networks that support learning. In: Proceedings of the 22nd ACM Conference on Economics and Computation, EC 2021, pp. 95–96 (2021)
4. Bahar, G., Arieli, I., Smorodinsky, R., Tennenholtz, M.: Multi-issue social learning. Math. Soc. Sci. **104**, 29–39 (2020)
5. Banerjee, A.V.: A simple model of herd behavior. Q. J. Econ. **107**(3), 797–817 (1992)
6. Bikhchandani, S., Hirshleifer, D., Welch, I.: A theory of fads, fashion, custom, and cultural change as informational cascades. J. Polit. Econ. **100**(5), 992–1026 (1992)
7. Chamley, C.: Rational Herds: Economic Models of Social Learning. Cambridge University Press (2004)
8. Erdős, P., Rényi, A.: On the evolution of random graphs. Publ. Math. Inst. Hung. Acad. Sci **5**(1), 17–60 (1960)
9. Golub, B., Sadler, E.: Learning in social networks. In: The Oxford Handbook of the Economics of Networks, pp. 504–542. Oxford University Press (2016)
10. Hofstad, R.V.D.: Random Graphs and Complex Networks. Cambridge Series in Statistical and Probabilistic Mathematics. Cambridge University Press (2016)
11. Laland, K.N.: Social learning strategies. Learn. Behav. **32**(1), 4–14 (2004)
12. Lu, K., Chong, J., Lu, M., Gao, J.: Enabling asymptotic truth learning in a social network. arXivpreprint: arXiv: 2410.04317 (2024)
13. Mobius, M., Rosenblat, T.: Social learning in economics. Annu. Rev. Econom. **6**(1), 827–847 (2014)
14. Monzón, I., Rapp, M.: Observational learning with position uncertainty. J. Econ. Theory **154**, 375–402 (2014)
15. Rossi, R.A., Ahmed, N.K.: The network data repository with interactive graph analytics and visualization. In: Proceedings of the Twenty-Ninth AAAI Conference on Artificial Intelligence, pp. 4292–4293 (2015)
16. Sgroi, D.: Optimizing information in the herd: Guinea pigs, profits, and welfare. Games Econ. Behav. **39**(1), 137–166 (2002)
17. Shoham, Y., Tennenholtz, M.: Emergent conventions in multi-agent systems: initial experimental results and observations (preliminary report). In: Proceedings of the Third International Conference on Principles of Knowledge Representation and Reasoning, KR 1992, pp. 225–231 (1992)
18. Smith, L.: Essays on dynamic models of equilibrium and learning. Ph.D. thesis, University of Chicago (1991)
19. Smith, L., Sørensen, P.: Pathological outcomes of observational learning. Econometrica **68**(2), 371–398 (2000)
20. Tsun, A.: Probability & Statistics with Applications to Computing (2020)
21. Welch, I.: Sequential sales, learning, and cascades. J. Finance **47**(2), 695–732 (1992)

Repeated Bidding with Dynamic Value

Benjamin Heymann, Alexandre Gilotte(✉), and Rémi Chan-Renous

Criteo AI Lab, Paris, France
a.gilotte@criteo.com

Abstract. We analyze a toy dynamic value model for auctions, where the buyer's valuation depends on the elapsed time since his last successful bid. That the current value depends on the outcome of the previous auctions is an overlooked aspect of repeated auctions in digital advertising. This dynamical nature stems from three facts: (a) ad slots are auctioned in real time, (b) an advertiser might repeatedly show banners to the same user, (c) the marginal value of winning an auction drops momentarily after a banner has been shown to the user. Facing dynamic values, the buyer must balance the immediate benefit of a current win against the potential decrease in future values. The toy model we introduce addresses a gap in the understanding of the implication of such dynamics. We demonstrate that traditional bidding strategies are suboptimal in this model and present an algorithmic method to derive the optimal bidding policy.

Keywords: repeated auctions · RTB · auto-bidding

1 Introduction

1.1 Repeated Auctions with Dynamic Value: A Planning Problem

This article analyzes a *repeated second price auction* where the buyer's value for the item depends on the time elapsed since their last successful bid. Said otherwise: an item value depends on the *age* of the last purchase of a similar item. Our main motivating application is real-time bidding for online advertising. Indeed, the rise of online marketplaces and digital advertising over the last decades has fuelled the study of *repeated auctions.* Our study relates to existing research in the realms of budget/ROI constraints, learning dynamics, and strategic interactions in auctions. But surprisingly, we seem to be entering unexplored territory when the buyer's utility for an item *depends on the previous auctions outcomes.* Our paper aims at addressing this gap in the literature by starting with the in-depth study of a minimal example.

We believe indeed that such a dependency on the previous auctions is an overlooked but frequent issue in online advertising. For example, it is widely believed that it can be beneficial to space the marketing interactions over time because of the user's *display fatigue.* Such belief implies the kind of dependency we propose to study. A key implication of this dependency on past auctions is

M. Mavronicolas et al. (Eds.): WINE 2024, LNCS 15534, pp. 548–563, 2026.
https://doi.org/10.1007/978-3-032-08560-3_31

that *winning the current auction changes the utility of future auctioned items.* Despite the prevalence of dynamic values in auctions, particularly in digital advertising, identifying the optimal bidding policy in these situations remains an unsolved challenge. This is even the case when the buyer disposes of an oracle perfectly predicting the utility of the item.

Figure 1 provides an example of value dynamics, visually representing how a bidder's valuation fluctuates over time. After an auction is won, the value of the next auctions drops and then increases back to its initial level as time goes by. Because winning an auction decreases the utility of winning the next auctions, it should be intuitive that the bidder should lower his bid to account for this future decrease. Bidding the true value (the *greedy* strategy) thus becomes suboptimal in this setting, even though the auction is second price. We illustrate this by showing in Fig. 2 the payoff of several linear scaling strategies. The graph confirms the intuition that the strategy of always bidding one's true value, represented by a scaling factor of 1, is suboptimal in our model.

This underscores the complexity introduced by the temporal dynamics of the bidder valuation.

Our findings notably contrast with the prevailing belief in the field that effective feature engineering alone can optimize bidding strategies. This intuition is flawed because it overlooks the underlying *planning problem.* This is well illustrated by this article's stylized model, since in what follows, the bidder has **oracle knowledge** of the item value, and we observe that bidding this value is suboptimal.[1] Considering the dynamic evolution of bidder valuation is crucial for truly optimizing bidding policies in a dynamic value context. Aside from the analysis of Algorithm 1, what motivated us to write this article was the need to dispel the idea —which seems widespread among practitioners– that feature engineering alone can solve the planning problem.

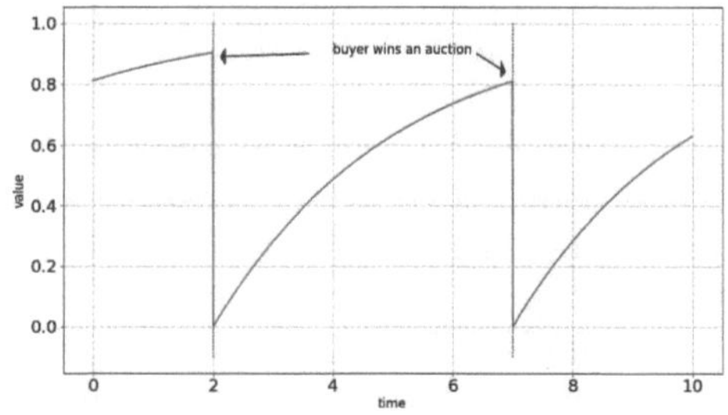

Fig. 1. Example of dynamical value. When the bidder wins an auction, his value drops brutally, and then progressively goes back to a maximum value. The shape displayed here is inspired by papers such as [7,14]. An optimal bidding strategy should account for this dynamical value phenomenon, which intensifies as the frequency of auctions over the timeline increases.

[1] We remind the reader that in a textbook second price setting, bidding the value is optimal.

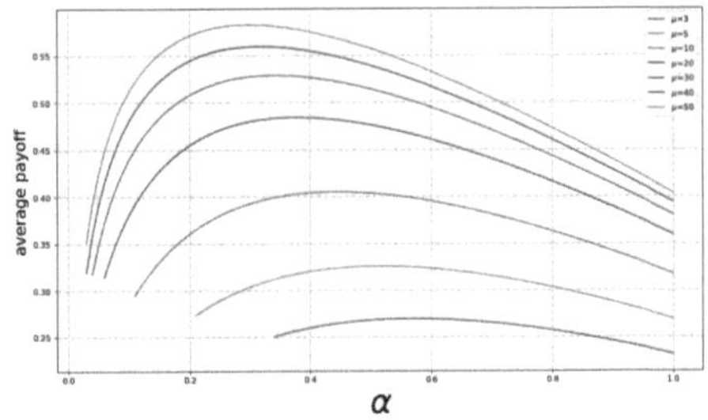

Fig. 2. Payoff $\lim_{\gamma \to 0} \gamma V_{b^\star,\gamma,\mu}(0)$ for different values of μ and α when the value as a function of the age is $k(t) = 1 - \frac{1}{1+t}$ and the cumulative distribution of the competition is $q(b) = \min(b, 1)$. The graph is obtained thanks to the closed form expression of Theorem 3. The greedy policy ($\alpha = 1$) is always beaten by $\alpha = 1/2$. For $\mu = 50$ we observe around 40% improvement between the greedy policy and a good choice of α. For greater value of μ, the opportunity cost is greater than 50%.

1.2 Display Advertising Auctions

In this section, we provide additional contextual details concerning our primary motivating use case, which revolves around display advertising auctions. The realm of display advertising offers a sophisticated mean to programmatically monetize content created by publishers. When a user reaches one of the publisher's pages, the right to show a display banner to the user is sold. The mechanism that elicits the deal is often an auction that takes place in real time, hence the name: Real-Time-Bidding (**RTB**). Real-Time-Bidding presents several practical challenges that have motivated a growing body of scientific work at the frontier of mathematics, economics, and computer sciences [11]. The buyer typically derives a valuation formula for the display opportunity of the shape [6,8,11] $Value\_Per\_Event \times \Pr(E|X)$, where $\Pr(E|X)$ is the probability that an event will take place in the future (such as click, sales... it depends) knowing what is available in the user's context (here denoted by X) and $Value\_Per\_Event$ is a multiplier that is independent of X. This multiplier often integrates business-related constraints such as budget [6] or ROI or cost per action. Such a valuation then becomes the input of a bidding module, tasked to find a bid that maximizes the immediate payoff (*i.e.* expected value minus cost). For a second-price auction, this bidding module simply returns the estimated value. For non-incentive compatible allocation mechanisms such as first-price auctions, methods to estimate the competition and adapt the bid accordingly have been proposed [29].

One specificity of display advertising is that bidders typically receive a sequence of auctions for the same user. Those auctions arise as the user browses the web and opens new pages, each new page triggering a few auctions that share a user's RTB identifier. However, it is usually not considered desirable to display too many similar ads to the same user in a short period of time, and a common practice consists in defining a *frequency capping* [11], which is a maximum number of ads which can be displayed to the same user for the same campaign on a predefined period of time (such as "maximum 10 displays per day per user").

With *frequency capping* activated, the bidder would thus temporarily stop bidding on a user once the maximum number of ads is reached, until the next period. The main rationale for applying such frequency capping is essentially that ads displayed to the same user usually have diminishing return. Those returns could even become negative, as it has been noted that too many similar ads in a short period of time result in a degraded user experience. Several studies also point to the usefulness of spreading the ads in time [9,26]. To deal with this decreasing return of the ads, another frequent method consists in directly modelling how the *past* displayed ads affect the value of the current auctioned ad opportunity. This can be done, for example, by keeping features measuring the *fatigue* such as the number of ads displayed to the user in a past period of time, or the time elapsed since the last ad displayed to this user, and using these fatigue features as predictors of the $\Pr(E|X)$ model. Several works point to a similar idea, such as [2,4,24].

However, while these features improve the immediate value estimation, they do not account for the impact of an ad on the future opportunities: they still miss the *planning problem*. In this work, we suppose that the value is perfectly estimated, to isolate this very planning problem. Indeed, one logical consequence of the diminishing return of ads is that winning an auction *now* impacts the value of the *next* opportunities for the same user. But while it is possible that the $\Pr(E|X)$ term of the valuation recognizes the fatigue effect, there is a hole in the literature on how the bidder should address this *dynamic value*.

The planning problem caused by a dynamic value is first recognized in [8] with a causal statistics perspective, but no solution is proposed as their counterfactual method falls short on estimating the impact of how the bid impacts future auctions costs. The authors in [20] also identify the planning issue we study in this paper. Their empirical solution is based on counterfactual estimation on a restricted set of options (it only tunes a small set of metaparameters and can only handle long-term fatigue). Despite these limitations, they report non-trivial performance improvements in a real production environment. Their results confirm the relevancy of studying this planning problem.

We recognize that today most RTB auctions are first-price. We decided to study the case of a second price auction instead; both for simplicity, and to emphasize that even in this simpler second price setting the problem is non-trivial. Although the article uses RTB auctions as a primary example, it is evident that the dynamic value phenomenon is present in various other online marketplaces.

1.3 Repeated Auctions

Various distinct challenges revolve around the concept of **repeated auctions**. First, **the existence of a budget** or an ROI constraint over a sequence of auctions couples all the individual bidding problems together. In the literature, such constraints are typically analyzed either by using a constraint relaxation or by solving an associated knapsack problem. The typical solution is then a scaling of the value with a well-chosen factor, that relates to the Lagrangian multiplier of the coupling constraint. See [6,10,12,17,19]. In this research stream, it is mostly the coupling of the bidding problems that is under scrutiny, and *repetition* is mostly about coupled optimization problems. Second, the level of bid can be seen as a decision that is repeated over time. The outcome (value and payment) depends on this decision. This setting can be studied with the **bandit framework**. See for instance [1,28]. In this research stream, this is mostly the explore-exploit tradeoff of the learning bidder that is under scrutiny, and *repetition* is mostly about learning. Third, there is a line of research focusing on the interactions between the seller and the multiple buyers. In this stream of work, *repetition* is mostly about playing a **repeated game** and mechanism design [3,5,13,16,18,21,25]. In particular, the question of how to design an auction that maximizes the sellers' revenue has attracted a lot of attention. The "dual" research question being how the buyer should adapt to the seller's strategy, researchers have investigated strategies that are robust to the other agents' choices. We introduce another perspective on *repeated auctions*, centered around the idea of **dynamical value**.

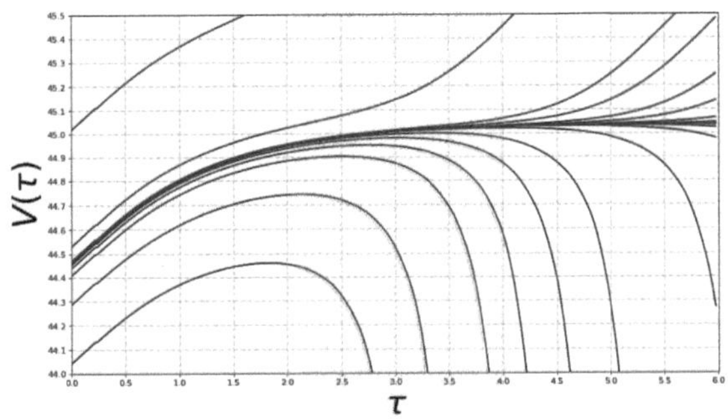

Fig. 3. An example of Algorithm 1 run. In red is the output of the algorithm, which is an estimate of the optimal value $V^\star(\tau)$. In blue are the iterates of the dichotomy on $V^\star(0)$. We can see that the iterates (a) do not cross and (b) explode as $\tau \to +\infty$. The output of the Algorithm can then be used to produce an optimal bid using Lemma 2. (Color figure online)

1.4 Contribution

The applied work [8] recognizes the difficulty of handling the coupling between present and future bidding decisions. We introduce a realistic minimal model to study this coupling, and frame the design of a bidding policy as a continuous time optimal control problem over an infinite horizon. We then present structuring properties of this control problem (Sect. 3), and introduce Algorithm 1 that iterates toward the optimal policy. A run of this Algorithm is displayed in Fig. 3. Such an algorithm allows us to estimate through Monte-Carlo simulations the economic loss of using the greedy policy (see for instance Table 1), that is, the economic cost of being oblivious to the planning problem and not accounting for future opportunities in the bidding policy design. To pinpoint what we believe is a distinctive technical characteristic of our approach: the convergence proof for Algorithm 1 relies on a dynamical system argument, more specifically, a perturbed version of the Cauchy-Lipschitz Theorem. We then take a look at shading policies because they constitute a natural class of simple approximations. We show that shading policies cannot be solutions to the optimal control problem, we also observe that, in the numerical experiments, they perform very well (see Fig. 4). While we do not have a guarantee for those performances, we show that asymptotically, the shading policies become optimal (Theorem 4).

We also derive a closed form expression of the buyer payoff in a specific case (Theorem 3). Such closed form expressions can be, we believe, useful to build intuition by leveraging a fully solved case. Overall, the study suggests that shading policies are a reasonable class of simple policies, if one wants to mitigate the economic loss of using a greedy policy in the presence of dynamic values of the type studied in the paper. Last, we mention that in the edge case where the buyer is interested in buying at most one item, we get a setting close in spirit to [23], who introduce a procurement version of the prophet problem, where the goal is to minimize the procurement cost. In what follows, we use the masculine pronoun *he* to refer to the buyer, and the feminine pronoun *she* for the seller. Proofs are provided in the supplementary material available on arxiv.

2 Model and Bidding Problem

This section introduces the stylized dynamic value environment under scrutiny. For the sake of simplicity, we suppose that all the items are similar, and we postulate that the value of an item depends solely on the time since the last won auction by the bidder of interest. We refer to this time as the system *age* thereafter. The section ends with Lemma 1, that characterizes the value function with a linear differential equation.

2.1 Model Primitives

An auctioneer receives items that she needs to sell immediately to a set of buyers, and we take the perspective of one of those bidders facing a stationary competition. The items arrivals follow a Poisson process of *intensity* μ. We denote by τ the *age*, which is the time that has elapsed since the last won auction.

In our model, the perceived value of an auctioned item to the bidder is uniquely determined by the variable τ, which represents the 'age' or time elapsed since the last auction won by the bidder. We denote this time-dependent value as $k(\tau)$, capturing the dynamic nature of the bidder's valuation over time. In our model, we assume k, representing the bidder's valuation, to be a non-decreasing and bounded function. This assumption reflects realistic scenarios where a bidder's interest in an item doesn't decrease over time, but has a natural upper limit. Figure 1 shows an example of how the value of buying a new item may evolve over a timeline. Every time an auction is won, the worth of an item drops, and then progressively returns to its nominal value.

We agree that postulating that the buyer's utility depends solely on the elapsed time does not capture the full picture of value dynamics. However, a tour of the literature [8,14] on display advertising shows that it is interesting enough to deserve an analysis. Furthermore, we deliberately chose to focus on a minimalistic model to clearly delineate what value dynamics can imply in terms of bidding behaviors.

For simplicity, our model assumes a 'second price' auction format, which used to be the dominant standard in digital advertising. For the sake of tractability and to focus on the dynamic valuation aspect, we assume that the competition is stationary. There are several reasonable ways of justifying this assumption: one is to consider that the price to beat is simply a reserve price, another one is to suppose that there are many buyers and that only a portion of them is called at each auction, yet another one is to simply suppose that the competition does not account for this time dependency, for example because it requires a too sophisticated data and machine learning pipeline. The prices-to-beat, which are the highest bids of the competition, are i.i.d. sampled from a known distribution of c.d.f. q. Said otherwise, $q(b)$ is the probability that the buyer wins with a bid equal to b. Finally, we assume that the sequence ends at a random time T_γ, where T_γ follows an exponential distribution of parameter γ, or equivalently we say that the buyer has a discount rate γ. In many statements, this γ is sent to 0, hence one can see this random end of time as a pure technical instrument. The bidder's goal is to maximize, in expectation, the sum of the values of the auctions he wins, minus the cost he pays, before the end of the sequence.

2.2 Linear Differential Equation

We note that this environment may be viewed as a continuous Markov process, where the state is fully summarized by the age τ. Because the state depends solely on the age τ, the bidder's policy can be fully described by a measurable function $\tau \longrightarrow b(\tau) \in \mathbb{R}^+$ specifying the bid when the age is τ. We denote by $\mathcal{B}$ the set of these bidding functions. We denote by $p(b)$ the average payment of the user when bidding b. Because the auction is second price, we have the relation $p(b) = q(b)b - \int_0^b q(t)dt$. Furthermore, we denote by $U(v,b) \stackrel{def}{=} q(b) \cdot v - p(b)$, the expected utility when bidding b and valuing v for winning, that is, U is the payoff of the corresponding classic, static second price auction. For a bidding function $b \in \mathcal{B}$, we denote by $V_{b,\gamma,\mu}(\tau)$ the expectation of the bidder's future revenue when the state is τ, or $V_b(\tau)$ for simplicity, when the context is clear. Let $T_1, T_2 \dots T_n \dots$ be the (random) times of the next auctions, $C_1 ... C_n ...$ be the (random) maximal bids in the competition at these times, and τ be the current age, then $V_b(\tau) = \mathbb{E}\sum_{i=1}^{\infty} e^{-\gamma T_i} \left(k(\tau(T_i)) - C_i\right) \mathbf{1}\{b(\tau(T_i)) > C_i\}$. We are ready to state Lemma 1.

Lemma 1. *For any bid policy $b \in \mathcal{B}$, V_b satisfies the differential equation*

$$V_b'(\tau) = -\mu \cdot U\big(k(\tau) + V_b(0) - V_b(\tau), b(\tau)\big) + \gamma V_b(\tau).$$

The proof of Lemma 1 relies on a dynamic programming argument applied to V_b. We observe that while the characterization in this lemma is a linear differential equation (U is affine in V), the initial condition is not known and appears in the dynamics.

3 Optimal Bid

This section derives some properties of the optimal bidding policy, as well as an algorithm to compute it. We define the optimal value function (or Bellman value), $V^\star(\tau)$, as the supremum of $V_b(\tau)$ over all possible bidding policies b, namely

$$V^\star(\tau) \stackrel{def}{=} \sup_{b \in \mathcal{B}} V_b(\tau) \ .$$

It is practical to introduce the notation $\pi(v)$ for the expected profit of a bidder of value and bid v in a one shot auction, namely

$$\pi(v) \stackrel{def}{=} U(v,v) = q(v)v - p(v).$$

This corresponds to the expected profit of the truth telling bid in the one shot auction. Lemma 2 provides a dynamic programming characterization of the Bellman value, and provides a relation between the Bellman value and the bid. The Bellman value is then proved to be the solution of a parameterized differential equation of unknown parameter (Lemma 3). Theorem 1 ensures that when k is concave, the optimal policy is increasing. Theorem 1 can be thought of as a sufficient condition for well-posedness: primitives for which the optimal bid would not be monotone could arguably be considered pathological from the econometric standpoint. For Sect. 3.3, that presents an algorithm for computing the optimal policy, we suppose that q is continuous, which implies that π is C^1.[2] Section 3 ends with some numerical applications, where we compare the optimal and the greedy policies.

3.1 Non-linear Differential Equation for $V^\star$

From Lemma 1, we get the following characterizations of $V^\star$ and of the optimal bid $b^\star$.

Lemma 2. *We have the relation*

$$V^\star(t) = \int_0^{+\infty} \mu e^{-(\mu+\gamma)t} \Big(\pi(k(t) + V^\star(0) - V^\star(t)) + V^\star(t)\Big) \mathrm{d}t \ ,$$

and the optimal bid $b^\star$ *satisfies the relation* $b^\star(\tau) = \max\left(0; k(\tau) + V^\star(0) - V^\star(\tau)\right)$.

[2] Indeed, by definition, $\pi(v) = q(v)v - p(v) = \int_0^v q(t)\mathrm{d}\ \mathrm{t}$.

This non-linear integral equation is too complicated to solve directly because its initial value $V^\star(0)$ also appears as a parameter of the dynamic. To clarify this point, we define $\Phi : \mathbb{R}\times\mathbb{R}_+\times\mathbb{R}_+ \to \mathbb{R}$ such that $\Phi(t,v,\lambda) = \gamma v - \mu\pi(k(t)+\lambda-v)$. We then reformulate Lemma 2 as a more usual Cauchy problem. It should be noted that in Lemma 3, the parameter y_0 is not given.

Lemma 3. *The value function $V^\star$ is the solution of the ordinary differential equation*

$$\begin{cases} \dot{Y}(t) = \Phi(t, Y(t), y_0) \\ Y(0) = y_0 \end{cases} \qquad (\mathcal{F}_{y_0})$$

for some $y_0 \in \mathbb{R}_+$.

3.2 Differential Equation for $b^\star$

In this section, we provide a reparameterization of the differential equation so that $b^\star$ becomes the solution. This parameterization requires additional assumptions on k.

Lemma 4. *If k is C^1-smooth and $b^\star(t) > 0$ for $t \in]t_1, t_2[$, then for $t \in]t_1, t_2[$, the optimal bid $b^\star$ satisfies the differential equation $\dot{b}(t) = \dot{k}(t) - \gamma(V^\star(0) + k(t)) + \mu\pi(b(t)) + \gamma b(t)$.*

Since the bidder's value $k(\tau)$ is an increasing function of the last purchase age τ, it is natural to expect that the optimal bid would also increase with τ. The following result confirms this intuition under the additional assumption that k is concave.

Theorem 1. *If the bidder's value k is concave, then $b^\star$ is increasing with τ, and strictly increasing on any interval where k strictly increases.*

We insist that this is (surprisingly) not true without the concavity assumption. In particular, Fig. 5 proposes a counter example where k is increasing but not concave.

3.3 Algorithm

Lemma 3 does not allow us to derive $V^\star$ from a Cauchy problem because the initial condition is also a parameter of the dynamics. Still, by the Cauchy-Lipschitz Theorem, the solution of the ordinary differential equation

$$\begin{cases} \dot{Y}(t) = \Phi(t, Y(t), \lambda) \\ Y(0) = y_0 \end{cases} \qquad (\mathcal{F}_{y_0,\lambda})$$

admits a unique maximal solution[3] $Z^{y_0,\lambda} : t \to Z^{y_0,\lambda}(t)$ for any $y_0 > 0$ and $\lambda > 0$. We also observe that the problem $\mathcal{F}_{y_0,y_0}$ is the same as $\mathcal{F}_{y_0}$ so that it

[3] It is a solution to the differential equation that cannot be extended any further.

makes sense to set $Z^v(t) = Z^{v,v}(t)$ for all $t \geq 0$. In the following, we assume that q is continuous, and thus π is C^1. This allows us to apply the differentiable dependency Theorem in Lemma 5, which is the key to our approach.

Lemma 5. *The differential of $v \to Z^v(t)$ is equal to*

$$e^{\gamma t+\mu \int_0^t q(u_s(v))\mathrm{d}s}\left(1 - \mu \int_0^t e^{-\gamma s-\mu \int_0^s q(u_l(v))\mathrm{d}l} q(u_s(v))\mathrm{d}s\right) ,$$

where $u_t(v) = k(t) + v - Z^v(t)$.

Lemma 5 is a sensitivity analysis statement that relates the initial condition of the differential equation with its solution. Its proof relies on a perturbed version of the Cauchy-Lipschitz Theorem from the dynamical system literature. We refer the reader to the supplementary material on arXiv for the full proof. Lemma 6 is a key technical step of the section, and is leveraged in the proof of Theorem 2.

Lemma 6. *If the probability $q(\cdot)$ of winning the auction given the bid is a continuous function of the bid, then the optimal value $v = V^*(0)$ is the unique value for which there exists finite a limit $\lim_{t\to\infty} Z^v(t)$*

Lemma 6 is well illustrated by Fig. 3, which shows trajectories $\tau \to Z^v(\tau)$ for different values of v. We see that all the numerical trajectories diverge for sufficiently large τ. This observation is the idea behind the next result. Indeed, we then use Lemma 6 to prove that a simple dichotomy on the parameter allows us to solve the bidding problem for repeated auctions with dynamic value. This dichotomy is formalized in Algorithm 1. Theorem 2 justifies this approach.

Theorem 2 (Algorithm). *Suppose q is continuous. Let $y^{(n)}$ be the iterations of Algorithm 1, then $||V^\star(0) - y^{(n)}(0)|| = O(1/2^n)$.*

Algorithm 1 performs a dichotomy on the value taken by the Bellman function at 0. It iterates on a guess based on the asymptotic behavior of the corresponding ODE solution. If the solution diverges to $-\infty$, then the guess needs to be increased; otherwise, the solution goes to $+\infty$ and the guess needs to be decreased.

3.4 Numerical Estimation of the Economic Cost of Being Greedy

Algorithm 1 is implemented in Python 3. We use SOLVE_IVP, an ODE solver from SCIPY [27] to numerically solve the ODE. We rely on the default Runge-Kutta method [15,22]. To compare the bidding policies, we compute the time average payoff over a long enough period. Through simulations comparing the optimal policy derived from our model to the traditional greedy policy, we observed significant variations in outcomes. These findings validate our model's efficacy and highlight the limitations of conventional strategies in dynamic valuation scenarios. We tested several values for μ, with $k(\tau) = 1 - e^{-\tau}$ and $k(\tau) = 1 - 1/(1 + \tau)$, and reported the results in Table 1. As expected, the optimality gap widens as μ increases, evolving from a barely noticeable effect to a significant double-digit influence.

Algorithm 1 Computing $V^\star(0)$

Require: (a, b, N), $a < b$, $\lim_{t\to\infty} Z^{a,a}(t) \to -\infty$, $\lim_{t\to\infty} Z^{b,b}(t) \to +\infty$
for $n \le N$ **do**
  $c \leftarrow (a+b)/2$
  Compute $Z^{c,c}$, solution of $\mathcal{F}_c$
  $y^{(n)} \leftarrow Z^{c,c}$
  if $t \to y^{(n)}(t)$ goes to $-\infty$ **then**
    $a \leftarrow c$
  else
    $b \leftarrow c$
  end if
end for

Table 1. Estimated value per time unit with the greedy and with the optimal bidder. $\gamma = 0.01$

μ	0.1	1	5	10	100
optimal	0.045	0.22	0.43	0.52	0.72
greedy	0.043	0.20	0.34	0.38	0.46

(a) with $k(t) = 1 - e^{-t}$

μ	0.1	1	5	10	50
optimal	0.035	0.16	0.32	0.40	0.69
greedy	0.035	0.14	0.26	0.31	0.44

(b) with $k(t) = 1 - 1/(1+t)$

4 Shading Strategies

A shading strategy consists in bidding a constant factor α (smaller than 1) of the value, *i.e.* $b(\tau) = \alpha k(\tau)$. This class of strategy is of great practical and theoretical interest because it is simple to implement and analyze. Moreover, as mentioned in the introduction, shading is used as a way to implement budget or ROI constraints. We use in this section the shorthand notation $V_\alpha = V_{\alpha k,\gamma,\mu}$. In this section, we are mostly interested in the case where $k(0) = 0$. We found it notable that in the settings we tested, the class of shading strategies performs very well. It can be proven, however, that it is not optimal in general. Indeed, by Lemma 4, the optimal bidding policy satisfies $\dot{b}(t) = \dot{k}(t) - \gamma(V^\star(0) + k(t)) + \mu\pi(b(t)) + \gamma b(t)$, which implies that a shading policy $b(t) = \alpha k(t)$ that would outperform all other strategies should satisfy $\alpha\dot{k}(t) = \dot{k}(t) - \gamma(V^\star(0) + k(t)) + \mu\pi(\alpha k(t)) + \gamma\alpha k(t)$, which in general does not hold.

4.1 Special Case

We mention that there is an instance of the problem where the time average payoff of a shading strategy can be computed explicitly.

Theorem 3. *If the competition is a uniform distribution on* $[0, 1]$*, if* $k(t) = 1 - 1/(1+t)$ *then for all* α *satisfying* $\mu\alpha > 1$

$$\lim_{\gamma\to 0} \gamma V_\alpha(0) = (1 - 1/2\alpha)(\mu\alpha)^2 \frac{\Gamma(\mu\alpha - 1, \mu\alpha)}{\Gamma(\mu\alpha + 1, \mu\alpha)},$$

where $\Gamma(\_,\_)$ is the upper incomplete gamma function.[4]

We use Theorem 3 to build a visualization of the average expected payoff, which we display in Fig. 2. We check that shading strategies outperform truthfulness in the presence of a value dynamic. We also check that the greater the arrival rate μ, the stronger the shading should be (the α multiplier close to zero).

4.2 Numerical Comparison

Figure 4 displays the proportion $V_\alpha/V^\star$ of the optimal value retrieved by bidding αk, for several values of the shading factor α and the opportunity arrival intensity μ, and two distinct concave functions k (and with γ was close to 0). We note that on all these settings, the best value of α gives results very close to the optimal value (it was higher than 99%). Another observation is that for smaller values of μ, the greedy bidder (*i.e.* $\alpha = 1$) performs nearly optimally. This observation is easily explained: a small μ means that there are few auctions per time unit; it is thus likely that the first auction only arrives a long time after 0, when the function $k(t)$ is already close to its supremum. This is not so different from the case when k is constant, for which the greedy bidder is optimal. On the other hand, for larger values of μ, the greedy bidder performs poorly, and the best shading factor decreases with the density of auctions μ.

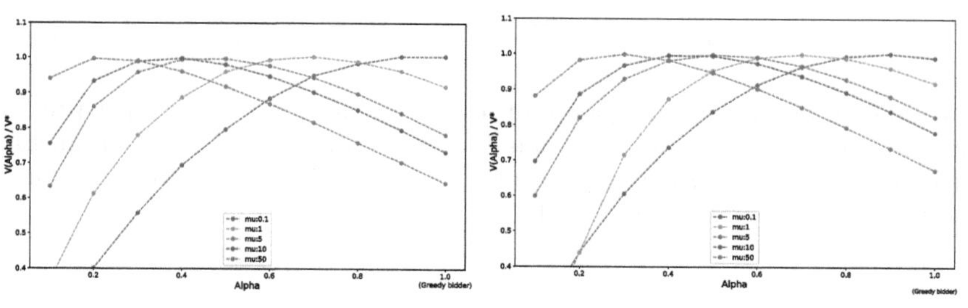

Fig. 4. Ratio $V_\alpha/V^\star$ as a function of α with $k(t) = 1-e^{-t}$ (left) and with $k(t) = 1-1/1+t$ (right). In these examples, there is always a shading close to optimality, although shading cannot be optimal.

4.3 Example with $k(\tau)$ Non Concave

We have seen in Lemma 2 that $b^\star$ is an increasing function when k is concave. We propose here an example where k is not concave to observe what can happen in this case. The function k we used is depicted in Fig. 5. It is a simple piece-wise linear function, making two steps. The optimal bids as a function of τ are plotted for different values of μ: we can observe that they are not monotonous.

[4] Defined as $\Gamma(s,x) = \int_x^\infty t^{s-1}e^{-t}\mathrm{d}t$.

The intuition for this counter-intuitive behavior is simple: since the function k increases quickly near 0, it is worth winning early because this generates a high value per time unit. Thereafter, the function stops increasing until the second step, where it increases sharply again. Just before the increase, it is better not to bid at all: the bidder should avoid resetting the state to 0 just before this strong increase. Figure 5 shows the value obtained when submitting a shaded bid $k(t) \times \alpha$, with the same two steps k function. We note that the best V_α here is quite sensibly smaller than $V^\star$. We also note that for large values of μ, V_α may have several local maxima. This example contradicts the intuition: $b^\star$ is *not* an increasing function of k. Indeed, if we removed the second step from the k function above, then we would get a concave function, and the optimal bid would be positive for all $t > 0$.

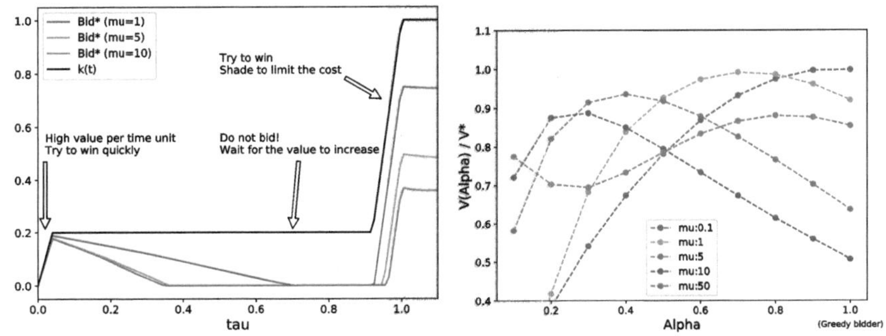

Fig. 5. Illustration of the consequence of having a non-concave k, the **purple** line is counterintuitive. (left) The bidder's value $k(\tau)$ and optimal bid $b^*(\tau)$ as a function of τ, for different values of μ (right). Optimal bids and V_α with the two steps function $k(t) = \min(5t, 0.2) + \max(0, \min(0.8, 10t - 9.2))$. (Color figure online)

4.4 Asymptotic Perspectives

Theorem 4 provides a complementary perspective on why shading policies might work well. Basically, as μ goes to $+\infty$, the best shading policy performs closer and closer to the optimal policy.

Theorem 4. *Suppose k concave and such that $k \in C^1$ and $k(0) = 0$. Suppose $q \in C^1$, and $\dot{q}(0) > 0$. Suppose $\mu_n \to +\infty$. There exist α_n such that*

$$\lim_n \lim_{\gamma \to 0} \gamma V_{\alpha_n k, \gamma, \mu_n}(0) = \lim_n \sup_{b \in \mathcal{B}} \lim_{\gamma \to 0} \gamma V_{b, \gamma, \mu_n}(0).$$

5 Conclusion

We are aware that the simplified model we studied here is quite far from the complexity of a real RTB environment. To get a more realistic system, we would need to account for inventories of different values, with different distributions of the competition, non-homogeneous parameters μ and γ... and the additional complexity that these parameters and the $k(t)$ function are not known by the bidder but must be estimated from the data. Additionally, expected value k could depend not only on the time of the last win, but also from other 'fatigue' features, such as the number of wins in a time window. All these complications are out of the scope of this paper,

We can, however, take away several lessons from this study: - First and foremost, having 'fatigue' variables in the model is not a sufficient condition to have a bidder correctly handling the fatigue - Second, we note that even in our simplified environment, the exact resolution is not trivial to find. We do not expect to be able to find an optimal solution in a real case. - Third, we note that α solutions are almost optimal in several limit cases, and seem to perform well in practice, at least for concave k. We believe a practical solution to the real problem would likely use this shape. - Finally, it may be worth keeping in mind that non-concave k function may lead to strongly non-intuitive behavior of the optimal policy. We note that non-concave here may happen in real cases from a design of the system: For example, if one decides to discount by half the value of display happening 'less than one hour after winning a previous display', the obtained k function is non-concave and looks a bit like the example in Fig. 4 (it would be even more pathological because it would be a discontinuous k). While our results do not extend trivially to more complicated real systems, it might still be better to avoid such non-concave k when it is possible.

To conclude, finding a robust method taking into account the impact on next auctions in a real environment is certainly the opportunity for future works.

References

1. Achddou, J., Cappé, O., Garivier, A.: Efficient algorithms for stochastic repeated second-price auctions. In: Algorithmic Learning Theory, pp. 99–150. PMLR (2021)
2. Agarwal, D., Chen, B.C., Elango, P.: Spatio-temporal models for estimating click-through rate. In: Proceedings of the 18th International Conference on World Wide Web, pp. 21–30 (2009)
3. Agrawal, S., Balkanski, E., Mirrokni, V., Sivan, B.: Robust repeated first price auctions. In: Proceedings of the 22nd ACM Conference on Economics and Computation, EC 2021, p. 4. Association for Computing Machinery, New York (2021). https://doi.org/10.1145/3465456.3467590
4. Aharon, M., et al.: Soft frequency capping for improved ad click prediction in yahoo gemini native. In: Proceedings of the 28th ACM International Conference on Information and Knowledge Management, pp. 2793–2801 (2019)
5. Amin, K., Rostamizadeh, A., Syed, U.: Repeated contextual auctions with strategic buyers. In: Advances in Neural Information Processing Systems, vol. 27 (2014)
6. Balseiro, S.R., Gur, Y.: Learning in repeated auctions with budgets: regret minimization and equilibrium. In: Proceedings of the 2017 ACM Conference on Economics and Computation, EC 2017, p. 609. Association for Computing Machinery, New York (2017). https://doi.org/10.1145/3033274.3084088

7. Bompaire, M., Désir, A., Heymann, B.: Fixed point label attribution for real-time bidding. CoRR abs/2012.01767 (2020). https://arxiv.org/abs/2012.01767
8. Bompaire, M., Gilotte, A., Heymann, B.: Causal models for real time bidding with repeated user interactions. In: Proceedings of the 27th ACM SIGKDD Conference on Knowledge Discovery & Data Mining, pp. 75–85 (2021)
9. Braun, M., Moe, W.W.: Online display advertising: modeling the effects of multiple creatives and individual impression histories. Mark. Sci. **32**(5), 753–767 (2013)
10. Castiglioni, M., Celli, A., Marchesi, A., Romano, G., Gatti, N.: A unifying framework for online optimization with long-term constraints. In: Oh, A.H., Agarwal, A., Belgrave, D., Cho, K. (eds.) Advances in Neural Information Processing Systems (2022). https://openreview.net/forum?id=DhHqObn2UW
11. Choi, H., Mela, C.F., Balseiro, S.R., Leary, A.: Online display advertising markets: a literature review and future directions. Inf. Syst. Res. **31**(2), 556–575 (2020)
12. Conitzer, V., Kroer, C., Sodomka, E., Stier-Moses, N.E.: Multiplicative pacing equilibria in auction markets. Oper. Res. **70**(2), 963–989 (2022). https://doi.org/10.1287/opre.2021.2167
13. Deng, Y., Mirrokni, V., Zhang, H.: Posted pricing and dynamic prior-independent mechanisms with value maximizers. In: Oh, A.H., Agarwal, A., Belgrave, D., Cho, K. (eds.) Advances in Neural Information Processing Systems (2022)
14. Diemert, E., Meynet, J., Galland, P., Lefortier, D.: Attribution modeling increases efficiency of bidding in display advertising. In: Proceedings of the ADKDD 2017. Association for Computing Machinery, New York (2017). https://doi.org/10.1145/3124749.3124752
15. Dormand, J.R., Prince, P.J.: A family of embedded runge-kutta formulae. J. Comput. Appl. Math. **6**(1), 19–26 (1980)
16. Drutsa, A.: Horizon-independent optimal pricing in repeated auctions with truthful and strategic buyers. In: Proceedings of the 26th International Conference on World Wide Web, WWW 2017, pp. 33–42. International World Wide Web Conferences Steering Committee, Republic and Canton of Geneva, CHE (2017). https://doi.org/10.1145/3038912.3052700
17. Gao, Y., Yang, K., Chen, Y., Liu, M., Karoui, N.E.: Bidding agent design in the linkedin ad marketplace. arXiv preprint arXiv:2202.12472 (2022)
18. Golrezaei, N., Javanmard, A., Mirrokni, V.: Dynamic incentive-aware learning: robust pricing in contextual auctions. In: Advances in Neural Information Processing Systems, vol. 32 (2019)
19. Heymann, B.: Cost per action constrained auctions. In: Proceedings of the 14th Workshop on the Economics of Networks, Systems and Computation, pp. 1–8 (2019)
20. Heymann, B., Chan-Renous, R., Gilotte, A.: A pragmatic policy learning approach to account for users' fatigue in repeated auctions (2024)
21. Kanoria, Y., Nazerzadeh, H.: Dynamic reserve prices for repeated auctions: learning from bids. In: Liu, T.Y., Qi, Q., Ye, Y. (eds.) Web and Internet Economics, p. 232. Springer, Cham (2014)
22. Lawrence, F.S.: Some practical runge-kutta formulas. Math. Comput. **46**, 135–150 (1986)
23. Livanos, V., Mehta, R.: Prophet inequalities for cost minimization. arXiv preprint arXiv:2209.07988 (2022)
24. Ma, H., Liu, X., Shen, Z.: User fatigue in online news recommendation. In: Proceedings of the 25th International Conference on World Wide Web, pp. 1363–1372 (2016)

25. Nedelec, T., Calauzènes, C., El Karoui, N., Perchet, V., et al.: Learning in repeated auctions. Found. Trends® Mach. Learn. **15**(3), 176–334 (2022)
26. Sahni, N.S.: Effect of temporal spacing between advertising exposures: evidence from online field experiments. Quant. Mark. Econ. **13**(3), 203–247 (2015)
27. Virtanen, P., et al.: SciPy 1.0: fundamental algorithms for scientific computing in python. Nat. Methods **17**, 261–272 (2020). https://doi.org/10.1038/s41592-019-0686-2
28. Weed, J., Perchet, V., Rigollet, P.: Online learning in repeated auctions. In: Conference on Learning Theory, pp. 1562–1583. PMLR (2016)
29. Zhou, T., et al.: An efficient deep distribution network for bid shading in first-price auctions. In: Proceedings of the 27th ACM SIGKDD Conference on Knowledge Discovery and Data Mining. ACM, New York (2021). https://doi.org/10.1145/3447548.3467167

A Fair Allocation is Approximately Optimal for Indivisible Chores, or Is It?

Bo Li, Ankang Sun(✉), and Shiji Xing

Department of Computing, Hong Kong Polytechnic University, Hong Kong, China
comp-bo.li@polyu.edu.hk, ankang.sun@polyu.edu.hk,
shi-ji.xing@connect.polyu.hk

Abstract. In this paper, we study the allocation of indivisible chores and consider the problem of finding a fair allocation that is approximately efficient. We shift our attention from the multiplicative approximation to the additive one. Our results are twofold, with (1) bounding how the optimal social cost escalates resulting from fairness requirements and (2) presenting the hardness of approximation for the problems of finding fair allocations with the minimum social cost. To quantify the escalation, we introduce *cost of fairness* (CoF) – an alternative to the price of fairness (PoF) – to bound the *difference* (v.s. *ratio* for PoF) between the optimal social cost with and without fairness constraints in the worst-case instance. We find that CoF is more informative than PoF for chores in the sense that the PoF is infinity regarding all EQX (equitable up to any item), EQ1 (equitable up to one item) and EF1 (envy-free up to one item), while the CoF is n regarding EQX and 1 regarding EQ1 and EF1, where n is the number of agents. For inapproximability, we present a detailed picture of hardness of approximation. We prove that finding the optimal EQX allocation within an additive approximation factor of n is NP-hard for any $n \geq 2$ where n is the number of agents and the cost functions are normalized to 1. For EQ1 and EF1, the problem is NP-hard when the additive factor is a constant and $n \geq 3$. When $n = 2$, we design additive approximation schemes for EQ1 and EF1.

Keywords: Fair Division · Cost of Fairness · Envy-freeness · Equitability

1 Introduction

The allocation of resources among multiple agents has long been a fundamental problem in various economic contexts [1,13,35,41]. Two orthogonal objectives, social welfare and individual fairness, used to be studied independently. A recent research trend focuses on understanding the trade-off between them [12,18,19].

The authors are ordered alphabetically. This work is funded by the Hong Kong SAR Research Grants Council (No. PolyU 15224823) and the Guangdong Basic and Applied Basic Research Foundation (No. 2023A1515010592).

M. Mavronicolas et al. (Eds.): WINE 2024, LNCS 15534, pp. 564–590, 2026.
https://doi.org/10.1007/978-3-032-08560-3_32

There are two types of problem. The first one is more quantitative, where we want to understand how much efficiency would be lost due to fairness constraints. The second one is more computational, where we want to design polynomial time algorithms to compute the optimal fair allocation, i.e., the fair allocation with maximum social welfare. In this work, we align with this research trend and address these two problems concerning the allocation of indivisible chores when agents incur costs to carry out their assigned chores.

Two widely studied comparison-based fairness notions are *equitability* (EQ) [13] and *envy-freeness* (EF) [24,42]. Informally, an allocation is EQ if all agents have the same cost and EF if no agent wants to exchange their bundles with any other agent's. When the items are not divisible, an EQ or EF allocation cannot be guaranteed to exist even in straightforward cases such that there is a single item with non-zero cost. Accordingly, since [16], most of the focus has been on their up to one relaxations. In particular, an allocation is *equitable up to one item* (EQ1) if every agent's cost excluding one item in her bundle is no greater than any other agent's cost, and is *envy-free up to one item* (EF1) if every agent's cost excluding one item in her bundle is no greater than her cost on any other agent's bundle. A stronger relaxation is up to any, which requires EQ and EF to be satisfied by each agent after the removal of any time in her bundle, and the resulting criteria are called EQX and EFX. It is known that EQX, EQ1 and EF1 allocations are guaranteed to exist [20,28,34], but the existence of EFX is still unknown.

To quantify the increase in social cost by restricting allocations to be fair, Caragiannis et al. [18,19] introduced the concept of *price of fairness* (PoF), which is the worst-case ratio between the optimal social cost with and without fairness constraints. Their original work focused on the unrelaxed fairness criteria EQ and EF, and excluded the instances which do not admit these fair allocations. Later, Bei et al. [11] and other followup works [5,32,33,37,38] studied the price of fairness regarding guaranteed fairness such as EQ1 and EF1. When the items are indivisible chores, it is known that there are instances whose optimal social cost is close to 0 but the social cost of any fair (in terms of all EQX, EQ1 and EF1) allocation is a constant, resulting in unbounded PoF. These results indicate that in the context of indivisible chores, fairness and efficiency are not compatible. Following these results, in this work, we want to understand the following problem.

If we could tolerate a constant increase in social cost, supposing that the total cost of items for every agent is normalized to 1, can we ensure the allocations are fair?

To investigate the above problem, we propose a new quantitative measure, the *cost of fairness* (CoF), which is the *difference*, in the worst-case instances, between the optimal social cost under fairness constraints and the optimal social cost without constraints. As we will prove in this work, CoF is more precise and revealing than PoF in terms of quantifying the efficiency loss. The known results that the PoF is infinity for all EQX, EQ1, and EF1 do not answer the above question nor provide insights into whether certain fairness criteria can be guaranteed while causing less social cost escalation than others. In stark contrast, the CoF is n regarding EQX and 1 regarding EQ1 and EF1. This exactly answers our problem: if we could tolerate a social cost loss of 1, we can ensure the allocations are EQ1 or EF1; however, EQX cannot be ensured.

We then take a computation perspective to study the previous problem. Among all fair allocations, we want to compute the one that is the most efficient. This problem has been extensively studied for indivisible goods, particularly under fairness notions of EF1 and EQ1 [3,6,14,37]. These results show that such a problem is NP-hard and is even hard to approximate. For indivisible chores, although the problem is proved to be NP-hard for EQ1 and for EQX in [37], the hardness of approximation is still unknown. Moreover, we are not aware of any prior work that studied the problem for EF1. Therefore, our paper also aims to fill the gap regarding the hardness of approximation. Inspired by the additivity nature of the cost of fairness, we investigate the computational complexity of finding an approximate solution within additive factors.

1.1 Our Contribution

In this work, we propose to study the *cost of fairness* for the allocation of indivisible chores and consider three guaranteed criteria, namely, EQX, EQ1, and EF1. On one hand, we give asymptotically tight bounds of the cost of fairness regarding all three fairness notions. On the other hand, we consider the computation problem of finding the fair allocations with minimum social cost. We show that all these problems are NP-hard, and are also hard to be approximated within either a constant or a linear additive factor. Our main results and their comparison to the price of fairness are shown in Table 1. If not stated otherwise, every agent's cost on all items is normalized to 1. In the following, we give a detailed introduction of our main results.

Table 1. Main results. PoF and CoF respectively refer to price of fairness and cost of fairness. Th.x points to Theorem x. The results on the price of fairness are from [37] and [38]. The parameter n is the number of agents and "$< \mathsf{OPT} + c$" means the problem of finding a fair allocation (EQX, EQ1, and EF1 respectively) with social cost smaller than the social cost of the optimal fair allocation plus c.

Fairness	PoF	CoF	Minimizing Social Cost
EQX	∞	n (Theorem 1)	$n \geq 2$: $< \mathsf{OPT} + n$ is NP-hard (Theorem 2)
EQ1	∞	1 (Theorem 3)	$n \geq 3$: $< \mathsf{OPT} + \frac{1}{2}$ is NP-hard (Theorem 5) $n = 2$: additive approximation scheme (Theorem 6)
EF1	∞	1 (Theorem 7)	$n \geq 3$: $< \mathsf{OPT} + \frac{n-1}{2n+2}$ is NP-hard (Theorem 8) $n = 2$: additive approximation scheme (Theorem 11)

Result 1. We start with EQX allocations, and our results in this part are negative. We first prove that the cost of fairness regarding EQX is n, which means that there are instances for which every EQX allocation results in a social cost as unfavorable as when each agent is assigned all the chores. This result further strengthens the hard instance designed in [37], where the optimal social cost is close to zero but the EQX allocation has a small constant cost resulting the price of fairness being unbounded. We then study the optimization problem of finding the EQX allocations with the minimum social cost. We prove that computing an approximate solution with an additive approximation factor smaller than n is NP-hard. As each agent's cost on all items is normalized to 1, our results are tight in the following two perspectives; (1) any EQX allocation (approximately) achieves the minimum social cost in the worst case instance, and (2) any EQX allocation achieves the best possible additive (polynomial time) approximation. if $P \neq NP$.

Result 2. We next consider EQ1 allocations and our results are significantly more positive. In sharp contrast to EQX, we first prove that every fair allocation instance admits an EQ1 allocation (which can be found in polynomial time) whose social cost is no greater than 1 – a single agent's total cost. This implies that the cost of EQ1 is upper bounded by 1. We also prove that the bound is asymptotically tight. This result is noteworthy because in the worst-case instance, the cost of an allocation can be as large as n. Therefore, the actual distance between the most efficient allocation and the most efficient EQ1 allocation can be relatively small, which cannot be concluded from the price of EQ, EQ1 or the price of EQX (which are all infinite as proved in [19] and [37]). We then study the optimization problem of finding the EQ1 allocations with the minimum social cost. This problem has been proved to be NP-hard, even when $n = 2$ [37]. We improve this result and show that for $n \geq 3$, there is no polynomial-time algorithm that ensures an additive approximation factor smaller than $\frac{1}{2}$ given $P \neq NP$. When $n = 2$, we complement the hardness results with an additive approximation scheme[1].

Result 3. We also study EF1 allocations. Similar to EQ1, we prove that the asymptotically tight bound of the cost of EF1 is 1. Moreover, the fair allocation with the matched social cost can be found in polynomial time. Note that the price of EF and that of EF1 are infinite as proved in [19] and [37]. Regarding the optimization problem, we prove that finding an EF1 allocation with the minimum social cost is NP-hard and moreover, when $n \geq 3$, there is no polynomial-time algorithm that ensures an additive approximation factor smaller than $\frac{n-1}{2n+2}$ given $P \neq NP$. Note that $\frac{n-1}{2n+2} \geq \frac{1}{4}$ and goes to $\frac{1}{2}$ when n goes to infinity. When $n = 2$, we also design an additive approximation scheme.

---

[1] The additive approximation scheme aims to compute a solution with an absolute error of at most ϵh in the objective where ϵ is the accuracy parameter and h is a predetermined fixed parameter. The additive approximation scheme has been applied to study load balancing problems [15].

It is worthwhile to mention that we en-route prove that for every optimization problem we consider, there is no polynomial-time algorithm having a bounded multiplicative approximation guarantee. This fact also argues in favor of using cost of fairness to investigate the fairness-efficiency trade-off for chores.

1.2 Related Work

The concept of the price of fairness draws inspiration from the extensively studied notions of the price of anarchy [31] and the price of stability [2] in game theory. Caragiannis et al. [19] first considered the price of EF, EQ and PROP (short for proportionality [35]) for divisible and indivisible goods and chores. Bei et al. [11] and followup works [5,32,33,37,38] propose to study the price of the up to one relaxation of these fairness notions. Barman et al. [5] also considered the price of approximate maximin share fairness, namely $\frac{1}{2}$-MMS. Some recent works study the price of fairness when items have graph structures [30,36,39,40]. In the graphical setting, the paradigm of quantifying welfare loss under constraints is also applied to the requirement of connected allocation to each agent, i.e., price of connectivity [9,10].

In addition to social welfare, Pareto optimality (PO) is also widely studied in resource allocation to measure the efficiency of an allocation. Caragiannis et al. [20] first observed that for indivisible goods, any allocation that maximizes the Nash welfare is EF1 and thus proved the existence of an EF1 and PO allocation. Later, Barman et al. [8] designed a pseudopolynomial time algorithm for this problem. It is still unknown whether an EF1 and PO allocation can be found in strongly polynomial time. For chores, the compatibility between EF1 and PO is largely unknown, except for several special cases [21,26,27,43]. A PROP1 and PO allocation exists and can be found in strongly polynomial time for goods, chores and even a mixture of them [4,7]. However, it is shown in [4] that a PROPX allocation may not exist for goods. Although PROPX allocations exist for chores, it may not be compatible with PO [4] when the excluded item can have zero cost. It is an open problem when the excluded item must have positive cost. Finally, EQ1 is compatible with PO for chores but not for goods, and EQX is not for both goods and chores [22,23]. It is NP-hard to decide whether an instance admits an EQX (or EQ1 for goods) and PO allocation.

2 Preliminaries

Given a positive integer k, denote by $[k]$ the set $\{1, \ldots, k\}$. There is a set of indivisible items $O = \{o_1, \ldots, o_m\}$ to be allocated to a set of agents $N = \{1, \ldots, n\}$. In this paper, we reserve n and m for the number of agents and items. Each agent $i \in N$ is associated with a disutility or cost function $c_i : 2^O \to \mathbb{R}_{\geq 0}$. For simplicity, we write $c_i(o)$ to refer $c_i(\{o\})$, i.e., the cost of o for agent i. It is assumed that the cost functions are normalized and additive, i.e., $c_i(O) = 1$ and for any $S \subseteq O$, $c_i(S) = \sum_{o \in S} c_i(o)$. An instance of the allocation problem is denoted by $\mathcal{I} = \langle N, O, \{c_i\}_{i \in N} \rangle$.

An allocation $A = (A_1, \ldots, A_n)$ is an ordered n-partition of O where for any $i \neq j$, $A_i \cap A_j = \emptyset$ and $\bigcup_{i \in N} A_i = O$. The *social cost* of an allocation is the total cost of agents. Given an allocation A, the social cost of A is $\mathsf{SC}(A) = \sum_{i \in [n]} c_i(A_i)$. The allocation with minimum social welfare without fairness constraints is called *optimal.* The allocation with the minimum social welfare among all allocations satisfying fairness criterion P is called an optimal P allocation.

In this paper, we consider two comparison-based fairness criteria, equitability (EQ) and envy-freeness (EF). Given an allocation $A = (A_1, \ldots, A_n)$, A is EQ if $v_i(A_i) = v_j(A_j)$ for all $i, j \in N$, and is EF if $v_i(A_i) \geq v_i(A_j)$ for all $i, j \in N$. Since an EQ or EF allocation is not guaranteed to exist, we consider their up to one relaxations.

Definition 1 (Equitable up to Any Item). *An allocation $A = (A_1, \ldots, A_n)$ is* equitable up to any item *(EQX) if for any pair of agents $i, j \in N$ and for every $o \in A_i$, $c_i(A_i \setminus \{o\}) \leq c_j(A_j)$.*

Definition 2 (Equitable up to One Item). *An allocation $A = (A_1, \ldots, A_n)$ is* equitable up to one item *(EQ1) if for any agents $i, j \in N$, there exists some $o \in A_i$ such that $c_i(A_i \setminus \{o\}) \leq c_j(A_j)$.*

Definition 3 (Envy-free up to One Item). *An allocation $A = (A_1, \ldots, A_n)$ is* envy-free up to one item *(EF1) if for any $i, j \in N$, there exists an item $o \in A_i$ such that $c_i(A_i \setminus \{o\}) \leq c_i(A_j)$.*

We now introduce the concept of *cost of fairness* (CoF). Informally, given a fairness criterion, the CoF is the supreme difference, among all instances, between the minimum social cost of all fair allocations and that of all allocations.

Definition 4. *For a given fairness criterion P, the cost of fairness (CoF) regarding P, also called the cost of P, is defined as*

$$\sup_{\mathcal{I}} \min_{A \in P(\mathcal{I})} \mathsf{SC}(A) - \mathsf{OPT}(\mathcal{I}),$$

where $P(\mathcal{I})$ is the set of allocations satisfying P and $\mathsf{OPT}(\mathcal{I})$ is the minimum social cost over all allocations of $\mathcal{I}$.

3 The Social Cost of EQX Allocations

Our results in this section are negative. We first prove that in the worst case example, any EQX allocation can be as bad as the "worst allocation" when every agent gets a cost of 1, while the social cost of optimal solution is close to 0, implying the cost of EQX being n. To surpass this obstacle, we consider the computation problem of finding the EQX allocation with the minimum cost. Unfortunately, we prove that for any $n \geq 2$, no polynomial time algorithm ensures an EQX allocation with social cost less than n, unless P = NP.

3.1 Bounding the Cost of EQX

We first bound the cost of EQX. All missing proofs can be founded in the full version of the paper.

Theorem 1. *For any $n \geq 2$, the cost of EQX is n.*

Proof (Proof Sketch). Since the social cost of any allocation is at most n and EQX allocation always exists [28], the cost of EQX is at most n.

As for the lower bound, let us consider an instance with n agents and $2n-1$ items. The items are divided into two categories, a set $\{o_1^1, \ldots, o_n^1\}$ of n items for the first category and a set $\{o_1^2, \ldots, o_{n-1}^2\}$ of $n-1$ items for the second category. For simple notations, let $O^1 = \{o_1^1, \ldots, o_n^1\}$ and $O^2 = \{o_1^2, \ldots, o_{n-1}^2\}$. The cost functions of agents are presented in Table 2 where K is a sufficiently large number, i.e., $K \gg n2^n$.

Table 2. The constructed instance for Theorem 1

$c_i(o)$		Items									
		o_1^1	o_2^1	o_3^1	......	o_n^1	o_1^2	o_2^2	o_3^2	......	o_{n-1}^2
Agents	1	K	1	1	......	1	n	$2n$	$4n$	......	$n \cdot 2^{(n-2)}$
	2	1	K	1	......	1	n	$2n$	$4n$	......	$n \cdot 2^{(n-2)}$
	3	1	1	K	......	1	n	$2n$	$4n$	......	$n \cdot 2^{(n-2)}$
	$\vdots$	$\vdots$	$\vdots$	$\vdots$	$\ddots$	$\vdots$	$\vdots$	$\vdots$	$\vdots$	$\ddots$	$\vdots$
	$\vdots$	$\vdots$	$\vdots$	$\vdots$	$\ddots$	$\vdots$	$\vdots$	$\vdots$	$\vdots$	$\ddots$	$\vdots$
	n	1	1	1	......	K	n	$2n$	$4n$	......	$n \cdot 2^{(n-2)}$

For any $i \in [n]$, the total cost of items for agent i is $K + n2^{n-1} - 1$ and hence cost functions are normalized. Note that by scaling, these cost functions can be converted to cost functions with total cost 1. For the ease of representation, throughout this paper, when presenting cost functions, we do not scale. By allocating each item to the agent having the least cost for it, one can verify that the social cost of optimal allocation is equal to $n2^{n-1}$. We then show that any EQX allocation has a social cost of at least nK, that is, every agent receives the item with cost K for her in any EQX allocation. The remaining proof is given by the contradiction and is given in the full version of the paper. For every possible allocation where some agents do not receive the item with cost K for her, we can prove that some agent violates the property of EQX.

3.2 Computing the Optimal EQX Allocations

We now consider the problem of minimizing social cost subject to EQX constraints, denoted as SC-EQX. Although SC-EQX is known to be NP-hard [37], the hardness of approximation has not been studied. Below we significantly strengthen the state-of-the-art result by proving an additive inapproximability factor of n.

To establish the hardness result, we provide a polynomial time reduction from the well-known PARTITION problem [25]: given a set $\{p_1, \dots, p_r\}$ of r integers whose sum is $2T$, can $[r]$ be partitioned into two indices sets I_1 and I_2 such that $\sum_{j \in I_1} p_j = \sum_{j \in I_2} p_j = T$?

Theorem 2. *For any $n \geq 2$, there is no polynomial time algorithm approximating SC-EQX within an additive factor smaller than n unless $P = NP$.*

Proof. We present the reduction from the PARTITION Problem. Given an arbitrary instance of PARTITION $\{p_1, \dots, p_r\}$, we construct a fair division $\mathcal{I}$ with n agents and $2n + r$ items. Items are divided into three categories, labelled by the superscript. Let $O^1 = \{o_1^1, \dots, o_n^1\}$ be the set of the n items belong to first category; $O^2 = \{o_1^2, \dots, o_r^2\}$ be the set of the r items belong to the second category; $O^3 = \{o_1^3, \dots, o_n^3\}$ be the set of n items belong to the third category. Agents' cost functions are described as follows:

- for $\{o_j^1\}$: every agent i has cost $K - x_i$ for o_i^1 and cost 0 for each of other items;
- for $\{o_j^2\}$: for every agent $i \in \{1, 2\}$, she has cost p_j for o_j^2; for every agent $i \geq 3$, her cost for every o_j^2 is equal to $10r^{i-2}n^{(n+i-2)}T$;
- for $\{o_j^3\}$: for any $i \in \{1, 2\}$, she has cost $4T$ for o_i^3 and cost $10n^{i-1}T$ for each of other items; for agent $i \geq 3$, she has cost $5T$ for o_i^3 and cost $10n^{i-1}T$ for each of other items.

In the construction, K is a sufficiently large number, satisfying $K \gg T \cdot n^{2n} \cdot r^n$ and for any i, $x_i = c_i(O^2 \cup O^3)$. The total cost of every agent is equal to K and hence cost functions are normalized.

In what follows, we will show that when the PARTITION instance has an answer "yes", there exists an EQX allocation with social cost $o(K)$. If PARTITION instance has "no" answer, then the social cost every EQX allocation is close to nK. The key ideas of the constructed fair instance are: in any EQX allocation, (i) if agent i receives her *big-item* o_i^1, then o_i^1 must be the only item in her bundle unless each of the other agents also receives their big-item, and (ii) if we require the total cost of the assignment of O^1 to be zero, then O^1 must be allocated to at least two agents, and moreover, their cost should be the same in order to satisfy EQX; note that when verifying EQX criterion, the item (if any) with cost 0 would be hypothetically removed.

Suppose we have a "yes" instance of PARTITION and let I_1 and I_2 be a solution. Now we consider allocation $A = (A_1, \dots, A_n)$ where

- A_1 contains $o_2^1, \dots, o_n^1$ and o_j^2 for all $j \in I_1$ and o_1^3;
- A_2 contains o_1^1 and o_j^2 for all $j \in I_2$ and o_2^3.
- for any $i \geq 3$, A_3 contains o_i^3 only.

Informally, in allocation A, the total cost of allocating O^1 is 0 and O^2 is allocated based on the solution of the PARTITION instance. It is not hard to verify that for any $i \in [n]$, $c_i(A_i) = 5T$. Hence, allocation A is EQX and has a social cost $5nT$, equal to the social cost of the optimal allocation.

For the other direction, suppose we have a "no" instance of PARTITION. We will prove that in any EQX allocation $B = (B_1, \dots, B_n)$, every agent i must receive her big-item o_i^1. Our first step is to show that in B, there exists at least one agent receiving her big-item. For a contradiction, assume no agent receives their big-item, which implies that the cost caused by the assignment of O^1 is 0 in allocation B. Let $N_1 = \{i : B_i \cap O^1 \neq \emptyset\}$. By the contradiction assumption, for any $i \in N_1$, $c_i(B_i \cap O^1) = 0$. Then it is not hard to verify that $|N_1| \geq 2$ because if one agent receives all o_j^1's, she would receive her big-item. For any $i, j \in N_1$, $c_i(B_i) = c_j(B_j)$ holds as when comparing EQX, item with cost 0 would be hypothetically removed. Moreover, for any $i \in N_1$, $c_i(B_i) > 0$ holds; If $N_1 = [n]$, then every agent need to have the same cost to satisfy EQX, and the above proposition holds trivially; If $N_1 \neq [n]$, then for every $j \in [n] \setminus N_1$, agent j can only receive one item in B due to EQX property and there will be items left unallocated.

Claim. Neither of agent 1 nor agent 2 is in N_1.

The above claim implies that every agent $i \in N_1$ satisfies $i \geq 3$. For any two agents $i, j \in N_1$ with $i < j$ (without loss of generality), the only possible way to achieve $c_i(B_i) = c_j(B_j) > 0$ is that: agents i, j do not receive items in O^2 and moreover $B_i \cap O^3 = \{o_i^3\}$ and $B_j \cap O^3 = \{o_j^3\}$. Note that for any $o \in O^2$, agent j has a cost $c_j(o) = 10r^{j-2}n^{n+j-2}T \geq 10r^{i-1}n^{n+i-1}T$, larger than the total cost of $O^2 \cup O^3$ for agent i. Thus, agent j receives no item from O^2, which then implies that agent i does not receive items from O^2 as for every $o \in O^2$, $c_2(o) = 10r^{i-2}n^{n+i-2}$, larger than the total cost of O^3 for agent j. Similarly, one can verify that in order to ensure $c_i(B_i) = c_j(B_j)$, agent i (resp. agent j) receives no item from $O^3 \setminus \{o_i^3\}$ (resp. $O^3 \setminus \{o_j^3\}$). Therefore, for every $i' \in N_1$, it must hold that $B_{i'} \cap O^3 = \{o_{i'}^3\}$ and moreover $c_{i'}(B_{i'}) = 5T$ as $i' \in N_1$. We then claim that for any $j \in [n]$, agent j receives exactly one item from O^3. This is clearly true for agents in N_1. As for $j \in [n] \setminus N_1$, if agent j receives more than two items from O^3, she violates EQX when compared to every agent $i \in N_1$; if j receives nothing from O^3, then another agent j' would receive at least two items from O^3 and thus violates EQX. Moreover for any $j \geq 3$ and $j \notin N_1$, it must hold that $B_j \cap O^2 = \emptyset$ as if $B_j \cap O^2 \neq \emptyset$, agent j would receive two items from $O^2 \cup O^3$ and hence violate EQX when compared to any agent $i \in N_1$; recall $c_i(B_i) = 5T$. Therefore, bundle O^2 must be entirely allocated between agents 1 and 2 in allocation B.

We now prove that given all the above-mentioned restrictions, allocation B can never be EQX. This is derived by the comparison among agents 1, 2 and agents in N_1. Thereafter fix $t \in N_1$. By the construction of N_1, there exists $o \in B_t$ such that $c_t(o) = 0$. Then, $\max_{o \in B_t} c_t(B_t \setminus \{o\}) = 5T$. Thus in order to make B be EQX, $c_i(B_i) \geq 5T$ must be true for all $i \in [2]$. For any $i \in [2]$, if agent i receives only o_i^3 from O^3, then due to "no" PARTITION instance, there must be an agent $l \in [2]$ such that $c_l(B_l) < 5T$, in which case allocation B is not EQX. Accordingly, at least one of agents $i \in [2]$ must receive some item o_j^3 with $j \neq i$. Without loss of generality, let agent 1 be such an agent. Then agent 1 only receives one item in allocation B as $c_t(B_t) = 5T$ and every o_j^3 with $j \neq 1$ yields a cost of $10T$ for agent 1. Accordingly, all o_j^2's would be allocated to agent 2. Recall that agent 2 also receives one item from O^3, then we have $\max_{o \in B_2} c_2(B_2 \setminus \{o\}) > 5T$, meaning that agent 2 violates EQX when compared to agent t. This yields the desired contradiction and thus allocation B can never be EQX if no agent receives their big-item. Therefore, there exists at least one agent receiving her big-item in allocation B. Let agent i_1 be one of such agents.

We next show that there exist at least two agents receiving their big-item in allocation B. For the sake of contradiction, suppose that agent i_1 is the only agent receiving their big-item. Then by the condition of EQX, it must hold that $|B_{i_1}| = 1$. We can indeed remove agent i_1 and $o_{i_1}^1$ and only consider the reduced instance with $n-1$ agents and $2n+r-1$ items as other agents never violate EQX when comparing to agent i_1. With a slight abuse of notations, let $N_1 = \{i : B_i \cap O^1 \neq \emptyset\}$ and similarly $|N_1| \geq 2$. Then by arguments similar to the proof of Claim, agents 1 and 2 are not in N_1; this holds no matter whether agent 1 or 2 is in the reduced instance. Then again, for any $i \in N_1$, $c_i(B_i) = 5T$ holds and moreover $B_i \cap O^3 = \{o_i^3\}$. For the reduced instance, the total number of agents not in N_1 is $n-1-|N_1|$ as the reduced instance contains $n-1$ agents. Since each agent $i \in N_1$ receives exactly one item from O^3 in allocation B, the total number of unallocated items from O^3 is $n-|N_1|$, and these $n-|N_1|$ items would be allocated to $n-1-|N_1|$ agents. Then there exists one agent receiving at least two items from O^3 and that agent violates EQX when compared to every agent $i \in N_1$, yielding the desired contradiction. Therefore in allocation B, there exist at least two agents receiving their big-item. We again eliminate two agents and their big-items and then consider the reduced instance with $n-2$ agents and $2r+n-2$ items. By similar arguments, we can prove that there exist at least three agents receiving their big-items in allocation B. We repeat this argument and can show that all n agents receive their big-items in allocation B. Therefore, the social cost of EQX allocation B is $nK - o(K)$.

By scaling the total cost of every agent to 1, we can claim that it is NP-hard to approximate SC-EQX within an additive term of

$$\frac{nK - o(K)}{K} \to n \text{ as } K \to +\infty,$$

which yields the desired inapproximation factor in the statement of Theorem 2.

For the reduction in the proof of Theorem 2, if the PARTITION instance has an answer "yes", there exists an EQX allocation with social cost $5nT$, while if the answer is "no", every EQX allocation has social cost close to nK. When K tends to infinity, the reduction also indicates no polynomial time algorithm can achieve a bounded multiplicative approximation factor for SC-EQX!

Corollary 1. *For any $n \geq 2$, there is no polynomial time algorithm having a bounded multiplicative approximation factor for SC-EQX unless $P = NP$.*

4 The Social Cost of EQ1 Allocations

In contrast to the negative results on EQX allocations, in this section we prove that we can always find an EQ1 allocation with social cost no more than 1 in polynomial time. We then prove that for any $n \geq 3$, the problem of computing the optimal EQ1 allocations does not admit a polynomial time algorithm with additive loss smaller than $\frac{1}{2}$ unless P = NP. When $n = 2$, we give an additive approximation scheme.

4.1 Bounding the Cost of EQ1

To bound the cost of EQ1, we first design an algorithm that returns an EQ1 allocation whose social cost is at most 1. The algorithm starts with the optimal solution S^* and implements a sequence of items reassignment. Given an intermediate allocation A, we execute items reassignment based on directed graph $G_A = (N, E_A)$ constructed as follows: each agent $i \in N$ forms a vertex of G_A and there is a directed edge or an arc $(i, j) \in E_A$ from i to j if $c_i(A_i \setminus \{o\}) > c_j(A_j)$ for all $o \in A_i$. In other words, an arc from i to j indicates that agent i violates EQ1 when compared to agent j. We classify agents into three categories based on their in- and out-degrees:

$$\begin{aligned} &N_0 \text{ contains all agents with zero in-degree and zero out-degree in } G_A;\\ &N_1 \text{ contains all agents with zero in-degree and positive out-degree in } G_A; \quad (1)\\ &N_2 \text{ contains all agents with positive in-degree in } G_A. \end{aligned}$$

The algorithm each time identifies an agent in N_1 and assigns one item from her bundle to the agent with the minimum cost in N_2. After each reassignment, we update G_A, N_0, N_1, N_2 and repeat the process until N_1 becomes empty. The formal description of the algorithm is shown in Algorithm 1.

We first present several lemmas and terminologies used in the below proof. Each iteration of the while loop in Algorithm 1 is called a round. We focus on two time frames of a given round t: *at the beginning of round* t refers to the moment right before the item reassignment happens; and *after round* t refers to the moment right after the item movement happens and the graph and corresponding sets get updated.

Observation 1. *Throughout the while loop of Algorithm 1, the minimum cost of an agent is non-decreasing.*

To see Observation 1, let us compare agents' cost before and after some round t. There exists exactly one agent (let's say agent i^*) whose cost is weakly decreased and moreover this agent belongs to N_1 at the beginning of round t. After round t, the cost of agent i^* should be strictly larger than the minimum cost of an agent in the allocation at the beginning of round t; otherwise right before round t, agent i^* satisfies EQ1 and should not be in N_1.

Algorithm 1: Computing EQ1 allocations with social cost ≤ 1

Input: An instance $\mathcal{I} = \langle N, O, \{c_i\}_{i\in N}\rangle$
Output: Allocation $A = (A_1, \ldots, A_n)$.
1: Compute the optimal allocation $S^* = (S_1^*, \ldots, S_n^*)$ by allocating each item to the agent with the smallest cost, where ties are broken arbitrarily.
2: Initialize $A_i \leftarrow S_i^*$ for all $i \in N$ and construct $G_A = (N, E_A)$ where there is a directed edge (i, j) for $i, j \in N$ if $c_i(A_i \setminus \{o\}) > c_j(A_j)$ for all $o \in A_i$. Define N_0, N_1 and N_2 as Equation (1).
3: **while** $N_1 \neq \emptyset$ **do**
4: Let $i^* \in \arg\max_{i\in N_1}(\min_{o\in A_i} c_i(A_i \setminus \{o\}))$ and arbitrarily pick $o' \in A_{i^*}$.
5: Let $j^* \in \arg\min_{j\in N_2} c_i(o')$ and make the reassignment $A_{i^*} \leftarrow A_{i^*} \setminus \{o'\}$ and $A_{j^*} \leftarrow A_{j^*} \cup \{o'\}$.
6: Update $G_A = (N, E_A)$, N_0, N_1 and N_2.
7: **end while**

Lemma 1. *For agent i, if $i \in N_0$ at the beginning of some round, then $i \in N_0$ holds in all future rounds.*

Lemma 2. *For agent i, if in G_{S^*}, the graph corresponding to the optimal allocation S^*, the in-degree of vertex i is 0, then the vertex i has in-degree 0 in G_A for every intermediate allocation A.*

Theorem 3. *Algorithm 1 runs in polynomial time and returns an EQ1 allocation with a social cost of at most 1.*

Proof. By the graph construction, $N_1 = \emptyset$ implies $N_2 = \emptyset$ and vice versa. As after each round, the cardinality of the union bundle of agents in N_2 is increased by 1, then the graph G_A varies once in at most every m rounds. Based on Lemmas 1 and 2, each time when G_A varies, we have three possible situations: (i) one agent moves from N_1 to N_0, (ii) one agent moves from N_2 to N_1 or to N_0, and (iii) both (i) and (ii) happen. As a consequence, N_1 and N_2 will be $\emptyset$ after G_A getting updated (at most) $2n$ times. Thus, Algorithm 1 terminates in $2nm$ rounds. The running time of executing a round is at most $O(mn)$ and therefore Algorithm 1 terminates in $O(m^2n^2)$ time.

The fact that returned allocation A is EQ1 directly follows from the construction of G_A and $N_1, N_2 = \emptyset$ at termination. We now bound the social cost of A. Let T be the last round of the algorithm, i.e., after round T, $N_1, N_2 = \emptyset$. Suppose N_2' is the set of vertex with positive in-degree at the beginning of round T and suppose $N_2' \ni k$. Denote by Q the set of items allocated to different agents in S^* and A. For any $o \in Q$, suppose $o \in A_j$, and then we have $c_j(o) \leq c_k(o)$ by Step 5 and the fact that k is in N_2 in every round $t = 1, ..., T$ (Lemmas 1 and 2). Accordingly, we have the following inequality,

$$\mathsf{SC}(A) \leq \sum_{o\in Q} c_k(o) + \sum_{i\in N} c_i(S_i^* \setminus Q) \leq \sum_{o\in Q} c_k(o) + \sum_{i\in N} c_k(S_i^* \setminus Q) = c_k(O) = 1,$$

where the second inequality transition is because S^* is the optimal allocation.

Theorem 4. *The cost of EQ1 is at most 1 and at least $1 - \frac{1}{n}$ for any $n \geq 2$.*

4.2 Computing the Optimal EQ1 Allocations

We now consider the optimization problem of minimizing social cost subject to EQ1 constraint, denoted by SC-EQ1. It is known that SC-EQ1 is NP-hard [37], where the reduction, however, does not suggest any hardness of approximation. We below show that SC-EQ1 is hard to approximate within some constant even when there are only three agents.

Theorem 5. *For any $n \geq 3$, there is no polynomial time algorithm approximating SC-EQ1 within an additive factor smaller than $\frac{1}{2}$ unless $P = NP$.*

Recall that by Theorem 3, Algorithm 1 can return an EQ1 allocation with social cost at most 1 in polynomial time. Thus, Algorithm 1 can approximate SC-EQ1 within an additive factor of 1, which complements the above negative result.

We remark that the reduction in the proof of Theorem 5 (see the full version) also implies that approximating SC-EQ1 within a multiplicative factor of $\frac{rK+n}{n+1}$ is NP-hard. As K can be arbitrarily large, we can then claim that no polynomial time algorithm can have bounded multiplicative approximation guarantee.

Corollary 2. *For $n \geq 3$, there is no polynomial-time algorithm that has a bounded multiplicative approximation factor for SC-EQ1 unless $P = NP$.*

As for the case of two agents, it is also known that SC-EQ1 is NP-hard [37]. We below present an additive approximation scheme for SC-EQ1 when $n = 2$. For the choice of parameter h, we follow the work of [17] and set h as the total cost of an agent for all items, i.e., $h = 1$. Denote by OPT^{EQ1} the social cost of the optimal EQ1 allocation.

Theorem 6. *Let $\epsilon > 0$ be an accuracy parameter. Given an instance of SC-EQ1 with two agents and m items, the algorithm runs in time $O(m2^{\frac{2}{\epsilon}})$ and computes an EQ1 allocation with a social cost of at most $\mathsf{OPT}^{EQ1} + \epsilon$.*

Our algorithm makes use of the PTAS idea for the classical scheduling problem of makespan minimization on identical machines [29]. We first partition the space of complete allocations $\mathcal{D}$ into q districts $\mathcal{D}^{(1)}, \mathcal{D}^{(2)}, \ldots, \mathcal{D}^{(q)}$ such that $\mathcal{D}^{(1)} \cup \cdots \cup \mathcal{D}^{(q)} = \mathcal{D}$ where q is polynomial in m and $\frac{1}{\epsilon}$. Then for each district $\mathcal{D}^{(l)}$, we compute a representative allocation that is EQ1 and has a social cost close to the optimal objective value in $\mathcal{D}^{(l)}$. At last, we pick the best representative among all districts.

We now present the algorithm in detail. Based on the accuracy parameter $\epsilon > 0$, we classify items into two categories as follows: for any $o \in O$,

it is a *big item* if $\exists i = 1, 2$ such that $c_i(o) > \epsilon$; it is a *small item* otherwise.

As the total cost of every agent is 1, the total number of big items is at most $\frac{2}{\epsilon}$. We define districts $\{D^{(l)}[i, o]\}$ as follows: two allocations F^1, F^2 lie in district $D^{(l)}[i, o]$ if F^1 allocates every big item to the same agent as F^2 and moreover item o is allocated to i in both F^1 and F^2.

Note that the requirement of the identical assignment of big items in F^1, F^2 does not violate the requirement of allocating item o to agent i no matter whether o is a big item or a small item. There are $2^{\frac{2}{\epsilon}}$ possible allocations of big items and at most m options for the choice of the specific o and two options for agent i. Thus, the total number of districts is at most $2m2^{\frac{2}{\epsilon}}$.

We next describe how to find the representative within each district. Consider a fixed district $D^{(l)}[1, o']$ (the district for agent 2 is symmetric). For every allocation in this district, assignments of big items and item o' have already been fixed while that of small items (possibly excepting o') may vary. Let $s_1, \dots, s_k$ be small items; note $k < m$. Let a_1 be the cost of agent 1 with respect to the fixed big items, excluding o' if o' is a big item, allocated to her and let a_2 be the cost of agent 2 with respect to the fixed big items allocated to her. We determine the assignment of small items based on the optimal solution of the following linear programming. In the LP, variable x_j refers to the fraction of s_j allocated to agent 1. The objective function is the total cost of small items and the first two constraints guarantee that (i) agent 1 receives less cost after removing o' and (ii) agent 2 receives less cost than agent 1. Note that we can get a complete (possibly fractional) allocation by extending the fixed partial assignment of big items and o' according to the solution of the LP. Moreover, the assignment of small items in those integral EQ1 allocations in district $\mathcal{D}^{(l)}[1, o']$, where o' is the largest cost item in the bundle of agent 1 and the cost of agent 1 is at least that of agent 2, corresponds to feasible solutions of the LP.

$$\begin{aligned}
\min\ & \textstyle\sum_{j=1}^{k} c_1(s_j)x_j + \sum_{j=1}^{k} c_2(s_j)(1-x_j) \\
\text{s.t.}\ & a_1 + \textstyle\sum_{j=1}^{k} c_1(s_j)x_j \le a_2 + \sum_{j=1}^{k} c_2(s_j)(1-x_j) \\
& a_2 + \textstyle\sum_{j=1}^{k} c_2(s_j)(1-x_j) \le a_1 + c_1(o') + \sum_{j=1}^{k} c_1(s_j)x_j \\
& 0 \le x_j \le 1, \quad j = 1, \dots, k.
\end{aligned}$$

We can compute in time polynomial to m the optimal solution $\{x_j^*\}$ of the LP. If $c_1(o') = 0$, the first two inequality constraint becomes an equality constraint. When $c_1(o') \neq 0$, the first two inequality constraints can not be fulfilled with equality at the same time. As a consequence, since the solution $\{x_j^*\}$ is a basic feasible solution of the underlying polyhedron in k-dimensional space, at least $k-1$ values x_j^* should be integral, i.e., at most one x_j^* is fractional. To find the representative in $D^{(l)}[1, o']$, we allocate small items as follows: item s_j is allocated to agent 1 if $x_j^* = 1$ and is allocated to agent 2 if $x_j^* = 0$. If every x_j^* is integral, then we find the representative. If $\{x_j^*\}$ is not an integral solution, as we have already argued, there is at most one value x_t^* being fractional. We next describe how to allocate item s_t. Define $E_1 = \{s_j : j \in [k] \text{ and } x_j^* = 1\}$ and $E_2 = \bigcup_{j \in [k]} s_j \setminus (E_1 \cup s_t)$, that is, E_1 (resp. E_2) refers to the set of small items determined to be allocated to agent 1 (resp. agent 2) based on $\{x_j^*\}$. Up to here, the assignment of every big item, item o' and small items excluding s_t has been determined; recall that o' is allocated to agent 1.

If $a_1 + c_1(E_1) \ge a_2 + c_2(E_2)$, then item s_t is allocated to agent 2. One can verify that this complete allocation is EQ1. If $a_1 + c_1(E_1) < a_2 + c_2(E_2)$, we compare $a_1 + c_1(E_1) + c_1(o')$ and $a_2 + c_2(E_2)$ and then allocate s_t to the agent with a smaller cost; if $a_1 + c_1(E_1) + c_1(o') > a_2 + c_2(E_2)$, we allocate s_t to agent 2 and otherwise allocate s_t to agent 1. The resulting complete allocation is also EQ1. The detailed proof of this part is in the full version.

We now have found the representative for each district and have shown that the representative allocation is EQ1. Suppose $B^* = (B_1^*, B_2^*)$ is the optimal EQ1 allocation and without loss of generality $c_1(B_1) \geq c_2(B_2)$. Moreover, item $o^* \in B_1$ satisfies $c_1(o^*) \geq c_1(o)$ for all $o \in B_1$. Then by the construction of districts, allocation B^* must lie in some district $\mathcal{D}^{(l^*)}[1, o^*]$. Let A^{l^*} be the representative of $\mathcal{D}^{(l^*)}[1, o^*]$ and z^{l^*} be the optimal objective value of the LP corresponding to $\mathcal{D}^{(l^*)}[1, o^*]$. Then we have

$$\mathsf{SC}(A^{l^*}) \leq a_1 + a_2 + z^{l^*} + \epsilon \leq \mathsf{SC}(B^*) + \epsilon.$$

The first inequality transition is because x_t^* is the only fractional value in $\{x_j^*\}$ and s_t is a small item. The second inequality transition is because the assignment of small items (possibly excluding o') of B corresponds to a feasible solution of LP. After visiting all the representatives, we pick the allocation with the minimum social cost. It is clear that the social cost of that allocation is no greater than $\mathsf{SC}(A^{l^*})$. Therefore, the algorithm runs in time $O(m2^{\frac{2}{\epsilon}})$ and computes an EQ1 allocation with a social cost of at most $\mathsf{OPT}^{EQ1} + \epsilon$.

5 The Social Cost of EF1 Allocations

Our last fairness criterion is EF1. As we will show the results are positive and similar as EQ1 allocations.

5.1 Bounding the Cost of EF1

Theorem 7. *The cost of EF1 is at most* 1 *and at least* $1 - \frac{1}{n}$.

The high-level idea of the proof of Theorem 7 is that we implement the round-robin algorithm on n different orderings of the agents. It is clear that all these n algorithms return EF1 allocations, and we further prove that one of them returns an allocation with a social cost of at most 1.

5.2 Computing the Optimal EF1 Allocations

Denote by SC-EF1 the optimization problem of minimizing social cost subject to EF1 constraint. We below first show that SC-EF1 is hard to approximate within a constant when $n \geq 3$. Then for the case of two agents, we present an additive approximation scheme.

Theorem 8. *For any $n \geq 3$, there is no polynomial time algorithm approximating SC-EF1 within an additive factor smaller than $\frac{n-1}{2n+2}$ unless $P = NP$.*

For the case of two agents, we also show the hardness of SC-EF1.

Theorem 9. *When $n = 2$, SC-EF1 is NP-hard.*

En-route, we also show that no polynomial time algorithm guarantees a bounded multiplicative approximation factor.

Theorem 10. *For $n \geq 4$, there is no polynomial time algorithm having a bounded multiplicative approximation factor for SC-EF1 unless $P = NP$.*

To consider the case of two agents and establish an additive approximation scheme for SC-EF1. In particular, we will establish an "equivalence" relationship between the allocations of goods and the allocations of chores when there are two agents. When allocating goods to two agents, an FPTAS (multiplicative approximation) for SW-EF1 exists [14]. We shall utilize their results and propose an additive approximation scheme for SC-EF1 by connecting SW-EF1 to our problem. Denote by OPT^{EF1} the social cost of the optimal EF1 allocation.

Theorem 11. *Let $\epsilon > 0$ be an accuracy parameter. Given an instance of SC-EF1 with two agents and m items, the algorithm runs in time $poly(m, \frac{1}{\epsilon})$ and computes an EF1 allocation with a social cost of at most $\mathsf{OPT}^{EF1} + \epsilon$.*

6 Conclusion

In this paper, we re-raised the discussion on the trade-off between fairness and system efficiency for the allocations of indivisible chores. We propose a new measurement, the cost of fairness, to quantify the social cost rise when requiring the allocations to be fair. We found that an EQ1 or EF1 allocation with social cost at most 1 can be computed in polynomial time. However, with three or more agents, finding an EF1 (or EQ1 respectively) allocation with social cost smaller than 1/2 (or 1/4 respectively) plus the social cost of the optimal EQ1 (or EF1 respectively) allocation is NP-hard. We complement these results with additive approximation schemes for the two-agent case. Regarding EQX, we prove that there exist instances in which any EQX allocation has social cost n and finding an allocation with cost smaller than n is NP-hard for any n.

Our work uncovers many open problems and future research directions. First, in the current paper, we did not consider share-based fairness notions such as PROP1, PROPX, and approximate MMS. The price of PROP1 and PROPX is shown to be n which is acceptable compared with infinity. It is proved in [32] that there is always a PROP1 and PROPX allocation with social cost of at most 1, which implies that the cost of PROP1 and PROPX is 1. We leave a complete study of share-based fairness notions for future study. Secondly, we only focused on the allocation of chores. It is interesting to see whether the cost of fairness brings new insights to the trade-off between fairness and efficiency for goods.

Appendix

A Missing Proofs in Section 4

A.1 Proof of Lemma 1

Proof (Proof of Lemma 1). For the sake of contradiction, assume that agent i has zero in- and out-degree at the beginning of round t while i has positive in- or out-degree after round t. Throughout the proof, let A', $G_{A'}$, N'_0, N'_1, N'_2 refer to the allocation, the graph and the corresponding sets of agents after round t; let A^b, G_{A^b}, N^b_0, N^b_1 and N^b_2 refer to the allocation, the graph and the corresponding sets of agents at the beginning of round t. By the construction, $i \in N^b_0$.

If $i \in N'_1$, by Observation 1, it holds that $\min_j c_j(A'_j) \geq \min_j c_j(A^b_j)$. As $i \in N^b_0$, agent i does not violate EQ1 in allocation A'_i and moreover $A^b_i = A'_i$. Combining these facts, the out-degree of i in $G_{A'}$ should be 0 and thus $i \notin N'_1$, deriving the desired contradiction.

If $i \in N'_2$, suppose that in $G_{A'}$ there is an arc (j, i). According to the reallocation rule, the appearance of arc (j, i) can only be caused by the fact that agent j takes some item o' in round t. Hence, $j \in N^b_2$ holds. Then we claim that $c_j(A^b_j) < c_i(A^b_i)$; otherwise, the vertex pointing to j in G_{A^b} should also point to i in G_{A^b}, making $i \in N^b_2$, a contradiction. For allocation A', we have

$$\min_{o \in A'_j} c_j(A'_j \setminus \{o\}) \leq c_j(A'_j \setminus \{o'\}) = c_j(A^b_j) < c_i(A^b_i) = c_i(A'_i),$$

which indicates that arc (j, i) does not exist in $G_{A'}$, deriving the desired contradiction.

A.2 Proof of Lemma 2

Proof (Proof of Lemma 2). For the sake of contradiction, suppose that some vertex i has in-degree 0 in G_{S^*} but has positive in-degree after some round t. Suppose that right after round t is the earliest moment such that vertex i has a positive in-degree. Throughout the proof, let A, G_A, N_0, N_1, N_2 refer to the allocation, the graph and the corresponding sets of agents after round t; let A^b, G_{A^b}, N^b_0, N^b_1, N^b_2 refer to the allocation, the graph and the corresponding sets of agents at the beginning of round t. Then the contradiction assumption implies that $i \notin N^b_2$ and $i \in N_2$. In the following, we prove no such an agent i exists.

As $i \notin N^b_2$, there are two possibilities: (i) $i \in N^b_1$, and (ii) $i \in N^b_0$. According to Lemma 1, it must hold that $i \in N^b_1$. We below split the proof by discussing whether an item is taken from the bundle of agent i in round t.

If an item is taken from the bundle of agent i in round t, i.e., $|A^b_i \setminus A_i| = 1$, then agent i is chosen by the Algorithm 1 at Step 4 in round t. Then for any $j \in N^b_1$ and any $o' \in A^b_j$, we have $c_j(A^b_j \setminus \{o'\}) \leq \min_{o \in A^b_i} c_i(A^b_i \setminus \{o\}) \leq c_i(A_i)$. Since for any $j \in N^b_1 \setminus \{i\}$, $A^b_j = A_j$ holds, then there is no arc (j, i) in graph G_A as every agent j satisfies EQ1 when compared to agent i in A. By Observation 1, there is no arc pointing from some vertex in N^b_0 to i. To show $i \in N_2$ never holds, the remaining is to prove that there is no arc (j, i) with $j \in N^b_2$ in G_A. Let us first prove the following claim.

Claim. For any $j \in N^b_2$, arc (i, j) exists in G_{A^b}.

Proof (Proof of Claim A.2*).* We first present properties satisfied by every intermediate graph $G_{A'}$. Let us now consider a fixed graph $G_{A'}$ and corresponding sets $\{N_i'\}$. There is no cycle in $G_{A'}$ as any arc (x, y) indicates $c_x(A_x') > c_y(A_y')$ and the existence of a cycle (containing some vertex x) implies $c_x(A_x') > c_x(A_x')$, a contradiction. Then consider any three vertices x, y, z. If there are arcs (x, y) and (y, z), then there exists an arc (x, z) as for any $o \in A_x'$, $c_x(A_x' \setminus \{o\}) > c_y(A_y') > c_z(A_z')$. Accordingly, for any $q \in N_2'$, there exists in $G_{A'}$ an arc (p, q) pointing from some $p \in N_1'$ to q. The reason is that vertex q is on some path as there is no cycle, and the starting vertex s of that path has in-degree 0 and positive out-degree and hence $s \in N_1'$. By the above argument, arc (s, q) exists in $G_{A'}$.

Now we are ready to prove Claim A.2. By above arguments, for any $j \in N_2^b$, there exists an arc (j', j) with $j' \in N_1^b$. By the construction of G_{A^b}, arc (j', j) implies $c_{j'}(A_{j'}^b \setminus \{o\}) > c_j(A_j^b)$ for all $o \in A_{j'}^b$. According to Step 4, agent i satisfies $\min_{o \in A_i^b} c_i(A_i^b \setminus \{o\}) \geq \min_{o \in A_q^b} c_q(A_q^b \setminus \{o\})$ for all $q \in N_1^b$, and hence for any $j \in N_2^b$, $\min_{o \in A_i^b \setminus \{o\}} c_i(A_i^b) > c_j(A_j^b)$ holds. Therefore for any $j \in N_2^b$, arc (i, j) exists in G_{A^b}.
For any $j \in N_2^b$, by Claim A.2, it is not hard to verify that she does not violate the condition of EQ1 in allocation A when compared to agent i. Hence, arc (j, i) does not exist in G_A.

We now consider the case where agent i is not chosen at Step 4 in round t, i.e., $A_i^b = A_i$. Let l be the agent receiving an item in round t. For any $j \neq l$, no arc (j, i) exists in G_A as $A_j = A_j^b$. As for agent l, it is clear that $l \in N_b^2$ and thus $c_i(A_i^b) > c_l(A_l^b)$ as $i \in N_1^b$; note if $c_i(A_i^b) \leq c_l(A_l^b)$, vertex i would have a positive in-degree in G_{A^b}, a contradiction. The inequality $c_i(A_i^b) > c_l(A_l^b)$ implies that agent l does not violate EQ1 in A when compared to agent i as $A_i = A_i^b$ and $|A_l \setminus A_l^b| = 1$. Therefore, there is no arc (l, i) in G_A.

A.3 Proof of Theorem 5

Proof (Proof of Theorem 5*).* We present the reduction from the PARTITION problem. Given an arbitrary instance of PARTITION, we construct a fair division instance $\mathcal{I}$ with n agents and a set $\{o_1, \ldots, o_{r+n-1}\}$ of $r + n - 1$ items. The cost functions of agents are shown in Table 3 where K is a sufficiently large number, i.e., $K \gg T$. For any $i \in [n]$, we have $\sum_{j=1}^{r+n-1} c_i(o_j) = (2r + \frac{n-1}{K})T$ and thus the cost functions are normalized.
Suppose we have a "yes" instance of PARTITION and let I_1 and I_2 be a solution. We now consider allocation $A = (A_1, \ldots, A_n)$ where $A_1 = \bigcup_{j \in I_1} o_j$, $A_2 = \bigcup_{j \in I_2} o_j$, $A_n = \{o_{r+1}, o_{r+2}\}$ and for any $3 \leq j \leq n - 1$, bundle A_j contains exactly one item from $\{o_{r+3}, \ldots, o_{r+n-1}\}$. It is not hard to verify that allocation A is EQ1 and achieves social cost $\mathsf{SC}(A) = \frac{(n+1)T}{K}$. Moreover, A is the optimal EQ1 allocation.

For the other direction, suppose we have a "no" instance of PARTITION. We first claim that in any EQ1 allocation B, the bundle of agent n does not include both o_{r+1} and o_{r+2}. Due to "no" instance, assigning

Table 3. The constructed fair division instance for Theorem 5

$c_i(o)$		Items										
		o_1	o_2	$\cdots$	o_r	o_{r+1}	o_{r+2}	o_{r+3}	o_{r+4}	$\cdots$	o_{r+n-1}
Agents	1	$\frac{p_1}{K}$	$\frac{p_2}{K}$	$\cdots$	$\frac{p_r}{K}$	rT	rT	$\frac{T}{K}$	$\frac{T}{K}$	$\cdots$	$\frac{T}{K}$
	2	$\frac{p_1}{K}$	$\frac{p_2}{K}$	$\cdots$	$\frac{p_r}{K}$	rT	rT	$\frac{T}{K}$	$\frac{T}{K}$	$\cdots$	$\frac{T}{K}$
	$\vdots$	$\vdots$	$\vdots$	$\ddots$	$\vdots$	$\vdots$	$\vdots$	$\vdots$	$\vdots$	$\ddots$	$\vdots$
	$n-1$	$\frac{p_1}{K}$	$\frac{p_2}{K}$	$\cdots$	$\frac{p_r}{K}$	rT	rT	$\frac{T}{K}$	$\frac{T}{K}$	$\cdots$	$\frac{T}{K}$
	n	$2T$	$2T$	$\cdots$	$2T$	$\frac{T}{K}$	$\frac{T}{K}$	$\frac{T}{K}$	$\frac{T}{K}$	$\cdots$	$\frac{T}{K}$

$\{o_1, \ldots, o_r, o_{r+3}, \ldots, o_{r+n-1}\}$ to the first $n-1$ agents makes one of them receive a cost less than $\frac{T}{K}$. If agent n receives $\{o_{r+1}, o_{r+2}\}$, then she violates the property of EQ1 when compared to the agent with the least cost. Thus in allocation B, at least one of o_{r+1}, o_{r+2} are assigned to first $n-1$ agents in allocation B. Accordingly for an EQ1 allocation B, its social cost $\mathsf{SC}(B) \geq (r + \frac{n}{K})T$.

By scaling the total cost of every agent to 1, we can claim that it is NP-hard to approximate SC-EQ1 within an additive factor of

$$\frac{r - \frac{1}{K}}{2r + \frac{n-1}{K}} \to \frac{1}{2} \text{ as } K \to +\infty,$$

which yields the desired inapproximation factor in the statement of Theorem 5.

B Missing Proofs in Section 5

B.1 Proof of Theorem 7

Proof (Proof of Theorem 7*).* We first consider the upper bound and show that there always exist EF1 allocations with social cost at most 1. Given an agent i with cost function c_i, suppose $c_i(o_{i_1}) \leq c_i(o_{i_2}) \leq \cdots \leq c_i(o_{i_m})$. We partition O into n bundles in a round-robin way, i.e., $M_k^i = \{o_i, o_{n+i}, o_{2n+i}, \ldots\}$ for $k = 1, \ldots, n$. It is clear that

$$c_i(M_1^i) + c_i(M_2^i) + \cdots + c_i(M_n^i) = 1, \text{ for any } i \in N. \tag{2}$$

Moreover, in a round-robin algorithm where agent i is the k-th to select items, her cost is no greater than $c_i(M_k^i)$. This is because in any round $l \geq 1$ of the algorithm, $l \cdot n + i - 1$ items have been allocated at the moment when it is i's turn to select an item and thus this item's cost cannot be greater than $c_i(o_{i_{l \cdot n+i}})$.

Next, consider n permutations of the agents:

$$\sigma_1 = (1, 2, \ldots n-1, n)$$
$$\sigma_2 = (2, 3, \ldots n, 1)$$
$$\vdots$$
$$\sigma_n = (n, 1, \ldots n-2, n-1).$$

Summing up the n equalities in Eq. (2) and rearranging the terms according to the permutations $\sigma_1, \ldots, \sigma_n$ gives

$$\sum_{k=1}^{n}\sum_{l=1}^{n} c_{\sigma_k(l)}(M_l^{\sigma_k(l)}) = n.$$

Thus

$$\min_{\sigma_1,\ldots,\sigma_n} \sum_{i=1}^{n} c_{\sigma_k(l)}(M_l^{\sigma_k(l)}) \leq 1.$$

Let $A^k = (A_1^k, \ldots, A_n^k)$ be the allocation returned by the round-robin algorithm where the agents are ordered by σ_k. As we have discussed, $c_{\sigma_k(l)}(A_l^k) \leq c_{\sigma_k(l)}(M_l^{\sigma_k(l)})$ for $l = 1, \ldots, n$. Therefore,

$$\min_{\sigma_1,\ldots,\sigma_n} \sum_{i=1}^{n} c_{\sigma_k(l)}(A_l^k) \leq 1.$$

That is, the allocation with the minimum cost among the n allocations returned by the n round-robin algorithms has social cost no greater than 1. Moreover, such an allocation can be found in polynomial time.

As for the lower bound, let us consider an instance with n agents and a set $\{o_1, \ldots, o_{K+1}\}$ of $K+1$ items, where K is a large number and divides n. The cost functions c_i's are as follows;

- for agent 1: $c_1(o_j) = \epsilon$ for all $j \leq K$ and $c_1(o_{K+1}) = 1 - K\epsilon$;
- for agent $i \geq 2$: $c_i(o_j) = \frac{1}{K}$ for all $j \leq K$ and $c_i(o_{K+1}) = 0$;

where $0 < \epsilon \ll \frac{1}{K}$.

In an optimal allocation $S^* = (S_1^*, \ldots, S_n^*)$, agent 1 receives $S_1^* = \{o_1, \ldots, o_K\}$ and agent 2 receives o_{K+1}, resulting in $\mathsf{SC}(S^*) = K\epsilon$. However, agent 1 violates EF1 in S^* when compared to agent $j \geq 3$. In the following, we will show that the social cost of the optimal EF1 allocation $B = (B_1, \ldots, B_n)$ is close to $1 - \frac{1}{n}$. It is not hard to verify that in the optimal EF1 allocation, o_{K+1} is not allocated to agent 1. As the cost functions of agents $2, \ldots, n$ are identical, we can assume that o_{K+1} is allocated to agent 2 in B due to symmetry. Then for every $i \in [n]$, let x_i refers to the number of the first K items allocated to agent i. Since $\epsilon \ll \frac{1}{K}$, the upper bound of x_1 implies the lower bound of the social cost of B. The EF1 condition among agents $2, \ldots, n$ requires that for any $i, j = 2, \ldots, n$, $|x_i - x_j| \leq 1$; EF1 condition among agents $1, 3, \ldots, n$ requires that for any $j = 3, \ldots, n$, $|x_1 - x_j| \leq 1$; EF1 condition between agents 1 and 2 requires $x_2 - x_1 \leq 1$. Given these constraints, we have $x_1 \leq \frac{K+n}{n}$ as $\sum_{j=1}^{n} x_j = K$. Then we have $\sum_{j=2}^{n} x_j \geq \frac{(n-1)K-n}{n}$ and thus $\mathsf{SC}(B) \geq \frac{n-1}{n} - \frac{1}{K}$. Therefore, the cost of fairness regarding to EF1 is at least $\frac{n-1}{n} - \frac{1}{K} - K\epsilon$, close to $1 - \frac{1}{n}$ when $\epsilon \ll \frac{1}{K} \ll \frac{1}{n}$.

B.2 Proof of Theorem 8

Proof (Proof of Theorem 8). We will reduce from the problem of *Partition-into-k-sets*: given a fixed number k and a set $\{p_1, \ldots, p_q\}$ of q positive integers whose sum is kT, can $[q]$ be partitioned into k indices sets $I_1, \ldots, I_k$ such that for every $j \in [k]$, $\sum_{j \in I_j} p_j = T$? It is easy to see that Partition-into-k-sets is NP-complete for every $k \geq 2$, via a polynomial time reduction from PARTITION. Starting from the PARTITION instance, we add $k-2$ integers each of value SUM/2 where SUM refers to the total value of all elements of the PARTITION instance. We have finished the construction of the Partition-into-k-sets instance. One can verify that the Partition-into-k-sets instance is a "yes" instance if and only if the PARTITION instance is a "yes" instance.

We present the reduction from the Partition-into-$(n-1)$-sets problem, known to be NP-complete based on the above argument. Given an arbitrary instance of Partition-into-$(n-1)$-sets, we construct a fair division instance $\mathcal{I}$ with n agents and a set $\{o_1, \ldots, o_{q+r}\}$ of $q+2$ items. The cost functions of agents are shown in Table 4 where K is a sufficiently large number, i.e., $K \gg T$. For any $i \in [n]$, we have $\sum_{j=1}^{q+2} c_i(o_j) = (\frac{n-1}{K} + 2)T$, and thus, cost functions are normalized.

Table 4. The constructed fair division instance in Theorem 8

$c_i(o)$		Items					
		o_1	o_2	$\cdots$	o_q	o_{q+1}	o_{q+2}
Agents	1	$\frac{p_1}{K}$	$\frac{p_2}{K}$	$\cdots$	$\frac{p_q}{K}$	T	T
	2	$\frac{p_1}{K}$	$\frac{p_2}{K}$	$\cdots$	$\frac{p_q}{K}$	T	T
	$\vdots$	$\vdots$	$\vdots$	$\ddots$	$\vdots$	$\vdots$	$\vdots$
	$n-1$	$\frac{p_1}{K}$	$\frac{p_2}{K}$	$\cdots$	$\frac{p_q}{K}$	T	T
	n	$\frac{2K+n-1}{Kn+K}p_1$	$\frac{2K+n-1}{Kn+K}p_2$	$\cdots$	$\frac{2K+n-1}{Kn+K}p_q$	$\frac{2K+n-1}{Kn+K}T$	$\frac{2K+n-1}{Kn+K}T$

Suppose we have a "yes" instance of Partition-into-$(n-1)$-sets. Let $I_1, \ldots, I_{n-1}$ be a solution. We now consider the allocation $A = (A_1, \ldots, A_n)$ where for any $i \in [n-1]$, $A_i = \bigcup_{j \in I_i} o_j$ and $A_n = \{o_{q+1}, o_{q+2}\}$. For every agent $i \in [n-1]$, it is not hard to verify that agent i does not envy other agents. For agent n, since $\min_{o \in A_n} c_n(A_n \setminus \{o\}) = \frac{2K+n-1}{Kn+K}T = c_n(A_i)$ for all $i \in [n-1]$, she also satisfies the condition of EF1 (and hence A is EF1). Moreover allocation A also achieves the minimum social cost $\frac{n-1}{K}T + \frac{4K+2n-2}{Kn+K}T$.

For the other direction, suppose we have "no" instance. We first claim that in an EF1 allocation, the bundle of agent n does not include both o_{q+1}, o_{q+2}. Let $\{S_1, \ldots, S_{n-1}\}$ be an arbitrary $(n-1)$-partition of $\{o_1, \ldots, o_q\}$. Due to "no" instance, $\min_{j\in[n-1]} c_n(S_j) < \frac{2K+n-1}{Kn+K}T$ holds. Accordingly, even when agent 3 receives only two items o_{r+1}, o_{r+2}, she violates the property of EF1 as $\min\{c_n(o_{q+1}), c_n(o_{q+2})\} > \min_{j\in[n-1]} c_n(S_j)$. Thus, in every EF1 allocation, at least one of o_{r+1}, o_{r+2} is not allocated to agent n and therefore the social cost of an EF1 allocation is at least $\frac{n-1}{K}T + \frac{2K+n-1}{Kn+K}T + T$.

By scaling the total cost of every agent to 1, we can claim that it is NP-hard to approximate SC-EF1 within an additive factor of

$$\frac{1 - \frac{2K+n-1}{Kn+K}}{\frac{n-1}{K} + 2} \to \frac{n-1}{2n+2} \text{ as } K \to +\infty,$$

which yields the desired inapproximation factor.

B.3 Proof of Theorem 9

Proof (Proof of Theorem 9). It suffices to prove NP-completeness of the decision version of SC-EF1: given a fair division instance and a threshold value W, does there exist an EF1 allocation with social cost at most W?

We present the reduction from the PARTITION problem. Given an arbitrary instance of PARTITION, we construct a fair division instance $\mathcal{I}$ of 2 agents and a set $\{o_1, \ldots, o_{r+3}\}$ of $r+3$ items. The cost functions of agents are shown in Table 5 where K is a sufficiently large number, i.e., $K \gg T$. The cost functions are normalized as for every $i \in [2]$, $\sum_{j=1}^{r+3} c_i(o_j) = 2T + 8K$. We define the threshold value W as $W = \frac{13}{2}K + 2T$.

Table 5. The constructed fair division instance in Theorem 9

$c_i(o)$		Items						
		o_1	o_2	$\cdots$	o_r	o_{r+1}	o_{r+2}	o_{r+3}
Agents	1	p_1	p_2	$\cdots$	p_r	$6K$	$2K$	0
	2	$\frac{p_1}{2}$	$\frac{p_2}{2}$	$\ldots$	$\frac{p_r}{2}$	$5K$	$\frac{3K+T}{2}$	$\frac{3K+T}{2}$

Suppose we have a "yes" PARTITION instance and let I_1 and I_2 be a solution. We now consider the allocation $A = (A_1, A_2)$ where $A_1 = \{o_{r+3}\} \cup \bigcup_{j\in I_1} o_j$ and $A_2 = \{o_{r+1}, o_{r+2}\} \bigcup_{j\in I_2} o_j$. As $c_1(A_1) < c_1(A_2)$ and $c_2(A_2 \setminus \{o_{r+1}\}) = c_2(A_1)$, allocation A is EF1 and moreover has social cost $\mathsf{SC}(A) = \frac{13}{2}K + 2T = W$.

We now prove the other direction. Suppose we have a "no" instance of PARTITION. Let $B = (B_1, B_2)$ be an allocation with $\mathsf{SC}(B) \le \frac{13}{2}K + 2T$. We will show that allocation B can never be EF1. Since $\mathsf{SC}(B) \leq \frac{13}{2}K + 2T$, it must hold that $o_{r+3} \in B_1$ and $o_{r+1}, o_{r+2} \in B_2$. Moreover as $\mathsf{SC}(B) \leq W$, agent 1 can receive a cost of at most T from the assignment of $o_1, \ldots, o_r$. Due to "no" PARTITION instance, we have $c_1(B_1 \cap \bigcup_{j\in[r]} o_j) < T$, implying $c_2(B_1) < \frac{3}{2}K + T$. Accordingly for agent 2, she receives cost larger than $\frac{T}{2}$ from the assignment of $o_1, \ldots, o_r$, and thus, $c_2(B_2 \setminus \{o_{r+1}\}) > \frac{3}{2}K + T > c_2(B_1)$, violating the condition of EF1. Therefore, no EF1 allocation can have social cost no greater than $W = \frac{13}{2}K + 2T$.

B.4 Proof of Theorem 10

Proof (Proof of Theorem 10*).* We present the reduction from the PARTITION problem. We will prove the statement for four agents and then explain how to extend the proof to any $n \geq 4$ agents. Given an arbitrary PARTITION instance, we construct a fair division instance $\mathcal{I}$ of 4 agents and a set $\{o_1, \ldots, o_{r+3}\}$ of $r + 3$ items. The cost functions of agents are shown in Table 6 where K is a sufficiently large number, i.e., $K \gg T$. Cost functions are normalized as for any $i =\in [4]$, $\sum_{j=1}^{r+3} c_i(o_j) = 3K + 2T$.

Table 6. The constructed fair division instance in Theorem 10

$c_i(o)$		Items						
		o_1	o_2	$\cdots$	o_r	o_{r+1}	o_{r+2}	o_{r+3}
Agents	1	p_1	p_2	$\cdots$	p_r	K	K	K
	2	p_1	p_2	$\cdots$	p_r	K	K	K
	3	p_1	p_2	$\cdots$	p_r	T	T	$3K - 2T$
	4	$\frac{K+2T}{r}$	$\frac{K+2T}{r}$	$\ldots$	$\frac{K+2T}{r}$	K	K	0

Suppose we have a "yes" PARTITION instance and let I_1 and I_2 be a solution. We now consider the allocation $A = (A_1, A_2, A_3, A_4)$ where $A_1 = \bigcup_{j\in I_1} o_j$, $A_2 = \bigcup_{j\in I_2} o_j$, $A_3 = \{o_{r+1}, o_{r+2}\}$ and $A_4 = \{o_{r+3}\}$. It is not hard to verify that A is an EF1 allocation and has social cost $\mathsf{SC}(A) = 4T$, the minimum social cost of an allocation.

We now prove the other direction. Suppose we have a "no" PARTITION instance. Let B be an EF1 allocation. We will prove that $\mathsf{SC}(B) = \Omega(K)$. For a contradiction, assume $\mathsf{SC}(B) = o(K)$. Then in allocation B, both o_{r+1}, o_{r+2} must be assigned to agent 3 and moreover o_{r+3} must be allocated to agent 4. Thus, we have $B_1 \cup B_2 \subseteq \bigcup_{j\in[r]} o_j$. Due to "no" PARTITION instance, we have $\min\{c_1(B_1), c_2(B_2)\} < T$, and hence, $\min_{o\in B_3} c_3(B_3 \setminus \{o\}) = T > \min\{c_3(B_1), c_3(B_2)\}$. Thus, agent 3 does not satisfy EF1. In other words, in an EF1 allocation, the union bundle of agents 1 and 2 should contain at least one item from $\{o_{r+1}, o_{r+2}, o_{r+3}\}$, implying social cost of at least K.

Then it is NP-hard to approximate SC-EF1 with a multiplicative factor of $\frac{K}{4T}$ that tends to $+\infty$ when K goes to $+\infty$. To extend to any constant $n \geq 4$ agents, one can add $n-4$ copies of agent 4.

B.5 Proof of Theorem 11

We next consider the case of two agents and establish an additive approximation scheme for SC-EF1. In particular, we will establish an "equivalence" relationship between the allocations of goods and the allocations of chores when there are two agents. We first introduce terminologies of the goods setting. In the allocations of goods, there is a set of O of items (yielding positive value) to be allocated to a set of agents N. Each agent $i \in N$ is associated with an additive valuation function v_i and $v_i(O) = 1$. To distinguish between instances for goods and instances for chores, we use $\mathcal{I}^+$ and A^+ (resp. $\mathcal{I}^-$ and A^-) to refer to the instance and the allocation for goods (resp. for chores). An allocation $A^+ = (A_1^+, \dots, A_n^+)$ of goods is EF1 if for any $i, j \in N$, there exists $o \in A_j^+$ such that $v_i(A_i^+) \geq v_i(A_j^+ \setminus \{o\})$. The *social welfare* of allocation A^+ is the total value of agents. Denote by $\mathsf{SW}(A^+)$ the social welfare of A^+ and by SW-EF1 the problem of maximizing social welfare subject to EF1 constraint for goods.

When allocating goods to two agents, an FPTAS (multiplicative approximation) for SW-EF1 exists [14]. We shall utilize their results and propose an additive approximation scheme for SC-EF1 by connecting SW-EF1 to our problem.

Lemma 3. *Given a goods instance $\mathcal{I}^+ = \langle \{1,2\}, O, \{v_i\} \rangle$, suppose $A^+ = (A_1^+, A_2^+)$ is an allocation maximizing social welfare subject to (goods) EF1 constraint. Considering the chores instance $\mathcal{I}^- = \langle \{1,2\}, O, \{c_i\} \rangle$ where $c_i(o) = v_i(o), \forall o \in O, \forall i = 1, 2$, allocation $A^- = (A_2^+, A_1^+)$ minimizes social cost subject to (chores) EF1 constraint.*

Proof. Let $B^+ = (B_1^+, B_2^+)$ be an arbitrary EF1 allocation of $\mathcal{I}^+$. Construct $B^- = (B_1^-, B_2^-)$, an allocation of instance $\mathcal{I}^-$, by letting $B_1^- = B_2^+$ and $B_2^- = B_1^+$. Then we have $\mathsf{SC}(B^-) = c_1(B_1^-) + c_2(B_2^-) = v_1(B_2^+) + v_2(B_1^+) = 2 - \mathsf{SW}(B^+)$; the second equality transition is because $c_i(o) = v_i(o), \forall o \in O, \forall i = 1, 2$ and the third equality transition is because $v_i(O) = 1$ for all i. We then prove that allocation B^- is also EF1. Since goods allocation B^+ is EF1, there exists $o' \in B_2^+$ such that $v_1(B_1^+) \geq v_1(B_2^+ \setminus \{o'\})$, equivalent to $c_1(B_1^+) \geq c_1(B_2^+ \setminus \{o'\})$ as $c_i(\cdot) = v_i(\cdot)$. The latter inequality translates to $c_1(B_1^- \setminus \{o'\}) \leq c_1(B_2^-)$ with $o' \in B_1^-$, indicating that agent 1 satisfies (chores) EF1 in allocation B^-. By similar arguments, agent 2 also satisfies EF1 and hence B^- is EF1. We remark that the reverse implication is also true, that is, given an arbitrary EF1 allocation B^- of $\mathcal{I}^-$, the allocation B^+ achieved by swapping bundles of two agents is EF1 and moreover $\mathsf{SC}(B^-) + \mathsf{SW}(B^+) = 2$.

The above argument proves that allocation A^- is EF1. The remaining is to show that A^- minimizes social cost among all EF1 allocations. For a contradiction, let $D^- = (D_1^-, D_2^-)$ be an EF1 allocation of $\mathcal{I}^-$ with social cost $\mathsf{SC}(D^-) <$

$\mathsf{SC}(A^-)$. Then consider allocation $D^+ = (D_2^-, D_1^-)$ of $\mathcal{I}^+$. By the above argument, D^+ is EF1. Note that $\mathsf{SW}(D^+) = 2 - \mathsf{SC}(D^-) > 2 - \mathsf{SC}(A^-) = \mathsf{SC}(A^+)$, violating the fact that A^+ maximizes social welfare among all EF1 allocations. Therefore, allocation A^- minimizes social cost among all EF1 allocations of $\mathcal{I}^-$.

Now we are ready to present the additive approximation scheme for SC-EF1. Denote by OPT^{EF1} the social cost of the optimal EF1 allocation.

Proof (Proof of Theorem 11). There is a known FPTAS algorithm for SW-EF1 when $n = 2$ [14], that is, for any instance $\mathcal{I}^+$ with m items and two agents and for $\epsilon' > 0$, their algorithm runs in time $poly(m, \frac{1}{\epsilon'})$ and computes an EF1 allocation A^+ with $\mathsf{SW}(A^+) \geq (1 - \epsilon')\mathsf{OPT}(\mathcal{I}^+)$ where $\mathsf{OPT}(\mathcal{I}^+)$ is the maximum social welfare of EF1 allocations for $\mathcal{I}^+$.

Given a chores instance $\mathcal{I}^- = \langle 2, O, \{c_i\}\rangle$, we construct a goods instance $\mathcal{I}^+ = \langle 2, O, \{v_i\}\rangle$ where $v_i(o) = c_i(o), \forall o \in O, \forall i = 1, 2$. Implement the FPTAS in [14] with instance $\mathcal{I}^+$ and parameter $\epsilon' = \frac{\epsilon}{2}$. Name the returned EF1 allocation B^+ that has social welfare $\mathsf{SW}(B^+) \geq (1 - \epsilon')\mathsf{OPT}(\mathcal{I}^+)$. Now let us consider allocation $B^- = (B_1^-, B_2^-)$ where $B_1^- = B_2^+$ and $B_2^- = B_1^+$. Then according to the proof of Lemma 3, allocation B^- is EF1 and has a social cost of $2 - \mathsf{SW}(B^+)$. Denote by $\mathsf{OPT}(\mathcal{I}^-)$ the minimum social cost of EF1 allocations for $\mathcal{I}^-$. Again by the proof of Lemma 3, we have $\mathsf{OPT}(\mathcal{I}^-) + \mathsf{OPT}(\mathcal{I}^+) = 2$. Thus, we have the following,

$$\begin{aligned}\mathsf{SC}(B^-) = 2 - \mathsf{SW}(B^+) &\leq 2 - (1 - \epsilon')\mathsf{OPT}(\mathcal{I}^+)\\ &= 2 - \mathsf{OPT}(\mathcal{I}^+) + \frac{\epsilon}{2} \cdot \mathsf{OPT}(\mathcal{I}^+)\\ &\leq \mathsf{OPT}(\mathcal{I}^-) + \epsilon,\end{aligned}$$

where the equality transition is due to $\epsilon' = \frac{\epsilon}{2}$ and the last inequality transition is due to $\mathsf{OPT}(\mathcal{I}^+) \leq 2$.

As for the running time, since converting $\mathcal{I}^-$ to $\mathcal{I}^+$ takes linear time, the total running time of our algorithm is $poly(m, \frac{1}{\epsilon})$.

References

1. Amanatidis, G., et al.: Fair division of indivisible goods: recent progress and open questions. Artif. Intell. **322**, 103965 (2023)
2. Anshelevich, E., Dasgupta, A., Kleinberg, J.M., Tardos, É., Wexler, T., Roughgarden, T.: The price of stability for network design with fair cost allocation. SIAM J. Comput. **38**(4), 1602–1623 (2008)
3. Aziz, H., Huang, X., Mattei, N., Segal-Halevi, E.: Computing welfare-maximizing fair allocations of indivisible goods. Eur. J. Oper. Res. **307**(2), 773–784 (2023)
4. Aziz, H., Moulin, H., Sandomirskiy, F.: A polynomial-time algorithm for computing a pareto optimal and almost proportional allocation. Oper. Res. Lett. **48**(5), 573–578 (2020)

5. Barman, S., Bhaskar, U., Shah, N.: Optimal bounds on the price of fairness for indivisible goods. In: Chen, X., Gravin, N., Hoefer, M., Mehta, R. (eds.) WINE 2020. LNCS, vol. 12495, pp. 356–369. Springer, Cham (2020). https://doi.org/10.1007/978-3-030-64946-3_25
6. Barman, S., Ghalme, G., Jain, S., Kulkarni, P., Narang, S.: Fair division of indivisible goods among strategic agents. In: AAMAS, pp. 1811–1813. International Foundation for Autonomous Agents and Multiagent Systems (2019)
7. Barman, S., Krishnamurthy, S.K.: On the proximity of markets with integral equilibria. In: AAAI, pp. 1748–1755. AAAI Press (2019)
8. Barman, S., Krishnamurthy, S.K., Vaish, R.: Finding fair and efficient allocations. In: EC, pp. 557–574. ACM (2018)
9. Bei, X., Igarashi, A., Lu, X., Suksompong, W.: The price of connectivity in fair division. SIAM J. Discret. Math. **36**(2), 1156–1186 (2022)
10. Bei, X., Lam, A., Lu, X., Suksompong, W.: Welfare loss in connected resource allocation. CoRR abs/2405.03467 (2024)
11. Bei, X., Lu, X., Manurangsi, P., Suksompong, W.: The price of fairness for indivisible goods. Theory Comput. Syst. **65**(7), 1069–1093 (2021)
12. Bertsimas, D., Farias, V.F., Trichakis, N.: The price of fairness. Oper. Res. **59**(1), 17–31 (2011)
13. Brams, S.J., Taylor, A.D.: Fair Division - From Cake-Cutting to Dispute Resolution. Cambridge University Press, Cambridge (1996)
14. Bu, X., Li, Z., Liu, S., Song, J., Tao, B.: On the complexity of maximizing social welfare within fair allocations of indivisible goods. CoRR abs/2205.14296 (2022)
15. Buchem, M., Rohwedder, L., Vredeveld, T., Wiese, A.: Additive approximation schemes for load balancing problems. In: 48th International Colloquium on Automata, Languages, and Programming, ICALP 2021. LIPIcs, vol. 198, pp. 42:1–42:17 (2021)
16. Budish, E.: The combinatorial assignment problem: approximate competitive equilibrium from equal incomes. J. Polit. Econ. **119**(6), 1061–1103 (2011)
17. Caragiannis, I., Ioannidis, S.D.: Computing envy-freeable allocations with limited subsidies. In: Feldman, M., Fu, H., Talgam-Cohen, I. (eds.) WINE 2021. LNCS, vol. 13112, pp. 522–539. Springer, Cham (2022). https://doi.org/10.1007/978-3-030-94676-0_29
18. Caragiannis, I., Kaklamanis, C., Kanellopoulos, P., Kyropoulou, M.: The efficiency of fair division. In: Leonardi, S. (ed.) WINE 2009. LNCS, vol. 5929, pp. 475–482. Springer, Heidelberg (2009). https://doi.org/10.1007/978-3-642-10841-9_45
19. Caragiannis, I., Kaklamanis, C., Kanellopoulos, P., Kyropoulou, M.: The efficiency of fair division. Theory Comput. Syst. **50**(4), 589–610 (2012)
20. Caragiannis, I., Kurokawa, D., Moulin, H., Procaccia, A.D., Shah, N., Wang, J.: The unreasonable fairness of maximum Nash welfare. ACM Trans. Econ. Comput. **7**(3), 12:1–12:32 (2019)
21. Ebadian, S., Peters, D., Shah, N.: How to fairly allocate easy and difficult chores. In: AAMAS, pp. 372–380. International Foundation for Autonomous Agents and Multiagent Systems (IFAAMAS) (2022)
22. Freeman, R., Sikdar, S., Vaish, R., Xia, L.: Equitable allocations of indivisible goods. In: IJCAI, pp. 280–286. ijcai.org (2019)
23. Freeman, R., Sikdar, S., Vaish, R., Xia, L.: Equitable allocations of indivisible chores. In: AAMAS, pp. 384–392. International Foundation for Autonomous Agents and Multiagent Systems (2020)
24. Gamow, G., Stern, M.: Puzzle-Math. Viking Press (1958)

25. Garey, M.R., Johnson, D.S.: Computers and Intractability: A Guide to the Theory of NP-Completeness. W. H. Freeman (1979)
26. Garg, J., Murhekar, A., Qin, J.: Fair and efficient allocations of chores under bivalued preferences. In: AAAI, pp. 5043–5050. AAAI Press (2022)
27. Garg, J., Murhekar, A., Qin, J.: Weighted EF1 and PO allocations with few types of agents or chores. CoRR abs/2402.17173 (2024)
28. Gourvès, L., Monnot, J., Tlilane, L.: Near fairness in matroids. In: ECAI 2014. Frontiers in Artificial Intelligence and Applications, vol. 263, pp. 393–398. IOS Press (2014)
29. Graham, R.L.: Bounds on multiprocessing timing anomalies. SIAM J. Appl. Math. **17**(2), 416–429 (1969)
30. Höhne, F., van Stee, R.: Allocating contiguous blocks of indivisible chores fairly. Inf. Comput. **281**, 104739 (2021)
31. Koutsoupias, E., Papadimitriou, C.: Worst-case equilibria. In: Meinel, C., Tison, S. (eds.) STACS 1999. LNCS, vol. 1563, pp. 404–413. Springer, Heidelberg (1999). https://doi.org/10.1007/3-540-49116-3_38
32. Li, B., Li, Y., Wu, X.: Almost (weighted) proportional allocations for indivisible chores. In: WWW 2022: The ACM Web Conference 2022, Virtual Event, Lyon, France, 25–29 April 2022, pp. 122–131. ACM (2022)
33. Li, Z., Liu, S., Lu, X., Tao, B., Tao, Y.: A complete landscape for the price of envy-freeness. In: AAMAS, pp. 1183–1191. International Foundation for Autonomous Agents and Multiagent Systems/ACM (2024)
34. Lipton, R.J., Markakis, E., Mossel, E., Saberi, A.: On approximately fair allocations of indivisible goods. In: Proceedings 5th ACM Conference on Electronic Commerce (EC-2004), New York, NY, USA, 17–20 May 2004, pp. 125–131. ACM (2004)
35. Steinhaus, H.: The problem of fair division. Econometrica **16**, 101–104 (1948)
36. Suksompong, W.: Fairly allocating contiguous blocks of indivisible items. Discret. Appl. Math. **260**, 227–236 (2019)
37. Sun, A., Chen, B., Doan, X.V.: Equitability and welfare maximization for allocating indivisible items. Auton. Agents Multi Agent Syst. **37**(1), 8 (2023)
38. Sun, A., Chen, B., Doan, X.V.: Fairness criteria for allocating indivisible chores: connections and efficiencies. Auton. Agents Multi Agent Syst. **37**(2), 39 (2023)
39. Sun, A., Li, B.: On the price of fairness in the connected discrete cake cutting problem. In: ECAI. Frontiers in Artificial Intelligence and Applications, vol. 372, pp. 2242–2249. IOS Press (2023)
40. Sun, A., Li, B.: Allocating contiguous blocks of indivisible chores fairly: revisited. In: AAMAS, pp. 1800–1808. International Foundation for Autonomous Agents and Multiagent Systems/ACM (2024)
41. Thomson, W.: Introduction to the theory of fair allocation. In: Handbook of Computational Social Choice, pp. 261–283. Cambridge University Press (2016)
42. Varian, H.R.: Equity, envy and efficiency. J. Econ. Theory **9**, 63–91 (1974)
43. Wu, X., Zhang, C., Zhou, S.: Weighted EF1 allocations for indivisible chores. In: EC, p. 1155. ACM (2023)

Edge Arrival Online Matching: The Power of Free Disposal on Acyclic Graphs

Tianle Jiang and Yuhao Zhang(✉)

Shanghai Jiao Tong University, Shanghai, China
{jiangtianle,zhang_yuhao}@sjtu.edu.cn

Abstract. Online matching is a fundamental problem in the study of online algorithms. We study the problem under a very general arrival model: the edge arrival model. Free disposal is an important notion in the online matching literature, which allows the algorithm to dispose of the previously matched edges. Without free disposal, we cannot achieve any bounded ratio, even with randomized algorithms, when edges are weighted.

Our paper focuses on clarifying the power of free disposal in both the unweighted and the weighted setting. As far as we know, it's still uncertain if free disposal can give us extra leverage to enhance the competitive ratio in the unweighted scenario, even in specific instances such as Growing Trees, where every new edge adds a new leaf to the graph. Our study serves as a valuable initial exploration of this open question. The results are listed as follows:

1. With free disposal, we improve the competitive ratio for unweighted online matching on Growing Trees from $5/9$ to $2/3 \approx 0.66$, and show that the ratio is tight. For Forests, a more general setting where the underlying graph is a forest and edges may arrive in arbitrary order, we improve the competitive ratio from $5/9$ to $5/8 = 0.625$.
2. Both the ratios of $2/3$ and 0.625 show a separation to the upper bound of the competitive ratio without free disposal on Growing Trees (0.5914). Therefore, we demonstrate the additional power of free disposal for the unweighted setting for the first time, at least in the special setting of Growing Trees and Forests.
3. We improve the competitive ratio for weighted online matching on Growing Trees from $1/3$ to $1/2$ using a very simple ordinal algorithm, and show that it is optimal among ordinal algorithms.

Keywords: Online Matching · Edge Arrival · Free Disposal

1 Introduction

Online matching stands as a central problem in the field of online algorithms because of its clean definition and broad applications. These problems come in

Granted by the National Natural Science Foundation of China (NSFC) under Grant No. 62102251.

M. Mavronicolas et al. (Eds.): WINE 2024, LNCS 15534, pp. 591–608, 2026.
https://doi.org/10.1007/978-3-032-08560-3_33

various forms with respect to the arrival models, edge weights, and objectives. In the seminal work of Karp, Vazirani, and Vazirani [24], they propose the Online Bipartite Matching Problem, where one side of the graph is known upfront, and on the other side, vertices arrive one by one with the information of their adjacent edges. The designed online algorithm needs to decide the matching of the arriving vertex immediately and irrevocably, with the objective of maximizing the total cardinality of the matching. We term this arrival model "one-sided vertex arrival" and the objective "unweighted". Regarding the arrival mode, we can further study it under a more generalized setting, such as the "general vertex arrival" [15,33], where all vertices in the graph become online, and the more general "edge arrival", where we receive edges one by one. On the other hand, considering the objective and the edge weight, we will have the special vertex-weighted variant [1,21], the edge-weighted variant, and the AdWords variant [17,23,26].

This paper focuses on the online matching problem under the edge arrival model and discusses both the unweighted and the weighted settings. For simplicity, we call the unweighted problem MCM (Max Cardinality Matching) and the weighted version MWM (Max Weight Matching). An optimal deterministic online algorithm for edge arrival MCM is Greedy, which is 0.5-competitive. Moreover, Gamlath et al. [15] showed that 0.5 is the best competitive ratio we can achieve, even for randomized and fractional algorithms.

If we associate edges with weight, the problem becomes harder. Without making further assumptions, we cannot achieve any bounded competitive ratio in edge arrival MWM, even when the graph is a star (it means that the hardness holds for almost all arrival models). To this end, the notion of "free disposal" (a.k.a., preemption) is introduced. In this setting, the algorithm is allowed to discard any previously matched edge, while a discarded edge cannot be re-matched later. Feigenbaum et al. [14] first studied the edge arrival MWM problem with free disposal and gave a 1/6-competitive deterministic algorithm. Later, the ratio was improved to $3 - 2\sqrt{2} \approx 0.1716$ by [25], and was proved to be optimal among all deterministic algorithms by [32]. Epstein et al. [12] presented a 0.1867-competitive randomized algorithm for edge arrival MWM.

Move back to the unweighted setting, can "free disposal" also be useful for MCM? The hard instance for the edge arrival setting by [15] does not hold when the algorithm can use the power of free disposal. The current best upper bound with disposal is 0.585 by Huang et al. [20], and the current best competitive ratio is only 0.5 by Greedy. The power of free disposal is still unknown, and whether there is an algorithm with a competitive ratio better than 0.5 in the edge arrival setting with free disposal remains open.

In conclusion, there are two main open questions in this line to clarify the power of free disposal in edge arrival online matching.

1. Can we beat 0.5 with free disposal in unweighted edge arrival online matching?
2. What is the best competitive ratio for weighted edge arrival online matching?

As a crucial starting point for answering the two open questions above, we focus on a special arrival model called Growing Trees introduced by Chiplunkar et al. [9]. In this setting, the input graph is a tree at every time point. In other words, each arriving edge "grows" a leaf to the tree. The algorithmic challenge is quite clear: "Should we benefit the new vertex by hurting an old one?" Although more challenges will be introduced when the arrival model becomes more general, it is worth clarifying the best competitive ratio we can achieve under this special setting, because it has been used to obtain some hardness results. Buchbinder et al. [7] constructed a growing tree as hard instance and showed a $\frac{2}{3+1/\phi^2} \approx 0.5914$ upper bound for edge arrival MCM, which also applies to vertex arrival MCM. Moreover, in edge arrival MWM, the $3-2\sqrt{2}$ tight hard instance for deterministic algorithms [32] is also a growing tree. Therefore, we believe that algorithmic improvements in this specialized model may offer valuable insights for the more general models.

Unfortunately, even for Growing Trees, the power of free disposal remains unclear in the literature. In the unweighted setting, Chiplunkar et al. [9] presented a $15/28 \approx 0.535$-competitive randomized algorithm for Growing Trees. Tirodkar et al. [31] generalized the Growing Trees model to Forests (allowing an arriving edge to merge two trees), analyzed the same algorithm as [9] and obtained a competitive ratio of $33/64 \approx 0.515$. Subsequently, Buchbinder et al. [7] presented a $5/9 \approx 0.555$-competitive algorithm for both settings that do not require the ability of free disposal, and no known algorithm with free disposal beats the 0.555 competitive ratio without free disposal. Therefore, it remains open whether there is a separation between the competitive ratio in the setting of "with free disposal" and the upper bound in the setting of "without free disposal".

In the weighted setting, the best randomized algorithm for Growing Trees MWM is 1/3-competitive [31], and there is still a huge gap between the lower and upper bound. In particular, we do not have a stronger upper bound than the unweighted setting.

1.1 Our Results

In this paper, we significantly improve the competitive ratio of MCM with free disposal on Growing Trees to $2/3 \approx 0.66$ and show that it is optimal among all randomized algorithms (even fractional). Remark that it beats the upper bound of 0.5914 in the setting without free disposal. Therefore, it shows a separation between the setting "with free disposal" and "without disposal" on Growing Trees. We also generalize our algorithm to the Forests setting and achieve a competitive ratio of $5/8 = 0.625 > 0.5914$, which also shows the strict power of "free disposal" on Forests.

Theorem 1. *There is a 2/3-competitive randomized algorithm for MCM on Growing Trees.*

Theorem 2. *No randomized (or even fractional) algorithm can achieve a competitive ratio better than 2/3 for MCM on Growing Trees.*

Theorem 3. *There is a 5/8-competitive randomized algorithm for MCM on Forests.*

As an extension of our techniques, we design a very simple 1/2-competitive algorithm for the edge arrival MWM problem on Growing Trees. Interestingly, our algorithm is "ordinal": it only makes comparisons between edge weights and does not need to known the exact values. Previously, research on online ordinal algorithms focused on the random order arrival model, such as various secretary problems [3,4,29] and the two-sided game of Googol [10,27]. We refer readers to [18] for more comprehensive results on this model. Therefore, it is rather surprising that it also applies to our setting with adversarial arrival order. We also prove that 1/2 is the optimal competitive ratio with respect to ordinal algorithms. Note that this part is completely deferred to Sect. 6 due to page limit.

Theorem 4. *There is a* 1/2*-competitive randomized ordinal algorithm for MWM on Growing Trees.*

Theorem 5. *No randomized (or even fractional) ordinal algorithm can achieve a competitive ratio better than* 1/2 *for MWM on Growing Trees.*

Our results and comparison with previous works are listed in Table 1. Note that all the competitive ratios here hold for randomized algorithms, and all hardness results hold for fractional algorithms and thus for randomized algorithms.

Table 1. Our results (marked in red).

		unweighted		weighted	
Settings	Free Disposal	algorithm	hardness	algorithm	hardness
General Graph	without FD	0.5	0.5 [15]	unbounded	unbounded
	with FD	0.5	0.585 [20]	0.1867 [12]	0.585 [20]
Forests	without FD	5/9 [7]	0.5914 [7]	unbounded	unbounded
	with FD	$5/9 \to 5/8$	2/3	0.1867 [12]	2/3
Growing Trees	without FD	5/9 [7]	0.5914 [7]	unbounded	unbounded
	with FD	$5/9 \to 2/3$	2/3	$1/3$ [31] $\to 1/2$	2/3

1.2 Overview of Techniques

The main technical highlights in our paper are a novel lossless rounding framework for solving this problem and an uncertain dual assignment technique for online primal-dual analysis.

All previous works in the Growing Trees and the Forests model [7,9,31] maintain a constant number of different matchings deterministically and consistently output each matching with a certain probability. However, this method has a drawback: since the multiple matchings are strongly correlated with each other, the complexity of the analysis grows as the number of matchings increases. For example, the algorithm for MCM on Growing Trees in [9] maintains 4 matchings, and although it is almost certain that maintaining more than 4 matchings can achieve better performance, it would be very hard to analyze.

Lossless Rounding Framework. We take a different approach to obtain improved results. We propose a novel framework for solving this problem. We first design a lossless rounding scheme, which shows that any fractional algorithm for Growing Trees or Forests that satisfies certain additional constraints can be rounded to a randomized integral algorithm with the same competitive ratio. A similar idea has been proposed by Buchbinder et al. [6] for online bipartite matching. Due to the simplicity of the tree structure, there is no correlation between vertices, so we only need a mild condition to promise a lossless rounding. For growing trees, no additional constraint is needed. The algorithm for Forests only needs to satisfy one additional constraint: if an arriving edge (u, v) merges two trees, let $x(u, v)$ denote the fraction we match to (u, v) and $x(u), x(v)$ denote the already matched fraction of u and v respectively, then the algorithm should satisfy $x(u, v) \leq (1 - x(u))(1 - x(v))$. This constraint allows us to successfully match (u, v) with probability $x(u, v)$, since if u and v belong to different trees when (u, v) arrives, the event that u is already matched and v is already matched are independent.

With the help of lossless rounding, we can focus on designing fractional algorithms, which makes the algorithmic approach conceptually cleaner. Especially for MCM on Growing Trees, we only use a 1-page proof to achieve the optimal competitive ratio of 2/3.

Uncertain Dual Assignment. Our design and analysis of the fractional algorithms mainly build on the online primal dual framework (refer to [5] for details). The main technical task of this framework is a carefully designed dual assignment scheme that maintains a set of dual variables online so that each edge is approximately "covered" (to keep the approximate dual feasibility).

In many previous works, the dual assignments are usually fixed after we make a matching decision [11,17,20,22]. For example, in the randomized primal dual analysis of the Ranking algorithm [11], when we match an offline vertex, the dual assignment is fixed immediately w.r.t. the rank of the offline vertex. However, in our work, we introduce a technique called uncertain dual assignment (for the Forests setting). When we match an edge, we can store some uncertain dual value on the edge, instead of assigning it to a fixed endpoint. This guarantees approximate dual feasibility at the current time. After some future event happens, we gain more knowledge of the graph structure and will then either fix the uncertain dual assignment to one endpoint or dispose of it to comply with the decrement in the primal solution. We note that this technique is not restricted to the tree structure, and may also be applied to other online problems.

1.3 Related Works

The MCM problem has been well studied under different arrival models besides edge arrival. As an extension of the one-sided online bipartite matching, Wang and Wong [33] introduced the general vertex arrival model, where vertices on both sides are revealed one at a time, together with edges to previously revealed neighbors. They presented a 0.526-competitive fractional algorithm. Gamlath et al. [15] presented a $1/2 + \Omega(1)$-competitive randomized algorithm, which essentially rounds online the fractional algorithm by [33]. The current best upper bound for this problem is 0.584 [30]. Other arrival models in literature include the Fully Online Matching [19,20,22,30] and the Online Windowed Matching [2], where each vertex has not only an arrival time but also a departure time.

Bullinger and Romen [8] studied the online coalition formation problem in the general vertex arrival model, where matching is a special case (each coalition has size 2). They gave a 1/6-competitive algorithm, which can actually be improved to $3 - 2\sqrt{2}$ in the same way as [25]. There is a substantial gap between general vertex arrival and one-sided arrival. For MWM with one sided arrival, greedy is 0.5-competitive, and Farbach et al. [13] designed a 0.5086-competitive randomized algorithm, beating the 0.5-barrier for the first time. Their algorithm was later improved to 0.519-competitive [16]. Besides free disposal, there are various other attempts to achieve a constant competitive ratio for online edge-weighted matching. For example, Yan [34] presented a 0.645-competitive algorithm for edge-weighted online stochastic matching, which does not need free disposal. This result is further improved to 0.650 [28].

There is relatively little research on lossless rounding for online matching problems. Devanur et al. [11] showed that lossless online rounding for arbitrary fractional bipartite matching is impossible. However, motivated by the fact that each randomized algorithm induces a fractional algorithm with the same competitive ratio, Buchbinder et al. [6] studies the necessary conditions for lossless rounding. They proposed a set of constraints such that any fractional algorithm that satisfies these constraints admits a lossless rounding scheme. Our rounding scheme differs from theirs in that we have to deal with free disposal.

2 Preliminaries

2.1 Problem Definitions

Edge Arrival Online Matching with Free Disposal. Consider an underlying graph $G = (V, E)$. Edges arrive one at a time, and once an edge arrives, the algorithm must immediately decide whether to match or discard that edge. We allow the algorithm to perform "free disposal": after an edge is matched, the algorithm can discard it at any time, possibly to make room for a newly revealed edge. However, if an edge is discarded in this way, it cannot be matched again afterward. In the MCM problem, the goal is to maximize the cardinality of the matching. In the MWM problem, each edge $e = (u, v)$ is associated with a weight that is also revealed upon its arrival, and is denoted by $w(e)$ or $w(u, v)$. The goal is to maximize the total weight of matched edges.

Growing Trees. Growing Trees is a special case of the edge arrival model. The underlying graph is a tree, and at any time, the edges that have arrived form a tree. In other words, each arriving edge "grows" a leaf to the tree. For convenience of analysis, we will assume that the tree is rooted at one of the endpoints of the first edge. Then, when a new edge (u, v) arrives, where only u is incident to some previous edge, we call u the parent of v, denoted as p_v. To be clear, in the context of Growing Trees, when we mention an edge (u, v), we always assume that u (the left vertex) is the parent vertex of v (the right vertex).

Forests. Forests is a slightly more general arrival model than Growing Trees. Now the underlying graph is a forest, and the edges may arrive in any order. In particular, an arriving edge can connect two trees and merge them into one larger tree.

2.2 Online Primal Dual Framework

The Online Primal Dual Framework has been widely used to analyze the competitive ratio of online matching algorithms [11,13,20,23]. Consider the LP of (edge-weighted) matching and its dual:

$$\begin{aligned} \max: \quad & \sum\nolimits_{(u,v)\in E} x(u,v) \cdot w(u,v) & \min: \quad & \sum\nolimits_{u\in V} \alpha(u) \\ \text{s.t.} \quad & \sum\nolimits_{v:(u,v)\in E} x(u,v) \le 1, \forall u \in V & \text{s.t.} \quad & \alpha(u) + \alpha(v) \ge w(u,v), \forall (u,v) \in E \\ & x(u,v) \ge 0, \forall (u,v) \in E & & \alpha(u) \ge 0, \forall u \in V \end{aligned}$$

Assume we have an algorithm for online matching that always maintains a feasible primal solution x.

Lemma 6. *The algorithm is c-competitive if there is a set of non-negative dual variables α that satisfy the following two conditions at all times:*

- *Reverse weak duality:* $P = \sum_{(u,v)\in E} x(u,v) \cdot w(u,v) \ge \sum_{u\in V} \alpha(u) = D$.
- *Approximate dual feasibility:* $\alpha(u) + \alpha(v) \ge c \cdot w(u,v), \forall (u,v) \in E$.

Following the framework, we can transform the task of proving the competitive ratio into constructing an approximate feasible dual solution.

3 Fractional Algorithms and A Loseless Rounding

3.1 Fractional Algorithm for Matching with Disposal

We use deterministic fractional algorithms as an intermediate step for designing randomized integral algorithms. For each edge $e = (u, v)$, let $x(e)$ or $x(u, v)$ denote the matched fraction of the edge. For each vertex u, let $x(u)$ denote the total matched fraction of its incident edges.

For each arriving edge $e = (u, v)$, the fractional algorithm should match a fraction to e under the matching constraint. That is, the algorithm should keep $x(u) \le 1$ and $x(v) \le 1$. With the ability to have free disposal, the algorithm can decrease the matched fraction of edges at any time. However, the matched fraction of an edge can only increase at its arrival time.

3.2 Lossless Randomized Rounding

Next, we present a rounding framework for Growing Trees and Forests. Under some mild conditions, we prove that the rounding is lossless: given a c-competitive fractional algorithm A_F, we can obtain a c-competitive randomized integral algorithm A_R.

When each edge arrives, A_F essentially performs the following two steps sequentially:

1. Dispose Step: lower the matched fraction of some previously arrived edges.
2. Matching Step: match some fraction to the arriving edge.

We treat the two steps separately. For a clear representation, we assume that the t-th edge arrives at time $2t-1$. The disposal step is performed at a time of $2t-1$, and the matching step is performed at time $2t$. We introduce the following notations for convenience:

- $M(e,t)$ denotes the event that edge e is matched by A_R right after time t.
- $M(u,t)$ denotes the event that vertex u is matched by A_R right after time t.
- $x_t(e)$ denotes the matched fraction of e in A_F right after time t.
- $x_t(u)$ denotes the total matched fraction of edges incident to u in A_F right after time t.

To ensure that the rounding is lossless, we maintain the following invariant in our rounding scheme: at any time t, for each edge e, $\mathbf{Pr}[M(e,t)] = x_t(e)$. If the invariant holds, the expected size of the matching maintained by A_R always equals the size of fractional matching maintained by A_F. Thus the competitive ratio is the same.

We will prove that this property is always held by induction. If the matched fraction of an edge does not change at time t, then the matching state of that edge does not change, and the property still holds trivially. Next, we consider three cases where we change the matched fraction of an edge.

Disposal Step. For an edge e, if $x_{t-1}(e) = \lambda$ and A_F disposes a δ-fraction from e when an edge arrives at time t, then A_R performs the following step:

- If e is matched, dispose it with probability $p = \frac{\delta}{\lambda}$. Otherwise, do nothing.

Therefore, if $\mathbf{Pr}[M(e,t-1)] = x_{t-1}(e) = \lambda$ by induction hypothesis, then

$$\begin{aligned}\mathbf{Pr}[M(e,t)] &= \mathbf{Pr}[M(e,t-1)] \cdot \mathbf{Pr}[e \text{ is not disposed}] \\ &= \lambda \cdot \left(1 - \frac{\delta}{\lambda}\right) = \lambda - \delta = x_t(e).\end{aligned}$$

Match an Edge that Extends a Leaf. In this case, an edge $e = (u,v)$ extends a leaf v to a tree, and A_F matches a γ-fraction to it at time t. Whether to match e depends on whether u is matched. Suppose $x_{t-1}(u) = \mu$, then A_R performs the following step:

– If u is not matched, match e with probability $\frac{\gamma}{1-\mu}$.

Any valid fractional algorithm guarantees $\mu+\gamma \leq 1$, so this probability does not exceed 1. Since at most one edge incident to u can be matched, the probability that u is matched is

$$\mathbf{Pr}[M(u,t-1)] = \sum_{e' \ni u} \mathbf{Pr}[M(e',t-1)] = \sum_{e' \ni u} x_{t-1}(e') = x_{t-1}(u).$$

Therefore the probability that e becomes matched at time t is

$$\begin{aligned}\mathbf{Pr}[M(e,t)] &= \mathbf{Pr}[\overline{M(u,t-1)}] \cdot \mathbf{Pr}[\text{match } e \text{ at time } t \mid \overline{M(u,t-1)}] \\ &= (1-\mu) \cdot \frac{\gamma}{1-\mu} = \gamma = x_t(e).\end{aligned}$$

Merge a Tree with Only One Edge with Another Tree. Suppose there is a tree with only one edge e', and the arriving edge e merges e' with another tree. In this case, we can assume that e arrives first and then e' arrives, both growing a leaf to the other tree, then calculate the matched fractions. To ensure that the desired fraction can always be achieved, we impose a trivial condition on the fractional algorithm: if an arriving edge is not incident to any previous edge (i.e. a new tree with one edge), the algorithm must match fraction 1 to the edge. The corresponding step in A_R goes as follows: first decide whether e and e' are matched (assuming e' arrives last), then dispose e' and match e if necessary.

Match an Edge that Merges Two Trees. In this case, the arriving edge $e = (u, v)$ merges two "non-trivial" trees (with at least 2 edges), and A_F matches a γ-fraction to it at time t. It only happens in the Forests setting. Whether to match e depends on whether u and v are matched. Following our rounding scheme, these two events are independent. Suppose $x_{t-1}(u) = \mu_1$ and $x_{t-1}(v) = \mu_2$, then A_R performs the following step:

– If u and v are both not matched, match e with probability $\frac{\gamma}{(1-\mu_1)(1-\mu_2)}$.

However, a valid fractional algorithm does not guarantee $\frac{\gamma}{(1-\mu_1)(1-\mu_2)} \leq 1$. Therefore, it should be an additional constraint in A_F. Combining with the independent condition, the probability that e becomes matched at time t is

$$\begin{aligned}\mathbf{Pr}[M(e,t)] &= \mathbf{Pr}[\overline{M(u,t-1)}] \cdot \mathbf{Pr}[\overline{M(u,t-1)}] \cdot \frac{\gamma}{(1-\mu_1)(1-\mu_2)} \\ &= (1-\mu_1) \cdot (1-\mu_2) \cdot \frac{\gamma}{(1-\mu_1)(1-\mu_2)} = \gamma = x_t(e).\end{aligned}$$

Finally, we formalize our rounding scheme as the following two lemmas:

Lemma 7. *For online matching with disposal on Growing Trees, if there is a c-competitive fractional algorithm, then there is a corresponding c-competitive randomized algorithm.*

Proof. When we consider growing trees, there is only a disposal step and a matching leaf step. We can keep the lossless invariant without any conditions.

Lemma 8. *For online matching with free disposal on Forests, if there is a c-competitive fractional algorithm with the following two conditions:*

1. *for any isolated edge e,* $x(e) = 1$*,*
2. *at the matching step t of a new edge* (u, v) *that merges two trees,*

$$\frac{x_t(u,v)}{(1 - x_{t-1}(u))(1 - x_{t-1}(v))} \leq 1,$$

then there is a corresponding c-competitive randomized algorithm.

Proof. Compared to growing trees, there may exist edges that connect two different trees. However, when the two conditions hold, we have also shown that $\mathbf{Pr}[M(e,t)] = x_t(e)$ can be satisfied.

4 Online MCM on Growing Trees

In this section, we first present a 2/3-competitive algorithm for MCM on Growing Trees, which proves Theorem 1. The proof of Theorem 2 is deferred to the full version of the paper.

4.1 A 2/3-Competitive Algorithm

We only present the fractional algorithm in this section, and by Lemma 7 we can obtain an equivalent randomized integral algorithm.

In our algorithm, edges can be divided into three types: unmatched, stable (1/3-matched), and unstable (2/3-matched or fully matched). The matched fraction of a stable edge will never change, while the matched fraction of an unstable edge may be decreased to 1/3 in the future. When an edge e arrives, if it is incident to another unstable edge e' then we make e' stable. After that we greedily match the maximum possible fraction to e. See Algorithm 1 for a formal description.

Algorithm 1: Fractional Algorithm for MCM on Growing Trees

When an edge $e = (u, v)$ arrives:
if *u has an incident edge $e' = (u, z)$ such that $x(e') \geq 2/3$* **then**
    Make u stable: $x(e') \leftarrow 1/3$;
end
$x(e) \leftarrow 1 - x(u)$;

We use the Online Primal Dual Framework to analyze the algorithm's competitive ratio. The following is a dual assignment rule that naturally satisfies reverse weak duality, and we will prove that approximate dual feasibility is satisfied for each edge under this rule.

Proposition 9 (Parent-only dual assignment rule). *If the matched fraction of an edge (u, v) is $x(u, v)$, where u is v's parent, then we assign $x(u, v)$ to $\alpha(u)$, while $\alpha(v)$ gets 0. In other words, for each non-leaf vertex u, we have $\alpha(u) = x(u) - x(p_u, u)$. (if u is the root, assume $x(p_u, u) = 0$)*

Lemma 10. *For each non-leaf vertex u, $\alpha(u) \geq 1/3$.*

Proof. Suppose the first child of u is v, then we match at least $2/3$ to (u, v) when it arrives. Then no matter what happens in the future, $x(u, v) \geq 1/3$ always holds. Therefore by Theorem 9 we have $\alpha(u) \geq x(u, v) \geq 1/3$.

Lemma 11. *For each non-leaf vertex u, if it is adjacent to at least one leaf, then $\alpha(u) \geq 2/3$.*

Proof. If u has only one child v, then before we greedily match (u, v), there is at most one stable edge incident to u. Therefore we match at least $2/3$ to (u, v), so $\alpha(u) = x(u, v) \geq 2/3$.

If u has at least two children, suppose the first two arriving children are v_1, v_2. Similar to the previous case, we greedily match at least $2/3$ to (u, v_1), and $x(u, v_1) \geq 1/3$ always holds afterward. Before we greedily match v_2, u has at most 2 stable incident edges, so we match at least $1/3$ to (u, v_2), and $x(u, v_2) \geq 1/3$ always holds afterward. Therefore we have $\alpha(u) \geq x(u, v_1) + x(u, v_2) \geq 2/3$.

Lemma 12. *Algorithm 1 is $2/3$-competitive.*

Proof. Consider the two constraints of the Online Primal Dual Framework:

- Reverse Weak Duality: $P = D$ is satisfied by Theorem 9.
- Approximate dual feasibility: for each edge (u, v), if v is not a leaf, then by Theorem 10, $\alpha(u) + \alpha(v) \geq 1/3 + 1/3 = 2/3$; otherwise if v is a leaf, then by Theorem 11, $\alpha(u) + \alpha(v) \geq 2/3 + 0 = 2/3$.

Therefore Algorithm 1 is $2/3$-competitive.

Finally, we can conclude the proof of Theorem 1.

Proof (Proof of Theorem 1). We apply the rounding from Sect. 3 to Algorithm 1, then by Theorem 12 and Theorem 7 the resulting randomized algorithm is $2/3$-competitive.

5 Online MCM on Forests

In this section, we present a $5/8$-competitive fractional algorithm for MCM with free disposal on Forests, which satisfies the conditions in Theorem 8. Due to page limit, we only present the algorithm and the invariants it aims to maintain and defer the detailed analysis to the full version.

This algorithm is based on the $2/3$-competitive algorithm for Growing Tree in the previous section, which works fine if no edges that merge two trees arrive. But problems start to occur when an arriving edge merges two trees. For example, consider two leaves u and v, following Algorithm 1 and the parent-only dual assignment rule, they can be both 2/3-matched with $\alpha(u) = \alpha(v) = 0$. However, if an edge $e = (u, v)$ arrives, from the dual perspective, we cannot afford to cover e because the primal solution can increase by at most 1/3. To overcome this challenge, we introduce a surplus dual assignment rule on each tree, which means we maintain

$$P = \sum_{(u,v)\in E} x(u,v) \geq \sum_{u\in V} \alpha(u) + C = D + C.$$

As a result, we target to prove $\mathsf{ALG} \geq c \cdot \mathsf{OPT} + C$. It can be done by carefully handling the root vertex because of the special property that the root does not have a parent. In our 5/8-competitive algorithm, we always maintain a 5/8-approximate dual solution on each tree, with a $C = 2/8$ amount of surplus as follows.

Invariant 1 (Surplus Invariant). *For each independent tree, we have a condition stronger than Reverse Weak Duality:* $P \geq D + 2/8$.

Note that the surplus does not affect the competitive ratio on each tree because it is only a constant. However, when we merge trees, we have two copies of surplus on each tree but only need to maintain one copy after merging. Therefore, we can use one copy of the surplus to ensure the approximate dual feasibility of the arriving edge.

Types of Edges. Before introducing our algorithm, we classify edges into three types. A *Growing Edge* is an edge that "grows" a leaf to a tree. For a growing edge (u, v), unless otherwise specified, we assume that u is already in the graph and v is a new vertex, and call u the parent of v. A *Non-Trivial Merging Edge* is an edge that merges two trees (with at least two edges). Conversely, if one of the trees only consists of one edge, the merging edge is called a *Trivial Merging Edge*. Note that we will view the trivial merging edges as growing edges but with a specific rule.

Types of Non-leaf Vertices. Corresponding to the edge types, we classify non-leaf vertices into two types. A non-leaf vertex is *type A* if all its incident edges with non-zero matched fractions are growing edges. Otherwise, a non-leaf vertex is *type B* if at least one of its incident (non-trivial) merging edges has a non-zero matched fraction. Our matching strategy and dual assignment rule always keep the type of a non-leaf vertex unchanged.

Invariant 2. *A non-leaf vertex is fixed to be type A or type B the first time it becomes a non-leaf vertex.*

5.1 Matching Rule for a Growing Edge to a Type A Vertex or a Leaf

In this section, we introduce how we match a growing edge to a type A vertex or a Leaf, which is almost aligned with Algorithm 1. To recall, the difference is that we need to carefully match edges incident to the root to gain a 2/8-surplus for dual assignments. In particular, here, the notion of "root" is slightly different from the Growing Tree setting.

For convenience, the matched fractions of each edge is always a multiple of 1/8, and we will always use 8 as the denominator when describing matched fractions. We call a growing edge e stable if $x(e) \leq 3/8$ and unstable if $x(e) \geq 4/8$. When a growing edge (u, v) arrives, we make each unstable edge adjacent to u (the parent) stable by lowering its matched fraction to 3/8. Notice that when u is the root, it may have two incident 4/8-matched edges. We present details in Algorithm 2.

Algorithm 2: Matching Rule for A Growing Edge to A Type A Vertex or A Leaf

When an growing edge $e = (u, v)$ arrives, and the parent u is currently type A or a Leaf;
if *e is not incident to any previous edge* **then**
    $x(e) \leftarrow 1$;
else if *e is incident to e', and e' is not incident to any previous edge* **then**
    $x(e') \leftarrow 4/8$;
    $x(e) \leftarrow 4/8$;
    regard the common vertex of e and e' as a "root".
else
    for *each growing edge $e' = (u, z)$ or $e' = (z, u)$ where z is also type A, such that $x(e') \geq 4/8$* **do**
        $x(e') \leftarrow 3/8$;
    end
    $x(e) \leftarrow 1 - x(u)$;
end
Update dual assignments using Theorem 13

Similar to the analysis of Algorithm 1, we use a parent-only dual assignment rule (Theorem 13) for each growing edge (also applies if u is type B). To precisely formalize the surplus, we need a specific rule for a root.

Proposition 13 (Parent-only Dual Assignment Rule). *The dual assignment rule for growing edges are:*

- *For each growing edge (u, v), where u is the parent, $\alpha(u)$ gets $x(u, v)$ from (u, v) and $\alpha(v)$ gets nothing.*
- *For the root vertex r, which must be type A, we remove 2/8 from its dual as the surplus. That is, $\alpha(r) = \sum_{u:(r,u)\in E} x(r, u) - 2/8$.*

The motivation of the matching strategy and the dual assignment rule is to maintain the following invariant for Type A vertices.

Invariant 3 (Type A Invariant). *For each type A vertex* u, $\alpha(u) \geq 3/8$, *and if* u *is adjacent to at least one leaf, then* $\alpha(u) \geq 5/8$.

Proposition 14. *A growing edge* e *(both to type A and to type B) is stable if* $x(e) \leq 3/8$. *The following invariants corresponding to the stable property should be kept.*

1. *If a growing edge* e *is stable, we will never decrease* $x(e)$ *in the future.*
2. *If a growing edge* e *is unstable, we may decrease* $x(e)$ *to make it stable. But we always have* $x(e) \geq 3/8$.

Remark for Trivial Merging Edges. The case that a tree to be merged has only two vertices is special: suppose a tree has only two vertices u_1, v_1. The arrival edge is (v_1, v_2), and v_2 is in another tree. We can view (v_1, v_2) arriving first as a growing edge, and then (u_1, v_1) arriving. Algorithm 2 shows that we have matched 1 to (u_1, v_1), so the matched fraction of (u_1, v_1) can be achieved by disposal.

5.2 Type B Vertices and Merging Edges

Before introducing specific rules, we introduce some basic invariants we plan to maintain for type B vertices. We define a stable invariant aligned with Invariant 14.

Invariant 4. *Generally speaking, a vertex becomes stable when it becomes type B. We have the following invariants:*

1. *The dual value of a type B vertex* u*, i.e.,* $\alpha(u)$*, never decreases.*
2. *All growing edges adjacent to a type B vertex are stable.*

On the other hand, we do not have the strong property that $\alpha(u) \geq 3/8$ as type A vertices. Instead, we introduce the following invariant.

Invariant 5. *For each type B vertex,* $\alpha(u) \geq 2/8$. *We call it safe if* $\alpha(u) + 1 - x(u) \geq 5/8$. *If* u *is unsafe, we have* $\alpha(u) + 1 - x(u) = 4/8$, *and there must be an adjacent edge* (u, v) *where we have an uncertain dual assignment of* $1/8$ *stored on* (u, v), *and* v *is also unsafe.*

Uncertain Dual Assignment. Here, the uncertain dual assignment is a special assignment rule when we handle merging edges. For example, we may match a $3/8$-fraction to a merging edge (u, v), but we do not fix the gain of $3/8$ to the dual of u or v immediately. There may be a $1/8$ amount of uncertain dual value temporarily stored on this edge. As a result, we may have $\alpha(u) = \alpha(v) = 2/8$ with an uncertain dual assignment of $1/8$ on (u, v). This edge still has the approximate dual feasibility because $2/8 + 2/8 + 1/8 = 5/8$.

We remark that the rule for growing edges to type A vertices never affects the dual value of type B vertices. So, Invariant 4 and Invariant 5 inductively hold by these rules.

Overview of Remaining Matching Rules. Due to space considerations, we leave a detailed discussion of other merging rules for the full version. Here we only give a brief overview.

First, when growing a leaf to a Type B vertex u, if u is safe, the rule is trivial; otherwise, by Invariant 5 there exists an edge incident to u that stores an *uncertain dual assignment*, and we utilize it to adjust the matching state and make u safe while preserving all required invariants.

Second, for a merging edge, we need to use the additional surplus of 2/8. Recall that both independent trees have one copy of 2/8-surplus before merging, and we only need to maintain one copy of 2/8-surplus after merging. In some easy cases, where we can satisfy approximate dual feasibility by only using the current dual variables and the surplus, such as matching two non-leaves, we do not match any portion of the edge, so that we mainatin every invariant.

However, in other cases, if we must match a non-zero fraction to the merging edge, it is non-trivial to mainatin all the invariants. In the hardest case, when an edge merges two leaves that are originally 3/8-matched and have dual equal to 0, we need to match at least 3/8 to maintain approximate dual feasibility $(3/8 + 2/8(\text{the surplus}) = 5/8)$. However, in this case, we cannot make them all safe (need 3/8-dual for each) by the 5/8 primal increment. To circumvent this problem, we store a 1/8-dual at the edge as the uncertain dual assignment and mark the two leaves unsafe. In the future, if an edge incident to an unsafe vertex arrives, we will find the edge that stores uncertain dual assignment and dispose of it (both the dual and the matching), thereby making both endpoints of the edge safe.

6 Online MWM on Growing Trees

6.1 A Simple Ordinal Algorithm

Following the general framework, we first maintain a fractional matching, which, for the sake of simplicity, only assigns 1/2-fractions to edges. If an edge has a 1/2-fraction, we call it half-matched for short. Decisions of disposal only depend on the comparative relationship between the weights of the two edges. Each vertex can have at most two half-matched incident edges, and when the third edge arrives, the algorithm will dispose of the edge with the lowest weight. The details are presented in Algorithm 3.

Unlike the analysis for MCM, this time, we do not maintain the dual variables or the matching. Instead, we show that it is always possible to construct a set of dual variables that satisfy the two constraints of the Online Primal Dual Framework.

Let U denote the set of non-leaf vertices. For each $u \in U$, define

$$h_u = \arg\max_{v \in child(u)} \{w(u, v)\},$$

i.e., the child of u with the largest edge weight. To break ties, choose the child that arrives the earliest. We refer to edges of the form (u, h_u) as heavy edges.

Algorithm 3: Fractional Algorithm for MWM on Growing Trees

```
When edge e = (u, v) arrives:
if m(u) ≤ 1/2 then
    x(e) ← 1/2;
else
    Suppose e1 = (u, x) and e2 = (u, y) are half-matched,
    Wlog. suppose w(e1) ≤ w(e2).
    if w(e) > w(e1) then
        x(e1) ← 0;
        x(e) ← 1/2;
    else
        x(e) ← 0;
    end
end
```

Lemma 15. *Let* M_F *be the subset of edges* $e \in E$ *such that* $x(e) = 1/2$*, then at any time,*

$$\sum_{e \in M_F} w(e) \geq \sum_{u \in U} w(u, h_u).$$

The proof is deferred to the full version of the paper.

Lemma 16. *Algorithm 3 is* $1/2$*-competitive.*

Proof. We assign $\alpha(u) = \frac{1}{2}w(u, h_u)$ for each $u \in U$, and for each leaf $v \notin U$, simply set $\alpha(v) = 0$. Consider the two constraints of the Online Primal Dual Framework:

1. Reverse weak duality: by Theorem 15,

$$P = \frac{1}{2} \sum_{e \in M_F} w(e) \geq \frac{1}{2} \sum_{u \in U} w(u, h_u) = D.$$

2. Approximate dual feasibility: for each edge (u, v) where u is the parent, we have

$$\alpha(u) + \alpha(v) \geq \alpha(u) = \frac{1}{2} \cdot w(u, h_u) \geq \frac{1}{2} \cdot w(u, v) .$$

Therefore Algorithm 3 is 1/2-competitive. The ratio of 1/2 is tight because we only match a 1/2-fraction to the first arriving edge.

Finally, we conclude the proof of Theorem 4.

Proof (Proof of Theorem 4). We apply the transformation from Sect. 3 to Algorithm 3, then by Theorem 16 and Theorem 7 the resulting randomized algorithm is 1/2-competitive.

6.2 Upper Bound for Ordinal Algorithms

Algorithm 3 is very simple, and we conjecture that better algorithms exist when making full use of the values of edge weights. Nevertheless, we are able to show that it is optimal among ordinal algorithms. The high-level idea of our proof is to construct two instances that are indistinguishable for an ordinal algorithm and show that any fractional ordinal algorithm cannot be simultaneously better than 1/2-competitive on both instances. The technical details are left for the full version.

Acknowledgement. Yuhao Zhang is supported by the National Natural Science Foundation of China (NSFC) under Grant No. 62102251.

References

1. Aggarwal, G., Goel, G., Karande, C., Mehta, A.: Online vertex-weighted bipartite matching and single-bid budgeted allocations. In: SODA, pp. 1253–1264. SIAM (2011)
2. Ashlagi, I., Burq, M., Dutta, C., Jaillet, P., Saberi, A., Sholley, C.: Edge-weighted online windowed matching. Math. Oper. Res. **48**(2), 999–1016 (2023)
3. Babaioff, M., Immorlica, N., Kempe, D., Kleinberg, R.: Matroid secretary problems. J. ACM **65**(6), 35:1–35:26 (2018)
4. Buchbinder, N., Jain, K., Singh, M.: Secretary problems via linear programming. Math. Oper. Res. **39**(1), 190–206 (2014)
5. Buchbinder, N., Naor, J.: The design of competitive online algorithms via a primal-dual approach. Found. Trends Theor. Comput. Sci. **3**(2–3), 93–263 (2009)
6. Buchbinder, N., Naor, J.S., Wajc, D.: Lossless online rounding for online bipartite matching (despite its impossibility). In: SODA, pp. 2030–2068. SIAM (2023)
7. Buchbinder, N., Segev, D., Tkach, Y.: Online algorithms for maximum cardinality matching with edge arrivals. Algorithmica **81**(5), 1781–1799 (2019)
8. Bullinger, M., Romen, R.: Online coalition formation under random arrival or coalition dissolution. In: ESA. LIPIcs, vol. 274, pp. 27:1–27:18. Schloss Dagstuhl - Leibniz-Zentrum für Informatik (2023)
9. Chiplunkar, A., Tirodkar, S., Vishwanathan, S.: On randomized algorithms for matching in the online preemptive model. In: Bansal, N., Finocchi, I. (eds.) ESA 2015. LNCS, vol. 9294, pp. 325–336. Springer, Heidelberg (2015). https://doi.org/10.1007/978-3-662-48350-3_28
10. Correa, J., Cristi, A., Epstein, B., Soto, J.A.: The two-sided game of googol. J. Mach. Learn. Res. **23**, 113:1–113:37 (2022)
11. Devanur, N.R., Jain, K., Kleinberg, R.D.: Randomized primal-dual analysis of RANKING for online bipartite matching. In: SODA, pp. 101–107. SIAM (2013)
12. Epstein, L., Levin, A., Segev, D., Weimann, O.: Improved bounds for online preemptive matching. In: STACS. LIPIcs, vol. 20, pp. 389–399. Schloss Dagstuhl - Leibniz-Zentrum für Informatik (2013)
13. Fahrbach, M., Huang, Z., Tao, R., Zadimoghaddam, M.: Edge-weighted online bipartite matching. In: FOCS, pp. 412–423. IEEE (2020)
14. Feigenbaum, J., Kannan, S., McGregor, A., Suri, S., Zhang, J.: On graph problems in a semi-streaming model. Theor. Comput. Sci. **348**(2–3), 207–216 (2005)

15. Gamlath, B., Kapralov, M., Maggiori, A., Svensson, O., Wajc, D.: Online matching with general arrivals. In: FOCS, pp. 26–37. IEEE Computer Society (2019)
16. Gao, R., He, Z., Huang, Z., Nie, Z., Yuan, B., Zhong, Y.: Improved online correlated selection. In: FOCS, pp. 1265–1276. IEEE (2021)
17. Goel, G., Mehta, A.: Online budgeted matching in random input models with applications to adwords. In: SODA, pp. 982–991. SIAM (2008)
18. Gravin, N., Sun, E., Tang, Z.G.: Online ordinal problems: Optimality of comparison-based algorithms and their cardinal complexity. In: FOCS, pp. 1863–1876. IEEE (2023)
19. Huang, Z., Kang, N., Tang, Z.G., Wu, X., Zhang, Y., Zhu, X.: How to match when all vertices arrive online. In: STOC, pp. 17–29. ACM (2018)
20. Huang, Z., Peng, B., Tang, Z.G., Tao, R., Wu, X., Zhang, Y.: Tight competitive ratios of classic matching algorithms in the fully online model. In: SODA, pp. 2875–2886. SIAM (2019)
21. Huang, Z., Tang, Z.G., Wu, X., Zhang, Y.: Online vertex-weighted bipartite matching: beating 1-1/e with random arrivals. ACM Trans. Algorithms **15**(3), 38:1–38:15 (2019)
22. Huang, Z., Tang, Z.G., Wu, X., Zhang, Y.: Fully online matching II: beating ranking and water-filling. In: FOCS, pp. 1380–1391. IEEE (2020)
23. Huang, Z., Zhang, Q., Zhang, Y.: Adwords in a panorama. In: FOCS, pp. 1416–1426. IEEE (2020)
24. Karp, R.M., Vazirani, U.V., Vazirani, V.V.: An optimal algorithm for on-line bipartite matching. In: STOC, pp. 352–358. ACM (1990)
25. McGregor, A.: Finding graph matchings in data streams. In: Chekuri, C., Jansen, K., Rolim, J.D.P., Trevisan, L. (eds.) APPROX/RANDOM -2005. LNCS, vol. 3624, pp. 170–181. Springer, Heidelberg (2005). https://doi.org/10.1007/11538462_15
26. Mehta, A., Saberi, A., Vazirani, U.V., Vazirani, V.V.: Adwords and generalized online matching. J. ACM **54**(5), 22 (2007)
27. Nuti, P., Vondrák, J.: Secretary problems: the power of a single sample. In: SODA, pp. 2015–2029. SIAM (2023)
28. Qiu, G., Feng, Y., Zhou, S., Wu, X.: Improved competitive ratio for edge-weighted online stochastic matching. In: WINE. LNCS, vol. 14413, pp. 527–544. Springer, Cham (2023)
29. Soto, J.A., Turkieltaub, A., Verdugo, V.: Strong algorithms for the ordinal matroid secretary problem. Math. Oper. Res. **46**(2), 642–673 (2021)
30. Tang, Z.G., Zhang, Y.: Improved bounds for fractional online matching problems. CoRR abs/2202.02948 (2022)
31. Tirodkar, S., Vishwanathan, S.: Maximum matching on trees in the online preemptive and the incremental dynamic graph models. In: Cao, Y., Chen, J. (eds.) COCOON 2017. LNCS, vol. 10392, pp. 504–515. Springer, Cham (2017). https://doi.org/10.1007/978-3-319-62389-4_42
32. Badanidiyuru Varadaraja, A.: Buyback problem - approximate matroid intersection with cancellation costs. In: Aceto, L., Henzinger, M., Sgall, J. (eds.) ICALP 2011. LNCS, vol. 6755, pp. 379–390. Springer, Heidelberg (2011). https://doi.org/10.1007/978-3-642-22006-7_32
33. Wang, Y., Wong, S.C.: Two-sided online bipartite matching and vertex cover: beating the greedy algorithm. In: Halldórsson, M.M., Iwama, K., Kobayashi, N., Speckmann, B. (eds.) ICALP 2015. LNCS, vol. 9134, pp. 1070–1081. Springer, Heidelberg (2015). https://doi.org/10.1007/978-3-662-47672-7_87
34. Yan, S.: Edge-weighted online stochastic matching: Beating 1-1/e. In: SODA, pp. 4631–4640. SIAM (2024)

The Impact of Generative Artificial Intelligence on Market Equilibrium: Evidence from a Natural Experiment

Kaichen Zhang[1], Zixuan Yuan[2], and Hui Xiong[1,3](✉)

[1] Thrust of Artificial Intelligence, The Hong Kong University of Science and Technology (Guangzhou), Guangzhou, China
kzhangbi@connect.ust.hk

[2] Thrust of Financial Technology, The Hong Kong University of Science and Technology (Guangzhou), Guangzhou, China
zixuanyuan@hkust-gz.edu.cn

[3] Department of Computer Science and Engineering, The Hong Kong University of Science and Technology, Hong Kong, China
xionghui@ust.hk

Abstract. Generative artificial intelligence (AI) exhibits the capability to generate creative content akin to human output with greater efficiency and reduced costs. This groundbreaking capability, however, has ignited a debate regarding its potential to displace human creators. In light of these discussions, this paper empirically investigates the impact of generative AI on market equilibrium, in the context of China's leading art outsourcing platform. We overcome the challenge of causal inference by identifying an unanticipated and sudden leak of an advanced image-generative AI as a natural experiment. This leak precipitated a notable reduction in the production costs of anime-style images compared to other genres, thereby providing a unique opportunity for difference-in-differences comparisons. Our analysis shows that the advent of generative AI led to a 64% reduction in average prices, yet it simultaneously spurred a 121% increase in order volume and a 56% increase in overall revenue. This growth is primarily driven by the rising demand for "low-end" personal orders, rather than commercial orders. Moreover, incumbent creators retain the majority of the market share and reap the most benefits of generative AI. Our research highlights the potential of generative AI to benefit all stakeholders across the platform economy, yielding both scholarly contributions and practical implications.

Keywords: Artificial Intelligence · Generative AI · Market Equilibrium · Online Platform · Platform Economy

1 Introduction

Generative artificial intelligence (AI) appears to propel our world closer to the realm of Cyberpunk—a futuristic milieu characterized by the coexistence of

M. Mavronicolas et al. (Eds.): WINE 2024, LNCS 15534, pp. 609–626, 2026.
https://doi.org/10.1007/978-3-032-08560-3_34

"high tech and low life": Generative AI demonstrates its ability to generate text, images, and videos that are similar to content created by human experts âĂŞ but in much less time, at a fraction of the cost, and with amazing creativity [43]. Such advances, however, have raised significant concerns about the potential for job displacement. One notable manifestation is the organized opposition of artists, who have taken to social media to voice their opposition through campaigns such as "NO to AI-generated images" [20].

However, the potential impact of generative AI remains controversial in academia. Distinguished AI scientists such as Turing Award winners Yoshua Bengio, Yann LeCun, and Geoffrey Hinton - recognized with what is often called the "Nobel Prize of computer science" - hold divergent views. While Bengio and Hinton have publicly expressed concern, warning of "the risk of extinction by AI", LeCun dismisses such fears as unfounded, citing his understanding of the limitation of current AI systems. This debate echoes the broader discourse found in the economics literature, which suggests that while information technology may carry the risk of rendering certain jobs obsolete [1,5,6,17,45], it may also benefit workers by increasing their productivity [7,28,29].

To this end, this paper investigates the impact of generative AI on market equilibrium, with empirical data from China's leading art outsourcing platform. Specifically, we seek to answer the following research questions: **RQ1**: In what ways does generative AI shape market equilibrium in terms of average price, order volume, and overall revenue? **RQ2**: What types of demand are eliminated or created by generative AI? **RQ3**: Who emerges as the primary suppliers in the market—are they incumbents or entrants utilizing generative AI?

The main challenge in assessing the impact of generative AI lies in achieving *causal inference*. This involves establishing a true cause-and-effect relationship between generative AI and the focal outcomes, rather than mere correlation. One straightforward approach is to contrast the outcome variable (e.g., the number of orders) after and before the emergence of generative AI. However, this method lacks the capacity for causal inference. For example, the demand for anime images could naturally fluctuate due to the seasonal effect of a new semester for students. This intrinsic variability could potentially bias our estimates of the impact of generative AI, leading to incorrect estimates. Another approach is randomized experiments (or A/B testing), which offer the most compelling evidence for causality. However, using experiments to study the impact of generative AI on markets is not feasible. Unlike individuals, markets cannot be randomly assigned to different groups, which limits the ability to implement generative AI effects only on the treated group while leaving the control group unaffected. Macro-level analyses involving countries, cities, and industries face similar limitations.

To overcome the challenge of causal inference, we identify an unanticipated and sudden leak of a highly proficient image-generative AI as a *natural experiment*, where markets inadvertently encounter experimental and control conditions, resembling a scenario wherein nature orchestrates an experiment.

Among the pioneer platforms in the field, NovelAI stands out as an early entrant, offering text-to-image generative AI services. These services are available

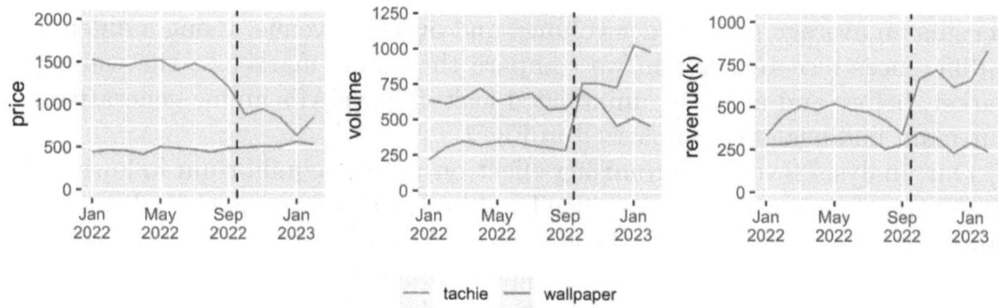

Fig. 1. Model-free evidence. The three graphs show the trajectories of average price, order volume, and overall revenue for the "wallpaper" and "tachie" (anime character paintings) markets. The dashed lines represent the leak moment of an advanced anime generative AI, which only affects only the "tachie" market. The post-leak landscape is intriguing: While the "tachie" market experiences a dip in price, it experiences a surge in both volume and revenue, marking a departure from the intuition of job replacement and illustrating the transformative potential of generative AI for market prosperity.

for a subscription fee ranging from \$10 to \$25 per month per user. Powered by Stable Diffusion [34], NovelAI's AI can quickly generate anime-style images in response to specific keyword-based requests. However, in early October 2022, hackers attacked NovelAI and leaked the AI model and training codes used by NovelAI. As a result, the leaked resources are available for download and local implementation, effectively reducing the cost of image generation to nearly zero.

The inadvertent leak, which originated in the Western world, has inadvertently reverberated throughout the Eastern world. Prior to October 2022, China lacked domestic providers of generative AI, while access to foreign alternatives was also impeded. After October 2022, however, the leaked AI, along with its accompanying tutorials, quickly spread throughout China's social media landscape. As a result, this leak has expeditiously transitioned China from a state of being unaffected by generative AI to one of being affected by it.

This leak served as an ideal natural experiment. First, China constitutes an environment where painting markets remained unaffected by generative AI prior to the leak due to policy restrictions, language restrictions, and payment restrictions. Second, the leaked AI specializes exclusively in generating anime-style images, thus affecting only the anime painting markets while leaving other painting styles unaffected, allowing for a comparison between treated and control units. Third, the unanticipated and sudden nature of the leak is exogenous. This feature ensures that the focal markets are unlikely to have been affected by unobserved confounding factors as stakeholders did not have time to react to the leak in advance. Thus, alternative explanations are effectively ruled out.

We collected 197,068 records from a leading art outsourcing platform in China from January 2022 to February 2023. This platform allows consumers to place orders with specifications and budgets, and artists as freelancers to fulfill them. The orders are segmented into distinct markets, such as anime, UI, live2d, and wallpaper. We transformed these order records into a market-week panel dataset, focusing on average price, order volume, and overall revenue. Using a difference-in-differences research design, we compare the anime market to the wallpaper market, before and after the unintended leak of NovelAI's anime generative AI. Our main findings are summarized in Fig. 1.

Our analysis reveals a promising effect on market equilibrium following the emergence of generative AI. Specifically, although this emergence leads to a 64% reduction in the average price, the generative AI also leads to a remarkable 121% increase in order volume and a consequent 56% increase in overall revenue. All effects are statistically significant at the 1% level. In addition, the effects are also economically significant, due to their substantial magnitude. The results suggest that generative AI can contribute to a more prosperous market, where each stakeholder benefits: consumers get cheaper services, and human creators and the platform get more orders and revenue.

We then examine what types of demand are created by generative AI. Orders on the platform are categorized into personal orders and commercial orders, with different licenses. We conduct a heterogeneous analysis, showing that the growth of generative AI is mainly driven by personal orders, rather than commercial orders. The "tachie" market achieved significantly more revenue through more personal orders at low prices, while the price, volume, and revenue of commercial orders are relatively stable. Our results indicate the potential of generative AI for business success by satisfying the "low-end" needs of individual users.

Finally, we examine who emerges as the primary suppliers in the market. Our results show that the incumbents, who registered before the leak, won 97% of the orders after the leak. As a result, they reap most of the benefits of increased order volume and revenue. The results rule out the possibility that new entrants using generative AI have taken over the market share of real human creators.

Our work provides valuable managerial implications. First, the increased order volume and overall revenue indicate more jobs and taxes, providing favorable economic evidence of generative AI for policymakers to make supportive AI policies. Second, generative AI mainly benefits incumbent creators, suggesting that they should welcome and incorporate generative AI into their workflow to further increase their productivity. Third, online platforms can conduct marketing campaigns and improve services specifically for individual users to better exploit the blue ocean driven by generative AI.

In conclusion, this paper makes the following contributions:

- To the best of our knowledge, we pioneer the estimation of the impact of generative AI on market equilibrium, making scholarly contributions to the growing literature on the economics of (generative) AI.
- We provide promising empirical evidence that generative AI can promote market prosperity and explain the reason. We also provide valuable managerial insights for various stakeholders.
- Our identified natural experiment facilitates causal inference for future research on generative AI—its effects on customer satisfaction, on employment, and beyond.

2 Related Literature

This paper adopts empirical research methods to study generative AI, which is related to the field of generative AI and the economics of AI.

2.1 Generative AI

Generative AI is defined as artificial intelligence (AI) algorithms that generate original outputs based on prompt inputs, which is distinguished from discriminative AI in that generative AI synthesizes the data it has been trained on and creates similar content created by human experts [43].

Generative AI has many applications, which their output format can categorize. For text, the applications include machine translation [3], code generation [38], natural language generation [11], etc. For images, the applications include text-to-image generation [31], style transfer [12], image translation [19], etc. For audio, the applications include text-to-speech [32], music composition [9], singing voice conversion [26], etc.

Generative AI's underlying generative models include Autoregressive Model (AR) [25], Generative Adversarial Network (GAN) [14], Normalizing Flow (Flow) [33], Variational Auto-Encoder (VAE) [22], and Denoising Diffusion Probabilistic Model (Diffusion) [16]. For example, OpenAI's ChatGPT adopts an autoregressive model [4] with Transformer architecture [41]. The focal generative AI in this paper, NovelAI's model, adopts stable diffusion [34].

Stable diffusion is a cutting-edge text-to-image deep learning model that generates images based on textual descriptions. Training Stable Diffusion takes two steps: initially, a large dataset of text-image pairs is used to teach the model the complex relationships between textual descriptions and visual content; subsequently, the model learns to iteratively refine images from noise, given these text inputs.

The technical foundations of stable diffusion are relevant to our work in the following ways. First, stable diffusion can only generate images from the same distribution of its training set. Since NovelAI's model is trained on anime images, it can only generate anime images, allowing for comparison across genres. Second, stable diffusion may struggle with maintaining fidelity to specific details in the text prompts, especially for complex or highly detailed descriptions, resulting in images that are somewhat generic or miss nuanced aspects of the desired output. This limitation can qualitatively explain why generative AI mainly affects low-end orders and cannot replace incumbent artists.

2.2 The Economics of AI

The economics of artificial intelligence (AI) refers to the study of how AI affects markets, organizations, and individuals, and how it changes the way businesses operate and make strategic decisions.

The main research method for studying the economics of AI is empirical research, which meticulously examines the impact of AI adoption on a focal area through the careful analysis of observational data and controlled experiments. Past research has examined the impact of AI on areas such as innovation [2], human resource management [39], sales [23], consumer behavior [37], international trade [13], education [30] and healthcare [44].

A stream of papers on the impact of AI on the labor market [10,42] is closely related to our work. These papers mainly focus on how AI affects the employment and wages of workers, and who is more or less likely to be affected by AI. The literature reveals two contradictory aspects: on the one hand, AI may carry the risk of job displacement [1,5,6,17,45]; on the other hand, AI may also benefit workers by increasing their productivity [7,28,29].

Our paper contributes to the literature on the economics of (generative) artificial intelligence. First, we pioneer the estimation of the impact of generative AI, the new era of AI whose effects are still unclear and understudied. Second, our research target, a marketplace, is unique and provides perspectives from both the supplier and consumer sides, providing more insights into the impact of generative AI. Third, our identified natural experiment facilitates future research on generative AI by identifying an ideal scenario for causal inference.

3 Empirical Strategy

In this section, we compare different empirical methods for estimating the causal effects of generative AI. We apply our analysis under the potential outcome framework [35], the most widely used framework for causal inference.

3.1 Pretest-Posttest Design and Experimental Design

Let us consider the case of analyzing a set of N units, labeled from 1 to N. For each unit i, there are two potential outcomes $\{Y_{1i}, Y_{0i}\}$ depending on whether the unit receives treatment or not ($D_i = \{1, 0\}$). The individual treatment effect for unit i is then defined as $Y_{1i} - Y_{0i}$. In this paper, the units are markets of different art genres, the treatment is the emergence of generative AI and the outcomes are market equilibrium variables, i.e., price, volume, and revenue.

The main challenge in making causal inferences is posed by the fact that there is only one potential outcome that can be observed for each unit. For treated units ($D_i = 1$), we can only observe Y_{1i}; for control units($D_i = 0$), we can only observe Y_{0i}. As a result, individual treatment effects can never be estimated unless strong assumptions about the data generation process are made. As an alternative, researchers estimate the average treatment effect (ATE, $E(Y_{1i} - Y_{0i})$) or the average treatment effect on the treated (ATT, $E(Y_{1i} - Y_{0i}|D_i = 1)$).

We further extend our demonstration to two time periods. In the pretest period ($t = 0$), all units are not treated, where we can observe $E(Y_0(t = 0)|D = 0)$ and $E(Y_0(t = 0)|D = 1)$. In the posttest period ($t = 1$), the treated units ($D = 1$) have received the treatment while the control units ($D = 0$) remain untreated. We can observe $E(Y_0(t = 1)|D = 0)$ and $E(Y_1(t = 1)|D = 1)$. Our goal is to estimate ATE ($E(Y_1(1)) - E(Y_0(1))$) or ATT ($E(Y_1(1)|D = 1) - E(Y_0(1)|D = 1)$).

Note that we still cannot observe another potential outcome ($E(Y_0(1)|D = 1)$ and $E(Y_1(1)|D = 0)$).

A straightforward yet unsophisticated method is the pretest-posttest design, which compares the outcome variable before and after the treatment. Its estimator is $E(Y_1(1)|D = 1) - E(Y_0(0)|D = 1)$. However, ATT equals $E(Y_1(1)|D = 1) - E(Y_0(1)|D = 1) = (E(Y_1(1)|D = 1) - E(Y_0(0)|D = 1)) - (E(Y_0(1)|D = 1) - E(Y_0(0)|D = 1))$, which indicates that the estimator of the pretest-posttest design is unbiased only if $E(Y_0(1)|D = 1) - E(Y_0(0)|D = 1) = 0$, i.e., the outcome variable does not change for the treated units if they do not receive the treatment.

However, the equation does not hold in the context of this study. A plausible explanation for this discrepancy could be the influence of seasonal effects. For example, the demand for anime images could naturally fluctuate after October 2022 due to the start of a new academic semester for students. This intrinsic variability could potentially skew our estimates of the impact of generative AI, leading to underestimates, overestimates, or even fundamentally incorrect estimates. As a result, it's clear that the pretest-posttest design is not an appropriate approach for this study.

Another method is randomized experiments, also called a/b tests in some literature. This method randomly assigns units to a treated group ($D = 1$) and a control group ($D = 0$), ensuring that the population of the treated group is the same as the population of the control group. So we have $E(Y_0(1)|D = 1) = E(Y_0(1)|D = 0)$. We can then derive ATT by subtracting $E(Y_1(1)|D = 1)$ from $E(Y_0(1)|D = 0)$, since we can observe both parts. Also, ATT is equal to ATE in randomized experiments. Thus, randomized experiments are considered the best method for causal inference.

However, randomized experiments are not feasible in this study. While the individual level of analysis allows for the division of subjects into treatment and control groups, the challenge arises at the market level of analysis. Unlike individuals, markets cannot be randomly assigned to different groups, which limits the ability to implement generative AI effects only on the treated group while leaving the control group unaffected. Macro-level analyses involving countries, cities, and industries face similar limitations. In these contexts, researchers have to gain valuable insights from observational data rather than from controlled interventions.

3.2 Difference-in-Differences Design

Difference-in-differences (DID) is arguably the most popular research design in the quantitative social sciences. It can be traced back to 1855 [36] and has also made great strides in recent years [15].

The DID estimator is $E(Y_1(1)|D = 1) - E(Y_0(0)|D = 1) - (E(Y_0(1)|D = 0) - E(Y_0(0)|D = 0))$. DID first calculates the differences between the pretest and posttest periods for each group. DID then takes the difference between these two differences as the estimator.

The DID design is based on the parallel trend assumption that the trends in the outcome variable for the treatment and control groups would have been parallel in the absence of the treatment. Formally, the DID estimator equals ATT if $E(Y_0(1)|D = 1) - E(Y_0(0)|D = 1) = E(Y_0(1)|D = 0) - E(Y_0(0)|D = 0)$. The parallel trend assumption can be deemed satisfied if the treatment and control groups exhibit comparable pre-treatment trends in the outcome variable.

The application of the difference-in-differences to examine the impact of generative AI on market equilibrium hinges on three pivotal conditions: an appropriate treatment that affects treated markets subsequent to the treatment, not prior to it; minimal impact on control markets; and the fulfillment of the parallel trend assumption between treated and control markets.

This study employs the difference-in-differences design, utilizing an unanticipated leak of an advanced generative AI as the treatment. Prior to the leak, the trends of the treated and control markets were stable and parallel, as illustrated in Fig. 1, which lends support to the validity of the parallel trend assumption.

4 Research Context and Data

4.1 The NovelAI "Experiment"

NovelAI is one of the first commercial websites offering generative AI services. NovelAI was initially launched on June 15, 2021, when it only offered AI story-writing services. After Stable Diffusion, a state-of-the-art text-to-image algorithm, was released on March 2, 2022, NovelAI began offering its text-to-image services. NovelAI put a lot of effort into collecting anime-style images and used them to train an anime-style generative AI model based on Stable Diffusion. NovelAI's text-to-image service was implemented on October 3, 2022 [27].

NovelAI's generative AI takes a series of prompts, or keywords, such as "masterpiece, magical girl, fantasy, caustic" as input. The generative AI will generate images according to the prompts. Figure 4 in the Appendix presents the official examples of NovelAI. Note that NovelAI's AI only generates anime-style images because it is trained on anime-style images.

NovelAI's generative AI can generate an image in a few seconds and generate multiple images simultaneously. The cost of generating an image depends on its quality and resolution size. NovelAI uses a monthly subscription scheme ranging from \$10 to \$25. \$10 users can generate approximately 200 images of normal quality per month. \$25 users can generate an unlimited number of images.

However, on October 6, 2022, a few days after NovelAI's text-to-image service was released, hackers attacked NovelAI's server and released its AI model and associated training code. Anyone can download the leaked material via a shared seed and implement the same AI locally, allowing people to generate an image in a few seconds for free.

This leak served as an ideal natural experiment for several reasons. First, China constitutes an environment where painting markets remained unaffected by generative AI before the leak. China did not yet have a local generative AI provider, and people cannot access foreign generative AI providers due to policy restrictions, language restrictions, and payment restrictions. Figure 2 shows the search trend from Baidu, the largest search engine in China. The trend verifies

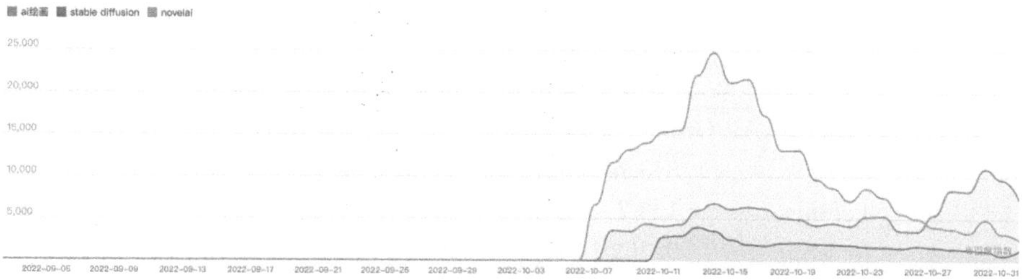

Fig. 2. Search Engine Trend of AI Painting (Blue), Stable Diffusion (Green), and NovelAI (Orange). (Color figure online)

that painting markets had not been affected by generative AI before the leak. Second, the leaked AI exclusively specializes in generating anime-style images, thereby exerting an impact solely on the anime painting markets while leaving other painting styles unaffected, allowing for a comparison between treated and control units. Third, the leak caused by hackers is unanticipated and sudden, which contributes to exogeneity. This attribute ensures that the focal markets are unlikely to have been affected by unobserved confounding factors since stakeholders had no time to react to the leak in advance. As a result, alternative explanations beyond the emergence of generative AI are effectively ruled out.

4.2 Data Description

We collect data from Mihuashi, the leading art outsourcing platform in China. Consumers, as demanders, can post an order on the focal platform, and artists, as producers, can express their interest in accepting the order. The consumer can then select one of the interested artists and enter into a contract with the selected artist. Once the artist has completed the task, the consumer makes the payment. The platform charges a commission of 5% of the price of the order.

An order on the focal platform is comprised of an art category, a price range, and a description. The art category includes "original character", "artwork", "avatar", "Q style", "tachie", "wallpaper", "live2d", "scenario", "cover", "poster", "UI", "character design", and "CG". The price range is defined by a lower and an upper bound, such as 50 to 200 yuan. The order description specifies the requirement, such as "a wallpaper for Vtuber live streaming background".

We collected data on the focal platform from January 2022 to February 2023, which contains 197,068 orders in total. We removed 408 outlier orders whose price is more than 70,000 yuan (about $10,000) because they are too expensive to become real orders. We then aggregate the data to the category-week level. For each category and each week, we calculate the average price (price), order volume (volume), and overall revenue (revenue). We calculate the average price and overall revenue based on the lower bound because the lower bound may better reflect the true willingness of consumers to accept their orders than the upper bound.[1]

[1] https://en.wiktionary.org/wiki/%E7%AB%8B%E3%81%A1%E7%B5%B5# Japanese.

Table 1. Summary Statistics

	Mean	SD	Min	Median	Max
Panel A: All Orders					
Price	843.7	460.7	334.6	628.0	2473.2
Volume	123.7	54.5	12.0	130.0	318.0
Revenue	92362.1	45096.0	6000.0	78950.0	264104.0
Price_ upper	2237.5	869.3	1212.2	1949.5	4689.1
Revenue_ upper	251996.5	103591.0	21550.0	230810.0	656074.0
Panel B: Completed Orders					
Price	885.9	472.3	327.9	721.5	2321.6
Volume	43.2	18.1	1.0	43.0	89.0
Revenue	34718.8	19238.0	500.0	30501.0	111170.0
Price_ upper	2266.1	901.7	1152.0	2010.9	5154.8
Revenue_ upper	92299.1	46196.5	1500.0	81650.0	286370.0

We select the category "tachie" as the treated market. The "tachie" is a Japanese term in the context of anime, manga, and video games that refers to the full-body standing drawing of a character. The "tachie" is precisely the style that NovelAI's AI produces. In contrast, we select the "wallpaper" as the control market. A control unit should be comparable to the treated unit; thus, 11live2d", "UI", and "CG" were excluded from consideration because they are beyond the scope of paintings. With regard to the remaining categories, they are likely to be partially affected by NovelAI's generative AI, because the generative AI may only be useful with careful prompt engineering and post-processing, which could result in an underestimation of the effects. The "wallpaper" may be the best control unit because it is unlikely that NovelAI's generative AI will generate elaborate wallpapers.

Table 1 presents the summary statistics for the category-week panel dataset in this paper. *Price* refers to the average price. *Volume* refers to order volume. *Revenue* refers to the overall revenue. The currency of *Price* and *Revenue* is Chinese yuan (7 yuan = $1). Also, we show the price and revenue based on the upper bound, which is *Price_ upper* and *Revenue_ upper*. Panel A presents the statistical data for all orders, while Panel B displays the statistical data for only completed orders.

5 Main Analysis

In this section, we answer the research question RQ1, by examining the impact of generative AI on market equilibrium in terms of average price, order volume, and overall revenue.

Table 2. The Impact of Generative AI on Market Equilibrium

Dependent Variables:	price	volume	revenue	log(price)	log(volume)	log(revenue)
Model:	(1)	(2)	(3)	(4)	(5)	(6)
Variables						
treated × after	−646.5***	140.3***	56,674.4***	−0.643***	1.21***	0.567***
	(2.0×10^{-12})	(4.9×10^{-13})	(1.7×10^{-10})	(1.8×10^{-15})	(3.8×10^{-15})	(1.5×10^{-15})
Fixed-effects						
category	Yes	Yes	Yes	Yes	Yes	Yes
week	Yes	Yes	Yes	Yes	Yes	Yes
Fit statistics						
Observations	126	126	126	126	126	126
R^2	0.88131	0.89730	0.84262	0.91405	0.93921	0.88433
Within R^2	0.49208	0.79015	0.36957	0.53003	0.83278	0.35453

Clustered (category) standard-errors in parentheses
*Signif. Codes: ***: 0.01, **: 0.05, *: 0.1*

We employ the standard multi-periods model specification of difference-in-differences with two-way fixed effects [8]:

$$y_{it} = \alpha_i + \delta_t + \beta \cdot treated_i after_t + \varepsilon_{it} \tag{1}$$

In this model, the variable y_{it} represents one of our outcome variables by category i in week t. $treated_i$ is a dummy variable indicating whether category i belongs to the treated group ($treated_i = 1$) or the control group ($treated_i = 0$). $after_t$ is also a dummy variable indicating whether week t is after the natural experiment ($after_t = 1$) or before the natural experiment ($after_t = 0$). The model includes a fixed effect δ_t for each week to control for observed and unobserved nonparametric time trends. Additionally, the model includes a fixed effect α_i for each category to account for time-invariant differences between categories. The parameter of interest is β, which captures the impact of generative AI on the focal outcome variable.

Table 2 shows the estimates for the change in price, volume, and revenue in columns (1) to (3), respectively. The coefficient estimate of price is -646, indicating generative AI lowered the average price by 646 yuan. Notably, the coefficient estimates of volume and revenue are 140 and 56,674, showing that generative AI increased the number of orders by 140 per week and increased the total revenue by 56,674 yuan per week. In addition, columns (4) to (6) take the logarithm of price, volume, and revenue and show their percentage change. The results show that price decreased by 64%, volume increased by 121%, and revenue increased by 56%. All effects are statistically significant at the 1% level.

Public concern about generative AI is often rooted in negative anticipation of the economic ramifications associated with its adoption. In particular, there is an understandable concern among various segments, including artists, that generative AI could lead to feelings of displacement, given the fear of rapid and unceremonious replacement after years of dedicated study and creative toil.

However, our empirical findings provide a counterintuitive insight, showing that contrary to initial impressions, generative AI can benefit all stakeholders across the platform economy. Consumers can fulfill their orders 64% cheaper than before, indicating an improvement in consumer welfare. Furthermore, artists receive 56% more revenue, as each monetary transaction completed by consumers is considered revenue for the artist. They receive 121% more orders, which underscores the increased opportunities that accelerate the evolution of their artistic endeavors by providing an invaluable platform for rapid prototyping and iterative refinement. Moreover, the platform's revenue also increases, as their commission is proportional to the total revenue. The substantial increase in order volume could further benefit the platform economy through the network effect [24,40].

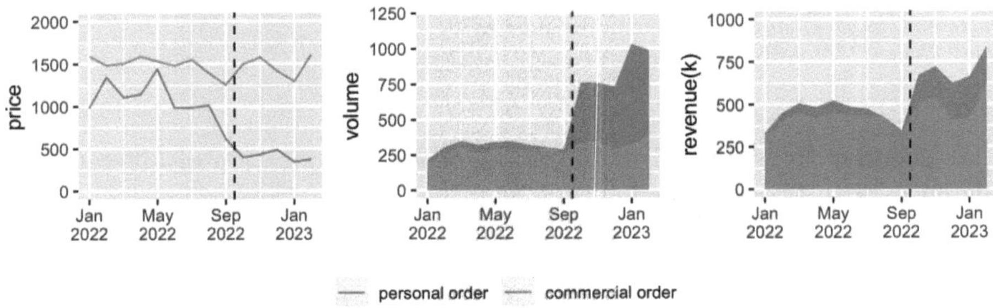

Fig. 3. Market Equilibrium of Personal and Commercial Orders in the "tachie" Market.

6 Demand Side Analysis

In this section, we answer research question RQ2, which aims to identify the specific demand growth that drives market prosperity[2]. To this end, the heterogeneous effects on order licenses are examined.

The orders on the focal platform are classified into two categories based on licenses: personal orders and commercial orders. The artwork of personal orders is for personal use only, whereas the artwork of commercial orders is permitted for commercial purposes. In particular, commercial orders are granted more extensive property rights than personal orders, such as distribution rights and rental rights.

Figure 3 shows the market equilibrium outcomes of personal orders and commercial orders in the "tachie" market. Prior to and following the advent of generative AI, the price, volume, and revenue of commercial orders remain relatively stable. In contrast, personal orders exhibit a substantial increase in volume and revenue with a corresponding decrease in price, which aligns with the overall "tachie" market. This suggests that the growth in the "tachie" market is primarily driven by personal orders.

[2] "Once you start thinking about growth, it's hard to think about anything else", said Robert Lucas, a Nobel Prize-winning economist.

Table 3. The Impact of Generative AI on Market Equilibrium - Heterogeneity by Licenses

Dependent Variables:	price	volume	revenue	log(price)	log(volume)	log(revenue)
Model:	(1)	(2)	(3)	(4)	(5)	(6)
Variables						
personal_treated × after	-746.0**	129.5***	41,244.4***	-0.9892***	2.953***	1.964***
	(17.61)	(1.520)	(195.6)	(0.0101)	(0.0162)	(0.0061)
commercial_treated × after	-50.57***	37.25***	21,115.6***	-0.0967***	0.3903***	0.2936***
	(1.36×10^{-12})	(2.62×10^{-13})	(1.03×10^{-10})	(2.27×10^{-15})	(6.77×10^{-15})	(4.43×10^{-15})
Fixed-effects						
art_category	Yes	Yes	Yes	Yes	Yes	Yes
year_week	Yes	Yes	Yes	Yes	Yes	Yes
Fit statistics						
Observations	185	185	185	185	185	185
R^2	0.78796	0.91291	0.82473	0.79054	0.92775	0.83418
Within R^2	0.59778	0.73866	0.77730	0.63074	0.85464	0.76433

Clustered (art_category) standard-errors in parentheses
*Signif. Codes: ***: 0.01, **: 0.05, *: 0.1*

Formally, we replace the term, *treated*, in Eq. 1 with *personal_treated* and *commercial_treated* to capture heterogeneous treatment effects for different order licenses:

$$y_{it} = \alpha_i + \delta_t + \beta_1 \cdot personal\_treated_i after_t + \beta_2 \cdot commercial\_treated_i after_t + \varepsilon_{it} \tag{2}$$

Table 3 presents the heterogeneous impact of Generative AI with different licenses. The results indicate that both personal and commercial orders exhibit a similar trend with regard to the primary average effects, namely a reduction in price and an increase in volume and revenue. Furthermore, the results demonstrate that the magnitude of the estimators for personal orders is notably higher than that of commercial orders, suggesting that generative AI benefits the "tachie" market by increasing the demand for personal orders with lower prices.

Our findings indicate that, as a blue ocean strategy [21], generative AI can achieve its success by fulfilling the "low-end" needs of individual users, which were previously unmet due to the high cost. In contrast, the requisite quality for commercial orders exceeds the capabilities of current generative AI. For example, there is a notable disparity between the demo images produced by NovelAI (Fig. 4 in Appendix) and successful commercial anime images (such as [18]) that feature considerably more intricate and detailed designs. Conversely, individual users may not require a very high level of quality for their orders. Generative AI can fulfill their demand directly or with a slight modification by human creators. Furthermore, generative AI may present potential copyright issues that limit its commercial applications. This is because the training of generative AI often involves the use of a substantial number of images collected from the Internet without copyright permission.

Table 4. The Impact of Generative AI on Market Equilibrium with Orders Completed by Incumbents

	Panel A: All Completed Orders			Panel B: Incumbent Completed Orders		
Dependent Variables:	log(price)	log(volume)	log(revenue)	log(price)	log(volume)	log(revenue)
Model:	(1)	(2)	(3)	(4)	(5)	(6)
Variables						
treated × after	-0.4915***	1.171***	0.6790***	-0.5127***	1.168***	0.6556***
	(1.33×10^{-15})	(3.51×10^{-15})	(2.02×10^{-15})	(1.5×10^{-15})	(3.1×10^{-15})	(1.92×10^{-15})
Fixed-effects						
art_category	Yes	Yes	Yes	Yes	Yes	Yes
year_week	Yes	Yes	Yes	Yes	Yes	Yes
Fit statistics						
Observations	126	126	126	126	126	126
R^2	0.79706	0.89378	0.73652	0.77257	0.89462	0.73014
Within R^2	0.21484	0.67848	0.19868	0.20654	0.67798	0.17595

Clustered (art_category) standard-errors in parentheses
*Signif. Codes: ***: 0.01, **: 0.05, *: 0.1*

7 Supply Side Analysis

Our previous findings indicate that generative AI has the potential to benefit artists by increasing their order volume and revenue. However, it remains unclear whether incumbents or new entrants utilizing generative AI will reap most of the benefits. In this section, we answer research question RQ3, by examining the incumbents in the market.

We focus on the completed orders in order to observe which artists won the orders. We define incumbents as the group of artists who registered prior to the leak, i.e., before October 2022. We will examine the market equilibrium before and after the leak with the orders completed by incumbents.

We employ the standard difference-in-differences in Eq. (1). Table 4 presents the estimation results. Panel A's dataset comprises all completed orders, whereas Panel B's dataset comprises only orders completed by incumbents. The results in Panel B are consistent with our main effects, indicating that incumbents benefit from generative AI by gaining more orders (116%) and more revenue (65%). Furthermore, the estimators in Panel B is close to those in Panel A, indicating that the majority of the orders were completed by incumbents, resulting in a similar sample for Panel A and B. According to our data, the incumbents won 97.41% of the orders and 97.96% of the revenue after the leak.

One potential explanation for this result is that the current generative AI is unable to fulfill orders without human artists. This is because the generation of content is an iterative process between consumers and artists, with artists required to make ongoing modifications in response to consumer feedback. The current generative AI, which outputs images, audio, and videos, lacks the capacity to make slight modifications independently due to the limitations of its underlying algorithms.

Our results rule out the possibility that new entrants using generative AI have taken over the market share of genuine human creators. In contrast, generative

AI serves to expand the market itself, with the incumbent creators deriving benefits from such expansion. The findings provide empirical evidence of a favorable economic consequence of generative AI for artists and suggest that they should embrace generative AI rather than resist it. The incorporation of generative AI into one's workflow has the potential to increase productivity and leverage the novelty and enhanced capabilities provided by generative AI to attract new customers who previously were not engaging with the art market, thereby "growing the pie".

8 Conclusions

This paper examines the impact of the advent of generative AI on market equilibrium, in terms of price, volume, and revenue. We identify an unanticipated and sudden leak of an advanced image-generative AI as a natural experiment, which facilitates difference-in-differences comparisons for causal inference. The results indicate that generative AI substantially increases order volume and overall revenue with a corresponding decrease in average price. The increase in volume and revenue is primarily driven by the rising demand for personal orders with lower prices, rather than expensive commercial orders. Moreover, the incumbent creators maintained the majority of the market share, reaping most of the benefits from generative AI.

This paper makes a significant scholarly contribution to the growing literature on the economics of AI, by providing one of the first empirical evidence on how generative AI affects the platform economy through diverse perspectives. Moreover, the findings also offer valuable practical insights. Creators should embrace generative AI rather than resist it to gain greater benefits from enhanced productivity. Online platforms can conduct marketing campaigns and improve services specifically for "low-end" personal orders, the emerging demand driven by generative AI. Policymakers may also implement supportive policies for generative AI to create more jobs and tax revenue.

It is our contention that generative AI is having and will continue to have a profound impact on our world. In the context of the marketplace, future research can explore how generative AI affects product quality, consumer satisfaction, and beyond through our identified natural experiment.

Acknowledgments. We would like to thank Reviewer HvRd, QbGb, and Area Chair Y8tc for their constructive feedback. This work was supported by the National Natural Science Foundation of China (Grant No.92370204), National Key R&D Program of China (Grant No.2023YFF0725001), Guangzhou-HKUST(GZ) Joint Funding Program (Grant No.2023A03J0008), Education Bureau of Guangzhou Municipality.

A Appendix

Fig. 4. An illustration of the style and quality of NovelAI's AI-generated images.

References

1. Acemoglu, D., Autor, D.: Skills, tasks and technologies: implications for employment and earnings. In: Handbook of labor economics, vol. 4, pp. 1043–1171. Elsevier (2011)
2. Allam, S.: The impact of artificial intelligence on innovation-an exploratory analysis. Int. J. Creative Res. Thoughts (IJCRT) ISSN pp. 2320–2882 (2016)
3. Bahdanau, D., Cho, K., Bengio, Y.: Neural machine translation by jointly learning to align and translate. arXiv preprint arXiv:1409.0473 (2014)
4. Brown, T., et al.: Language models are few-shot learners. Adv. Neural. Inf. Process. Syst. **33**, 1877–1901 (2020)
5. Brynjolfsson, E., Hui, X., Liu, M.: Does machine translation affect international trade? Evidence from a large digital platform. Manage. Sci. **65**(12), 5449–5460 (2019)
6. Brynjolfsson, E., Mitchell, T., Rock, D.: What can machines learn and what does it mean for occupations and the economy? In: AEA Papers and Proceedings, vol. 108, pp. 43–47. American Economic Association 2014 Broadway, Suite 305, Nashville, TN 37203 (2018)
7. Brynjolfsson, E., Rock, D., Syverson, C.: Artificial intelligence and the modern productivity paradox. Econ. Artif. Intell. Agenda **23**, 23–57 (2019)
8. Callaway, B., Sant Anna, P.H.: Difference-in-differences with multiple time periods. J. Econometrics **225**(2), 200–230 (2021)
9. Fernández, J.D., Vico, F.: AI methods in algorithmic composition: a comprehensive survey. J. Artif. Intell. Res. **48**, 513–582 (2013)
10. Frank, M.R., et al.: Toward understanding the impact of artificial intelligence on labor. Proc. Natl. Acad. Sci. **116**(14), 6531–6539 (2019)

11. Gatt, A., Krahmer, E.: Survey of the state of the art in natural language generation: core tasks, applications and evaluation. J. Artif. Intell. Res. **61**, 65–170 (2018)
12. Gatys, L.A., Ecker, A.S., Bethge, M.: Image style transfer using convolutional neural networks. In: Proceedings of the IEEE Conference on Computer Vision and Pattern Recognition, pp. 2414–2423 (2016)
13. Goldfarb, A., Trefler, D., et al.: Artificial intelligence and international trade. Econ. Artif. Intell.: Agenda 463–492 (2019)
14. Goodfellow, I., et al.: Generative adversarial nets. In: Advances in Neural Information Processing Systems, vol. 27 (2014)
15. Goodman-Bacon, A.: Difference-in-differences with variation in treatment timing. J. Econometrics **225**(2), 254–277 (2021)
16. Ho, J., Jain, A., Abbeel, P.: Denoising diffusion probabilistic models. Adv. Neural. Inf. Process. Syst. **33**, 6840–6851 (2020)
17. Horton, J.J., Tambe, P., Wharton, U.P.: The death of a technical skill. Unpublished Manuscript, p. 299 (2019)
18. Hoyoverse: Genshin impact characters (2024). https://genshin.hoyoverse.com/en/character/liyue?char=12
19. Isola, P., Zhu, J.Y., Zhou, T., Efros, A.A.: Image-to-image translation with conditional adversarial networks. In: Proceedings of the IEEE Conference on Computer Vision and Pattern Recognition, pp. 1125–1134 (2017)
20. Jonas, J.: No to AI-generated images (2022). https://twitter.com/JoJoesArt/status/1603071104528023553
21. Kim, W.C., Mauborgne, R.A.: Blue ocean strategy, expanded edition: how to create uncontested market space and make the competition irrelevant. Harvard Business Review Press (2014)
22. Kingma, D.P., Welling, M.: Auto-encoding variational bayes. arXiv preprint arXiv:1312.6114 (2013)
23. Luo, X., Qin, M.S., Fang, Z., Qu, Z.: Artificial intelligence coaches for sales agents: caveats and solutions. J. Mark. **85**(2), 14–32 (2021)
24. McIntyre, D.P., Srinivasan, A.: Networks, platforms, and strategy: emerging views and next steps. Strateg. Manag. J. **38**(1), 141–160 (2017)
25. Model, A.: System identification and. In: Instrumentation, Control and Automation of Water and Wastewater Treatment and Transport Systems: Proceedings of the 5th IAWPRC Workshop Held in Yokohama and Kyoto, Japan, 26 July–3 August 1990, p. 121. Elsevier (2013)
26. Mohammadi, S.H., Kain, A.: An overview of voice conversion systems. Speech Commun. **88**, 65–82 (2017)
27. NovelAI: Image generation announcement (2022). https://blog.novelai.net/image-generation-announcement-807b3cf0afec
28. Noy, S., Zhang, W.: Experimental evidence on the productivity effects of generative artificial intelligence. Science **381**(6654), 187–192 (2023)
29. Peng, S., Kalliamvakou, E., Cihon, P., Demirer, M.: The impact of AI on developer productivity: evidence from github copilot. arXiv preprint arXiv:2302.06590 (2023)
30. Popenici, S.A.D., Kerr, S.: Exploring the impact of artificial intelligence on teaching and learning in higher education. Res. Pract. Technol. Enhanc. Learn. **12**(1), 1–13 (2017). https://doi.org/10.1186/s41039-017-0062-8
31. Ramesh, A., et al.: Zero-shot text-to-image generation. In: International Conference on Machine Learning, pp. 8821–8831. PMLR (2021)
32. Ren, Y., et al.: Fastspeech: fast, robust and controllable text to speech. In: Advances in Neural Information Processing Systems, vol. 32 (2019)

33. Rezende, D., Mohamed, S.: Variational inference with normalizing flows. In: International Conference on Machine Learning, pp. 1530–1538. PMLR (2015)
34. Rombach, R., Blattmann, A., Lorenz, D., Esser, P., Ommer, B.: High-resolution image synthesis with latent diffusion models. In: Proceedings of the IEEE/CVF Conference on Computer Vision and Pattern Recognition, pp. 10684–10695 (2022)
35. Rubin, D.B.: Causal inference using potential outcomes: design, modeling, decisions. J. Am. Stat. Assoc. **100**(469), 322–331 (2005)
36. Snow, J.: On the mode of communication of cholera. Edinb. Med. J. **1**(7), 668 (1856)
37. Sun, C., Shi, Z.J., Liu, X., Ghose, A., Li, X., Xiong, F.: The effect of voice AI on consumer purchase and search behavior. NYU Stern Sch. Bus. (2019)
38. Svyatkovskiy, A., Deng, S.K., Fu, S., Sundaresan, N.: Intellicode compose: code generation using transformer. In: Proceedings of the 28th ACM Joint Meeting on European Software Engineering Conference and Symposium on the Foundations of Software Engineering, pp. 1433–1443 (2020)
39. Tambe, P., Cappelli, P., Yakubovich, V.: Artificial intelligence in human resources management: challenges and a path forward. Calif. Manage. Rev. **61**(4), 15–42 (2019)
40. Uzzi, B.: The sources and consequences of embeddedness for the economic performance of organizations: the network effect. Am. Sociol. Rev. 674–698 (1996)
41. Vaswani, A., et al.: Attention is all you need. In: Advances in Neural Information Processing Systems, vol. 30 (2017)
42. Webb, M.: The impact of artificial intelligence on the labor market. Available at SSRN 3482150 (2019)
43. Wessel, M., Adam, M., Benlian, A., Thies, F.: Generative AI and its transformative value for digital platforms
44. Wolff, J., Pauling, J., Keck, A., Baumbach, J.: The economic impact of artificial intelligence in health care: systematic review. J. Med. Internet Res. **22**(2), e16866 (2020)
45. Yilmaz, E.D., Naumovska, I., Aggarwal, V.A.: AI-driven labor substitution: evidence from Google translate and chatgpt (2023)

Extended Abstracts

Generalized Principal-Agency: Contracts, Information, Games and Beyond

Jiarui Gan[1], Minbiao Han[2], Jibang Wu[2](✉), and Haifeng Xu[2]

[1] University of Oxford, Oxford OX1 2JD, UK
jiarui.gan@cs.ox.ac.uk
[2] The University of Chicago, Chicago, IL 60637, USA
{minbiaohan,wujibang,haifengxu}@uchicago.edu

Abstract. This paper introduces a generalized principal-agent model where agents have private information (type) and make private decisions (action), both of which are unobservable to the principal. The model unifies several important design problems in economics as special cases including contract design, information design, and Stackelberg games. We first extend the revelation principle to this general model, based on which a polynomial-time algorithm is then derived for computing the optimal mechanism for the principal. This algorithm not only implies new efficient solutions simultaneously for all the aforementioned special cases but also significantly simplifies previously known algorithms designed for special cases. We then turn to constrained economic design in our general principal-agent model, motivated by similar recent interest in specific problems such as constrained contract design. Interestingly, in contrast to the above *unification*, our results here illustrate the other facet of *diversity* among different principal-agent design problems and demonstrate how their different structures can lead to different complexities: some are tractable whereas others are APX-hard. A full version of this paper can be found at https://arxiv.org/abs/2209.01146.

Keywords: Principal-agent Problem · Information Design · Contract Design · Stackelberg Game

M. Mavronicolas et al. (Eds.): WINE 2024, LNCS 15534, p. 629, 2026.
https://doi.org/10.1007/978-3-032-08560-3

Interconnected Conflict

Marcin Dziubiński[1(✉)], Sanjeev Goyal[2], and Junjie Zhou[3]

[1] Institute of Informatics, University of Warsaw, Warsaw, Poland
m.dziubinski@mimuw.edu.pl
[2] University of Cambridge & NYUAD, Abu Dhabi, United Arab Emirates
sg472@cam.ac.uk
[3] School of Economics and Management, Tsinghua University, Beijing, China

We consider a problem of conflict with multiple battlefields with the possibility of spillovers of investments between the battlefields. There are two parties facing a number of battles. Each battle has a value, common to the two parties, and an investment on a battle may have an indirect effect on other battles. The degree of this effect varies across the battles. The parties take these investment externalities into account when choosing their investment levels. The probability of winning the conflict at a given battlefield is determined by a contest success function (CSF) and depends on the amount of direct investment to the battlefield *plus* the amount of indirect investment spilling over from other battlefields. We study Nash equilibrium of the conflict game and in particular how the network of spillovers affects the allocation of resources and the outcomes for the two parties.

We start with a preliminary observation that Nash equilibrium exists and is (generically) unique. Let us define the effective investment in a battle as the sum of direct investment in that battle plus the investment in neighbouring battles. We show that in equilibrium each battlefield receives a positive effective investment from both players. However, in general, only a subset of battlefields receive positive direct investment. The set of battlefields receiving zero investment is the same for each of the two players.

We then turn to the effects of networks on allocations and earnings and obtain a *network invariance result*: equilibrium payoffs, equilibrium total expenditure, and equilibrium probabilities of winning individual battlefields are independent of the network of spillovers. This means that as we move across networks, the players adjust their equilibrium investment strategies to keep these quantities unchanged.

Marcin Dziubiński's work was supported by the Polish National Science Centre through grant 2018/29/B/ST6/00174.

M. Mavronicolas et al. (Eds.): WINE 2024, LNCS 15534, pp. 630–631, 2026.
https://doi.org/10.1007/978-3-032-08560-3

In equilibrium, individual efforts vary across nodes in a network. Our final result provides a characterization of allocations across battles using a network index. The vector of equilibrium investment levels is obtained by scaling this index by a factor that depends on the costs of investments and the CSF. The network index is a solution to a non-linear optimization problem defined only by the network of spillovers. We show that the index satisfies neighbourhood inclusion, a minimal property for a network index to be a network centrality.

Full version of the paper is available at:

http://www.mimuw.edu.pl/~amosild/interconnectedconflict-full.pdf

Gradual Matching with Affirmative Action

Kriti Manocha[1] and Bertan Turhan[2(✉)]

[1] Shiv Nadar Institute of Eminance, Greater Noida, India
kritimanocha18@gmail.com
[2] Iowa State University, Ames, USA
bertan@iastate.edu

Abstract. We examine multi-period matching problems with complex affirmative action constraints, motivated by admissions to technical colleges in India. It extends the framework of [2] to settings where institutions' choice rules incorporate diversity constraints and are updated consistently across periods. Students are allowed to update their ROLs in each stage. It is crucial in such a setting that each student is matched to a program that is weakly better than their match in the previous stage, with respect to their current-stage ROLs (*monotone gradual mechanism*). We show that students cannot be worse off in subsequent stages if and only if the preference update rules are *regular*. We introduce a "backward-looking" notion of stability for gradual matching mechanisms adjusted for affirmative action constraints (gradual stability). We show that a mechanism is gradually stable if and only if it is monotone and the stage mechanism is stable.

We provide theoretical validation for the claims made by [1] that the Multi-Round Deferred Acceptance (MRDA)(The MRDA mechanism is a version of the iterative deferred acceptance algorithm introduced in [3] and [4].) they introduced improves students' welfare across rounds. This admissions process implement a complex affirmative action policy with both vertical reservations (for Scheduled Castes, Scheduled Tribes, Other Backward Classes, and Economically Weaker Sections) and horizontal reservations (for women within each vertical category). Institutions' choice rules are modeled using Position-Specific Priorities (PSP) structure. The MRDA mechanism allows candidates to update their choices in each round by selecting options like "freeze," "float," "slide," or "withdraw/reject." The paper proves that the MRDA mechanism is both monotone and gradually stable.

The full paper is available at https://sites.google.com/site/bertanturhan/

References

1. Baswana, S., Chakrabarti, P.P., Chandran, S., Kanoria, Y., Patange, U.: Centralized admissions for engineering colleges in India. INFORMS J. Appl. Analytics **45**(5), 338–354 (2019)

M. Mavronicolas et al. (Eds.): WINE 2024, LNCS 15534, pp. 632–633, 2026.
https://doi.org/10.1007/978-3-032-08560-3

2. Haeringer, G., Iehlé, V.: Gradual college admission. J. Econ. Theory **198**, 105378 (2021)
3. Manjunath, V., Turhan, B.: Two school systems, one district: what to do when a unified admissions process is impossible. Games Econ. Behav. **95**, 25–40 (2016)
4. Turhan, B.: Welfare and incentives in partitioned school choice markets. Games Econ. Behav. **113**, 199–208 (2019)

Market Design for Capacity Sharing in Networks

Saurabh Amin[1], Patrick Jaillet[1], Haripriya Pulyassary[2], and Manxi Wu[2(✉)]

[1] Laboratory for Information and Decision Systems, MIT, 77 Massachusetts Ave, Cambridge, MA 02139, USA
{amins,jaillet}@mit.edu

[2] School of Operations Research and Information Engineering, Cornell University, Rhodes Hall, Ithaca, NY 14850, USA
{hp297,manxiwu}@cornell.edu

In transportation and logistics networks, inefficiencies arise from congestion due to the limited capacity of infrastructure. This paper introduces a market-based approach that incentivizes agents to form coalitions and share network resources. The market sets prices for each unit of capacity on network edges. Agents form coalition to select a path to use, and split the total cost of using that path among themselves. A market equilibrium is defined by edge prices, coalition formation and agent payments that satisfy individual rationality, stability, budget balance, and market clearing.

Our main results are summarized in Table 1, which shows the conditions for equilibrium existence based on network topology and agents' preferences. We demonstrate that a market equilibrium is guaranteed to exist in series-parallel networks with homogeneous capacity-sharing disutilities. In our framework, market equilibrium existence is closely tied to zero integrality gap of the linear relaxation of the integer resource sharing program. This connection allows us to show that in series-parallel networks, an integer route flow computed by greedy algorithm is the flow induced by the optimal resource sharing regardless of agents' preferences and coalition formation. Then, the efficient coalition formation and payment mechanisms relies on the homogeneous resource sharing disutility parameter. For non-series-parallel networks or agents with general preferences, more complex pricing mechanisms, such as path-based pricing, are required. We also show the existence of market equilibrium that is equivalent to the VCG mechanism, and extend all results to multi-period settings.

(The full paper can be found at https://arxiv.org/abs/2307.03994)

M. Mavronicolas et al. (Eds.): WINE 2024, LNCS 15534, pp. 634–635, 2026.
https://doi.org/10.1007/978-3-032-08560-3

Table 1. Conditions for market equilibrium existence

	Series-parallel network	**Non-series-parallel network**
Homogeneous capacity sharing disutility	Edge-based pricing same for all agents (Theorem 2)	Path-based pricing same for all agents (Proposition 4)
General preferences	Edge-based and population-specific pricing (Proposition 4)	Path-based and population-specific pricing (Proposition 4)

Mechanism Design for Exchange Markets

Yusen Zheng[1,2](✉), Yukun Cheng[3](✉), Chenyang Xu[4](✉), and Xiaotie Deng[1,2](✉)

[1] CFCS, School of Computer Science, Peking University, Beijing 100871, China
[2] PKU-WUHAN Institute for Artificial Intelligence, Wuhan 430075, China
yusen@stu.pku.edu.cn , xiaotie@pku.edu.cn
[3] School of Business Jiangnan University, Wuxi 214122, China
ykcheng@amss.ac.cn
[4] Software Engineering Institute, East China Normal University, Shanghai, China
cyxu@sei.ecnu.edu.cn

Abstract. Exchange markets are a significant type of market economy, in which each agent holds a budget and certain (divisible) resources available for trading. Most research on equilibrium in exchange economies is based on an environment of completely free competition. However, the orderly operation of markets in reality also relies on effective economic regulatory mechanisms. This paper initiates the study of the mechanism design problem in exchange markets, exploring the potential to establish truthful market rules and mechanisms. This task poses a significant challenge as unlike auctioneers in auction design, the mechanism designer in exchange markets lacks centralized authority to fully control the allocation of resources.

In this paper, the mechanism design problem is formalized as a two-stage game. In stage 1, agents submit their private information to the manager, who then formulates market trading rules based on the submitted information. In stage 2, agents are free to engage in transactions within these rules, ultimately reaching an equilibrium. We generalize the concept of liquid welfare from classical budget-feasible auctions and use *market liquid welfare* as a measure to evaluate the performance of the designed mechanism. Moreover, an extra concept called *profitability* is introduced to assess whether the market is money-making (profitable) or money-losing (unprofitable). Our goal is to design a truthful mechanism that achieves an (approximate) optimal welfare while minimizing unprofitability as much as possible. Two mechanisms for the problem are proposed. The first one guarantees truthfulness and profitability while approaching an approximation ratio of 1/2 in large markets. The second one is also truthful and achieves 1/2 approximation in general markets

Full version of the paper can be found at https://arxiv.org/abs/2410.07023. This work is supported by Wuhan East Lake High-Tech Development Zone (also known as the Optics Valley of China, or OVC) National Comprehensive Experimental Base for Governance of Intelligent Society, NSFCs (No. 12471339, No. 62302166 and No. 62172012), and the Key Laboratory of Interdisciplinary Research of Computation and Economics (SUFE), Ministry of Education.

M. Mavronicolas et al. (Eds.): WINE 2024, LNCS 15534, pp. 636–637, 2026.
https://doi.org/10.1007/978-3-032-08560-3

but incurs bounded unprofitability. Our aim is for both mechanisms to provide valuable insights into the truthful market design problem.

Keywords: Exchange Market · Budget Constraints · Truthfulness

Information Acquisition in Fragmented Markets

Xian Wu[1] and Mengjia Xia[2(✉)]

[1] University of Wisconsin–Madison, Madison, WI 53706, USA
xwu394@wisc.edu
[2] University of Pennsylvania, Philadelphia, PA 19104, USA
xiax@sas.upenn.edu

Abstract. This paper analyzes how market fragmentation affects traders' information choices and market outcomes. We built a model of a decentralized market with costly information acquisition based on a double-auction model. We show that the information acquisition is lower than the efficient level due to the externality of price informativeness and market power. We find that market fragmentation leads to lower information acquisition. However, fragmentation can be welfare-improving if traders' value correlations are sufficiently heterogeneous.

The full version can be found at https://tinyurl.com/WuXiaWINE24

Keywords: Costly Information Acquisition · Market Fragmentation · Market Power · Welfare

M. Mavronicolas et al. (Eds.): WINE 2024, LNCS 15534, p. 638, 2026.
https://doi.org/10.1007/978-3-032-08560-3

Grace Period Is All You Need: Individual Fairness Without Revenue Loss in Revenue Management

Patrick Jaillet, Chara Podimata, and Zijie Zhou(✉)

Massachusetts Institute of Technology, Cambridge, MA 02139, USA
zhou98@mit.edu

Abstract. Imagine you saw a friend purchasing an item at a store on social media and receiving a discount. Would your friend's discount make you feel unfairly treated by the store if you do not get a discount yourself? And would you be less willing to purchase from that store again in the future? Based on a large-scale online survey that we ran on Prolific, it turns out that the answers to the above questions are *positive.* Our findings suggest: (i) Customers who perceive "unfair" treatment before shopping are less likely to buy. (ii) A customer who has bought from a store in the past and then experiences unfair treatment is likely to decrease their purchasing from that store in the future. (iii) People view recent unfairness as worse than distant events.

Motivated by these findings, in this work, we propose a notion of *individual fairness* in online revenue management and an algorithmic module (called *"Grace Period"*) that can be embedded in traditional revenue management algorithms and guarantee individual fairness. Specifically, we show how to embed the Grace Period in five common revenue management algorithms, including Deterministic Linear Programming with Probabilistic Assignment (DLP-PA), Resolving Deterministic Linear Programming with Probabilistic Assignment (RDLP-PA), Static Bid Price Control (S-BPC), Booking Limit (BL), and Nesting (NESTING), thus covering both *stochastic* and *adversarial* customer arrival settings.

Based on the different structure of these algorithms, we employ different methods to incorporate the grace period design into them. For example, in DLP-PA, each type i customer is accepted with certain probability p_i, leading to unfairness because consecutive type i customers are treated differently with a large probability $p_i(1-p_i)$. To address this, we avoid random decisions for individual arrivals. Instead, we divide the timeline into segments, accepting a random number of customers in a first-come-first-serve manner per segment. Grace periods *inside* each segment ensure fairness when decisions change from acceptance to rejection within each segment; grace periods *between each segment* guarantee fairness when decisions change from rejection to acceptance between the segments.

Embedding the Grace Period does *not* incur additional regret for any of these algorithms, thus indicating that, in an asymptotic regime, there

M. Mavronicolas et al. (Eds.): WINE 2024, LNCS 15534, pp. 639–640, 2026.
https://doi.org/10.1007/978-3-032-08560-3

is *no* tradeoff between a seller maximizing their revenue and guaranteeing that customers feel fairly treated.

The full paper can be found in: https://arxiv.org/abs/2402.08533

Keywords: Individual Fairness · Revenue Management · Online Algorithms

Mechanism Design with Delegated Bidding

Gagan Aggarwal[1], Marios Mertzanidis[2], Alexandros Psomas[2(✉)], and Di Wang[1]

[1] Google Research, Mountain View, CA, USA
{gagana,wadi}@google.com
[2] Purdue University, West Lafayette, IN, USA
{mmertzan,apsomas}@purdue.edu

Abstract. We consider the problem of a designer who wants to allocate resources to representatives, that then distribute the resources they receive among the individuals they represent. Motivated by the way Feeding America, one of the largest U.S. charities, allocates donations to food banks, which then further distribute the donations to food-insecure individuals, we focus on mechanisms that use artificial currencies. We compare auctions through the lens of the Price of Anarchy, with respect to three canonical welfare objectives: utilitarian social welfare (sum of individuals' utilities), Nash social welfare (product of individuals' utilities), and egalitarian social welfare (minimum of individuals' utilities). We prove strong lower bounds on the Price of Anarchy of all auctions that allocate each item to the highest bidder (regardless of how the artificial currency budgets are distributed), subject to a mild technical constraint; this includes Feeding America's current auction, the First-Price auction. In sharp contrast, our main result shows that adapting the classic Trading Post mechanism of Shapley and Shubik to this setting, and coupled with Feeding America's choice of budget rule (each representative gets an amount of artificial currency equal to the number of individuals it represents), achieves a small Price of Anarchy for all generalized p-mean objectives (which include the three canonical objectives as special cases) simultaneously. Our bound on the Price of Anarchy of the Trading Post mechanism depends on ℓ: the product of the rank and the "incoherence" of the underlying valuation matrix, which together capture a notion of how "spread out" the values of a matrix are. This notion has been extremely influential in the matrix completion literature, and, to the best of our knowledge, has never been used in auction theory prior to our work. Perhaps surprisingly, we prove that the dependence on ℓ is necessary: the Price of Anarchy of the Trading Post mechanism is $\Omega(\sqrt{\ell})$.

A full version of this paper appears at: https://arxiv.org/abs/2409.19087

Keywords: Delegation · Mechanism design · Artificial currencies

M. Mavronicolas et al. (Eds.): WINE 2024, LNCS 15534, p. 641, 2026.
https://doi.org/10.1007/978-3-032-08560-3

Truthful Budget Aggregation: Beyond Moving-Phantom Mechanisms

Mark de Berg[1], Rupert Freeman[2], Ulrike Schmidt-Kraepelin[1], and Markus Utke[1(✉)]

[1] TU Eindhoven, Groene Loper 3, 5612 AE Eindhoven, The Netherlands
m.t.d.berg@tue.nl , u.schmidt.kraepelin@tue.nl, m.utke@tue.nl

[2] Darden School of Business, University of Virginia, 100 Darden Boulevard, Charlottesville, VA 22903, USA
FreemanR@darden.virginia.edu

Abstract. We study a budget-aggregation setting in which a number of voters report their ideal distribution of a budget over a set of alternatives, and a mechanism aggregates these reports into an allocation. Ideally, such mechanisms are truthful, i.e., voters should not be incentivized to misreport their preferences. For the case of two alternatives, the set of mechanisms that are truthful and additionally meet a range of basic desiderata (anonymity, neutrality, and continuity) exactly coincides with the so-called moving-phantom mechanisms, but whether this space is richer for more alternatives was repeatedly stated as an open question. We answer this question in the affirmative by presenting a class of truthful mechanisms that are not moving-phantoms but satisfy the three properties. Since moving-phantom mechanisms can only provide limited fairness guarantees (measured as the worst-case distance to a fair share solution), one motivation for broadening the class of truthful mechanisms is the hope for improved fairness guarantees. We dispel this hope by showing that lower bounds holding for the class of moving-phantom mechanisms extend to all truthful, anonymous, neutral, and continuous mechanisms.

Full paper: https://arxiv.org/abs/2405.20303

Keywords: Participatory Budgeting · Social Choice · Truthfulness

Acknowledgements. Part of this research was carried out while Ulrike Schmidt-Kraepelin was supported by the National Science Foundation under Grant No. DMS-1928930 and by the Alfred P. Sloan Foundation under grant G-2021-16778 while being in residence at the Simons Laufer Mathematical Sciences Institute (formerly MSRI) in Berkeley, California, during the Fall 2023 semester.

We also thank Felix Brandt, Matthias Greger, and Warut Suksompong for interesting discussions on the topic.

Disclosure of Interests. The authors have no competing interests to declare that are relevant to the content of this article.

M. Mavronicolas et al. (Eds.): WINE 2024, LNCS 15534, p. 642, 2026.
https://doi.org/10.1007/978-3-032-08560-3

When Should You Offer an Upgrade: Online Upgrading Mechanisms for Resource Allocation

Patrick Jaillet[1], Chara Podimata[1], Andrew Vakhutinsky[2], and Zijie Zhou[1](✉)

[1] Massachusetts Institute of Technology, Cambridge, USA
zhou98@mit.edu
[2] Oracle Lab, Cambridge, USA

Abstract. In this work, we study an upgrading scheme for online resource allocation problems. We work in a sequential setting, where at each round a request for a resource arrives and the decision-maker has to decide whether to accept it (and thus, offer the resource) or reject it. The resources are ordered in terms of their value. If the decision-maker decides to accept the request, they can offer an *upgrade-for-a-fee* to the next more valuable resource. This fee is dynamically decided based on the currently available resources. After the upgrade-for-a-fee option is presented to the requester, they can either accept it, get upgraded, and pay the additional fee, or reject it and maintain their originally allocated resource.

We take the perspective of the decision-maker and wish to design upgrading mechanisms in a way that simultaneously maximizes revenue and minimizes underutilization of resources. Both of these desiderata are encapsulated in a notion of *regret*, which is the asymptotic gap between the online algorithm and the hindsight offline benchmark when the time horizon T scales up. It is hard to establish a benchmark for the online upgrading problem since at every period we not only have to decide whether to accept or reject a request but also decide on an upgrade price if we do decide to accept. To navigate this challenge, we propose a "hybrid programming" approach which corresponds to an upper bound of the hindsight optimal expected revenue. The "hybrid programming" interpolates between a *deterministic* and a *hindsight optimum* counterpart, which draws intuition from approaches in price- and quantity- based revenue management respectively.

We present a fast algorithm called DynUp-n that achieves $O(n \log T)$ regret under n types of resources. At a high level, Algorithm DynUp-n determines the upgrading price at each time period t by solving the aforementioned hybrid programming with the remaining availability of each resource and remaining expected demand for each type of resource at t. Although the hybrid programming is a non-convex optimization problem, we show that we can either translate it into a convex optimization or obtain a closed-form solution to it. Finally, we implemented our algorithm utilizing data akin to the hospitality industry and estimated our upgrading mechanism would increase the annual revenue by over 17%.

The full paper can be found in: https://arxiv.org/abs/2402.08804

M. Mavronicolas et al. (Eds.): WINE 2024, LNCS 15534, pp. 643–644, 2026.
https://doi.org/10.1007/978-3-032-08560-3

Keywords: Online Upgrading · Revenue Management · Dynamic Pricing · Online Non-convex Optimization

Infrequent Resolving Algorithm for Online Linear Programming

Guokai Li, Zizhuo Wang(✉), and Jingwei Zhang

School of Data Science, The Chinese University of Hong Kong, Shenzhen, Guangdong 518172, People's Republic of China

guokaili@link.cuhk.edu.cn , wangzizhuo@cuhk.edu.cn, zhangjingwei@cuhk.edu.cn

Abstract. Online linear programming (OLP) has gained significant attention from both researchers and practitioners due to its extensive applications, such as online auction, network revenue management and advertising. Existing OLP algorithms fall into two categories: LP-based algorithms and LP-free algorithms. The former one typically guarantees better performance, even offering a constant regret, but requires solving a large number of LPs, which could be computationally expensive. In contrast, LP-free algorithm only requires first-order computations but induces a worse performance, lacking a constant regret bound. In this work, we study the case where the inputs are drawn from an unknown finite-support distribution, and bridge the gap between these two extremes by proposing an algorithm that achieves a constant regret while solving LPs only $\mathcal{O}(\log\log T)$ times over the time horizon T. Moreover, when we are allowed to solve LPs only M times, we propose an algorithm that can guarantee an $\mathcal{O}\left(T^{(1/2+\epsilon)^{M-1}}\right)$ regret. Furthermore, when the arrival probabilities are known at the beginning, our algorithm can guarantee a constant regret by solving LPs $\mathcal{O}(\log\log T)$ times, and an $\mathcal{O}\left(T^{(1/2+\epsilon)^{M}}\right)$ regret by solving LPs only M times. Numerical experiments are conducted to demonstrate the efficiency of the proposed algorithms.

Keywords: online linear programming · network revenue management · resolving

---

Link to the full paper: https://arxiv.org/abs/2408.00465.

Zizhuo Wang's research is partially supported by the National Natural Science Foundation of China (NSFC) [Grant NSFC-72394361, 72425013], and the Guangdong Key Lab of Mathematical Foundations for Artificial Intelligence. Jingwei Zhang is partly supported by the National Natural Science Foundation of China (NSFC) [Grant NSFC-72394361] and the Guangdong Provincial Key Laboratory of Mathematical Foundations for Artificial Intelligence (2023B1212010001).

M. Mavronicolas et al. (Eds.): WINE 2024, LNCS 15534, p. 645, 2026.
https://doi.org/10.1007/978-3-032-08560-3

When Does Additional Information Lead to Longer Travel Time in Multi-origin-destination Networks?

Xujin Chen[1,2], Xiaodong Hu[1,2], Xinqi Jing[1,2(✉)], and Zhongzheng Tang[3,4(✉)]

[1] Academy of Mathematics and Systems Science, Chinese Academy of Sciences, Beijing 100190, China

[2] School of Mathematical Sciences, University of Chinese Academy of Sciences, Beijing 100049, China

[3] Key Laboratory of Mathematics and Information Networks (Beijing University of Posts and Telecommunications), Ministry of Education, Beijing, China

[4] School of Science, Beijing University of Posts and Telecommunications, Beijing 100876, China

{xchen,xdhu,jingxinqi,tangzhongzheng}@amss.ac.cn

Abstract. The *informational Braess' paradox* (IBP) illustrates a counterintuitive phenomenon in network traffic where disseminating information about available paths to *certain* travelers paradoxically results in increased travel time for *these* individuals. IBP extends the original Braess' paradox by relaxing the assumption that all travelers have identical and complete information about the network. For non-atomic selfish travelers, Acemoglu et al. (Operations Research, 2018) characterized *IBP-free* undirected networks with a single origin-destination (OD) pair in which IBP never occurs. While they provided a sufficient condition for IBP-freeness in networks with multiple OD pairs, a necessary and sufficient condition remained an open question. In this paper, we resolve this question by establishing the following complete characterization: an undirected network with multiple OD pairs is IBP-free if and only if its *constituent subnetworks* (each associated with a single OD pair) satisfy the following conditions:

(i) each constituent subnetwork is a series of linearly independent networks;

(ii) each pair of constituent subnetworks either is edge-disjoint or their common blocks either are cycles or share identical sets of origin and destination.

The full paper can be found at https://drive.google.com/file/d/1B-nFU6clUMOFbF2uwC7EpeIm6B5oeuRR/view

Xujin Chen and Xinqi Jing were supported by the Chinese Academy of Sciences [Grant XDA27000000] and National Natural Science Foundation of China [Grant 12331014, 72192804, and 71988101]. Zhongzheng Tang was supported by National Natural Science Foundation of China [Grant 12101069].

M. Mavronicolas et al. (Eds.): WINE 2024, LNCS 15534, pp. 646–647, 2026.
https://doi.org/10.1007/978-3-032-08560-3

Keywords: Non-atomic selfish routing · Informational Braess' paradox · Wardrop equilibrium · Multi-origin-destination networks

Computing Most Equitable Voting Rules

Lirong Xia(✉)

Rutgers University-New Brunswick and DIMACS, Piscataway, NJ, USA
lirong.xia@rutgers.edu

Abstract. How to design fair and (computationally) efficient voting rules is a central challenge in Computational Social Choice. In this paper, we aim at designing efficient algorithms for computing *most equitable rules* for large classes of preferences and decisions, which optimally satisfy two fundamental fairness/equity axioms: *anonymity* (every voter being treated equally) and *neutrality* (every alternative being treated equally). By revealing a natural connection to the graph isomorphism problem and leveraging recent breakthroughs by Babai [2019], we design quasipolynomial-time algorithms that compute *most equitable rules with verifications*, which also compute verifications about whether anonymity and neutrality are satisfied at the input profile. Further extending this approach, we propose the *canonical-labeling tie-breaking*, which runs in quasipolynomial-time and optimally breaks ties to preserve anonymity and neutrality. As for the complexity lower bound, we prove that even computing verifications for most equitable rules is GI-complete (i.e., as hard as the graph isomorphism problem), and sometimes GA-complete (i.e., as hard as the graph automorphism problem), for many commonly studied combinations of preferences and decisions. To the best of our knowledge, these are the first problems in computational social choice that are known to be complete in the class GI or GA.

This work is supported by NSF #2450124, 2313136, 2303000, 2106983
Full version of this paper: https://arxiv.org/abs/2410.04179

Keywords: Social choice · anonymity · neutrality · most equitable rules · graph isomorphism

M. Mavronicolas et al. (Eds.): WINE 2024, LNCS 15534, p. 648, 2026.
https://doi.org/10.1007/978-3-032-08560-3

Optimism in the Face of Ambiguity Principle for Multi-Armed Bandits

Mengmeng Li[1(✉)], Daniel Kuhn[1], and Bahar Taşkesen[2]

[1] Risk Analytics and Optimization Chair, EPFL, Route Cantonale, 1015 Lausanne, Switzerland
mengmeng.li@epfl.ch , daniel.kuhn@epfl.ch

[2] Booth School of Business, 5807 S Woodlawn Ave, Chicago, IL 60637, USA
bahar.taskesen@chicagobooth.edu
https://rao.epfl.ch

Abstract. Follow-The-Regularized-Leader (FTRL) algorithms often enjoy optimal regret for adversarial as well as stochastic bandit problems and allow for a streamlined analysis. However, FTRL algorithms require the solution of an optimization problem in every iteration and are thus computationally challenging. In contrast, Follow-The-Perturbed-Leader (FTPL) algorithms achieve computational efficiency by perturbing the estimates of the rewards of the arms, but their regret analysis is cumbersome. We propose a new FTPL algorithm that generates optimal policies for both adversarial and stochastic multi-armed bandits. Similar to FTRL, our algorithm admits a unified regret analysis, and similar to FTPL, it offers low computational costs. Unlike existing FTPL algorithms that rely on independent additive disturbances governed by a *known* distribution, we allow for disturbances governed by an *ambiguous* distribution that is only known to belong to a given set and propose a principle of optimism in the face of ambiguity. Consequently, our framework generalizes existing FTPL algorithms. It also encapsulates a broad range of FTRL methods as special cases, including several optimal ones, which appears to be impossible with current FTPL methods. Finally, we use techniques from discrete choice theory to devise an efficient bisection algorithm for computing the optimistic arm sampling probabilities. This algorithm is up to 10^4 times faster than standard FTRL algorithms that solve an optimization problem in every iteration. Our results not only settle existing conjectures but also provide new insights into the impact of perturbations by mapping FTRL to FTPL (ML and DK are supported by the Swiss National Science Foundation under the NCCR *Automation*, grant agreement 51NF40_180545. The full paper is available at https://arxiv.org/pdf/2409.20440.)

Keywords: Bandits · Discrete choice models · Online learning

M. Mavronicolas et al. (Eds.): WINE 2024, LNCS 15534, p. 649, 2026.
https://doi.org/10.1007/978-3-032-08560-3

Packing a Knapsack with Items Owned by Strategic Agents

Javier Cembrano[1(✉)], Max Klimm[2], and Martin Knaack[2]

[1] Department of Algorithms and Complexity, Max Planck Institut für Informatik, Saarbrücken, Germany
jcembran@mpi-inf.mpg.de

[2] Institute for Mathematics, Technische Universität Berlin, Berlin, Germany
klimm@math.tu-berlin.de, knaack@math.tu-berlin.de

Abstract. This paper considers a scenario where a mechanism designer is interested in selecting items with maximum total value under a knapsack constraint. The items, however, are controlled by strategic agents who aim to maximize the total value of their items in the knapsack. This is a natural setting, e.g., when agencies select projects for funding, companies select products for sale in their shops, or hospitals schedule MRI scans for the day. A mechanism governing the packing of the knapsack is strategyproof if no agent can benefit from hiding items controlled by them to the mechanism. We are interested in mechanisms that are strategyproof and α-approximate in the sense that they always approximate the maximum value of the knapsack by a factor of $\alpha \in [0, 1]$. First, we give a deterministic mechanism that is $\frac{1}{3}$-approximate. For the special case where all items have unit density, we design a $\frac{1}{\phi}$-approximate mechanism where $1/\phi \approx 0.618$ is the inverse of the golden ratio. This result is tight as we show that no deterministic strategyproof mechanism with a better approximation exists. We further give randomized mechanisms with approximation guarantees of $1/2$ for the general case and $2/3$ for the case of unit densities. For both cases, no strategyproof mechanism can achieve an approximation guarantee better than $1/(5\phi - 7) \approx 0.917$.

A full version can be found on arXiv:2410.06080.

Keywords: Mechanisms without money · Knapsack

M. Mavronicolas et al. (Eds.): WINE 2024, LNCS 15534, p. 650, 2026.
https://doi.org/10.1007/978-3-032-08560-3

Data-Scarce Identification of Game Dynamics via Sum-of-Squares Optimization

Iosif Sakos(✉), Antonios Varvitsiotis, and Georgios Piliouras

Singapore University of Technology and Design, 8 Somapah Road, Singapore 487372, Singapore

iosif_sakos@sutd.edu.sg , antonios@sutd.edu.sg, georgios@sutd.edu.sg

Understanding how players adjust their strategies in games, based on their experience, is a crucial tool for policymakers. It enables them to forecast the system's eventual behavior, exert control over the system, and evaluate counterfactual scenarios. The modern approach of dynamical system discovery relies on vast amounts of available measurements, coupled with powerful machine learning algorithms and inexpensive parallel processing power. Nevertheless, techniques that require access to massive data sets have limited applicability in settings where data are scarce, and expensive to acquire, or in time-critical applications where there is not enough time to collect data.

In this work, we introduce the Side-Information Assisted Regression (SIAR) framework, designed to identify game dynamics in multiplayer normal-form games only using data from a short run of a single system trajectory. By design, our framework allows the swift integration of a wide range of constraints, which are native to game-theoretic applications, such as positive correlation between the players' directions of motion and payoffs, game symmetries, a.o. SIAR relies on sum-of-squares optimization, and the construction of a hierarchy of approximations that provably converge to the true dynamics of the system.

We test the SIAR framework in a wide range of classic benchmarks including: 1) periodic replicator dynamics in matching pennies games; 2) chaotic replicator dynamics in ϵ-perturbed rock-paper-scissors games; and 3) equilibrium selection prediction in numerous game dynamics in multiplayer atomic congestion games with up to hundreds of pure, attracting Nash equilibria. Thus, our methodology can also be used to provide insights into two fundamental problems of game theory: equilibrium selection and non-equilibrating (possibly chaotic) dynamics. In each of the above settings, our method showcases near perfect system recovery strongly outperforming previous benchmarks. The complete manuscript can be found in https://doi.org/10.48550/arXiv.2307.06640

Acknowledgements. This research is supported by the MOE Tier 2 grant (MOE-T2EP20223-0018) and by the National Research Foundation, Singapore, A*STAR under the Quantum Engineering Programme (NRF2021-QEP2-02-P05), and DSO National Laboratories under its AI Singapore Program (AISG Award No: AISG2-RP-2020-016).

M. Mavronicolas et al. (Eds.): WINE 2024, LNCS 15534, p. 651, 2026.
https://doi.org/10.1007/978-3-032-08560-3

Continuous-Time Best-Response and Related Dynamics in Tullock Contests with Convex Costs

Edith Elkind, Abheek Ghosh(✉), and Paul W. Goldberg

Department of Computer Science, University of Oxford, Oxford, UK
edith.elkind@cs.ox.ac.uk , abheek.ghosh@cs.ox.ac.uk,
paul.goldberg@cs.ox.ac.uk

Abstract. Tullock contests model real-life scenarios that range from competition among proof-of-work blockchain miners to rent-seeking and lobbying activities. We show that continuous-time best-response dynamics in Tullock contests with convex costs converges to the unique equilibrium using Lyapunov-style arguments. We then use this result to provide an algorithm for computing an approximate equilibrium. We also establish convergence of related discrete-time dynamics, e.g., fictitious play. These results indicate that the equilibrium is a reliable predictor of the agents' behavior in these games.

Full paper is available at https://arxiv.org/abs/2402.08541

Keywords: Contests · Best-Response Dynamics · Fictitious Play

This research was funded in part by EPSRC Grant EP/X040461/1. For the purpose of Open Access, the author has applied a CC BY public copyright licence to any Author Accepted Manuscript (AAM) version arising from this submission.

M. Mavronicolas et al. (Eds.): WINE 2024, LNCS 15534, p. 652, 2026.
https://doi.org/10.1007/978-3-032-08560-3

When to Push Ads: Optimal Mobile Ad Campaign Strategy Under Markov Customer Dynamics

Guokai Li, Pin Gao, and Zizhuo Wang(✉)

School of Data Science, The Chinese University of Hong Kong, Shenzhen, Guangdong 518172, People's Republic of China
guokaili@link.cuhk.edu.cn , gaopin@cuhk.edu.cn, wangzizhuo@cuhk.edu.cn

Abstract. We investigate a seller's optimal advertising campaign strategy targeting customers who interact with the seller over time. We model customers' engagement by a continuous-time Markov chain with two states, active and inactive. While in the active state, customers make purchases according to a Poisson process; in contrast, customers in the inactive state make no purchases. The seller can use advertisements to activate customers and the activation probability may be affected by customers' fatigue to promotion. The objective of the seller is to maximize the long-term average expected profit by designing an optimal ad campaign strategy based on customers' purchase histories. For a single customer without budget constraints, we demonstrate the optimality of a triple-threshold policy based on the elapsed time since the last purchase or ad campaign. Moreover, we find that the seller tends to push ads earlier to customers with high purchase rates and low recapture rates (i.e., low transition rates from an inactive state to an active state). Surprisingly, we also find that the seller should push ads earlier to customers with intermediate churn rates compared to those with small or large churn rates. When confronted with a budget constraint on the overall ad cost, we suggest first solving a relaxed problem and then implementing a separate triple-threshold policy for each customer with a specified budget, and establish the asymptotic optimality of this policy. In addition, we explore the case with strategic customer reactions.

Keywords: mobile ad campaign · Markov customer dynamics · revenue management

---

Link to the full paper: https://papers.ssrn.com/sol3/papers.cfm?abstract_id=4477931.

Pin Gao's research is supported by the National Natural Science Foundation of China [Grants 72201234 and 72192805] and the Research Grant Council of Hong Kong (the Collaborative Research Fund) [Grant C6032-21G], The Guangdong Key Lab of Mathematical Foundations for Artificial Intelligence. Zizhuo Wang's research is partially supported by the National Natural Science Foundation of China (NSFC) [Grant NSFC-72394361, 72425013], and the Guangdong Key Lab of Mathematical Foundations for Artificial Intelligence.

M. Mavronicolas et al. (Eds.): WINE 2024, LNCS 15534, p. 653, 2026.
https://doi.org/10.1007/978-3-032-08560-3

Continuous Social Networks

Julián Chitiva[1(✉)] and Xavier Venel[2]

[1] HEC Paris, 1 Rue de la Libération, 78351 Jouy-en-Josas, France
julian.chitiva@hec.edu
[2] Luiss University, Viale Romania 32, 00197 Rome, Italy
xvenel@luiss.it

Abstract. The dynamics of opinions play a crucial role in shaping economic outcomes and decision-making processes. Motivated by this, we synthesize the theoretical frameworks from the model of DeGroot, DiKernel Theory, and the models of lobby competition into an integrated approach that comprehensively captures the dynamics of opinion formation in large economic systems. This work contributes to the growing strand of literature in economics that uses DiKernels to study the interactions between non-atomic agents in large economic environments and characterizes the equilibrium sets of these games.

The main contribution of this paper is the extension of the model of DeGroot to networks with a continuum of agents. The seminal model of DeGroot offers a solid foundation for social learning; nevertheless, it faces computational restrictions when considering large networks. Our extension tackles these limitations by studying these dynamic economic interactions through their continuous approximation. Our first result states that under Lipschitz conditions on the DiKernel, the continuous DeGroot model is the limit case of the discrete one.

We propose some applications of our model both in non-strategic and strategic setups. First, we define a canonical method to reduce the dimensionality of networks by quantifying the error one is making when considering aggregation of agents in groups. Second, we focus on the problem of competition between two opposing lobbies that aim to influence economic decision-making in a society through strategic campaigns. We propose a static game between lobbies in which they exert influence over a continuum of non-strategic agents to transform their initial opinion before the agents update it *à la* DeGroot. Our second result states sufficient conditions for the existence of the Nash Equilibrium. Finally, our third result studies the conditions under which the Nash Equilibria of the game are ε-Nash Equilibria of the discretization of the game.

The full version is available at https://arxiv.org/abs/2407.11710

Julián Chitiva gratefully acknowledges the support of Hi!PARIS and the ANR ("Investissements d'Avenir programme") under grant ANR-18-EURE-0005/EUR DATA EFM. This work was partially done when Julián Chitiva was visiting LUISS University in Rome. Xavier Venel's work was partially supported by the MIUR PRIN projects "Learning in Markets and Society" (2022EKNE5K) and "Supply Chain Disruptions, Financial losses and their prevention" (PNRR P2022XT8C8).

M. Mavronicolas et al. (Eds.): WINE 2024, LNCS 15534, pp. 654–655, 2026.
https://doi.org/10.1007/978-3-032-08560-3

Keywords: Continuous DeGroot · Dynamic of Opinions · Lobby Competition · Dimensionality Reduction · DiKernels · Graphons

Simultaneous vs Sequential: Optimal Assortment Recommendation in Multi-store Retailing

Yicheng Liu[1(✉)], Xiao Alison Chen[2], Yan Liu[3], and Zizhuo Wang[1]

[1] School of Data Science, The Chinese University of Hong Kong (CUHK-Shenzhen), Shenzhen, China
yichengliu@link.cuhk.edu.cn , wangzizhuo@cuhk.edu.cn
[2] Peter T. Paul College of Business and Economics, University of New Hampshire, Durham, NH, USA
alison.chen@unh.edu
[3] Department of Logistics and Maritime Studies, Faculty of Business, Hong Kong Polytechnic University, Kowloon, Hong Kong
yan.y.liu@polyu.edu.hk

Abstract. We study a multi-store assortment planning problem in which a seller centrally decides the assortment of products in each store. Each store is visited by a certain fraction of customers. Upon arrival, customers observe the products offered in the current store. The seller can either 1) show products offered in other stores to customers simultaneously (called simultaneous offering strategy) or 2) show products offered in other stores to customers if they decide not to purchase in the current store (called sequential offering strategy). Customers may incur a disutility if they choose products from other stores. We study the optimal assortment planning problem under each of these two strategies and compare their performance. We show that under the simultaneous strategy, the optimal assortment for each store is revenue-ordered, and the seller cannot do better than if it operates each store separately. In contrast, adopting the sequential strategy can significantly increase the seller's revenue. In particular, when customers' disutilities of purchasing from other stores are universally homogeneous, the optimal assortment under the sequential offering strategy is revenue-ordered for each store, and the store with a lower (higher, resp.) demand should offer a larger (smaller, resp.) assortment to facilitate the sequential selling process. We then compare the two offering strategies in terms of the seller's revenue and customer surplus and analyze the impact of parameters on the revenue comparison. We conclude that the sequential offering strategy may be a better strategy for sellers in practice.

Full paper link: https://papers.ssrn.com/sol3/papers.cfm?abstract_id=4592172
Zizhuo Wang's research is partially supported by the National Natural Science Foundation of China (NSFC) [Grant NSFC-72394361, 72425013], and the Guangdong Key Lab of Mathematical Foundations for Artificial Intelligence.

M. Mavronicolas et al. (Eds.): WINE 2024, LNCS 15534, pp. 656–657, 2026.
https://doi.org/10.1007/978-3-032-08560-3

Keywords: Multi-store retailing · Assortment optimization · Simultaneous/Sequential offering strategy

Churning While Experimenting: Maximizing User Engagement in Recommendation Platforms

Michael L. Hamilton[1(✉)] and Raghav Singal[2]

[1] University of Pittsburgh, Pittsburgh, PA 15213, USA
mhamilton@katz.pitt.edu

[2] Dartmouth College, Hanover, NH 03755, USA
singal@dartmouth.edu

Abstract. Online media platforms rely on long-term user engagement to generate revenue. The primary operational lever they control is content recommendation, which determines what content should be suggested to each user. However, due to the constantly evolving supply of content and the heterogeneity in user behavior, achieving optimal recommendations is challenging. It necessitates a delicate balance between experimentation (to gauge the effectiveness of new content) and exploitation (recommending high-quality existing content).

Motivated by a real-world dataset, we present a novel framework for the platform recommendation problem with the aim of maximizing long-term user engagement. Our framework comprises a *behavioral* model and a *prescriptive* model. The behavioral component is used to model the period-over-period heterogeneous behavior of users and is grounded in the existing RFM-style models to capture customer lifetime value. The prescriptive component is used to model the dynamic optimization problem faced by the platform, with the experimentation policy being the key lever. Our main contribution is to analytically characterize the structure of the optimal experimentation policy. Doing so allows us to extract important managerial insights as we establish the myopic policy of targeting users based on their churn sensitivity is long-term optimal under business-relevant scenarios.

A full version is available at: https://papers.ssrn.com/sol3/papers.cfm?abstract_id=3871915

M. Mavronicolas et al. (Eds.): WINE 2024, LNCS 15534, p. 658, 2026.
https://doi.org/10.1007/978-3-032-08560-3

Auctioning with Strategically Reticent Bidders

Jibang Wu[1(✉)], Ashwinkumar Badanidiyuru[2], and Haifeng Xu[1]

[1] The University of Chicago, Chicago, IL 60637, USA
{wujibang,haifengxu}@uchicago.edu
[2] Google, Mountain View, CA 94043, USA
ashwinkumarbv@google.com

Abstract. We propose and study a novel mechanism design setup where each bidder holds two kinds of private information: (1) *type variable*, which can be misreported; (2) *information variable*, which the bidder may want to conceal or partially reveal, but *not* to misreport. We refer to bidders with such behaviors as *strategically reticent bidders*. Among others, one direct motivation of our model is the ad auction in which many ad platforms today elicit from bidders not only their private value per conversion but also their private information about Internet users (e.g., user activities on the advertisers' websites) in order to improve the platform's estimates of conversion rates. We show that in this new setup, it is still possible to design mechanisms that are both *Incentive and Information Compatible* (IIC). We develop two different black-box transformations, which convert any mechanism $\mathcal{M}$ for classic bidders to a mechanism $\bar{\mathcal{M}}$ for strategically reticent bidders, based on either outcome of expectation or expectation of outcome, respectively. We identify properties of $\mathcal{M}$ such that the resultant mechanism $\bar{\mathcal{M}}$ is IIC. Interestingly, as corollaries of these results, we show that running VCG with bidders' *expected values* maximizes welfare, whereas the mechanism using *expected outcome* of Myerson's auction maximizes revenue. Finally, we study how regulating auctioneer's usage of bidders' information can lead to more robust mechanisms. A full version of this paper can be found at https://arxiv.org/abs/2109.04888.

Keywords: Auction Design · Information Design · Data Sharing

M. Mavronicolas et al. (Eds.): WINE 2024, LNCS 15534, p. 659, 2026.
https://doi.org/10.1007/978-3-032-08560-3

Regulating Discriminatory Pricing in the Presence of Tacit Collusion

Zongsen Yang[1], Xiao Lei[2(✉)], and Pin Gao[1]

[1] School of Data Science, The Chinese University of Hong Kong, Shenzhen, Guangdong 518172, People's Republic of China
zongsenyang@link.cuhk.edu.cn , gaopin@cuhk.edu.cn
[2] HKU Business School, The University of Hong Kong, Hong Kong SAR, China
xlei@hku.hk

Abstract. Price-setting algorithms have facilitated widespread awareness-based price discrimination, wherein firms charge high prices to customers unaware of alternative choices and low prices to those in competitive markets. This unethical behavior has increased customer complaints and prompted policymakers to enact fairness regulations in response. However, while limiting price discrimination may improve consumer welfare under genuine competition, it also affects firms' incentives to form tacit collusion, another regulatory concern arising from the proliferation of pricing algorithms. We develop an analytical model to examine the interplay between fairness regulation and tacit collusion, and discuss its impact on consumer welfare and policymaking. Firms utilize customers' product unawareness to implement price discrimination and decide whether to collude by comparing profits from collusion and deviation. We then explore the consequences of price fairness regulation on the sustainability of tacit collusion. For homogeneous products, fairness regulation can substantially weaken collusion, potentially rendering it unattainable. However, for differentiated products, strict fairness inadvertently supports collusive behavior, harming consumer welfare. In this case, mild fairness permitting moderate price differentiation can prevent market collusion and optimize welfare. Our key results are valid under various robustness checks such as incorporating Q-learning or using alternative demand models and punishment strategies. We also propose a novel randomized approach to tame fairness-induced collusion. Overall, our study emphasizes the importance of a nuanced approach to regulating discriminatory pricing in the presence of tacit collusion.

The paper is accepted by The 20th Conference on Web and Internet Economics (WINE2024). Here is the link to the full paper: https://ssrn.com/abstract=4633784. Xiao Lei's research is financially supported by the Hong Kong Research Grants Council (RGC) funding (RGC Early Career Scheme 2023/24) [Grant 27503123]. Pin Gao's research is supported by the National Natural Science Foundation of China [Grants 72201234 and 72192805] and the Research Grant Council of Hong Kong (the Collaborative Research Fund) [Grant C6032-21G], The Guangdong Key Lab of Mathematical Foundations for Artificial Intelligence.

M. Mavronicolas et al. (Eds.): WINE 2024, LNCS 15534, pp. 660–661, 2026.
https://doi.org/10.1007/978-3-032-08560-3

Keywords: Price Discrimination · Algorithmic Collusion · Dynamic Competition · Fairness

Robust Optimal Selling Mechanism Under Non-linear Utility

Pin Gao[1], Yicheng Liu[1(✉)], Shixin Wang[2], Zhen Wang[3], and Zizhuo Wang[1]

[1] School of Data Science, The Chinese University of Hong Kong (CUHK-Shenzhen), Shenzhen, China
gaopin@cuhk.edu.cn, yichengliu@link.cuhk.edu.cn , wangzizhuo@cuhk.edu.cn
[2] CUHK Business School, The Chinese University of Hong Kong, Hong Kong SAR, China
shixinwang@cuhk.edu.hk
[3] School of Information Management and Engineering, The Shanghai University of Finance and Economics (SUFE), Shanghai, China
wangzhen@sufe.edu.cn

Abstract. We consider a robust mechanism design problem in which a seller wishes to choose a selling mechanism to maximize the revenue obtained from a buyer with private valuation information and a nonlinear utility function. The objective is to provide a profit guarantee for the seller across all possible valuation distributions whose mean and variance are consistent with the prior information. When the utility function is of a quadratic form and the coefficient of variation (i.e., the ratio between the standard deviation and the mean) is low, we find that a quadratic pricing mechanism yields the optimal revenue guarantee. In this case, the worst-case valuation distribution follows a generalized Pareto distribution. However, when the coefficient of variation is high, a unit price mechanism is optimal. We further extend our analysis to general elasticity utility functions and investigate how the optimal worst-case revenue varies with elasticity and coefficient of variation. Furthermore, given the widespread implementation of a two-part tariff mechanism in real-world scenarios, we also study the optimal two-part tariff mechanism. We show that under the quadratic utility form, the optimal two-part tariff pricing strategy achieves at least 80% of the optimal worst-case revenue when

Full paper link: https://papers.ssrn.com/sol3/papers.cfm?abstract_id=4731184. Pin Gao's research is supported by the National Natural Science Foundation of China [Grants 72201234 and 72192805] and the Research Grant Council of Hong Kong (the Collaborative Research Fund) [Grant C6032-21G], The Guangdong Key Lab of Mathematical Foundations for Artificial Intelligence. Shixin Wang's research is supported by the National Natural Science Foundation of China [Grant 72394395]. Zhen Wang's research is partially supported by the National Science Foundation of China [Grant 72301235] and the Major Program of the National Natural Science Foundation of China [Grant NSFC-72192830, 72192832]. Zizhuo Wang's research is partially supported by the National Natural Science Foundation of China (NSFC) [Grant NSFC-72394361, 72425013], and the Guangdong Key Lab of Mathematical Foundations for Artificial Intelligence.

M. Mavronicolas et al. (Eds.): WINE 2024, LNCS 15534, pp. 662–663, 2026.
https://doi.org/10.1007/978-3-032-08560-3

compared with the optimal robust mechanism. We also develop corresponding performance guarantees when the elasticity coefficient is less than two. However, when the elasticity coefficient exceeds two, the competitive ratio of the two-part pricing scheme goes to zero as the variance of valuation grows to infinity.

Keywords: Robust mechanism design · Moment condition · Non-linear utility

Computing Competitive Equilibrium for Chores: Linear Convergence and Lightweight Iteration

He Chen[1], Chonghe Jiang[2], and Anthony Man-Cho So[3](✉)

[1] The Chinese University of Hong Kong, Hong Kong SAR, China
hechen@link.cuhk.edu.hk
[2] The Chinese University of Hong Kong, Hong Kong SAR, China
chjiang@link.cuhk.edu.hk
[3] The Chinese University of Hong Kong, Hong Kong SAR, China
manchoso@se.cuhk.edu.hk

Abstract. Competitive equilibrium (CE) for chores has recently attracted significant attention, with many algorithms proposed to approximately compute it. However, existing algorithms either lack iterate convergence guarantees to an exact CE or require solving high-dimensional linear or quadratic programming subproblems. This paper overcomes these issues by proposing a novel unconstrained difference-of-convex formulation, whose stationary points correspond precisely to the CE for chores. We show that the new formulation possesses the local error bound property and the Kurdyka-Łojasiewicz property with an exponent of 1/2. Consequently, we present the first algorithm whose iterates provably converge linearly to an exact CE for chores. Furthermore, by exploiting the max structure within our formulation and applying smoothing techniques, we develop a subproblem-free algorithm that finds an approximate CE in polynomial time. Numerical experiments demonstrate that the proposed algorithms outperform the state-of-the-art method.

Keywords: Competitive equilibrium for chores · Linear convergence · Lightweight method · Error bound

Anthony Man-Cho So was supported in part by the Hong Kong Research Grants Council (RGC) General Research Fund (GRF) project CUHK 14204823.The full paper is available at http://arxiv.org/abs/2410.04036.

M. Mavronicolas et al. (Eds.): WINE 2024, LNCS 15534, p. 664, 2026.
https://doi.org/10.1007/978-3-032-08560-3

How Much Should You Pay for Restaking Security?

Tarun Chitra[1] and Mallesh M. Pai[2,3](✉)

[1] Gauntlet, New York, NY, USA
[2] Rice University, Houston, TX, USA
[3] Consensys Inc, New York, NY, USA
mallesh.pai@consensys.net

Abstract. Restaking protocols have aggregated billions of dollars of security by utilizing token incentives and payments. A natural question to ask is: How much security do restaked services *really* need to purchase? To answer this question, we expand a model of Durvasula and Roughgarden (DR) that includes incentives and an expanded threat model consisting of strategic attackers and users. Our model shows that an adversary with a strictly submodular profit combined with strategic node operators who respond to incentives can avoid the large-scale cascading failures of (DR). We utilize our model to construct an approximation algorithm for choosing token-based incentives that achieve a given security level against adversaries who are bounded in the number of services they can simultaneously attack. Our results suggest that incentivized restaking protocols can be secure with proper incentive management.

The full paper is available at https://arxiv.org/abs/2408.00928

Keywords: Blockchain · staking · risk management

M. Mavronicolas et al. (Eds.): WINE 2024, LNCS 15534, p. 665, 2026.
https://doi.org/10.1007/978-3-032-08560-3

Static Pricing for Online Selection Problem and Its Variants (Extended Abstract)

Bo Sun[1(✉)], Hossein Nekouyan Jazi[2], Xiaoqi Tan[2], and Raouf Boutaba[1]

[1] University of Waterloo, Waterloo N2L 3G1, Canada
{bo.sun,rboutaba}@uwaterloo.ca

[2] University of Alberta, Edmonton T6G 2R3, Canada
{nekouyan,xiaoqi.tan}@ualberta.ca

Abstract. This paper studies an online selection problem where a seller seeks to sequentially sell multiple copies of an item to arriving buyers. We consider an adversarial setting, making no assumptions about buyers' valuations for the items, except acknowledging a finite support. The adversarial online selection problem has been studied through posted pricing mechanisms, where the seller sets a unit price for the item before each buyer's arrival, and buyers decide to purchase an item if their utility is non-negative. Nearly all prior work relies on dynamic pricing strategies, which set different prices for different buyers to make the best use of pricing power. However, dynamic pricing raises additional concerns in practice, as it introduces price discrimination, i.e., selling the same item at different prices to buyers arriving at different times. In contrast, static pricing, which maintains a fixed price for all buyers, is considered a fairer pricing scheme and is easier to implement in practice.

In this paper, we focus on a class of static pricing algorithms that sample a price from a predetermined distribution and sell items to buyers whose valuations exceed the sampled price. Our work shows that this simple static pricing strategy can achieve strong guarantees comparable to the best-known dynamic pricing algorithms. Specifically, we design the optimal static pricing algorithms for the adversarial online selection problem and its two important variants: (i) the online assignment problem, which involves selecting from multiple supply sources; and (ii) the online selection problem with convex cost, where each unit of the item incurs a non-decreasing marginal production cost.

The static pricing algorithms we propose can even achieve the optimal competitive ratios among all online algorithms for both the online selection problem and the online assignment problem. To obtain these results, we introduce an economics-based approach to the competitive analysis of static pricing algorithms and develop a novel representative function-based method to derive lower bounds. We expect these approaches to be useful in related problems, such as online matching (Extended abstract. Full version appears as arXiv:2410.07378.).

This work is supported by the NSERC Discovery Grant RGPIN-2019-

M. Mavronicolas et al. (Eds.): WINE 2024, LNCS 15534, pp. 666–667, 2026.
https://doi.org/10.1007/978-3-032-08560-3

06587, Alberta Machine Intelligence Institute (Amii), Alberta Major Innovation Fund, and NSERC Discovery Grant RGPIN-2022-03646.)

Optimal Guarantees for Online Selection Over Time

Sebastian Perez-Salazar[1,2] and Victor Verdugo[3,4(✉)]

[1] Department of Computational Applied Mathematics and Operations Research, Rice University, Houston, USA
sperez@rice.edu

[2] Ken Kennedy Institute, Rice University, Houston, USA

[3] Institute for Mathematical and Computational Engineering, Pontificia Universidad Católica de Chile, Santiago, Chile

[4] Department of Industrial and Systems Engineering, Pontificia Universidad Católica de Chile, Santiago, Chile
victor.verdugo@uc.cl

Abstract. Prophet inequalities are a cornerstone in optimal stopping and online decision-making. Traditionally, they involve the sequential observation of n non-negative independent random variables and face irrevocable accept-or-reject choices. The goal is to provide policies that provide a good approximation ratio against the optimal offline solution that can access all the values upfront—the so-called prophet value. In the *prophet inequality over time problem* (POT), the decision-maker can commit to an accepted value for τ units of time, during which no new values can be accepted. In the IID setting, Abels et al. [1] identified the optimal dynamic programming policy and provided lower and upper bounds on the approximation ratio against the prophet.

In this work, we provide best possible worst-case approximation ratios in the IID setting of POT for single-threshold algorithms and the optimal dynamic programming policy. We show a single-threshold algorithm that achieves an approximation ratio of $(1 + e^{-2})/2 \approx 0.567$, and we prove that no single-threshold algorithm can surpass this guarantee. With our techniques, we can analyze simple algorithms using k thresholds and show that with $k = 3$ it is possible to get an approximation ratio larger than ≈ 0.602. Then, for each n, we prove it is possible to compute the tight worst-case approximation ratio of the optimal dynamic programming policy for instances with n values by solving a convex optimization program. A limit analysis of the first-order optimality conditions yields a nonlinear differential equation showing that the optimal dynamic programming policy's asymptotic worst-case approximation ratio is ≈ 0.618. Finally, we extend the discussion to adversarial settings and show an optimal worst-case approximation ratio of ≈ 0.162 when the values are streamed in random order.

The full version is available at [https://arxiv.org/abs/2408.11224

M. Mavronicolas et al. (Eds.): WINE 2024, LNCS 15534, pp. 668–669, 2026.
https://doi.org/10.1007/978-3-032-08560-3

Reference

1. Abels, A., Pitschmann, E., Schmand, D.: Prophet inequalities over time. In: ACM Conference on Economics and Computation (EC), pp. 1–20 (2023)

Impartial Selection Under Combinatorial Constraints

Javier Cembrano[1(✉)], Max Klimm[2], and Arturo Merino[3]

[1] Department of Algorithms and Complexity, Max Planck Institut für Informatik, Saarbrücken, Germany
jcembran@mpi-inf.mpg.de

[2] Institute for Mathematics, Technische Universität Berlin, Berlin, Germany
klimm@math.tu-berlin.de

[3] Institute of Engineering Sciences, Universidad de O'Higgins, Rancagua, Chile
arturo.merino@uoh.cl

Abstract. Impartial selection problems are concerned with the selection of one or more agents from a set based on mutual nominations from within the set. To avoid strategic nominations of the agents, the axiom of impartiality requires that the selection of each agent is independent of the nominations cast by that agent. This paper initiates the study of impartial selection problems where the nominations are weighted and the set of agents that can be selected is restricted by a combinatorial constraint. We call a selection mechanism α-optimal if, for every instance, the ratio between the total sum of weighted nominations of the selected set and that of the best feasible set of agents is at least α. We show that a natural extension of a mechanism studied for the selection of a single agent remains impartial and $\frac{1}{4}$-optimal for general independence systems, and we generalize upper bounds from the selection of multiple agents by parameterizing them by the girth of the independence system. We then focus on independence systems defined by knapsack and matroid constraints, giving impartial mechanisms that exploit a greedy order of the agents and achieve approximation ratios of $\frac{1}{3}$ and $\frac{1}{2}$, respectively, when agents cast a single nomination. For graphic matroids, we further devise an impartial and $\frac{1}{3}$-optimal mechanism for an arbitrary number of unweighted nominations.

A full version can be found on arXiv:2409.20477.

Keywords: Impartial selection · Mechanism design without money · Social choice

---

J. Cembrano and M. Klimm — Research was supported by the Deutsche Forschungsgemeinschaft (DFG, German Research Foundation) under project number 431465007.
A. Merino — This work is part of the project TIPEA that has received funding from the European Research Council (ERC) under the European Unions Horizon 2020 research and innovation programme (grant agreement No. 850979).

M. Mavronicolas et al. (Eds.): WINE 2024, LNCS 15534, p. 670, 2026.
https://doi.org/10.1007/978-3-032-08560-3

Splitting Guarantees for Prophet Inequalities via Nonlinear Systems

Johannes Brustle[1], Sebastian Perez-Salazar[2,3](✉), and Victor Verdugo[4,5]

[1] Department of Computer and Systems Sciences, Sapienza University of Rome, Rome, Italy
brustle@diag.uniroma1.it
[2] Department of Computational Applied Mathematics and Operations Research, Rice University, Houston, USA
[3] Ken Kennedy Institute, Rice University, Houston, USA
sperez@rice.edu
[4] Institute for Mathematical and Computational Engineering, Pontificia Universidad Católica de Chile, Santiago, Chile
victor.verdugo@uc.cl
[5] Department of Industrial and Systems Engineering, Pontificia Universidad Católica de Chile, Santiago, Chile

Abstract. The prophet inequality is one of the cornerstone problems in optimal stopping theory and has become a crucial tool for designing sequential algorithms in Bayesian settings. In the i.i.d. k-selection prophet inequality problem, we sequentially observe n non-negative random values sampled from a known distribution. Each time, a decision is made to accept or reject the value, and under the constraint of accepting at most k items. For $k = 1$, Hill and Kertz [Ann. Probab. 1982] provided an upper bound on the worst-case approximation ratio that was later matched by an algorithm of Correa et al. [Math. Oper. Res. 2021]. The worst-case tight approximation ratio for $k = 1$ is computed by studying a differential equation that naturally appears when analyzing the optimal dynamic programming policy. A similar result for $k > 1$ has remained elusive.

In this work, we introduce a nonlinear system of differential equations for the i.i.d. k-selection prophet inequality that generalizes Hill and Kertz's equation when $k = 1$. Our nonlinear system is defined by k constants that determine its functional structure, and their summation provides a lower bound on the optimal policy's asymptotic approximation ratio for the i.i.d. k-selection prophet inequality. To obtain this result, we introduce for every k an infinite-dimensional linear programming formulation that fully characterizes the worst-case tight approximation ratio of the k-selection prophet inequality problem for every n, and then we follow a dual-fitting approach to link with our nonlinear system for sufficiently large values of n. We use our new lower bounds to establish a tight approximation ratio for the stochastic sequential assignment problem in the non-negative regime.

The full version is available at [https://arxiv.org/abs/2406.17767

M. Mavronicolas et al. (Eds.): WINE 2024, LNCS 15534, p. 671, 2026.
https://doi.org/10.1007/978-3-032-08560-3

How to Implement Reservation Policy in India

Orhan Aygün[1] and Bertan Turhan[2](✉)

[1] Boğaziçi University, Bebek, Istanbul, Turkey
orhan.aygun@bogazici.edu.tr
[2] Iowa State University, Ames, IA, USA
bertan@iastate.edu

Abstract. We analyze India's affirmative action program, focusing on two landmark Supreme Court cases: (1) Indra Sawhney vs. Union of India (1992) which established reservations as "hard" constraints, and (2) Ashoka Kumar Thakur vs. Union of India—henceforth AKT—(2008) in which the Court advocated for "soft" reservations for Other Backward Classes (OBC). We demonstrates that these two judgments are incompatible through axiomatic analysis and proposes a family of assignment rules that aim to minimize deviations from the established judicial norms. Moreover, the paper characterizes a particular assignment rule in this family—the Forward Transfer (FT) Choice Rule—as merit-maximizing. The analysis is extended to a centralized marketplace framework, characterizing deferred acceptance mechanisms under some of the introduced assignment rules.

We formalize three key principles from Indra Sawhney (1992)—over-and-above principle, within-category fairness, and quota-filling subject to eligibility—as formal axioms (similar to [1]) and provide a formal proof to show that these three axioms uniquely characterize the choice rule introduced in Indra Sawhney (1992). AKT (2008) proposes a modification to the eligibility criteria for OBC positions by mandating that OBC cutoff score cannot be more than 10 points lower than that of the open category. This stipulation may lead to vacancies in the OBC category. The AKT judgment recommends that any vacant positions in the OBC category should revert to the open category. This paper formulates the new eligibility criteria introduced in AKT and the transfer of unfilled OBC positions to the open category as formal axioms. We show that the over-and-above and within-category fairness principles established in Indra Sawhney (1992) are incompatible with the new quota-filling mechanism subject to the eligibility and de-reservation axioms from AKT (2008). We introduce a family of choice rules that satisfies the over and above, within-category fairness, quota filling subject to eligibility principles from Indra Sawhney (1992) and the de-reservation principle from AKT (2008). Choice rules in this family are parameterized by *transfer functions* that determine where in the precedence order each unfilled OBC position is filled. This family includes the Backward Transfer (BT) choice function introduced in [1] and the Forward Transfer (FT) choice rule. The main result of the paper characterizes the FT choice rule as merit-maximizing

M. Mavronicolas et al. (Eds.): WINE 2024, LNCS 15534, pp. 672–673, 2026.
https://doi.org/10.1007/978-3-032-08560-3

in this family.

The full paper is available at https://sites.google.com/site/bertanturhan/

Reference

1. Aygün, O., Turhan, B.: How to de-reserve reserves: admissions to technical colleges in India. Manage. Sci. **69**, 5695–6415 (2022)

Informational Size in School Choice

Di Feng[1] and Yun Liu[2(✉)]

[1] Dongbei University of Finance and Economics, 116025 Dalian, China
dfeng@dufe.edu.cn
[2] Shandong University, 250100 Jinan, China
yunliucer@sdu.edu.cn

Abstract. Inspired by [1], this paper introduces a novel measurement of *informational size* to school choice problems. Informational size measures the *informativeness* of a matching mechanism through quantifying the maximal preference and priority rankings required to secure a specific student-school assignment across all possible matching problems. Our analysis yields two main insights regarding four prominent matching mechanisms: Deferred Acceptance (DA), Top Trading Cycles (TTC), Serial Dictatorship (SD), and Immediate Acceptance (IA). First, we find that the three strategy-proof mechanisms (DA, TTC, SD) require more information, i.e., with a larger informational size, compared to IA. This result highlights a previously omitted advantage of IA in term of its lower information demand, which partially explains its popularity in real-world matching problems where acquiring information is both peculiarly and cognitively costly. Second, when the market involves at least four students, DA requires more information than TTC. The issue of comparing between TTC and DA has puzzled researchers both theoretically and experimentally. Our result responds to this issue from an informational perspective: in experiments with relatively fewer students, agents tend to prefer DA over TTC as DA requires fewer information to secure one's allocation in all problems, while the opposite is true when the market size increases.

Keywords: Matching · Informational size · School choice · Immediate acceptance · Deferred acceptance · Top trading cycles · Serial dictatorship

---

Part of this research was conducted during Di Feng's residency at the Simons Laufer Mathematical Sciences Institute (formerly MSRI) in Berkeley, California, in the Fall 2023 semester, supported by the National Science Foundation under Grant No. DMS-1928930 and by the Alfred P. Sloan Foundation under grant G-2021-16778. Yun Liu acknowledges financial support from the Ministry of Education Project of Youth Fund of Humanities and Social Science (No. 23YJC790090), and the Shandong Provincial Natural Science Foundation (No. ZR2023MG009). The full version of the paper is available at https://arxiv.org/abs/2407.11273.

M. Mavronicolas et al. (Eds.): WINE 2024, LNCS 15534, pp. 674–675, 2026.
https://doi.org/10.1007/978-3-032-08560-3

Reference

1. Mount, K., Reiter, S.: The informational size of message spaces. J. Econ. Theory **8**(2), 161–192 (1974)

Author Index

M. Mavronicolas et al. (Eds.): WINE 2024, LNCS 15534, pp. 677–679, 2026.
https://doi.org/10.1007/978-3-032-08560-3

MIX
Papier aus verantwortungsvollen Quellen
Paper from responsible sources
FSC® C105338

If you have any concerns about our products,
you can contact us on
ProductSafety@springernature.com

In case Publisher is established outside the EU,
the EU authorized representative is:
Springer Nature Customer Service Center GmbH
Europaplatz 3, 69115 Heidelberg, Germany

Printed by Libri Plureos GmbH
in Hamburg, Germany